COMPUTER AND COMPUTING TECHNOLOGIES IN AGRICULTURE II, VOLUME 2

IFIP – The International Federation for Information Processing

IFIP was founded in 1960 under the auspices of UNESCO, following the First World Computer Congress held in Paris the previous year. An umbrella organization for societies working in information processing, IFIP's aim is two-fold: to support information processing within its member countries and to encourage technology transfer to developing nations. As its mission statement clearly states,

> *IFIP's mission is to be the leading, truly international, apolitical organization which encourages and assists in the development, exploitation and application of information technology for the benefit of all people.*

IFIP is a non-profitmaking organization, run almost solely by 2500 volunteers. It operates through a number of technical committees, which organize events and publications. IFIP's events range from an international congress to local seminars, but the most important are:

• The IFIP World Computer Congress, held every second year;
• Open conferences;
• Working conferences.

The flagship event is the IFIP World Computer Congress, at which both invited and contributed papers are presented. Contributed papers are rigorously refereed and the rejection rate is high.

As with the Congress, participation in the open conferences is open to all and papers may be invited or submitted. Again, submitted papers are stringently refereed.

The working conferences are structured differently. They are usually run by a working group and attendance is small and by invitation only. Their purpose is to create an atmosphere conducive to innovation and development. Refereeing is less rigorous and papers are subjected to extensive group discussion.

Publications arising from IFIP events vary. The papers presented at the IFIP World Computer Congress and at open conferences are published as conference proceedings, while the results of the working conferences are often published as collections of selected and edited papers.

Any national society whose primary activity is in information may apply to become a full member of IFIP, although full membership is restricted to one society per country. Full members are entitled to vote at the annual General Assembly, National societies preferring a less committed involvement may apply for associate or corresponding membership. Associate members enjoy the same benefits as full members, but without voting rights. Corresponding members are not represented in IFIP bodies. Affiliated membership is open to non-national societies, and individual and honorary membership schemes are also offered.

COMPUTER AND COMPUTING TECHNOLOGIES IN AGRICULTURE II, VOLUME 2

The Second IFIP International Conference on Computer and Computing Technologies in Agriculture(CCTA2008), October 18-20, 2008, Beijing, China

Edited by

Daoliang Li
China Agricultural University
China

Chunjiang Zhao
National Engineering Research Center
for Information Technology in Agriculture
China

Computer and Computing Technologies in Agriculture II, Volume 2

Edited by Daoliang Li and Chunjiang Zhao

p. cm. (IFIP International Federation for Information Processing, a Springer Series in Computer Science)

ISSN: 1571-5736 / 1861-2288 (Internet)
ISBN: 978-1-4419-5494-7
eISBN: 978-1-4419-0211-5

Printed on acid-free paper

Printed in the United States of America.

9 8 7 6 5 4 3 2 1

springer.com

Contents

Foreword

The papers in this volume comprise the refereed proceedings of the Second IFIP International Conference on Computer and Computing Technologies in Agriculture (CCTA2008), in Beijing, China, 2008.

The conference on the Second IFIP International Conference on Computer and Computing Technologies in Agriculture (CCTA 2008) is cooperatively sponsored and organized by the China Agricultural University (CAU), the National Engineering Research Center for Information Technology in Agriculture (NERCITA), the Chinese Society of Agricultural Engineering (CSAE) , International Federation for Information Processing (IFIP), Beijing Society for Information Technology in Agriculture, China and Beijing Research Center for Agro-products Test and Farmland Inspection, China. The related departments of China's central government bodies like: Ministry of Science and Technology, Ministry of Industry and Information Technology, Ministry of Education and the Beijing Municipal Natural Science Foundation, Beijing Academy of Agricultural and Forestry Sciences, etc. have greatly contributed and supported to this event. The conference is as good platform to bring together scientists and researchers, agronomists and information engineers, extension servers and entrepreneurs from a range of disciplines concerned with impact of Information technology for sustainable agriculture and rural development. The representatives of all the supporting organizations, a group of invited speakers, experts and researchers from more than 15 countries, such as: the Netherlands, Spain, Portugal, Mexico, Germany, Greece, Australia, Estonia, Japan, Korea, India, Iran, Nigeria, Brazil, China, etc. are gathering Beijing to review the new advancement of Information and Communication Technology (ICT) applications for sustainable agriculture and food quality and safety control, to present new research findings, and to look for the new challenges and opportunities in the future.

Information technology, the convergence of computing and communication technologies, has had an enormous impact on all aspects of socio-economic development and human life in the past 30 years. Powered by the unprecedented and continuous advances in microelectronics and photonics, the power and capacity of our expanding information infrastructure has risen exponentially, while simultaneously its cost has fallen also exponentially. At least for the foreseeable future, the exponential pace of technology improvement is likely to be continued. The modern ICT is playing increasingly important roles in every facet of agricultural and biological system improvement. While traditional sectors of agricultural technology are being constantly updated, the new sectors, such as biological informatics, information network services, information & knowledge-based precision farming system are bringing new concepts and contents into the agricultural & food chain management. The arrival of new requirements for agricultural system sustainability is accompanied by greater challenge in our profession. The goals for farming productivity, resources conservation & environmental sustainability require to develop intelligent equipment, technologies & services in extension of ICT for agriculture. The automated data acquisition is the fit way to provide spatial and temporal high-resolutions and safe documentations. The huge amount of raw data needs to be processed by a easy-to-use and safe data processing systems. A well-founded documentation will be the base of many agricultural applications in the future. A web based data management and information system are able to provide safety and effective information management for the farmers-avoiding problems with local installed software, time and costs. To promote ICT for agriculture, we need "Simplicity Theory", that is to find the simplest method to solving real problems in farming management. To develop a low-cost with high technologies are the future of innovation activities of ICT engineers for agriculture.

The main subjects of this conference are:

■ Exploitation of the strategic problems on ICT for agricultural resources, environment & production system management, web-based technology & agro-information and knowledge service system;

■ Spatial information technologies (GPS, GIS, RS) for agriculture, modeling of resources, ecological and biological systems; Precision Agriculture; advanced sensors and instrumentation for farm use; & process automation; expert system and knowledge system & DSS development;

■ Applied software development for farm users and macro management;

■ Intelligent & virtual technology for agriculture, knowledge dissemination and remote education, etc.

More than 432 academic manuscripts have been received by this organizing committee. After review process by a group of experts, 244 English papers are accepted and published by Spring IFIP US. Taking this opportunity, We would like to express our gratefulness to the hard word by all the contributors and members of Academic Committee.

Finally, we would like to extend the most earnest gratitude to our organizers, College of Information and Electrical Engineering (CAU), EU-China Centre for Information & Communication Technologies (CAU), also to Beijing Fu-Chi Technology Co., Ltd., all members and colleagues of our preparatory committee, for their generous efforts, hard work and precious time!

This is the Second series of conferences dedicated to real-world applications of computer and computing technologies in agriculture around the world. The wide range and importance of these applications are clearly indicated by the papers in this volume. Both are likely to increase still further as time goes by and we intend to reflect these developments in our future conferences.

Daoliang LI

Chunjiang Zhao

Co-Chairs of CCTA2008

Organizing Committee

Co-Chairs

Prof. Daoliang Li
China Agricultural University, China
Director of EU-China Center for Information & Communication technologies in Agriculture

Prof. Chunjiang Zhao
Director of National Engineering Research Center for Information Technology in Agriculture, China

Members [in alpha order]

Baozhu Yang, Professor of National Engineering Research Center for Information Technology in Agriculture, China

Dehai Zhu, Professor of College of Information and Electrical Engineering, China Agricultural University, China

Haijian Ye, Professor of College of Information and Electrical Engineering, China Agricultural University, China

Jianing Cai, Official of Department of International Cooperation, Ministry of Science and technology, China

Ju Ming, Official of Department of science and technology, Chinese Ministry of Education, China

Qingshui Liu, Secretary-general of China Agricultural University Library, China

Rengang Yang, Professor of College of Information and Electrical Engineering, China Agricultural University, China

Renjie Dong, Professor of Office of International Relations, China Agricultural University, China

Songhuai Du, Professor of College of Information and Electrical Engineering, China Agricultural University, China

Wanlin Gao, Professor of College of Information and Electrical Engineering, China Agricultural University, China

Weizhe Feng, Professor of International College at Beijing, China Agricultural University, China

Xinting Yang, Associate Professor of National Engineering Research Center for Information Technology in Agriculture, China

Program Committee

Chair

Maohua Wang

Professor of China Agricultural University, Academician of Chinese Academy of Engineering, China

Members [in alpha order]

Baoguo Li, Professor of College of Resources and Environmental Sciences, China Agricultural University , China

Béatrice Balvay, Professor of Institut de l'Elevage, France

Benhai Xiong, Professor of Institute of Animal Science, Chinese Academy of Agricultural Sciences, China

Chunjiang Zhao, Professor of National Engineering Research Center for Information Technology in Agriculture, China

Daoliang Li, Professor of College of Information and Electrical Engineering, China Agricultural University, China

Deepa Thiagarajan, Doctor of Michigan State University, USA

Dehai Zhu, Professor of College of Information and Electrical Engineering, China Agricultural University , China

Fangquan Mei, Professor of Agricultural Information Institute, Chinese Academy of Agricultural Sciences, China

Fanlun Xiong, Professor of Hefei Institute of Intelligent Machines, Chinese Academy of Sciences

Fazhong Jin, Professor of Center for Agro-food Quality & Safety, Ministry of Agriculture, China

Fernando Bienvenido, Professor of Universidad de Almeria, Spain

Gang Liu, Professor of College of Information and Electrical Engineering, China Agricultural University , China

Guohui Gan, Professor of Institute of Geographic Sciences and Natural Resources, Chinese Academy of Sciences, China

Guomin Zhou, Professor of Agricultural Information Institute, Chinese Academy of Agricultural Sciences, China

Heinz-W. Dehne, Professor of University of Bonn, Germany

Jihua Wang, Professor of National Engineering Research Center for Information Technology in Agriculture , China

Jinsheng Ni, Doctor of Beijing Oriental TITAN Technology Co., LTD, China
Joanna Kulczycka, Doctor of Polish Academy of Sciences Mineral and Energy Economy Research Institute, Poland
João Cannas da Silva, Vice President of European College of Bovine Health Management
K.C. Ting, Professor of University of Illinois at Urbana-Champaign
Koji Sugahara , Professor of National Agricultural Research Center, NARO, Japan
Kostas Komnitsas, Professor of Technical University of Crete, Greece
Liangyu Chen, Professor of Rural Technology Development Center , Ministry of Science & Technology , China
Louise Marguin, Professor of Institut de l'Elevage, France
Max Bramer, Professor of University of Portsmoth , UK
Michele Genovese, Director of Unit Specific International Cooperation Activities, International Cooperation Directorate, DG Research , UK
Minzan Li, Professor of College of Information and Electrical Engineering, China Agricultural University , China
Nick Sigrimis, Professor of Agricultural University of Athens , Greece
Nigel Hall, Professor of Harper Adams University College , England
Rohani J. Widodo, Professor of Maranatha Christian University , Indonesia
Shihong Liu, Professor of Agricultural Information Institute, Chinese Academy of Agricultural Sciences, China
Theodoros Varzakas, Doctor of Technological Educational Institution of Kalamata , Greece
Weixing Cao, Professor of Nanjing Agricultural University , China
Xiwen Luo, Professor of South China Agricultural University, China
Yanqing Duan , Professor of University of Bedfordshire, UK
Yenu Wan, Professor of Taiwan Chung Hsing University , China
Yeping Zhu, Professor of Agricultural Information Institute, Chinese Academy of Agricultural Sciences, China
Yibin Ying, Professor of Zhejiang University , China
Yiming Wang, Professor of College of Information and Electrical Engineering , China Agricultural University , China
Yud-Ren Chen, Professor of Instrumentation and Sensing Laboratory, Department of Agriculture, USA
Yuguo Kang, Professor of China Cotton Association, China
Zetian Fu, Professor of China Agricultural University , China

Secretariat

Secretary-general

Baoji Wang , China Agricultural University, China
Jihua Wang, National Engineering Research Center for Information Technology in Agriculture, China
Liwei Zhang , China Agricultural University, China

Secretaries

Bin Xing, China Agricultural University, China
Chengxian Yu, China Agricultural University, China
Dongjun Wang, China Agricultural University, China
Liying Xu, China Agricultural University, China
Miao Gao, National Engineering Research Center for Information Technology in Agriculture, China
Ming Li, National Engineering Research Center for Information Technology in Agriculture, China
Ming Yin, National Engineering Research Center for Information Technology in Agriculture, China
Rui Guo, China Agricultural University, China
Xiaochen Zou, China Agricultural University, China
Xiaohong Du, National Engineering Research Center for Information Technology in Agriculture, China
Xin Qiang, China Agricultural University, China
Xiuna Zhu, China Agricultural University, China
Yanjun Zhang, China Agricultural University, China
Yingyi Chen, China Agricultural University, China
Zhenglu Tao, China Agricultural University, China

A STATISTICALLY DEPENDENT APPROACH FOR THE MONTHLY RAINFALL FORECAST FROM ONE POINT OBSERVATIONS

J. Pucheta[1,*], D. Patiño[2], B. Kuchen[2]

[1] *LIMAC, Departments of Electronic Engineering and Electrical Engineering, National University of Córdoba, Vélez Sarsfield ave. 1611, ARGENTINA X5016GCA.*

[2] *Institute of Automatics, Faculty of Engineering, National University of San Juan, Lib. San Martín ave., 1109, ARGENTINA J5400ARL.*

[*] *Corresponding author, Address: LIMAC, Departments of Electronic Engineering and Electrical Engineering, National University of Córdoba, Vélez Sarsfield ave. 1611, ARGENTINA. X5016GCA. Email: julian.pucheta@gmail.com.*

Abstract: In this work an adaptive linear filter model in a autoregressive moving average (ARMA) topology for forecasting time series is presented. The time series are composed by observations of the accumulative rainfall every month during several years. The learning rule used to adjust the filter coefficients is mainly based on the gradient-descendent method. In function of the long and short term stochastic dependence of the time series, we propose an on-line heuristic law to set the training process and to modify the filter topology. The input patterns for the predictor filter are the values of the time series after applying a time-delay operator. Hence, the filter's output will tend to approximate the current value available from the data series. The approach is tested over a time series obtained from measures of the monthly accumulative rainfall from La Perla, Córdoba, Argentina. The performance of the presented approach is shown by forecasting the following 18 months from a hypothetical actual time for four time series of 102 data length.

Keywords: Adaptive filter, rainfall forecasting, Autoregressive moving average

---

Please use the following format when citing this chapter:

Pucheta, J., Patino, D. and Kuchen, B., 2009, in IFIP International Federation for Information Processing, Volume 294, *Computer and Computing Technologies in Agriculture II, Volume 2*, eds. D. Li, Z. Chunjiang, (Boston: Springer), pp. 787–798.

1. INTRODUCTION

This work presents an approach to the future rainfall water availability problem for agricultural purposes. There are several approaches based on non linear autoregressive moving average filters that face the rainfall forecast problem for water availability by taking an ensemble of measurement points (Liu and Lee, 1999; Masulli *et al.*, 2001). Here, the proposed approach is based on the classical linear autoregressive filter moving average using time lagged feedforward approach, by considering the historical data from one geographical point. One of the motivations for this study follows the closed-loop control scheme (Pucheta et al., 2007a) where the controller considers future conditions for the control law's design as shown Fig. 1. In that scheme the controller takes into account the actual state of the crop by a state observer and the monthly accumulative rainfall. However, this paper presents only the controller portion concerning with the rainfall forecast. The controller design is inspired on the one presented in (Pucheta et al., 2007a).

The main contribution of this work lies on the tuning process and filter structure, which employs the gradient descendent rule and considers the long and short term stochastic dependence of passed values of the time series to adjust at each time-stage the number of patterns, the number of iterations, and the length of the tapped-delay line, in function of the Hurst's value (H) of the time series. According to the stochastic characteristics of each series, H can be greater or smaller than 0.5, which means that each series tends to present long or short term dependence, respectively. In order to adjust the design parameters and see the performance of the proposed prediction model, sinusoidal and square signals are used. Then, the predictor filter is applied to the monthly accumulative rainfall from La Perla -Córdoba, Argentina- as the time series to forecast the next 18 values given a historical data set.

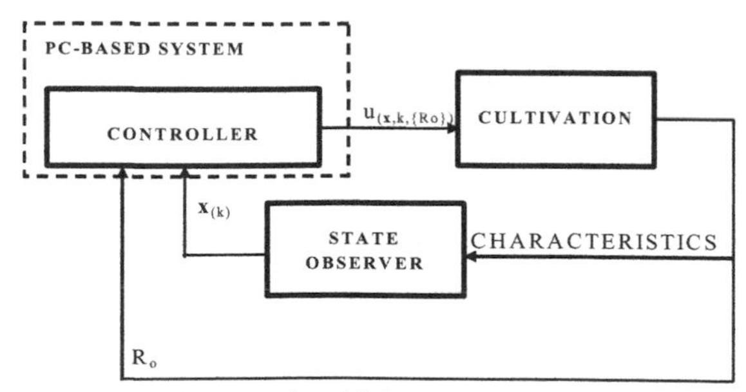

Fig. 1. PC-based control approach, which considers an accumulative rainfall R_o.

1.1 Overview on Fractional Brownian motion

In this work the Hurst's value is used in the learning process to modify on-line the number of patterns and number of iterations presented. The H parameter is useful for the definition of the Fractional Brownian Motion (fBm). The fBm is defined in the pioneering work by Mandelbrot (1983), through its stochastic representation

$$B_H(t) = \frac{1}{\Gamma\left(H + \frac{1}{2}\right)} \left(\int_{-\infty}^{0} \left((t-s)^{H-\frac{1}{2}} - (-s)^{H-\frac{1}{2}} \right) dB(s) + \int_{0}^{t} (t-s)^{H-\frac{1}{2}} dB(s) \right) \tag{1}$$

where, $\Gamma(\cdot)$ represents the Gamma function

$$\Gamma(\alpha) = \int_0^{\infty} x^{\alpha-1} e^{-x} dx, \tag{2}$$

and $0<H<1$ is called the Hurst parameter. The integrator B is a stochastic process, ordinary Brownian motion. Note, that B is recovered by taking $H=1/2$ in (1). Here, it is assumed that B is defined on some probability space (Ω, F, P), where Ω, F and P are the sample space, the sigma algebra (event space) and the probability measure, respectively. So, an fBm is a continuous-time Gaussian process depending on the so-called Hurst parameter $0<H<1$. It generalizes the ordinary Brownian motion corresponding to $H=0.5$, and whose derivative is the white noise. The fBm is self-similar in distribution and the variance of the increments is given by

$$Var\left(B_H(t) - B_H(s)\right) = v\left|t-s\right|^{2H} \tag{3}$$

where, v is a positive constant. This special form of the variance of the increments suggests various ways to estimate the parameter H. In fact, there are different methods for computing the parameter H associated to Brownian Motion (Dieker, 2004). In this work, the algorithm uses a wavelet-based method for estimating H from a trace path of the fBm with parameter H (Abry *et al.*, 2003; Dieker, 2004). Three trace path from fBm with different values of H are shown in Fig. 2, where can be noted the difference in the velocity and the amount of its increments.

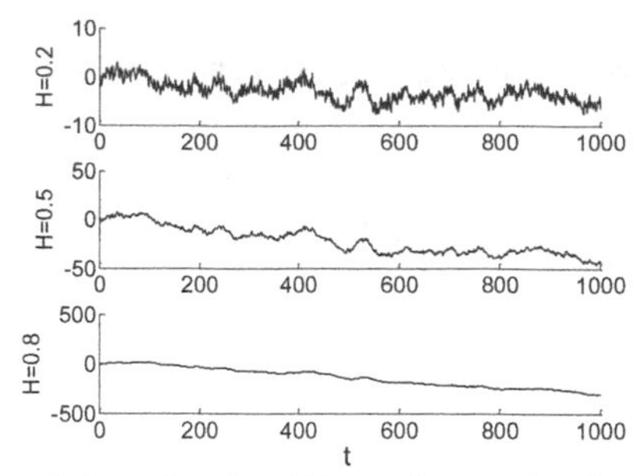

Fig. 2. Three sample path from fractional Brownian motion for three values of H.

2. PROBLEM STATEMENT

The classical prediction problem may be formulated as follow. Given past values of a process that are uniformly spaced in time, as shown by *x(n-T), x(n-2T), . . . , x(n-mT)*, where T is the sampling period and m is the prediction order, it is desired to predict the present value x(n) of such process. Therefore, obtain the best prediction (in some sense) of the present values from a random (or pseudo-random) time series is desired.

The predictor system may be implemented using an ARMA linear filter. Here, the model follows the classic linear schemme (Ljung, 1999). The linear model structure is self tuned in such a way that smaller the prediction error is (in a statistical sense), the better the filter serves as model of the underlying physical process responsible for generating the data. In this work, time lagged feedforward scheme are used. Thus, the present value of the time series is used as the desired response for the adaptive filter, and the past values of the signal supply as input of the adaptive filter. Then, the adaptive filter output will be the one-step prediction signal. In Fig. 3 the block diagram of the linear prediction scheme based on a ARMA filter is shown. Here, a prediction device is designed such that starting from a given sequence $\{x_n\}$ at time n corresponding to a time serics it can be obtained the best prediction $\{x_e\}$ for the following 18 values sequence. Hence, it is proposed a predictor filter with an input vector l_x, which is obtained by applying the delay operator, Z^{-1}, to the sequence $\{x_n\}$. Then, the filter output will generate x_e as the next value, that will be equal to the present value x_n. So, the prediction error at time k can be evaluated as

$$e(k) = x_n(k) - x_e(k),$$

which is used for the learning rule to adjust the filter's coefficients. The coefficients of the filter are adjusted on-line in the learning process, by considcring a criterion that modifies at each time-stage the number of patterns, the number of iterations, and the length of the tapped-delay line, in function of the Hurst's value (H) calculated from the time series. According to the stochastic behavior of the series, H can be greater or smaller than 0.5, which means that the series tends to present long or short term dependence, respectively. A similar algorithm was presented in (Pucheta *et al.*, 2007b).

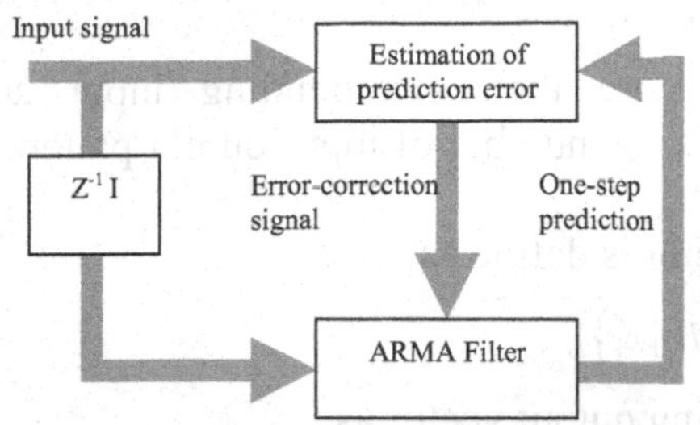

Fig. 3. Block diagram of the linear prediction.

3. PROPOSED APPROACH FOR PREDICTION

3.1 Autoregressive Linear Model

Now, a linear autoregressive filter model (Haykin, 1999; Ljung, 1991) is proposed. The filter used is a time lagged feedforward type. The filter topology consists of one input with l_x taps, and one output. The rule used in the tuning process is based on the standard descendent gradient (Ljung, 1991). The tuning rule modifies the number of patterns and the number of iterations at each time-stage according to the Hurst's parameter H, which gives short and long term dependence of the sequence $\{x_n\}$ or —from a practical point of view, it gives the ruggedness of the time series. In order to predict the sequence $\{x_e\}$ one-step ahead, the first delay taken off from the tapped-line x_n is used as input. Therefore, the output prediction can be denoted by

$$x_e(n+1) = F_p\left(Z^{-1}I(\{x_n\})\right) \tag{4}$$

where, F_p is the nonlinear predictor filter operator, and $x_e(n+1)$ the output prediction at n+1.

3.2 The Proposed Learning Process

The filter's coefficients are tuned by means of the gradient-descendent rule in a batch scheme, which in turn considers the long and short term stochastic dependence of the time series measured by the Hurst's parameter H. The proposed learning process consists on changing both the number of patterns and the number of iterations in function of the parameter H for each corresponding time series. Here, the tuning process is performed using a batch model. In this case the update of the coefficients is being performed after the presentation of all tuning examples, which forms an epoch. The pairs of the used input-output patterns are

$$(x_i, y_i) \quad i = 1,2,...., N_p \tag{5}$$

where, x_i and y_i are the corresponding input and output pattern respectively, and N_p is the number of input-output patterns presented at each epoch.

Here, the input vector is define as

$$X_i = Z^{-1} I(\{x_i\}), \tag{6}$$

and its corresponding output vector as

$$Y_i = x_i, \tag{7}$$

Furthermore, the index i is within the range of N_p given by

$$l_x \le N_p \le 4 \cdot l_x,$$

where, l_x is the dimension of the input vector.

In addition, through each epoch the number of iterations performed it is given by

$$l_x \le i_t \le 4 l_x .$$

The proposed criterion to modify the pair (i_t, N_p) is given by the statistical dependence of the time series $\{x_n\}$, supposing that is an fBm. The dependence is evaluated by the Hurst's parameter H, which is computed using a wavelet-based method (Abry *et al.*, 2003).

Then, a heuristic adjustment for the pair (i_t, N_p) in function of H according to the membership functions shown in Fig. 4 is proposed. Finally, the number of inputs of the filter is tuned —that is the length of tapped-delay line, according to the following heuristic criterion. After the training process is completed, both sequences —$\{x_n\}$ and $\{\{x_n\},\{x_e\}\}$, should have the same H parameter. The error between $H(\{x_n\})$ and $H(\{\{x_n\},\{x_e\}\})$ is used for tuning the value of l_x, by mean of a well-known PID scheme. Thus, l_x is updated by,

$$l_x = l_x + round\left(K_P \cdot e_H + K_D \cdot \dot{e}_H\right). \tag{8}$$

where K_P, K_D are constants, and e_H is defined as

$$e_H = H(\{x_n\}) - H(\{\{x_n\},\{x_e\}\}). \tag{9}$$

Thus, the objective is to forecast the time series in such a way that the predicted time series and the data time series present the same H.

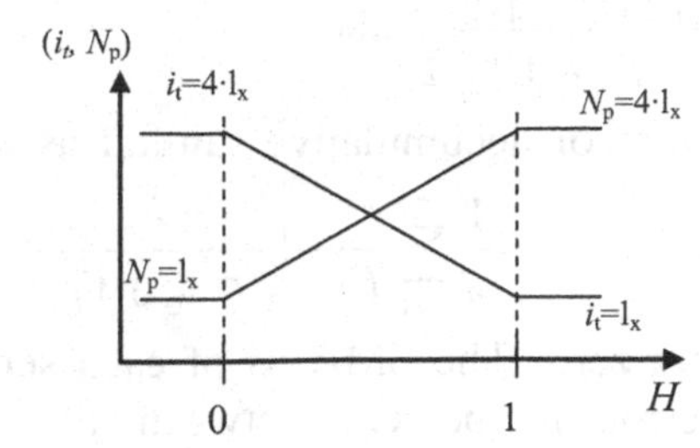

Fig. 4. Heuristic adjustment of (i_t, N_p) in terms of H.

4. MAIN RESULTS

4.1 Set-up of Filter and Tuning Process

Table 1. Initial condition of the tuning algorithm.

Variable	Initial Condition
l_x	12
i_t	$5 \cdot l_x$
H	0.5
η	0.1

The initial conditions for the filter and tuning algorithm are shown in Table 1. Note that the initial number of iteration is set in function of the input number. The variable η is the step-size of the gradient descendent method, used for tuning the filter parameter. These initial conditions of the tuning algorithm were used for forecasting the monthly accumulative rainfall time series, whose sizes have a length of 102 values each.

4.2 Preliminary Results Using Test Time Series

In order to test the proposed design procedure of the linear predictor, an experiment with sinusoidal and square signals was carried out. The performance of the filter is evaluated using the mean Symmetric Mean Absolute Percent Error (SMAPE):

$$SMAPE_S = \frac{1}{n}\sum_{t=1}^{n}\frac{|X_t - F_t|}{(X_t + F_t + 0.1)/2}\cdot 100 \qquad (10)$$

where, t is the time observation, n is the (Data series) test set size, s each time series, X_t and F_t are the actual and the forecast time series values at time t respectively. The shortcoming arise when the denominator is 0 in

$$SMAPE_s = \frac{1}{n}\sum_{t=1}^{n} \frac{|X_t - F_t|}{(X_t + F_t + 0.1)/2} \cdot 100$$ (10), so we opt to include the value 0.1 mm of accumulative rainfall as a dummy minimum only for evaluating

$$SMAPE_s = \frac{1}{n}\sum_{t=1}^{n} \frac{|X_t - F_t|}{(X_t + F_t + 0.1)/2} \cdot 100 \quad (10)$$

to avoid the division by zero. The SMAPE of each series s calculates the symmetric absolute error in percent between the actual X_t and its corresponding forecast F_t value, across all observations t of the test set of size n for each time series s. Fig. 5 (a) shows the filter response, when it forecasts the 18 future values for a sinusoidal time series. The used sine time series has a period T=0.48 s, and it is sampled at T_0=0.05 s. The initial length of the tapped-delay line was set-up at 12 taps and at the end of the tuning process was equal to 12, given that Eq $l_x = l_x + round\left(K_P \cdot e_H + K_D \cdot \dot{e}_H\right)$. (8) does not apply because H equals 1 for all the trials. For a square time series, Fig. 5 (b) presents the forecasted 18 values. Here the value of H, across for the complete time series $\{x_n\}$ and $\{x_e\}$, differs at a 5%. To improve the forecasting performance of the H-dependent filter, it is used as initial condition l_x= 15, in order to increase H of the $\{x_e\}$. The new results are shown in Fig. 5 (c) and Fig. 5 (d), where the percentage is declined in the order of 1%.

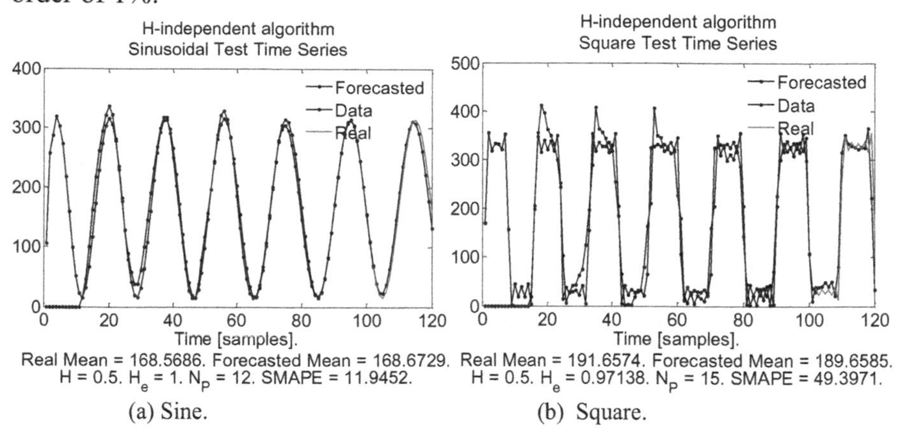

(a) Sine. (b) Square.

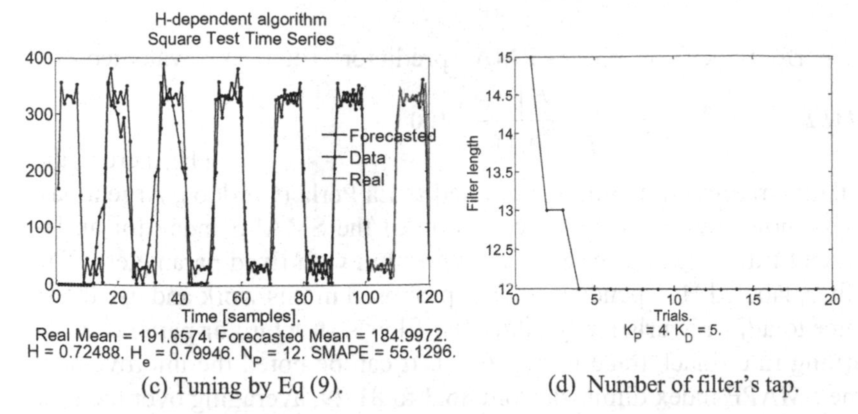

(c) Tuning by Eq (9). (d) Number of filter's tap.

Fig. 5. Algorithm performance on test series.

4.3 Prediction Results for the Rainfall Time Series

Each one of the time series are composed by observations from the monthly rainfall, which over ten years are yields 120 values. However, the 18 last values where used to validate the performance of the prediction system. So, 102 values forms the Data set, and the Forecasted set are 120, and the Real data are 18 values. The Data was obtained along 4 decades, which is the laps from January, 1962 to December, 1971, from Jan 1972 to Dec 1981, and so on up to Dec 2001. Obtained results are show in Fig. 6.

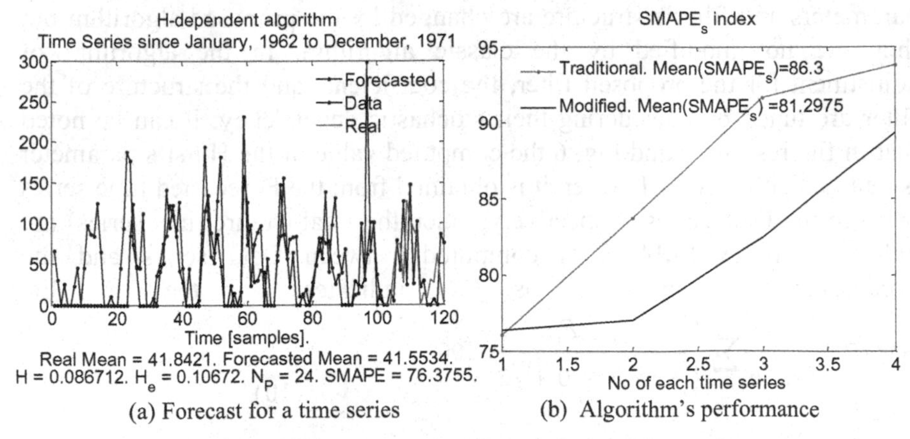

(a) Forecast for a time series (b) Algorithm's performance

Fig. 6. Performance on the prediction of rainfall time series.

4.4 Comparative Results

The performance of the ARMA predictor filter is evaluated by

$$SMAPE_s = \frac{1}{n}\sum_{t=1}^{n}\frac{|X_t - F_t|}{(X_t + F_t + 0.1)/2}\cdot 100$$

Eq. (10) across the accumulative rainfall time series measured at La Perla (Córdoba, Argentina). Fig. 6 (b) shows by red trace the evolution of the SMAPE index for an H-independent filter, which uses a tuning algorithm with fixed parameters. The other filter, labeled H-dependent filter is proposed in this work and use the H parameter to adjust heuristically either the filter's structure or parameters of the learning rule -black trace in Fig. 6 (b). It can be noted the improvement since the SMAPE index diminish from 86.3 to 81.29, averaging over the four time series.

5. DISCUSSION

The evaluation of the obtained results has been realized by comparing the performance of the proposed filter against that of the classic filter, both based on linear scheme. Although the difference between both filters only resides in the adjustment algorithm, the coefficients that each filter has perform different behaviors. In the four analyzed cases, the generation of 18 future values from 102 present values was made by each algorithm. The same initial parameters were used for each algorithm, although such parameters and filter's structure are changed by the proposed algorithm but they are not modified by the classic algorithm. In the algorithm of adjustment for the proposed filter, the coefficients and the structure of the filter are tuned by considering their stochastic dependency. It can be noted that in figures Fig. 5 and Fig. 6 the computed value of the Hurst's parameter is denoted either H_e or H when it is obtained from the Forecasted time series or from the Data series, respectively, since the Real (future time series) are unknown. Index SMAPE is computed between Data series and the Forecasted one, as indicates the Ec.

$$SMAPE_s = \frac{1}{n}\sum_{t=1}^{n}\frac{|X_t - F_t|}{(X_t + F_t + 0.1)/2}\cdot 100 \qquad (10).$$

6. CONCLUSION

In this work a statistically dependent linear filter for forecasting time series has been presented. The tuning rule proposed to adjust the filter coefficient is based on the standard gradient-descendent method. Furthermore, in function of the long and short term stochastic dependence of the time series evaluated by the Hurst parameter H, an on-line heuristic adaptive law was proposed to update the filer topology at each time-stage, which is the number of input taps, the number of patterns and the number of iterations. The main result shows a good performance of the predictor system applied to the accumulative rainfall time series when the observations are taken from a single point, due to similar roughness for both the original and the forecasted time series, evaluated by H and He respectively was obtained. These results encourage one to go on working with this new tuning algorithm, applying to other filter models (such as non linear autoregressive moving average), due to the time series generated by humans interaction presents short and long term stochastic dependence.

ACKNOWLEDGMENTS

This work was supported by the National University of Córdoba (Secyt-UNC 69/08), the National University of San Juan (UNSJ), the National Agency for Scientific and Technological Promotion (ANPCyT) under grant PAV-TIC-076 and PICT/04 25423. The authors like to thank the help from Carlos Bossio (Coop. Huinca Renancó), Ronald del Águila (LIADE).

REFERENCES

Abry, P.; P. Flandrin, M.S. Taqqu, D. Veitch, Self-similarity and long-range dependence through the wavelet lens, Theory and applications of long-range dependence, Birkhäuser, 2003, pp. 527-556.

Dieker, T., Simulation of fractional Brownian motion, MSc theses, University of Twente, Amsterdam, The Netherlands, 2004.

Haykin, S, Neural Networks: A comprehensive Foudation, 2nd Edition, Prentice Hall, 1999.

Liu, J.N.K.; Lee, R.S.T., Rainfall forecasting from multiple point sources using neural networks, Systems, Man, and Cybernetics, 1999. IEEE SMC '99 Conference Proceedings. 1999 IEEE International Conference on, Vol.3, Iss., 1999, Pages:429-434 vol.3.

Ljung, Lennart. System Identification Theory for the user. 2nd Ed, Prentice Hall. 1999.

Mandelbrot, B. B., The Fractal Geometry of Nature, Freeman, San Francisco, CA., 1983.

Masulli, F., Baratta, D., Cicione, G., Studer, L., Daily Rainfall Forecasting using an Ensemble Technique based on Singular Spectrum Analysis, in Proceedings of the International Joint Conference on Neural Networks IJCNN 01, pp. 263-268, vol. 1, IEEE, Piscataway, NJ, USA, 2001.

Pucheta, J., Patiño, H., Schugurensky, C., Fullana, R., Kuchen, B., Optimal Control Based-Neurocontroller to Guide the Crop Growth under Perturbations, Dynamics Of Continuous, Discrete And Impulsive Systems Special Volume Advances in Neural Networks-Theory and Applications. DCDIS A Supplement, Advances in Neural Networks, Watam Press, Vol. 14(S1) 2007a, pp. 618—623.

Pucheta, J., Patiño, H.D. and B. Kuchen, Neural Networks-Based Time Series Prediction Using Long and Short Term Dependence in the Learning Process, In proc. of the 2007 International Symposium on Forecasting, 24th to 27th of June 2007 Marriott Marquis Times Square, New York, 2007b.

THE RESEARCH OF VERTICAL SEARCH ENGINE FOR AGRICULTURE

Weiying Li, Yan Zhao[*], Bo Liu, Qiang Li

College of Information Science and Technology, Agriculture University of HeBei, Baoding 071001,China

[*] *Corresponding author, Address: College of Information Science and Technology, Agriculture University of HeBei, Baoding 071001,China, Tel: +86-312-7526424, Email: leeweiying@163.com*

Abstract: Following rapid expansion of huge Agriculture information body on the Web, the efficient Agriculture information gathering on specified top becomes more and more important in search engine research. Through the statement of the developing trend of search engine and sharing agriculture information resource, this paper discusses the necessity of building search engine for agriculture information. The author clarifies the working principles of professional search engine for agriculture and finally analyses the improvement of searching technique of agriculture and proposes a model for agriculture - focused search.

Keywords: search engine, agriculture

1. INTRODUCTION

The internet in China has enormously promoted the development of agriculture. The amount of agriculture information networks in China has reached to huge numbers, which has shown a marked efficacy in terms of agriculture information communication and technological achievements communication. With the substantial increase of the information resource in the network, the phenomenon of information overload has been paid attention. It has become a problem that needs urgent solution how to obtain the web page that contains the information the user needs in an effective and accurate way.

Please use the following format when citing this chapter:

Li, W., Zhao, Y., Liu, B. and Li, Q., 2009, in IFIP International Federation for Information Processing, Volume 294, *Computer and Computing Technologies in Agriculture II, Volume 2*, eds. D. Li, Z. Chunjiang, (Boston: Springer), pp. 799–803.

Using the Vertical Search Engine is the most effective way to settle this problem. The vertical search engine directed towards agriculture searches the agriculture information of the website appropriately to make it possible that the information can be searched efficiently. With the exponential increasing of the information on Internet, there is a great deal of information waste. It becomes an important side that to provide the high quality and a modest number of query result.

2. CURRENT STATUS AND DEVELOPMENT TREND

Most of the Vertical Search Engines are lying in the phase of scientific research. The portals have appeared that faces some fields after making use of the searched result and after the professional person's processing. The research about the subject-based searching engine is getting hotter overseas, but in our country it is at the first time step.

The actual Vertical Search Engine adopts two kinds of technologies list below:

One is based upon the content which is the extension of the traditional information retrieval technologies. Its main way is to establish a word list in connection with the subject in the Search Engine. The crawler of Search Engine makes index from web according to the word list in the Search Engine. The complexity of establishing the word list is quite differently according to the different systems.

The other is based upon the analysis of interlink. Some scholars consider that the interlink age between websites is very similar as the traditional citation Index. The relationship between websites can be found out through analyzing the interlink age. The lots of Web pages can be easily classified according to the relationship of quoting because the details of websites that are referenced and references are interrelated in content.

3. KEY PROBLEMS IN TECHNOLOGY

The Vertical Search Engine that faces agriculture has its property. The four key technologies of it are listed below.

(1) The target-oriented, high real-time and manageable technique for webpage collecting: The Vertical Search Engine that faces agriculture has its professional requirement and target. It only collects the websites of partial source. The quantity of web pages is moderate. But the all sidedness and depth of collecting is demanding and the priority of collecting the dynamic web pages is highly required. The technologies of website collecting should

control the objective and range; support the deep collecting and the complicated dynamic web pages collecting according to needs in practice. The more target-oriented, Real-timing and more inclined of management of the collecting technology is needed, as well as the shorter cycle of information refreshing and the prompter acquiring for message.

(2) The web page analyzing technology of structured data: Because of the particularity of the Vertical Search Engine that faces agriculture, the period, source, and the other metadata are needed as well as the specifically content in web pages. For the purpose of providing preferable and more valuable searching service in the Vertical Search Engine, it demands that the author, the theme, the region, the name of the institute and product and the specialized shoptalk must be extracted.

(3) full-text index and combined retrieval technology: Because of the higher requirements in the professional information and useful value, the Vertical Search Engine that faces agriculture can support full-text searching and precise searching and can provide many different ways of sorting. In addition, the combined retrieval which is structured or unstructured should be support according to need, such as combined retrieval based on the author, the content and the classification.

(4) Intelligent zed text mining technology: The Vertical Search Engine that faces agriculture takes the structured data as the basic components. The Search Engine can furnish the more valuable service more accurately based on the integrating of structured data and full-text data. The entire structured information extracting goes through the process from analyzing to handling. In accordance with the above requirements, the Vertical Search Engine has the function of intelligent processing, such as automatic classification, automatic indexing, automatic decomposition and text mining. They are leading technology in the areas of Vertical Search and message processing.

4. DESIGN PROPOSAL

4.1 Technology pathway

(1)According to the characteristic of agricultural information's distributing and the practical requirements of the customer and based on thorough investigation, after understanding and comparing of the significant addition in the same field, put forward the overall framework of the platform.

(2)Specify the platform further combining the object-oriented technology to determine the specific pattern of the project.

(3)In connection with the concrete issue that is faced with, such as how to increase the crawl speed, the system's resource constraints, the precision of web page classification, the resolve of HTML file, provide effectual solutions.

(4) Reduction to practice and form a platform faced agriculture for information searching and sharing. Measure and test the multiple parameters further, improve and optimize them to form a user-friendly Search Engine, which meets the user's requirement in the response speed, recall ratio and precision ratio.

4.2 System architecture

The construction drawing is list below:

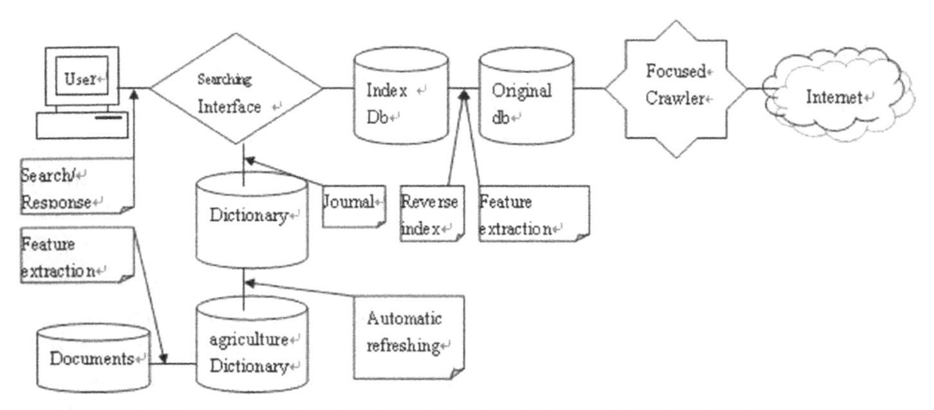

Figure. 1. The construction drawing of the Vertical Search Engine that faces agriculture

4.3 Technical innovations

(1)The system has the modular construction and the coupling is relaxed. It makes reference to the characteristic of Internet search engine and takes the file manipulation as an individual module.

(2)The system is developed by JAVA Language, so some code can be reuse availably.

(3)The object-oriented ideas are drawn into during the design process which is convenient for the developer to exchange with each other. The code is reconstructed ceaselessly during the coding session, which makes it high efficiency and its readability is raised too.

(4) The XML document is used to transit information between modules when the system is integrated.

(5) Many hash tables are used to raise the system's performance.

5. CONCLUSIONS

The primary focus about search engine has varies from how to find more information to how to find the exact and useful information. The precision ratio becomes the first priority for many of the search engine. The Vertical Search Engine that faces agriculture seeks the thematic information appropriately to make the user search what they want effectively. It can affect the development of search engine deeply.

REFERENCES

Chen H. Machine learning for information retrieval: neural networks, symbolic learning, and genetic algorithms [J].J. Am. Soc. Inf Sci. . 1995.46(3): 194-216

WANG Zheng, WANG Qing, WANG Ding-wei.Design of Domain-specific Search Model for Meta-search Engine on Internet, JOURNAL OF SYSTEM SIMULATION, 2008, 20(5)

CFD SIMULATION TO THE FLOW FIELD OF VENTURI INJECTOR

Xingfa Huang*, Guangyong Li, Miao Wang

College of Water Conservancy and Civil Engineering, China Agricultural University, Beijing 100083, China

* *Corresponding author, Address: P. O. Box 104, China Agricultural University, 17 Qinghua donglu, Beijing,100083,P.R. China, Tel:+86-10-62737874, Fax:+86-10-62736273, Email: huangxingfa@cau.edu.cn*

Abstract: Venturi injector is widely used in fertigation system due to its obvious advantages such as cheap and robust system without mobile pieces, simple structure, convenient to operation, stable performance, needless of external energy for operation etc. At present, the hydraulic parameters such as suction capacity (injection rate) for the most of the Venturi injectors produced domestically are not very desirable. In this paper, CFD (Computational Fluid Dynamics) method was used to simulate the inner flow field of the Venturi injectors, and the relationships among the structure parameters (i.e., throat length L, throat diameter D, slot diameter Da) and suction capacity q, and the optimal structure sizes of the Venturi injector were analyzed. The results show that when the inlet pressure and the slot position are kept unchanged as the sample one, the suction capacity of Venturi injector increases with the decrease of throat diameter D and throat length L, and the increase of slot diameter Da; while keeping the slot diameter Da, throat diameter D and throat length L unchanged, the suction capacity of Venturi injector q increases with the increase of inlet pressure P. The optimal combination of the structural parameters in this size was selected as follows: throat diameter D=8mm, slot diameter Da=18.5mm, and throat length L=14mm. In this case, the suction capacity of the Venturi injector q=1.203m3/h. The results can provide theoretic support for domestic Venturi injector research, design and manufacturing.

Key words: venturi injector, suction capacity (injection rate), inlet pressure, structure parameters, numerical simulation

---

Please use the following format when citing this chapter:

Huang, X., Li, G. and Wang, M., 2009, in IFIP International Federation for Information Processing, Volume 294, *Computer and Computing Technologies in Agriculture II, Volume 2*, eds. D. Li, Z. Chunjiang, (Boston: Springer), pp. 805–815.

1. INTRODUCTION

The application of chemical products in pressurized irrigation systems presents evident advantages. This technique, called chemigation, is widely used in sprinkler irrigation and drip irrigation systems (Johnson et al, 1986). The fertirrigation system needs an injector. Venturi injector is the one that widely used in the chemigation system in small and medium size farms (Manzano Juárez J. et al, 2005). The pressure drop through a venturi must be sufficient to create a negative pressure (vacuum) as measured relative to atmospheric pressure. Under these conditions the fluid from the tank will flow into the injector.

As the rapid development of agricultural modernization in China, it is of importance for sprinkler and micro irrigation system to apply in the field crops, facility agriculture, industrialized agriculture and landscape etc (Zhao Jingcheng et al, 1999). Fertigation in drip irrigation system is the optimal selection for fertilizing perennial crops in root zone, which is also the important measure to obtain high yields, good quality and efficient use of water and fertilizer. According to the investigation of Irrigation Training and Research Center, Science & Engineering College of California, U.S.A., in normal case with fertigation in drip system, 25% fertilizer can be saved, and more labors and energies be saved (Li jiusheng et al, 2003; 2004). At present, the common fertilizing equipments in China include self-pressured injecting device, pressure difference injecting device, fertigation pump, fertilizer container, Venturi injector, jet fertigation device etc.(Shen Xuemin, 2000).

Venturi injector is widely used in fertigation system due to its obvious advantages such as cheap system, simple structure, robust system without mobile pieces, convenient to operation, labor saving, stable performance, needless of external energy for operation etc. (Manzano Juárez J. et al, 2005; Shen Xuemin, 2001; Sha Yi et al, 1995). Venturi injector is one of the basic types of injector in subsurface drip irrigation system (Graham Harris, 2005)

Some researches on performance of Venturi injector in irrigation system were carried out. Better fertilizer distribution in the greenhouse experiment was obtained with Venturi one (Bracy R P et al, 2003). The effect of chemical temperature change on the injection flow rate of a Venturi injector was evaluated by Yuan Z et al (2000). Manzano Juárez J et al (2005) carried out the study on hydraulic modeling of Venturi injector using CFD, with its predictions for the total pressuer losses had a small error, less than 2%.

The study of Venturi injector in irrigation system in China began at late 1970's, when advanced Western micro irrigation system came to China. There were not series products then and up to now, and mostly relied on importing from foreign companies. As the rapid development of drip irrigation and fertigation technologies, reliable performance of Venturi injector was required. This requirement leads to the further research on Venturi injector. But at present, most studies mainly focus on theoretic analyses and experimental research on its hydraulic performances such as discharge, suction capacity (injection rate), pressure at throat, outlet water velocity etc., which needs a huge amount of experiments (Sha Yi et al, 1995; Shen Xuemin, 2001; Li Baijun et al, 2001). As the development of computer and simulation technologies, numerical simulation can do the similar things with its great advantages such as high efficiency, low cost, wide suitability to various variables etc.. It is widely extended and used in different areas (Chen Zuobing et al, 2005). In this article, The CFD (Computational Fluid Dynamics) numerical calculation method was used to simulate the performance of Venturi injector.

2. DESIGN AND METHODS

2.1 Structure and parameters of Venturi Injector

A sample of Venturi injector was designed and made before numerical simulation, as shown at Figure 1. For this sample Venturi injector, its inlet diameter is d_1=30mm, diameter of suction pipe is d_2=18mm, the slope of inlet and out let of throat is 17.5°, 4°, respectively. The throat diameter is D=12mm, and its length is L=27mm; The slot is at the top of suction pipe, and its diameter is D_a=17mm. After the sample was made, its performance test was carried out. The experiment condition was set as follows: inlet pressure was P=0.21MPa, outlet pressure was 0. Then the measured inlet discharge was Q_1=6.27m^3/h, outlet discharge was Q_2=6.95 m^3/h, while the suction capacity was q=0.68m^3/h.

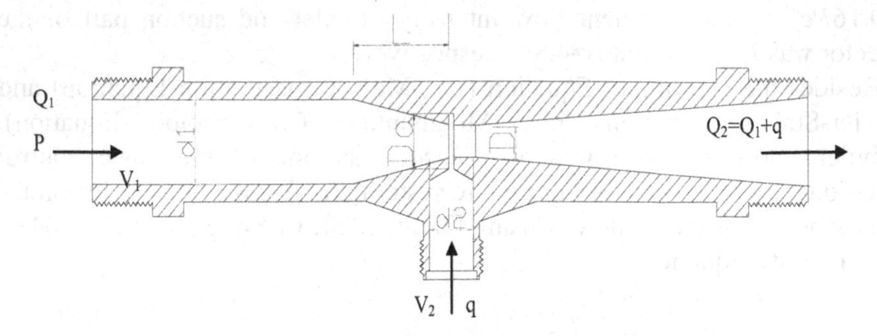

Fig. 1 Schematic diagram of Venturi injector

(L is the throat length; D is the throat diameter; D_a is the slot diameter; d_1, d_2 are the inlet diameter and suction pipe diameter, respectively; P is the inlet water pressure; Q_1, Q_2, q are the discharges from inlet, outlet, suction part; v_1, v_2 are the flow rates at inlet and suction part, respectively)

2.2 CFD Models of Venturi Injector and Its Validation

2.2.1 CFD Models

In this case, one of the CFD software, FLUENT 6.1.22 was used to simulate the performance of Venturi injector. It applies Definite Volume Method to transfer the differential equation to algebraical equation, and the approach of First Order Windward was used. For the solution of pressure – velocity coupled dispersed equations, typical SIMPLE algorithm was used (Chen Zuobing *et al*, 2005), in which the convergent value that is the difference between two variable values of conjoint alternation was less than 10^{-4}. The three dimensional geometric model of the Venturi injector sample was established by using the software of GAMBIT 2.04 , in which hexahedron grid with its size of 1mm was used to divide its grids for most parts, except for the slot part, which non structure grid with its size of 1mm was used. The total number of grids was 162138.

The inlet flow rate of the sample Venturi injector was: $v_1=4Q_1/(\pi d_1^2)=4*(6.27/3600)/(\pi*0.03^2)= 2.46$ m/s ; suction flow rate was : $v_2=4q/(\pi d_2^2)=4*(0.68/3600)/(\pi*0.018^2)= 0.742$ m/s. From above data, and taking $\gamma=0.0101 cm^2/s$, Renault values of inlet and suction part of the injector can be calculated, and they are $Re_1= v_1d_1/\gamma =(2.46*100)*(30/10)/0.0101= 73069 > 2000$, $Re_2= v_2d_2/\gamma = (0.742*100)*(18/10)/0.0101=13224 > 2000$, respectively. Thus, the water flow status of these two parts was considered as turbulence. So the flow status of the sample injector was regarded as unconstringent turbulent flow, and the $\kappa-\varepsilon$ turbulent flow model was used then (Li Baijun *et al*, 2001). By calculating the turbulent flow intensity, $I=0.16Re^{-0.125}$, the turbulent flow intensity of inlet and suction part of the injector was I_1=3.9% and I_2=4.8%, respectively.

Besides the Continuity Equation (i.e., Mass Conservation Equation) and Navier-Stokes Equation (i.e., Monmentum Conservation Equation), turbulent flow equation was also needed as one of the basic control equations. In the Descartes Coordinate system, the forms of the basic control equations were list as follows (Wang Fujun, 2004; Li Yongxin, et al, 2004):

Continuity Equation

$$\frac{\partial u}{\partial x}+\frac{\partial v}{\partial y}+\frac{\partial w}{\partial z}=0 \tag{1}$$

Navier-Stokes Equation

$$\frac{\partial(\rho u)}{\partial t} + \nabla \cdot (\rho u \mathrm{u}) = -\frac{\partial p}{\partial x} + \mu \nabla^2 u + F_x$$
$$\frac{\partial(\rho v)}{\partial t} + \nabla \cdot (\rho v \mathrm{u}) = -\frac{\partial p}{\partial y} + \mu \nabla^2 v + F_y \quad (2)$$
$$\frac{\partial(\rho w)}{\partial t} + \nabla \cdot (\rho w \mathrm{u}) = -\frac{\partial p}{\partial z} + \mu \nabla^2 w + F_z$$

Where, t is time; $\boldsymbol{u}$ is flow rate vector, and u, v, w are the tree sub values of $\boldsymbol{u}$ in directions of x, y, z; ρ and μ are the water density and dynamical viscosity coefficient, respectively; P is the water pressure on the micro flow unit; F_x, F_y, F_z are the mass forces of the micro flow unit in the directions of x, y, z. If the mass force was just gravity, and the direction of z was vertical upward, in this case, we can get that $F_x = 0$, $F_y = 0$, $F_z = -\rho g$.

In standard k-ε turbulent flow model, the expression of turbulent flow kinetic energy k and dissipation ratioε was listed as follows (Wang Fujun, 2004; Li Yongxin et al, 2004):

$$k = \frac{1}{2}(\overline{u'^2} + \overline{v'^2} + \overline{w'^2}) \qquad \varepsilon = \frac{\mu}{\rho}\overline{(\frac{\partial u_i'}{\partial x_k})(\frac{\partial u_i'}{\partial x_k})}$$

Where, u', v', w' are the fluctuation values of flow rate in the three directions of x, y, z.

In the standard k-ε model, k andε are two unkown variables. The corresponding control equation is (Chen Zuobing *et al*, 2005):

$$\frac{\partial}{\partial t}(\rho k) + \frac{\partial}{\partial x_i}(\rho k u_i) = \frac{\partial}{\partial x_j}[(\mu + \frac{\mu_t}{\sigma_k})\frac{\partial k}{\partial x_j}] + G_k - \rho\varepsilon \quad (3)$$

$$\frac{\partial}{\partial t}(\rho \varepsilon) + \frac{\partial}{\partial x_i}(\rho \varepsilon u_i) = \frac{\partial}{\partial x_j}[(\mu + \frac{\mu_t}{\sigma_\varepsilon})\frac{\partial \varepsilon}{\partial x_j}] + \frac{C_{1\varepsilon}\varepsilon}{k}G_k - C_{2\varepsilon}\rho\frac{\varepsilon^2}{k} \quad (4)$$

Where, μ_t is turbulent viscosity efficient; G_k is the turbulent kinetic energy k produced by average velocity gradient; σ_k, σ_ε, $C_{1\varepsilon}$, $C_{2\varepsilon}$ are constants. The meaning of other variables are the same as previous mentioned. The detailed expression of the equation can be seen in literature of Wang Fujun (2004).

By numerical simulation to the flow field of the sample Venturi injector, its performance such as suction capacity can be obtained.

2.2.2 Validation of the CFD Models

The simulation was carried out according to the above model, the structure size of the sample Venturi injector as at Figure 1, and its running parameters. The simulated suction capacity was 0.720 m^3/h, while the actual measured suction capacity was q=0.68m^3/h. The error between these two values was 5.88%, which showed that the model and its solution approach are reasonable. This model can be applied to simulation analyses of practical Venturi injector performance.

3. RESULTS AND DISCUSSION

3.1 Effects of Key Parameters of Venturi Injector on Suction Capacity

3.1.1 Effects of throat diameter and slot diameter on suction capacity

In this case, throat diameter and slot diameter were changed, while other structural parameters and inlet pressure kept unchanged. The simulation results are showed at Figure 2, which show that the correlation between the suction capacity and slot diameter under different throat diameters are quite well. When fixed slot diameter D_a, the suction capacity of the sample injector decreases as the throat diameter D is increased; While fixed throat diameter D, the suction capacity of the sample injector increases as the slot diameter D_a is increased. The effect of throat diameter D on suction capacity is much more sensitive than that of slot diameter D on it.

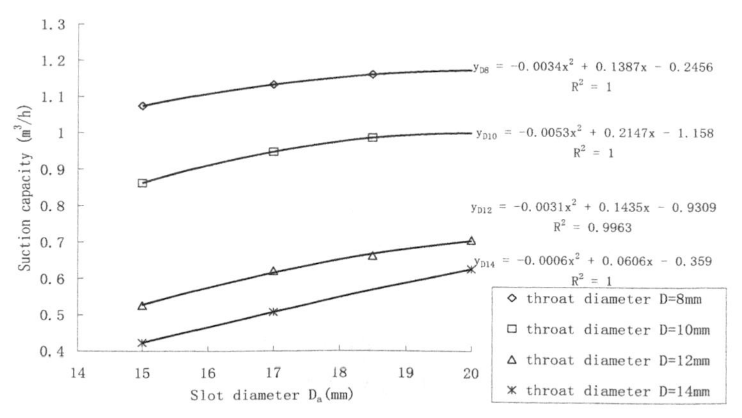

Fig. 2 Suction capacity changing curves under different slot diameters

3.1.2 Effects of throat length and slot diameter on suction capacity

In this case, throat length and slot diameter were changed, while other structural parameters and inlet pressure kept unchanged. The simulation results are showed at Table 2 and Figure 3.

The results from Figure 3 show that when fixed slot diameter D_a, the suction capacity of the sample injector decreases as the throat length L is increased; While fixed throat length L, the suction capacity of the sample injector increases as the slot diameter D_a is increased. The effect of slot diameter D_a on suction capacity is much more sensitive than that of throat length L on it.

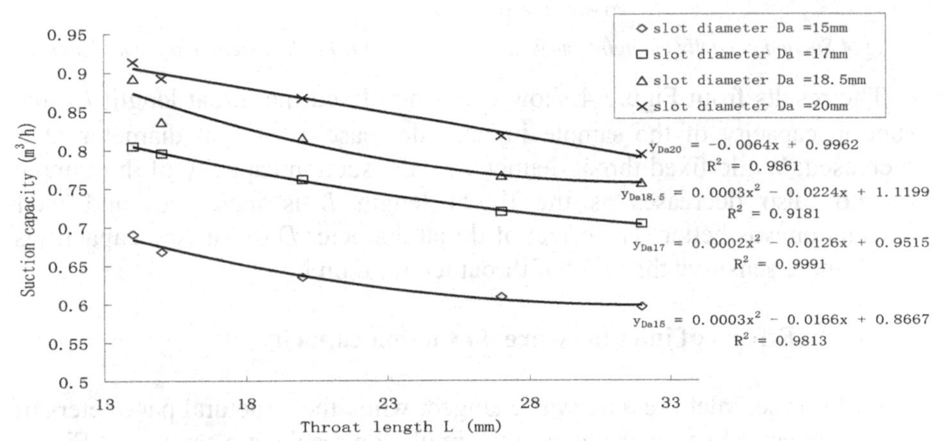

Fig. 3 Suction capacity changing curves under different slot diameters and throat length

3.1.3 Effects of throat diameter and throat length on suction capacity

In this case, throat length L and throat diameter D were changed, while other structural parameters and inlet pressure kept unchanged. The simulation results are showed at Figure 4.

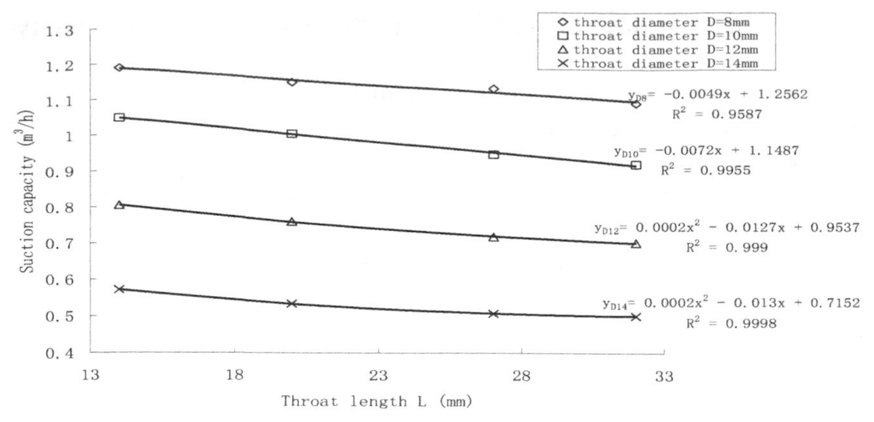

Fig. 4 Suction capacity changing curve under different throat diameters and throat lengths

The results from Figure 4 show that when fixed the throat length L, the suction capacity of the sample injector decreases as throat diameter D is increased; While fixed throat diameter D, the suction capacity of the sample injector also decreases as the throat length L is increased, and their correlation was better. The effect of throat diameter D on suction capacity is much more sensitive than that of throat length L on it.

3.1.4 Effects of inlet pressure on suction capacity

In this case, inlet pressure was changed, while the structural parameters of sample injector kept unchanged. The simulation results are showed at Figure 5, which show that the suction capacity of Venturi injector increases as its inlet pressure P increases.

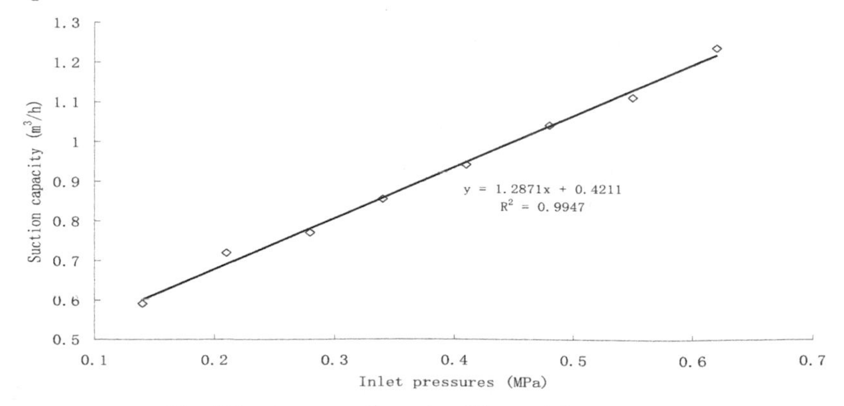

Fig. 5 Suction capacity under different inlet pressures

As mentioned above, the main sensitive factors to affect suction capacity of Venturi injector are throat diameter D, slot diameter D_a, and throat length L. The order of its sensitivity from strong to weak is as follows: first throat diameter D, then slot diameter D_a, and last throat length L.

3.2 Full Factors Experiment for Suction Capacity of Venturi Injector with Main Structural Parameters

In order to analyze the regulation of the effects of different factors on suction capacity of Venturi injector, Full Factors Simulation Experiment were carried out based on previous simulation. In this experiment period, slot diameter D_a was selected as 15mm, 17mm and 18.5mm; throat diameter D was selected as 8mm, 10mm and 12mm; while throat length L was selected as 14mm, 20mm and 27mm. The layout of experiment and the results were shown as Table 1.

Table 1 Experimental results with all 3 factors by 3 levels

Treatments	Q_1 (m^3/h)	q (m^3/h)	Q_2 (m^3/h)
D_a=15, D=8, L=14	5.709	1.137	6.846
D_a=15, D=8, L=20	5.709	1.097	6.806
D_a=15, D=8, L=27	5.709	1.073	6.782
D_a=15, D=10, L=14	5.831	0.951	6.782
D_a=15, D=10, L=20	5.831	0.908	6.739
D_a=15, D=10, L=27	5.831	0.862	6.693
D_a=15, D=12, L=14	5.902	0.691	6.593
D_a=15, D=12, L=20	5.902	0.635	6.537
D_a=15, D=12, L=27	5.902	0.609	6.511
D_a=17, D=8, L=14	5.709	1.189	6.898
D_a=17, D=8, L=20	5.709	1.152	6.861
D_a=17, D=8, L=27	5.709	1.135	6.844
D_a=17, D=10, L=14	5.831	1.049	6.880
D_a=17, D=10, L=20	5.831	1.007	6.838
D_a=17, D=10, L=27	5.831	0.950	6.781
D_a=17, D=12, L=14	5.902	0.805	6.707
D_a=17, D=12, L=20	5.902	0.762	6.664
D_a=17, D=12, L=27	5.902	0.720	6.622
D_a=18.5, D=8, L=14	5.709	1.203	6.912
D_a=18.5, D=8, L=20	5.709	1.178	6.887
D_a=18.5, D=8, L=27	5.709	1.163	6.872
D_a=18.5, D=10, L=14	5.831	1.076	6.907
D_a=18.5, D=10, L=20	5.831	1.036	6.867
D_a=18.5, D=10, L=27	5.831	0.988	6.819
D_a=18.5, D=12, L=14	5.902	0.893	6.795
D_a=18.5, D=12, L=20	5.902	0.816	6.718
D_a=18.5, D=12, L=27	5.902	0.767	6.669

The results from Table 1 show that, The suction capacity under all treatments with throat diameter D=8mm was much better than that under other treatments, and this verified the conclusion that throat diameter D is the most sensitive factor affecting the suction capacity of Venturi injector. The optimal combination of the structural parameters was selected asfollows:

throat diameter D=8mm, slot diameter D_a =18.5mm, and throat length L=14mm. In this case, the suction capacity of the Venturi injector was q=1.203m^3/h.

4. CONCLUSION

The following conclusions can be drawn by numerical simulation analysis to the effect of 30mm Venturi injector structural parameters on its suction capacity: 1) As keeping other parameters unchanged, the suction capacity of Venturi injector increases with the decrease of throat diameter, or the decrease of throat length, or increase of the slot diameter, or increase of the inlet water pressure. It also decreases with the slot position to the right; 2) The results of Full Factors Experiments show that the most obvious affected factor to the suction capacity is the throat diameter; 3) The optimal structural parameters of Venturi injector with inlet diameter of 30mm are D=8mm, D_a =18.5mm, L=14mm. In this case, its suction capacity is q=1. 203m^3/h.

REFERENCES

Bhattarai S P, Huber S, Midmore D J. Aerated subsurface irrigation water gives growth and yield benefits to Zucchini, vegetable soybean and cotton in heavy clay soils. Annals of Applied Biology, 2004,144 (3): 285-298

Bracy R P, Parish R L, Rosendale RM. Fertigation uniformity affected by injector type. Hort Technology, 2003, 13 (1): 103-105

Chen Zuobing, Dou Haijian, Chen Siwei, et al. Numerical research on flow field of Venturi Tube. China Cement, 2005, (4): 61-63 (in Chinese)

Harris G A. Sub-surface drip irrigation – Advantages and Limitations, DPI&F Note, Brisbane, 2005

He Shaohua, Wen Zhuqing, Lou Tao. Experiment Design and Data Process. Beijing: National Defence Scientech University Press, 2002: 67-69 (in Chinese)

Jiusheng Li, R.E.Yoder, J.Zhang. Simulation of nitrate distribution under drip irrigation using artificial neural networks. Irrigation Science, 2004, 23: 29-37

Johnson A W, Young J R, Thereadgill E D, et al. Chemigation for crop production management. Plant Disease. 1986, 70 (11) : 998-1004

Li Baijun, Mao Hanping, Li Kai. Study on the parallel connected Venturi injectors and their parameters selection. Drainage and Irrigation Machinery. 2001, 19(1): 42-45 (in Chinese)

Li jiusheng, Zhang jianjun, Xue Kezong. Drip Irrigation Principals for Fertigation and Its Applications. Beijing: Beijing Agricultual Scientech Press, 2003: 10-24 (in Chinese)

Li Yongxin, Li Guangyong, Qiu Xiangyu, et al. Numerical Simulation to the inner flow field of labyrinth type drip with CFD models. 6th National Micro Irrigation Congress. 2004: 275-282 (in Chinese)

Manzano Juárez J, Palau Salvador G. Hydraulic Modeling of Venturi Injector by Means of CFD. 2005 ASAE Annual International Meeting, paper No. 052070, Tampa Convention Center, Tampa, Florida, U.S.A., 17 - 20 July 2005

Sha Yi, Hou Sujuan. Experimental study on parallel connected Venturi injectors. Irrigation and Drainage Machinery, 1995, (2): 37-39 (in Chinese)

Shen Xuemin, Feng Jun, Zhang Xuejun. Performance research on Venturi injector in sprinkler irrigation system. Water Saving Irrigation, 2001, (1): 20-21 (in Chinese)

Shen Xuemin. Introduction to the 100PS-1 Venturi Injector in Sprinkler Irrigation System. Water Saving Irrigation, 2000, (11): 14-15 (in Chinese)

Wang Fujun. Analyze to CFD(Computational Fluid Dynamics). Beijing: Tsinghua University Press, 2004: 1-259 (in Chinese)

Yuan Z, Choi C Y, Waller P M, Colaizzi P. Effects of liquid temperature and viscosity on Venturi injectors. Transactions of the ASAE, 2000, 43 (6): 1441-1447

Yuan Zhifa, Zhou Jingyu. Experiment Design and Analyze. Beijing: Higher Education Press, 2000: 292-296 (in Chinese)

Zhao Jingcheng, Ren Xiaoli. Sprinkler Engineering Technology. Beijing: China Water Conservation and Hydraulic Electricity Power Publish Press, 1999(in Chinese)

APPLICATION OF STATISTIC EXPERIMENTAL DESIGN TO ASSESS THE EFFECT OF GAMMA-IRRADIATION PRE-TREATMENT ON THE DRYING CHARACTERISTICS AND QUALITIES OF WHEAT

Yong Yu[1,*] , Jun Wang[1]
[1] *Department of Biosystems Engineering, Zhejiang University, Hangzhou, 31002*
[*] *Corresponding author, Address: Department of Biosystems Engineering, Zhejiang University, Hangzhou, 31002, Zhejiang Province, P. R. China, Tel: +86-571-86971053, Email: yyu_zju@yahoo.com.cn*

Abstract: Wheat, pretreated by 60Co gamma irradiation, was dried by hot-air with irradiation dosage 0-3 kGy, drying temperature 40-60 ℃, and initial moisture contents 19-25% (drying basis). The drying characteristics and dried qualities of wheat were evaluated based on drying time, average dehydration rate, wet gluten content (WGC), moisture content of wet gluten (MCWG)and titratable acidity (TA). A quadratic rotation-orthogonal composite experimental design, with three variables (at five levels) and five response functions, and analysis method were employed to study the effect of three variables on the individual response functions. The five response functions (drying time, average dehydration rate, WGC, MCWG, TA) correlated with these variables by second order polynomials consisting of linear, quadratic and interaction terms. A high correlation coefficient indicated the suitability of the second order polynomial to predict these response functions. The linear, interaction and quadratic effects of three variables on the five response functions were all studied.

Keywords: statistic experimental design, gamma irradiation, drying, quality, wheat

---

Please use the following format when citing this chapter:

Yu, Y. and Wang, J., 2009, in IFIP International Federation for Information Processing, Volume 294, *Computer and Computing Technologies in Agriculture II, Volume 2*, eds. D. Li, Z. Chunjiang, (Boston: Springer), pp. 817–829.

1. INTRODUCTION

wheat was one of the major cereals and harvested usually at high moisture content to minimize shattering losses. Moisture content was one of the most important factors affecting the quality of wheat during storage and subsequent handling. The moisture content of wheat at harvesting was as high as 20-35% (drying basis). Drying was necessary to prevent quality during storage. The current drying method used widely was hot-air drying. The drawback of this method was a long drying time and the controlling of drying conditions (e.g. increasing drying temperature) will result in the quality of dried wheat. It is necessary that a new method be applied for not only high dehydration rate but also quality of wheat.

Technology of gamma irradiation had long been employed to decontaminate and to extend the shelf life of food. 1-3 kGy dose was efficient for bacterial decontamination and elimination of potential pathogens and improve the keeping quality of fruit, vegetable (Bidawid, et al., 2000), condiment (Kamat, et al., 2003), and beef (Formanek, et al., 2003); a low dose of 50 Gy and 150 Gy could completely prevent and control the apple maggot (Hallman, 2004) and Hawaii's fruit flies (Follett, 2004), etc.

Irradiation, as a pre-treatment method before drying, was studied about its influence on the drying characteristics of apple and potato slices (Wang, et al., 2002). The interior tissue structure of apple would be changed and injured by ^{60}Co gamma ray irradiation, which would bring about different drying characteristics and affect dehydration rate and qualities of dried products. L-value, Vc content, and rehydration ratio of dried samples were greatly affected by irradiation dose (Wang, et al., 2003). Dehydration rate was increased and Vc content was decreased with increasing dose, and L-values of dried product after irradiation was greater than under non-irradiation.

The qualities would change after dried rough rice and wheat was irradiated (Wu, et al., 2002). Apparent amylose content (ACC) was reduced and gel consistency (GC) was improved with increasing dose (0-12 kGy). Four major parameters of RVA profile, peak viscosity, hot pasting viscosity, cool pasting viscosity, and setback viscosity, were considerably decreased with increasing dose. The viscosity of rice irradiated were reduced by 25%, 50%, 65%, 72% and 74% for Local Black barley cultivar, while, in Shoaa cultivar the reductions were 15%, 30%, 52%, 69% and 67% at 10, 50, 100, 150 and 200 kGy, respectively (Al-Kaisey, et al., 2002).

It can be found that, irradiation, as a pre-treatment method, can increase dehydration rate of apple and potato and change the qualities of wheat that was irradiated after drying. It was necessary to studies the effect of gamma radiation on drying characteristics and quality of wheat that was irradiated before drying, and to find a new wheat drying method. The previous researches showed that the drying time, average dehydration rate, WGC,

MCWG and TA of wheat were signifcant effected by gamma irradiation (Yu et al., 2005; 2006; 2007).

In this research, a quadratic rotation-orthogonal composite experimental design, with three variables (at five levels each) and five response functions (drying time, average dehydration rate, WGC, MCWG, TA), and analysis method were employ to study the effect of the three variables on the individual response functions.

2. MATERIALS AND METHODS

2.1 Wheat

Wheat (Zhenong 1) harvested in June, 2004, from the experimental farm of Agronomy, Zhejiang University, was used for this experiment immediately after harvest. The initial moisture content was determined by drying five samples of wheat at 105 ℃ in a constant temperature oven till weight of the samples became constant (GB/5497-85, National Standard of China), and 25 % (dry basis).

Before irradiation and drying experiments, samples were air-dried in natural condition to different initial moisture content respectively (Table 1). The sample mass for each drying test was 250 g. Samples of wheat were stored in a refrigerator at 5 ℃ and warmed to room temperature before each tests.

Table 1. Quadratic rotation-orthogonal composite experimental design in coded and actual level of variables

Exp. No.	Irradiation dosage		Drying temperature		Initial moisture content	
	Coded level x_1	Actual level X_1 (kGy)	Coded level x_2	Actual level X_2 (%, d.b.)	Coded level x_3	Actual level X_3 (℃)
1	1	2.4	1	56	1	23.8
2	1	2.4	1	56	-1	20.2
3	1	2.4	-1	44	1	23.8
4	1	2.4	-1	44	-1	20.2
5	-1	0.6	1	56	1	23.8
6	-1	0.6	1	56	-1	20.2
7	-1	0.6	-1	44	1	23.8
8	-1	0.6	-1	44	-1	20.2
9	1.682	3	0	50	0	22
10	-1.682	0	0	50	0	22
11	0	1.5	1.682	60	0	22
12	0	1.5	-1.682	40	0	22
13	0	1.5	0	50	1.682	25
14	0	1.5	0	50	-1.682	19
15*	0	1.5	0	50	0	22

* Experiment 15 was repeated nine times.

2.2 Experimental procedure

The wheat was irradiated by ^{60}Co γ-ray in the Institute of Nuclear-agriculture Sciences, Zhejiang University. The doses were controlled respectively (Table 1), and the dose rate was 1 kGy/h.

Irradiated samples were symmetrically placed in a sifter. Drying experiments were conducted at the above different doses, air temperatures, and initial moisture contents (Table 1). Air velocity was kept at 0.5±0.1 m/s. The samples were dried until it reached a final moisture content of 14.5 ±0.1 % (dry base), which represented the safe moisture value for grain storage. All the drying tests were replicated three times.

The experimental setup consisted of a drying chamber. For each experiment, the setup was allowed to run with a dummy sample till the desired drying conditions attained steady state. The dummy sample container was then replaced quickly with the actual sample container. The sample container was periodically removed and was weighed to generate the moisture loss data on an electric-balance having a weighing capacity of 500 g with an accuracy ± 0. 01 g. While weighing the sample, a dummy sample container was placed in the drying chamber so as not to disturb the state drying conditions. Thus the moisture content at any stage during drying was determined once the initial moisture content was known, and the accuracy of moisture content was ± 0.1 % based on the accuracy of electronic balance.

After drying experiment, wheat samples was milled into flour by milling with a Quadrumat Junior laboratory mill (Brabender OHG, Duisberg, Germany). After separating wheat bran, wheat flour extraction rate ranged from 70% to 80%. The wheat flour samples were used for wet gluten content (WGC), moisture content of wet gluten (MCWG) and titratable acidity (TA).

2.3 Response functions

2.3.1 Drying time

Drying time was the total time for the samples drying from its initial moisture content to the same final moisture content (14.5 ±0.02 %, drying base).

2.3.2 Average dehydration rate

Average dehydration rate was calculated with the value of initial moisture content (%) minus final moisture content (%) and divided by drying time (h). To increase the average dehydration rate was also one of the objectives of improving drying method.

2.3.3 Gluten content

Wet gluten yield were determined by the machine washing Method (GB/5506-85, National Standard of China). Paste was forming with 25.00±0.01 g flour (W) mixed 12.5±0.1 mL of water, followed by washing for 10 min with water at a flow rate of 50–60 mL/min on a special 88-μm sieve using a Perten Glutomatic Gluten Index machine (Perten Instruments AB, S-141 05 Huddinge, Sweden) to wash away any remaining starch or coarse particles. Afterwards, the wet gluten piece was centrifuged at 6000 rpm for 1 min on a special 600-μm metallic sieve using a Perten Centrifuge 2015 machine (Perten Instruments AB, S-141 05 Huddinge, Sweden) and the wet gluten yield was obtained (W1). The wet gluten samples was cut into twelve pieces, placed in a tin and dried for 30 min at 155 ℃ using a special Perten Glutork 2020 dryer. After cooling in a desiccator, weight of dry gluten (W2) was recorded. The gluten indexes of wet gluten content (WGC) (Eq. 1), dry gluten content (DGC) (Eq. 2) and moisture content of wet gluten (MCWG) (Eq. 3) was obtained. Tests were carried out in triplicates for each sample.

$$WGC = \frac{W_1}{W} \times 100\% \tag{1}$$

$$MCWG = \frac{W_1 - W_2}{W_2} \times 100\% \tag{2}$$

$$DGC = \frac{W_2}{W} \times 100\% \tag{3}$$

2.3.4 Titratable acidity

The titratable acidity was expressed as sodium hydroxide required to neutralize the acids in a 100.00±0.01 g sample, using phenolphthalein as an indicator (AOAC, 1990) with the unit mg NaOH/100g. Triplicates were carried out for each sample.

2.4 Experimental design and analysis of data

The method employed was a three-variable (five levels of each variable), quadratic rotation-orthogonal composite experimental design with nine replications at the centre points (0, 0, 0), in coded levels of variables (-1.682, -1, 0, 1, 1.682) (Akhnazarova, et al., 1982). In three independent variables in controlling were: X1 (irradiation dosage), X2 (drying temperature), X3 (initial moisture content).

The experimental design in the actual (X) and coded (x) levels of variables was shown in Table 1. The response function y, i.e., drying time, average dehydration rate, WGC, MCWG, TA was approximated by a second degree polynomial (Eq. 4) with linear, quadratic and interaction effects (in coded level of variables) using the method of least squares (Little, et al., 1978).

$$y_k = b_0 + \sum_{i=1}^{n} b_i x_i + \sum_{i=1}^{n} \sum_{\substack{j=1 \\ i \le j}}^{n} b_{ij} x_i x_j \tag{4}$$

The number of variables was denoted by n, and i, j, k were integers. The coefficients of the polynomials were represented by b0, bi, bij; when i < j, bij represents the interaction effects of the variables xi and xj. Analysis of variance (ANOVA) was conducted using the SAS software (SAS, 1999). Response surface graphs were obtained from the regression equations in actual level of variables, keeping the response function on the Z axes with X and Y axes representing the two independent variables while keeping the other variable at its optimum point.

3. RESULTS AND DISCUSSION

The experimental results on the effect of the three variables (irradiation dosage, initial moisture content and drying temperature) on the five response functions or targeted parameters (drying time, average dehydration rate, WGC, MCWG, TA) were detected and the condensed analysis of variances (ANOVA, in coded level of variables) table was shown in Table 2 for all the five response functions. The coefficients of the second order polynomial (in actual level of variables) were cited in Table 3 for case in using them to predict the response functions. The response surfaccs, with X and Y axes representing two of the most influential variables while keeping another variable constant at their centre point (Fig. 1-5) (other response surface graphs were omitted), were presented to aid in visualizing the effect of the variables.

3.1 Drying time

A high correlation coefficient ($r=0.971$, $p \leqslant 0.01$) indicated the suitability of the second order polynomial to predict the drying time (Y1) (Table 2). Among the variables, total lincar effect ($p \leqslant 0.01$) dominated the total quadratic effect ($p \leqslant 0.01$), and the total interaction effect ($p \leqslant 0.05$). Similar those conclusions could be draw from the other functions except average dehydration rate. Of the individual variables, drying temperature had the maximum negative effect ($p \leqslant 0.01$) on drying time followed by the positive effect of initial moisture content ($p \leqslant 0.01$) and the negative effect of irradiation dosage ($p \leqslant 0.01$). On the contrary, the quadratic effect of drying temperature was significant ($p \leqslant 0.01$) having a positive effect. Among the various interactions, the effect of drying temperature × initial moisture content was marginally significant ($p \leqslant 0.05$), having negative effect.

Table 2. Experimental results for the response functions

Source of variations	Y_1		Y_2		Y_3		Y_4		Y_5	
	Coefficient of polynomial	F-Value	Coefficient of polynomial	F-Value	Coefficient of polynomial	F-Value	Coefficient of polynomial	F-Value	Coefficient of polynomial	F-Value
Constant	**170.526**	**-**	**2.995**	**-**	**26.171**	**-**	**154.246**	**-**	**17.646**	**-**
x_1	**-22.583**	**20.763*****	**0.361**	**27.4448*****	**-1.526**	**169.391*****	**-16.241**	**180.350*****	**1.470**	**575.925*****
x_2	**-68.879**	**193.164*****	**1.097**	**253.361*****	**0.191**	**2.655 NS**	**2.787**	**5.312****	**0.233**	**14.491*****
x_3	**28.058**	**32.052*****	**0.257**	**13.906*****	**-0.328**	**7.834****	**-3.946**	**10.646*****	**0.486**	**62.930*****
x_1^2	**1.122**	**0.050 NS**	**0.057**	**0.767 NS**	**0.163**	**2.039 NS**	**2.094**	**3.210***	**-0.096**	**2.577 NS**
x_2^2	**18.800**	**16.792*****	**0.098**	**2.306 NS**	**0.994**	**83.252*****	**10.173**	**81.852*****	**-0.527**	**85.618*****
x_3^2	**-5.065**	**1.282 NS**	**0.036**	**0.298 NS**	**0.181**	**2.534 NS**	**2.235**	**3.681***	**-0.103**	**2.995NS**
x_1x_2	**8.125**	**1.575 NS**	**0.011**	**0.016 NS**	**-0.025**	**0.027 NS**	**-0.025**	**0.000 NS**	**0.000**	**0.000 NS**
x_1x_3	**4.375**	**0.457 NS**	**-0.068**	**0.571 NS**	**0.025**	**0.027 NS**	**0.025**	**0.000 NS**	**0.003**	**0.001 NS**
x_2x_3	**-15.625**	**5.823****	**0.254**	**7.931 ****	**0.000**	**0.000NS**	**-0.025**	**0.000 NS**	**0.003**	**0.001 NS**
TLE	**-**	**245.978*****	**-**	**294.711*****	**-**	**179.880*****	**-**	**196.308*****	**-**	**653.346*****
TQE	**-**	**18.124*****	**-**	**3.370 NS**	**-**	**87.825*****	**-**	**88.743*****	**-**	**91.190*****
TIE	**-**	**7.854****	**-**	**8.518****	**-**	**0.054 NS**	**-**	**0.001 NS**	**-**	**0.003 NS**
r	**0.971*****	**-**	**0.963*****	**-**	**0.983*****	**-**	**0.978*****	**-**	**0.939*****	**-**

[a]*Variables: x_1- irradiation dosage, x_2- drying temperature, x_3- initial moisture content. TLE: total linear effect; TQE: total quadratic effect; TIE: total interaction effect; NS: non-significant at $p > 0.10$. *Significant at $p \leq 0.10$. ** Significant at $p \leq 0.05$. ***Significant at $p \leq 0.01$; Response functions: Y_1 = drying time, Y_2 = average dehydration rate, Y_3 = wet gluten content, Y_4 = moisture content of wet gluten, Y_5 = titratable acidity.*

The response surface of effect of drying temperature and initial moisture content on drying time was shown in Fig. 1 based on the equation in Table 3. It could be found that drying time was reduced with the increasing drying temperature and decreasing initial moisture content. The effect of drying temperature was more significant than initial moisture content.

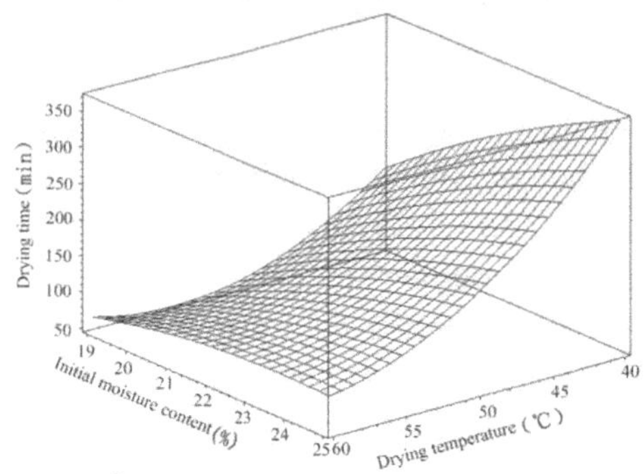

Fig. 1:. Drying time as a function of drying temperature and initial moisture content (irradiation dosage at 1.5 kGy)

Table 3. Regression equations for the response functions in the actual level of variables[a]

$Y_1 =$	$-398.2078 - 163.8928X_1 - 34.1298X_2 + 152.6592X_3 + 1.3854X_1^2 + 0.5222X_2^2 - 1.5633X_3^2 + 1.5046X_1X_2 + 2.7006X_1X_3 - 1.4468\ X_2X_3$
$Y_2 =$	$27.0300 + 1.0089X_1 - 0.6084X_2 - 1.4569X_3 + 0.0702X_1^2 + 0.0027X_2^2 + 0.0111X_3^2 + 0.0021X_1X_2 - 0.0042X_1X_3 + 0.0235\ X_2X_3$
$Y_3 =$	$119.4950 - 2.4084X_1 - 2.5562X_2 - 2.6635X_3 + 0.2016X_1^2 + 0.0276X_2^2 + 0.0559X_3^2 - 0.0046X_1X_2 + 0.0154\ X_1X_3$
$Y_4 =$	$1250.0800 - 25.9078X_1 - 27.7346X_2 - 32.4554X_3 + 2.5849X_1^2 + 0.2826X_2^2 + 0.6899X_3^2 - 0.0046X_1X_2 + 0.0154X_1X_3 - 0.0023\ X_2X_3$
$Y_5 =$	$-41.8064 + 1.6476X_1 + 1.4513X_2 + 1.5244X_3 - 0.1179X_1^2 - 0.0146X_2^2 - 0.0317X_3^2 + 0.0154X_1X_3 + 0.0023\ X_2X_3$

[a]*Variables: X_1 = irradiation dosage (kGy), X_2 = drying temperature (℃), X_3 = initial moisture content (%); Response functions: Y_1 = drying time (min), Y_2 = average dehydration rate (%/h), Y_3 = wet gluten content (%), Y_4 = moisture content of wet gluten (%), Y_5 = titratable acidity (mg NaOH/100g).*

3.2 Average dehydration rate

The ANOVA table (Table 2) for average dehydration rate (Y2) showed a high correlation coefficient ($r=0.963$, $p\leqslant 0.01$) which indicates the suitability of the second order polynomial to predict average dehydration rate. Among the variables, total linear effect ($p\leqslant 0.01$) dominated the total interaction effect ($p \leqslant 0.05$), and the total quadratic effect ($p > 0.1$). Among the individual variables, drying temperature had the most positive effect ($p\leqslant 0.01$) on average dehydration rate followed by the positive effect of irradiation dosage ($p\leqslant 0.01$) and the positive effect of initial moisture content ($p\leqslant 0.01$). The quadratic effect of all the variables had no significant effect ($p>0.1$) on average dehydration rate. Among their interactions, the effect of drying temperature × initial moisture content was marginally significant ($p\leqslant 0.05$), having positive effect.

The response surface of effect of drying temperature and irradiation dosage on average dehydration rate was presented in Fig. 2 based on the equation in Table 3. It was clearly that, as irradiation dosage and drying temperature increased, average dehydration rate was increased. The effect of drying temperature was more significant than irradiation dosage.

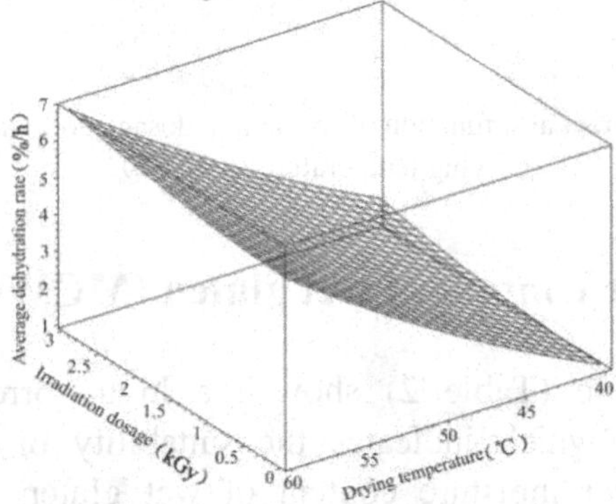

Fig. 2: Average dehydration rate as a function of drying temperature and irradiation dosage (initial moisture content at 22 %)

3.3 Wet gluten content (WGC)

A high correlation coefficient ($r=0.983$, $p\leq0.01$) which indicates the suitability of the second order polynomial to predict wet gluten content (Y3) was shown by the ANOVA table (Table 2). All the various interactions had no significant effects ($p>0.1$) on wet gluten content. Irradiation dosage possessed the maximum influence (negative effect) ($p\leq0.01$) on wet gluten content followed by initial moisure content ($p\leq0.05$). The quadratic effect of drying temperature was significant ($p\leq0.01$), having a positive effect and other quadratic effects had no significant effects ($p>0.1$).

The response surface of effect of irradiation dosage and initial moisture content on WGC was shown in Fig. 3 based on the equation in Table 3. It could be found that WGC was reduced with the increasing irradiation dosage and reduced followed by increased with the increasing initial moisture content. The effect of irradiation dosage was more significant than initial moisture content.

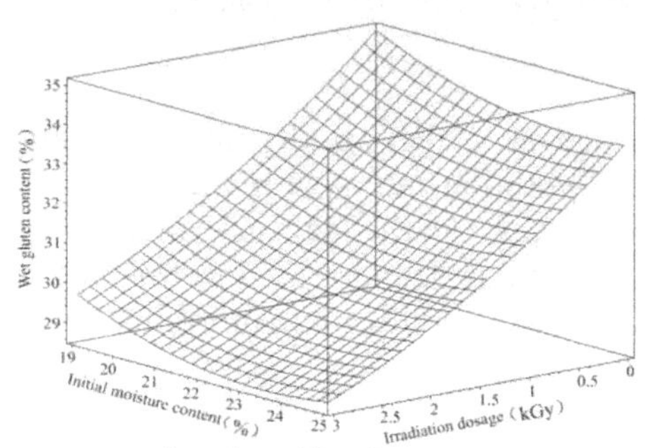

Fig. 3: Wet gluten content as a function of irradiation dosage and initial moisture content (drying temperature at 50 ℃)

3.4 Moisture content of wet gluten (MCWG)

The ANOVA table (Table 2) showed a high correlation coefficient ($r=0.978$, $p \leqslant 0.01$) which indicates the suitability of the second order polynomial to predict moisture content of wet gluten (Y4). Among the individual variables, irradiation dosage had the maximum negative effect ($p \leqslant 0.01$) followed by the negative effect of initial moisture content ($p \leqslant 0.01$) and the positive effect of drying temperature ($p \leqslant 0.05$) on MCWG. Among the quadratic effect of variables, drying temperature had the maximum positive effect ($p \leqslant 0.01$) followed by the positive effect of initial moisture content ($p \leqslant 0.10$) and the positive effect of irradiation dosage ($p \leqslant 0.10$) on MCWG. And all the various interactions had no significant effects ($p > 0.1$) on MCWG.

The response surface of effect of irradiation dosage and initial moisture content on MCWG was shown in Fig. 4 based on the equation in Table 3. It could be found that MCWG was decreased with the increasing irradiation dosage and decreased followed by increased with the increasing initial moisture content. The effect of irradiation dosage was more significant than initial moisture content.

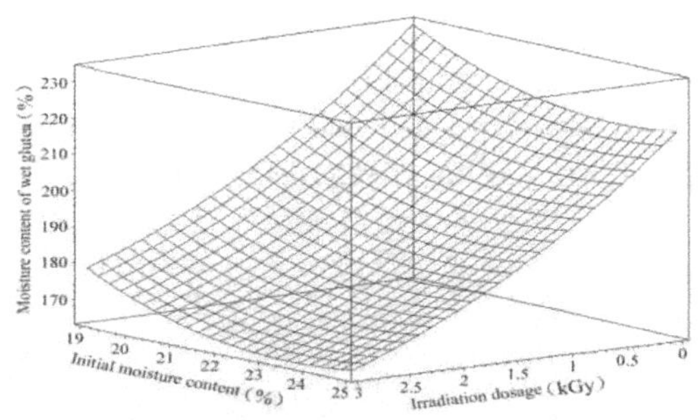

Fig. 4: Moisture content of wet gluten as a function of irradiation dosage and initial moisture content (drying temperature at 50 ℃)

3.5 Titratable acidity (TA)

A high correlation coefficient (r=0.971, $p \leqslant 0.01$) which indicates the suitability of the second order polynomial to predict titratable acidity (Y5) was shown by the ANOVA table (Table 2). Among the individual variables, irradiation dosage had the maximum positive effect ($p \leqslant 0.01$) followed by the positive effect of initial moisture content ($p \leqslant 0.01$) and the positive effect of drying temperature ($p \leqslant 0.01$) on TA. The quadratic effect of drying temperature had the maximum positive effect ($p \leqslant 0.01$) and other quadraic effects and all the various interactions had no significant effects ($p > 0.1$) on TA.

The response surface of effect of irradiation dosage and initial moisture content on TA was presented in Fig. 5 based on the equation in Table 3. It could be found that TA was increased with the increasing irradiation dosage and initial moisture content. The effect of irradiation dosage was more significant than drying temperature.

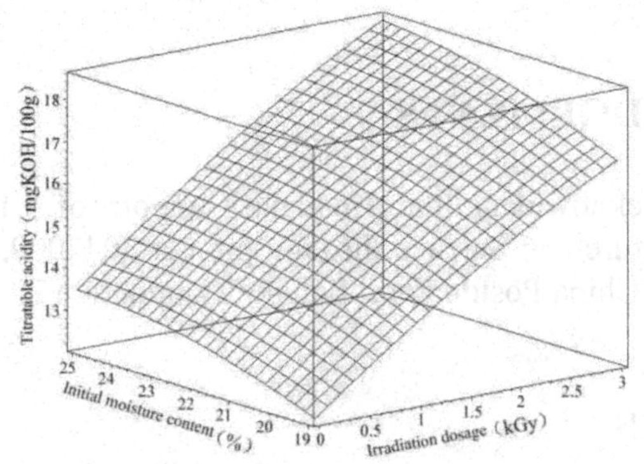

Fig. 5: titratable acidity as a function of irradiation dosage and initial moisture content (drying temperature at 50 ℃)

3.6 Effect of variables

Among the variables, irradiation dosage had most significant linear effect on the five response functions (drying time, average dehydration rate, WGC, MCWG, TA) followed by initial moisture content and drying temperature. An increase in irradiation dosage had a negative effect on drying time and a positive effect on average dehydration rate. These effects improved drying characteristics, representing in shortening drying time and increasing average dehydration rate. It was assumably because of the damages and changes of tissue and structure of wheat grain caused by gamma irradiation

(Yu, et al., 2006). An increase in irradiation dosage, had a negative effect on WGC and MCWG and a positive effect on TA. It was assumably because of the breakage or cleavage of long chains in amylopection caused by gamma irradiation (Wu et al., 2002).

4. CONCLUSIONS

The drying time, average dehydration rate, WGC, MCWG and TA were sensitive to irradiation dosage, drying temperature, and initial moisture content. Among the variables, drying temperature is most dominating factor on drying characteristics and irradiation dosage is most dominating factor on the quality of wheat flour.

Irradiation dosage had negative effect on drying time, WGC and MCWG, and positive effect on average dehydration rate and TA. Drying temperature had negative effect on drying time, and positive effect on average dehydration rate WGC, MCWG and TA. Initial moisture content had negative effect on WGC and MCWG, and positive effect on drying time, average dehydration rate and TA.

ACKNOWLEDGEMENTS

The authors acknowledge the financial support of Chinese National Foundation of Nature and Science through project 3047000, and the project was supported by China Postdoctoral Science Foundation 20060400320.

REFERENCES

Akhnazarova, S., Kafarov, V. Experiment optimization in chemistry and chemical engineering. Moscow: Mir. 1982, 151-240

Al-Kaisey, M.T., Mohammed, M.A., Alwan, A.H., Mohammed, M.H. The effect of gamma irradiation on the viscosity of two barley cultivars for broiler chicks. Radiation Physics and Chemistry, 2002, 63: 295-297

AOAC. Official method of analysis 14th ed.. Association of Official Analytical Chemists, Washington DC, USA. 1990

Bidawid, S., Farber, J. M., Sattar, S.A. Inactivation of hepatitis A virus (HAV) in fruits and vegetables by gamma irradiation. International Journal of Food Microbiology 2000, 57: 91-97

Follett, P.A. Irradiation to control insects in fruits and vegetables for export from Hawaii. Radiation Physics and Chemistry, 2004, 71: 163-166

Formanek, Z., Lynch, A., Galvin, K., Farkas, J., Kerry, J.P. Combined effects of irradiation and the use of natural antioxidants on the shelf~life stability of overwrapped minced beef. Meat Science, 2003, 63: 433-440

Hallman, G.J. Irradiation disinfestation of apple maggot (Diptera : Tephritidae) in hypoxic and low~temperature storage. Journal of economic entomology, 2004, 97: 1245-1248

Kamat, A., Pingulkar, K., Bhushan, B., Gholap, A., Thomas, P. Potential application of low dose gamma irradiation to improve the microbiological safety of fresh coriander leaves. Food Control, 2003, 14: 529-537

Little, T.M., Hills, F.J. Agricultural experimentation: design and analysis. New York: John Wiley, 1978, 247-266

Wang, J., Chao, Y. Drying characteristics of irradiated apple slices. Journal of Food Engineering, 2002. 52: 83-88

Wang, J., Chao, Y. Effect of gamma irradiation on quality of dried potato. Radiation Physics and Chemistry, 2003, 66: 293-297

Wu, D. X., Shu, Q.Y., Wang, Z.H., Xia, Y.W. Effect of gamma irradiation on starch viscosity and physicochemical properties of different rice. Radiation Physics and Chemistry, 2002, 65: 79-86

Yu Y, Wang J. Effect of gamma-ray irradiation on drying characteristics of wheat. Biosyst. Eng., 2006, 95:219-225

Yu Y, Wang J. Effect of γ irradiation pre-treatment on drying characteristics and qualities of rice. Radiat. Phys. Chem., 2005, 74:378-383

Yu Y, Wang J. Effect of γ-irradiation treatment before drying on qualities of dried rice. J. Food Eng., 2007, 78:529-536

APPLICATION OF JAVA TECHNOLOGY IN THE REGIONAL COMPARATIVE ADVANTAGE ANALYSIS SYSTEM OF MAIN GRAIN IN CHINA

Xue Yan, Yeping Zhu *

Agricultural Information Institute of Chinese Academy of Agricultural Sciences 100081

* *Corresponding author:ZHU Ye ping Address: Agricultural Information Institute of Chinese Academy of Agricultural Sciences 100081 Tel: +86-010-82103120, Fax: +86-010-82103120, Email: zhuyp@mail.caas.net.cn*

Abstract: Since no comparative advantage analysis system developed for agricultural products in China counties, this paper discussed how to use comparative advantage theory and Java language to design regional main grain comparative advantage analysis system. The system applies comparative advantage index calculation method to rapidly provide users with relevant data and analysis of grain production comparative advantage in difference region .For future extension, It not only ensures system and data safety but also makes the system run on internet and any computer with Java Virtual Machine.

Keywords: Region, Comparative advantage, Index, Java

1. INTRODUCTION

According to the theory of comparative advantage, any country or region, whether developed, developing or underdeveloped in agriculture,has agricultural products with comparative advantages due to regional features and difference; the only thing is that regional agricultural products judged to have advantages may not have shown real comparative advantage in production. At the same time, comparative advantage is not still but ever changing with the Chinese and international markets. (Ma Huilan, 2004). This reveals the importance to improve international competitiveness of agricultural products by optimizing regional agricultural structure based on

Please use the following format when citing this chapter:

Yan, X. and Zhu, Y., 2009, in IFIP International Federation for Information Processing, Volume 294, *Computer and Computing Technologies in Agriculture II, Volume 2*, eds. D. Li, Z. Chunjiang, (Boston: Springer), pp. 831–839.

regional agricultural development and regional comparative advantage analysis of agricultural products in China. Similarly, for the Chinese market, it should be a practical choice to analyze the regional comparative advantage of agricultural products and meet the domestic demands by obtaining more cheap products through regional labor division and trade. (Zhao Bolin, 2002). Therefore, using comprehensive comparative indices, one of the methods for quantitative analysis of comparative indices, and Java, the object-based programming language, this paper designs a regional comparative advantage analysis system of main grain based on rural economic data of counties, offering users a tool for quick access to comprehensive comparative indices of main grain in counties as well as relevant information analysis.

The reason why Java is selected as the language for developing the regional comparative advantage analysis system is, with features of cross-platform, security, object-based, multi-threading, universal design and virtual machine, and based on full consideration of system design and functional requirements, Java meets the following system requirements:

The system shall have scalable functions for further development. The system shall be able to operate online. The system security shall be emphasized as data of counties are not made public.

2. MAIN FUNCTIONS OF THE SYSTEM

The system comprises information maintenance subsystem, information query subsystem, calculation and display subsystem, comparison and analysis subsystem, and help subsystem.

The information maintenance subsystem provides input and maintenance of basic system data. Use “Add” to input a single data record, “Maintain” to modify or delete existing data in the system, and “Import from Excel” to add data from existing county-level main grain statistical database and national main grain statistical database to the system.

The information query subsystem supports query by year or by variety. When “Query by Year” is used, and year, province and county are confirmed, the system will display the area and per unit yield of all varieties of a certain year in some counties; when “Query by Variety” is used, and variety, year, province and county are confirmed, the system will display the area and per unit yield of a certain variety (rice, wheat, maize and soybean are available) in some years and some counties.

The index calculation subsystem calculates the scale advantage index, efficiency advantage index and comprehensive comparative advantage index of each variety every year according to the input data in the system. The calculation results are saved in the system in the form of SQL Server datasheet.

The comparison and analysis subsystem, playing a key role in analyzing comparative advantage of main grain in a region, has four functions: "Index ordering" sorts indices of a certain year at national, provincial or county level; "Inter-region comparison" selects counties of different provinces to compare the indices of the same variety in the same year, displaying results in table or column; "Intra-region comparison" selects counties of the same province to compare indices of one or more varieties in the same year, displaying results in table or column; "Chronological comparison" selects a county to compare indices of different varieties in successive years, displaying results in table or curve.

In addition, the system provides Help function for the users' convenience.

3. SYSTEM ARCHITECTURE

The system is developed by Struts and Hibernate. Struts is used to create presentation and control layers in Model-View-Controller (MVC), while Hibernate is used for model layer and data persistence. See Figure 1 for the system architecture.

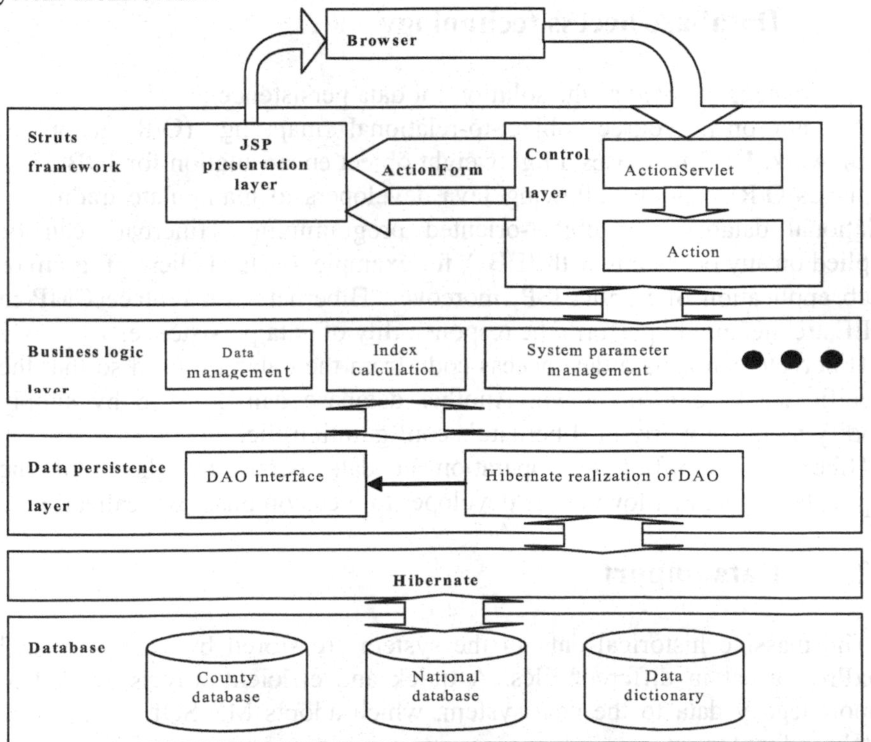

Fig.1: System Architecture.

In Model-View-Controller, Model represents business logic (realized by JavaBean and EJB components) of the application and data persistence; View is the presentation layer (generated by JSP page) of the application; Controller controls the processing of the application (usually one Servlet. The MVC design pattern separates the application logic, processing and display logic into distinct components.

Struts, like other projects of Apache, is an open source project. As an excellent MVC framework, it provides invisible underpinnings for MVC system development, and mainly uses Servlet, JSP and custom tag library.

The MVC structure is attached with great importance for separating modules of the business logic layer from the presentation layer. The DAO mode and Hibernate framework are adopted at data persistence layer, isolating data persistence from actual applications, so that only modification of DAO and Hibernate configuration will be required to change data storage solution in the future.

4. TECHNOLOGY REALIZATION

4.1 Database access technology

Hibernate is adopted as the solution for data persistence.

As an open source object-to-relational mapping (O/R mapping) framework, Hibernate uses a lightweight object encapsulation for JDBC and achieves O/R mapping, allowing Java developers to manipulate traditional relational database by object-oriented programming. Hibernate can be applied on any occasions with JDBC, for example, on Java client program or Web application of Servlet/JSP; moreover, Hibernate can replace CMP in J2EE architecture to perform the responsibility of data persistence.

Hibernate can isolate data access code from the database used so that the specific database is unknown. Another database can be used by simply modifying the property in Hibernate's configuration file.

Hibernate simplifies the operation at data persistence layer of the application system, allowing the developer to focus on business realization.

4.2 Data import

The massive historical data in the system are stored by year in Visual FoxPro format in different files. A quick and efficient way is needed to import legacy data to the new system, which adopts MS SQL Server2000 database for storage.

To fit Java language, a VFP data file should be converted into an Excel file and processed by POI tool in Apache Jakata project.

The file system API of POI, an open source tool, uses pure Java language to realize OLE2 compound document format, while HSSF API allows read/write of Excel file by Java POI.

Specifically, class ExcelFileParser in net.edu.caas.argsys. poi package is responsible for realizing the function. The class has four methods including two parserArgDataCountyFile() and parserArgDataCountyFile(), and the external data file has two formats – provincial data and national data. The two parserArgDataCountyFile() methods in ExcelFileParser are for analyzing provincial data files, while the two parserArgDataNationalFile() methods are for national data files.

4.3 Data table display and export

As a lot of data tables are generated in the system, display-tag open source tool is adopted to solve problems in display, including paging, ordering, data export, etc. Paging and ordering of tables are realized using tags of the tool in JSP page. See Figure 2 for codes.

```
<display:table name="sessionScope.nodeList" class="its" pagesize="15" sort="list" export="true">
   <display:column property="areaId" title="地区代码" sortable="true" headerClass="sortable" />
   <display:column property="areaName" title="地区名称" sortable="true" headerClass="sortable" />
   <display:setProperty name="basic.empty.showtable" value="true" />

</display:table>
```

Fig.2: Program Codes for Table Paging and Ordering.

In addition, display-tag provides another package – org.displaytag.export to solve the problem with data export. The package can help exporting data in the table to an Excel file according to the table format.

4.4 Figure display

After index calculation, the system can compare and analyze the indices in three ways, “Inter-region comparison”, “Intra-region comparison” and “Chronological comparison”. The results need to be presented in figures so as to show the comparison and change of indices in a more intuitive way.

Cewolf tag library is adopted to display figures in the system. Cewolf, a JSP tag library developed with JfreeChart tool, can be used inside a Servlet/JSP based web application to embed complex graphical charts of all kinds (e.g. line, pie, bar chart, etc.) into a web page. It provides a full featured tag library to define all properties of the chart (colors, strokes, legend, etc.). Thus the JSP which embeds the chart is not polluted with any java code.

Figure 3 shows the codes of cewolf tags in the page of inter-region comparison results

```
<cewolf:chart id="line" title="地区间比较结果" type="verticalbar" xaxislabel="地区" yaxislabel="指数">
    <cewolf:data>
        <cewolf:producer id="cpBtwAreasDataSet" usecache="false"/>
    </cewolf:data>
</cewolf:chart>
<p>
    <cewolf:img chartid="line" renderer="cewolf" width="500" height="400" />
<P>
```

Fig.3: Codes of cewolf tag.

5. DISPLAY OF SYSTEM FUNCTIONS

5.1 Information Maintenance Subsystem

It features Add information, Maintain information, and Import from Excel. See Figure 4.

The interface displays in four pages the area and per unit yield of rice, wheat, maize, soybean and all grains of Chaoyang district, Beijing in 1995-2004. On the interface the user can query, modify or delete data. Figure 4 shows the first page.

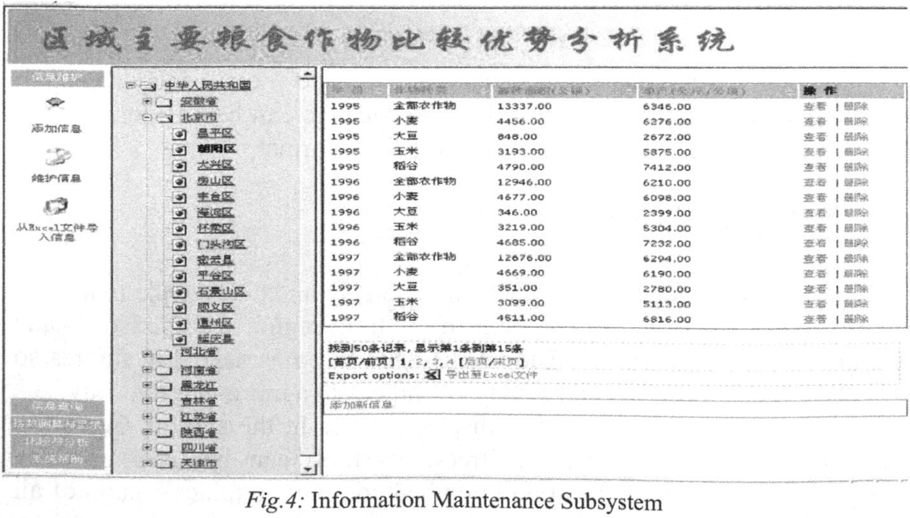

区域主要粮食作物比较优势分析系统

[illegible]	[illegible]	[illegible]	[illegible]	操作
1995	全部农作物	13337.00	6346.00	查看 | 删除
1995	小麦	4456.00	6276.00	查看 | 删除
1995	大豆	848.00	2672.00	查看 | 删除
1995	玉米	3193.00	5875.00	查看 | 删除
1995	稻谷	4790.00	7412.00	查看 | 删除
1996	全部农作物	12946.00	6210.00	查看 | 删除
1996	小麦	4677.00	6098.00	查看 | 删除
1996	大豆	346.00	2399.00	查看 | 删除
1996	玉米	3219.00	5304.00	查看 | 删除
1996	稻谷	4685.00	7232.00	查看 | 删除
1997	全部农作物	12676.00	6294.00	查看 | 删除
1997	小麦	4669.00	6190.00	查看 | 删除
1997	大豆	351.00	2780.00	查看 | 删除
1997	玉米	3099.00	5113.00	查看 | 删除
1997	稻谷	4511.00	6816.00	查看 | 删除

找到50条记录，显示第1条到第15条
[首页/前页] 1, 2, 3, 4 [后页/末页]
Export options: 导出至Excel文件

添加新信息

Fig.4: Information Maintenance Subsystem

5.2 Information Query Subsystem

It displays the area and per unit yield of rice, wheat, maize and soybean of counties by year and by variety. See figure 5.

区域主要粮食作物比较优势分析系统

信息维护
信息查询
按年份查询
按品种查询

查询结果
2004年数据如下:

110114	昌平区	9251.0	3602.0	1224.0	4336.0	852.0	1741.0	852.0	3657.0	143.0	5902.0
110105	朝阳区	767.0	6166.0	131.0	5267.0	170.0	2229.0	170.0	5379.0	0.0	0.0
110115	大兴区	27048.0	5320.0	9673.0	5429.0	827.0	2732.0	827.0	5773.0	0.0	0.0
110111	房山区	26180.0	3961.0	7259.0	4587.0	2422.0	1867.0	2422.0	4563.0	130.0	6162.0
110106	丰台区	785.0	4706.0	51.0	4157.0	30.0	2267.0	30.0	3025.0	0.0	0.0
110108	海淀区	1463.0	5200.0	111.0	3505.0	436.0	2232.0	436.0	5273.0	298.0	6403.0
110116	怀柔区	11086.0	3564.0	1193.0	5267.0	1866.0	2245.0	1866.0	3601.0	13.0	6231.0
110109	门头沟区	2153.0	1974.0	23.0	2304.0	570.0	761.0	570.0	1869.0	2.0	2000.0
110228	密云县	13903.0	3166.0	845.0	4780.0	0.0	0.0	0.0	3392.0	0.0	0.0
110117	平谷区	9071.0	3653.0	1532.0	5364.0	724.0	2441.0	724.0	3661.0	0.0	0.0

找到10条记录，显示所有记录
1
Export options: 导出至Excel文件

返回查询

Fig.5: Information Query Subsystem.

5.3 Comparison and Analysis Subsystem

The subsystem features index ordering, inter-region comparison, intra-region comparison, chronological comparison and other functions. It can display the analysis result in form or in graph. See Figure 6.

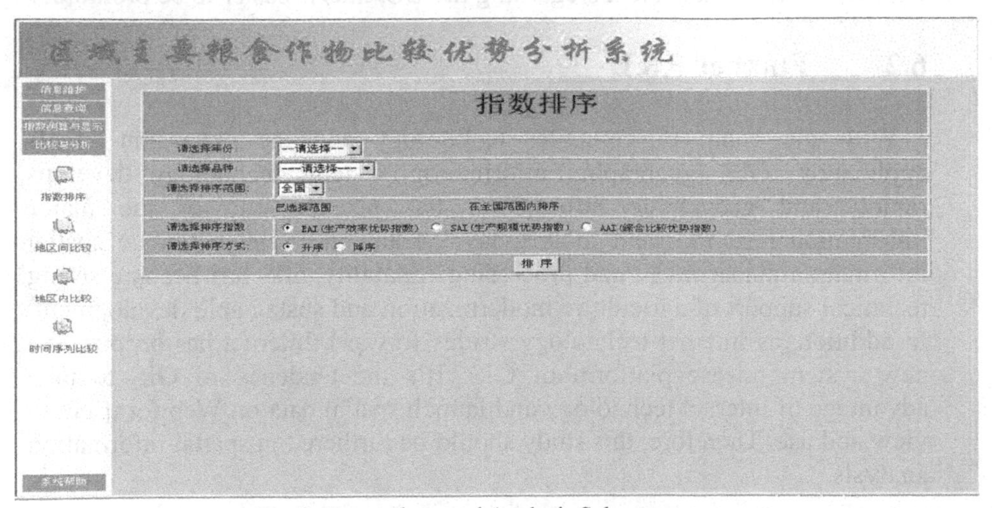

Fig.6: Comparison and Analysis Subsystem

6. SUMMARY

6.1 Conclusion

The "Regional Comparative Advantage Analysis System of Main Grain" is a software tool designed for managers and researchers. It builds main grain statistic databases at national, provincial and county level respectively. With functions of information maintenance and query, the system realizes the comparative advantage analysis of main grain production in counties using comprehensive comparative indices. Main achievements of the research include:

Comparative advantage theory, information technology and model technology are used to scientifically evaluate and analyze the advantages and disadvantages of counties in main grain production, providing supporting information for the structural adjustment and arrangement of agricultural production of counties in China.

Since there's few comparative advantage analysis and research on agricultural products of counties in China using information technology, the design and development of the system have made up the gap.

The system realizes the comparative advantage analysis of main grain in counties throughout China. The regional information analysis given by the system can basically reveal the production features and status of the regional grain.

By adopting Java language, one of the best enterprise and mobile application development platforms in the world, the system not only realizes online operation, but also can run on any computing equipment installed with Java virtual machine disregarding the OS, thus is easier to be promoted.

6.2 Further R&D

With the boom of computer technology and the expansion of its application, GIS (geographic information system) technology develops rapidly and geographic information has become one of the major information resources and tools today. Meanwhile, due to unique spatial information management and processing capability, GIS has become strong technical support of agriculture modernization and sustainable development. In addition, as Internet technology strides forward, Internet has become the new system release platform of GIS. It's the tendency of GIS to take advantage of Internet technology and launch spatial data on Web for users to view and use. Therefore, this study should be furthered in spatial information analysis.

6.2.1 To realize GIS-based rural economic information management and analysis of counties

The regional comparative advantage analysis system of main grain mainly realizes the geographic comparison and analysis of grain production in counties, so its information features spatial nature; meanwhile, traditional data management systems and methods can no longer meet the requirement for processing the ever-increasing agricultural economic information, especially mass spatial data. Therefore, as another way of information exchange and sharing, the system meets the tendency of leveraging GIS to analyze and process volume geographic data.

6.2.2 To combine with WEBGIS

WEBGIS is one of the tendencies of GIS. We will mange, analyze and convey agricultural economic information resources through Internet, and improve the system into WEBGIS software that integrates map processing and release.

ACKNOWLEDGEMENTS

The work was funded by National Scientific and Technical Supporting Programs Funded by Ministry of Science and Technology of China (2006BAD10A06, 2006BAD10A12).

REFERENCES

Ma Huilan. Theory of Comparative Advantage of Regional Agricultural Products [Doctoral degree dissertation of Xinjiang Agricultural University], Xinjiang: Xinjiang Agricultural University, 2004.

Zhao Bolin. A Study of the Countermeasures of China's Regional Agriculture after Joining WTO, Journal of Shanxi University of Finance & Economics, 2002, 24 (1): 41～44.

FINITE ELEMENT ANALYSIS OF SINGLE WHEAT MECHANICAL RESPONSE TO WIND AND RAIN LOADS

Li Liang, Yuming Guo [*]

College of Engineering $ Technology, Shanxi Agricultural University,Taigu, Shanxi Province, P. R. China 030801

[*] *Corresponding author, Address: College of Engineering $ Technology, Shanxi Agricultural University,Taigu,030801, Shanxi Province, P. R. China, Tel: +86-354-6288906, Fax: +86-354-6288906, Email: guoyuming99@sina.com*

Abstract: One variety of wheat in the breeding process was chosen to determine the wheat morphological traits and biomechanical properties. ANSYS was used to build the mechanical model of wheat to wind load and the dynamic response of wheat to wind load was simulated. The maximum Von Mises stress is obtained by the powerful calculation function of ANSYS. And the changing stress and displacement of each node and finite element in the process of simulation can be output through displacement nephogram and stress nephogram. The load support capability can be evaluated and to predict the wheat lodging. It is concluded that computer simulation technology has unique advantages such as convenient and efficient in simulating mechanical response of wheat stalk under wind and rain load. Especially it is possible to apply various load types on model and the deformation process can be observed simultaneously.

Keywords: finite element analysis, mechanical response, wheat stalk, wind and rain loads

1. INTRODUCTION

Crop lodging is one of the most important factors affecting wheat yield. Many researchers have studied the wheat lodging resistance for years. (1) Building mechanical models of crops are a popular method to evaluate the wheat lodging resistance capability by analyzing the mechanical response of

Please use the following format when citing this chapter:

Liang, L. and Guo, Y., 2009, in IFIP International Federation for Information Processing, Volume 294, *Computer and Computing Technologies in Agriculture II, Volume 2*, eds. D. Li, Z. Chunjiang, (Boston: Springer), pp. 841–846.

wheat stalk under wind. Some theoretical models were introduced to analyze the natural frequencies for cereals and trees under wind (C.J. Baker, 1995, S. E. T. Saunderson, et al, 1999, Yuan Zhihua, et al. 2002). However, most of these models are calculated by pure theory and the calculating process is extremely complicated. Sometimes, only the numerical solution is obtained by software for numerical mathematics (Hanns-Christop Spatz, Olga Seck, 2002). (2) Field tests of wheat were carried out in a portable wind tunnel (M. Sterling, et al., 2003), which could simulate the mechanical response of wheat close to the natural situation, but the testing and control system of wind tunnel is difficult to design and very costly.

With the development and application of computer technology in agriculture, more agricultural problems have been solved by computer. For example, in the field of simulating crop growth, many results are obtained. A peach crop production computer simulation model has been built to predict the fruits growth (T. M. DeJong, et al. 1990). Two models were utilized to simulate the effects of CO_2 and ozone on growth and developmental processes of spring wheat in response to climatic conditions (F. Ewert, et al. 1999). Methods for estimating potential evaporation and potential transpiration among crops are presented to help predict the soil moisture (Huang Guanhua, 1995).

A mechanical model of wheat under wind and rain loads built by ANSYS software is introduced in this paper. And mechanical responses of wheat are simulated. This method is convenient for applying different type of loads on model. The change of stress and displacement during loading can be shown clearly with stress and displacement nephograms. Besides, the maximum Von Mises stress is calculated simultaneously to predict the dangerous position.

2. FINITE ELEMENT MODEL OF SINGLE WHEAT

2.1 Element type and material attributes

In the process of building wheat finite element model, Beam188 (defined in ANSYS) shown in Fig.1 is chosen to be the element type.

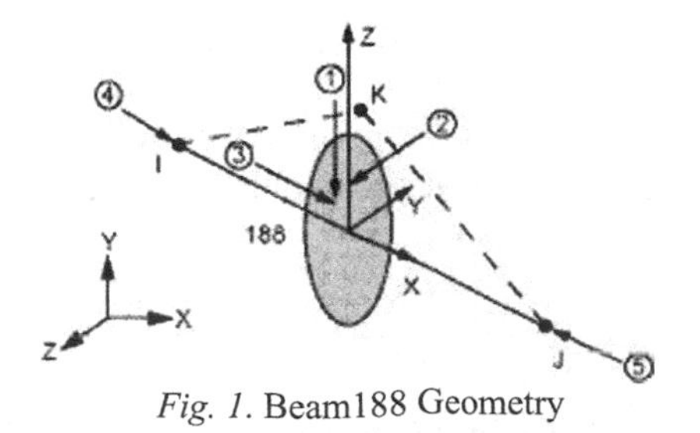

Fig. 1. Beam188 Geometry

It is suitable for analyzing slender to moderately stubby/thick beam structure. It is based on Timoshenko beam theory, and also considered shear deformation effects. A cross-section associated with this element type can be a built-up section referencing more than one material and the cross-section characteristics of beam can be shown clearly.

Due to different material of each wheat internode, morphological traits and biomechanical properties of wheat variety in the breeding process (named ShanNong040121-135) were tested during the wheat mature stage at June16th, 2007. Mean of the tested indexes are listed in Table1.

Table 1. Mean of wheat morphological traits and biomechanical properties

Internode number	Internode Distance (mm)	diameter (mm)	wall thickness (mm)	Ear weight (g)	Young's modulus (MPa)	Bending strength (MPa)	Bending rigidity (Nmm^2)
1	78.49	3.63	0.83		2168.62	27.18	16203.73
2	118.69	3.60	0.54		2228.98	19.85	14601.44
3	139.80	3.68	0.45	3.89	1429.63	15.32	8925.47
4	176.76	3.58	0.37		1458.55	13.84	7256.45
5	284.86	2.96	0.43		1973.65	13.39	5464.96

Biomaterials including wheat stalk have more complicated properties than other engineering materials. It is strictly a viscoelastoplastic material. However, it is difficult to test the related index so that the stalk material in this paper is supposed as linear isotropic elasticity. The parameters of material properties are defined according to Table 1 from the bottom to the top of the stalk. Poisson's ratio is taking as 0.413 determined by our test.

2.2 Finite element model

Wheat stalk is simplified as cantilever beam of variable rigidity with an apical load. It is negligible mass of the stalk (Hanns-Christof Spatz and Olga Speck, 2002). The beam cross-section is defined as circular hollow section and the value is determined according to table1. Wind load is applied as pressure on the beam which is calculated by the formula as follows:

$$\omega = \frac{\upsilon^2}{1600} \qquad (1)$$

Where: υ is wind speed.

The apical load is ear weight applied on the stalk as concentrated force.

Large displacement static is chosen as the analysis type. When rain and wind loads are applied together, rain load is simplified as concentrated force applied on the top of stalk. The finite element model is shown in Fig. 2.

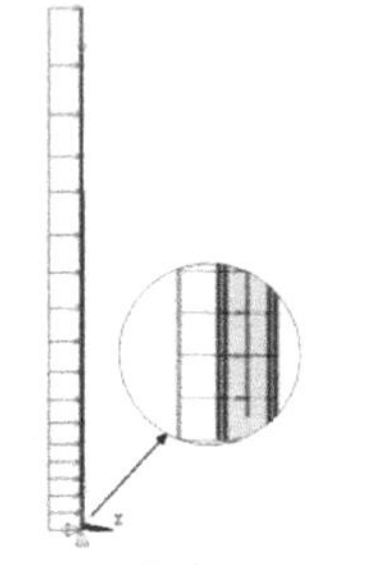

Fig.2. Stalk mechanical model

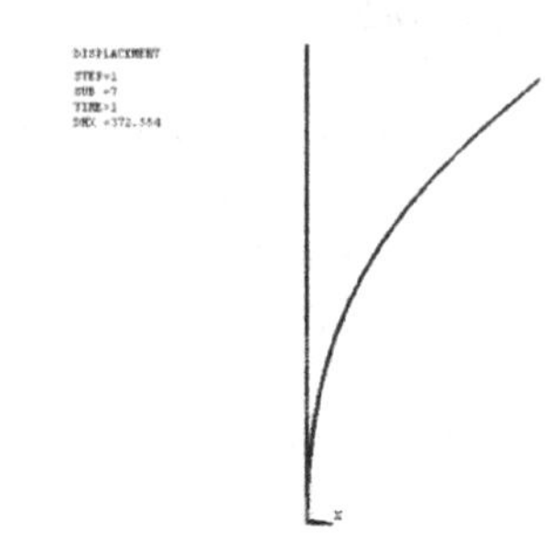

Fig. 3. Stalk deformed shape at wind speed of 3.4m/s

3. RESULTS AND ANALYSIS

Various displacements and stresses for single stalk under different loads were obtained by changing load type and size. Fig. 3 shows the deformed shape of wheat stalk under wind load. Meanwhile, the software can calculate the value of displacement and stress for each node and finite element. The results are presented as displacement and stress nephogram. It indicates that the maximum Von Mises stress of wheat stalk under wind load occurs at the second internode.

(1) Wind load acts alone. At wind speed of 3.4m/s, the maximum displacement down the wind direction is 353.634mm, the vertical maximum displacement is 119.467mm and the maximum von mises stress is 4.52MPa. When wind speed reaches to 19.3m/s, the top of wheat stalk is close to ground and the maximum Von Mises stress is 16.996MPa, which is lower than the bending strength of wheat stalk. Therefore, it will be returned to its original position. Von Mises stress nephograms for single wheat under above wind speeds are shown in Fig. 4.

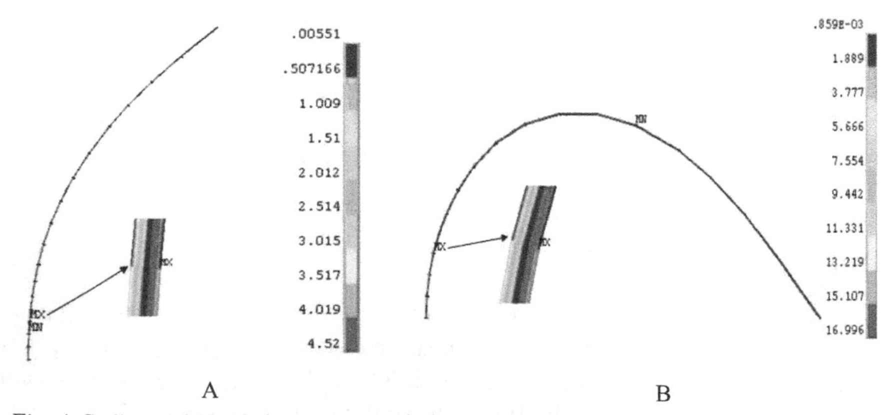

A B

Fig. 4: Stalk nodal Von Mises stress at wind speed of 3.4m/s (A) and 19.3m/s (B) (MPa)

(2) Both wind and rain load are considered. Assume that ear weight adds rain load of 0.01N. The maximum displacement in horizontal direction is 525.412mm, and in vertical direction is 331.093mm when wind speed is 3.4 m/s. The maximum Von Mises stress is 7.945MPa as shown in Fig. 5 (A). The maximum Von Mises stress of wheat increases 75.8% compared to the situation when wind load acts alone. The vertical displacement of wheat stalk reaches the largest point at wind speed of 17.8m/s, and the Von Mises stress is 16.886MPa as shown in Fig. 5 (B). Because of rainfall, the field becomes wet and soft. At that moment, both stem and root lodging might be occurred.

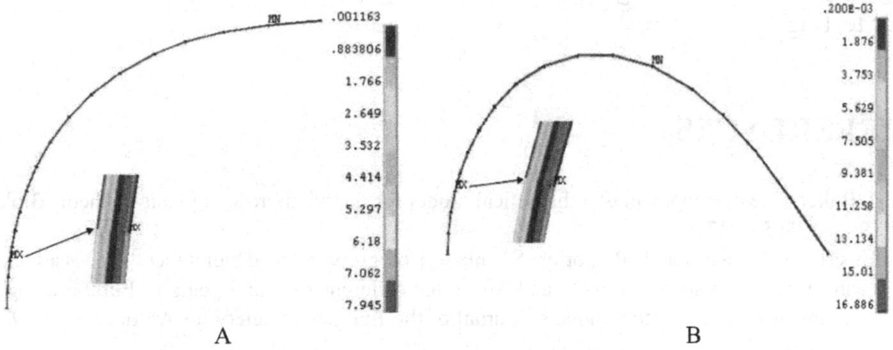

Fig. 5: Stalk von mises stress (MPa) nephogram considered rain weight at wind speed of 3.4m/s (A) and 17.8m/s (B)

4. DISCUSSION

Displacement and stress responses for single wheat under different load are simulated by ANSYS software. It is concluded that

(1) When wind load acts on wheat alone, lodging is difficult to occur. If both wind and rain load are considered simultaneously, the vertical displacement of wheat stalk is already reaches the utmost at wind speed of 17.8m/s with ear added 1g rainfall amount. Lodging is easier to happen.

(2) At the same wind speed of 3.4m/s, the vertical displacement of wheat stalk increases 177.1% and Von Mises stress increased 75.8% considered rain load, compared to the wind load acts alone.

(3) Related analysis above indicates that computer simulation method has unique advantages on analyzing the mechanical response of wheat stalk under various wind and rain load. It is convenient and efficient to help predict the wheat lodging. And the powerful calculation function is useful to evaluate the load support capability and present the changeable trend of displacement and stress. Therefore, fine results can be obtained in deformation analysis. However, in the process of building finite element

model, the material properties of wheat stalk is assumed as linear isotropic viscoplastoelastic, which is an ideal status. The applying types of wind and rain loads still need to do further study. These problems will be studied accompanied by verifying the model with experiment and improving it.

ACKNOWLEDGEMENT

Funding for this research is provided by Doctoral Fund of Ministry of Education of P. R. China. The authors would like to thank the college of agriculture in Shanxi Agricultural University (P. R. China) to provide wheat for testing.

REFERENCES

C. J. Baker. The development of a theoretical model for the windthrow of plants. J. theor. Biol. 1995, 175: 355-372

F. Ewert, M. Oijen, van. J. R. Porter, Simulation of growth and development processes of spring wheat in response to CO2 and ozone for different sites and years in Europe using mechanistic crop simulation models. Journal of the European Society for Agronomy, 1999, 10(3-4): 231-247

Hanns-Christof Spatz, Olga Speck. Oscillation Frequencies of Tapered Plant Stems. American Journal of Botany, 2002, 89(1): 1-11

Huang Guanhua, Shen Rongkai, Zhang Yufang, et al. Simulating Evaporation and Transpiration and Forecasting Soil Moisture Regime under Conditions of Crop Growth, Journal of Wuhan University of hydraulic and electric engineering, 1995, 28(5):481-487 (in Chinese)

M. Sterling, C.J. Baker, P.M. Berry, A. Wade. An experimental investigation of the lodging of wheat, Agricultural and Forest Meteorology, 2003, 119(3-4):149-145

S. E. T. Saunderson, A. H. England, C. J. Baker. A Dynamic Model of the Behavior of Sitka Spruce in High Winds, J. theor. Biol. 1999, 200(3): 245-259

T. M. DeJong, R. S. Johnson, S. P. Castagnoli, Computer simulation of the carbohydrate economy of peach crop growth. Acta horticulturae, 1990, (276): 97-104

Yuan Zhihua, Li Yundong, Chen Heshun. Dynamic model and vibration analysis of wheat or sort, Hennan Science, 2002, 20 (1):11-13. (in Chinese)

DESIGN AND IMPLEMENTATION OF DYNAMIC KNOWLEDGE MODEL FOR SUITABLE VARIETY SELECTION AND SOWING TIME DETERMINATION OF SOYBEAN

Xiangliang Wu [1,2] , Julin Gao [1,*] , Yvdong Zhao [2] , Lijun Li [1] , Man Li [1] , Zhigang Wang [1]

[1] *Agricultural institute of Inner Mongolian Agriculture University,huhhot 010010,China;*

[2] *Inner Mongolian animal husbandry hall information center;*

[*] *Corresponding author, Address: Gao Julin(1964-),male, professor, tutor of doctoral candidates, agricultural college of inner Mongolia university, direction of study:crop optimizing cultivation an decision system, E_mail:gaojulin@yahoo.com.cn*

Abstract: Using knowledge engineering principle and mathematical modeling technique, the model was established, it was the dynamic knowledge model of suiting soybean varieties, variety selection and variety sowing time determination through aggregating the results of existing research and analysising Experimental data. the knowledge model was verified through six different ecological points of Inner Mongolia and different varieties data. The results show that the model has a better decision-making and universal application.

Keywords: soybean, model, expert system, variety selection, sowing time

1. INTRODUCTION

Soybean variety selection is the guarantee of obtaining high quality and high yield. It is mainly through variety introduction test and experience of expert in production, it lacks quantitative indicators and increases the variety updating time. The study determines the suitability, variety selection and

Please use the following format when citing this chapter:

Wu, X., Gao, J., Zhao, Y., Li, L., Li, M. and Wang, Z., 2009, in IFIP International Federation for Information Processing, Volume 294, *Computer and Computing Technologies in Agriculture II, Volume 2*, eds. D. Li, Z. Chunjiang, (Boston: Springer), pp. 847–852.

variety sowing time determination of soybean,the result of study guides soybean production.

2. MATERIALS AND METHODS

Looking up a large number of documents and consulting the expert of soybean cultivation, these are the basis.determine.then thinking the relationship between characteristics of variety kinds and ecological environment, in the end ,the model is established through the summary, inducing and refining the study data of the soybean cultivation principles and technology. Further do mathematical statistical analysis on the basis of obtaining test data and transform the collected conceptual model to quantitative mathematical model which can solve problem.

3. MODEL DESIGN

3.1 The model of soybean suitability

The suitability of soybean is the primary considered problem in plan design before planting. It is on the basis of the perennial weather condition (≥ 10 ℃ active accumulated temperature) of the decision point, such as light and water and so on.

The effective accumulated temperature can show temperature condition of the local growing season. It determines the suitability of planting soybean Considering the local precipitation and irrigation.

Whether the decision point can plant soybean or not can be described quantitatively by equation (1). SGP means the possibility of soybean planting. When SGP is equal to one, it means that it can plant soybean ,but when SGP is equal to zero, it means the field can't plant soybean. JW、JS、GS、KH respectively means the active accumulated temperature that is higher than ten centigrade, annual rainfall(mm), whether have irrigation condition or not(1 means have irrigation condition) and whether select drought-resistant varieties or not.

$$SGP = \begin{cases} 1 & JW \geq 1900 \quad \text{and} \quad JS \geq 400 \\ 1 & JW \geq 1900 \quad \text{and} \quad 400 \geq JS \geq 350 \quad \text{and} \quad KH = 1 \\ 1 & JW \geq 1900 \quad \text{and} \quad JS \leq 350 \quad \text{and} \quad GS = 1 \\ 0 & \text{Non - coincidenc} \quad \text{e above} \end{cases} \qquad (1)$$

Temperature directly impact on the whole growth process of soybean which is from emergence, flowering to maturity. When the activities temperature that is higher than ten centigrade is lower than 1900 centigrade, the soybean will be difficult to be mature or instable of production. The study shows that when the soybean yield is higher 2625kg·hm-2,the water requirement is more than 400mm.The system regulates(Mao Hongxia et al.2007,Li Yunge et al.2001, Xu Shuqin et al.2001) that when the precipitation of growth period is between 350mm and 400mm, it need to choose drought-resistant variety under non-irrigation condition. The variety drought-resistant evaluation can be seen in the table six. When the precipitation is less than 350mm in the growth period, the field can't plant soybean under non-irrigation condition.

3.2 The model of soybean variety selection

Yield is the most concerned problem by user. The soybean variety selection can be described quantitatively by equation (2). It usually needs that the confidence degree of yield is more than 80 percent.

$$ZXD = \begin{cases} PY/STY & PY < STY \text{ and } SYQ \leq WSQ - 12 \\ 1.0 & PY \geq STY \text{ and } SYQ \leq WSQ - 12 \end{cases} \quad (2)$$

ZXD、PY、SYT respectively means confidence degree of variety need, yield potential of variety and target yield. When the confidence is more than 80 percent, it will be possible to be chosen. SYQ is the growth period of variety and WSQ is the frost-free period of decision point.

Spring soybean is mainly cultivated in the north of China. Selection of longer growth period variety will obtain high yield easily but also be threated by early frost. Central Meteorological Station regulations show that an early frost ahead of six days is normal, but ahead of seven to eleven days it is partial early. When ahead of more than twelve days, it is special early frost, etc have analyzed the earliest frost of the northeastern region between 1967 and 1997(Yang Keming et al.1999). The result shows that early frost in extremely early years is nineteen percent and is twenty-nine percent in partial early years. Partial and normal year is about respectively half each other.

Therefore, in order to use the regional yield-increasing potential and reduce the influence of frost, the model requires that the frost-free period of decision point should be twelve days more than the selected variety.

Confidence degree is the credibility of chosing result by system and can be expressed by equation (3).

$$ZXD = 1 - \prod_{i=1}^{N} \frac{\left| X_{iy} - X_{ix} \right|}{X_{ix}} \quad (3)$$

ZXD is the confidence degree, Xix and Xiy mean the predicted value and actual value of some indictor of model in turn.

3.3 Model of soybean sowing time

Not only Soybean sowing time have direct effect on growth and development, but also maturity and yield have significantly effect to the quality of soybean(Han TianFu,2004). In general, oil content of spring sowing soybean is higher than the summer sowing and the summer sowing soybeans is higher than the autumn sowing(Ren XiuRong et al.2005, Xu HaiTao et al.2007). The trend of the protein content is opposite. Therefore, the proper sowing time benefit to improve the yield and quality of high oil soybean.

When the average soil temperature from 0cm to 5cm passes the temperature of from 8℃ to10℃ stably the soybean can be sowed. The system identify T as 10℃.when under film-covering condition identify T as 8 ℃ (Zhou Baoku et al.2004).Despite existing regional inter-annual temperature difference, it is stable in general. The system is used to predict the future years' climate situation by perennial climate. The identification of perennial climate can refer to the method of Gao Liangzhi. According to the years' average value input by user and included month average temperature, month average highest and lowest temperature and month sunshine hours, use harmonic analysis and generate daily climate data(average temperature, the maximum , Low, etc.) automatically. Use the equation (4) during the process of sowing time identification:

$$Tdi \geq T \quad (i = 1,2,3,4,5) \tag{4}$$

Tdi means continuous few days average temperature of decision place, T means the suitable temperature for sowing. d1—d5 can be as the suitable sowing timc of soybean which meet the equation.

4. MODEL TEST

The model of suitability soybean is feasible in principle because it is the application of the integration, analysis and refining of the existing research. So the target of model test is variety selection and sowing time identification. On the basis of the situation of the decision place, requirement of user and predicted yield, select suitable variety from the varieties resources database of soybcan using variety selection knowledge model and give confidence degree. Analyze the variety selection model by six different ecological sites of HuLunBeiEr. Table 1 shows the suitable planting variety which knowledge model select for user on the basis of their input demand.

Table 1 Variety selection determined by knowledge model for six different eco-sites during normal climatic year

Decision site	frost-free season	Accumulative Temperatures(℃)	whether or not irrigating	expected yield (kg·hm^{-2})	name of cultivars	confidence degree
HuLunBeiEr Agricultural Institute	100～112	2380	yes	3150	Jiangdou 9	0.965
					Jingmodou 1	0.893
					Mengdou 12	0.845
ARong Banner	90—120	2150	yes	300	Mengdou 11	0.983
					Hefeng 40	0.893
MoLiDaWa Banner	110—120	2250	yes	3300	Dongnong 434	0.921
					Heihe 750	0.912
ZaLaiTe Banner	125—135	2760	yes	3600	Dongnong 434	0.991
					Dongnong 163	0.904
WengNiuTe Banner	130—158	3071.7	no	2100	Dongnong 9031	0.952
					Zhongzuo 983	0.876
SaiHan Urban in huhhot	125—131	2847.9	yes	3450	Jiyu 47	0.962
					Jiyu 56	0.923

Simulating the main soybean planting areas of Inner Mongolia by the model of sowing time , the results are following in table 2. it shows that the model guides local soybean production.

Table2 Suitable sowing-date of soybean in inner mongolia

Area	planting date	
	prediction	Practical(2006)
Northern HuLuBeiEr	May 12	May 14
Southern HuLuBeiEr	May 7	May 8
XingAn,TongLiao,ChiFeng	May 3	May 6
HuHHot	April 29	April 28

5. CONCLUSION AND DISCUSSION

5.1 Using the knowledge engineering principles and mathematical model techniques, we have established the model of variety selection and sowing time identification , the model is refined and analyzed the soybean growing and management indictor and variety type, environmental factors between the basic relations and quantitative algorithm.

5.2 Lack of soybean grain quality quantitative description in the model and need to be modified and improved in future study.

REFERENCES

Han TianFu. Technique guideline in Soybean's Plant of Good Quality and High Production[M].beijing: China Agricultural Science and Technology press,2005.

Li YunGe,Gao YuFeng,Chai YongZhao,etc. the modulus water requirements study of Soybean spray irrigation[J].soybean bulletin. 2001(5):8-9.

Mao HongXia,Zhang ChangFu,He LinWang. The influence of different irrigation to soybean yield and quality under spray irrigation [J]. Xinjiang Farmland Reclamation Science & Technology. 2007(6): 35-36.

Ren XiuRong,Xu HaiTao,Wu DeKe.etc.the affect of different sowing time and climatic condition to soybean grain quality and main traits[J].soybean science. 2005(1): 71-73.

Xu HaiTao,Xu Bo,Wang YouHua.the affact of different sowing time and good fertilization to high oil soybean yield and quality[J].shanxi agicultural science, 2007, 35(5): 51-53.

Yang KeMing,Chen XiuFeng,Wang DongSheng,etc.the climatic characteristic of early frost of north-east region[J].meteorology. 1999(6):13-18.

Yang QingKai, Ning HaiLong,Xu YanLi,etc.the affect of successive and alternate stubble to soybean chemical quality under different ecological condition.soybean science. 2001,20 (3):187-190.

Zhou Baoku,Zhang XiLin,Huan LiHai,etc.the early study result of film corvered to soybean yield increasing.heilongjian agicultural science 2004,(3):6-8.

CONTENT-BASED IMAGE RETRIEVAL USING SALIENT BOUNDARY AND CENTROID-RADII MODEL

Qing Wang[1], Haijian Ye[1,*], Yan Wang[2], Hua Zhang[1]

[1] *College of Information and Electrical Engineering, China Agricultural University, No. 17, Tsinghua East Road, Haidian, Beijing, 100083, China*

[2] *Ocean College, Hebei Agricultural University, No. 52, East Part of Hebei Street, Qinhuangdao City, Hebei Province, 066003, China*

[*] *Corresponding author, Address: P. O. Box 147, College of Information and Electrical Engineering, China Agricultural University, No. 17 Tsinghua East Road, Haidian, Beijing, 100083, China, Tel: +86-10-62737188, Fax: +86-10-62736401, Email: hjye@cau.edu.cn*

Abstract: In view of the instability and low efficiency of the present image retrieval method, especially for simple image comparison with some salient shapes, a new image retrieval algorithm based on salient closed boundary is presented. Firstly, the Canny operator is performed to detect edges. Secondly, the ratio contour is used to extract the most salient closed boundary of some shape from the image. Finally, the similarities are measured by feature vector of the salient closed boundary based on the centroid-radii model. Preliminary experimental results demonstrate that the proposed method is quite suitable for many professional image retrieval systems and has a good performance in both retrieval efficiency and effectiveness.

Key words: content based image retrieval, salient closed boundary, radio contour, centroid-radii model

1. INTRODUCTION

Due to the fast development of digital imaging and networking technologies, Content-Based Image Retrieval (CBIR) has emerged as an important area of research in computer visual and multimedia computing. CBIR retrieves images from the database similar to the query image by the

---

Please use the following format when citing this chapter:

Wang, Q., Ye, H., Wang, Y. and Zhang, H., 2009, in IFIP International Federation for Information Processing, Volume 294, *Computer and Computing Technologies in Agriculture II, Volume 2*, eds. D. Li, Z. Chunjiang, (Boston: Springer), pp. 853–860.

image characteristics such as color, texture, shape or any combination of these features. According to the human vision system theory, people tend to evaluate or understand image by its shape, namely shape-based image retrieval, which is an intuitive and useful way to make use of image's information. To find out the image shape's features close enough to the actual perceptive trait of human eyes is not a simple task. Therefore shape-based image retrieval is one of the most challenging subjects within content-based image retrieval.

As key technology of image retrieval system, image retrieval method is the main content of study on the systematic development of image retrieval system. Indeed it doesn't merely affect the results of image retrieval directly but also determines the working efficiency and some other performances of the whole system to a great extent. Therefore, it has important significance in image retrieval algorithm's study. With the quick growth in the number of image information, it is urgent to effectively retrieve a great deal image information.

Yet, we find there are massive images with some salient shapes in, especially in a variety of professional fields such like the monitoring of diseases, insect pests and crop condition monitoring in the agricultural field, as well as the footprint comparing in the process of investigating and solve a case and so on. This kind of images is quite common. The common characteristics of those images are they have outstanding simplicity compared to general scene image, their content are usually comparatively simple and have shapes with salient boundary.

According to the characteristics of this kind of images, this paper proposes a new retrieval algorithm for images based on the shape's salient closed boundary. The basic idea of this algorithm is to extract the most salient closed boundary of some shape from the image by the ratio contour, and then the similarities are measured by the selected boundary's shape descriptor (feature vector) which is based on centroid-radii model. Experimental results demonstrate that the proposed method has a good performancc in both retrieval efficiency and effectiveness.

2. EDGE EXTRACTION

2.1 Edge detection

Edge detection, as the primary condition of the image features extraction, is the first step of all kinds of image segmentation based on edge. The features extraction of the shape can be accurate and meaningful only with a flawless edge detection of the image.

In this paper, we use the Canny operator, which is a relatively new edge detection operator with good detection performance and has been widely applied in many scopes, to extract edges from image called "fragments". The Canny algorithm is an optimal edge detection method based on a specific mathematical model for edges. According to the edge detecting theory and algorithm, Canny operator has been compared with Sobel operator, Prewitt operator, Robert operator and LoG operator through edge-detection experiments by programming (Bai X, 2007). The experimental results indicate that the edge-detection performance of Canny operator within a professional image retrieval system is better than other edge operators, especially under the situation of noises of images the Canny operator has both differencing and smoothing effects and the later is useful to reduce noise in resulting fragments.

2.2 Salient closed boundary extraction (Wang S et al., 2005)

All the so-called edges gotten from edge detection by Canny algorithm in fact are only a set of boundary fragments, not the meaningful boundaries what we really want. In order to represent an image, a boundary of the shape with the largest saliency is needed. Shape is usually a polygonal or irregular region with closed boundary composed of multiple straight or curve lines. Ratio contour (RC) is adopted to extract salient closed boundaries from noisy images.

The ratio-contour algorithm encodes the Gestalt laws of closure, proximity and continuity in the boundary-saliency measure based on the relative gap length and average curvature when connecting fragments to form a closed boundary. Closure requires that the boundary be a cycle. Proximity requires the gap between two neighboring fragments to be small. Continuity requires the resulting boundary to be smooth. In RC, all the arcs are constructed to be smooth curve segments to achieve good boundary continuity. Let $q(t)=(x(t),y(t))$, $t\in[0,L(B)]$ be the arc-length parameterized representation of a valid closed boundary B formed by alternately connecting a sequence of arcs and fragments, where *L(B)* is the boundary length. We know that q(L(B))=q(0) as the boundary is closed.

In RC, the boundary-cost $R(B)$ (negatively related to the boundary saliency/desirability, e.g., $\phi(B)=e^{-R(B)}$) is defined as:

$$R(B)=\frac{\int_0^{L(B)}\left[\sigma(t)+\lambda k^2(t)\right]dt}{L(B)} \tag{1}$$

where $\sigma(t)=1$ if $q(t)$ is on a gap-filling arc and $\sigma(t)=0$, otherwise. $k(t)$ is the curvature of the boundary at $q(t)$.

Since all the fragments are of zero length, they have no direct contribution to this cost. In the numerator of (1), the first term $\int_0^{L(B)} \sigma(t)dt$ makes it biased to a boundary with longer detected arcs and shorter gap-filling arcs. This reflects the preference of better proximity.

The second term $\int_0^{L(B)} k^2(t)dt$ reflects the favor of smoother boundaries, or better continuity. The denominator normalizes the cost by the boundary length *L(B)* to avoid a bias to shorter boundaries. λ >0 is a regularization factor that balances the proximity and continuity in the cost function. From (1), boundary-cost *R(B)* associate the length and the curvature of all gaps of boundary B. So the most salient closed boundary is the boundary with minimum boundary-cost, that is an alternate cycle with minimum cycle ratio namely Minimum Ratio Alternate (MRA) cycle.

The ratio contour algorithm for finding an MRA cycle in a solid-dashed (SD) graph G consists of three steps:

1). Setting the weight and length of solid edges in G to zero. Since the weight and length of any edge in the constructed SD graph G are usually nonzero, so in this step we transform the edge weights and lengths in G so that all solid edges have zero weight and length without changing the MRA cycle.

2). Detecting negative-weight alternate (NWA) cycles. By this step we reduce the problem of finding an MRA cycle in an SD graph G to the problem of finding an alternate cycle with negative total edge weight in the same graph G by searching for an appropriate transformation of the edge weights in G that preserves the MRA cycle but where the MRA cycle has a cycle ratio of zero.

3). Finding minimum weight perfect matchings (MWPM) to get the MRA cycle. The problem of detecting an NWA cycle in an SD graph where all the solid edges have zero weight and length can be reduced to the problem of finding an MWPM in the same graph. A perfect matching in G denotes a subgraph of G that contains all the vertices in G, but where each vertex only has one incident edge.

Fig. 1 illustrates the operation of the most-salient closed boundary extraction implementation on a sample image of black cutworm from the agricultural pests image database.

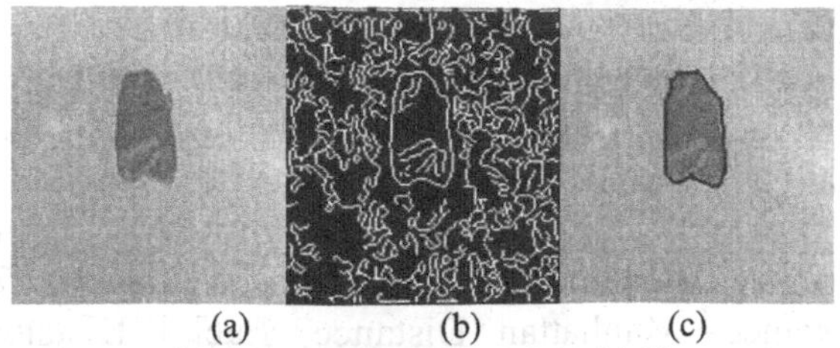

Fig. 1. (a) A sample image; (b) Output of edge detection; (c) The extracted salient boundary using RC.

3. FEATURE DESCRIPTION BASED ON THE SALIENT CLOSED BOUNDARY

In this paper, we adopt the centroid-radii model (CRM) (K. L. Tan et al., 2000) to represent the most-salient closed boundary feature. In the centroid-radii model, lengths of a shape's radii from its centroid at regular intervals are captured as the shape's descriptor (see Fig. 2). More formally, let θ be the regular interval (measured in degrees) between radii. Then, the number of intervals is given by $k = \lfloor 360 / \theta \rfloor$. Furthermore, we suppose that the intervals are taken anticlockwise starting from the x-axis direction. Thus, the shape descriptor can be represented as a vector $(l_0, l_\theta, l_{2\theta}, ..., l_{(k-1)\theta})$ where $l_{i\theta}$ $(0 \leqslant i \leqslant (k-1))$ is the $(i+1)$th radius from the centroid to the boundary of the shape. With sufficient number of radii, dissimilar shapes can be differentiated from each other.

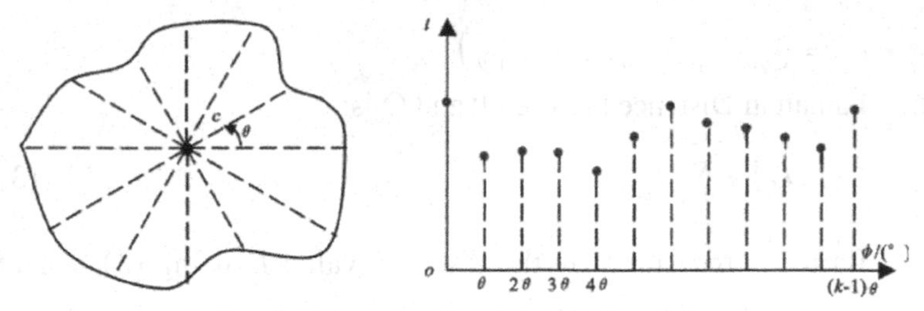

Fig. 2. The centroid-radii modeling of shape.

4. SIMILARITY MEASURE

Similarity measure is a key step to achieve good results of an image retrieval system. There are various formulas for distance calculation between the feature vectors and the result of the similarity measure is closely linked to which kind formula is adopted. A fixed image similarity measure cannot meet the need to adapt to different focuses of attention of different users.

It could be argued that we should choose different calculation methods for different feature vectors. There are many methods of image similarity measurement such as all kinds of distance measures, statistics methods and non-geometry similarity measure ways. Nowadays, commonly used formulas for distance calculation in content based image retrieval include: Minkowsky Distance, Manhattan Distance, Euclid Distance, Weighted Euclid Distance, Mahalanobis Distance, Quadratic Distance and they all have their respective appropriate application situations.

Considering the characteristics of the image we deal with and the retrieval efficiency of the image retrieval system, this paper adopts Manhattan Distance (2) to calculate the similarity between the query image and the images in the database.

$$D(X,Y) = \sum_{i=1}^{n} |x_i - y_i| \tag{2}$$

In (2), x_i and y_i are corresponding elements of feature vectors participated in matching, n is the number of elements of the feature vector. $D(X,Y)$ is the similarity distance. Here the smaller the similarity distance, the better the similarity. In order to embody the different importance of the different feature for psychological recognizing and matching, the Manhattan Distance could be adjusted by weighting parameter.

Suppose P is the query image, Q is an arbitrary image in image database, the centroid-radii vector of the salient shape boundary of P and Q are respectively:

$$L_p = (l_{p0}, l_{p\theta}, l_{p2\theta}, \ldots, l_{p(k-1)\theta})$$

$$L_q = (l_{q0}, l_{q\theta}, l_{q2\theta}, \ldots, l_{q(k-1)\theta})$$

then the Manhattan Distance between P and Q is:

$$D(P,Q) = \sum_{i=0}^{k-1} |l_{pi\theta} - l_{qi\theta}| \tag{3}$$

The similarity is measured by the distance value D using (3), where $l_{pi\theta}$ and $l_{qi\theta}$ is the centroid-radii of both P and Q. k as the number of the centroid-radii is determined by θ the regular interval between radii, and θ can be determined by the experiments or relevance feedback.

5. PRELIMINARY EXPERIMENTAL RESULTS

In this section, we demonstrate the performance of our method using an experiment. The aim of this experiment is to evaluate our algorithm and compare it with the feature vector based retrieval image algorithm. For our results to be relevant, we used the same hardware environment running on the Windows XP operating system. Preliminary experiments were done on agricultural pests image database. Precision and recall are used to evaluate the efficiency and effectiveness of the algorithm. Precision is the ratio of the number of images related to the query image retrieved by the system to the total number of images retrieved automatically by the system. Recall is the ratio of the number of images related to the query image retrieved by the system to the number of all images related to the query image in the database.

Fig. 3 shows the 10 sample images used as query images to be retrieved in the database. Fig. 4 shows a graph illustrating retrieval precision and recall of the two methods used in image retrieval system on agricultural pests image database.

Fig. 3. The sample images used to be retrieved in the database.

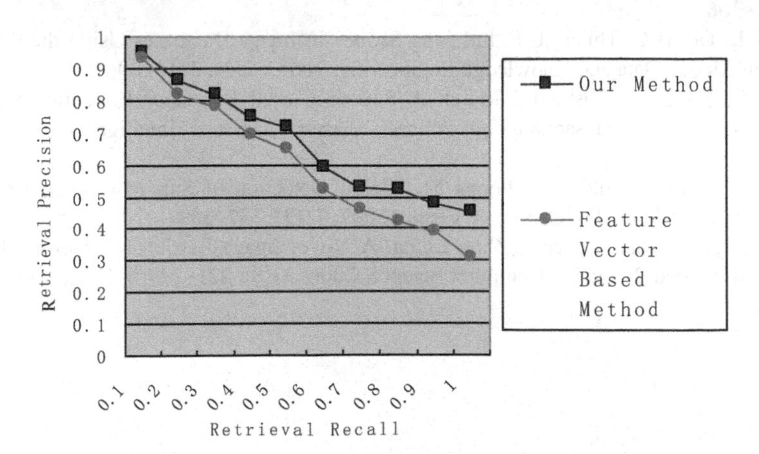

Fig. 4. Retrieval precision and recall of the two methods.

As be seen from Fig. 4, the preliminary experimental results show that the proposed retrieval method has better retrieval precision and recall. Moreover, there are two factors to be worth notice for precision and recall statistics. One is the structure of the image database and the other is the subjective judgment of human's eyes for image similarity.

6. CONCLUSION

With our new image retrieval algorithm, based on shape's salient closed boundary, we are aiming at a flawless image retrieval of relatively simple images including some shapes with salient boundary. Compared to the conventional image retrieval methods, a simple shape feature representation focused on salient closed boundary has been used in this algorithm. Similarities were measured by feature vector of the salient closed boundary based on the centroid-radii model. Experimental results show that our method used in many professional image retrieval systems has a better performance in terms of efficiency and effectiveness.

REFERENCES

Bai Xue. Study on the Algorithm of Footprint Comparing [D]. Beijing: China Agricultural University, 2007, 18~23(in Chinese)

Gondra I, Heisterkamp D R. Content-based Image Retrieval with the Normalized Information Distance. Computer Vision and Image Understanding, 2008

Han Jun Wei, Guo Lei. A Shape-Based Image Retrieval Method Using Salient Edges. Image Communication, 2003, 18: 141~156

Kong Xiaodong, Luo Qingshan, Zeng Guihua, Lee M H. A New Shape Descriptor Based on Centroid–Radii Model and Wavelet Transform. Optics Communications, 2007, 273: 362~366

Tan K L, Ooi B C, Thiang L F. Indexing Shapes in Image Databases Using the Centroid - Radii Model. Data and Knowledge Engineering, 2000, 32(3): 271~289

Wang S, Kubota T, Siskind J M, et al. Salient Closed Boundary Extraction with Ratio Contour. IEEE Transactions on Pattern Analysis Machine Intelligence, 2005, 27(4): 546~561

Wang S, Stahl J, Bailey A, Dropps M. Global Detection of Salient Convex Boundaries. International Journal of Computer Vision, 2007, 71(3): 337~359

Zeng Zhiyong, Zhang Xuejun, Zhou Lihua. A Novel Image Retrieval Algorithm Based on Salient Closed Boundary. Computer Science, 2006, 33(8): 221~224(in Chinese)

OILSEED RAPE PLANTING AREA EXTRACTION BY SUPPORT VECTOR MACHINE USING LANDSAT TM DATA

Yuan Wang [1,2] , Jingfeng Huang [1,3] , Xiuzhen Wang [4] , Fumin Wang [1,3] , Zhanyu Liu [1,3] , Junfeng Xu [1,3*]

[1] *Institute of Agricultural Remote Sensing and Information Technology, Zhejiang University, 310029, Hangzhou, P. R. China*

[2] *Ministry of Education Key Laboratory of Environmental Remediation and Ecological Health, Zhejiang University, 310029, Hangzhou, P. R. China*

[3] *Key Laboratory of Agricultural Remote Sensing and Information System, Zhejiang Province*

[4] *Zhejiang Meteorological Institute, 310004, Hangzhou P. R. China*

* *Ccorrespondence Author, Address: Institute of Agricultural Remote Sensing and Information Technology, Zhejiang University, 310029, Hangzhou, P. R. China, Tel: +86(0)571 8697 1830, Fax: 86-571-86971831, E-mail: xjfl1@zju.edu.cn*

Abstract: One parametric classify (Maximum likelihood classify, MLC for short) and two non-parametric classifiers (Adaptive resonance theory mappings and Support vector machines, ARTMAP and SVM for short) were presented in this study. Base on the confusion matrix and the pixels fuzzy analysis, the non-parametric classifier may be a more preferable approach than the parametric classifier for some remote sensing applications and deserves further investigation. The ARTMAP classify represent much best than the rest of classify, especially for the grade of pure pixel (90-100% pureness), Kappa coefficients and overall accuracy were nearly 100%. The higher pureness the pixels were, the better classification accuracy was got.

Keywords: planting area extraction, SVM, ARTMAP, neural networks, oilseed rape, confusion matrix

---

Please use the following format when citing this chapter:

Wang, Y., Huang, J., Wang, X., Wang, F., Liu, Z. and Xu, J., 2009, in IFIP International Federation for Information Processing, Volume 294, *Computer and Computing Technologies in Agriculture II, Volume 2*, eds. D. Li, Z. Chunjiang, (Boston: Springer), pp. 861–870.

1. INTRODUCTION

Spectral classifiers use only the spectral information of the pixel to be classified. Spectral classifiers are split into two main categories: (*a*) parametric classifiers, if they assume the existence of an underlying probability distribution of the data, and (*b*) non-parametric classifiers, if they do not assume anything about the probability distribution. When classifying a multi-spectral image by using a spectral classifier such as the classical maximum likelihood classifier(MLC), the resulting thematic map will usually give the overall impression of a 'noisy' classification. The thematic map that they obtain is composed by many isolated labeled pixels inside patches of other classes giving the overall impression of a noisy classification. This negative effect is more evident when there is overlapping between the training sets in the spectral space (Cortijo *et al.*, 1995). When this occurs the traditional quadratic ML classifier and the linear classifiers are not able to obtain a high accuracy classification due to the fixed-form decision boundaries they impose.

In most classification applications the assumption that the forms of the underlying density functions are known is unrealistic. The common parametric forms rarely fit the densities actually encountered in practice. For instance, the parametric models manage unimodal densities, whereas many practical problems present multi-modal densities (Duda and Hart, 1973). The only available information is the training set and the classification rules must be built only from it with no additional assumptions.

In this study, One popular used parametric classify, Maximum likelihood classify (MLC), andtwo non-parametric classifiers were selected, support vector machine (SVM) (Cristianini and Shawe-Taylor, 2000) and ARTMAP (Vapnik, 1995) neural network. We shall show that alternative spectral classifiers may improve the accuracy of the classifications as they are able to define a wider set of decision boundaries between the clusters in the representation space, thus reducing the classification error rate. In this work, we perform a comparative study of different spectral classifiers, when the available training samples are highly overlapping in the representation space.

2. MATERIALS AND METHODS

2.1 Study area and data acquisition

An area of 5km×5km located around Haiyan country within the Zhejiang province, China was selected. Mapping urban and sub-urban land covers is also recognized as an important but difficult task (Barnsley and Barr, 1996). In order to obtain the ground reference data, a precision tracking of the study area feature was got using a viliami grade GPS Receiver (Trimble GeoXT) in April to May, 2004, and created the detail feature vector map of the study area (Fig. 1). There were a variety of urban thematic and topographic features at the site. The main crops of this season were rape and wheat. The other crops were sporadic, blend in a mixed type. According to the area and attributes of the study area, the district features of the ground vector map category into residential area, rape, wheat, road, tree, water, grass and the other crops. Then transferred the features map to 1-meter, 3-meter and 30-meter grid data. 1-meter grid data use for remote sensing images for simulation, 3-meter data for the pixel fuzzy analysis, got the pixel percentage of the 30-meterdata. 30-meter grid data as a ground reference data of TM image.

Landsat 5 TM data of the site were used. All analyses were based on the data acquired in six of the TM's six wavebands. These were TM bands 1, 2, 3, 4, 5 and 7. The Landsat 5 TM data was acquired in April 5th, 2004; correspond to the ground reference data, and can be safely assumed free of significant temporal differences in land cover. For the purpose of this study, the following five classes appropriate to the scene defined: residential area (built-up land), rape, wheat, road (road land including road, bridge, barren land and sand road), tree (including fruit trees and mulberry), water (water bodies, including lakes ponds, and channel), grass (including grassland, grass road, bush, vegetable and wasteland) and the other crops (other crops, including soybean, horsebean, leek, cotton and corn).

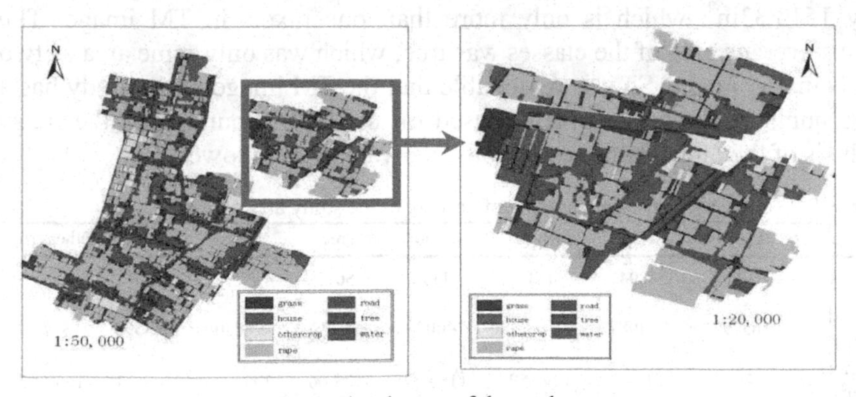

Fig.1: Sketch map of the study areas

2.2 Derivation of fuzzy ground data

Mixed pixel problem has a great influence for remote sensing image classification accuracy. In order to analysis the impact of mixed-pixel, pixel purity data should be got. This study used 3-meter grid data, degrade to 30-meter grid data, so have 100 3-meter lattices in one 30meter lattice. Calculating 3-meter lattice numbers of every ground features in the 30-meter lattice, got the percentage of every features in the pixels, expressed as the pixel features purity. Finally, the fuzzy ground data derived from the pixel features purity were stored as an eight-band image, one for each class.

Data handling and analysis were made in ENVI 4.2, ERDAS IMAGINE 8.7, ARCGIS 9.0, MATLAB 7.0 (Mathworks, USA), and LIBSVM 2.8.3 (Chang and Lin, 2001).

3. RESULTS

3.1 Pixel Fuzzy Analysis

In order to have a summary recognition of the study area, there was a brief statistic shown in Table 1.Among all 8 classes of the study area, grass is the largest class which have 33.68% of the area, followed by rape, tree, wheat, water, residential area, road and other crop, rape and wheat are the mainly crop of the area. Grass has the most parcels counts, the average area was only 1513.33m^2, which is only more than one pixels in TM image. The largest average area of the classes was tree, which was only same area as two pixels in TM image. So it is reasonable that the TM image of the study has a large number of mixture pixels. Based on the GPS ground data, the fuzzy analysis of the 30m spatial resolution data was given followed.

Table 1. Average area and compactness of parcels in the study area

Class	Residential area	Rape	Wheat	Road	Tree	Water	Grass	Other crop
Count	125	894	151	117	501	132	1020	85
Summation (m^2)	165797	1340877	194880	140023	870772	194820	1543598	131786
Average (m^2)	1326.37	1499.86	1290.59	1196.77	1738.06	1475.9	1513.33	1550.42
Percentage (%)	3.62	29.26	4.25	3.06	19.00	4.25	33.68	2.88
Standard deviation	883.39	870.39	846.41	935.35	857.19	870.75	863.10	822.59
Variance	780381.10	757582.01	716411.37	874887.51	734782.58	758213.56	744953.07	676654.36

The fuzziness of TM image can be calculated from the proportion of one class in overall pixels that contain this class in the 30m ground reference data. 7 grades was ranked in this study, 100%, 90-99%, 80-89%, 60-79%, 40-59%, 20-39% and 1-19%, the result shown in Table 2. It was clear showed that in the pixels that contain class rape, the pure pixels only have 15.86%, the pixels contain 1-19% of rape have 18.02% overall, 11.51%, 11.35%, 15.54%, 13.87% and 13.87% for the 90-99%, 80-89%, 60-79%, 40-59% and 20-39% pixels respectively. The residential area has 17.27% pure pixels, which was the most in all class, the least class was grass, and only have 1.53%. Otherwise, the road and grass have 66.70% and 65.13% of 1-19% pixels.

Table 2. Percent of one object in pixels contained the object in reference map with 30m resolution

Grade	Residential area	Rape	Wheat	Road	Tree	Water	Grass	Other crop
100%	17.27	15.86	3.66	2.75	4.23	3.59	1.53	12.22
90-99%	9.35	11.51	2.48	2.66	2.96	6.34	1.02	4.52
80-89%	6.06	11.35	4.05	3.20	4.12	4.79	0.82	4.75
60-79%	12.36	15.54	8.62	3.37	8.88	12.68	2.61	8.82
40-59%	13.65	13.87	14.23	4.88	14.23	17.76	6.73	10.18
20-39%	16.22	13.87	26.89	16.43	23.48	19.80	21.97	16.74
1-19%	25.10	18.02	40.08	66.70	42.09	35.02	65.31	42.76
Total	100	100	100	100	100	100	100	100

3.2 Classification evaluation

Because of the few pure pixels in TM image, the train data of the three classifies used the pixels which was more than 90% purity of all pixels, the total of train data was 2465, the number of train data for residential area, rape, wheat, road, tree, water, grass and the other crops was 558, 1358, 47, 61, 136, 141, 90 and 74 respectively. The inputs of the classifies were the six wavebands of the TM image(bands 1, 2, 3, 4, 5 and 7), and the out put of the classifies was the code of each classes(1, 2, 3, 4, 5, 6, 7 and 8 for residential area, rape, wheat, road, tree, water, grass and the other crops, respectively). Maximum likelihood classifies (MLC) was automatic run in software ENVI. The SVM classify selected RBF kernel, and the train epoch was 10000. In the ARTMAP classify, three alertness index was selected for train, 0.10, 0.75 and 0.90, finally used 0.75 as the alertness index because of the fine classification result.

The accuracy of each classification method is assessed based on its confusion matrix. Recall that the entry of *i*th row and *j*th column of this matrix is the number of sample pixels from the *j*th class that have classified as belonging to the *i*th class. Various indicators are derived from this matrix, such as overall accuracy, producer's accuracy, user's accuracy and Kappa coefficients. The confusion matrixes of each classification were showed in Table 3-5. The overall accuracy and Kappa coefficients denoted that ARTMAP has the highest value among the three classifies, followed by SVM, MLC has the lowest. Compared producer's accuracy and user's accuracy of different classifies; results showed that ARTMAP was not always the highest. The highest user's accuracy of wheat, road, tree and grass were ARTMAP, residential area and rape for MLC, water and other crop for SVM. The highest producer's accuracy of residential area, water, grass and other crop were ARTMAP, wheat, road and tree for MLC, rape for SVM. It is obvious that different classifies has differentia for different classes.

Table 3. The confusion matrix of classification by MLC

Class	Ground reference data									
	Residential area	Rape	Wheat	Road	Tree	Water	Grass	Other crop	Total	User accuracy
Residential area	440	136	7	13	75	85	39	13	808	54.46
Rape	175	2150	60	13	129	62	117	52	2758	77.96
Wheat	17	125	115	0	13	12	23	2	307	37.46
Road	71	102	2	63	19	40	27	7	331	19.03
Tree	197	234	12	7	250	87	67	20	874	28.60
Water	57	88	2	15	36	233	33	6	470	49.57
Grass	101	254	29	11	71	29	68	15	578	11.76
Other crop	131	179	5	13	28	49	21	60	486	12.35
Total	1189	3268	232	135	621	597	395	175	6612	3379
Prod.accuracy	37.01	65.79	49.57	46.67	40.26	39.03	17.22	21.67	34.21*	51.10**

* Kappa coefficients, ** overall accuracy

Table 4. The confusion matrix of classification by SVM

Class	Ground reference data									
	Residential area	Rape	Wheat	Road	Tree	Water	Grass	Other crop	Total	User Accuracy
Residential area	760	319	15	71	200	224	95	49	1733	43.85
Rape	378	2818	155	80	265	157	243	98	4194	67.19
Wheat	0	28	58	0	1	1	3	0	91	63.74
Road	1	6	0	14	2	0	5	0	28	50.00
Tree	29	52	1	4	134	21	21	6	268	50.00
Water	19	36	1	10	13	189	19	2	289	65.40
Grass	2	7	2	7	2	1	3	1	25	12.00
Other crop	6	6	0	4	4	5	2	21	48	43.75
Total	1195	3272	232	190	621	598	391	177	6676	3997
Prod.accuracy	63.60	86.12	25.00	7.37	21.58	31.61	0.77	11.86	36.99*	59.87**

Table 5. The confusion matrix of classification by ARTMAP

Class	Ground reference data									
	Residential area	Rape	Wheat	Road	Tree	Water	Grass	Other crop	Total	User accuracy
Residential area	817	241	15	43	134	152	84	23	1509	54.14
Rape	246	2727	108	67	197	128	141	55	3669	74.33
Wheat	3	24	79	0	2	3	4	0	115	68.70
Road	6	12	1	43	1	5	5	0	73	58.90
Tree	39	75	6	5	211	24	18	5	383	55.09
Water	38	85	3	13	38	257	31	7	472	54.45
Grass	22	59	18	7	27	11	102	5	251	40.64
Other crop	22	44	2	12	11	18	5	81	195	41.54
Total	1193	3267	232	190	621	598	390	176	6667	4317
Prod.accuracy	68.48	83.47	34.05	22.63	33.98	42.98	26.15	46.02	47.63*	64.75**

* Kappa coefficients, ** overall accuracy

The overall accuracy and Kappa coefficients description previous only evaluated the total accuracy of classification; producer's accuracy and user's accuracy were discrete and always disagree with each other. In order to get much better classification evaluation, a new Classification evaluation index—"precision product" was used in this study, in was the product of producer's accuracy and user's accuracy, the higher of precision product of one class, the higher classification accuracy of this class. Fig. 2 showed the precision product of various classifiers. Rape has the highest classification accuracy for each classifies, which followed by residential area, water, wheat, tree, other crop and road, grass has the lowest classification accuracy. In the three classifies, the ARTMAP has the highest classification accuracy for each class.

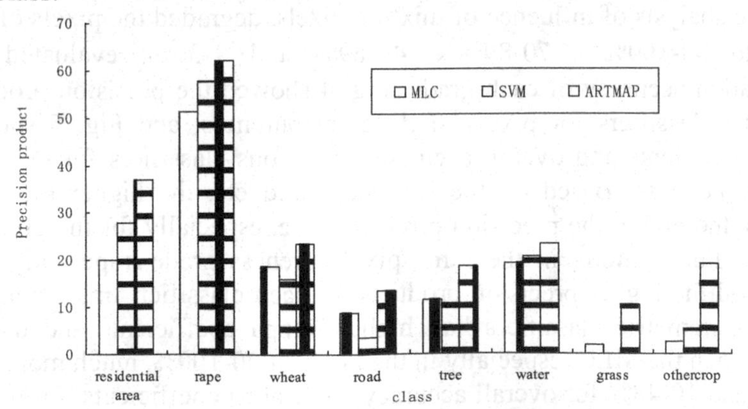

Fig. 2. Precision product of various classifiers

The non-parametric classifiers have better classification accuracy than the parametric classifiers. Non-parametric classifiers still have a low value of classification accuracy index. The ARTMAP has the highest overall accuracy, which was only 64.87%; it may be influence by the mixture pixels (Ju et al., 2005). It is necessary to analysis the influence of the mixture pixels for the classification. Fig. 3 was showed the Classification results of various classifiers.

Fig.3. Classification results of various classifiers

3.3 The influence of mixture pixels for Classification

For the analysis of influence of mixture pixels, degraded the pixels of TM image to 90-100% 、 70-89% 、 40-69% and 1-39%, evaluated the classification accuracy of each grade. Fig. 4 showed the precision products of various classifiers for pixels of different pureness, and Fig. 5 showed Kappa coefficients and overall accuracy of various classifiers for pixels of different pureness. Based on the precision products, the higher the pixel pureness, the higher the precision products have, especially for the grade of 90-100%. Furthermore, in the same pixel pureness grade, rape and water always had the higher precision products for each classifier. In general, the two non-parametric classifiers had higher Kappa coefficients and overall accuracy than the MLC, especially in the grade of 90-100%, much more than 11-33% and 10-48% for overall accuracy and Kappa coefficients. From Fig. 4 and 5, it can be concluded that the ARTMAP was the much fine classify, especially in the grade of 90-100%, Kappa coefficients and overall accuracy were nearly 100%, and it's an excellent result.

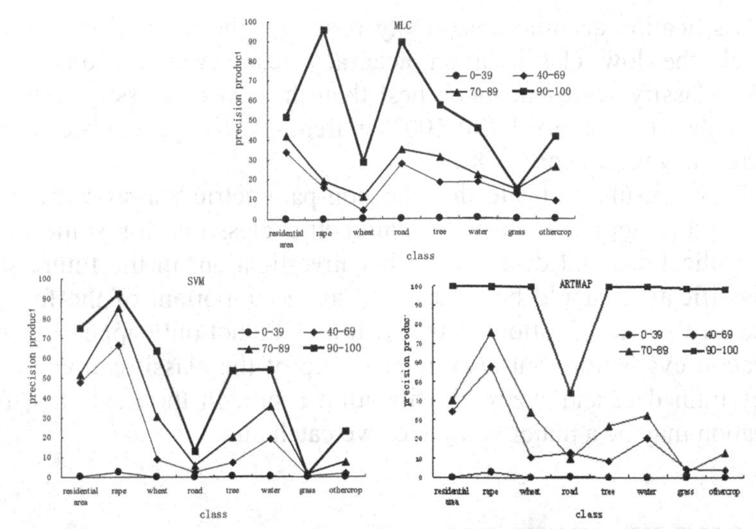

Fig. 4. Precision product of various classifiers for pixels of different

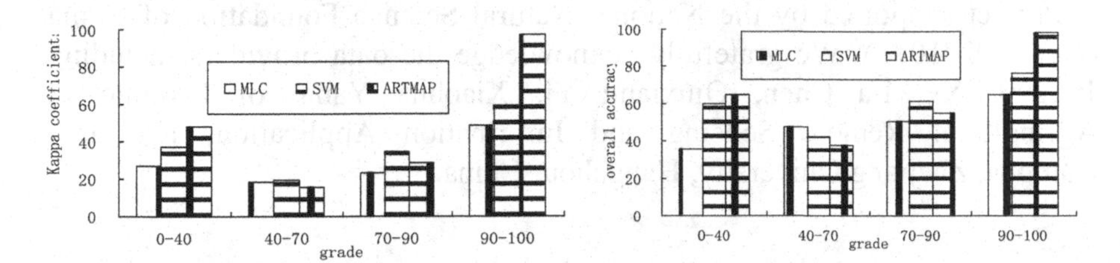

Fig. 5. Kappa coefficients and overall accuracy of various classifiers for pixels of different pureness

4. CONCLUSIONS

Supervised classification is a technique used commonly in the analysis of remotely sensed data. In the course of a classification many assumptions are made. To avoid problematic assumptions about the datasets made by conventional statistical classification algorithms, much attention has recently focused on non-parametric classifiers. This was illustrated for classifications with SVM and ARTMAP neural networks. For both types of non-parametric classifiers, the classification accuracy was prior to the parametric classifier—MLC, whereas the classification evaluation of both non-parametric classifiers were also not so much satisfaction.

There were a majority of mixture pixels of the TM image in the study area; it was also the representation for the agriculture region. The influence of the mixture pixels was the primary problem for the classification result and accuracy evaluation. The more fuzziness of the remote sensing data, the worse classification accuracy has. Only based on the spectral data and the pure pixel, the low classification accuracy result was reasonable. The ARTMAP classify represent much best than the rest of classify, especially for the grade of pure pixel (90-100% pureness), Kappa coefficients and overall accuracy were nearly 100%.

Overall, the results indicate that the non-parametric classifier may be a more preferable approach than the parametric classifier for some remote sensing applications and deserves further investigation. In the future study, fuzzy classification should be focused on as the important of the fuzziness influence for the classification accuracy. It may be not only appearing in the classification evaluation, but also in each step of the classification, such as the fuzzy train data and fuzzy classification result. In the next step, fuzzy classification may be a major study area we catch out.

ACKNOWLEDGEMENTS

Project supported by the National Natural Science Foundation of China (Nos. 40271078). We gratefully acknowledge the data providers including Junfeng Xv, La Chen, Qiuxiang Yi, Xiaohua Yang, of Institute of Agricultural Remote Sensing and Information Application, Huajiachi Campus, Zhejiang University, Hangzhou, China.

REFERENCES

C. C. Chang, C. J. Lin. LIBSVM : a library for support vector machines, 2001. Software available at http://www.csie.ntu.edu.tw/~cjlin/libsvm.

F. J. Cortijo. A Comparative Study of Classification Methods for Multispectral Images. PhD thesis, DECSAI, Universidad de Granada, Spain. 1995. Available at http://decsai.ugr.es/~cb/entrada_tesis.html, electronic edition.

M. J. Barnsley, S. L. Barr. Inferring urban land use from satellite sensor images using kernel-based spatial reclassi. Cation, Photogrammetric Engineering and Remote Sensing, 1996,62: 949-958.

N. Cristianini, J. Shawe-Taylor. An Introduction to Support Vector Machines. Cambridge University Press, Cambridge, UK, 2000.

R.O. Duda, P.E. Hart. Pattern Classification and Scene Analysis (New York: John Wiley & Sons), 1973.

V. Vapnik. The Nature of Statistical Learning Theory. Springer-Verlag, New York. 1995.

ASSESSING THE IMPACT OF CLIMATE CHANGE ON SOIL WATER BALANCE IN THE LOESS PLATEAU OF CHINA

Zhi Li [1, 2,*] , Wenzhao Liu [2], Xunchang Zhang [3]

[1] *College of Resources and Environmental Science, Northwest Sci-Tech University of Agriculture and Forestry, Yangling Shaanxi 712100, China;*

[2] *Institute of Soil and Water Conservation, Chinese Academy of Sciences and Ministry of Water Resources, Yangling Shaanxi 712100, China;*

[3] *USDA-ARS Grazinglands Research Laboratory, El Reno, OK, USA.*

[*] *Corresponding author, Address: College of Resources and Environmental Science, Northwest Sci-Tech University of Agriculture and Forestry, Yangling Shaanxi 712100, China; Tel: +86-29-87011683, Email: lizhibox@126.com*

Abstract: Soil water balance has response to climate change and evaluation of soil water change is one of the most important items of climate change impact assessment. GCM outputs under three scenarios were statistically downscaled during 2010～2039 to simulate the potential change of soil water balance in Wangdonggou watershed on the Loess Plateau with WEPP model. GCM predicted a 1.8 to 17.5% increase in annual precipitation, 0.5 to 0.9 ℃ rises in maximum temperature, 2.0 to 2.3 ℃ rise in minimum temperature for the region. Plant transpiration will mainly change from April to June and soil evaporation mainly changed during July to September. Percent increases under climate changes, as averaged for each emissions scenario and slope, ranged from -5 to19% for crop transpiration, -4 to 4% for soil moisture, -7 to 7% for soil evaporation, 6.5 to 44.1% for wheat grain yield, 26.3 to 41.7% for maize yield. Climate change will affect soil water balance significantly and some countermeasures are necessary.

Keywords: climate change; soil water balance; GCM, WEPP, CLIGEN

Please use the following format when citing this chapter:

Li, Z., Liu, W. and Zhang, X., 2009, in IFIP International Federation for Information Processing, Volume 294, *Computer and Computing Technologies in Agriculture II, Volume 2*, eds. D. Li, Z. Chunjiang, (Boston: Springer), pp. 871–880.

1. INTRODUCTION

As the main constituent of terrestrial ecosystem, the functions and processes of soil changes in response to global climate change. Soil water reserve is one of the main sources of water that can be utilized by vegetation. The potential change of soil water induced by climate change may cause great change to ecological environment and agricultural production. As unique geomorphic units with fragile ecological environment, the potential changes of soil water in future on the Loess Plateau need to be assessed.

The current researches about impact of climate change on soil water are carried out mainly through integration of ecological models with future climate scenarios (Huszar et al., 1999; Mehrotra, 1999; Naden and Watts, 2001; Pan et al., 2001; Ramos and Mulligan, 2005; Zhang and Liu, 2005). The climate change scenarios used in these studies falls in two groups, one is from synthesis and the other is from GCM output. The former was derived according to some laws of climate change; the latter was the most popular method of developing climate change scenarios for GCM.

When GCMs are used to assess the impacts of climate change, two major obstacles exist in the site-specific impacts assessment. These obstacles are spatial and temporal scale mismatches between coarse resolution projections of GCMs and fine resolution data requirements of agricultural systems models(Hansen and Indeje, 2004). Various methods are developed to bridge the spatiotemporal gaps, which can be divided into two kinds: dynamic and empirical (statistical) approaches. Dynamic downscaling is used to achieve higher spatial resolutions by nesting Regional Climate Models (RCM) within GCM output fields. RCM output is computationally costly and quantitatively heavy (Solman and Nunez, 1999), and is only available for limited regions. Statistical technique derives statistical relationships between observed local climatic variables (predictands) and large scale GCM output (predictors) using regression-type methods, and applies this relationship to GCM future output to get the information of future climate change. Statistical methods are easily implemented and can be calibrated to local situation, therefore, statistical methods are frequently used to downscale GCM projections to finer spatiotemporal scales. A diverse range of statistical downscaling techniques has been developed. Those techniques in principle fall in three categories: weather generators, transfer functions, and weather typing schemes (Chen, 2000; Wilby et al., 1998).

At present, the research about impacts of climate change on soil water in China is mainly theoretical discussion rather than quantitative studies. When quantification was carried out, large scale GCM outputs were usually used directly. GCM cannot take into account the difference of regional condition, therefore, using GCM grid outputs directly to assess site-specific impact of climate change is not reliable(von Storch, 1995). These studies contributed to the development of impact assessment of climate change on soil, however, more detailed climate scenarios needed to be developed to carry out climate change impact. Owing to the limited use of RCM, empirical approach should be discussed further.

The objectives of this study were to (i) spatially downscale GCM grid output with a new statistical approach to Wangdonggou watershed on the Loess Plateau and temporally downscale GCM monthly output to daily series data using CLIGEN; (ii) evaluate the potential impacts of HadCM3-projected climate changes during 2010-2039 under A2a, and B2a, and GGa1 on soil water balance (Soil water, soil evaporation, plant transpiration and percolation). Results would provide reference information for ecological environment construction and agricultural production in the region.

2. MATERIALS AND METHODS

2.1 Site description

Wangdonggou watershed is located at Changwu County, Shaanxi Province on the Loess Plateau (107°40′30″ ~ 107°42′30″E, 35°12′16″ ~ 35°16′00″N). The elevation is 946~1226 m above sea level. The prevailing landform is loessial tableland and gullyland, covering 35% and 65% of the watershed respectively. The loess is more than 100-m thick on the tableland. The soil is predominantly silt loam with silt content greater than 50% (two soil series: Huangmiantu and Heilutu). The averaged annual precipitation is 582.3 mm, with 52.8% falling in July through September. The mean annual temperature is 9.2 ℃. The common regional cropping system is a three-year rotation of winter wheat–winter wheat–spring maize. Rainfed agriculture is the dominant production system.

2.2 Data sources

Two group data are needed, viz. the CGM grid output where Wangdonggou watershed is located and the measured data of Wangdonggou watershed. Three GCM scenarios of A2a, B2a, and GGa1 from HadCM3 and three meteorologic variables (precipitation, maximum and minimum temperature) of each scenario were used. Three scenarios all include the hindcasts of 1957-2001 and the projections of 2010-2039, the measured data are from Changwu weather bureau, including daily series of precipitation, maximum and minimum temperature, wind speed and direction et al.

2.3 Generating climate scenarios

2.3.1 Spatially downscaling GCM grid outputs

A new statistical approach proposed by Zhang (2005) was used to spatially downscale. The approach emphasizes the parity of probability distributions between measured monthly quantities and downscaled GCM-projections while it relaxes the prerequisite of strong correlationships between local measurements (predictands) and GCM output (predictors). The procedures are as following:

The GCM hindcast monthly precipitation from 1957 to 2001 was used as the control, and the historical monthly precipitation between 1957 and 2001 as the baseline. For each calendar month, the ranked observational monthly precipitation (Y-axis) was plotted with the ranked GCM-projected precipitation (i.e., paired by their ranks or corresponding quartiles of the observed vs. projected monthly precipitation, also called qq-plot). A simple univariate linear and a nonlinear function were fitted to each plot to obtain transfer functions for each month.

For impact assessment, those transfer functions were further used to downscale 2010-2039 monthly precipitation at the native GCM scale to those at the Wangdonggou watershed under the premise that the transfer functions developed under the present climate are applicable to the changed climate. For each calendar month, the nonlinear function was used to transform the projected monthly precipitation values that were within the range in which the nonlinear function was fitted, while the linear function was used for the values outside the range. The use of linear functions for the out-of-range values is to generate conservative, first order approximations. The downscaled monthly precipitation values, which represent the future monthly precipitation distribution at the Wangdonggou watershed, were then used to calculate monthly mean and variance of the changed climate for the location. Those calculated mean and variance of the downscaled monthly precipitation were used in the temporal downscaling method to generate daily weather series of the changed climate at the Wangdonggou watershed as presented in the following section.

Likewise, the GCM-projected monthly maximum and minimum temperatures were downscaled spatially in the same manner as was for monthly precipitation. Mean temperature shifts as well as variance ratios between the downscaled monthly GCM projections of 2010-2039 and the local monthly measurements of 1957-2001 were calculated for each month and were further used in temporal downscaling.

2.3.2 Temporally downscaling GCM monthly outputs

Measured daily weather data of 1957-2001 at Wangdonggou watershed were used to estimate the baseline CLIGEN input parameters, which were subsequently adjusted for the relative changes to generate the changed climate scenario for the target station. The precipitation-related baseline parameters including Pw/w, Pw/d, mean and variance of daily precipitation of wet days were adjusted as follows. For each month, future transitional probabilities of precipitation were estimated for projected monthly means from linear relationships developed using historical transitional probability and monthly precipitation at the Changwu station. The projected monthly means were obtained by multiplying mean ratios of GCM-projected monthly precipitation between 2010-2039 and 1957-2001 by the baseline monthly precipitation means measured during 1957-2001 at the Changwu station. The mean daily precipitation per wet day, which is a CLIGEN input parameter, was analytically computed using the adjusted transitional probabilities, projected monthly mean, and number of days in the month. New variances of daily precipitation under climate change, which is another input parameter for CLIGEN, was approximated by multiplying the baseline variances derived from the daily station records by the monthly variance ratios between the target and control periods under the assumptions that transitional probabilities and autocorrelation of daily precipitation in both baseline and changed climates are similar.

Projected mean maximum and minimum temperature shifts were directly added to the corresponding baseline means. Adjusted daily temperature variances were obtained by multiplying the baseline temperature variances by the calculated variance ratios. All new parameter values were then input into CLIGEN, and 100 years of daily weather data were generated for each of three emission scenarios.

2.4 WEPP calibration

Measured soil, climate, crop management information, surface runoff, and sediment yield from 1988 to 1992 were used to calibrate soil erodibility parameters of the Water Erosion Prediction Project (WEPP) model (v2004.7), which was modified to incorporate the effect of elevated CO2 on plant growth and evapotranspiration. Based on the above emissions scenarios, CO2 concentration by the year 2025 would increase to 592 ppmv (parts per million by volume) for A2a, 416 ppmv for B2a, and 445 ppmv for GGa1. Two field runoff plots and two cropping systems were selected. One runoff plot is 20.1 m long by 5 m wide with a 5° slope, the other is 20.3 m long by, 5 m wide with a 10° slope. A common regional three-year rotation of wheat-wheat-maize was selected. In the simulation under the baseline climate condition, winter wheat was planted on September 23 and harvested on June 27 of the following year; and maize was planted on April 15 and harvested on September 22. However, under the changed climates, wheat was planted 3 days later and harvested 3 days earlier; and maize was planted 3 days earlier and harvested 3 days earlier to accommodate the increased temperature. It should be noted that overall mean of measured storm duration at Changwu station was 2.88 times that of CLIGEN-generated storm durations. To adjust this bias, a factor of 2.88 was multiplied to CLIGEN-generated storm durations and relative peak intensities in both baseline and changed climates.

3. RESULTS AND DISCUSSIONS

3.1 Projected climate change

Precipitation of Wangdonggou watershed from 2010~2039 can be projected by downscaling (Table 1, Fig. 1). Projected mean annual precipitation during 2010~2039 compared with 1957~2001 would increase by 10.8, 80.6 and 101.4 mm (equivalent to 1.8, 13.9 and 17.5% increase) respectively, for the A2a, B2a and GGal scenario. The projected percent increases in precipitation varied with emission scenarios. In general, three emission scenarios projected more precipitation increases in the spring and summer than in the winter and fall (Fig. 1). The most change occurred in May and from July to September, during which precipitation increased for B2a and GGal. Precipitation of A2a decreased greatly in July and August while increased rapidly during September and October.

Table 1. Averaged annual climate perturbations between 1957～2001 and 2010～2039.

Emissions Scenario	P Change (%)	Tmax Shift (°C)	Tmin Shift (°C)
A2a	1.8	0.9	2.3
B2a	13.9	0.5	2.1
GGal	17.5	0.8	2.0

Compared with 1957~2001, projected mean annual temperature during 2010~2039 would increase by 0.9, 0.5 and 0.8 °C for maximum temperature and increase by 2.3, 2.1 and 2.0 °C for minimum temperature for the A2a, B2a and GGal scenario, respectively. In general, the projected temperatures increased for each scenario; however, the increase of maximum temperature was less than that of the minimum temperature, which was consistent with the results of National Assessment Report of Climate Change (Ding et al., 2007). The monthly distributions of temperature are shown in Fig. 2.

Generally, there were two peaks of temperature increase: one in the spring and the other in the winter, which would mean warmer winter in 2010~2039.

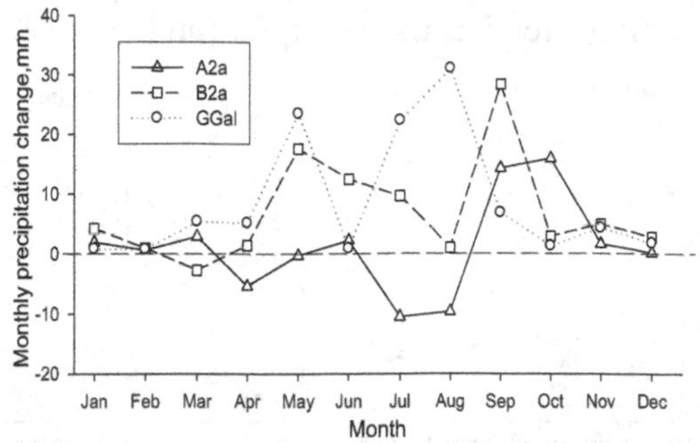

Fig.1: Predicted change in precipitation between 1957~2001 and 2010~2039 under three emission scenarios

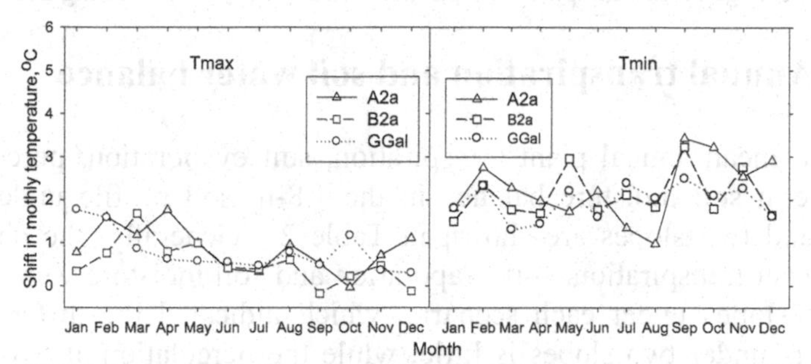

Fig. 2: Predicted change in maximum and minimum temperature between 1957~2001 and 2010~2039 under three emission scenarios

3.2 Monthly change of plant transpiration and soil evaporation

Seasonal patterns of plant transpiration and soil evaporation are shown in Fig. 3. The transpiration mainly occurred from April to June, and the soil evaporation mainly occurred from July to September, which was possibly the integrated effect of precipitation change and crop growth. From April to June are the growing periods of winter wheat from disjointing stage to heading stage, during which water consumption of wheat transpiration increases rapidly and the amount of transpiration accounts for more than 80% of the total ET. The water consumption of transpiration from April to June is mainly from precipitation that varies with the three climate scenarios in 2010~2039: (i) Precipitation changes little during April, which leads to little change in plant transpiration; (ii) Precipitation decreases for A2a and increases for B2a and GGal during May, and change of transpiration exhibits the similar trend; (iii) Precipitation and transpiration increases generally for three scenarios. From July to September is the fallow period of Wangdonggou fields when soil evaporation reaches the peak. As the soil water supply is from precipitation, the difference of soil evaporation depends also on the precipitation change. Monthly soil evaporation from July to September has the similar trend as the precipitation (Fig. 1 & 3).

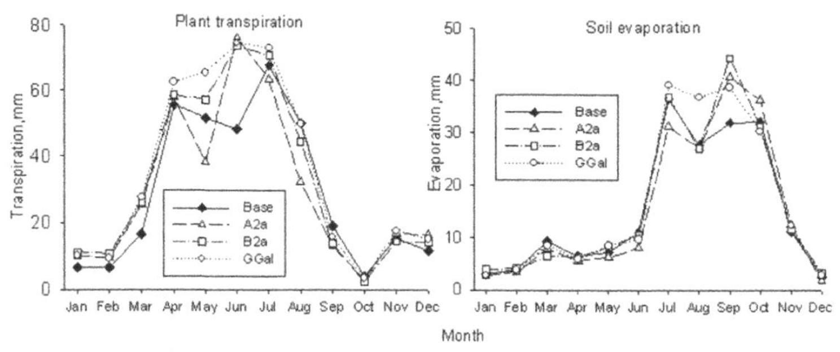

Fig. 3: The change of monthly plant transpiration and soil evaporation during 2010-2039

3.3 Annual transpiration and soil water balance

Predicted mean annual plant transpiration, soil evaporation, percolation, and long-term soil moisture balance in the 1.8-m soil profile under three scenarios and two slopes are shown in Table 2. Generally, the changing trends of plant transpiration, soil evaporation and soil moisture were similar across two slopes under each scenario, which indicated the difference of these factors under two slopes is little; while the percolation at two slopes has different changing trends, with greater percolation at 5°slope.

Table 2. Response of ET and soil water balance to three scenarios under conventional tillage

Scenario(CO_2 ,ppmv)		Baseline(350)		A2a(592)		B2a(416)		GGal(445)	
Slope		5°	10°	5°	10°	5°	10°	5°	10°
Transpiration	Depth(mm)	353	348	339	332	396	389	419	413
	Change (%)	0	0	-4	-5	12	12	19	19
Soil moisture	Depth(mm)	301	297	289	285	305	300	312	308
	Change (%)	0	0	-4	-4	1	1	4	4
Evaporation	Depth(mm)	183	179	171	166	187	182	196	192
	Change (%)	0	0	-7	-7	2	2	7	7
Percolation	Depth(mm)	0.58	0.58	0.87	0.76	0.58	0.58	0.83	0.71
	Change (%)	0	0	50	32	0	0	43	23
Wheat	Yield($t \cdot ha^{-1}$)	2.9	2.8	2.9	2.8	3.5	3.4	4.1	3.9
	Change(%)	0	0	-1	-1	23	22	41	41
Maize	Yield($t \cdot ha^{-1}$)	7	6.8	8	7.8	8.6	8.3	9.6	9.5
	Change(%)	0	0	15	14	23	22	38	38

Compared with the baseline conditions, plant transpiration changed from -5 to 19% across all emissions scenarios and slopes. The percent increases averaged across all slopes were -4.5% for A2a, 12% for B2a, and 19% for GGa1. The change of plant transpiration is the integrated effect of precipitation change and crop growth, for example, A2a predicted the least precipitation and crop yields, which led to the decrease of plant transpiration. The GGal scenario predicted the most precipitation increase and crop yields, which resulted in the most transpiration. Simulated soil moisture changed from -4 to 4%, compared with the baseline condition, A2a decreased while B2a and GGal increased, and GGal has the most increase. Soil evaporation has the similar trend as soil moisture, changed from -7 to 7%, simply because soil evaporation was mainly limited by soil water supply rather than by evaporative demand in the study region. Compared with the baseline condition, the deep percolation loss of A2a and GGal increased greatly while the B2a changed little.

4. CONCLUSION

With the new statistical approach, GCM grid outputs of 2010~2039 were downscaled to Wangdonggou watershed, three scenarios of A2a, B2a and GGal were developed to drive WEPP model to simulate the potential change of soil water balance under 5° and 10° slopes. The Hadley Centre model (HadCM3) predicted a 1.8 to 17.5% increase in annual precipitation, 0.5 ℃ to 0.9 ℃ rise in maximum temperature, and 2.0 ℃ to 2.3 ℃ rise in minimum temperature for the region, the increase of minimum temperature were more than maximum temperature, and increase of temperature mainly happened in spring and winter.

Compared with the baseline condition, the change trends of soil water balance under two slopes are similar. Change of plant transpiration and soil evaporation mainly occurred from April to June and from July to September, respectively. Compared with the baseline condition, predicted percent change under climate change, ranged from -5 to 19% for plant transpiration, -4 to 4% for soil moisture, -7 to 7% for soil evaporation, A2a decreased while B2a and GGal increased for these factors. Compared with the baseline condition, deep percolation loss of A2a increased the most, GGal intermediate and B2a changed little. The above changes are mainly caused by the precipitation change and crop growth. For example, the transpiration mainly occurs from April to June when the water supply is mostly from precipitation. The transpiration changed differently according to the precipitation change. The least increase of precipitation and yields for A2a led to the decrease of transpiration, while the most increase of precipitation and yields for GGal caused the most increase of transpiration. The above results suggest that climate change would influence soil water balance significantly and some countermeasures must be taken to mitigate the adverse effect of climate change.

ACKNOWLEDGEMENTS

This study was funded by National Natural Science Foundation of China (No. 40640420061) and the Outstanding Overseas Chinese Scholars Fund of Chinese Academy of Sciences (No. 2005-2-3).

REFERENCES

Chen, D., 2000. A monthly circulation climatology for Sweden and its application to a winter temperature case study. International Journal of Climatology, 20: 1067-1076.

Ding, Y. et al., 2007. China's National Assessment Report on Climate Change (I): Climate change in China and the Future Trend. Advances in Climate Change Research, 2(1): 3-8.

Hansen, J.W., Indeje, M., 2004. Linking dynamic seasonal climate forecasts with crop simulation for maize yield prediction in semi-arid Kenya. Agric. For. Meteo, 125: 143-157.

Huszar, T., Mika, J., Loczy, D., Molnar, K. and Kertesz, A., 1999. Climate change and soil moisture: A case study. Phys. Chem. Earth, 24(10): 905-912.

Mehrotra, R., 1999. Sensitivity of runoff, soil moisture and reservoir design to climate change in central Indian River basins. Climatic Change, 42(4): 725-757.

Naden, P.S. and Watts, C.D., 2001. Estimating climate-induced change in soil moisture at the landscape scale: An application to five areas of ecological interest in the U.K. Climatic Change, 49(4): 411-440.

Pan, Z., Arrit, R.W., Gutowski W, Jr. and Takle, E.S., 2001. Soil moisture in a regional climate model: Simulation and projection. Geophys. Res. Lett., 28(15): 2947-2950.

Ramos, M.C. and Mulligan, M., 2005. Spatial modelling of the impact of climate variability on the annual soil moisture regime in a mechanized Mediterranean vineyard. J Hydrol, 306(1-4): 287-301.

Solman, S. and Nunez, M., 1999. Local estimates of global climate change: a statistical downscaling approach. Int. J. Climatol, 19: 835-861.

von Storch, H., 1995. Inconsistencies at the interface of climate impacts studies and global climate research. Meteorologie Zeitschrift, NF4: 72–80.

Wilby, R.L. et al., 1998. Statistical downscaling of general circulation model output: A comparison of methods. Water Resources Research, 34: 2995-3008.

Zhang, X.C., 2005. Spatial downscaling of global climate model output for site-specific assessment of crop production and soil erosion. Agr. Forest. Meteorol., 135(1-4): 215-229

Zhang, X.C. and Liu, W.Z., 2005. Simulating potential response of hydrology, soil erosion, and crop productivity to climate change in Changwu tableland region on the Loess Plateau of China. Agr. Forest. Meteorol., 131: 127-142.

RESEARCH ON TRANSITIONAL FLOW CHARACTERISTICS OF LABYRINTH-CHANNEL EMITTER

Wanhua Zhao*, Jun Zhang, Yiping Tang, Zhengying Wei, Bingheng Lu
[1] *State Key Laboratory for Manufacturing Systems Engineering, Xi'an Jiaotong University, Xi'an, Shaanxi Province, P. R. China 710049*
* *Corresponding author, Address: Institute of Advanced Manufacturing Technology, School of Mechanical Engineering, Xi'an Jiaotong University, Xi'an 710049, Shaanxi Province, P. R. China, Tel:+86-29-82665575, Fax:+86-29-82660114, Email: whzhao@mail.xjtu.edu.cn*

Abstract: A physical model for flow characteristics analysis of labyrinth-channel emitter is reconstructed by Reverse Engineering from an injection molded part, and both laminar flow and turbulence models are adopted to simulate the flow state under the condition of low Reynolds numbers. According to the distribution of separation and reattachment points, the onset of transition from laminar to turbulent flow in labyrinth channels occurs at a range of Re=250~300. Furthermore, a visualization system of the flow field inside the labyrinth channels is also established and two kinds of tracers are used in the experiments. The experiment of tracing particles verifies the calculated flow field distribution, and another experiment using dyeing liquor showed the critical Reynolds number characterizing the transition, which is reasonably consistent with numerical simulation results. The critical Reynolds number obtained shows the fact that the flow inside this emitter is turbulent under the pressures of 40~150 kPa.

Keywords: emitter, labyrinth channel, visualization, transitional flow

1. INTRODUCTION

Emitter, one of the key parts in drip irrigation system, is designed to let out the pressurized water in pipes to drop into the soil slowly and uniformly through energy dispassion by its internal structure. The structure has a great

Please use the following format when citing this chapter:

Zhao, W., Zhang, J., Tang, Y., Wei, Z. and Lu, B., 2009, in IFIP International Federation for Information Processing, Volume 294, *Computer and Computing Technologies in Agriculture II, Volume 2*, eds. D. Li, Z. Chunjiang, (Boston: Springer), pp. 881–890.

effect on the hydraulic and anti-clogging performance of emitters, especially for those non-compensating emitters with labyrinth channels. Therefore in recent years, many scholars have carried out the research on the relationship between emitter's structure and performance (macro performance (Wang, et al., 2003; Yao, et al., 2003) and micro performance (Wei, et al., 2006; Zhang, et al., 2007)). In the macro scope, the available experiments such as on flow rate uniformity and relationship between flow rates and pressure heads have been standardized. While, in the micro scope, numerical methods are usually employed due to the intricacy and complexity of emitters. Thus, simulation accuracy mainly depends on the mathematics model. Labyrinth channels, a special structure, differ greatly from the straight channels in terms of their flow characteristics. It is very important for selecting a right mathematical model to explore the transitional flow in the labyrinth channels.

Hence, this paper aims to conduct a numerical study on the transitional flow of an emitter in use, and a flow visualization system is constructed and the flow characteristics can be visualized in labyrinth channels. Experimental results can be used to guide the selection of mathematical model as well as the accuracy verification of numerical simulations.

2. NUMERICAL ANALYSIS ON THE FLOW TRANSITION CHARACTERISTICS

2.1 Physical model and grids system

A kind of labyrinth-channel emitter assembled together with the external pipe system which can be used to carry out the experiments of hydraulic performance is considered. Fig.1 shows the structure stripped from the pipe. In order to measure its dimensions, an optical microscope (VH-8000) and a laser profilometer (Song, et al., 2003) were employed respectively to get the horizontal geometries and the channel depth. The filter grids are considered and the outlet is lengthened a little to obtain a stable flow field, the physical model for numerical analysis is shown in Fig.2.

Fig.1: An injection molded emitter with labyrinth channels

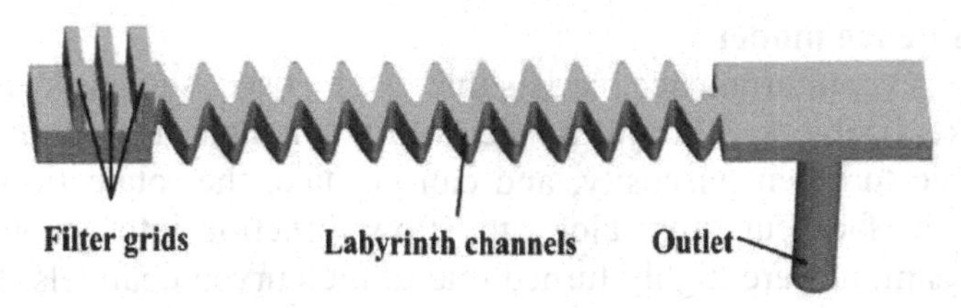

Fig.2: Physical model used for numerical simulations

Considering the sharp turns of the channel boundary, to get high quality grids, physical model was divided into three parts, filter grids, labyrinth channels and outlet. The filter grids and outlet were meshed with structured hexahedron grids, while the labyrinth channels can be meshed with pyramidal grids for their complexity. The regions near the wall and corners were given a finer mesh to simulate the flow field with great velocity gradient. The whole computational domain employed in the computation contained 245,782 control volume cells. From an investigation on the grid independence on the numerical solutions, little influence on the final results was found when using even finer grids.

2.2 Mathematical model

The fluid in the emitter is water, as a result, it is assumed to be viscous, steady, incompressible at room temperature. The fluid gravity and the surface roughness of channel wall are considered while ignoring the surface tension. For the similar structure used in other areas, many scholars have performed a lot of numerical and experimental studies (Nishimura, et al., 1984; Nishimura, et al., 1990) and they all found that the transition from laminar to turbulent flow occurred at a low Reynolds number, e.g. in a range of *Re*=300~500. Therefore in this paper, both laminar flow and turbulence models are applied to explore the flow characteristics under low *Re* values in all cases.

(a) Laminar flow model

The equations of mass conservation and momentum conservation can be expressed as follows:

$$\frac{\partial u_j}{\partial x_j} = 0 \tag{1}$$

$$\rho \frac{\partial}{\partial x_i}\left(u_i u_j\right) = -\frac{\partial p}{\partial x_j} + \frac{\partial}{\partial x_i}\left[\mu\left(\frac{\partial u_i}{\partial x_j} + \frac{\partial u_j}{\partial x_i}\right)\right] \tag{2}$$

Where ρ (kg/m^3) is the fluid density, u_j is the velocity vector in the jth direction.

(b) Turbulence model

There are several turbulence models that can be used, such as standard k-ε, RNG k-ε, Realizable k-ε, but all of them only calculate the turbulent stress with isotropic turbulent viscosity, and can not take the rotate flow and the variation of surface curvature along the flow direction into account. Here, the flow streamlines are highly turned due to the curved channels, Reynolds Stress Model (RSM) (Mohammad, et al., 2002) is adopted which can accounts for the effects of streamline curvature, swirl, rotation, and rapid changes in strain rate in a more rigorous manner than one-equation and two-equation models. It is a more advanced turbulence model because the turbulent stresses are directly related to the mean velocity gradient. The governing equations involve 12 equations such as Eq. (1), Eq. (2), k equation, ε equation and the Reynolds stress transport equations (Tao, 2001).

2.3 Boundary conditions and numerical methods

At the inlet of the channel, a uniform fluid velocity may be specified as a boundary condition for the momentum equations according to the flow rate through the emitter under each working condition, this fluid velocity can be obtained from the hydraulic performance experiment of emitters. The outlet condition was treated as standard atmosphere. Turbulence intensity and hydraulic diameter were used to define the turbulence parameters both in inlet and outlet planes. On the wall surface, a value of 0.01 mm was defined as the roughness and the velocity was considered to be zero (non-slip condition), the standard log-law wall function was used in this paper to bridge the near-wall linear sublayer, where acute changes in the fluid velocity are expected. The governing equations were solved by using the SIMPLE algorithm with a segregated solver and a second-order upwind scheme. The CFD program Fluent6.2® was used for the calculations.

2.4 Results and discussion

Fig.3 shows the flow state of different Reynolds numbers in the same channel unit using laminar flow model. Due to the sharp turns (α=31°), a small vortex is developed at the corner of the channel even when Re=1 (Fig.3a), while the streamlines are still symmetrical about the central plane of channel unit, which indicates the inertia force is less dominant. As the Re increases to 40 shown in Fig.3b, the inertia force also increases and then a phase shift appears between flow streamlines and channel boundary. Moreover, the vortex slowly shifts to the mainstream area and becomes larger. When Re=80 as shown in Fig.3c, the first vortex arrives at the middle of the channel, which leaves space for the second vortex. Subsequently, with the development of the second vortex, flow three dimensionality appears as a result of oscillations of the shear-layer.

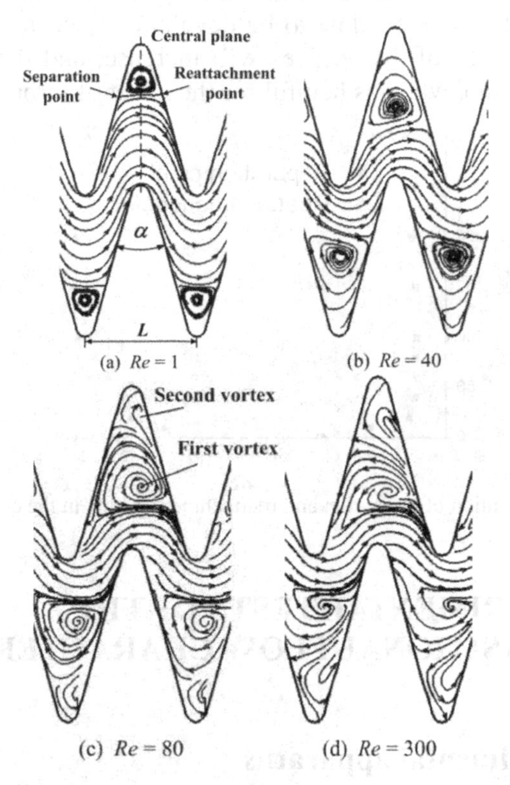

Fig.3: Streamlines in the same channel unit under different Reynolds numbers

Before the fluid reaches the central plane of each unit, the flow separation is developed because of the higher pressure near the outer edge. After the fluid passes through the central plane, the flow reattachment appears due to the opposite pressure gradient plotted in Fig.3a. Fig.4 shows the distribution of flow separation and reattachment points under various Reynolds numbers. In the beginning, the distance between two points become longer as *Re* increases. The flow separation points begin to be stationary when *Re* is more than 40, while the reattachment points tend to be stable at *Re*=100. It can also be concluded from the distribution of the two points that the vortex firstly moves to down left and then to down right corners with the development of *Re*. When *Re* ranges from 250 to 300, the distance between the two points becomes shorter. According to the experimental results performed by Rush (Rush, et al., 1999), the vortex shrinks when the flow transits from laminar to turbulent flow. Thereby, we can also conclude that the transition to turbulence occurs when *Re*=250-300 for this labyrinth channels. This phenomenon can be partly explained by the theory of pipe flow. The pressure near the outer edge is higher than near the inner edge for the centrifugal force acting on the fluid in labyrinth channels, so it is easier for the vortex and secondary flow to be developed than in straight channel. The intensity and size of the vortices will increase, and they could appear early in this condition, which is helpful for the onset of turbulence.

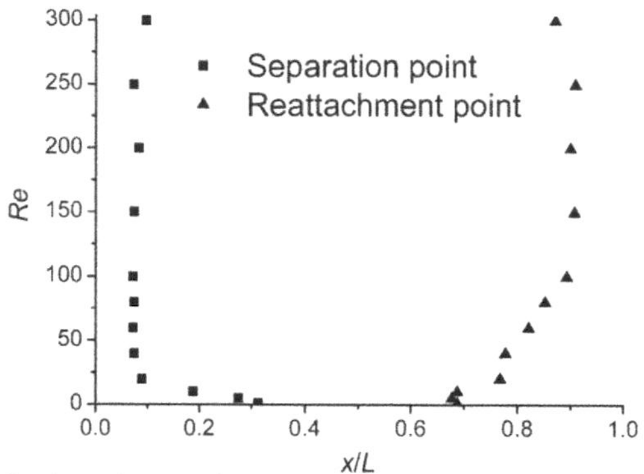

Fig.4: Distribution of separation and reattachment points in the channel unit

3. EXPERIMENTAL TESTS OF THE TRANSITIONAL FLOW CHARACTERISTICS

3.1 Experimental apparatus

The experimental apparatus is shown in Fig.5. A self-priming pump (25WZ-45) is employed as the driven source and a filter with 200 mesh is followed to deal with the large particles in the water. Plexiglass with a thickness of 1 mm is used to make the test model of labyrinth channels because of its excellent transparence. The whole model is consist of three parts, the middle is the contour of the labyrinth channels which is fabricated by CNC carving machine based on 2-D geometry dimensions, and the former and the back are used to fasten and seal the middle part with 8 bolts, as shown in Fig.6. The actual dimensions of the machined model were measured by the optical microscope (VH-8000) and the results show great consistence with the design structure. The high speed video camera (TroubleShooter1000, Fastec Imaging Inc.) was employed to record the flow state. The water passing through labyrinth channels is weighted by an electronic scale, thus the relative *Re* can be calculated.

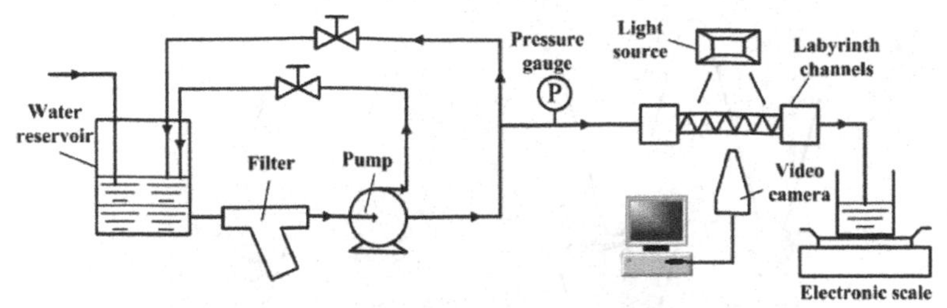

Fig.5: Schematic of the visualization system for the fluid in labyrinth channels

Fig.6: The test model of labyrinth channels fabricated in plexiglass

3.2 Flow visualization methods

Two methods were employed to carry out the experiments. One is the marked particle tracking method, the aluminum powder with a diameter of 5-15 μm was used as the marked particles and its concentration for imaging processing was calculated by the method in the literature (Fan, 2002). In order to prevent the flocculation of particles, a little alcohol was added into the fluid. The other is dyeing liquor method. The ink, used as the dyeing liquor, was filled into an injector with a cubage of 1ml which was fixed at the entrance of labyrinth channels and was parallel with the flow direction.

Considering the influence of dyeing liquor on the observability of labyrinth channels, it is necessary to perform the marked particle tracking experiment firstly and then the dyeing liquor experiment.

3.3 Experimental results and discussion

Fig.7 shows the flow field distribution at *Re*=213, where (a) is from the numerical analysis using laminar flow model, (b) is from the experiment. Except a little noise, the velocity vectors from experiments are consistent with those of numerical analysis. Those noises may be incurred by the error of image processing or the light source which cannot produce the slice light.

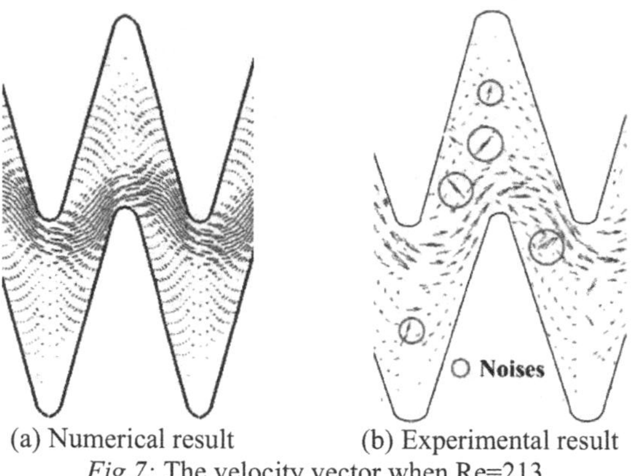

(a) Numerical result (b) Experimental result

Fig.7: The velocity vector when Re=213

The channels in different flow regimes are shown in Fig.8, it indicates the flow states in the same three channel units at different Reynolds numbers. The flow streamlines are still symmetrical about the central plane at low *Re* value such as *Re*=22 (Fig.8a), it can be seen from the width of dyeing liquor that the flow mainly flows along the inside of each unit corner. As Reynolds number increases to 57 (Fig.8b), the flow streamlines shown by the dyeing liquor become narrow, which indicates that the area of mainstream regime is shrunk, in addition, a phase shift appears from here. With the development of *Re* to 93-131, the streamlines begin to diffuse to non-core flow regime as shown in Fig.8c and Fig.8d, the fluid in core and non-core flow regimes intermixed. At *Re*=277 (Fig.8e), the mainstream is getting obscure while at the corner of the channel, the dyeing liquor concentration is increasing. When the flow speed continues to increase (Fig.8f), it becomes difficult to track the flow by dyeing liquor method because of the rapidly decreased definition of dyeing liquor after transition to turbulence which causes a complete mixing with the fluid. Therefore, the present results are consistent with those reported in numerical analysis.

The *Re* is 692 when the emitter is under the pressure of 40 kPa which can be obtained from the experiment of the relationship between pressure heads and flow rates. For this emitter, thereby, the fluid inside the channels has been in the state of turbulence either under the working condition (100 kPa) or in a range of 40-160 kPa that is the experimental scope of relationship between pressure heads and flow rates. As a result, it is more accurate to employ turbulence model to simulate the flow characteristics inside the labyrinth channels.

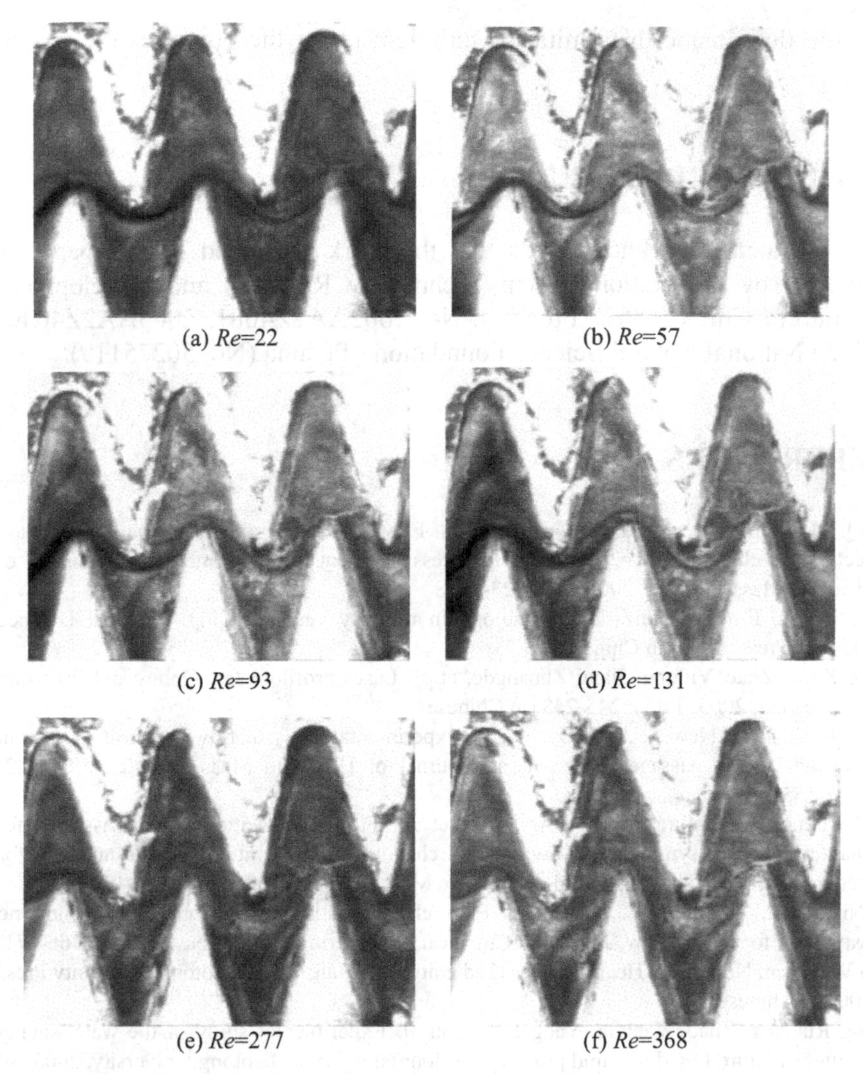

(a) *Re*=22 (b) *Re*=57

(c) *Re*=93 (d) *Re*=131

(e) *Re*=277 (f) *Re*=368

Fig.8: The fluid field visualization by dyeing liquor method

4. CONCLUSION

The transitional flow characteristics in a labyrinth-channel emitter in engineering use were analyzed by numerical simulations and experiments. The results show that various vortexes are developed with the increase of Reynolds numbers for such a labyrinth channel with sharp turns. The transition from laminar to turbulent flow occurs well in advance in the range of *Re*=250-300, which is consistent with the experimental result using dyeing liquor that the critical *Re* is about 277. The result also shows the fact that the flow inside this emitter is turbulent under the pressures of 40~150 kPa.

ACKNOWLEDGEMENTS

It is gratefully acknowledged that the work presented in this paper is supported by the National High Technology Research and Development Program of China ("863" Program, No. 2002AA2Z4081, 2005AA2Z4040) and the National Natural Science Foundation of China (No. 50275119).

REFERENCES

A. Q. Mohammad, Y J Jang, H C Chen, et al. Flow and heat transfer in rotating two-pass rectangular channels(AR=2) by Reynolds stress turbulence model. International Journal of Heat and Mass Transfer, 2002, 45: 1823-1838

Fan Jichuan. Flow visualization technology in recently years. Beijing: National Defence Industry Press, 2002 (in Chinese)

Song Kang, Zhao Yulong, Jiang Zhuangde, et al. Laser profilometer. Optics and Precision Engineering, 2003, 11(3): 245-248 (in Chinese)

T. A. Rush, T. A. Newell, A. M. Jacobi. An experimental study of flow and heat transfer in sinusoidal wavy passages. International Journal of Heat and Mass Transfer, 1999, 42: 1541-1553

T. Nishimura, S. Murakami, S. Arakawa, et al. Flow observations and mass transfer characteristics in symmetrical wavy-walled channels at moderate Reynolds numbers for steady flow. International Journal of Heat and Mass Transfer, 1990, 33: 835-845

T. Nishimura, Y. Ohori, Y. Kawamura. Flow characteristics in a channel with symmetric wavy wall for steady flow. Journal of Chemical Engineering of JAPAN, 1984, 17: 466-471

Tao Wenquan. Numerical Heat Transfer (2nd edition). Xi'an: Xi'an Jiaotong University Press. 2001(in Chinese)

Wang Ruihuan, Zhao Wanhua, Yang Laixia, et al. Experimental study of the water-saving emitter structure based on rapid prototyping. Journal of Xi'an Jiaotong University, 2003, 37: 542-545 (in Chinese)

Wei Qingsong, Shi Yusheng, Dong Wenchu, et al. Study on hydraulic performance of drip emitters by computational fluid dynamics. Agricultural Water Management, 2006, 84: 130-136

Yao Bin, Liu Zhifeng, Zhang Jianping. Study on the influence of channel length on dripper hydraulic performance. Water Saving Irrigation, 2003, (5): 38-39 (in Chinese)

Zhang Jun, Zhao Wanhua, Lu Bingheng et al. Numerical and experimental study on hydraulic performance of emitters with arc labyrinth channels. Computers and Electronics in Agriculture, 2007, 56:120-129

CHINA-US BORDER EFFECT OF AGRICULTURAL TRADE USING GRAVITY MODEL

Haixia Zhu[1,2,*], Haiying Gu[1]

[1] *Antai School of Economics and Management, Shanghai Jiao Tong University, Shanghai, P. R. China 200052*

[2] *School of Foreign Languages, Shanghai Maritime University, Shanghai, P. R. China 200135*

[*] *Corresponding author, Address: School of Foreign Languages, Shanghai Maritime University, Shanghai, P. R. China 200135, Tel: +86-21-58855200, Fax: +86-21-58854605, Email: haixiazhu@126.com*

Abstract: Two border effect models capturing the characteristics of agricultural trade based on the Gravity Model are put forward to make a study on China-US border effects in the sector of agricultural trade. Different methods are used to check the robustness of the models. Applying the panel data covering 1987-2005, the empirical results show that the border effects of China-US exist with a great magnitude and tend to drop over time. They differ greatly from the direction of the international trade. Linear model in log-form can well explain the border effects of China-US agricultural trade.

Keywords: border effect, Gravity Model, China-US, agricultural trade

1. INTRODUCTION

The Gravity Model has been widely and successfully used to explain international trade flows for several decades. In particular, the literature on the effects of national borders on trade has adopted this model for investigating the relative volumes of internal versus external trade. The term "border effect" or "home bias" refers to the extent to which volume of domestic trade exceeds the volume of international trade. In other words,

Please use the following format when citing this chapter:

Zhu, H. and Gu, H., 2009, in IFIP International Federation for Information Processing, Volume 294, *Computer and Computing Technologies in Agriculture II, Volume 2*, eds. D. Li, Z. Chunjiang, (Boston: Springer), pp. 891–900.

two different countries trade much less with each other than do two regions within one country, taking into account income, size and distance. Many economists believe that national borders represent large and mostly unidentified barrier to trade and reveal existence. Since the study of McCallum (1995) where it was found that inter-provincial trade in Canada is 22 times as large as Canada's international trade with United States, there has been growing research effort done to measure and understand trade border effect in order to achieve world or regional integration.

There are two main ways used so far in the vast empirical studies for estimation of border effect using the Gravity Model. The first one calculates it by comparing intra-national and international trade data, as have been done by McCallum. Another sort of border effect literature instead of measuring tries to explain why national borders have so significant trade deterring effect and to find out whether there are any policy instruments to influence them (Evans, 2003).

However, agricultural trade liberalization has lagged far behind even though agricultural trade plays a crucial role in the total merchandise international trade. Some have paid attention to border effect of agricultural trade and made conclusion that border effect of agricultural trade is also significantly large among developed countries (Furtan & van Melle, 2004; Olper & Raimondi, 2005; Paiva, 2005). Besides, the border effect studies have been focusing more on advanced developed countries or regions than on developing countries. Thus this paper tries to study China-US (two world major agricultural countries) border effect of agricultural trade from different approaches based on the Gravity Model.

2. MODEL SPECIFICATION

This paper mainly deals with the agricultural products, so more variables with specific importance for agricultural activity need to be included among the main explanatory variables in the basic model used by McCallum (1995): each region's share of agricultural product in GDP and rural population density. In the specification used here, a country's imports of agricultural products from a partner country depends on the size of the countries' respective economies, their land areas, the physical distance between and several dummy variables capturing agricultural characteristics. Thus the equation estimated in log form has the following specification:

$$\begin{aligned} X_{ijt} = \alpha &+ \beta_1 Y_{it} + \beta_2 Y_{jt} + \sigma D_{ij} + \beta_3 y_{it} + \beta_4 y_{jt} \\ &+ \beta_5 SAY_{it} + \beta_6 SAY_{jt} + \beta_7 land_{it} + \beta_8 land_{jt} \\ &+ \beta_9 den_{it} + \beta_{10} den_{jt} + \delta B_{ijt} + \mu_{ijt} \end{aligned} \tag{1}$$

where the gravity equation is estimated in log-near form. Adopting standard notation, the independent variable Xijt denotes the real US dollar amount of agricultural products imported from i to j in year t. Yit and Yjt are the total agricultural products of importer i and exporter j respectively. Dij is the economic distance between importer and exporter. yit and yjt are the region's real per capita agricultural GDP in year t. SAYit and SAYjt indicate the importer's share of agriculture in GDP in year t. landit and landjt denote the farm land area of importer and exporter. denit and denjt are rural population density of importer and exporter. Bijt is the dummy variable at in year t, taking the value of 1 for inter-provincial trade, 0 otherwise and uij is the standard classical error term.

However, equation (1) can't reflect the real bilateral trade flow between each Chinese province and US state since it ignores the direction of the trade. So the dummy is further divided into BEijt and BIijt to denote Chinese agricultural export to US and its agricultural import from US respectively:

$$\begin{aligned} X_{ijt} &= \alpha + \beta_1 Y_{it} + \beta_2 Y_{jt} + \sigma D_{ij} + \beta_3 y_{it} + \beta_4 y_{jt} \\ &+ \beta_5 SAY_{it} + \beta_6 SAY_{jt} + \beta_7 land_{it} + \beta_8 land_{jt} \\ &+ \beta_9 den_{it} + \beta_{10} den_{jt} + \delta_1 \mathrm{BE}_{ijt} + \delta_2 \mathrm{BI}_{ijt} + \mu_{ijt} \end{aligned} \tag{2}$$

3. EMPIRICAL STUDY

The estimation was done through ordinary least squares to explore the magnitude of China-US border effect in the field of agricultural trade. Different methods are used to make a comparatively comprehensive study on this issue by estimating equation (1) and (2) respectively with six methods: method (1) just makes regular regression of the models; method (2) uses Y_i+Y_j as the weight in the regression to check the heteroscedasticity of the models; method (3) used GDP and per capita GDP to replace total agricultural product and total agricultural product per capita as some scholars did in their study of agricultural trade; methods (4) and (5) respectively challenge the two main ways to calculate the key distance variable of the model-Wei (1996) and Nitsch (2000) & Leamer (1997) (short as NL method); the last method adds the square of distance to the basic specification to check the nonlinear effects of distance on agricultural trade.

3.1 Data

This paper tries to investigate the bilateral trade between China and US every 5 year from 1987 to 2005. Agricultural products are defined according to HS classification in accordance with Monthly Statistical Report of Chinese Agricultural Import and Export. In order to learn the trade pattern between them, first, each country is divided into different regions according to the geographical places and economic development: China is divided into eight regions, namely Northeast region, Beijing & Tianjin, East Coast, South Coast, Middle China, Southwest region, Northwest region and North Coast according to the book entitled "Multi-Regional Input-Output Model for China" (2005). Because of the great differences in Chinese development before 1995, China is divided into seven regions (Shicunzhenyi & Wang, 2007). Likewise, U.S. is divided into nine regions according to US Census Bureau, namely New England, Middle Atlantic, East North Central, West North Central, South Atlantic, East South Central, West South Central, Mountain region and Pacific region.

The data for each Chinese region including total agricultural product, population, GDP and land area are taken from the Bureau of Statistics of China, China Yearbook of Statistics, each province's Statistical Yearbook, statistical websites, agricultural websites and Yearbook of Chinese Cities' Statistics. Each region's land area is taken from Chinese Resources and Environment Database and each province's statistical website. The data for each US state are from the 2007 Statistical Abstract-National Data Book, state fact express of US Census Bureau. Bilateral trade data is taken from the Office of Trade and Industry Information, Manufacturing and Services, International Trade Administration of US Department of Commerce, and FAO. The bilateral trade data between each Chinese region is taken from the I-O table provided by the National Center of Information (2005). As no data on bilateral trade flows between Chinese provinces and US states, it is necessary to proceed to some adjustments to reconcile the model with the degree of aggregation of the available trade data.

Distance variable plays an important role in the model, and any error in data processing can bias the last regression result, so this paper applies Head & Mayer (2000)'s method in dealing with both international and intra-national distances. We use the Great Circle Formula to calculate the economic distances between each pair of regions using the population of each region's provincial capitals as the weight.

3.2 Results and discussion

In the estimation of China-US border effects, four questions are to be answered: How big were border effects between China and US in agricultural trade? How did these border effects change over time? To what extent did the Chinese import from US differ from its export to US? How did the independent variables influence the dependent variable?

3.2.1 The analysis of different variables

The regression results of equation (1) in different sample years are shown in Table1.-Table 5. with standard errors in the parentheses and summarized afterwards. Each model for each method includes 836 observations in total.

Table 1. Regression results of equation (1)- 1987

	(1)	(2)	(3)	(4)	(5)	(6)
α	-15.028 (0.176)	-12.820 (0.254)	-17.358 (0.084)	-15.916 (0.774)	-24.520 (0.603)	-15.028 (0.176)
Y_{it}	0.966 (0.482)	0.793 (0.492)	1.043 (0.481)	0.966 (0.482)	1.022 (0.481)	0.966 (0.482)
Y_{jt}	1.263 (0.482)	1.146 (0.493)	1.352 (0.481)	1.263 (0.482)	1.326 (0.481)	1.263 (0.482)
D_{ij}	-0.831 (0.710)	-0.920 (0.691)	-0.770 (0.704)	-0.831 (0.710)	-0.442 (0.935)	-0.415 (0.355)
y_{it}	0.898 (0.683)	0.741 (0.675)	1.065 (0.671)	0.898 (0.683)	1.135 (0.747)	0.898 (0.683)
y_{jt}	2.083 (0.687)	1.886 (0.679)	2.237 (0.675)	2.083 (0.687)	2.376 (0.765)	2.083 (0.687)
SAY_{it}	-0.547 (0.787)	-0.539 (0.788)	1.373 (0.779)	-0.547 (0.787)	-0.687 (0.805)	-0.547 (0.787)
SAY_{jt}	1.209 (0.787)	1.238 (0.787)	2.178 (0.783)	1.209 (0.787)	1.342 (0.804)	1.209 (0.787)
$land_{it}$	-0.314 (0.407)	0.384 (0.408)	0.277 (0.404)	0.314 (0.407)	0.275 (0.410)	-0.314 (0.407)
$land_{jt}$	-0.207 (0.406)	-0.126 (0.408)	-0.247 (0.404)	-0.207 (0.406)	-0.227 (0.409)	-0.207 (0.406)
den_{it}	-0.054 (0.783)	-0.254 (0.772)	-0.025 (0.775)	-0.054 (0.783)	-0.173 (0.793)	-0.054 (0.783)
den_{jt}	-0.163 (0.782)	-0.343 (0.771)	-0.133 (0.775)	-0.163 (0.782)	-0.260 (0.790)	-0.163 (0.782)
B_{ijt}	-4.470 (0.391)	-4.395 (0.393)	-5.074 (0.906)	-3.582 (0.084)	-7.569 (0.508)	-4.470 (0.391)
R^2	0.515	0.538	0.524	0.515	0.511	0.515
D.W.	1.032	1.006	1.037	1.032	1.022	1.032

Table 2. Regression results of equation (1)- 1992

	(1)	(2)	(3)	(4)	(5)	(6)
α	-12.611 (0.666)	-12.027 (0.752)	-13.283 (0.825)	-12.680 (0.522)	-17.449 (0.148)	-12.611 (0.666)
Y_{it}	1.353 (0.437)	1.216 (0.450)	1.402 (0.447)	1.353 (0.437)	1.376 (0.431)	1.353 (0.437)
Y_{jt}	1.114 (0.436)	1.051 (0.449)	1.168 (0.446)	1.114 (0.436)	1.120 (0.431)	1.114 (0.436)
D_{ij}	-0.148 (0.631)	-0.197 (0.616)	-0.078 (0.638)	-0.148 (0.631)	-1.295 (0.812)	-0.074 (0.316)
y_{it}	2.117 (0.557)	2.019 (0.556)	2.090 (0.565)	2.117 (0.557)	2.484 (0.595)	2.117 (0.557)
y_{jt}	2.203 (0.559)	2.092 (0.558)	2.185 (0.568)	2.203 (0.559)	2.617 (0.605)	2.203 (0.559)
SAY_{it}	-1.410 (0.709)	-1.460 (0.714)	2.040 (0.704)	-1.410 (0.709)	-1.606 (0.713)	-1.410 (0.709)
SAY_{jt}	1.625 (0.709)	1.690 (0.714)	1.687 (0.706)	1.625 (0.709)	1.807 (0.711)	1.625 (0.709)
$land_{it}$	-0.145 (0.348)	-0.092 (0.351)	-0.089 (0.348)	-0.145 (0.348)	-0.195 (0.346)	-0.145 (0.348)
$land_{jt}$	0.236 (0.347)	0.285 (0.350)	0.286 (0.347)	0.236 (0.347)	0.228 (0.345)	0.236 (0.347)
den_{it}	-0.096 (0.692)	-0.244 (0.688)	-0.050 (0.697)	-0.096 (0.692)	-0.295 (0.694)	-0.096 (0.692)
den_{jt}	-0.420 (0.691)	-0.563 (0.687)	-0.375 (0.697)	-0.420 (0.691)	-0.605 (0.693)	-0.420 (0.691)
B_{ijt}	-3.910 (0.882)	-3.949 (0.889)	-3.970 (0.906)	-3.831 (0.084)	-5.735 (0.280)	-3.910 (0882)
R^2	0.612	0.625	0.607	0.612	0.618	0.612
D.W.	1.310	1.348	1.268	1.368	1.468	1.310

Table 3. Regression results of equation (1)- 1997

	(1)	(2)	(3)	(4)	(5)	(6)
α	-12.045	-11.167	-13.089	-12.358	-19.727	-12.045
	(0.043)	(0.096)	(0.413)	(0.875)	(0.037)	(0.043)
Y_{it}	0.983	0.974	0.981	0.983	1.002	0.983
	(0.349)	(0.353)	(0.353)	(0.349)	(0.348)	(0.349)
Y_{jt}	0.933	0.941	0.975	0.933	0.952	0.933
	(0.349)	(0.353)	(0.352)	(0.349)	(0.348)	(0.349)
D_{ij}	-0.503	-0.539	-0.455	-0.503	-0.963	-0.251
	(0.575)	(0.578)	(0.576)	(0.575)	(0.732)	(0.287)
y_{it}	2.455	2.160	2.496	2.455	2.847	2.455
	(0.486)	(0.486)	(0.491)	(0.486)	(0.539)	(0.486)
y_{jt}	2.108	1.871	2.184	2.108	2.499	2.108
	(0.486)	(0.486)	(0.491)	(0.486)	(0.539)	(0.486)
SAY_{it}	-0.215	-0.341	3.194	-0.215	-0.551	-0.215
	(0.629)	(0.625)	(0.661)	(0.629)	(0.663)	(0.629)
SAY_{jt}	0.939	1.009	2.142	0.939	1.275	0.939
	(0.629)	(0.625)	(0.663)	(0.629)	(0.663)	(0.629)
$land_{it}$	-0.861	-0.749	-0.807	-0.861	-0.836	-0.861
	(0.340)	(0.339)	(0.338)	(0.340)	(0.339)	(0.340)
$land_{jt}$	-0.039	0.037	-0.031	-0.039	-0.014	-0.039
	(0.340)	(0.339)	(0.340)	(0.340)	(0.339)	(0.340)
den_{it}	-0.725	-0.729	-0.760	-0.725	-0.879	-0.725
	(0.550)	(0.548)	(0.549)	(0.550)	(0.549)	(0.550)
den_{jt}	-0.509	-0.580	-0.545	-0.509	-0.663	-0.509
	(0.550)	(0.548)	(0.550)	(0.550)	(0.549)	(0.550)
B_{ijt}	-3.110	-3.140	-3.308	-2.797	-5.651	-3.110
	(0.103)	(0.104)	(0.117)	(0.366)	(0.669)	(0.103)
R^2	0.605	0.641	0.605	0.605	0.607	0.605
D.W.	1.032	1.289	1.149	1.148	1.258	1.032

Table 4. Regression results of equation (1)- 2002

	(1)	(2)	(3)	(4)	(5)	(6)
α	-8.237	-7.688	-7.902	-8.323	-12.436	-8.237
	(0.684)	(0.708)	(0.718)	(0.583)	(0.185)	(0.684)
Y_{it}	0.881	0.895	0.833	0.881	0.920	0.881
	(0.361)	(0.367)	(0.365)	(0.361)	(0.359)	(0.361)
Y_{jt}	0.859	0.873	0.817	0.859	0.898	0.859
	(0.361)	(0.367)	(0.364)	(0.361)	(0.359)	(0.361)
D_{ij}	-0.247	-0.276	-0.267	-0.247	-1.180	-0.124
	(0.584)	(0.587)	(0.589)	(0.584)	(0.732)	(0.292)
y_{it}	2.585	2.254	2.546	2.585	2.969	2.585
	(0.462)	(0.460)	(0.474)	(0.462)	(0.503)	(0.462)
y_{jt}	1.966	1.705	1.894	1.966	2.350	1.966
	(0.462)	(0.460)	(0.685)	(0.462)	(0.503)	(0.462)
SAY_{it}	-0.770	-0.895	2.655	-0.770	-1.078	-0.770
	(0.592)	(0.587)	(0.685)	(0.592)	(0.614)	(0.592)
SAY_{jt}	0.903	0.970	1.891	0.903	1.211	0.903
	(0.592)	(0.587)	(0.686)	(0.592)	(0.614)	(0.592)
$land_{it}$	-0.529	-0.418	-0.490	-0.529	-0.517	-0.529
	(0.332)	(0.331)	(0.328)	(0.332)	(0.330)	(0.332)
$land_{jt}$	0.053	0.121	0.070	0.053	0.065	0.053
	(0.332)	(0.331)	(0.329)	(0.332)	(0.330)	(0.332)
den_{it}	-0.741	-0.696	-0.777	-0.741	-0.908	-0.741
	(0.582)	(0.582)	(0.588)	(0.582)	(0.579)	(0.582)
den_{jt}	-0.550	-0.567	-0.582	-0.550	-0.716	-0.550
	(0.582)	(0.582)	(0.588)	(0.582)	(0.579)	(0.582)
B_{ijt}	-2.321	-2.308	-2.094	-2.234	-3.738	-2.321
	(0.564)	(0.566)	(0.557)	(0.723)	(0.866)	(0.564)
R^2	0.618	0.651	0.612	0.618	0.623	0.618
D.W.	1.022	1.381	1.138	1.382	1.378	1.022

Table 5. Regression results of equation (1)- 2005

	(1)	(2)	(3)	(4)	(5)	(6)
α	-10.105 (0.056)	-10.095 (0.101)	-10.554 (0.117)	-10.129 (0.998)	-12.149 (0.186)	-10.105 (0.056)
Y_{it}	1.493 (0.232)	1.480 (0.238)	1.532 (0.321)	1.493 (0.232)	1.418 (0.231)	1.493 (0.232)
Y_{jt}	1.315 (0.307)	1.375 (0.311)	1.289 (0.320)	1.315 (0.307)	1.334 (0.303)	1.315 (0.307)
D_{ij}	-0.117 (0.488)	-0.102 (0.489)	-0.024 (0.510)	-0.117 (0.488)	-1.307 (0.596)	-0.058 (0.244)
y_{it}	2.056 (0.414)	1.917 (0.406)	2.618 (0.421)	2.056 (0.414)	2.389 (0.431)	2.056 (0.414)
y_{jt}	2.246 (0.405)	2.122 (0.396)	2.416 (0.421)	2.246 (0.405)	2.563 (0.420)	2.246 (0.405)
SAY_{it}	-2.377 (0.272)	-2.384 (0.278)	1.779 (0.547)	-2.377 (0.272)	-2.474 (0.271)	-2.377 (0.272)
SAY_{jt}	2.583 (0.278)	2.624 (0.280)	0.951 (0.548)	2.583 (0.278)	2.719 (0.280)	2.583 (0.278)
$land_{it}$	-0.640 (0.232)	-0.712 (0.231)	-0.136 (0.280)	-0.640 (0.232)	-0.640 (0.229)	-0.640 (0.232)
$land_{jt}$	0.917 (0.269)	0.976 (0.267)	1.010 (0.280)	0.917 (0.269)	0.903 (0.266)	0.917 (0.269)
den_{it}	-1.475 (0.382)	-1.523 (0.386)	-1.854 (0.403)	-1.475 (0.382)	-1.814 (0.398)	-1.475 (0.382)
den_{jt}	-1.918 (0.388)	-1.974 (0.386)	-2.122 (0.403)	-1.918 (0.388)	-2.268 (0.404)	-1.918 (0.388)
B_{ijt}	-1.942 (0.237)	-1.951 (0.240)	-2.033 (0.247)	-1.918 (0.321)	-2.721 (0.363)	-1.942 (0.237)
R^2	0.705	0.745	0.681	0.707	0.714	0.705
D.W.	1.181	1.190	1.111	1.176	1.190	1.181

It can be found that the coefficients of the same parameter do not differ too much in each year, indicating that this revised model is well-organized. To look closely, we can see that the last method using squared distance tends to share the same results with the regular regression except the coefficients of the distance are cut half. The result of method (5) shows a greater difference than that of the rest, probably because of the wrong assumption of the distance calculation in NL method, the assumption that the customers are randomly distributed in a round circle.

The signs and the magnitudes of the coefficients in all sample years are in line with most estimates of the impact of distance and economic size on agricultural trade found in the literature: Imports of agricultural products are negatively affected by the distance between the trading pairs and positively affected by the size of their economies. Five distance coefficients tend to decrease in absolute value, showing the decline of the effect of distance on bilateral trade. The agricultural GDP or GDP of both importers and exporters positively affect the bilateral trade. The estimates also show that the bilateral trade seems to be positively affected a little more by the economy of the importer than by that of the exporter, probably reflecting the fact that the agricultural demand has a greater influence in the market. In addition, a higher share of agricultural product in GDP is related to higher exports of agricultural goods as expected and a higher share of agricultural product in

GDP tends to result lower imports of agricultural goods. The quantity of agricultural products a region imports seems to be inversely related to its land area. But land area of the exporting region is not found consistent in both specifications. Maybe it's because the greater land area of the exporting region might be due to its own great demand in agricultural products. Rural population density is also found to be inversely related to the bilateral trade. The greater number of the farms in one region results in more demands of agricultural products, so less trade is resulted.

Finally, all dummies in five tables are below zero indicating that border creates an invisible barrier to bilateral agricultural trade and that China-US agricultural trade is also possibly influenced by the other trade partners in the world. The regression results of equation (2) using the same six methods show similar results and are omitted here for space consideration.

3.2.2 The magnitude and evolution of border effects

Border effects of each year can be calculated from the anti-log of the dummy coefficients. The evolution of the bilateral border effects, the China export border effect to US and China import border effect from US are better seen in Fig.1 to Fig.3 respectively.

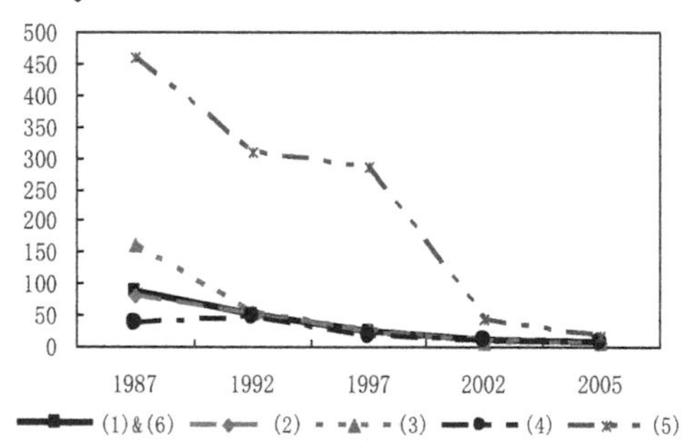

Fig. 1: China-US border effect over time

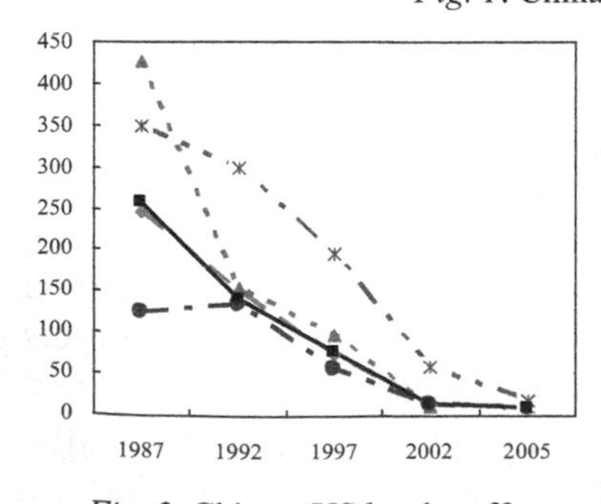

Fig. 2: China→US border effect

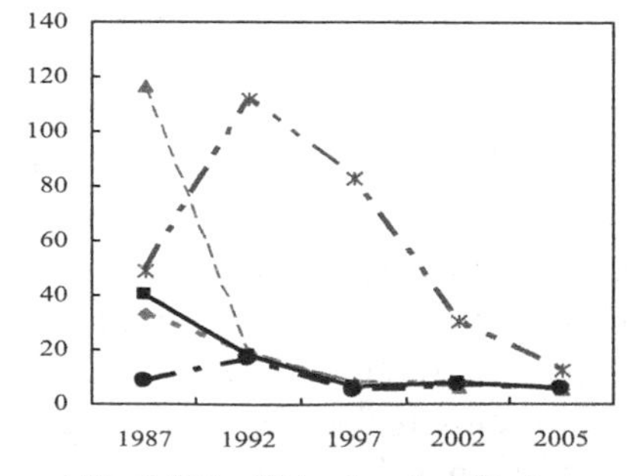

Fig.3: US→China border effect

First, the estimation of equation (1) using six methods all indicate the great magnitude of China-US border effects in agricultural trade. The border effects without considering the direction of the trade range from 36 to 160 in 1987 and drop quickly to 7 -15 in 2005 with the exception of method (5). In average, this means one Chinese region tends to trade agricultural products around 87 times more with another Chinese region than with an identical US region in 1987. In 2005 it drops to 7. Different methods of distance calculation only affect the border effect of data 1987, while all the other methods except method (5) show a similar magnitude of border effect over 1992-2005. The third method using total economy development named GDP tends to overestimate border effect while Wei's method tends to underestimate it.

Second, border effects are also significant in value with the decreasing trend when the direction of trade is taken into consideration. Also method (5) shows a different result. On the one hand, the Chinese agricultural export border effects to US are amazingly large in value, ranging from 426 to 125 in 1987 and from year 1992 it drops quickly from 142 to 9 in year 2005 in average. This shows that more Chinese agricultural products have been exported to US than to the other Chinese regions due to the government's new policy in agriculture. On the other hand, the Chinese agricultural import border effects from US remain comparatively stable ranging only from 6 to 40 during the two decades. The smaller magnitudes of the import effect than the export effect shows that in the international agricultural market with US, Chinese regions import more agricultural products from US than from other domestic regions, but export more to domestic regions than to US. This fact leads to Chinese large deficit in agricultural trade with US.

4. CONCLUSION

This paper tries to explore the border effects of China-US agricultural trade based on the Gravity Model. Two models are put forward and six methods are used to make an empirical study on the magnitudes, evolution of the border effects and on the relationship of variables involved in the model applying the panel data covering 1987-2005. Results indicated that the main determinants of trade in a gravity framework- economic size and distance-have the same influence on agricultural trade as they have on total merchandise trade. As for new variables specific to agricultural trade, the estimates show that a higher share of agriculture in GDP and a lower rural population density are associated with higher bilateral trade flows of agricultural products. Distance variable tends to have a decreasing negative influence on bilateral agricultural trade, indicating the fewer barriers to the trade. Logarithmic linear equations can explain the border effects of China-US agricultural trade. The method of using total economy development tends to overestimate all border effects of China-US agricultural trade while Wei's method of calculating distance tends to underestimate them.

The border effects of China-US agricultural trade are significantly large no mater whether the direction of trade is considered. The impact of borders on both sides has declined over time. It is found that the Chinese import border effect in agricultural trade is more stable and smaller than its export border effect to US. China needs to establish appropriate policies to promote China's agricultural trade to US to ease the burden of deficit in this area.

ACKNOWLEDGEMENTS

Funding for this research was provided by National Natural Science Foundation of China under grant 70473057. The first author is grateful to the Shanghai Maritime University for providing her with pursuing a PhD degree at the Shanghai Jiao Tong University.

REFERENCES

A. Olper, V. Raimondi. Access to OECD agricultural market: a gravity border effect approach, in the 99th seminar of the EAAE (European Association of Agricultural Economists), 2005, 24-27

C. Paiva. Assessing protectionism and subsidies in agriculture: a gravity approach, IMF working paper, 2005, no. 21

C.L. Evans. The economic significance of national border effects, American Economic Review, 2003, 93(4): 1291-1312

Center of National Information of China. Multi-Regional Input-Output Model for China, Beijing: Publishing House of Social Science and Literature, 2005 (in Chinese)

E. Leamer. Access to western markets, and eastern effort levels. in S. Zecchini, Lessons from the Economic Transition: Central and Eastern European the 1990s, Dor-drecht; Boston: Kluwer Academic Publishers, 1997

J. E. Anderson, E. V. van Wincoop. Gravity with gravitas: a solution to the border puzzle, The American Economic Review, 2003, 93: 170-192

J. McCallum. National borders matter: Canada-U.S. regional trade patterns, American Economic Review, 1995, 85(3) 615-623

K. Head, T. Mayer. Non-Europe: the magnitude and causes of market fragmentation in the EU, Weltwirtschaftliches Archive, 2000, 136(2): 285-314

Shicunzhenyi, Wang Huijing. Interregional Input-Output Analysis of the Chinese Economy, Beijing: Publishing House of Chemical Industry, 2007 (in Chinese)

V. Nitsch. National borders and international trade: evidence from European Union, Canadian Journal of Economics, 2000, 22: 1091-1105

W. H. Furtan, M. Blain. van Melle. Canada's agricultural trade in north America: do national borders matter?, Review of Agricultural Economics, 2004, 26(3): 317–331

Wei Shangjin. Intra-national versus international trade: how stubborn are nations in global integration?, NBER Working Paper, 1996, no. 5531

METHOD FOR ACHIEVING IRREGULAR BOUNDARY AREA FOR COMPLETE FLUIDIC SPRINKLER

Junping Liu*, Shouqi Yuan, Hong Li, Xingye Zhu
Engineering Research Centre for Fluid Machinery ,Jiangsu University, Jiangsu , Province, P. R. China 212013
* *Corresponding author, Address: Engineering Research Centre for Fluid Machinery, Jiangsu University, Jiangsu 212013, Jiangsu Province, P. R. China, Tel: +13914555629, Email: liujunping401@hotmail.com*

Abstract: For resolving the problem of sprinkle repeated, overtaken and went beyond in irrigation, it is important to research the approach of irregular boundary area. The equation of range and flow for achieving the square and triangle spray were deduced. Pressure is proportional to range. Specific method of changing the sectional area was put forward for achieving square and triangle spray. Adopted MATLAB language editor to analyzing the theoretical relation and emulate for achieving square and triangle spray. The experiments of theory pressure were carried out. The results showed that the experimental value were consistent with the theoretical value.

Keywords: theoretical relation; complete fluidic sprinkler; simulation; experiment

1. INTRODUCTION

As water supplies become limited, agricultural water use needs to become more efficient to maintain current productivity levels. Sprinkler irrigation has some problem such as sprinkle repeated, overtaken and went beyond and so on. It is important to find some way to fulfill variable irrigation. The shape of the irrigation field is decided by the variety of the range of the sprinkler (H. Sourell et al., 2003; Li J et al., 1995). Recent developments in variable irrigation were attempting to solve this issue. Han W et al. (2007)

Please use the following format when citing this chapter:

Liu, J., Yuan, S., Li, H. and Zhu, X., 2009, in IFIP International Federation for Information Processing, Volume 294, *Computer and Computing Technologies in Agriculture II, Volume 2*, eds. D. Li, Z. Chunjiang, (Boston: Springer), pp. 901–908.

have deduced theatrical range equation between rotational speed, flow and range. The field of the irrigation was square and draft star. Han X et al. (2005) obtain the curve of amendment pressure and angle of rotation through experiments. Zhang S et al. (2001) put forward a mechanics approach for changing the range of the sprinkler. Meng Q et al. (2003) carried through the feasibility analysis for sprinkler achieving irregular boundary area by mathematical principle. The theatrical studies on realized method of variable irrigation above afford us the theatrical guide. They are important references for the structural design.

But they were not put forward the method and validate for the parameter relation. Through the theatrical analysis and simulation for range, flow and pressure, the method for achieving the square and triangle spray area were put forward. And the experiments by complete fluidic sprinkler were carried out. The results indicate the correctness of the theatrical analysis.

2. SPRINKLER WORKING PARAMETER RELATION EQUATIONS

The spray process of tradition rotary type is the whole-circle spray. As shown in Fig. 1.

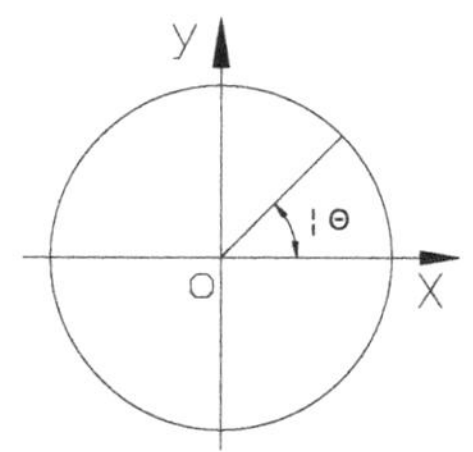

Fig 1: Sketch map of the whole-circle spray

The sprinkler is at O point. When the sprinkler revolting form axis X to Y with the angle θ , thc spray area is:

$$\Delta S = \pi R^2 \frac{\theta}{2\pi} = \frac{1}{2} R^2 \theta = \frac{1}{2} R^2 \omega t \qquad (1)$$

In which Q ——capacity of spray in unit time (m3/h). The major factors of sprinkler flow are working pressure and the size of injecting nozzlc. The Q will enlarge with the pressure.

For a sprinkler, the structure parameters are fixed value. When the sprinkler is working, the purees, wind velocity and the rotational speed affect the range. At the scope of working pressure, the pulverization will increase and the range wills not increase.

Calculation sprinkler irrigation intensity is the capacity of spray in unit time and area. The calculation sprinkler irrigation intensity will calculate as the formula:

$$\rho = \frac{1000\ Q}{S} \tag{2}$$

In which ρ —— calculation sprinkler irrigation intensity (mm/h). $\rho = \bar{h}/t$, $\bar{h}$ ——average sprinkler irrigation water (mm) S——area of spray with the Q (m2).

Put the formula (1) into(2):

$$\frac{\bar{h}}{t} = \frac{1000\ Q}{1/2R^2\omega t} \tag{3}$$

After simplification, the working equation of sprinkler is as follow:

$$Q = \frac{1}{2000}\bar{h}\omega R^2 \tag{4}$$

In a sprinkle irrigation system, $\bar{h}$ is fixed value for the crop requirement. At the scope of working pressure, the rotational speed changed a little. The relation ship between the flow and range is: $Q_1/Q_2 = R_1^2/R_2^2$, in an other word, flow is proportional to the square of range. In which Q_1, Q_2, R_1 and R_2——flow and range in different working conditions.

3. BOUNDARY EQUATION OF SQUARE AND TRIANGLE

In the system of sprinklers combination, there are many methods to layout sprinklers. The spray area of a sprinkler is square or triangle can increase the water usage and the uniformity of the spray. The sprinkle repeated, overtaken and went beyond in irrigation reduced obviously.

The control area of spray for square and triangle is reduced than the whole-circle. But according to the point of view of precision irrigation, the overlap rate and overtake rate reduced obviously. So that, it is necessary to study the variety orderliness of theory parameters relation of square and triangle spray area, which were carried out by complete fluidic sprinkler.

When the sprinkler achieving square spray area, the form is shown in figure 3, the sprinkler is at O point, set OA=R0=1, OB=R. The range of sprinkler changed for four periods. The angle is α from OA to OB, the range reduced gradually. There are four peak value of range in one period.

The range changed fast at the max value and slows at the min. The form of spray area is shown in Fig. 2.

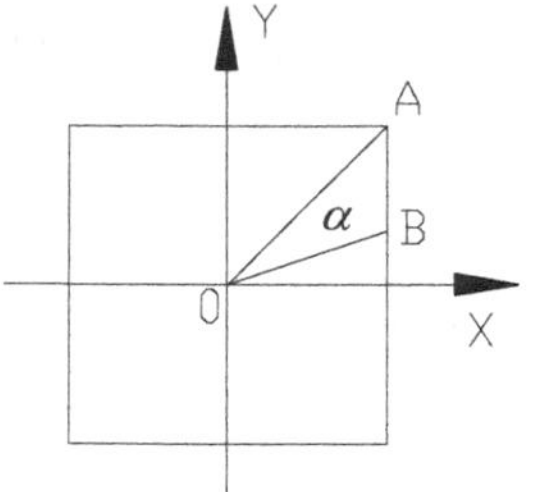

Fig 2: Sketch map of square spray

The boundary function for square is as follows:

$$R = \begin{cases} \dfrac{R_0}{\sqrt{2}\cos(\pi/4-\alpha)} & 0 \le \alpha \le \dfrac{\pi}{2} \\ \dfrac{R_0}{\sqrt{2}\cos(3\pi/4-\alpha)} & \dfrac{\pi}{2} \le \alpha \le \pi \\ \dfrac{R_0}{\sqrt{2}\cos(5\pi/4-\alpha)} & \pi \le \alpha \le \dfrac{3\pi}{2} \\ \dfrac{R_0}{\sqrt{2}\cos(7\pi/4-\alpha)} & \dfrac{3\pi}{2} \le \alpha \le 2\pi \end{cases} \tag{5}$$

In one period, the solution of the equation is the area boundary of square.

When the sprinkler achieving triangle spray area, the form is shown in Fig. 4, the sprinkler is at O point, set $OA=R_0=1$, $OB=R$. The range of sprinkler changed for three periods. The angle is β from OA to OB. There are three peak value of range in one period. The range changed fast than square. The form of spray area is shown in Fig. 3.

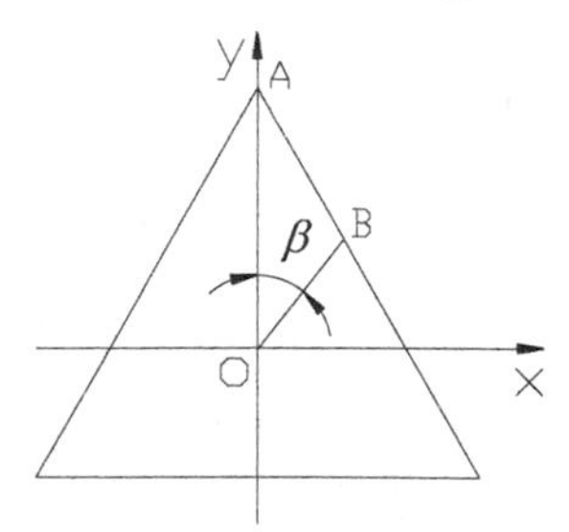

Fig 3: Sketch map of triangle spray

The boundary function for triangle is as follows:

$$R = \begin{cases} R_0 / 2\cos(\frac{\pi}{3} - \beta) & 0 \le \beta \le \frac{2\pi}{3} \\ R_0 / 2\cos(\pi - \beta) & \frac{2\pi}{3} \le \beta \le \frac{4\pi}{3} \\ R_0 / 2\cos(\frac{5\pi}{3} - \beta) & \frac{4\pi}{3} \le \beta \le 2\pi \end{cases} \quad (6)$$

In one period, the equation solution is the area boundary of triangle.

4. METHODS FOR ACHIEVING IRREGULAR BOUNDARY AREA

There are many factors for the range of sprinkler. For example, elevation of sprinkler, inlet velocity of flow, and the inlet velocity of flow which were effected by flow and pressure. The experiments for PXH30 complete fluidic sprinkler indicated that the prominence factor is pressure. For achieving the square and triangle range variety, the inlet pressure can be modified.

Because of the deficient of fix height, the range of sprinkler is different. But it doesn't affect the spray form. So that, the studies ignore the fix height of sprinkler, other parameters hold the line. The equation of sprinkler range is:

$$R = 2\varphi^2 P \sin 2\alpha \quad (7)$$

In which φ ——speed coefficient, P——sprinkler working pressure, α ——elevation of sprinkler.

After simplification: $\frac{R_1}{R_2} = \frac{P_1}{P_2}$

MATLAB language has so powerful figure disposal function. The simulation was carried out for range of square and triangle through MATLAB. The parameters relation became more visualize. The curves of square and triangle range variety were edited through MATLAB.

According to the relation of range and pressure in one period, the pressure variety cures for achieving square and triangle are shown in Fig. 4.

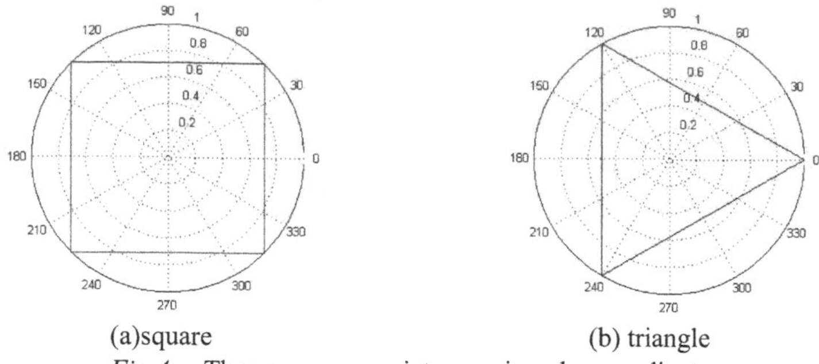

(a)square (b) triangle

Fig 4: Theory purees variety cure in polar coordinate

As is shown in Fig. 4, the pressure variety achieved four and three periods, appeared four and three peak value.

Above the analysis for parameters relation for achieving irregular boundary area, the main methods for achieving irregular boundary area were change the inlet pressure and elevation of sprinkler. And for changing the inlet pressure, the inlet section area can be changed.

There is:

$$v = \varphi\sqrt{2gp} \tag{8}$$

$$v = \frac{Q}{A} \tag{9}$$

Put the formula (9) to (8): $$A = \frac{Q}{\varphi\sqrt{2gp}} \tag{10}$$

After simplification: $$\frac{A_1}{A_2} = \frac{Q_1}{Q_2}\cdot\sqrt{\frac{p_2}{p_1}} = \frac{R_1^2}{R_2^2}\sqrt{\frac{p_2}{p_1}} = \frac{R_1^{1.5}}{R_2^{1.5}}$$

The cures of inlet section area of sprinkler are shown in Fig. 5.

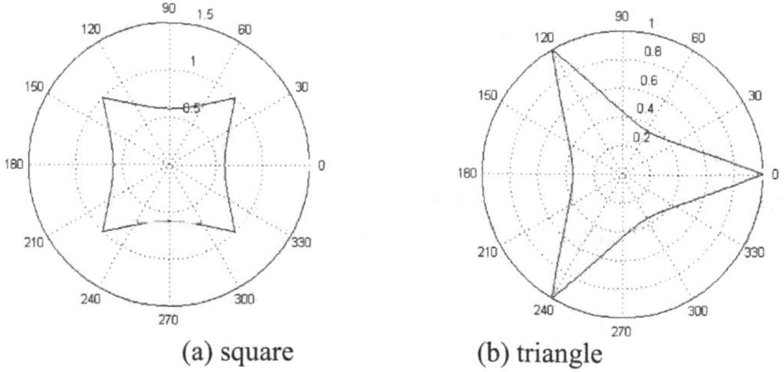

(a) square (b) triangle

Fig 5: Theory inlet section area variety cures in polar coordinate

It can be seen from the cures that, when achieving the square and triangle areas, the trend of inlet section area and the range are consistent, and the scope is different. The simulation was carried out for inlet section areas of square and triangle through MATLAB.

For the complete fluidic sprinkler, it can add a pressure control to modify the inlet section area. The control can adopt movement pieces to achieve. Form of movement can accord section variety cures.

5. CONCLUSIONS

For resolving the problem of sprinkle repeated, overtaken and went beyond in irrigation, it is important to research the approach of irregular boundary area.

(1)The equation of range and flow for achieving the square and triangle spray were deduced. Flow is proportional to the square of range. Pressure is proportional to range.

(2)Adopted MATLAB language editor to analyzing the theoretical relation and emulate for achieving square and triangle spray.

(3)The experiments results showed that the experimental value were consistent with the theoretical value.

For changing the inlet section area of complete fluidic sprinkler, the main issue is that the driving force can not overcome friction. So that, the theory parameters can instructs the structure design of irregular boundary area sprinkler. The complete fluidic sprinkler will analysis and design farther.

ACKNOWLEDGEMENTS

Funding for this research was provided by the National High-Tech Program Grant No.2006AA100211. (P. R. China).

REFERENCES

H. Sourell, J. M. Faci, and E. Playán. .Performance of Rotating Spray Plate Sprinklers in Indoor Experiments. J. Irrig. Drain. Eng., 2003.129.(5):376～380.

Han Xin, HAO Pei-ye. Analysis of Spraying Mechanism of a Whirl Sprinkler with Square Spray Field.Transactions of the CSAM,2005,36(3): 40～44. (in Chinese)

Han, Wenting; Fen, Hao; Wu, Pute; Yang, Qing. Evaluation of sprinkler irrigation uniformity by double interpolation using cubic splines. ASABE Annual International Meeting, Technical Papers, 2007,(6);1-7.

Li,J.and H.Kawano.Simulating water-drop movenent from noncircular sprinkler nozzles. Journal of Irrigation.and.DrainageEngineering,1995,121(2):152～158.

Zhang She-qi, LIU Shu-ming, HAN Wei-sheng. The hydromechanics way to change the nozzle sprinkling orbit . Jour. of Northwest Sci2Tech Univ. of Agri and For. (N at. Sci. Ed.),2001,29(4):118～121. (in Chinese)

Meng Qin-qian, WANG Jian, CAI Jang-bi. The research on realizing non-circular spraying distr ict of sprinkle. Jour. of Northwest Sci2Tech Univ. of Agri and For. (N at. Sci. Ed.),,2003,(4):145～148. (in Chinese)

IRRIGATION UNIFORMITY WITH COMPLETE FLUIDIC SPRINKLER IN NO-WIND CONDITIONS

Xingye Zhu [*] , Shouqi Yuan, Hong Li, Junping Liu

Technical and Research Center of Fluid Machinery Engineering, Jiangsu University, Zhenjiang, Jiangsu Province, P.R. China 212013

[*] *Corresponding author, Address: Technical and Research Center of Fluid Machinery Engineering, Jiangsu University, Zhenjiang 212013, Jiangsu Province, P.R. China, Tel: +86-511-88782121, E-mail:xingye488@hotmail.com*

Abstract: The Complete fluidic sprinkler was originally created by China. Irrigation uniformity in no-wind conditions was studied in this paper. Radial water distribution for complete fluidic sprinkler type 10 was got by experiment. MATLAB was used to establish computation program to change the radial data into net data. Three-dimensional water distribution figures for single sprinkler and combined sprinklers were figured out after the calculation of MATLAB. Combined uniformity coefficient of the complete fluidic sprinkler type 10 was simulated in combined spacing coefficient form 1 to 1.8.The maximal uniformity coefficient was 80.5% or 88.2% for rectangle or triangular combination respectively. Combined uniformity surpasses 75% when combined spacing coefficient was 1 to 1.7 in rectangle combination and 1.4 to 1.7 in triangular combination. A case study shows that the MATLAB is reliable for simulating water distribution in sprinkler irrigation.

Keywords: complete fluidic sprinkler, uniformity coefficient, combined spacing, interpolation

1. INTRODUCTION

Complete fluidic sprinkler is a water-saving sprinkler which is based on the wall-attachment effect. There were several modal fluidic sprinklers developed in China (Yuan et al. 2006). But the knowledge about the water

Please use the following format when citing this chapter:

Zhu, X., Yuan, S., Li, H. and Liu, J., 2009, in IFIP International Federation for Information Processing, Volume 294, *Computer and Computing Technologies in Agriculture II, Volume 2*, eds. D. Li, Z. Chunjiang, (Boston: Springer), pp. 909–917.

application of the fluidic sprinkler is quite limited. Most sprinkler irrigation systems require a minimum value of water distribution uniformity [Christiansen's Coefficient of Uniformity (CU)≥80%] (Keller and Bliesner,1990). Low values of CU are usually indicators of a faulty combination of the number and size of nozzles, working pressure and spacing of sprinklers. To determine Christiansen's Coefficient of Uniformity (Christiansen, 1942) and other parameters characterizing surface water distribution, we need to know the application rate caught in a grid of cans within the wetted area. The procedures to determine sprinkler water distribution can be grouped into three types: 1. Apply the catch can grid to the existing irrigation system: evaluation of the system (Merriam and Keller, 1978; Merriam et al., 1980). 2. Place a catch can grid around a single sprinkler head in no-wind conditions and established the corresponding overlapping for any sprinkler spacing (Solomon, 1979). 3. Reduce the catch cans grid to a single-leg in a radial pattern, in no-wind and with high relative humidity conditions. The application rate can be calculated by rotating the radial pattern around the sprinkler (Vories and von Bemuth, 1986). The first procedure describes working conditions of an existing irrigation system. The second has thc advantage of identifying the entire distribution pattern of the sprinkler, as well as uniformity parameters under any irrigation spacing. The third has the advantage of controlling all factors in the process, especially sprinkler water distribution, thus allowing us to establish comparisons between different sprinklers. The objective of this study was to provide an interpolation algorithm to prepare distribution maps of water depth from catch-can test data for a complete fluid sprinkler irrigation system. Uniformity parameter (CU) are then calculated from the distribution maps of application depth. This method assumes that water application depth is a continuous variable.

2. LABORATORY PROCEDURES

Performing the experiments in an indoor facility ensures a radial water distribution and avoids drift and losses (Sourell et al. 2003). Measurements were made in the Indoor Sprinkling Laboratory at Jiangsu University. It is a circular indoor laboratory with a diameter of 44m. Fig.1 presents the condition of the laboratory. The affection of wind and other natural factors were excluded and test results were made exact and reliable. The test of point irrigated intensity is the technical sticking point and most difficult for the whole testing system. If every point is tested manually, the testing efficiency would have been low and the reliability of data would not have been as accurate. So developing an auto-testing system was an important guarantee for exploitation of new-typed sprinkler(Li 1996, Li et al. 1998).

Jiangsu University rebuilt the testing system and changed it from a centralizing system to a total line distribution system based on the RS485. Fig.2 showed a diagram of this new testing system. The technique of using rain-collection as an implement was installed in the laboratory and connected to the computer in the control room using the RS485 assembled line. It reduced cable lengths and link workload considerably. The system was made more compact and reliable. The flow discharged from the sprinkler was determined by an auto-select system using software we had developed. Fig.1 showed that testing points for the water were distributed radially in the testing field. The collectors were 20cm in diameter and 60cm high, spacing every 1 m and there were 40 points in all. The flow rate was measured using a flow-computer which was accurate to +_0.5 percent over the entire flow rate range in the irrigated room.

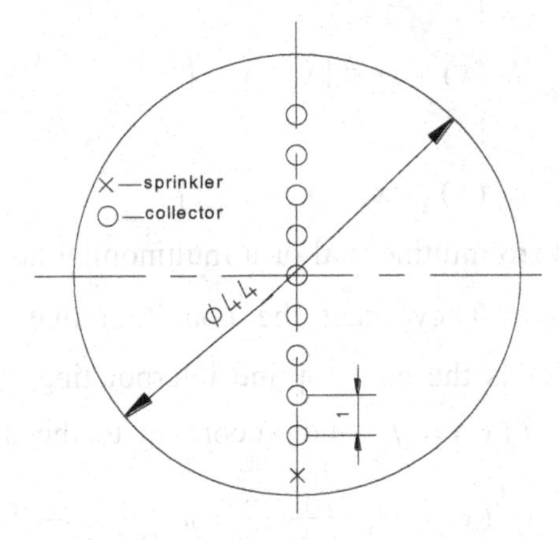

Fig.1: Testing layout for uniformity tests

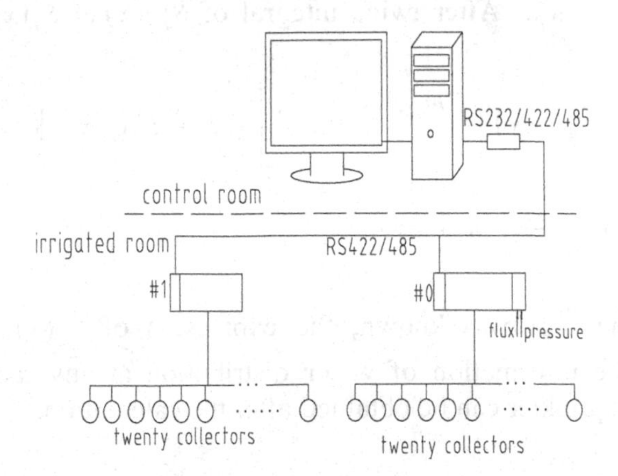

Fig.2: Auto-testing system based on total line distribution

3. MATHEMATICAL MODEL

Radial data of water distribution was changed into net data for the complete fluidic sprinkler. It is the main step of analyzing uniformity coefficient and drawing three-dimensional water distribution. The water distribution can be supposed to be the same in all directions of the complete fluidic sprinkler. The depth of net point depends on the distance away from the sprinkler. It can be got using interpolation method from data that checked indeed. The model of stick insert function was established as follows:

f[x] was supposed to be a continued function in the limited of [a,b]. Some basic points were given in [a,b]. a=x1<x2<...<xn+1=b. On the assumption that $S(x)$ is a twice continuously differentiable function as:

$$S(x) = \begin{cases} S_1(x) & x \in [x_1, \ x_2] \\ \vdots & \\ S_i(x) & x \in [x_i, \ x_{i+1}] \\ \vdots & \\ S_n(x) & x \in [x_n, \ x_{n+1}] \end{cases}$$

where $S_i(x)$ is a zero multinomial or a multinomial no higher than thrice i =1, 2, 3, ..., n. They meet the condition that $S_i(x) = f(x_j)$, j =1, ..., n+1. $S(x)$ is the cubic spline interpolating function of $f(x)$. In that $m_i = S''(x)$, $f(x_i) = f_i$ and according to the definition of cubic spline, we can get $S_i''(x) = m_i \dfrac{x_{i+1} - x}{h_i} + m_{i+1} \dfrac{x - x_i}{h_i}$ $x \in [x_i, \ x_{i+1}]$, where $h_i = x_{i+1} - x_i$. After twice integral of $S_i''(x)$, $S_i(x)$ was got as follow:

$$S_i(x) = h_i\left[\frac{m_i}{6}(x_{i+1} - x)^3 + \frac{m_{i+1}}{6}(x - x_i)^3\right] + f_i + f[x_i, x_{i+1}](x - x_i) - \frac{h_i^2}{6}\left[(m_{i+1} - m_i)\frac{x - x_i}{h_i} + m_i\right].$$

When m_i and m_{i+1} was known, the expression of $S_i(x)$ was wholly determined. The information of water distribution at any radial direction away from the sprinkler can be obtained after the calculation.

4. RESULTS AND DISCUSSION

Water distribution is an important index of sprinkler characteristics. It is also the main gist of irrigation programming design. Tests were conducted adopting the national standards for irrigated sprinkler JB/T 7867-1997. The factors for water distribution include flow, point irrigated intensity and range under relevant pressure. Complete fluidic sprinkler type 10 was chosen in the experiments. The working pressure was 0.25MPa and the flow rate was 0.84m3/h. Fig.3 showed the water distributions of the sprinkler. MATLAB has a wonderful numerical calculation function. It can constitute the interpolating function and fulfill the calculation automatically. The program of drawing water distribution using MATLAB was established. Fig.4 represents the three-dimensional water distribution of this sprinkler. As can be seen from Fig.4, Water distribution of any net point around complete fluidic sprinkler can be got easily and conveniently. Water distribution by sprinkling was made more intuitionistic. It was supplied as a method expressing water distribution effectiveness more apparent.

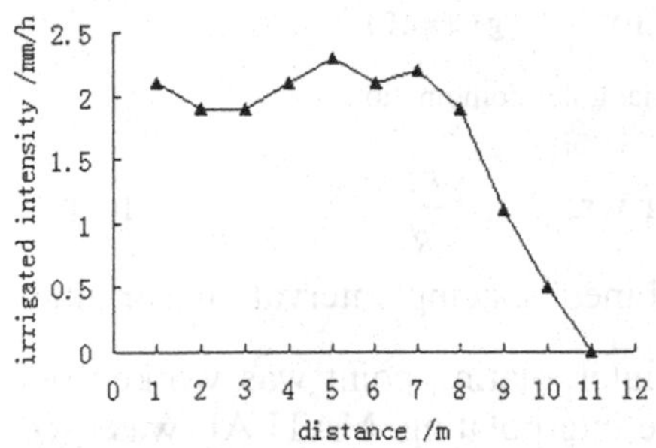

Fig.3: Water distribution of the sprinkler

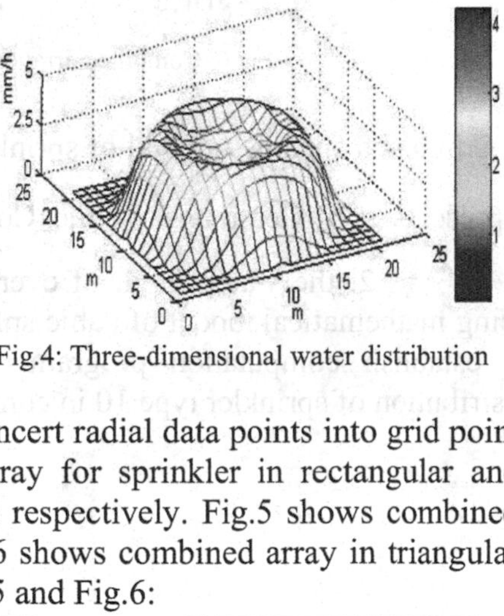

Fig.4: Three-dimensional water distribution

Double interpolation is needed to concert radial data points into grid point data (Han et al. 2007). Combined array for sprinkler in rectangular and triangular combination were discussed respectively. Fig.5 shows combined array in rectangular combination. Fig.6 shows combined array in triangular combination. As can be seen from Fig.5 and Fig.6:

$$l_{ma} = \sqrt{x^2 + y^2}, \quad l_{mb} = \sqrt{(x - r_a)^2 + y^2}, \quad l_{mc} = \sqrt{(x - r_a)^2 + (y - r_b)^2},$$

$$l_{md} = \sqrt{x^2 + (y - r_b)^2} \quad l_{me} = \sqrt{x^2 + y^2}, \quad l_{mf} = \sqrt{(x - r_a)^2 + y^2},$$

$$l_{mg} = \sqrt{(x - 2r_a)^2 + y^2}, \quad l_{mh} = \sqrt{(x - \frac{3r_a}{2})^2 + (y - r_b)^2},$$

$$l_{mi} = \sqrt{(x - \frac{r_a}{2})^2 + (y - r_b)^2}$$

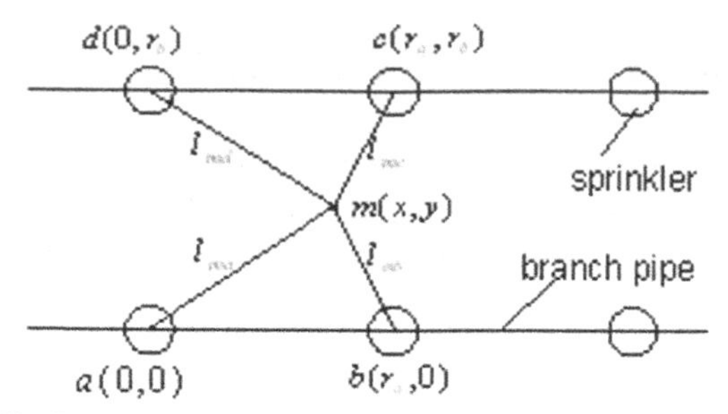

Fig.5: Combined array in rectangular combination

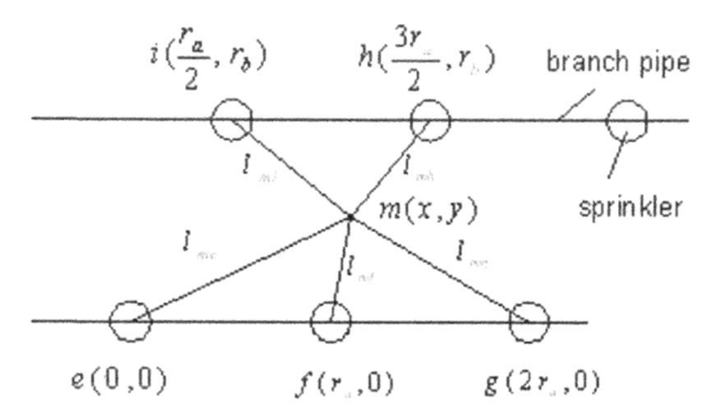

Fig.6: Combined array in triangular combination

Combined spacing interval of sprinkler was $k_a = \frac{r_a}{R}$, $k_b = \frac{r_b}{R}$, where R represents sprinkler range. While Combined spacing interval of sprinkler $k_a = k_b = 1.2$, the water depth of every interpolating point was worked out using mathematical model of cubic spline interpolating. MATLAB was used to establish computation program. Fig 7 was three-dimensional water distribution of sprinkler type 10 in combined irrigation.

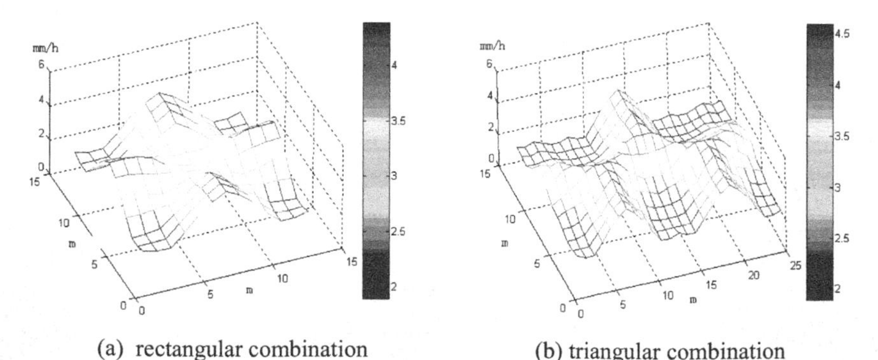

(a) rectangular combination (b) triangular combination

Fig.7: Three-dimensional water distribution of sprinkler type 10 in combined irrigation

As can be seen from Fig.7, water distribution was uniform all around the irrigated control area in rectangular combination. The maximal irrigated intensity was about 4mm/h in the middle of two sprinklers. The nearer any sprinkler was, the less water distribution of combined irrigation was. Water was triangular distribution all around the irrigated area when sprinklers work in triangular combination. The maximal irrigated intensity was about 4.5mm/h. Combined water distributions were similar for both rectangular and triangular combination. But water distribution face was more flat for combined array in rectangular combination than combined array in triangular combination.

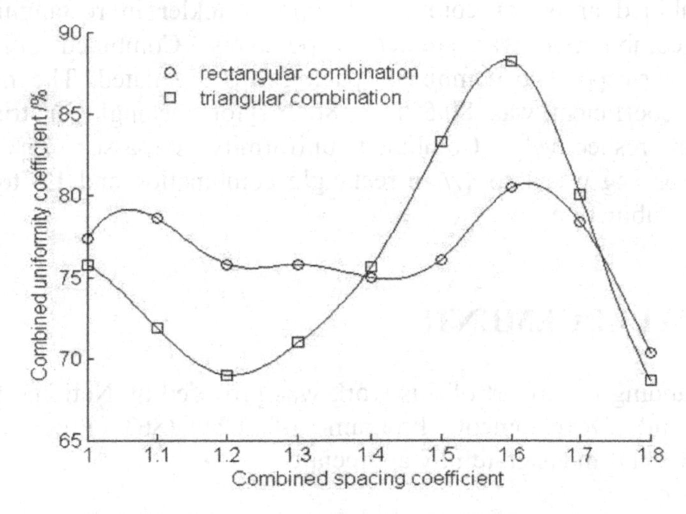

Fig.8: Simulated curves of combined irrigation coefficient

After the definition of water distribution for single sprinkler, the main factors in affecting combined irrigated uniform were combined manner and combined spacing. Under the condition that combined manner was rectangular or triangular, combined spacing coefficient $k = k_a = k_b$ was chosen by 1.1、1.2、1.3、1.4、1.5、1.6、1.7、1.8 respectively. Simulation program was established. Fig.8 was the uniformity coefficient simulated curve of combined irrigation. As can be seen from the figure, the curve of uniformity coefficient is more flat in rectangle than in triangular. It means that uniformity coefficient was stable as the change of combined spacing coefficient in rectangle combination. Uniformity coefficient changed rapidly as the change of combined spacing coefficient in triangular combination. The maximal uniformity coefficient was 80.5% or 88.2% for rectangle or triangular combination respectively. When combined spacing was 1 to 1.7 in rectangle combination and 1.4 to 1.7 in triangular combination, combined uniformity coefficient of the complete fluidic sprinkler surpasses 75%, which reached the irrigated standard.

5. CONCLUSION

(1) Complete fluidic sprinkler type 10 was chosen to study the irrigation uniformity in no-wind conditions. The radial water distribution was got by experiment in the Indoor Sprinkling Laboratory at Jiangsu University.

(2) Radial data of water distribution was changed into net data using cubic spline interpolating. MATLAB was used to establish computation program Three-dimensional water distribution of this sprinkler was figured out. It supplied a software platform to study the complete fluidic sprinkler.

(3) Combined array for complete fluidic sprinkler in rectangular and triangular combination was studied respectively. Combined uniformity coefficient correspond to combined spacing was simulated. The maximal uniformity coefficient was 80.5% or 88.2% for rectangle or triangular combination respectively. Combined uniformity surpasses 75% when combined spacing was 1 to 1.7 in rectangle combination and 1.4 to 1.7 in triangular combination.

ACKNOWLEDGEMENTS

Partial funding in support of this work was provided by National Hi-tech Research and Development Program of China(863 Program, No. 2006AA100211) and is gratefully appreciated.

REFERENCES

Christiansen, J.E. (1942). "Irrigation by sprinkling." Bulletin 670, California Agricultural Experiment Station, University of California, Berkeley, California.

Faci, J.M., Salvador, R., Playan, E., and Sourell, H. (2001). "Comparison of fixed and rotating spray plate sprinklers." J. Irrig. Drain. Eng., 127(4), 224-233.

Han, W.T., Fen, H., Wu, P.T., Yang, Q., and Chen, X.W. (2007). "Evaluation of sprinkler irrigation uniformity by double interpolation using cubic splines." Effective utilization of agricultural soil & water resources and protection of environment: 250-255.

Hendawi,M., Molle,B., Folton,C., and Granier, J. (2005). "Measurement accuracy analysis of sprinkler irrigation rainfall in relation to collector shape." J. Irrig. Drain. Eng., 131(5), 477-483.

Keller, J., and Bliesner, R.D. (1990). "Sprinkler and trickle irrigation." New York, N.Y.: Van Nostrand Reinhold.

Li, J., and Kawano, H. (1995). "Simulating water-drop movement from noncircular sprinkler nozzles." J. Irrig. Drain. Eng., 121(2), 152-158.

Li,J., and Hiroshi, K.,(1998). "Sprinkler performance as affected by nozzle inner contraction angle." Irrig. Sci.,18, 63-66.

Louie, M.J., and Selker, J.S. (2000). "Sprinkler head maintenance effects on water application uniformity." J. Irrig. Drain. Eng., 126(3), 142-148.

Merriam, J.L., and Keller, J. (1978). "Farm irrigation system evaluation." A guide for management. Logan, Utah: Utah State University.

Merriam, J.L., Shearer, M.N., and Burt, C.M. (1980). "Evaluating irrigation system and practices." In design and operation of Farm Irrigation System, ed. M.E. Jensen, 721-760. Joseph, St., Mich.: ASAE.

Solomon, K. (1979). "Variability of sprinkler coefficient of uniformity test results." Trans ASAE, 22(5), 1078-1080, 1086.

Sourell, H., Faci, J. M., and Playán,E. (2003). "Performance of rotating spray plate sprinklers in indoor experiments." J. Irrig. Drain. Eng., 129 (5), 376-380.

Tarjuelo, J.M., Montero, J., Valiente, M., Honrubia, F.T., and Ortiz, J. (1999). "Irrigation uniformity with medium size sprinklers. Part I: characterization of water distribution in no-wind conditions." Trans ASAE, 42(3), 665-675.

Vories, E., and von Bernuth, R.D. (1986). "Single nozzle sprinkler performance in wind." Transactions of the ASAE 29(5):1325-1330.

Yuan Shouqi, Zhu Xingye, Li Hong, and Ren Zhiyuaan. (2006). "Effects of complete fluidic sprinkler on hydraulic characteristics based on some important geometrical parameters." Trans CSAE， 22(10), 113-116. (in Chinese)

STRUCTURAL BREAKS AND THE RELATIONSHIP BETWEEN SOYBEAN AND CORN FUTURES PRICES ON THE DALIAN COMMODITY EXCHANGE OF CHINA

Rufang Wang[1] ,Yonghong Du[2,*] ,Jian Wang[2]
[1] *School of Economics, Beijing Wuzi University, Beijng, P. R. China 101149*
[2] *School of Economics, Nankai University, Tianjing, P. R. China 300071*
[*] *Corresponding author, Address: School of Economics, Nankai University, Tianjing, P. R. China 300071, Tel: +86-13910121099, Fax: +86-10-89534320, Email: Yonghongdu@sina.com*

Abstract: Co-movement between futures prices can arise when commodities are substitutes. Using Johansen's co-integration procedure, we fail to find a significant long-run link between soybean and corn prices on the Dalian Commodity Exchange of China. This relationship is re-examined using Johansen's co-integration procedure that permits structural breaks. Results show evidence of co-integration and hence price discovery. There is a significant break in July 2007 by reason of rare drought in China's main soybean producing areas. The soybean–corn futures market is perfectly integrated, and the soybean price Granger-causes the corn price. Modeling structural breaks in price relationships appears important.

Key words: structural break, co-integration, agricultural commodity futures price, price discovery

1. INTRODUCTION

Co-movement between futures prices occurs when two or more prices move together in the long run and the discovery of one price provides valuable information about others. Commonality between commodity futures prices arises particularly when the commodities are substitutes on the

---

Please use the following format when citing this chapter:

Wang, R., Du, Y. and Wang, J., 2009, in IFIP International Federation for Information Processing, Volume 294, *Computer and Computing Technologies in Agriculture II, Volume 2*, eds. D. Li, Z. Chunjiang, (Boston: Springer), pp. 919–926.

supply and/or demand sides, but other reasons include the common impact of changes in macroeconomic variables and speculator behavior (Pindyck et al., 1990).

Malliaris and Urrutia (1996) examine price discovery on the Chicago Board of Trade (CBOT) by using bivariate single-equation models and residual-based tests of co-integration. Using daily data for 1981–1991, they examine long-run linkages between six related agricultural commodities—corn, wheat, oats, soybean, soybean meal, and soybean oil—where cross-price elasticities of supply and demand are nonzero. The findings show interdependencies between pairwise prices and hence price discovery. Dawson and White (2002) extend this analysis by examining interdependencies between futures prices on the London International Financial Futures Exchange (Euronext.LIFFE) using Johansen's co-integration procedure and daily data for 1991–2000. Findings show no long-run relationships between the prices of barley, cocoa, coffee, sugar, and wheat. The result is particularly surprising because barley and wheat are close substitutes. If futures prices change, processors can hedge prices and change mixes. Dawson, Sanjuán and White (2006) re-examine the long-run relationship between feed barley and feed wheat futures prices on the Euronext.LIFFE using the recent co-integration procedure of Johansen, Mosconi, and Nielsen (2000) where structural breaks are allowed at known points in time and weekly data for 1996–2002. Results show evidence of co-integration and hence price discovery. There is a significant break in October 2000 following Common Agricultural Policy intervention price reductions, the barley–wheat futures market is perfectly integrated, and the barley price Granger-causes the wheat price. Modeling structural breaks in price relationships appears important. The failure to model structural breaks may have led to Dawson and White's counterintuitive conclusion that barley and wheat futures prices are unrelated.

Our aim is to examine the long-run relationship between soybean and corn futures prices on the Dalian Commodity Exchange and hence price discovery, using daily data from 1 July 2005 to 20 June 2008. The price discovery hypothesis is tested here using Johansen's co-integration procedure without breaks and the Johansen's co-integration procedure where structural breaks are allowed at known points in time. The breaks are identified from Perron's unit root test. Perfect price integration is tested where a given percentage change in one price is mirrored in the other; weak exogeneity tests examine Granger-causality; and impulse response functions explore dynamics.

2. EMPIRICAL METHOD

We utilized a vector autoregressive (VAR) model that does not require specifying the causal order prior to estimation

$$Y_t = \mu + \sum_{i=1}^{k} A_i Y_{t-i} + \varepsilon_t \qquad (1)$$

where Yt = [pb,t pc,t]' and pb,t and pc,t are futures prices of soybean and corn in logarithms, μ and Ai are matrices of parameters, k is the lag length, and εt are error terms. Many price series are nonstationary and regressions between such data are generally spurious. Nonstationarity price series are typically integrated of order one, I(1), and must be first differenced to render them stationary, I(0). Where two series are I(1) and move together so that their linear combination is stationary, they are co-integrated, a meaningful long-run equilibrium exists, and the problem of spurious regression does not arise. The Dickey–Fuller test is commonly used to test for unit roots (or nonstationarity) but when a series is subject to a deterministic trend and an exogenous shock causes a structural break, it tends to underreject. Therefore, we follow Perron and test the null of a unit root with a structural break and the alternative is stationarity around a broken level.

If both prices are I(1), the test for co-integration is a test of long-run equilibrium. Johansen's standard test is based on the estimation of the VAR model in (1) transformed into its vector error-correction model (VECM) form. This test is inappropriate however, if structural breaks within the individual series occur, either at different times or at the same time and do not cancel each other out. Johansen, Mosconi, and Nielsen generalize it by admitting up to two predetermined breaks. The sample is divided into q periods and j denotes each period. The VECM form of (1) where breaks are admitted is

$$\Delta Y_t = \alpha\beta Y_{t-1} + \gamma E_t + \sum_{i=1}^{k-1} \Gamma_i \Delta Y_{t-i} + \sum_{i=1}^{k}\sum_{j=2}^{q} \kappa_{j,i} D_{j,t-i} + \varepsilon_t \qquad (2)$$

Where Δ is the difference operator; Et is a vector of q dummy variables Et = [E1t E2t,. . .,Eqt]' with Ej,t =1 (j = 1,. . .,q) if observation t behind the first observation of the jth period and zero otherwise; Dj,t−i (j = 1,. . .,q and i = 1,. . .,k) is an impulse dummy that equals unity if observation t is the ith observation of the jth period and are included to facilitate estimation.

The short-run parameters are γ of order (2 × q), Γi of order (2 × 2) for i = 1,. . .,k−1, and κj,i of order (2 × 1) for j = 1,. . .,q and i = 1,. . .,k. The innovations ε t is a (2 × 1) vector of white noise residuals. The long-run drift parameters are α= [αb ,αc]′ is a vector of adjustment parameters, and β= [βb , βc] are the long-run coefficients in the co-integration vector. The co-integration hypothesis is formulated by testing the rank (r) of $\pi = \alpha\beta$ by the trace statistic.

Given the co-integration rank, we test two further restrictions on the cointegration space. First, we test that the soybean– corn futures market is perfectly integrated where a 1% increase in one price leads to a 1% increase in the other, and the null is

$$H0: \beta b = -\beta c = 1 \quad (3)$$

Second, we test for weak exogeneity where the null for pb,t for example is

$$H0 : \alpha b = 0. \quad (4)$$

In the case of r = 1, a test of weak exogeneity is one of Granger noncausality, and if for example, the nulls that $\alpha b = 0$ is rejected and $\alpha c = 0$ is not rejected, pb,t Granger-causes pc,t. For each test, the log-likelihood ratio is χ_1^2.

Finally, impulse response functions (Lütkepohl, 1993) are estimated to explore the dynamics of price adjustment following a shock to each price. Since deviations from the long-run equilibrium are stationary, any shock to the system generates time paths that eventually return to the equilibrium. Each impulse response shows the effect on one price to a shock in the other. When co-integration exists, impulse responses are generally permanent.

3. DATA AND RESULTS

Daily rollover nominal closing prices of futures contracts for soybean and corn are used from 1 July 2005 to 20 June 2008, giving 726 observations (Dalian Commodity Exchange). We get missing observations by interpolation. Figure 1 shows that both prices have trended upwards, peaking in March 2008 at ¥5495/tonne for soybean and in May 2008 at ¥1981/tonne for corn. The minima occurred at ¥2370/tonne for soybean in augest 2006, and ¥1193/tonne for corn in August 2005. Of particular interest is the trend chang in late 2007.

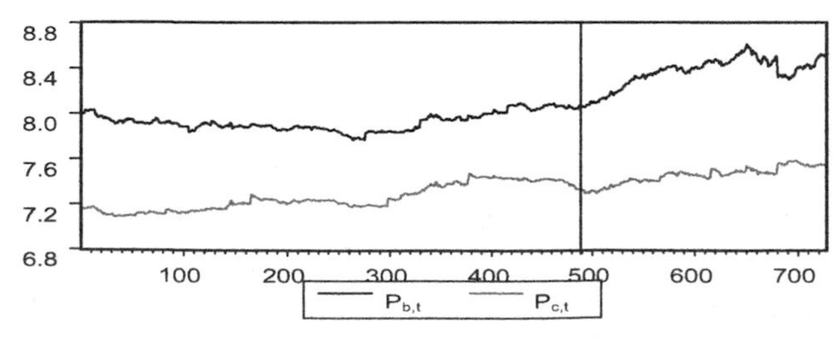

Figure 1. Soybean and corn prices

Following Perron (1997), we examined the nonstationarity and location of a structural break in each price series. The test is performed by estimating the breakpoint that minimizes the t-statistic on the parameter associated with the change in intercept. The number of lags, with a maximum of six, is chosen following Said and Dickey's (1984) general-to-specific method where sequential t-statistic on the parameter associated with ΔYt-i are used at the 10% significance level. The test statistics are −3.68 for pb,t and −3.70 for pc,t (critical value at the 95% confidence level: −4.84). Thus, the null of a unit root is not rejected for both prices and each series is I(1). There is a break in each series, on 13 July, 2007 for pb,t and on 17 May 2007 for pc,t. Because of the U.S. soybean acreage substantially reduced, coupled with China's main soybean producing areas of Heilongjiang historical experience rare drought , the generally upward trend of soybean prices changed sharply in July 2007. Corn acreage substantially increased imposed a reduction in the prices for corn on March, 2007.

Co-integration between pb,t and pc,t is tested without breaks in Models 1. In Models 2 implicit in (2), co-integration between pb,t and pc,t is tested allowing for breaks on 17 May, 2007 and on 13 July, 2007. In Models 3, co-integration between pb,t and pc,t is tested allowing for a single break on 17 May, 2007. In Models 4, co-integration between pb,t and pc,t is tested allowing for a single break on 13 July, 2007. The Akaike Information Criterion (AIC) is used to determine the number of lags and k = 1 lags for both models. The trace statistics is used to test for co-integration (table 1). No co-integration vector is found (r = 0) in Model 1 and one co-integration vector is found (r = 1) in Model 2 which includes two change in the level. Allowing for a single break, no co-integration vector is found (r = 0) in Model 3 and one co-integration vector is found (r = 1) in Model 4. Thus, there is the long-run equilibrium relationship between pb,t and pc,t , a single break appears in the vector error-correction model.

Table 1. Co-integration tests (trace statistics)

	Model 1	Model 2	Model 3	Model 4
Breaks		13.07.2007 17.05.2007	17.05.2007	13.07.2007
$H_0: r = 0;$	5.60	15.5*	9.12	16.85*
$H_1: r \geq 1$	(0.74)	(0.05)	(0.35)	(0.03)
$H_0: r = 1;$	0.13	1.46	0.13	3.38
$H_1: r \geq 2$	(0.72)	(0.23)	(0.72)	(0.07)

Notes: 1.The values in parentheses denotes the sgnificance level.
2. *denotes rejection of the hypothesis at the 0.05 level

Hypothesis tests are conducted:

(1) the null that βb =−βc =1 is not rejected (χ^2_1 =1.99, p-value: 0.16) and the soybean–corn futures market is perfectly integrated;

(2) the null that $\alpha b = 0$ is rejected (χ_1^2 =4.36, p-value: 0.04) while the null that $\alpha c = 0$ is not rejected (χ_1^2 =2.49, p-value: 0.11). Thus, pb,t is weakly exogenous, which implies that pb,t Granger-causes pc,t.

Normalizing the co-integrating vector on pc,t and imposing the restriction that $\beta b = -\beta c = 1$ gives

$$Pc,t = pb,t - 0.749 + \varepsilon t \quad (5)$$

the vector error-correction model is

$$\Delta pb,t = -0.019ecmt - 0.078\Delta pb,t\text{-}1 - 0.083\Delta pc,t\text{-}1 - 0.001 + 0.006Et - 0.009Dt\text{-}1 + 0.023Dt\text{-}2$$

$$\Delta pc,t = -0.0001ecmt - 0.058\Delta pb,t\text{-}1 - 0.001\Delta pc,t\text{-}1 + 0.0003 + 0.001Et - 0.009Dt\text{-}1 + 0.018Dt\text{-}2$$

Where $ecmt = \Delta pb,t\text{-}1 - \Delta pc,t\text{-}1 - 0.749$ and the single break is on 13 July, 2007. the βb coefficient in an identified co-integrating relationship such as Equation (5) is the ceteris paribus long-run elasticity, and a 1% increase in pb,t leads to a 1% increase in pc,t. To explore dynamics more fully, (orthogonal) impulse response functions are calculated. These show the impact on one price of a one standard error increase in the other as both adjust back to long-run equilibrium and are illustrated in figure 2 with 90% confidence intervals. The shock in pc,t has no significant long-run effect on pb,t, thereby substantiating the corresponding weak exogeneity test. The shock in pb,t leads to a significant and permanent increase of 0.2% in pc,t in the long run with full adjustment before long-run equilibrium is restored.

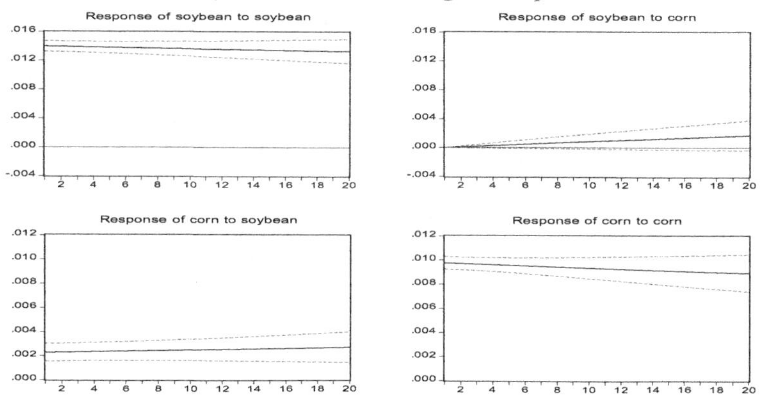

Figure 2. Impulse responses

4. CONCLUSIONS

This paper examines the long-run relationship between the futures prices of soybean and corn on the Dalian Commodity Exchange using daily data from 1 July 2005 to 20 June 2008. Unlike previous studies that examine futures price relationships, we used Johansen's co-integration procedure that allows for structural breaks at known points in time. The results show evidence of co-integration: a long-run relationship exists between soybean and corn futures prices, and the hypothesis of price discovery is not rejected. This substantiates the results of Malliaris and Urrutia's examination of price discovery on the CBOT for corn, wheat, oats, soybean, soybean meal, and soybean oil.

Using Perron's unit root test, there is a break in each price series: in May 2007 for corn and in July 2007 for soybean. Modeling these two breaks in Johansen's co-integration procedure that permits structural breaks shows that the latter is significant whereas the former is not. The significant break in July, 2007 permanently affected the link between the two prices resulting. Intuitively, there appear to be two explanations for this break in the soybean–corn futures price relationship. Because the U.S. soybean acreage substantially reduced, coupled with China's main soybean producing areas of Heilongjiang historical experience rare drought, domestic and international futures market surged, domestic soybean spot market is also good with the support of many continue to rise significantly. The resulting structural break in the soybean–corn futures price relationship demonstrates the importance of modeling structural breaks. Failure to do so can lead to biased estimates and inappropriate conclusions. Hypothesis tests show that the soybean price is weakly exogenous while the corn price is not. Thus, the soybean price Granger-causes the corn price and not vice versa, and causality is unidirectional. This is perhaps surprising: a priori, we expect bidirectional causality where markets respond to common information about demand and supply. Our results suggest that the soybean futures price incorporates such information while the corn price adjustment follows but not vice versa. There appear to be three explanations. First, soybean is harvested around three weeks before corn and has a price-formation role. Second, soybean and corn are close substitutes where the high protein content favors soybean. Third, the futures market for soybean and corn on the Dalian Commodity Exchange is inefficient because information is not incorporated rapidly into either price due to thin markets, limited speculative involvement, or a relatively small number of traders or small volumes. The implication for traders, that the soybean price Granger-causes the corn price, is that price discovery exists on the Dalian Commodity Exchange of China, and since one price may be used to predict another, consistent profits can be generated. Estimates of long-run coefficients in the co-integrating vector imply that the soybean–corn futures market is perfectly integrated and a 1% increase in the soybean price leads, ceteris paribus, to a 1% increase in the corn price. Impulse responses show the impacts on one price of a one standard error increase in the other as both adjust back to long-run equilibrium. A shock in the corn price has no significant impact on the soybean price. In contrast, a 1% shock in the soybean price leads to a significant and permanent 0.2%

increase in the corn price. This supports the conclusion that the soybean price is weakly exogenous. The permanent change in the corn price in response to the change in the soybean price may indicate that part of the industry responds to soybean price changes by permanently changing the amount of corn in formulations. We chose to examine the link between soybean and corn futures prices because commonality between commodity futures prices arises particularly when the commodities are substitutes on the supply and/or demand side. Commonality can also arise because of common impacts of changes in macroeconomic variables and speculator behavior. Analysis of these other bivariate relationships, which are not reported, suggests the result here does generalize: there is evidence of co-integration in all relationships when breaks are included, and at least one break is significant in each co-integrating vector. Thus, there is wide price discovery on the Dalian Commodity Exchange. Our findings suggest that it is important to model structural breaks when examining the relationship between futures prices and a failure to do so may lead to the erroneous conclusion that price discovery is absent.

ACKNOWLEDGEMENTS

Funding for this research was provided by the research funding base of Beijing Wuzi University (P. R. China).

REFERENCES

Dawson, P.J., Ana I.Sanjuán, and B. White. Structural Breaks and the Relationship between Barley and Wheat Futures Prices on the London International Financial Futures Exchange." J. Review of Agricultural Economics, 2006,28 (12):585–594.

Dawson, P.J., and B. White. "Interdependencies between Agricultural Commodity Futures Prices on the LIFFE." J. Futures Markets, 2002,22 (6):269–80.

Johansen S., R. Mosconi, and B. Nielsen. "Cointegration Analysis in the Presence of Structural Breaks in the Deterministic Trend," J. Econom., 2000, 3 (12):216–49.

Johansen, S. "Statistical Analysis of Cointegration Vectors." J. Econ. Dyn. Control 12, (June 1988):231–54.

Lütkepohl, H. Introduction to Multiple Time Series Analysis. Berlin: Springer-Verlag, 1993:43–56.

Malliaris, A.G., and J.L. Urrutia. "Linkages between Agricultural Commodity Futures Contracts." J. Futures Markets, 1996,16 (8):595–609.

Perron, P. "Further Evidence on Breaking Trend Functions in Macroeconomic Variables," J. Econometrics, 1997 ,80(10):355–85.

Pindyck, R.S., and J.J. Rotemberg. "The Excess Co-Movement of Commodity Prices," Econ. J. 1990, 100(12):1173–89.

Said, S.E., and D.A. Dickey. "Testing for Unit Roots in Autoregressive-Moving Average Models of Unknown Order," Biometrika, 1984, 71(12):599–607 .

DESIGN OF MEASURE AND CONTROL SYSTEM FOR PRECISION PESTICIDE DEPLOYING DYNAMIC SIMULATING DEVICE

Yong Liang [1,*] , Pingzeng Liu [1] , Lu Wang [1] , Jiping Liu [2] , Lang Wang [2] , Lei Han [3] , Xinxin Yang [4]

[1] *School of Information Science and Engineering, Shandong Agricultural University, Taian Shandong Province, China 271018*

[2] *Chinese Academy of Surveying and Mapping, Beijing 100039, CHINA;*

[3] *School of economics and management, Shandong Agricultural University, Taian Shandong Province, China 271018*

[4] *School of foreign languages, Shandong Agricultural University, Taian Shandong Province, China 271018*

Abstract: A measure and control system for precision deploying pesticide simulating equipment is designed in order to study pesticide deployment technology. The system can simulate every state of practical pesticide deployment, and carry through precise, simultaneous measure to every factor affecting pesticide deployment effects. The hardware and software incorporates a structural design of modularization. The system is divided into many different function modules of hardware and software, and exploder corresponding modules. The modules' interfaces are uniformly defined, which is convenient for module connection, enhancement of system's universality, explodes efficiency and systemic reliability, and make the program's characteristics easily extended and easy maintained. Some relevant hardware and software modules can be adapted to other measures and control systems easily. The paper introduces the design of special numeric control system, the main module of information acquisition system and the speed acquisition module in order to explain the design process of the module.

Keywords: precision pesticide deploying, measure and control, numerical control system, function module, Information measure terminal.

Please use the following format when citing this chapter:

Liang, Y., Liu, P., Wang, L., Liu, J., Wang, L., Han, L. and Yang, X., 2009, in IFIP International Federation for Information Processing, Volume 294, *Computer and Computing Technologies in Agriculture II, Volume 2*, eds. D. Li, Z. Chunjiang, (Boston: Springer), pp. 927–935.

1. INTRODUCTION

The resulting pollution from pesticide over use is increasingly become health threat. It is a main factor breaking the ecological balance and hindering the sustainable development of agriculture .The out-dated pesticide deployment technology leads to poor efficiency use of pesticide and it is the main factor of pesticide pollution. Employing pesticide is a complex process affected by many factors which affect and restrict each other. The factor affecting pesticide deployment efficiency has been accurately researched in order to study new pesticide technology and equipment. The precision pesticide deployment simulation equipment is designed in order to conquer the restriction of natural conditions. The system can simulate every state of practical pesticide deployment, adjust and carry through precise, simultaneous measures to every factor affecting pesticide deployment. The measure and control system for pesticide deployment pesticide dynamic simulation device is developed in order to enhance the precision and efficiency of simulating device.

2. THE ANALYSIS OF PRECISION EMPLOYING PESTICIDE DYNAMIC SIMULATION DEVICE DESIGN

The precision pesticide deployment dynamic simulating system includes a moving simulated prototype, static fog distributing test-bed and wind fed simulation equipment. The simulated prototype is the main part of the simulation system. It simulated the movement of signal nozzle or grope nozzles through measurement of speed flux and pressure. The fog distribution is measured on the static fog distributing test-bed. It is realized by measuring the fog weight of fog collection groove which is measured by 30 precise subminiature weight sensors. The step less regulation of wind direction and speed is realized by simulated wind feed. So it is convenient for research on the affect of wind feed effects on pesticide deployment. The wind speed is gathered by the wind speed sensor.

The mission of system is controlling pesticide deployment prototype, at the same time collecting and disposing of data. The dynamic simulation precision deployment process includes automatic run of prototype, on-off control of pump and control of liquid spray. The measure parameters include the speed of prototype, temperature, humidity, wind speed, the distribution of fog, flux and so on. The system measures each factor accurately and simultaneously. After pretreatment the data is transferred wirelessly to numeric control system and then transferred to industrial computer. The control mechanism of the industrial computer is executed by the numeric control system. Each control function is achieved on keyboard of numerical control system. Thinking about the character of precision pesticide deployment dynamic simulation device and functional request of measure and control system, the whole configuration of precision pesticide deployment dynamic simulation system is designed as fig 1 (Liu Pengzeng,et al., 2006).

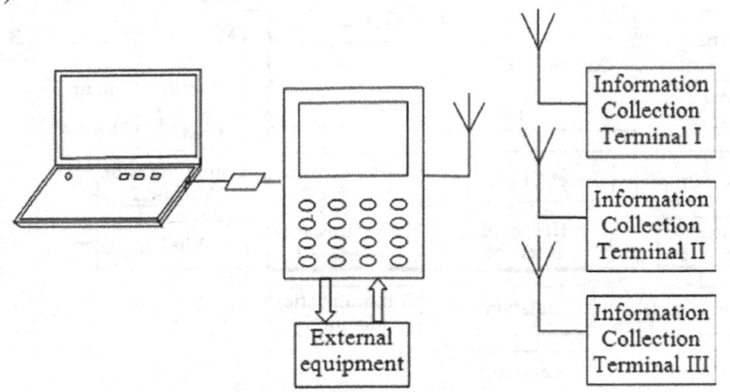

Figure 1: The Structure of Measure and Control System for Precision Pesticide Deployment Dynamic Simulation Device

3. THE MODULARIZATION DESIGN OF NUMERIC CONTROL SYSTEM

In order to enhance the system's universality, efficiency of exploder and systemic reliability, and lower system cost the design of hardware incorporates a structural design of modularization in the numeric control system and information acquisition terminal. Independence, integration and exchange of module function is the directional thinking of module partition. Function is more centralized, is worse, in contrast, when function is dispersive, more parts leads bad reliability of system. The modules' interfaces are uniform defined, which is convenient for the connection of different modules.

The basic principle of module partition:

(1)Integration .Each module is a unit which has integrated function.

(2)Independence. The function of each module is independent of each other.

(3)Exchange. When define module, the currency of module is needful on the basic of monished PCB area.

(4)Singularity. The definition of interface must be united, convenient and reliable. The singularity and convenience of interface is a very important in the process of PCB.

The whole structure of numeric control system is designed as fig2 on the thinking of module partition (Guo Shifu et al.,2007;Li Quan et al.,2007;Yang Shuming et al.,2007).

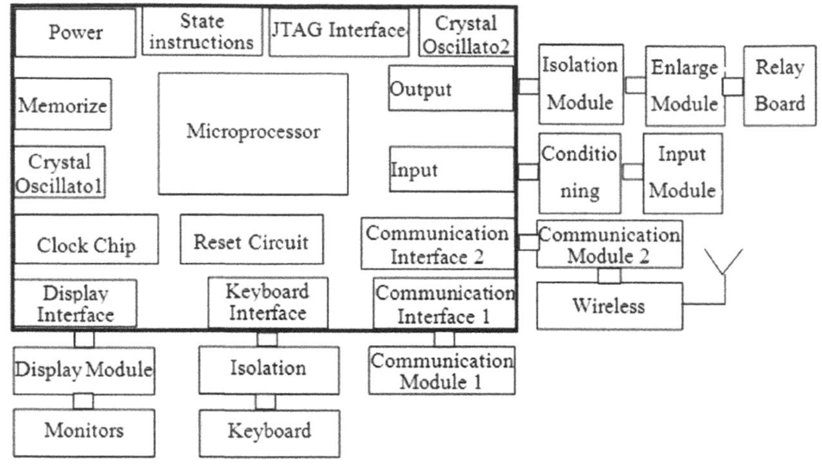

Figure 2: Structure of the NC System Module

The unit the MCU on is the main board of numeric control system which includes some basic collocation. The MCU is the core of the system, the capability of the whole system is decided by MCU. The MSP430F149 whose power is especially low and function is strong is used in the system. AT24C08 is used as serial memorizer which can memorize some basic system information. A 32.760k oscillator of low frequency is chosen as the first oscillator which suits situation of low power. A 8MHZ oscillator is chosen as the second oscillator which is good at disposal of high speed information. The display interface joint the P6 I/O port of MCU. It joins exterior display through isolated module by the way of serial port or parallel port. The OCMJ4*8C LCD of serial communication can show characters and figures. The keyboard joins the P2 I/O port which has the interruption function. It joins exterior keyboard through isolated module. There are two serial ports of MSP430F149, the first port communicate with IPC through communication module, the second port communicate with wireless module through communication module which transport wirelessly. It also can join USB transformation interface through parallel port, so it can join IPC through USB. The output signal is magnified after being isolated, and then it drives a subminiature relay. The connection of relay drive contactor to control exterior equipment such as exterior electromotor. The exterior status messages are produced through the input port.

The signal of module's two ports is isolated by the isolation module.

There are united defined interface on the two ports of module, which can join each other expediently. The digital isolation between peripheral equipment and MCU, peripheral equipment and peripheral equipment, and between MCUs can implement by the isolation module.The module can isolate eight channels signal ,so we can make our choice according the situation less then eight channels.

The first communication module joins MCU and the IPC. Two different communication modules are designed to join different interface of computer, so system are more flexible and current. One is RS485-RS232 communication module which joins MCU and the RS232 serial interface of IPC expediently. The other is USB conversion interface communication module which joins MCU and USB interface of IPC expediently. A TTL-RS485 communication module which can transform TTL to RS485 is designed to join MCU and peripheral equipment of RS485 bus.

The amplifier module mainly amplifies the drive signal of the relay and isolates them. There are eight channels output signals, they can mostly drive the relay and peripheral equipment. We can make our choice according the number of relays. We need more amplifier module and relay module when there are more peripheral equipments.

4. THE DESIGN OF INFORMATION ACQUISITION TERMINAL

The design of information acquisition terminal incorporates structural design of modularization. The main module of information acquisition terminal and the speed acquisition module are designed as function unit respectively.

4.1 The design of information acquisition terminal main module

The main module of information acquisition terminal is designed according the request and character of common information acquisition system. The main module of information acquisition terminal is designed as fig 3.In the process of information acquisition main module low power is a very important guideline, whatever choice such as CPU or peripheral equipment comply the guideline strictly. The MSP430F149 of very low power is chosen and then the system can use battery as power supply, so the main module of information acquisition can be applied more widely. The SD2303AP is chosen as clock chip, it is high accuracy real-time clock chip which has built-in oscillator and two-way serial interface. It can promise the clock accuracy as the ± 5 ppms. The function of clock accuracy numeral adjustment can correct the deviation of clock over a very wide range. The typical power of the chip is 0.5uA(VDD=3.0v).It can export year, month, date, week, hour, minute, second and operate on 1.8 to 5.5 volts. The interface of display is designed in the main module of information acquisition. When acquired information needs to be shown, we join the display through the module of display interface to realize the real-time display.

4.2 The design of speed acquisition module

The speed acquisition module is designed to acquire the speed of the prototype. There are many ways of speed acquisition. The system acquires the speed by photoelectric encoder. The speed of simulation prototype is calculated by testing the number of pulse acquired in unit time, so the function of speed acquisition module is acquiring the number of pulses thrown out by encoder in unit time (MA Dongtao et al., 2005).

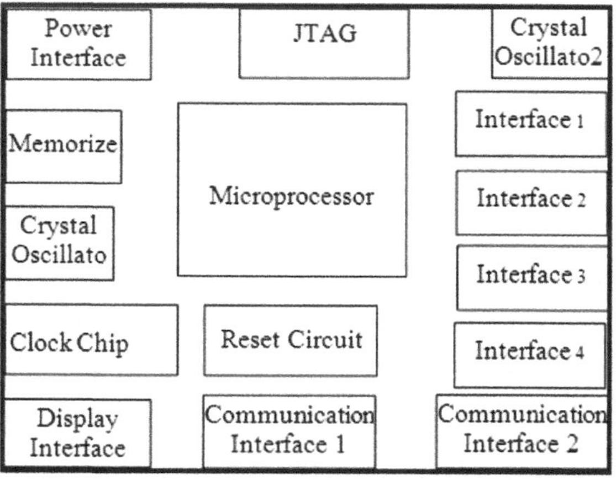

Figure 3: Structure of Information Collection Main Module

In order to enhance the system's universality, three circuits of pulse signal acquisition are designed in the module.The module can receive pulse signal of pulse encoder as well as on-off signal of Hall element. When the voltage of exterior pulse changed we can change the resistance to adjust. When we connect the photoelectric encoder externally one circuit acquire the number of pulses, the other circuit joins the other phase of the photoelectric encoder's output and use it to distinguish the direction. When the direction is positive the number of pulse is increased by one, by contrast, the number of pulses reduces by one. Accordingly, the analysis of function the chart of speed acquisition module is shown on the fig 4.

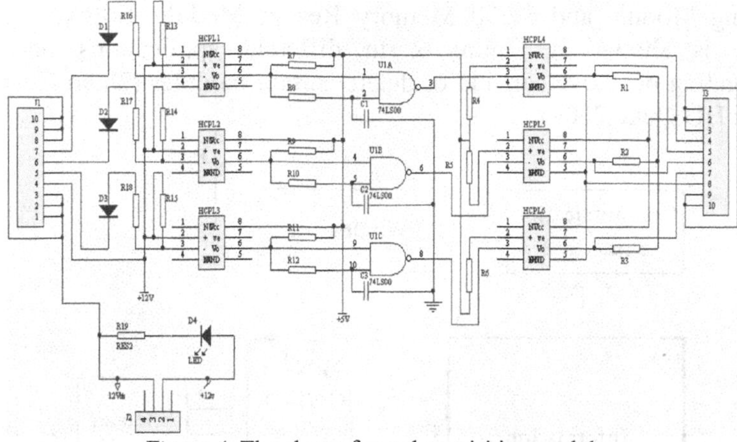

Figure 4: The chart of speed acquisition module

As the chart of speed acquisition module shown on the basis of pulse counting the module makes use the technology of isolation, divorcement of voltage, divorcement of current and differential coefficient to achieve anti-jamming and a muffler effect. The two ports of module are joined by traversing order. It is convenient to connect and assure the security of the join. When the voltage of input pulse changed we can change the divorcement of voltage and divorcement of resistance to fit new voltage.

4.3 The design of speed acquisition terminal

The function of speed acquisition terminal is to acquire the information of speed and then transmit to numeric control system wirelessly. After pretreatment of numeric control system information, it is transmitted to the management computer. It is easy to design the system using the information acquisition main module, speed acquisition module, communication interface module and transceiver module. The char of speed acquisition terminal is shown on the fig 5.

5. MODULAR DESIGN OF SOFTWARE SYSTEM

System software design also fully adopted the structure ideas, the software will be divided into several relatively independent of the functional modules, so, the procedures are easily extended, and easy to maintain. We designed many modules as follows: Initialization Module, Temperature and Humidity Detection Modules, LCD Display Modules, Keyboard Module, Communications Module, Speed Detection (pulse count) Modules, Data-processing Module and Serial Memory Reader Module. Software system structure is shown in Figure 6. In different applications under the corresponding need, we can add or delete certain modules (Duan Guiping et al.,2007; HU Dake,2001).

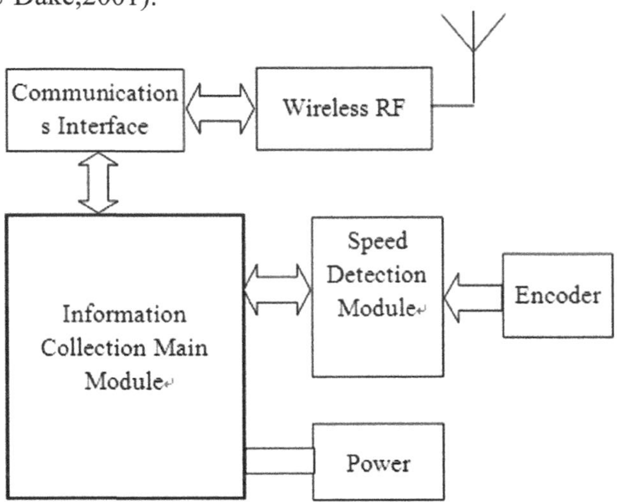

Figure 5: Terminal Velocity Structure of the Acquisition

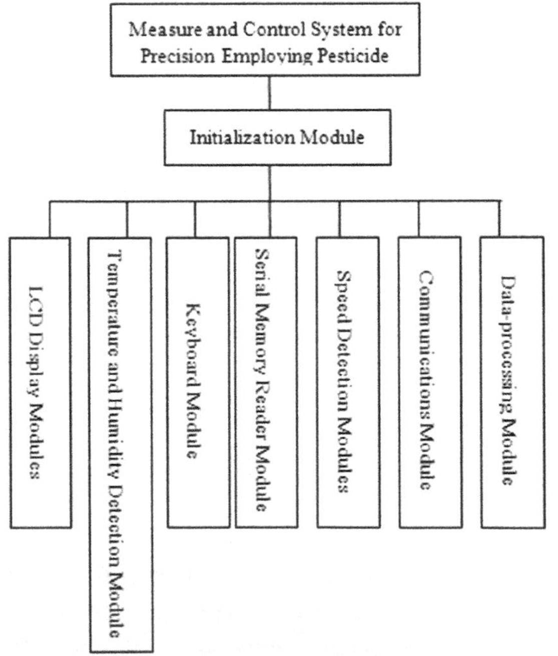

Figure 6: Software System Structure

6. CONCLUSIONS

The development of the measure and control system for precision pesticide deployment simulation equipment has created the conditions for Precision Measurement of the relevant factors about precision pesticide deployment. The system motherboard and modules developed during the design of NC system can be easily used to control other occasions, the information collection at terminal modules and related modules can be easily used for further information analysis. At the same time, in conjunction with the development of the software modules we can easily assemble various measurement and control system, which is very effective for developing small-batch control systems. The design of Measurement and Control System modular architecture has effectively improved the efficiency of the design of the system, improved the system of universality and reliability and easy maintenance. After extensive testing at the scene, it showed that the system is stable and reliable, easy to maintain, and better to meet the requirements of the measure and control for precision pesticide deployment simulation equipment.

REFERENCES

DUAN Guiping, SUN Yong, CHAI Xiaoli. Design of CPCI Common Processor Module Based on PCI6254[J]. Computer Engineering, 2007,33(14):236-238

GUO SHifu, MA SHuyuan, WU Pingdong. Pulse Signal Test System Based on ZigBee Wireless Sensor Networks[J]. Computer Application Study, 2007, 24(4):258-260

HU Dake. Ultra-low-power 16-bit Microcontroller of FLASH-MSP430 Series[M]. BeiJing: Beijing University of Aeronautics and Astronautics Press, 2001

LIU Pengzeng, DING Weimin, XUE Xinyu. Development of Test Bed Used for Precision Test of Nozzles' Comprehensive Performance[J]. Jiangsu University Journal(Natural Science), 2006, 27(5):388-391

LIU Quan, LIU Hong. Research and Implementation of Communication System in Digital Car[J]. Micro-computer Application, 2007,28(7):763-766

MA Dongtao, YAN Jianguo, LIU Gequn. Design of Unmanned Helicopter Data Editing System[J]. Computer Measurement and Control, 2005,13(4):389-391

YANG SHuming, YANG Qing, YANG Liangliang. Development of Speed Measuring Instrument Based on Infrared Ray[J]. Micro-Computer Information, 2007,23(7-2):72-73, 173

THE CROP DISEASE AND PEST WARNING AND PREDICTION SYSTEM

Juhua Luo[1,2] , Wenjiang Huang[1,*] , Jihua Wang[1] , Chaoling Wei[2]
[1] *National Engineering Research Center For Information Technology In Agriculture, Beijing, P. R. China 100097*
[2] *Resources and Environment College of Anhui Agricultural University, Hefei ,Anhui Province, P. R. China 230036*
[*] *Corresponding author, Address: National Engineering Research Center for Information Technology In Agriculture, Beijing 100097, P. R. China, Tel: +86-10-51503647, Fax: +86-10-51503750, Email: yellowstar0618@163.com*

Abstract: The aim of this study was to establish the warning and prediction system for crop diseases and pests based on SuperMap IS. NET geographic information system (GIS), which was developed by Supermap company. In this system, the author used GIS and remote sensing (RS) technology. The system could transform data information into a geographical information map to show the occurrence degree and distribution on various diseases and pests. This paper described mainly warning flow, database design and the main functions of the system. Finally, the system realized successfully the warning of the wheat stripe rust in Xifeng region of Qingyang city in Gansu province in 2002, and the prediction result was satisfactory. It indicated that we could classify and predict diseases and pests, and select right time and technology to control the diseases and pests by this GIS system.

Key words: disease and pest; geographic information system (GIS); Warning and prediction system

1. INTRODUCTION

The crop diseases and insect pests were all important biological hazard in agricultural production, and for many years, they restricted seriously the agricultural sustainable development. Statistics from the UN Food and

---

Please use the following format when citing this chapter:

Luo, J., Huang, W., Wang, J. and Wei, C., 2009, in IFIP International Federation for Information Processing, Volume 294, *Computer and Computing Technologies in Agriculture II, Volume 2*, eds. D. Li, Z. Chunjiang, (Boston: Springer), pp. 937–945.

Agriculture Organization showed the world grain yield lost 10% because of pests and 14% because of diseases for many years. At the same time, the world cotton yield lost 16% because of pests and 14% because of diseases. China is a big agricultural country in the world, so the lose because of diseases and pests were approximately equivalent with the above statistic (Huang Muyi et al, 2003). It was more than important to predict the occurrence and development of diseases and insect pests by different prediction methods (Zeng Shimai,2005). According to prediction result, decision maker s and users could make correct prevention standards and proper treatment measures in order to obtain the maximum economic benefits on the condition of minimum capital investment .

The paper introduced a warning system of diseases and pests which was established based on GIS (geographic information system) and RS (remote sense). And the establishment of the system involved lots of objects, including agronomy , soil science and meteorology, and so on. Besides using the kinds of functions with GIS, the remote sense was also developed in the warning system to obtain fast and real-time the information of diseases and insects, which improved the accuracy of early warning result to a certain extent.

In summary, the establishment of the warning system of diseases and pests could provide scientific basis for and prediction scientific managing farm land.

2. THE ESTABLISHMENT OF THE WARNING SYSTEM

2.1 The basic goals of the warning system

The warning on diseases and pests was to predict the occurrence condition and the trend of development on diseases and pests in certain range and period, and made some strategy to control diseases and pests according to the pre-warning results (Wang Shuhai et al, 2005).

The basic goals of the warning system were as follows:

1) By using the warning system, data information could be transformed into a geographical information map to show the occurrence and distribution on variety of diseases and pests by using visualization and spatial analysis with GIS.

2) The warning system could obtain fast and real-time the information of diseases and insects by using RS, and finally it could accurately monitor the occurrence and predict the development of the diseases and pests in large area (Jiang Jinbao et al, 2008).

3) Some standards and measures could be made for the investigation and controlling diseases and pests according to the pre-warning results of the warning system.

2.2 The design and realization the warning system of crops diseases and pests

2.2.1 Design route

At first, it was important to choose feasible GIS platform and the prediction model on diseases and pests after requirement analysis. Second, database was established according to parameter of model, and function modules of the system were also designed. Finally, the warning system was established and published after debugging the system (Feng Jiangfan et al,2006).

2.2.2 Platform selection

SuperMap IS. NET was chosen as the system development platform, the reasons were as follows:

1) Components of design could be easily managed.

2) The Multi-source data could be integrated and the massive image could be quickly accessed to.

3) Server was clustering; it was with a high degree of flexibility

4) In the platform, Client and server belonged multi-level cache structure, which could support a variety of map engine work together.

2.2.3 Database design

1) Attribute database

The attribute database of the system was composed with monitoring data, basic data on diseases and pests , national meteorological observation data and the data table about latitude and longitude of meteorological observation site, and so on (Wang Minghong et al,2006).

Monitoring data on diseases and pests: field investigation data, monitoring data using remote sensing, field experiments data, and so on.

Basic data on diseases and pests: diseases and pests species data, damage characteristics data, control and preventive methods data, and so on.

National meteorological observation data: temperature, moisture, rainfall and sunshine number, and so on.

2) Spatial databases

The spatial database of the warning was composed of basic map, thematic map and warning information map.

Basic map: national administrative division map, digital elevation map.

Thematic map was got by basic map according to some goal, which included weather map, remote sensing imagery and crop division map.

Warning information map was obtained by spatial analysis on the basic of few thematic maps and analysis result of prediction model, by which the system could show the occurrence and distribution on variety of diseases and pests. And ultimately, some effective guidance can be given to prevent and control the diseases and pests according the warning information map.

2.2.4 Basic functions of the system

The functions of warning system was composed of special functions and general functions which were offered by GIS platform, including the translating, mitigating and amplificating graphics, and adding and deleting map layer, and so on. While the special functions were as follows:

1) Retrieval of diseases and insect pests

Users could search all kinds of knowledge about diseases and pests, for example, latin name and English name of disease and pest, harm symptom and epidemic law on diseases and pests and control methods, and so on (Zhou Qiang,et al,2004).

2) Diagnosis of diseases and pests

Users could obtain some papers on warning results and prediction methods after the operations of systems by various data and prediction models of diseases and pests.

3) Model base management

Administrators could add, delete, or modify the models in the system by Model base management (Liu Shuhua et al, 2003).

4) Monitoring and evaluation using remote sensing

The function was realized by the module of pests monitoring. Multiple and multi-temporal remote data was applicated in the module, and assessment model based on RS was call in the module (SONG Y H et al,1993). By the function module, users could develop monitoring and assessment the diseases and pests. And related remote sensing image s could also were call from spatial database and displayed for users.

5) Visualization of warning result

By making visualization and spatial analysis with GIS, the system could transform prediction result and emergence grade into a geographical information map to display visually the occurrence and distribution on variety of diseases and pests. And ultimately, some effective guidance could be given by the system to prevent and control the diseases and pests (Si Lili et al, 2006).

2.2.5 Warning flow

The warning system was established on the basic on a database which was composed of attribute database and spatial database and a model base including diseases models and pests prediction models. At first, the result which was obtained by prediction model form model base and data from database was displayed in the GIS platform and was transformed into information diagram, meanwhile, some thematic maps from spatial database were overlaid on the information map. Finally, the system could offer a clearly electronic information map which could show the occurrence and distribution on variety of diseases and pests (Fig.1).

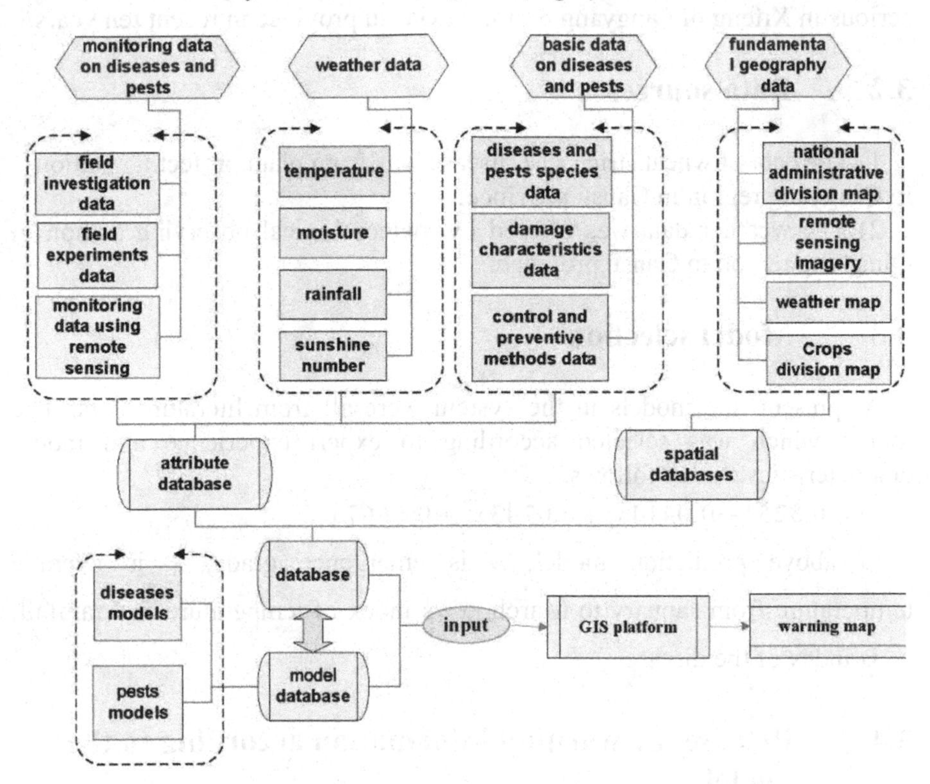

Fig.1. Flow chart of warning system for diseases and pests

3. THE APPLICATION INSTANCE OF THE WARNING SYSTEM

The process how to realize the function of the warning system was displayed by taking the wheat stripe rust disease in Xifeng of Qingyang region in Gansu province as an example.

3.1 Summary of study area

The wheat stripe rust disease was one of the most serious diseases in China. Xifeng of Qingyang region in Gansu province was taken as an example to display the process how to realize the function of the warning system.

It was reported that the occurrence of the wheat stripe rust disease was mildly severe and severe symptoms in 1986,1989,1990,1991,2002,2003, outbreak in 1985, moderate in 1993, and mild or none occurrence in other years. The wheat stripe rust disease was intermittent disease, but the occurrence was continuous and frequent, and becoming more and more serious in Xifeng of Qingyang region in Gansu province in recent ten years.

3.2 Data sources

1) The data of wheat stripe rust disease was from plant protecting station in Qingyang region in Gansu province;

2) The weather data was offered by meteorological observing station in Qingyang region in Gansu province;

3.3 Model selection

At present, the models in the system were all from literature data. The model which was selected according to expert experience and model characteristics was as follows:

$$y = 0.3251 + 0.0414x_1 + 0.0243x_2 + 0.0667x_3$$

In above prediction model, y is emergence grade, x_1 is average temperature from January to March, x_2 is index of temperature and rainfall, x_3 is index of the disease.

3.4 Release the warning information according to the model

3.4.1 Grading standard on the stripe rust wheat

Six degrees was adopted to evaluate the emergence of the wheat tripe rust, none, mild, lower than moderate, moderate, more than moderate, outbreak, and were correspondingly marked as 0, 1, 2, 3, 4, 5 and identified as white, blue, green, yellow, orange, red (table 1).

Table 1 Grading standards for occurrence degree of wheat stripe rust disease

emergence grade	occurrence degree	warning color
zone	none	white
one	mild	blue
two	lower than moderate	green
three	moderate	yellow
four	more than moderate	orange
five	outbreak	red

3.4.2 The pre-warning results on study area

1) System interface

The left page of the warning system concluded the field of user ' logging and operational area of prediction, while the data about the warning was display on the right of system interface with victor diagram.

2) The operating process of the warning on the wheat stripe rust disease

At first, the users could enter the main interface of the system by login; second, users could enter the main interface of diseases prediction by clicking the diseases prediction model. Finally, used could develop step-by-step operation in the operational area of prediction according.

3) The pre-warning results on study area

By the above operation, users could call,, in the model from the database of the system, and got the warning result by calculation. Finally, the pre-warning results were displayed on the right of the main interface by different color by electrical information map, at the same time, the warning results were also display by character form on the left of main interface (Fig.2).

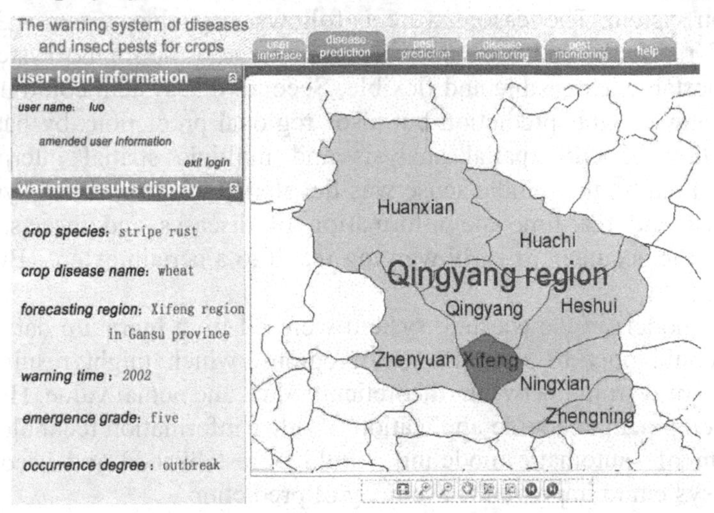

Fig.2. Pre-warning results of wheat stripe rust disease in Xifeng county of Qingyang city in 2002

It was displayed that the emergence grade of the wheat stripe rust was five and occurrence degree was outbreak in Xifeng region of Gansu province. The fact that the warning results coincided well with report results from plant protecting station of Gansu province showed that this warning system and prediction method were all feasible.

4. RESULT

A warning and information system for diseases and pests of crops was established base on Geography Information System (GIS). The occurrence and damage of major diseases and pests of crops could be monitored and forecasted in real time by using the system, based on the forecasting model .And the system could submit warning maps with 6 colors and words information to the users. By using the system, we could standardize the forecasting data collection, transmit information through network and forecasting results viewable. Users could use it to classify and predict diseases and pests, also could use it to select fitting time and technology to control.

5. DISCUSSION

1) With the development of all kinds of subjects and lots of technology, many warning systems had been established at home and abroad. However, the system which was based on GIS was different from others traditional prediction system. The reasons were as follows: first, this system was based on .NET platform, and the interface was friendly , so the system was stable , portable, extensible and flexible. Second, this system could not only realize single point prediction but also regional prediction by basing on the function of GIS spatial analysis and multiple spatial interpolation methods. Finally, the remote sense was developed in the warning system to obtain fast and real-time the information of diseases and insects, which improved the accuracy of early warning result to a certain extent (BONE, S et al,2005).

2) The models in the warning system were all from literature data, so the models could only be used in some regions, which might result in the deviation of warning between theoretical value and actual value. However, with development and wide application of other information technology, the subsystem of automatic modeling could be established and used in the warning system to improve the accuracy of prediction.

ACKNOWLEDGEMENTS

This work was subsidized by the National High Tech R&D Program of China (2006AA10Z203, 2007AA10Z201), National Natural Science Foundation of China (40701119), and funds (2006BAD10A01, 2007BAH 12B02). The authors are grateful to Mr.Weiguo Li, and Mrs. Hong Chang for data collection.

REFERENCES

C. BONE, S. DRAGICEVIC and A. ROBERTS. Integrating high resolution remote sensing, GIS and fuzzy set theory for identifying susceptibility areas of forest insect infestations[J]. International Journal of Remote Sensing. 2005,26(21) :4809–4828

Feng Jiangfan, Ying Xuewei, Zhang Hong, et al. Forewarning System of Blue Algae for Taihu Lake Based on GIS. Environmental Science & Technology. 2006.29(9) :59-62

Huang Muyi, Wang Jihua, et al.Hyperspectral character of stripe rust on winter wheat and monitoring by remote sensing. Journal of Agricultural Engineering Research.2003,19(6):154-158

Jiang Jinbao, Chen Yunhao, Huang Wenjiang, et al. Hyperspectral estimation models for LTN content of winter wheat canopy under stripe rust stress. Journal of Agricultural Engineering Research. 2008,24 (1): 34-39

Liu Shuhua. Yang Xiaohong, et al. Decision support system for crop diseases and insect pests prevention and control based on GIS. Journal of Agricultural Engineering Research 2003,19(4):148-150

Si Lili, Cao Keqiang, et al. Establishment of a real-time monitoring and forecasting system on main crop diseases and pests of China based on GIS. Journal of Plant Protection. 2006,33(3):282-286

SONG Y H,HEONG K L.Use of geographical Information system in analyzing large area distribution and dispersal of rice insects in South Korea[J].Journalof Applied Entomolite.1993,32(3):307-316

Wang Minghong, Jin Xiaohua, Liu Qian. The establishment of a long-distance warning and information system for major diseases and pests of corps in Beijing and its application. China Plant Protection. 2006,7(26):5-10

Wang Shuhai, et al. Establishment of space information database of prediction systems of Dendrolimus sp. Ased on GIS. Journal of Liaoning Forestry Science & Technology. 2005,4(13) : 35-36

Zeng Shimai. Maco-phytopathology. Chinese Agricultural Technology Publishing House. 2005:182-184

Zhou Qiang, Zhang Runjie. Studies of W ebGIS-Based Warning Information System of Planthoppers Disaster. ACTA Universitatis Sunyatseni. 2004,30(21):64-66

KNOWLEDGE ACQUISITION AND REPRESENTATION OF THE GENERAL AUXILIARY DIAGNOSIS SYSTEM FOR COMMON DISEASE OF ANIMAL

Jianhua Xiao, Hongbin Wang [*], Ru Zhang, Peixian Luan, Lin Li, Danning Xu

College of Veterinary Medicine, NorthEast Agricultural University, Harbin,Heilongjiang Province, P. R. China, 150030

* *Corresponding author, Address: College of Veterinary Medicine, NorthEast Agricultural University, Harbin,150030, Heilongjiang Province, P. R. China, Tel: +86-451-55191940, Fax: +86-451-55190470, Email: neau1940@yahoo.com.cn*

Abstract: The main methods of diagnosis for animal is symptom, generally the end diagnosis conclusion can be get by experiment after diagnosis by symptoms, for this reason, the symptom, epidemic character and pathological changes are most important factors in diagnosis. Based on the level of diagnosis in china, the knowledge acquisition and representation methods were stated in this papers, the animal disease diagnosis knowledge was acquired from experts in animal disease and books, the knowledge was represented in rule, and one factor is assigned to each condition of a rule, representing its significance in drawing the conclusion. This results in better representation, and facilitates knowledge acquisition and maintenance.

Key words: diagnosis, knowledge acquisition, knowledge representation

1. INTRODUCTION

Computer-based methods are increasingly used to improve the efficacy and effects of veterinary medical services. In medical diagnosis, expert system is mainly used for performing diagnoses based on symptom of patient, since they can naturally represent the way expert's reason. Diagnosis

Please use the following format when citing this chapter:

Xiao, J., Wang, H., Zhang, R., Luan, P., Li, L. and Xu, D., 2009, in IFIP International Federation for Information Processing, Volume 294, *Computer and Computing Technologies in Agriculture II, Volume 2*, eds. D. Li, Z. Chunjiang, (Boston: Springer), pp. 947–952.

of animal diseases is greatly facilitated by the symptoms and epidemic characters but not by experiment or pathological change just as pig, chicken etc. In this paper, knowledge acquisition and representation of the general auxiliary diagnosis system for common disease of animal was presented.

2. KNOWLEDGE ACQUISITION

Generally, the data of patient animal can be distinguished in three types: epidemic symptoms, pathological changes and laboratory results. Symptoms are those detected by a physical examination of the patient, like e.g. the existence and the kind of a pain etc. pathological changes are those detected by ptomatopsia and check by microscope. Laboratory results are those detected via laboratory tests, like e.g. blood tests etc. the knowledge for diagnosis disease is the inference from patient data to disease. In another words, what disease the animal get if it has one symptom. This kind of knowledge alike declarative knowledge very much, but not belong to it. Declarative knowledge refers to knowing that something is true or false. It is concerned with knowledge expressed in the form of declarative statements such as "sky is blue". But the diagnosis knowledge for disease can not be affirmed directly (Joseph C, 2005). This type of knowledge is fuzzy and represented by Fuzzy technique better than traditional two-valued logic. For this reason, the diagnosis knowledge for animal diseases should be obtained from expert by analysis but not by statistics. The knowledge in this system obtained from expert and books about animal disease. The knowledge from book was arranged as the follow form by some postgraduates at begins:

If there is one symptom alike "lameness" then there is cf possibility That the animals get foot- and-mouth disease

Then this statement was filled into one table as below:

Table.1 the form of knowledge when Acquisition

Patient data	disease	cf
lameness	foot-and-mouth	
Cough	pneumonia	
Frequent micturition	nephritis	
...	...	

And then the confidence of this knowledge was determined by expert. In fact, it is very difficult to analyze the possibility of this symptom result in this disease. And it is not useful if the knowledge were obtained in this rambling form. Because one symptom may signify many diseases, and one disease may show many symptoms. Therefore, if the confidence of knowledge be analyzed leave other knowledge alone, the confidence of this knowledge would separate with that knowledge which has close relation with symptom or disease in this knowledge. The relations among different knowledge (see figure.1) were disregarded consequently.

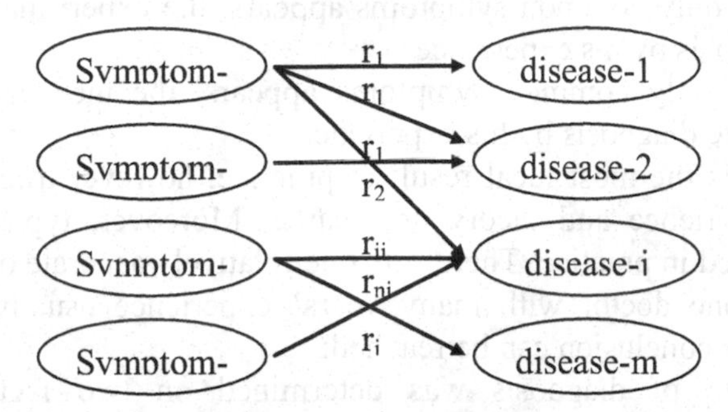

Figure.1 the inner relation among different knowledge for diagnosing of animal disease

r_{ij} represent the rule between *symptom-i* and *disease-j*. if symptom-1 appeared in one animal then disease-1, disease-2, disease-j can be occurred. On the contrary, if disease-I occurred then symptom-1, symptom-i and symptom-n can be appeared. Therefore when the confidence of r_{11} will be analyzed, the confidence of r_{12}, r_{1n}, r_{ij},r_{nj} should be analyze in same time. Although one disease may show many symptoms, but these symptoms can be divided into two types: common symptom and typical symptom.

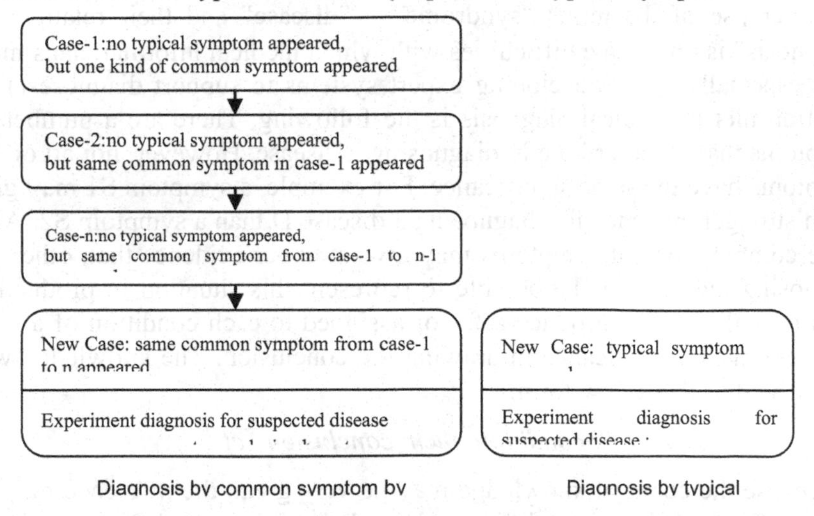

Fig.2 The different diagnosis ways for by typical symptom and common symptom

Generally, there are 4 kinds of results may occur when diagnose by inexperienced doctor and expert in practice:

Situation 1: the typical symptom appears, the expert make most accurate diagnosis at once

Situation 2: the typical symptom appears, the inexperienced doctor relatively accurate diagnosis

Situation 3: only common symptoms appears, the expert make relatively accurate diagnosis by his experience

Situation 4: only common symptoms appears, the inexperience doctor make inaccurate diagnosis by his experience

Situation 1 is the most ideal result in practice, however true expert with plenty of experience and theory are scarce. Moreover, typical symptom seldom appeared in practice. Therefore, one relatively accurate diagnosis can be made by one doctor with many years' experience usually. By above discussion, one conclusion can be reached:

The accurate of diagnosis was determined on two factors: typical symptom and experience. If the typical symptom do not appeared, the expert can make one relatively accurate by his many experiences on same case (see fig.2).

3. KNOWLEDGE REPRESENTATION

Representing medical knowledge is a highly complex endeavor. The improper use of the terms "syndrome", "disease" and their relations to "diagnosis" is one of the difficulties with which medical informaticians must deal, especially when developing expert systems to support diagnoses(). A situation met in medical diagnosis is the following. There are a number of symptoms that all contribute in diagnosing a disease. However, not all of the symptoms have the same significance. For example, a symptom S1 may give much stronger evidence for diagnosing a disease D than a symptom S2. Also, some combinations of symptoms may give stronger evidence than others in diagnosing the disease. To be able to represent this situation in production rules formalism, we introduce a factor assigned to each condition of a rule, representing its significance in drawing the conclusion. The knowledge was represented as the follow form:

If condition then conclusion cf

Because the count of knowledge may be very great, the inference may be very difficult. Otherwise, different knowledge plays a different role in inference. The knowledge based on typical symptoms play a determined role in diagnosis, but common symptom play a not important role. If one typical symptom appears then there is a great possibility that one conclusion can be make, while some symptom appears can not make any conclusion, in another words, these symptom have a little relation with one disease, for example: anorexia. Any disease may result in the decrease of appetite. Therefore the symptoms as anorexia have no help to diagnosis generally. However, anorexia may be very important to diagnosis diseases of digestive system, so one symptom has different role for different diseases. Based on above ideas, a five-grade score was used in this expert system. They are certain not, little possible, possible, very possible, certain. The significance factor *sf* takes value from these five grades. The principle of take value for one knowledge is list in table.2.

Tab.2 the principle of take value for one knowledge

score	means	The principle of take value
0	Certain not	The disease have no relation with this symptom certainly
1-4	Little possible	The disease have little relation with this symptom, but this symptom does not signify this disease
5	Possible	The disease have relation with this symptom, if this symptom appears then it is likely that this disease has occurred, but not very certain
6-7	Very possible	If this symptom appears then the possibility for this disease occurred is very great
8-10	certain	If this symptom appears then this disease occurred on the whole

The above principles reflect the relation between symptom and disease to a great extent. But when acquire knowledge in above principles, a lot of disease may corresponding with one symptom, and one disease may corresponding many symptom. The result may be when one symptom be input a lot of disease may output. Moreover, the system runs slowly. In fact, the essence of disease was not embodied in above principles. Generally diseases of animal can be divided into two categories: infective diseases and noninfective diseases. the symptoms of infective diseases appear in many system, and the symptoms of noninfective diseases appear in few system. Moreover, there was no difference among symptoms just as wet cough, dry cough, and painful cough in essence. The difference among those symptoms is degree. Otherwise, when animal cough, it may wheeze, nose running, even lose breath in the same time. And the symptoms as those are only the typical symptoms of infection disease of respiratory system but not others. Furthermore, by symptoms like cough, wheeze etc. concrete disease can not be inferred. For above reasons, a more complex knowledge representation form was design to solve above problem. The new representation form is:

If *condition (class)* then *conclusion cf is-Typical*

The “class” represents the relation among different condition for one disease. Class is take value from “X1, X2, Xn; Y1, Y2, Yn; Z1, Z2, Zn”, n<10. X Y, Z means this condition belong to group X, group Y and group Z respectively. 1, 2, ···, n means any class which include 1,2, ,n, be selected the sum weight will be add 1, 2, , n respectively. "is-Typical" take value from "is" or "not", if "is" then means the condition in this knowledge is a typical symptom. Therefore by this condition one conclusion can be making certainly.

4. CONCLUSION

The accurate was relative low in veterinary medical. There was more problem not solved in veterinary medical than human medical. Therefore one human doctor will get much knowledge than one veterinary doctor with same time and energy. Maybe the veterinary medicine has developed to the same level as human medicine in some area. But the level was relative low in most area. There are more than 100 kinds of animal disease occurred usually. These diseases belong to 8 systems. Every disease will show several symptoms. Therefore, it is a more complex procedure of diagnosis for animal disease than disease of human being. The knowledge representation methods in rule are one simple but effective ways.

ACKNOWLEDGEMENTS

Thanks to surgical department faculty working office of the Northeast Agricultural universdity, as well as Professor Liu Yun, associate professor Gao Li by the help regarding to this research.

REFERENCES

C Joseph, Giarratano, D Gary. Riley. Expert Systems Principles and Programming, 4th, Edition. Thomson Learning Ltd, 2005.

S.V. Ellam, M.N. Maisen. A Knowledge-based System to Assist in the Diagnosis of Thyroid Disease from a Radioisotope Scan, in Pretschner D.P. and Urrutia B., (Eds), Knowledge-based systems to aid medical image analysis, vol.1, Commission of the European Community, 1990.

THE DEVELOPMENT OF A GENERAL AUXILIARY DIAGNOSIS SYSTEM FOR COMMON DISEASE OF ANIMAL

Jianhua Xiao, Hongbin Wang*, Ru Zhang, Peixian Luan, Lin Li, Danning Xu

College of Veterinary Medicine, NorthEast Agricultural University, Harbin,Heilongjiang Province, P. R. China, 150030

* *Corresponding author, Address: College of Veterinary Medicine, NorthEast Agricultural University, Harbin,150030, Heilongjiang Province, P. R. China, Tel: +86-451-55191940, Fax: +86-451-55190470, Email: neau1940@yahoo.com.cn*

Abstract: In order to development one expert system for animal disease in china, and this expert system can help veterinary surgeon diagnose all kinds of disease of animal. The design of an intelligent medical system for diagnosis of animal diseases is presented in this paper. The system comprises three major parts: a disease case management system (DCMS), a Knowledge management system (KMS) and an Expert System (ES). The DCMS is used to manipulate patient data include all kinds of data about the animal and the symptom, diagnosis result etc. The KMS is used to acquire knowledge from disease cases and manipulate knowledge by human. The ES is used to perform diagnosis. The program is designed in N-layers system; they are data layer, security layer, business layer, appearance layer, and user interface. When diagnosis, user can select some symptoms in system group by system. One conclusion with three possibilities (final diagnosis result, suspect diagnosis result, and no diagnosis result) is output. By diagnosis some times, one most possible result can be get. By application, this system can increased the accurate of diagnosis to some extent, but the statistics result was not compute now.

Key words: expert system, diagnosis,

Please use the following format when citing this chapter:

Xiao, J., Wang, H., Zhang, R., Luan, P., Li, L. and Xu, D., 2009, in IFIP International Federation for Information Processing, Volume 294, *Computer and Computing Technologies in Agriculture II, Volume 2*, eds. D. Li, Z. Chunjiang, (Boston: Springer), pp. 953–958.

1. INTRODUCTION

Computer-based methods are increasingly used to improve the efficient and effects of veterinary medical services. Those methods include both conventional techniques, such as database management systems(DBMSs), and Expert system (ES) techniques. From born of techniques, medical diagnosis and management has been a very active field(Silvia Alayon, 2007; Sengur, A, 2007; Xuewei Wang, 2004; I. Hatzilygeroudis,1994). In medical diagnosis, database management system are used for storing, retrieving and generally manipulating patient data, whereas expert system are mainly used for performing diagnoses based on symptom of patient, since they can naturally represent the way experts reason. The structure of the paper is as follows.

2. SYSTEM DESCRIPTION

2.1 The structure of system

To develop one expert system for disease diagnosis, two matters needs to be done: collection of knowledge by typical symptom and the converting of experience into knowledge. There is only one difficult to be overcome. That is the latter of two works. The experiences were obtained from many cases by common symptoms. This experience can be converted into computer easily. But its adaptivity is limited by a lot of reasons. The experience obtained form one region may not applicable to another region. The experience obtained at past may not applicable to now. The disease is different in different region and time. Therefore it is very difficult to develop one ideal expert system. One general expert system for common disease of animal based on typical knowledge and common knowledge origin from cases was design and partly developed in this study. The structure of this expert system is show in figure.1.

It consists of three major parts: a disease case management system (DCMS), a Knowledge management system (KMS) and an Expert System (ES). The DCMS is used to manipulate patient data include all kinds of data about the animal and the symptom, diagnosis result, prescriptions, treatment results of disease case. After final diagnoses have been made, the patient animal cases are stored in the disease cases database. The KMS is used to acquire knowledge from disease cases and manipulate knowledge by human. The ES is used to perform diagnosis. The Knowledge Base (KB) contains the heuristic knowledge based on common symptom and knowledge based on typical symptom for diagnosing diseases. An augmented knowledge representation (KR) formalism based on production rules(see figure 2), the most widely employed KR formalism by ESs [3], is used. The Working Database (WDB) contains the case specific data, that is temp patient data, partial conclusions, data given by the user, and any other information relevant to the case under consideration. Patient data are transferred from the PDB. The Inference Engine (IE) uses the available knowledge to draw conclusions and make diagnoses. The Knowledge acquisition facility (KAF) is used for rule training. Finally, the UI performs a number of functions. The user can manipulate patient data and the cases. Also, the user can interact with the IE to start a diagnosis process as well as use the KAF to perform rule training.

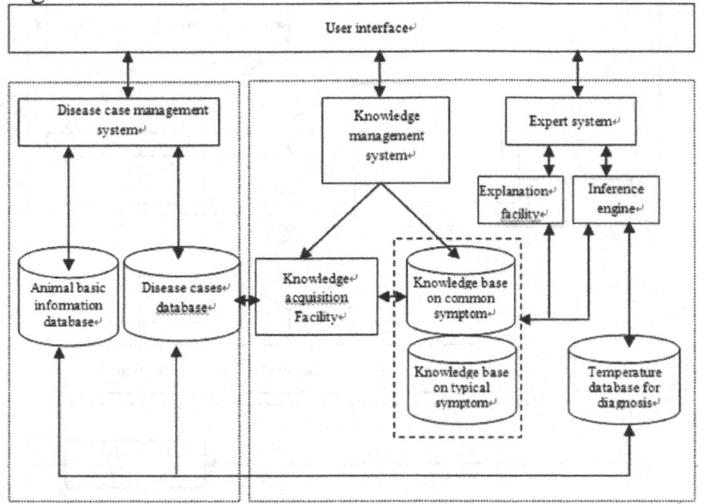

Fig.1: The structure of animal disease diagnosis expert system

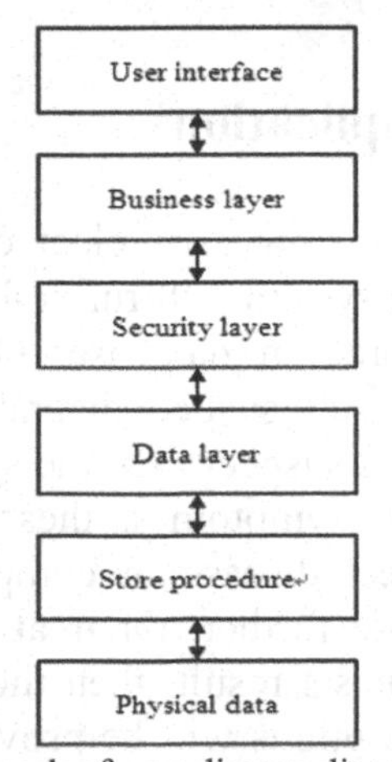

Fig. 2: The framework of cow disease diagnosis expert system

2.2 The system structure

N-layers system was used in this program, they are data layer, security layer, business layer, appearance layer, and user interface.Data layer is one levis component that return data set to up-layer by communication with store procedure directly. Security layer take responsibility to validation role of user, password, user id. Business layer is one buffer area, it can protect the resources requested from client, and increasing flexibility significantly. Compare with traditionary three-layer structure, this structure has high security, high efficiency, and significant flexibility (see figure.3).

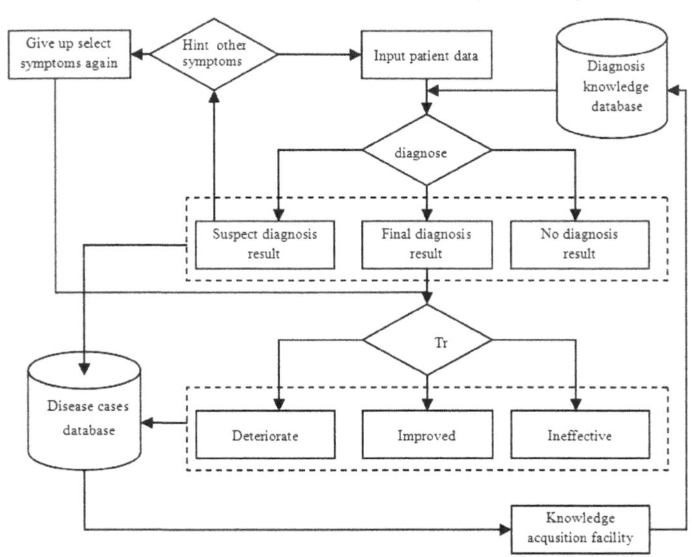

Fig. 3: the flow-sheet of operation for expert system

2.3 flow-sheet of application

Based on the symptom the user can select corresponding symptom in system group by system. The system will run inference procedure and make diagnosis by knowledge stored in database. One conclusion with three possibilities (final diagnosis result, suspect diagnosis result, and no diagnosis result) is output. If the result is suspect then the system will return user with a question: if the animal have symptom as these:", the system will list the possible symptoms by suspect disease in computer. The user can select symptom again. And diagnose further. Or treat as has got final diagnosis result. If there were no diagnosis result, then the system stops. If the treat procedure is run, some prescriptions will be provided and user can treat the animal with these prescriptions. These methods are adequate generally. The diagnosis result and treatment result will all be stored in disease cases database finally, the user must input the evaluation information for this diagnosis and treatment so as to train the knowledge in database. The knowledge acquisition facility will excavate knowledge from disease cases database.

3. CONCLUSION

The differential diagnosis can be complete from one class of disease in medical of human by doctor special in one subject. Under this condition, the degree of complexity will decreased greatly. Diagnosis will not be transfer to another doctor with different specialty until a well-pleasing result be produced. However, it is not so lucky that the doctor must give the ultimate answer of disease but not transfer to others because there were no one doctor that faced to one subject technically in veterinary more often. Any veterinary doctor must prescribe medicines whatever the disease belong to. There are more than 150 kinds of animal disease occurred usually. These diseases belong to 8 systems. Every disease will show several symptoms. Therefore, it is a more complex procedure of diagnosis for animal disease than disease of human being. Generally, common diseases of animal can be divided into infectious and noninfectious, noninfectious disease can be further divided into nervous system disease, digestive system, respiratory system, urinary system, circulation system, urinary system etc. diseases belong to one system have same symptoms, moreover these symptoms belong to one system generally. For example, respiratory system diseases have same symptoms such as cough, breathe heavily, respiratory sound etc. these symptoms belong to respiratory either. However, infectious diseases have more complex symptoms than infectious disease. Many symptoms belong to different system will show in one infectious diseases. Therefore, it is so difficult to make one decision of which one disease the animal can contract that any doctor is cautious without reliable experiment data or long year's experience. The diagnosis of disease is a very complex procedure especially in veterinary. The accurate of diagnosis depend on the knowledge of physician to some extent. But there was a great of difference between medicine and veterinary medicine. One physician may have plenty of experience in his profession because he will diagnose a great deal of case belong to his domain. He can study deeply in his profession either. But one veterinary can only meet a few cases in one domain and cases in every domain. Otherwise, the case can not be diagnosed to a very accurate level for short of precise and advanced equipment. The accurate was relative low in veterinary medical. There was more problem not solved in veterinary medical than human medical. Therefore one human doctor will get much knowledge than one veterinary doctor with same time and energy. Maybe the veterinary medicine has developed to the same level as human medicine in some area. But the level was relative low in most area. Based on above reason, a general expert system is more suitable than a lot of expert system special for one domain in veterinary practice.

ACKNOWLEDGEMENTS

Thanks to surgical department faculty working office of the Northeast Agricultural universdity, as well as Professor Liu Yun, associate professor Gao Li by the help regarding to this research.

REFERENCES

I. Hatzilygeroudis, P. J. Vassilakos, A. Tsakalidis. An Intelligent Medical System for Diagnosis of Bone Diseases, Proceedings of the 1st International Conference on Medical Physics and Biomedical Engineering (MPBE'94), Nicosia, Cyprus, May 1994, Vol. I, 148-152.

Sengur, A, & Turkoglu, I., A hybrid method based on artificial immune system and fuzzy k-NN algorithm for diagnosis of heart valve diseases , Expert Systems with Applications (2007), doi:10.1016/j.eswa.2007.08.003

Silvia Alayon, Richard Robertson , Simon K. Warfield, et al. A fuzzy system for helping medical diagnosis of malformations of cortical development. Journal of Biomedical Informatics, 40 (2007) 221 - 235

Xuewei Wang, Haibin Qu, Ping Liu, Yiyu Cheng. A self-learning expert system for diagnosis in traditional Chinese medicine, Expert Systems with Applications 26 (2004) 557–566

THE RELATIONSHIP BETWEEN MONTH DISEASE INCIDENCE RATE AND CLIMATIC FACTOR OF CLASSICAL SWINE FEVER

Hongbin Wang, Danning Xu, Jianhua Xiao, Ru Zhang, Jing Dong

College of Veterinary Medicine, NorthEast Agricultural University, Harbin,Heilongjiang Province, P. R. China, 150030

* *Corresponding author, Address: College of Veterinary Medicine, NorthEast Agricultural University, Harbin,150030, Heilongjiang Province, P. R. China, Tel: +86-451-55191940, Fax: +86-451-55190470, Email: neau1940@yahoo.com.cn*

Abstract: The Swine Fever is a kind of acute, highly infective epidemic disease of animals; it is name as Classical Swine Fever (CSF) by World animal Health organization. Meteorological factors such as temperature, air pressure and rainfall affect the epidemic of CSF significantly through intermediary agent and CSF viral directly. However there is significant difference among different region for mode of effects. Accordingly, the analyze must adopt different methods. The dependability between incidence rate each month of CSF and meteorological factors from 1999 to 2004 was analyzed in this paper. The function of meteorological factors on CSF was explored and internal law was expected to be discovered. The correlation between the incidence rate of Swine Fever and meteorological factors, thus the foundation analysis of the early warning and the decision-making was made, the result indicated that the incidence rate of CSF has negative correlation with temperature, rainfall, cloudage; relative humidity has positive correlation with disease; for air pressure, except average air pressure of one month, other air pressure factors have positive correlation with disease; for wind speed, except Difference among moths of wind speed and average temperature of one month. have positive correlation with disease, other wind speed factors has negative correlation with disease.

Key words: e-government, knowledge management, frameworks, e-governance

Please use the following format when citing this chapter:

Wang, H., Xu, D., Xiao, J., Zhang, R. and Dong, J., 2009, in IFIP International Federation for Information Processing, Volume 294, *Computer and Computing Technologies in Agriculture II, Volume 2*, eds. D. Li, Z. Chunjiang, (Boston: Springer), pp. 959–966.

1. DATA

1.1 Regions were studied

5 regions which incidence rate is relatively high were selected by epidemic situation. There was no different regularity among those regions for count of pigs and density of cultivation. The data is random distribution(see table 1).

Table 1 the cultivation situation of pig in 5 selected region

Region	Count of pigs (ten thousand)	Ranking in nation	Densityfor raising (head/km^2)	Ranking in nation
A	2817.00	fifth	119.01	eleventh
B	1081.33	eighteenth	23.82	twentieth
C	2544.76	seventh	64.59	Twenty-one
D	1853.43	twelfth	105.31	sixteenth
E	1937.14	eleventh	138.76	tenth

1.2 Epidemic situation data of CSF

The CSF epidemic situation data from 1999 to 2004 was collect from china statistics year book. The data include count of all alive pigs by the end of year. The incidence rate was calculated as count of morbidity for CSF/count of all pigs by end of year. The calculation formula as below:

$$\text{Incidence rate permonth} = \frac{\text{Total count of new cases in one month}}{\text{The total count of pigs in same month}} \times K$$

(*K:coefficient, unit:* 1/ten thousands pigs)

1.3 Meteorology data

576 meteorology data be obtained for each province in 6 years(see table2).

Table 2 Synopsis of meteorology factors

Atmosphere factors	Simplified character	Atmosphere factors	Simplified character
Average air temperature per month	t	Average air pressure per month	a
Max air temperature per month	t_{max}	Min air temperature per month	t_{min}
Average relative humidity per month	h	Total cloudage per month	c
Average wind speed per month	w	Total rainfall per month	p

2. METHOD OF ANALYSIS

2.1 The treatment of meteorology factors

By treat further, more variables were created. Every factor was analyzed independently, and new variables were listed in table 3.

Table3 new variables of meteorological factors

variable	Index	variable	Index
Difference among moths of temperature	*d*	average temperature of one month.	*jp*
The same month meteorology value in correspondence with disease incidence rate	Footnote 0	The previous month meteorology value in correspondence with disease incidence rate	Footnote -1
The second month before outbreak's meteorology value in correspondence with disease incidence rate	Footnote -2	The meteorological factor mean value of the same month and the previous month	Footnote -01
The meteorological factor mean value of the same month and the two months before outbreak	Footnote -02	The meteorological factor mean value of the two months before outbreak	Footnote -12
Meteorological factor differential value of the same month and the previous month	Footnote 0-1		

Regarding the rainfall amount, subscript-01、subscript -02、subscript -12 are on behalf of total quantity, but is not the mean value.

Difference among moths of temperature was define as difference between max average temperature and min、average temperature of one month. was define as deviation from the mean between average temperature situation of one period and many years(Abeku TA, 2002), it was defined as difference between this month of this year and average value of 6 years. By this methods, 71 meteorology factors were produced and represent as x1-x71, those new factors were list in table 4.

Table 4 Contrasting form of variable forms of meteorological factors

Variable name	Actual value	Variable name	Actual value	Variable name	Actual value	Variable name	Actual value	Variable name	Actual value
x_1	a_0	x_{16}	c_{-01}	x_{31}	t_{minjp}	x_{46}	t_{-2}	x_{61}	w_{-12}
x_2	t_0	x_{17}	w_{-01}	x_{32}	h_{jp}	x_{47}	t_{max-2}	x_{62}	p_{-12}
x_3	t_{max0}	x_{18}	p_{-01}	x_{33}	c_{jp}	x_{48}	t_{min-2}	x_{63}	a_{0-1}
x_4	t_{min0}	x_{19}	a_{-02}	x_{34}	w_{jp}	x_{49}	d_{-2}	x_{64}	t_{0-1}
x_5	d_0	x_{20}	t_{-02}	x_{35}	p_{jp}	x_{50}	h_{-2}	x_{65}	t_{max0-1}
x_6	h_0	x_{21}	t_{max-02}	x_{36}	a_{-1}	x_{51}	c_{-2}	x_{66}	t_{min0-1}
x_7	c_0	x_{22}	t_{min-02}	x_{37}	t_{-1}	x_{52}	w_{-2}	x_{67}	d_{0-1}
x_8	w_0	x_{23}	d_{-02}	x_{38}	t_{max-1}	x_{53}	p_{-2}	x_{68}	h_{0-1}
x_9	p_0	x_{24}	h_{-02}	x_{39}	t_{min-1}	x_{54}	a_{-12}	x_{69}	c_{0-1}
x_{10}	a_{-01}	x_{25}	c_{-02}	x_{40}	d_{-1}	x_{55}	t_{-12}	x_{70}	w_{0-1}
x_{11}	t_{-01}	x_{26}	w_{-02}	x_{41}	h_{-1}	x_{56}	t_{max-12}	x_{71}	p_{0-1}
x_{12}	t_{max-01}	x_{27}	p_{-02}	x_{42}	c_{-1}	x_{57}	t_{min-12}		
x_{13}	t_{min-01}	x_{28}	a_{jp}	x_{43}	w_{-1}	x_{58}	d_{-12}		
x_{14}	d_{-01}	x_{29}	t_{jp}	x_{44}	p_{-1}	x_{59}	h_{-12}		
x_{15}	h_{-01}	x_{30}	t_{maxjp}	x_{45}	a_{-2}	x_{60}	c_{-12}		

(1)footnote-1、-2、-12 represent that this factor will affect disease persistently.

(2)footnote-01、-02 represent the effect of average level of these meteorological factors on disease in recent period.

(3)footnote 0 represent those factors that affect the disease at present. (4) footnote 0-1、*jp* represent those factors that affect disease when changed.

2.2 Statistical analysis methods

The spearman's interclass correlation method was adopted in this study to analyze the relationship between meteorological factors and incidence rate of CSF each month, and hypothesis test of coefficient correlation was done either. The judge method of coefficient correlation r_s was: r_s range from -1 to 1, if $|r_s|>0.95$, there is significant correlation between meteorological factor and disease; if $|r_s|\geqslant 0.8$ indicate correlated highly; if $0.5\leqslant|r_s|<0.8$ indicate correlated moderately. If $0.3\leqslant|r_s|<0.5$, indicate lowly correlated. If $|r_s|<0.3$ indicate there was little correlation between factors and disease. If the r_s is positive, there is direct correlation between meteorological factor and disease, conversely negative correlated. r_s Supposition examination: H_0: $\rho_s=0$; H_1: $\rho_s\neq 0$; $\alpha=0.05$。

The SPSS13.0 statistical analysis software was used to process, and data file which contains the month disease incidence rate and various meteorological factor was established. The command procedure was: [Analyze]→[Correlate]→[Bivariate Correlations],and choose Spearman rank correlation coefficient analysis, and other options are default.

3. RESULT

3.1 Incidence rate of CSF in five regions

The computation of the month disease incidence rate extreme value computation sees the table 5.

Table 5 Extreme number of monthly incidence

incidence (1/10 thousand heads)	A	B	C	D	E
max value of Month incidence rate	1.383	1.726	2.017	1.985	4.143
min value of Month incidence rate	0.003	0.016	0.006	0.006	0.024
meanvalue of Month incidence rate	0.392	0.290	0.359	0.355	0.715

The tendency shows that incidence rate in autumn and winter higher than spring and summer. By incidence rate of every year, there emerge two peak rates in 2000 and 2002 respectively. However, the total tendency had been decreased in recent 3 years. See table 6, table 7 and Fig. 1, Fig.2.

Table 6 The calculation of monthly incidence

Incidence (1/10 thousand heads)	A	B	C	D	E
January	0.414	0.495	0.432	0.499	1.256
February	0.470	0.308	0.529	0.497	0.927
March	0.401	0.230	0.341	0.281	0.393
April	0.364	0.298	0.181	0.210	0.321
May	0.325	0.095	0.128	0.136	0.189
June	0.339	0.192	0.270	0.242	0.112
July	0.273	0.243	0.401	0.273	0.596
August	0.377	0.264	0.264	0.386	0.632
September	0.428	0.270	0.269	0.468	0.831
October	0.392	0.261	0.580	0.527	1.667
November	0.354	0.365	0.523	0.411	0.782
December	0.572	0.464	0.389	0.332	0.871

Table 7 The calculation of annual incidence

annual incidence (1/10 thousand heads)	A	B	C	D	E
1999	0.081	0.181	0.238	0.235	1.248
2000	0.328	0.376	0.469	0.217	0.736
2001	0.223	0.194	0.165	0.146	0.668
2002	0.767	0.732	0.853	1.019	1.228
2003	0.742	0.177	0.340	0.441	0.252
2004	0.213	0.083	0.088	0.074	0.157

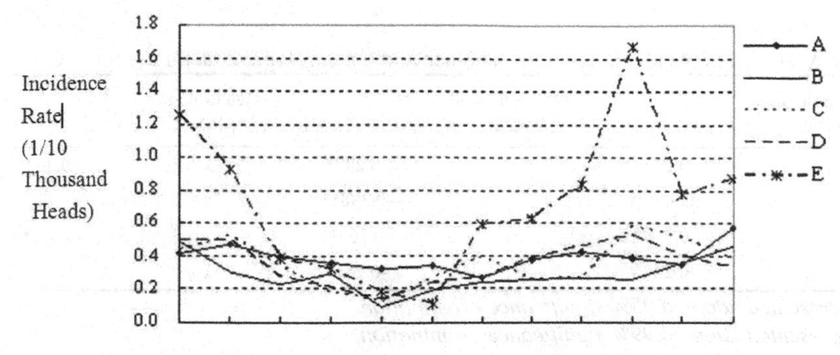

Fig.1 The scheme of monthly incidence

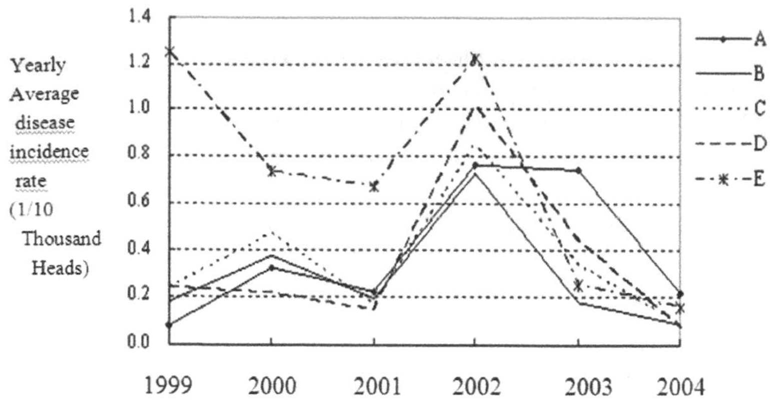

Fig.2 The incidence of CSF each month and the incidence each year from 1999 to 2004.

3.2 Analysis result calculated by Spearman's interclass correlation

The coefficient correlation r_s between 71 meteorological factors and incidence rate of CSF in these 5 regions calculated by Spearman's interclass correlation are show in table 8, 9, 10 and 11.

Table8 Analysis results about monthly incidence and meteorological factor in region B

Meteorological factor	r_s	Meteorological factor	r_s	Meteorological factor	r_s	Meteorological factor	r_s
x_1	0.357**	x_7	-0.361**	x_{12}	-0.252*	x_{36}	0.281*
x_2	-0.292*	x_9	-0.298*	x_{16}	-0.280*	x_{54}	0.251*
x_3	-0.291*	x_{10}	0.357**	x_{19}	0.312**		
x_4	-0.293**	x_{11}	-0.247*	x_{26}	-0.236*		

* *represented adopted 95% significance examination,*
** *represented adopted 99% significance examination.*

Table9 Analysis results about monthly incidence and meteorological factor in region C

Meteorological factor	r_s	Meteorological factor	r_s	Meteorological factor	r_s
$x2$	-0.245*	x_{26}	-0.297*	x_{43}	-0.302*
$x3$	-0.311**	x_{28}	-0.388**	x_{52}	-0.377**
$x12$	-0.280*	x_{32}	0.268*	x_{61}	-0.357**
$x21$	-0.247*	x_{38}	-0.251*	x_{70}	0.315**

* *represented adopted 95% significance examination,*
** *represented adopted 99% significance examination.*

Table10 Analysis results about monthly incidence and meteorological factor in region D

meteorological factor	r_s	meteorological factor	r_s
x_{15}	0.291*	x_{50}	0.300*
x_{24}	0.338**	x_{59}	0.313**
x_{41}	0.238*		

* *represented adopted 95% significance examination,*
** *represented adopted 99% significance examination.*

Table 11 Analysis results about monthly incidence and meteorological factor in region E

meteorological factor	r_s	meteorological factor	r_s	meteorological factor	r_s	meteorological factor	r_s
x_1	0.360**	x_9	-0.337**	x_{23}	-0.337**	x_{63}	0.359**
x_2	-0.290*	x_{10}	0.254*	x_{34}	0.327**	x_{64}	-0.412**
x_3	-0.309**	x_{14}	-0.330**	x_{40}	-0.256*	x_{65}	-0.369**
x_4	-0.287*	x_{18}	-0.250*	sx_{58}	-0.299*	x_{66}	-0.440**

* *represented adopted 95% significance examination,*
** *represented adopted 99% significance examination.*

A region: there was no associatively between average incidence rate each month and 71 meteorological factors. The result show that meteorological factors has little influence on epidemic of CSF in this Region (Artois M, 2002)

B region: there was associatively between incidence rate each month and 14 factors: (1) 5 of 31 temperature factors have associatively with disease, they are t_0、t_{max0}、t_{min0}、t_{-01}、t_{max-01}, all these factors are inverse correlation with disease. The effects of temperature on incidence rate each month mainly reflect in instantaneity and average level, and the continued effect is little. (2) 5 in 8 air pressure factors are direct relative with disease. They are a_0、a_{-01}、a_{-02}、a_{-1}、a_{-12}, the effect of air pressure on CSF is very great that instantaneity, average level and continued effects exist at the same time; (3)p_0 in rainfall has negative correlation with incidence rate, this result show that p0 affect the disease immediately.(4)the w_{-02} in wind speed factors has negative correlation with incidence rate, the continued function of wind speed affected incident rate. (5) 2 of 8 total cloudage factors affect incidence rate of CSF in an negative correlation ways, they are c_0 and c_{-01}.

C region: there are correlation between incidence rate of CSF and 12 meteorological factors.(1) t_0、t_{max0}、t_{max-01}、t_{max-02}、t_{max-1}, has negatively correlation with epidemic of CSF. Among these 5 factors, average max temperature affect incidence rate most significant. The survival rate of CSFV decreased in high temperature situation.(2)5 factors include w_{-02}、w_{-1}、w_{-2}、w_{-12}、w_{0-1} have correlation with incidence rate of CSF. Among of these factors, none but Difference among moths of wind speed have positive correlation with disease. This result shows that the change of wind speed increased the incidence rate of CSF. (3) the incidence rate of CSF has negative correlation with a_{jp} and positive correlation with h_{jp}, and indicate that the change of these two factors affect the disease.

D region: only relative humidity factor has positive correlation with epidemic of CSF. These five factors are h_{-01}、h_{-02}、h_{-1}、h_{-2}、h_{-12}. this result show that the incidence rate of CSF would increased in moist environment. A continuous and hysteretic effect of these factors would make on disease.

E region: 16 factors have correlation with incidence rate of CSF. (1)10 factors include t_0、t_{max0}、t_{min0}、t_{0-1}、t_{max0-1}、t_{min0-1}、d_{-01}、d_{-02}、d_{-1}、d_{-12} in 31 temperature factors have negative correlation with disease. And the real-time and change of temperature affect incidence rate of CSF significantly. (2)3 factors including a_0、a_{-01}、a_{0-1} in 8 air pressure factors as positive correlation with incidence. The real-time and recent average air pressures affect disease greatly. (3)there are 2 in rainfall factors has negative correlation with disease. These correlations reflected as instantaneity and total level. (4)w_{jp} has positive correlation with disease. This result shows that change of wind speed in same time can affect incidence rate of CSF.

4. CONCLUSION

With result of correlation analysis, some conclusions can be arriving at: (1)incidence rate of CSF has negative correlation with temperature, rainfall, cloudage. (2)relative humidity has positive correlation with disease.(3)for air pressure, except average air pressure of one month, other air pressure factors have positive correlation with disease. (4) For wind speed, except Difference among moths of wind speed and average temperature of one month. Have positive correlation with disease, other wind speed factors has negative correlation with disease.

REFERENCES

M Artois, KR Depner, V Guberti, Classical swine fever (hog cholera) in wild boar in Europe. Rev Sci Tech.2002,21(2):287～303

TA Abeku, SJ Vlas, G Borsboom, et al. Forecasting malaria incidence from historical morbidity patterns in epidemic-prone areas od Ethiopia: a simple seasonal adjustment method performs best. Trop Med Int Heath.2002. 7(10):851～857

A STUDY ON THE METHOD OF IMAGE PROCESSING AND FEATURE EXTRACTION FOR CUCUMBER DISEASED

Youwen Tian [1,*] , Yan Niu [1] ,Tianlai Li [2]

[1] *Department of Information and Electric Engineering, Shenyang Agricultural University, Shenyang, Liaoning Province, P. R. China 110161*

[2] *Department of Horticulture, Shenyang Agricultural University,Shenyang, Liaoning Province, P. R. China 110161*

[*] *Corresponding author, Address: Department of Information and Electric Engineering, Shenyang Agricultural University ,Shenyang 110161, Liaoning Province, P. R. China, Tel: +86-24-88487129, Fax: +86-24-88487122, Email: youwen_tian10@163.com*

Abstract: In order to improve the recognition accuracy of cucumber diseased image, a new method based on image preprocessing and extraction of texture feature had been proposed, which making full use of color information included in cucumber image. At first, vector median filter for color images was applied to remove noise, and then chromaticity moments were extracted as texture feature. Experimental results show that vector median filter for color images can remove the noise of diseased leaf with efficiency, and the adoption of chromaticity moments as texture feature of color image is simple and rapid, which had good recognition effect.

Keywords: chromaticity moments, vector median filter, cucumber disease, image processing

1. INTRODUCTION

The recognition of cucumber disease based on image processing has been paid attention to and studied by more and more researchers (Zhanwu Peng,2007; Xiuli Si,2006).While image processing and texture feature are the key factors of recognition in diseased cucumber(Dong Ren et al.,2007; Ran Li,2008). At present most of methods by image processing and extraction of

---

Please use the following format when citing this chapter:

Tian, Y., Niu, Y. and Li, T., 2009, in IFIP International Federation for Information Processing, Volume 294, *Computer and Computing Technologies in Agriculture II, Volume 2*, eds. D. Li, Z. Chunjiang, (Boston: Springer), pp. 967–972.

feature were based on gray images with losing abundant color information of cucumber disease image, which affected the successive recognition results. The method of image preprocessing by vector median filter and extraction of chromaticity moments as texture feature was proposed, which was proved a good basis for recognition of cucumber disease.

2. PREPROCESSING OF CUCUMBER DISEASE IMAGE

In processing of color image of cucumber disease, caused by collective device, environment et al., blurry edge of disease spots and spots of disease leaves always occurred. Vector median filter had been taken to enhancing image, stressing some useful information and getting rid of or weakening harmful information in this research.

The method of vector median filter for color image, takes the average of all vectors X_i in given window to get average vector $\overline{X}$, and calculates the distance between X_i(in given window) and $\overline{X}$, then takes the nearest vector as the output value for window çenter pixel .It supposes that a color image express by $X_i=[R_i, G_i, B_i]$ (i=1,2， …， N,N is pixel),for given window there are s vector sets, that is $X=\{X_i\}$(i=1,2， …， s),according to the definition, the algorithm of median vector for this window as follows:

(1)Calculate average vector of the window and the distance S_i'

$$\bar{r} = \sum_{i=1}^{s} r_i / s \qquad \bar{g} = \sum_{i=1}^{s} g_i / s \qquad \bar{b} = \sum_{i=1}^{s} b_i / s \tag{1}$$

$$S_i' = \left\| x_i - \bar{x} \right\| \tag{2}$$

(2)Compare the value of S_i', select the smallest one as S_{min}'

(3)X_{min} corresponding to S_{min}' is the median vector value, which will be used to replace window center pixel vector.

The application of vector median filter for color image had been taken to enhancing cucumber disease image, as shown Fig.1. The experimental result indicates that vector median filter for color image can not only remove spot and noise but keep detail of edge without increasing new color. The major reason is that center pixel is replaced by min value of the nearest vector, rather than synthetic vector by R, G, B.

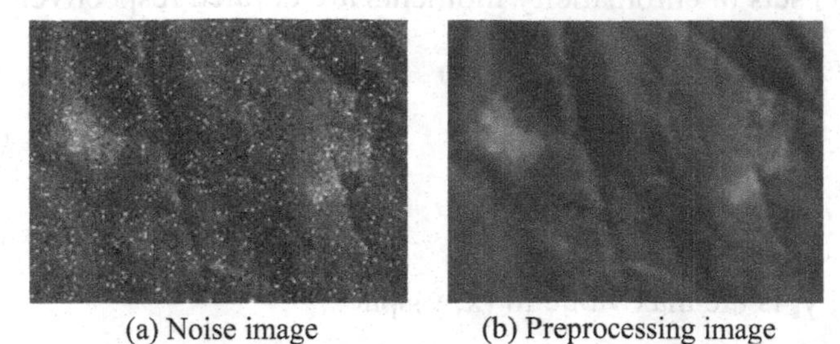

(a) Noise image (b) Preprocessing image

Fig.1. Experimental result of median filter for color image

3. EXTRACTION OF CUCUMBER DISEASE FEATURE

As the images of diseased cucumber shown, various diseased varieties have different texture and color. Hence the color texture feature of diseased cucumber was extracted for study, which became the main basis of successive recognition of cucumber disease.

3.1 Chromaticity moments

In this research, the selection of texture feature of color image for cucumber disease was based on chromatic concept of CIE XYZ color space. In CIE XYZ color space, every pixel has a couple chroma (x, y) derived from the XYZ color space as follows:

$$x = \frac{X}{X+Y+Z} \qquad y = \frac{Y}{X+Y+Z} \tag{3}$$

The chromatic diagram is, thus, a two dimensional representation of an image where each pixel produces a pair of (x, y) values. The trace of chromaticity diagram for an image is defined as follow:

$$T(x,y) = \begin{cases} 1 & if\ (x,y) \\ 0 & others \end{cases} \tag{4}$$

Two-dimensional distribution (i.e., histogram) of the chromaticity diagram is defined as:

$$D(x, y)=n_k \tag{5}$$

where: n_k is the producing number of(x, y)

Now (m,1) sets of chromaticity moments are defined respectively, as:

$$M_T(m,l)=\sum_{x=1}^{x_s}\sum_{y=1}^{y_s}x^m y^l T(x,y) \quad (6)$$

(6)

$$M_D(m,l)=\sum_{x=1}^{x_s}\sum_{y=1}^{y_s}x^m y^l D(x,y) \quad (7)$$

where: x_s, y_s is the max value in (x, y)space.

3.2 The analysis and comparison of different chromaticity moments

Images with 120*60 pixels can totally reflect the diseased texture feature with certain representativeness, therefore, fifty sub images with 120*160 pixels had been selected as training sample from every types of disease texture image. Then the chromaticity moments sets for every texture image such as CM30, CM03, CM33, CM50, CM05, CM55 had been calculated. CM30 is expressed by $M_T(1,0)$, $M_T(0,1)$, $M_T(1,1)$ as feature vector, CM03 by $M_D(1,0)$, $M_D(0,1)$, $M_D(1,1)$; CM33 by $M_D(1,0)$, $M_D(0,1)$, $M_D(1,1)$, $M_T(1,0)$, $M_T(0,1)$, $M_T(1,1)$; CM50 by $M_T(1,0)$, $M_T(0,1)$, $M_T(1,1)$, $M_T(2,1)$, $M_T(1,2)$; CM05 by $M_D(1,0)$, $M_D(0,1)$, $M_D(1,1)$, $M_D(2,1)$, $M_D(1,2)$; CM55 by $M_D(1,0)$, $M_D(0,1)$, $M_D(1,1)$, $M_D(2,1)$, $M_D(1,2)$,$M_T(1,0)$,$M_T(0,1)$, $M_T(1,1)$, $M_T(2,1)$, $M_T(1,2)$.

Then ten sub images with 120*160 pixels had been selected as testing samples from every type of diseased texture image. The recognition experiment had been done by SVM with radial basic kernel function where studying parameter c=10,ζ=0.001,σ=1/3. The different chromaticity moments sets such as CM30, CM50, CM03, CM05, CM33, CM55 had been selected as feature vector, recognition results shown as Tabel1.

Tabel1. Recognition performance of the CM method

Types	The numbers of SVM	Correct rate of recognition	Running time（ms）
CM30	93	70%	47
CM50	85	70%	47
CM03	83	80%	47
CM05	85	83.3%	62
CM33	68	86.7%	47
CM55	75	90%	47

It can be concluded that when only selecting M_T as the feature vector, recognition results had bad performance, and the correct rate of recognition was no more than 70%.When adding M_D as feature vector, the correct rate of recognition on testing sample improved. For M_T only expressed if a chromatic value was existing or not, texture image with same (m, l)order may had different textures, while M_D included distribution characteristic of chromaticity, and similar color texture images had same M_D. There are great differences between different color texture images in M_D. The correct rate of recognition of using M_T or M_D alone was lower than the two sets of moments combined. More over, CM55 had easy calculation and good recognition performance. So that selected chromaticity moments sets CM55 as feature vector was ideal.

3.3 Analysis of chromaticity moments invariance

Rotation, translation, enlargement, reduction had been done to cucumber texture images shown as Fig.2. Ten sub images with 120*160 pixels were selected as testing sample from every transformed images, while training sample were same to previous experiment. The recognition experiment had been done by SVM with radial basic kernel function, where $c=10,\zeta=0.001,\sigma=1/3$.The recognition results were shown as Table2.

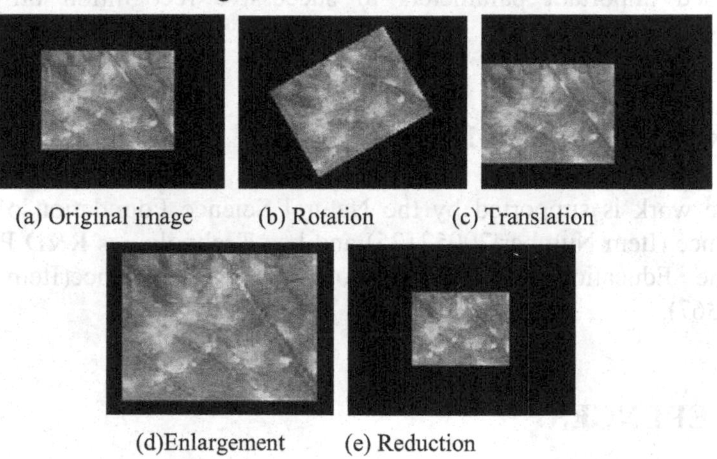

Fig.2. Transform and scaling image of cucumber disease

The data of column 2 to 4 of Tabel2 was recognition result. The right of "/" stands for the test sample numbers of every kind of disease, and the left side stands for the numbers of being recognized correctly.

Seen from Tabel2, the recognition results changed little when previous deformation occurring in images. M_T, M_D were only related to the color and color distribution of images, which having no reflected to the position of color. Thus, color and color distribution of images had no change when rotation, translation, enlargement occurring, that is, the recognition results had no change; while reduction occurring, color distribution had few differences which resulted in little change in recognition result.

Tabel2.Recognition about invariance of chromaticity moments

Transformed types	Disease types		
	Power mildew	Downy mildew	Black spot
rotation	10/10	8/10	9/10
translation	10/10	8/10	9/10
enlargement	10/10	8/10	9/10
reduction	9/10	7/10	9/10

4. CONCLUSION

This paper presents a method for enhancing the diseased plant image by vector median filter, extracting the texture feature vector of color image by chromaticity moments in color space CIE XYZ. The analysis of diseased cucumber experiment shows that vector median filter for color image have fast calculation, and the effect of filter is ideal. Extraction of the texture feature vector of color image by chromaticity moments is fast, which provided important parameters to successive recognition on cucumber disease.

ACKNOWLEDGEMENTS

The work is supported by the Natural Science Foundation of Liaoning Province (Item Number:20052125) and key Technologies R&D Programme of the Educational Department of Liaoning Province(Item Number: 2005367).

REFERENCES

Dong Ren, Haiye Yu, Xiaojun Qiao. Research on Cucumber Disease Recognition in Greenhouse Based on Support Vector Machine, Journal of Agricultural Mechanization Research, 2007, 3:25-27(in Chinese)

Ran Li. Preprocessing of Leaf Image Based on Mathematical Morphology, Agriculture Network Information, 2008, 1:43-45(in Chinese)

Xiuli Si. A Study of Image Processing and Identifying Technology on Cucumber Disease. Jilin Agricultural University Master's degree paper, 2006, 1-4 (in Chinese)

Zhanwu Peng. Research on Cucumber Disease Identification Based on Image Processing and Pattern Recognition Technology. Jilin Agricultural University Master's degree paper, 2007, 1-6 (in Chinese)

STUDY ON EARLY-WARNING SYSTEM OF COTTON PRODUCTION IN HEBEI PROVINCE

Runqing Zhang*, Teng Ma

School of Economics and Business, Agricultural University of Hebei, Baoding, Hebei Province, P.R China, 071000

* *Corresponding author, Address: School of Economics and Business, Agricultural University of Hebei, Baoding, 071000, Hebei Province, P. R. China, Tel: 13012078786, Email: runqingzhang@126.com*

Abstract: Cotton production plays an important role in Hebei. It straightly influences cotton farmers' life, agricultural production and national economic development as well. In recent years, due to cotton production frequently fluctuating, two situations, "difficult selling cotton" and "difficult buying cotton" have alternately occurred, and brought disadvantages to producers, businesses and national finance. Therefore, it is very crucial to research the early warning of cotton production for solving the problem of cotton production's frequent fluctuation and ensuring the cotton industry's sustainable development. This paper founds a signal lamp model of early warning through employing time-difference correlation analysis method to select early-warning indicators and statistical analysis method associated with empirical analysis to determine early-warning limits. Finally, it not only obtained warning conditions of cotton production from 1993 to 2006 and forecast 2007's condition, but also put forward corresponding countermeasures to prevent cotton production from fluctuating. Furthermore, an early-warning software of cotton production is completed through computer programming on the basis of the early warning model above.

Key words: Cotton Production; Early-warning Model; Early-warning Software

1. INTRODUCTION

Hebei is one of largest cotton production provinces in China. In 2006, the actual acreage of cotton planted reached 623,100 hectares, 11.82% of the

---

Please use the following format when citing this chapter:

Zhang, R. and Ma, T., 2009, in IFIP International Federation for Information Processing, Volume 294, *Computer and Computing Technologies in Agriculture II, Volume 2*, eds. D. Li, Z. Chunjiang, (Boston: Springer), pp. 973–980.

national cotton sown area; and cotton ginned 631,000 tons, about 10% of the country. Since 1990s, cotton production has shown a significant fluctuation. In 1991, the sown area of cotton arrived at the highest level, 955,200 hectares. But from then on, it dropped down to 1999's 266,600 hectares and after then gradually rose year by year to 2006's 623,100 hectares. Cotton total yield decreased from 570,800 tons of 1990 to 222,600 tons of 1999 and then gradually increased to 2006's 631,000 tons (Hebei Rural Statistical Yearbook, 1991-2007). Cotton production fluctuating not only influences related industries' development but seriously restricts the economy in rural areas and the growth of farmers' income. Therefore, it is very imperative to take research on the early warning of cotton production in order to efficiently monitor and manage cotton production, and further feedback timely abnormal symptoms to relevant departments for making decisions to ensure cotton production normal running.

2. THE FOUNDATION OF EARLY-WARNING INDICATORS

2.1 Selecting the Referential Indicator

The final result of cotton production is cotton total field. In the whole process of cotton production, when cotton total yield is unsatisfactory, even if most indicators are good, that can't represent cotton production lying in safety. So this paper takes cotton total yield's loop ratio as the referential indicator, which can eliminate its long-term trend and seasonal factor.

2.2 Selecting and Classifying Other Indicators

Selecting early-warning indicators should abide by the following principles (Bai Jiyun, 2006):

(1) Synchronousness or antecedence. Early-warning indicators' changes should be coincident with cotton total yield or earlier, and sensitively represent the trend of cotton production.

(2) Relevance. That means indicators should have a direct and inherent relation with cotton production.

(3) Importance. That requires selecting indicators should refer to the agricultural economic theory and can represent cotton production's main aspects.

Employing the method of time-difference correlation analysis to calculate relevant data about cotton production in Hebei from 1990 to 2006, we eventually determine 8 early-warning indicators, involving 3 synchronous indicators and 5 antecedent indicators (see Table 1).

Table 1 Results of Selecting Early-warning Indicators

NO.	Indictors	2-Period Lagging	1-Period Lagging	Synch-ronous	1-Period Leading	2-Period Leading	Period Relationship
1	Loop ratio of cotton unit yield	-0.37	-0.4	0.69	0.11	-0.63	Synch ronous
2	Loop ratio of cotton sown area	-0.39	0.18	0.79	-0.26	-0.23	Synch ronous
3	Loop ratio of total output value of faming, forestry, animal husbandry and fishery	-0.41	0.09	0.43	-0.14	-0.36	Synch ronous
4	Ratio of production cost between grain and cotton	-0.33	-0.45	0.032	0.46	-0.21	1-Period Leading
5	Loop ratio of cotton import volume of China	0.1	0.09	0.3	0.41	-0.2	1-Period Leading
6	Loop ratio of household consumption of Hebei province	-0.18	0.3	0.17	-0.09	-0.33	2-Period Leading
7	Loop ratio of per capita GDP of Hebei	-0.11	-0.03	0.03	0.07	-0.39	2-Period Leading
8	Loop ratio of fixed assets investment of Hebei rural residents	-0.5	-0.03	0.42	-0.3	-0.81	2-Period Leading

2.3 Determining Every Indicator's Limits

Determining early-warning limits is a key step to set up an early warning model. What kind of warning condition will be forecast much depends on how early-warning limits are determined.

First of all, considering conditions of cotton production in past 16 years, the paper classifies them into 5 sorts—"Severely Cold", "Slightly Cold", "Normal", "Slightly Hot" and "Severely Hot", in which " Severely Cold" means cotton production in extreme depression and it is quite difficult to recover; "Slightly Cold" means cotton production may become normal or depressive in the near future; "Normal" means cotton production in a healthy condition and cotton market is running well; "Slightly Hot" means cotton production runs still stably with the output growing a little fast and may turn to "Severely Hot" or "Normal" in a short period; "Severely Hot" means cotton output grows so fast that supply and demand are in extreme imbalance, together with the common phenomenon of being difficult to sell cotton.

Secondly, determining early-warning indicator's limits should follow two principles below: For one thing, it should be based on every indicator's historical data to set the centerline of fluctuation as the normal field's center and then according to the probability that indicators appear in different fields to get the basic limits—mathematic critical points. For another, when data are not enough or it lies in a long-term abnormal condition, unsuitable values must be eliminated by relevant theories and empirical judgments.

Concrete measures are as follows: First, make "●", green lamp, represent "Normal". It should be in the middle and its probability is between 40% and 60%. Here, we choose the probability of 50%. Second, make "●", red lamp and "●", blue lamp respectively represent two extreme fields "Severely Hot" and "Severely Cold", and their probabilities are generally fixed 10% each. Last, " ", yellow lamp and " " light blue lamp are used to represent " Slightly Hot" and "Slightly Cold", two relative stable fields. Each of their probabilities is 15%, a little more than the extreme fields' (Gong Yingying, 2005).

Every early-warning indicator's limits are as follows:

Table 2 Limits of Early-warning Indicators

Indicators	Blue	Light Blue	Green	Yellow	Red
Loop ratio of cotton total yield	0~0.78	0.78~1.05	1.05~1.33	1.33~2.03	2.03~∞
Loop ratio of cotton sown area	0~0.6	0.6~0.85	0.85~1.24	1.24~1.39	1.39~∞
Loop ratio of cotton unit yield	0~0.72	0.72~0.98	0.98~1.16	1.16~1.42	1.42~∞
Loop ratio of total output value of faming, forestry, animal husbandry and fishery	0~1.02	1.02~1.08	1.08~1.22	1.22~1.49	1.49~∞
Ratio of production cost between grain and cotton	∞~1.64	1.64~1.47	1.47~1.30	1.30~1.10	1.10~0
Loop ratio of cotton import volume of China	0~0.45	0.45~1.26	1.26~2.83	2.83~5.39	5.39~∞
Loop ratio of household consumption of Hebei province	0~1.04	1.04~1.09	1.09~1.14	1.14~1.24	1.24~∞
Loop ratio of per capita GDP of Hebei	0~1.07	1.07~1.14	1.14~1.23	1.23~1.31	1.31~∞
Loop ratio of fixed assets investment of Hebei rural residents	0~0.88	0.88~1.03	1.03~1.29	1.29~1.74	1.74~∞
States	Severely Cold	Slightly Cold	Normal	Slightly Hot	Severely Hot

2.4 Determining the Composite Early-warning Limits

After getting each indicator's limits, we need to fix the composite early-warning limits. Firstly, we set every signal different scores. Here, "Red Lamp" is given 5 scores, "Yellow Lamp" 4 scores, "Green Lamp" 3 scores, "Light Blue Lamp" 2 scores and "Blue Lamp" only 1 score. And then, we make 36 scores, 80% of the full scores calculated by $5 \times M$ (M is the number of indicators, and here it is 9) as the boundary between "Red Lamp" and "Yellow Lamp"; 31.5 scores, 70%, and 22.5 scores, 50% respectively as two boundaries between "Yellow Lamp" and "Green Lamp" and between "Green Lamp" and "Light Blue Lamp"; 18 scores, 40% as the boundary between "Light Blue Lamp" and "Blue Lamp". Table 3 below shows each year's early-warning condition from 1993 to 2006.

Table 3 Cotton Production State 1993-2006 in Hebei Province

Year	Loop ratio of cotton total yield	*Loop ratio of cotton sown area*	Loop ratio of cotton unit yield	Loop ratio of total output value of faming, forestry, animal husbandry and fishery	Ratio of production cost between grain and cotton	Loop ratio of cotton import volume of China	Loop ratio of household consumption of Hebei province	Loop ratio of per capita GDP of Hebei	Loop ratio of fixed assets investment of Hebei rural residents	Total Points	State
1993	1	1	3	3	3	1	2	3	5	22	Light Blue
1994	4	4	5	5	5	4	3	3	1	34	Yellow
1995	2	3	2	4	4	4	4	4	3	30	Green
1996	1	2	3	3	3	2	4	4	4	26	Green
1997	2	3	3	3	3	2	5	4	4	29	Green
1998	3	2	4	2	3	2	3	3	3	25	Green
1999	2	2	2	1	3	1	3	2	4	20	Light Blue
2000	4	3	4	1	4	1	1	1	2	21	Light Blue
2001	4	4	3	3	4	3	2	1	2	26	Green
2002	2	3	2	2	2	3	3	2	3	22	Light Blue
2003	3	4	2	3	2	5	3	2	3	27	Green
2004	3	3	3	3	1	3	3	2	2	23	Green
2005	2	3	2	3	2	2	3	3	2	22	Light Blue
2006	3	3	3	2	2	3	3	4	3	26	Green

Note: Because leading-2 periods indicators' loop ratios are involved from 1991 the early-warning condition can only be given from 1993.

3. DEVELOPING EARLY-WARNING SOFTWARE OF COTTON PRODUCTION

After analyzing relevant theoretical issues on the early warning of cotton production, this chapter will introduce the development of an early-warning software to operate corresponding functions.

3.1 General Design of Software

Java and SQL Sever 2000 are used to compile this software together. Java is an object-oriented programming language with features of safety, simplicity, easy usage and cross-platform. Microsoft SQL Server 2000 is a comprehensive, integrated data management and analysis software that enables organizations to reliably manage mission-critical information and confidently run increasingly complex business applications. At the beginning of developing software, the object-oriented programming method and form technique of Java is used to generate GUI——Graphics User Interface. All of these make the main menu very friendly and easy to operate (Zhou Haijun, 2006).

This software provides the following 4 functions: (1) User Management——add users, delete users and rcset password (2) Data management——input and edit every indicator's data and calculate their loop rates for eliminating the season factor of time series (3) Graphics analysis——show curve diagrams of cotton total yield and sown area (4) Early-warning signal system——analyze processed data and display their individual early-warning signals and the composite condition.

3.2 Software Introduction

On the basis of the above explanation of the software's design and function, we develop the cotton production early-warning system by computer software programming.

Log on the software and enter the main interface (See Fig.1). Four functions of the main interface are shown, consisting of "Indicator", "Early Warning", "System Management" and "Help".

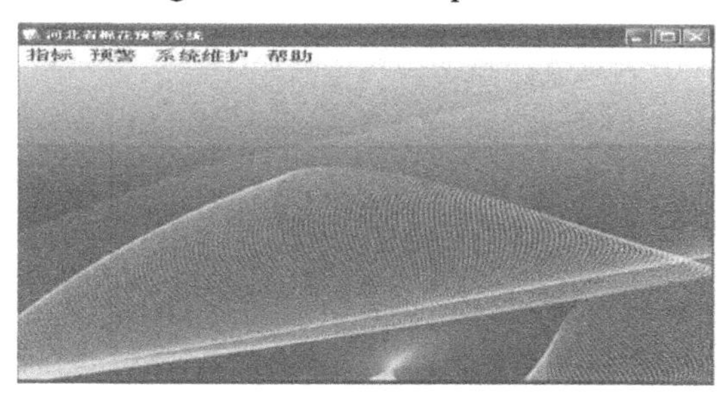

Fig.1 Main Interface of Early-Warning Software

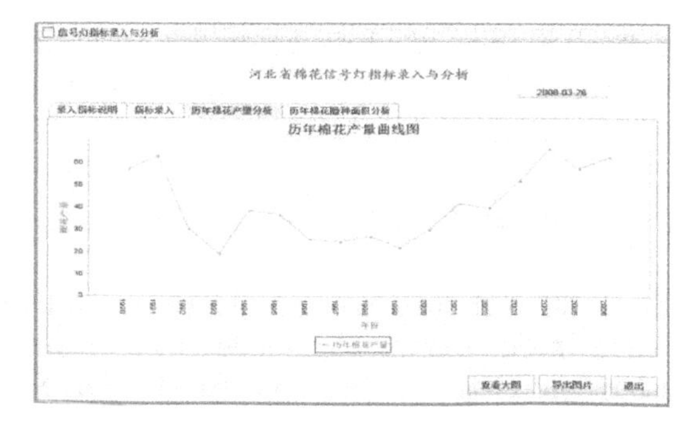

Fig.2 Analysis of Cotton Total Yield

After analyzing indicators' data inputted, Fig.2 displays the drawing function of the software on cotton total yield.

Finally, we can easily and straightly get early-warning results on cotton production from 1993 to 2006, shown in Fig.3.

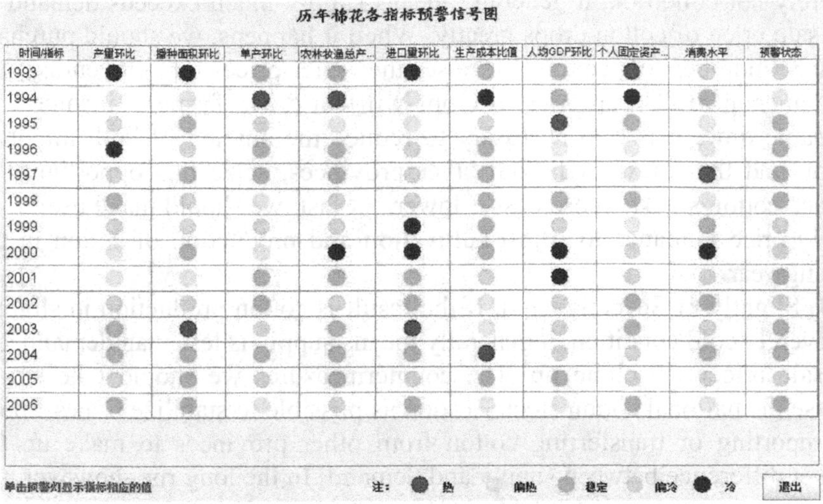

Fig.3 Signal Chart of Early-Warning Software

4. CONCLUSION

From the investigation in early 2007 by Hebei Survey Organization of National Bureau of Statistics, 2007cotton production in Hebei Province shows three characters——all of cotton sown area, unit yield and total yield will have some increases compared with 2006, among which cotton sown area is 623,100 hectares, unit yield 1018.5 kg / ha and total field expected to be 661,000 tons. On the basis of 3 synchronous indicators' values and 5 indicators with leading one or two period, we can infer that the whole scores of 8 indicators added is 22 scores when neglecting the indicator of total output value of faming, forestry, animal husbandry and fishery. Because its score scale is 1 to 5, the whole score scale in 2007 can be inferred to be 23 to 27, embraced in the range of "Green Lamp", 22.5 to 31.5. Thus, we can draw a conclusion that the early-warning condition of cotton production in 2007 will lie in the field of "Green Lamp", being normal.

In accordance with cotton production's features, we should take corresponding countermeasures for different early-warning conditions (Tan Yanwen, 2005).

(1) Normal. If the early-warning result is cotton production in a normal condition, what we should do is to steady cotton production, deal well with breeding, marketing, producing, circulating and processing and so forth, in order to raise the whole competent ability of cotton industry, which is a long-term task.

(2) Slightly or Severely Hot. If the result is cotton production in slightly or severely hot condition, it generally means supply much exceeds demand so that sale price of cotton drops greatly. When it happens, we should purchase a large number of cotton to raise the sale price for preventing the phenomenon of "Cheap price of crops will cut down farmers' income". At the same time, it is also necessary to reduce the amount of both imported cotton and the transferred from other provinces, seize this opportunity to replace cotton stock, whose cost is lower. At last, we should make use of the timely price guidance to adjust cultivation and production of cotton in the coming year.

(3) Slightly or Severely Cold. If the result is cotton production in slightly or severely cold condition, it generally means supply is less than demand and the sale price will climb up. The countermeasures we should take are to release the national cotton stock as soon as possible to stabilize prices, speed up importing or transferring cotton from other provinces to make up the market difference between supply and demand. In the long run, however, the main countermeasure is to stably improve the capacity of cotton production for guaranteeing the cotton supply.

All in all, the above are just some general countermeasures under different early-warning conditions. In practice, we should inquire into all kinds of conditions, find out the main reason and then take corresponding measures. Furthermore, sometimes it is necessary to make a comprehensive analysis and diagnosis for early-warning conditions.

REFERENCES

Bai Jiyun. The Forecast and Early Warning Research and Concrete Evidence Analysis about Production of Soybean in Hei Longjiang Province, Northeast Agricultural University, 2006

Gong Yingying. Research on Economic Supervising and Early Warning System Based on Prosperity Index, Wuhan University of Technology, 2005

Hebei Economic Yearbook 1991-2006. Beijing: China Statistics Publishing House, 2007

Hebei Rural Statistical Yearbook 1991-2007. Beijing: China Statistics Publishing House, 2007

Tan Yanwen. Study on Fluctuation of Cotton Production in China. Beijing: China Economy Press, 2005

Zhou Haijun. Java Programming. Beijing: China Railway Publishing House, 2006

REVIEW OF APPLICATION OF MATHEMATICAL MORPHOLOGY IN CROP DISEASE RECOGNITION

Zhihua Diao [1], Chunjiang Zhao [2,*], Gang Wu [1], Xiaojun Qiao [2]

[1] *Department of Automation, University of Science and Technology of China, HeFei, AnHui Province, P.R. China 230027*

[2] *National Engineering Research Center for Information Technology in Agriculture, Beijing, P.R. China 100097*

[*] *Corresponding author, Address: Shuguang Huayuan Middle Road 11#, National Engineering Research Center for Information Technology in Agriculture, Beijing, China, 100097, Tel: +86-10-51503411, Fax: +86-10-51503449, Email: zhaocj@nercita.org.cn*

Abstract: Mathematical morphology is a non-linear image processing method with two-dimensional convolution operation, including binary morphology, gray-level morphology and color morphology. Erosion, dilation, opening operation and closing operation are the basis of mathematical morphology. Mathematical morphology can be used for edge detection, image segmentation, noise elimination, feature extraction and other image processing problems. It has been widely used in the field of image processing. Based on the current progress, this thesis gives a comprehensive expatiation on the mathematical morphology classification and application of crop disease recognition. In the end, open problems and the further research of mathematical morphology are discussed.

Keywords: binary morphology, gray-level morphology, color morphology, erosion, dilation, crop disease

1. INTRODUCTION

Mathematical morphology is a new theory and method, which is used in the field of digital image processing and recognition. Its mathematical background and language is set theory, which has self-contained

---

Please use the following format when citing this chapter:

Diao, Z., Zhao, C., Wu, G. and Qiao, X., 2009, in IFIP International Federation for Information Processing, Volume 294, *Computer and Computing Technologies in Agriculture II, Volume 2*, eds. D. Li, Z. Chunjiang, (Boston: Springer), pp. 981–990.

mathematical background. Mathematical morphology was born in 1964, which is firstly proposed by the Ph. D student J. Serra and his academic advisor G. Mathorn. They proposed the "hit/miss Transformation" and they introduced the expression of morphology on the theory level firstly and established particle analysis method. In 1968, they found the "Fontainebleau Mathematical Morphology research institute". Based on the hard work of the researchers in this Institute and other country's researchers, Mathematical morphology is gradually developed and self-contained. In 1970s, with the commercial applications of grain analyzer and the publication of Mathern's "random set and integral set", the development of Mathematical Morphology is focus on gray-level aspects. In 1982, after publish of J.Serra's "image analysis and mathematical morphology", Mathematical morphology is world-wide known. Subsequently, Mathematical morphology was developed vigorously. Because the algorithm of Mathematical morphology has naturally parallel realizing structure, which realizes morphology analysis and parallel process algorithms, the method could be realized by hardware easily, which improves the image process speed and analysis.

Mathematical morphology is found on the self-contained mathematical theory and its ideas and methods have great effect on image process theory and technology and have been used in many subjects' image process and analysis. Moreover, the application of Mathematical morphology in agriculture field has also achieved great improvement. The application is focus on crop disease recognition, including wheat, cotton, vegetable, etc. In this paper, we mainly summarize the applications of Mathematical morphology in agriculture field and discuss open problems and further research.

2. MATHEMATICAL MORPHOLOGY CLASSIFICATION

Through people's efforts, mathematical morphology used in binary image originally was extended to gray-level image and made a rapid progress of the theory and applied study of gray-level morphology by right of the umbrella theory. Recently, the study of the mathematical morphology placed emphasis on the color morphology and made some achievements. According to the expression and display format of the object of study for morphology, this paper divided the mathematical morphology into binary morphology, gray-level morphology and color morphology.

2.1 Binary morphology

Mathematical morphology put forward by Maheron and Serra studied the binary image and was called binary morphology. The morphological transform of binary image in mathematical morphology was a process for sets. The essential of the morphological operator is the interaction between the sets expressed the object and shape and structure element, the shape of the structure element decide the shape information of the signal extracted by the operation. The morphological image processing is the set operation of the moving a structure element in the image and then transforming or combining between structure element and binary image. The basic morphological operations are erosion and dilation.

In the morphological operation, structure element is the most basic and important conception, which plays the role of the wave filtering in the signal process. If $B(x)$ expresses the structure element, for the every dot x of the work space E , the erosion and dilation are defined respectively as:

Erosion: $$X = E \odot B = \{x : B(x) \subset E\} \quad (1)$$

Dilation: $$Y = E \oplus B = \{y : B(y) \cap E \neq \Phi\} \quad (2)$$

From the definition and mathematical expression, binary morphological dilation and erosion can translate into the logical operation of the set, and the algorithm is easy. Because of the facility of parallel processing and hardware realization, binary image can be operated in multiple ways, such as edge detection, image segmentation, thinning, feature extraction, figure analyzing. However, in different condition, the selection of the structure element and the corresponding algorithm is different, different structure element and algorithm is designed for different aim image. The size of the structure element and the selection of the shape will influence the result of image morphological operation. (Huang et al. 2003) adopted round, triangle, square and other basic geometric figure as structure element and eroded the binary image for some times, they detached the section of the hexapod by the segmenting method of filtering image with morphological template at last. The result showed that the segmenting algorithm could get the better effect and established all right basis for extracting the figure character of the disease image. (Bouaynaya et al. 2008) founded the operator of space-variant mathematical morphology in Euclidean space and presented geometric structure element based on space variable, the simulated result illuminated the theory and the huge potentials in many kinds of image processing application.

2.2 Gray-level morphology

Gray-level morphology is the natural extension of binary morphology for gray-level image, which operating object is not any more set but image function. For gray-level morphology, the intersection and union operation used in binary morphology are replaced by maximum and minimum operation. The erosion and dilation process of gray-level image can be calculated directly from the gray-level function of image and structure element. If $g(x, y)$ expresses the structure element, for one dot $f(x, y)$ in the image, the erosion and dilation are defined respectively as:

Erosion:

$$(f \odot g)(x, y) = \min_{i,j} \left\{ f(x-i, \quad y-j) - g(-i, \quad -j) \right\} \tag{3}$$

Dilation:

$$(f \oplus g)(x, y) = \max_{i,j} \left\{ f(x-i, \quad y-j) + g(i, \quad j) \right\} \tag{4}$$

In order to apply the gray-level morphological operation in reality, some scholars proposed many improved algorithms. (Kang et al. 2006) proposed an extended definition of mathematical morphology for the problem that although edge detection methods based on classical morphology has good ability of noise elimination, it could not reflect whole edge characters. And they proposed an edge detection method based on extended mathematical morphology. The simulation result implied that this method not only eliminates noise effectively, but also the edge image by detecting has good edge characters. (Bouaynaya et al. 2008) proposed spatially-variant mathematical morphology, and gave out the geometrical concept of structure function. Simulation results showed the potential power of this theory in image analysis and computer vision applications.

2.3 Color morphology (Fan et al., 2007)

The research of morphology in color image processing area is not that much. Although some scholars have presented some morphology method used in color image, most of them only consider each vector individually, neglecting the relationship between the vectors. It is an effective and reasonable research approach to process the color pixel through vector methods describing relation among each vector. Simultaneously, the research on the morphology transformation in the HIS color space can reflect its relationship with the gray-level morphology.

For the color image $\left\{V(x);\quad x \in X,\quad X \subset D_V\right\}$ in HIS space, where D_V is the image domain in RGB color space, erosion and dilation in color morphology for the structure element B are as followed:

Erosion: $V \Theta_c B = \max\left\{V(x+y); (x+y) \in D_V, y \in B\right\}$ (5)

Dilation: $V \oplus_c B = \min\left\{V(x-y); (x-y) \in D_V, y \in B\right\}$ (6)

In recent years, a lot of scholars have dedicated on the research of color morphology. (Zhang et al. 2006) proposed an edge detecting method based on mathematical morphology. In this method, the image is pre-processed, and then the gradient transformation is done through mathematical morphology. At last, the edges are detected by the edge detecting method based on statistics. The method eliminates the shadow edge caused by the illumination, extracts the contour of objects directly, and has some effect on background noise suppression. (Leozoray et al. 2007) present a new graph-based ordering of color vectors for mathematical morphology purposes. An attractive property of the proposed ordering is its color space independence. A complete graph is analyzed to construct an ordering of color vectors by finding a Hamiltonian path in a two-step algorithm. This method can be used in any color picture.

3. MATHEMATICAL MORPHOLOGY APPLICATION

The basic idea of mathematical morphology and its method could be used in any aspects in the area of picture processing. With the development of computers, picture processing, pattern recognition and computer vision, mathematical morphology is developing quickly, and the application area is becoming vaster. Especially in the area of the crop disease recognition, many great results were achieved. In the existing software system, there are many implementations of mathematical morphology. The mathematical morphology is applied in the many areas, such as edge detection, image segmentation, noise elimination, feature extraction, and etc.

3.1 Edge detection

Mathematical morphology depicts and analyzes image from the angle of set, makes geometrical transformation for the target objects through a " probe " set (structural element) in order to outburst the required information. Along with the continuously development and improvement of mathematical morphology theory, mathematical morphology has gotten extensive research and application in image edge detection. Compared with the traditional image edge detection algorithm (Sobel operator, Prewitt operator et al.), morphology has unique advantage in image edge detection, and obtains better effect. Morphological method applied in image edge detection can keep preferably the image detail characters, and solves the coordinative problem of edge detection precision and anti-noise performance. (Zhou 2005) firstly made gray-level processing for color image, then used mathematical morphology method for edge detection, where the structure element was 3*3 square template. This method could solve preferably the problems of noise elimination and edge detection of pests in stored grain. (Kang et al. 2006) proposed an extended edge detection method of mathematical morphology in order to solve the problem of not perfectly reflecting the whole edge characters of the classical mathematical morphology. Selection definition of distance operator was given and the concept of multi-resolution analysis was applied in the extended morphological method. Results indicated that this method has good edge detection performance. (Hu et al. 2006) proposed a multi-direction edge detection algorithm based on fuzzy morphological method. The combination of multi-direction characters and fuzzy image characters was introduced to mathematical morphology. Then mathematical morphology was used for edge detection. Results showed that this algorithm could detect the edge successfully, and had better effect than other traditional edge detection algorithms.

3.2 Image segmentation

In the research of image and its application process, people often are interested in certain parts of image. The image segmentation refers to the technology that the image is divided into each characteristic region and withdraws the interest goal. Here the characteristics may be the pixel's gray, color, texture and so on. The image edge segmentation algorithm based on mathematical morphology uses mathematical morphology transformation to divide the complicated target X to the simple subset which a scrics of docs not intersect mutually " X_1, $X_2,\ldots,X_N$ " ,where

$$X = \sum_{i=1}^{N} X_i \tag{6}$$

The segmentation process for goal X can be achieved according to following method: firstly extracts the biggest inscribed circle X_1 of X , then calculates $X - X_1$, and extracts the biggest inscribed circle X_2 of $X - X_1$,..., respectively, until the set which finally obtained is the null set.

At present some scholars has already applied mathematics morphology to the agricultural crop disease image segmentation. (Huang et al. 2003) proposed a segmentation method that uses morphological template to filter color image, and applied this method for colorized digital image segmentation of vermin in cropper foodstuff. The result indicated that, this segmentation method can obtain good effect. (Xue et al. 2006) firstly use mathematical morphology to segment image in three two-dimensional color subspaces, then make two-dimensional histogram, and implements watershed algorithm respectively for the three two-dimensional histograms, lastly achieves final image segmentation by region splitting-merging process. This method is faster and less memory than the method applied in three-dimensional space, has good segmentation result. (Huang 2007) applies hole filling, erosion, dilation, opening and closing operation of mathematical morphology to extract the whole lesion areas of Phalaenopsis seedling, and obtains good effect.

3.3 Noise elimination

In pre-processing images, it is indispensable to eliminate the noise. The combination of opening and closing operations forms a morphological noise filter.

Concerning the binary image, noise includes noise block around the objects and noise hole inside the objects. The noise block will be eliminated by opening operation to operate A with structure element B. The noise hole will be eliminated by close operator to operate A with structure element B. However, these operations are related to the selection of structure element. When the noise block and the noise hole are smaller than the structural elements, the noise could be eliminated successfully. When they are larger than the structural elements, the noise can not be eliminated. (Jianga et al. 2008) used the erosion and dilation operators of mathematical morphology to filter the selected Apple image, and the structure element is a 3×3 structure.

Concerning the gray-scale image, to eliminate the noise is to smooth it morphologically. Actually, we commonly use opening operation to eliminate bright details which are smaller than structure element and keep the overall value of the gray-scale and large bright region invariable. And we use closing operation to eliminate dark details which are smaller than structure element and keep the overall value of the gray-scale and large dark region invariable. With the combination of these two operations, we can get the goal of eliminating noises both in bright and dark areas. (Li 2008) presented a pre-processing method of plant leaf image based on mathematical morphology, and had very good results by using opening and closing operations of mathematical morphology to eliminate the isolated noise points and fill the internal holes of the leaves. (Pina et al. 2006) first did the closing operation to the image of olive trees with an isotropic structuring element and then did the erosion operation. The noise points can be eliminated successfully.

3.4 Feature extraction

In general, the feature extraction is a transformation, which maps or translates the samples from high-dimensional space to low-dimensional space in order to decrease the dimensional degree. In agriculture disease recognition application, the features such as color, texture, shape have been widely used. By using the mathematical morphology, it will extract not only the disease texture features such as energy, entropy, moment of inertia, but also the disease shape feature like perimeter, area, degree of rotundity, length to width ratio. (Huang 2007) has applied the same method to Phalaenopsis seedling diseases and obtained features like centre coordinate, area, degree of rotundity. (Qian et al. 2003) had extracted the statistical features of framework image sequence by means of framework reconstructing algorithm of mathematical morphology, which provided the basis for further identification. (JR et al. 2003) has proposed a feature extraction method base on morphological operator for the iris identification, resulting that the system has lower complexity and lower storage requires. (Zheng et al. 2007) have used the mathematical morphology to achieve the four shape features of cotton by using the 3×3 square matrix template as the structure element in processing.

4. OPEN PROBLEMS AND FURTHER RESEARCH

The mathematical morphology, used as a powerful tool to analyze and depict the texture and shape signature, has developed rapidly. And the mathematical morphology-based image process system has attracted great attention world widely (Wu et al., 2003). Particularly, the agriculture disease intelligently recognition has drawn great enthusiasms from researchers both home and abroad. So far the mathematical morphology has been applied to the real-system. However, there could be many improvements in real application. So the open problems and further research could be concluded as follows:

(1) The morphology actually is a two-dimensional convolution operation. And the operation speed will be very low in case of the high-dimensional

grey morphology, color morphology or the image morphology. So it is not fit to deal with the real-time process required system.

(2)Selection of structure element plays a decisional role in the morphology operation. However until now there hasn't a standard to the structure elements' selection.

(3)The combination of morphology with the neutral network, the wavelet could be developed further and it could improve the currently used image process method.

(4) Different structure elements can have different effects on the morphology operation' results. And the color image process based on the multi-structure elements could be one of the future researches.

So the development of fast speed algorithms for grey-level and color morphology, the design of the structure elements select standard, and the improvements of the morphology's generality will be the problems for the mathematical morphology. Also the adaptation of the morphology operation must be enhanced. By using the advanced achievements in mathematical morphology, its application to the agriculture disease recognition can enrich and develop the morphology-based image processing method. Meanwhile, the research of color morphology theory and the color morphology operator's application will be the further researches.

ACKNOWLEDGEMENTS

Funding for this research was provided by the National High Technology Research and Development Program of China (863 Program, Grant 2007AA10Z237). The first author is grateful to the University of Science and Technology of China for providing him with pursuing a PhD degree at the National Engineering Research Center for Information Technology in Agriculture.

REFERENCES

Deng Tingquan, Dai Qionghai. Representation Theorem of Gray-scale Mathematical Morphology, Computer Engineering,2005,31(15): 1-3(in Chinese)

Fan Linan, Zhang Guangyuan, Han Xiaowei. Image Processing and Pattern Recognition, Science Press,2007,3(in Chinese)

Guo Jun, Pan Shen, Hu Xiaojian. Edge Detection in Tobacco Leaf Image Based on Grayscale Morphology, Computer Engineering,2007,33(21): 163-165(in Chinese)

Hu Dong, Tian Xianzhong. A Multi-directions Algorithm for Edge Detection Based on Fuzzy Mathematical Morphology, Proceedings of the 16th International Conference on Artificial Reality and Telexistence- Workshops(ICAT2006), 2006

Huang Xiaoyan, Guo Yong, Zhao Taifei. Segmentation Method Based on Mathematical Morphology for Colorized Digital Image of Vermin in Cropper Foodstuff,2003,11(6): 467-469(in Chinese)

J. D. M. JA, J. MAYER. Image Feature Extraction for application of Biometric Identification of Iris-A Morphological Approach, Proceedings of the XVI Brazilian Symposium on Computer Graphics and Image Processing (SIBGRAPI 2003),2003

Joe-Air Jiang, Hsiang-Yun Chang, Ke-Han Wu, et al. An adaptive image segmentation algorithm for X-ray quarantine inspection of selected fruits, Computers and electronics in agriculture,2008,60: 190-200

Kang Huaiqi, Shi Caicheng, Zhao Baojun, et al. A method of edge detection based on extended mathematical morphology, Optical Technique,2006,32(4): 634-638(in Chinese)

Kuo Yihuang. Application of artificial neural network for detecting Phalaenopsis seedling diseases using color and texture features, Computers and electronics in agriculture,2007,57: 3-11

Li Ran. Preprocessing of leaf image based on mathematical morphology, Agriculture Network Information,2008,1: 43-45(in Chinese)

N. Bouaynaya, D. Schonfeld. Theoretical Foundations of Spatially-Variant Mathematical Morphology Part II: Gray-Level Images, IEEE Transactions on pattern analysis and machine intelligence,2008,30(5): 837-850

N. Bouaynaya, M. Charif-Chefchaouni, D. Schonfeld. Theoretical Foundations of Spatially-Variant Mathematical Morphology Part I: Binary Images, IEEE Transactions on pattern analysis and machine intelligence,2008,30(5): 823-836

O. Lezoray, A. Elmoataz, C. Meurie. Mathematical Morphology in any color space, 14th International Conference of Image Analysis and Processing-Workshops(ICIAPW 2007), 2007

P. Pina, T. Barata, L. Bandeira. Morphological recognition of the spatial patterns of olive trees, The 18th International Conference on Pattern Recognition (ICPR 2006), 2006

Qian Wei, Chen Wei, Bai Shilei, et al. Image Recognition Method for Pathological Changes Based on Morphology and SVM, Journal of Image and Graphics,2003,8(10):1201-1204(in Chinese)

Wang Shuwen, Yan Chengxin, Zhang Tianxu, et al. Application of Mathematical Morphology in Image Processing, Computer engineering and application,2004,32: 89-92(in Chinese)

Wu Dan， Liu Xiuguo， Shang Jianga. The Application and Prospect of Mathematical Morphology in Image Processing and Analysis, Journal of engineering graphics,2003,2: 120-125(in Chinese)

Xue Heru, Ma Shuoshi, Pei Xichun. Color Image Segmentation Based on Mathematical Morphology and Fusion, Journal of Image and Graphics,2006,11(12): 1764-1768(in Chinese)

Zhang Dongfang, Wang Xiangzhou. Image Edge Processing based on Mathematical-morphology, Microcomputer Information,2006,22(8): 186-188(in Chinese)

Zheng Shicha, Mao Hanping, Hu Bo, et al. Morphological feature extraction for cotton disease recognition by machine vision, Microcomputer Information,2007,23(4): 290-292(in Chinese)

Zhou Long. Investigate on image's edge detection of pests in stored grain Based on Mathematical morphology, Microcomputer Information,2005,21(3):224-225(in Chinese)

DESIGN AND DEVELOPMENT OF COUNTY-LEVEL INFORMATION MANAGEMENT SYSTEM FOR DISEASES AND PESTS OF HEBEI PROVINCE ON GIS

Xiaoyan Cheng[1] , Xiaoli Zhang[1,*] , Fangyi Xie[1]

[1] *Key Laboratory for Silviculture and Conservation, Ministry of Education, Beijing Forestry University, 100083 , P. R. China*

[*] *Corresponding author, Address: Key Laboratory for Silviculture and Conservation, Ministry of Education, Beijing Forestry University, 100083 , P. R. China, Tel: +86-13681272638, Email:zhang-xl@263.net*

Abstract: To get a more convenience work in forest application, GIS and information system is used in forestry. GIS technology is used to build an informational management system of forest disease. For the practical requirement, the system is implemented by PDA which works outside to help completing the data collection. The major function of the system is input and output of the forest disease data and processing the report which is based on the criteria report and the assistant function of GIS. This article is aim to discuss about the theory, the process and the critical points of the information system. Besides the general information management system, GIS and PDA is introduced into the diseases system, which could combine the map and the attribute information and realize inventory data reform by PDA. The system is developed with VB and SuperMap Object (SuperMap Company).

Key words: forest diseases, second development, Design, development, PDA, Geographic Information System

1. INTRODUCTION

The forest diseases and pests are the main factors of the forest health. With the resource of the forest is more and more deficiency, monitoring the

---

Please use the following format when citing this chapter:

Cheng, X., Zhang, X. and Xie, F., 2009, in IFIP International Federation for Information Processing, Volume 294, *Computer and Computing Technologies in Agriculture II, Volume 2*, eds. D. Li, Z. Chunjiang, (Boston: Springer), pp. 991–1000.

diseases and pests is becoming important, and it is more important to the economist and ecologist. With the development of the software and hardware, the information management of diseases and pests has a bright future. Be under the reality, we use GIS technology as the second development platform to manage the diseases and pests of HEBEI province and give some decision for the manager by the system. We use VB as the development language and SuperMap Object (SuperMap Company) as the second develop plat in order to provide management and decision-making supports for the managers.

2. THE NECESSARY AND FEASIBLE

2.1 Necessary of developing the system

Although the information is widely used in every industry, the application in the forestry is not extensive yet. There is a large distance between the resource management, monitor technology, information level and modern development of forest. It is known as: (1) The information circulation is not fluency. The inquirers must go to the sample plot and have the record of subplot, which is more time-wasting. (2) A lot of data is not shared. With the increased data, the usage is not enough without the information. (3) There are many mistakes when calculating the report form. The forest farm staff often makes mistakes with the confusion concepts which are not satisfying the time effectiveness. Especially in some county-level forest prevention station, report is usually delayed. So it is more important to develop an information management system.

The purpose of the system is collecting field data through PDA, and then transferring it to the Server which collect the data and calculate various data without mistakes.

2.2 Feasible of developing the system

After understanding the process of serious data collecting and calculating in the forest prevention station, also with a clear idea of the data organization, we choose VB as the development language and SuperMap Object 5.2 (SuperMap Company) as the second development platform to develop the system to fit the practical demand.

3. THE SYSTEM DESIGN

The system is developed with the unity data standard based on VB and SuperMap Object 5.2 XML format is used by PDA which collects the field data and then input it into the system. The system accepts the XML data and input them into database. The system has several functions, include system setup, data management, data query, view edit.

3.1 Database designing

Database is aim to initialize the system and output the report form. Above all, setting up several standard sample plots, and surveying every detail data of them is the most important function of it. When there are some data of the standard sample plots, the next step is surveying the other plots and summarizing them as the survey table based on the sample plots data. So the tables of the database include the same diseases level table, survey table, standard record table. And there are other tables which separated by survey way, include pests tree number table, pests ratio table, insect gall table and so on. The other related tables are: the prevent table, the quarter total table, administrative division table, pests code table, diseases code table, rodent damage table and so on.

On the needs of the practical usage, the staff needs to survey every aspect of disease. Different survey way is suggested because of the different disease: including pests' tree number table, pests' ratio table, and insect gall table and so on.

The system need other auxiliary tables include administrative code table, disease code table, pests code table, rodent damage table, user-password table, preventing happen table and quarter total table. These are the entire database which used in setting up the system and output report.

3.2 The data interface

Subplot is considered as the fundamental unit in the survey, which every data belongs to is recorded in the survey table. The species of the disease and the same disease level plot or other practical data are recorded in PDA. As the format of XML, the data of PDA is introduced into the system, and the report is exported. Input the data of PDA and filling the survey table by manual, with the prevent table and the data summary are the core text of the system.

The inquirer surveys every subplot to collect the field data by PDA which has the same table with the system. The system has the data-processing interface and the result could be seen by the inquirer.

The table of the system could be filled out by the inquirer where there is no PDA. After finishing the necessary data, such as the number of the harmed trees, the degree of the diseases, and the other index, the national report could be output. The same disease level table, survey table and standard record table are designed in the system which needs to fill in. These data are connected in the database by programming and are calculated to fit the need of the report.

As a result of the system, we can obtain the required report while calculating the index of several data. The inquire function of the report is supported by the month, which could be printed according to the demands.

3.3 The auxiliary function of GIS

After connecting GIS and information management with VB and SuperMap Object 5.2, the auxiliary function is implemented, such as measuring the area and connecting the map together with the attribute tables. And of course, converting the data format is allowed.

The damage degree of every year and every month of the main disease could be inquired and showed on the map, with different pests and different diseases level.

4. THE DETAILED DESIGN AND THE KEY TECHNIQUES

4.1 Designing of the system interface and the main function

For the sake of the practical application, initialization of the system includes several parameters. When the area is definitely choose, the report will be output with the chosen area ID and other information automatically. The other operations are based on the chosen area (Fig. 1 and Fig. 2).

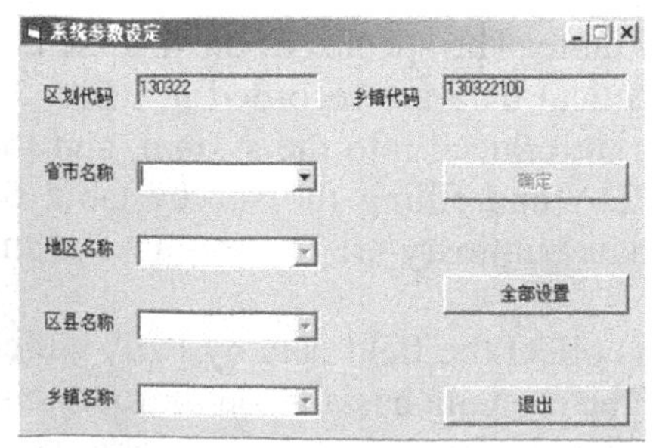

Fig. 2. The initial parameters

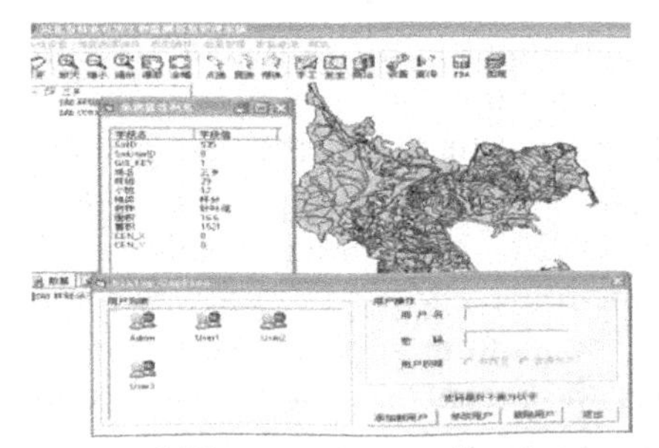

Fig. 2. User-management interface

GIS had been introduced into the information management system of the diseases and pests, which could be visualized when inquiring the situation of the diseases. The function included: format reforming of the area, input the map, the attribute of the object in the map, mapping the thematic maps of diseases degree. The decision of preventing diseases could be helped to make on the intuitive map. Other GIS function as zoom out, zoom in, roam, full width were included (Fig. 3 and Fig. 4).

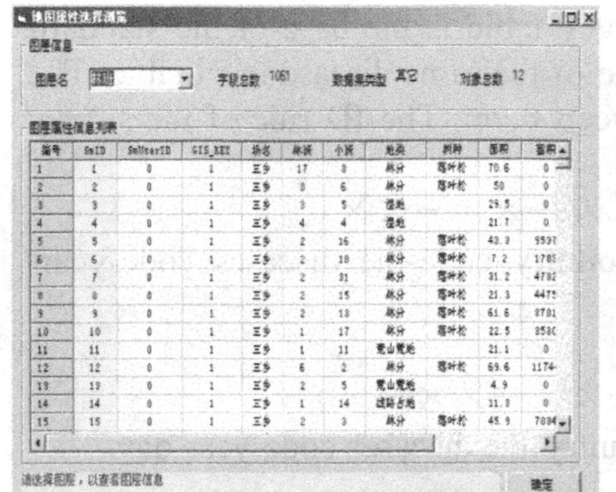

Fig. 3. The attribute of the subplot of the map

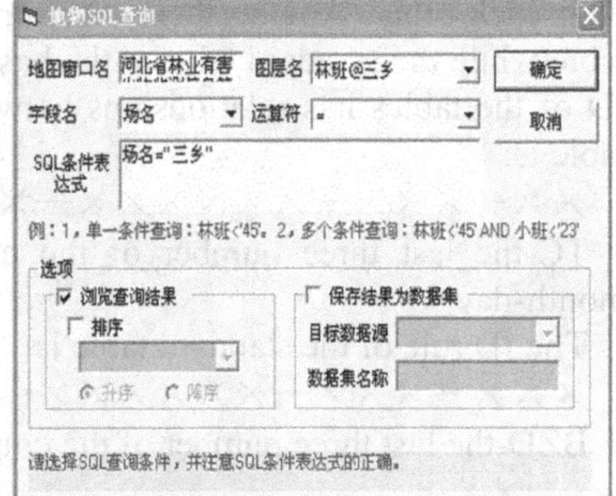

Fig. 4. SQL inquiring of subplot

Another important function is checking the area and the degree of the diseases through clicking the map. SQL inquiring of the condition after writing the ID of the subplot or the degree of the diseases, and the result is showed on the map (Fig. 5).

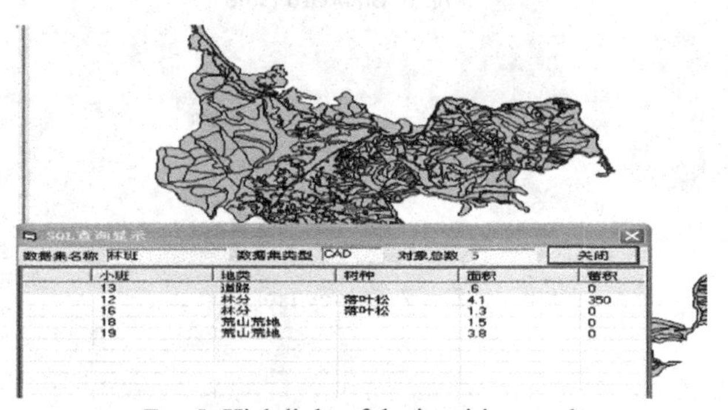

Fig. 5. High light of the inquiring result

As explained before, the data collected by PDA, which have three important tables in it is useful. Survey table (Fig. 7) include: survey table ID, inquirer, county code, name of county, diseases code, diseases name, generations, survey way, units, time and so on. The standard table (Fig. 6) include: the standard table ID, the survey table ID, diseases code, the own compartment, the own subplot, the degree. The same diseases level table (Fig. 8) included: same level table ID, the connected standard table ID, the own subplot of the forest, the own subplot, the area of the subplot, the preventing way.

When the inquirer starts a new survey, a new survey table is coming forth. The inquirer takes some typical subplot as the standard sample which will be the base of the other survey data such as the level of the damage. The same disease level table is filled of the data in which subplot has the same condition with the standard sample. The tables of the database have relations with each other. Among them, the survey table is the base of the standard table while the standard table is the base of the same disease level table. The ID of the tables has relationships between them. The ID rule of the survey table is:

××—×××—××××××—××—××—××

TC-the last three number of the country code-the diseases code-year-month-day

The ID rule of the standard table is:

×××—×××—××××××—××—××—××

BZD-the last three number of the country-the diseases code-year-day

标准地记录表

标准地记录表ID	踏查记录表ID	病虫代码	所在林班号	所在小班号	发生程度	添加新字段
BZD3224160017521	TC32241600175	416001	2	1	轻	
BZD3224160017911	TC32241600179	416001	1	1	中	
BZD3224160017923	TC32241600179	416001	2	3	重	
BZD322473011710	TC322473011710	473011	2	1	重	
BZD3224730117111	TC32247301171	473011	1	1	重	
BZD3224730117112	TC32247301171	473011	1	2	轻	
BZD3224730117611	TC32247301176	473011	1	1	轻	
BZD3224730117613	TC32247301176	473011	1	3	中	
BZD3224810037311	TC32248100373	481003	1	1	中	
BZD3224810037713	TC32248100377	481003	1	3	中	
		0				

Fig. 6. Standard table

踏查记录表

踏查记录表ID	乡镇代码	乡镇名称	病虫害代码	病虫害名称	世代/次	调查方法代	计量
TC32241600175	322	12	416001	松突圆蚧	1	1	万亩
TC32241600179	322	12	416001	松突圆蚧	2	1	万亩
TC32247301171	322	12	473011	松毛虫	1	1	万亩
TC322473011710	322	12	473011	松毛虫	3	1	万亩
TC32247301176	322	12	473011	松毛虫	2	1	万亩
TC32248100373	322	12	481003	美国白蛾	1	1	万亩
TC32248100377	322	12	481003	美国白蛾	2	1	万亩
	0				0	0	万亩

Fig. 7. Survey table

编号	标准地记录	林班号	小班号	小班面积	地物编号	防治方法	添加新字段
1	BZD32247301	1	2	11.6	Sm12		
2	BZD32247301	1	3	9.8	Sm13		
3	BZD32247301	1	1	10.1	Sm11		
4	BZD32247301	1	4	7.2	Sm14		
5	BZD32248100	1	3	9.8	Sm13		
6	BZD32248100	1	1	10.1	Sm11		
7	BZD32241600	2	1	13.2	Sm21		
8	BZD32241600	1	3	9.8	Sm13		
9	BZD32241600	2	3	9.2	Sm23		
10	BZD32241600	2	2	14.6	Sm22		
11	BZD32247301	1	3	9.8	Sm13		
12	BZD32247301	2	3	9.2	Sm23		
13	BZD32247301	3	1	14.1	Sm31		
14	BZD32247301	1	1	10.1	Sm11		
15	BZD32248100	1	3	9.8	Sm13		
16	BZD32248100	2	3	9.2	Sm23		
17	BZD32248100	1	4	7.2	Sm14		
18	BZD32241600	1	1	10.1	Sm11		
19	BZD32241600	1	4	7.2	Sm14		
20	BZD32241600	3	1	14.1	Sm31		

第 1 项(共 27 项)

Fig. 8. Same diseases level table

The connection of the tables is more important in this system. The information could be got outside or completed inside. When there is no PDA data could be input the system, the table could be filled out by hand through add or browse survey interface (Fig. 9). In this interface, the record of the survey data could be edited and browsed but could not new one. In the survey table management interface (Fig. 10), the record could be edited or deleted in a number. When the record of the standard table is selected in the right table, the related record of the survey table will be showed in left table (Fig. 10). The survey table ID is the primary key of the survey table while it is the foreign key of the standard table. Some of the survey table record will be also as the record of the standard table, and after ensuring the degree of the diseases, the other records will take it as the standard record. Those which record have the same degree with the standard record will be filled in the same diseases level table.

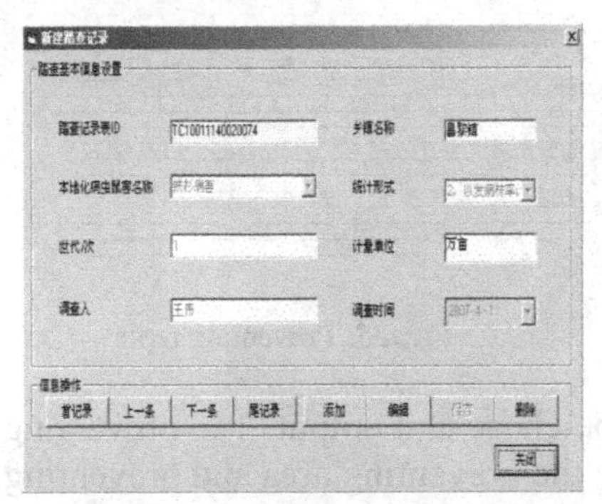

Fig. 9. Add or browse survey interface

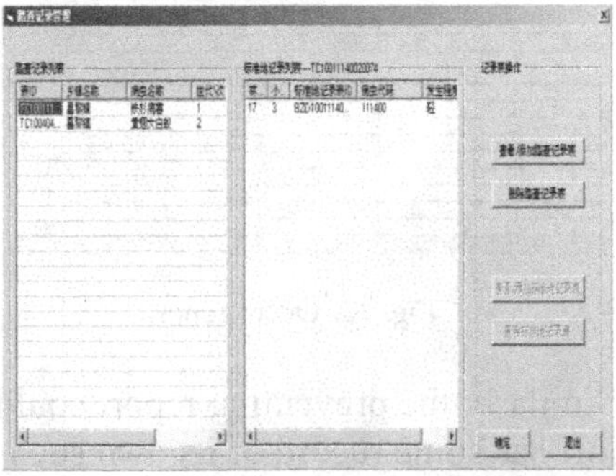

Fig. 10. Survey table management interface

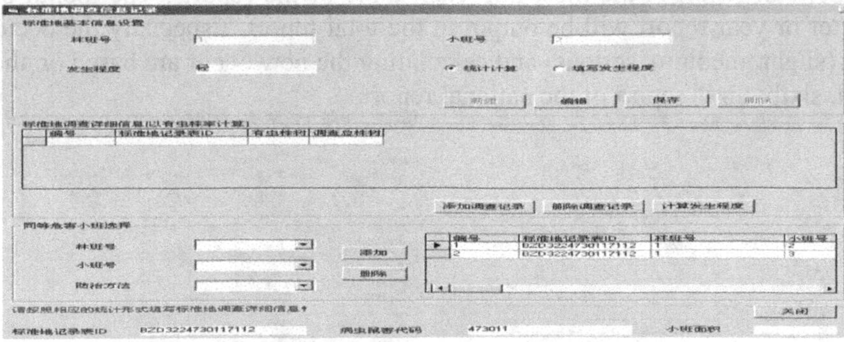

Fig. 11. Standard table management

The standard table management interface (Fig. 11) includes the basic information of the standard sample, the detail information of the standard sample and same diseases level. The detail information includes the kind of the diseases in the standard sample and the survey way, the total number of the area which will do goods to calculating the percentage of the diseases. According to the same preventing way of the same disease, the number of subplot is filled into the same diseases level table and use the same survey way. Or we could input the record through clicking the map. These three tables are the core of the database and have a key effect of report and other operation.

The report could be output after completing the necessary data of the table. As one of the final task of the system, the report include: occur report (Fig. 12), preventing report (Fig. 13), and total report.

The month report (Fig. 14) has been output with the name of the area, the name of the pests, generations, the host tree area, the distribution area and occur area, the new occur area and so on. The format is according to the national system and allowed to be printed in EXECL.

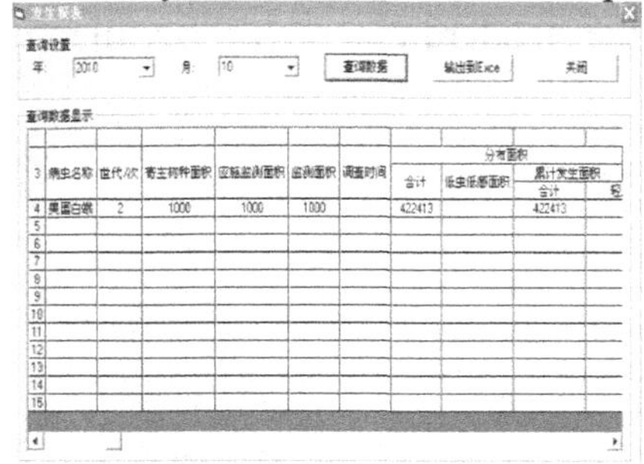

Fig. 12. Occur report

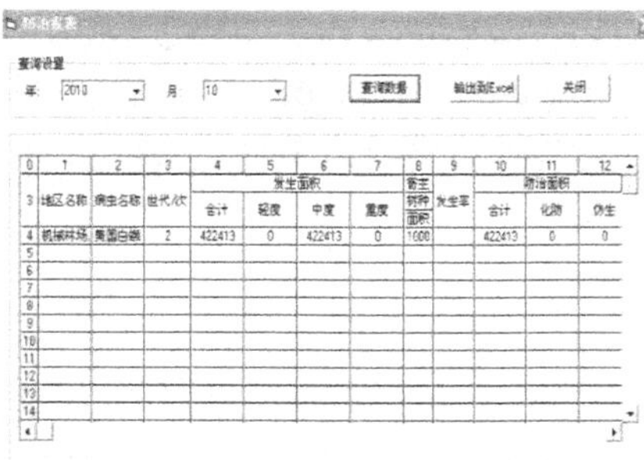

Fig. 13. Preventing report

Similarly the preventing report could be chose and output. The preventing report has some record of its own such as the preventing area and preventing way.

The total report is the most important table of the reports. The month or quarter or year report will be output in the total report. Especially the occur area (slight, medium, serious) and calculating the new occur are based on the third, sixth, ninth report of the national report.

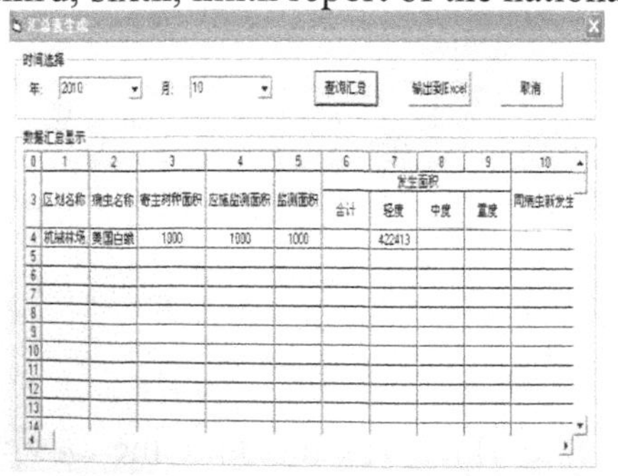

Fig. 14. The month report

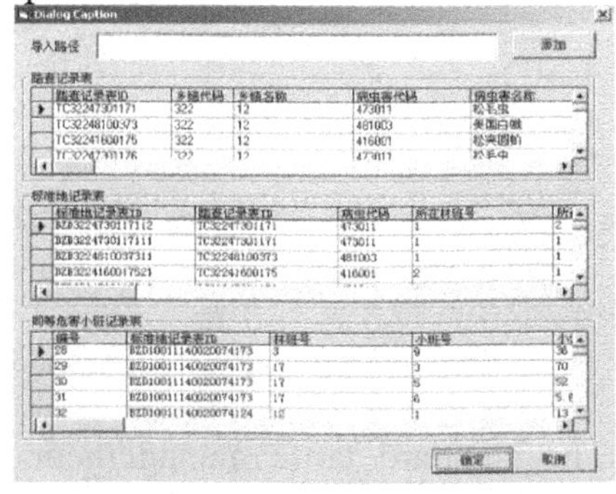

Fig. 15. Input the data into the system

4.2 The key technology in the system

One of the important innovations is the use of PDA, which saves labor and money. Inquirer input the data into the PDA then lead it to the system which has the same interface with PDA (Fig. 15). When input into the system, the data of PDA is as a record in the database of the system. Then the data will be as the record used in the data management function.

Besides amount of the data, the tables have complicated relations between each other such as mathematics relations. The input interface of the system should be consistent with database and result in the output report.

5. CONCLUSION AND PROSPECT

The report data is complicated while the basic work is easy, especially some confusion concepts which lead to the wrong result. When fill the table by hand, some index are usually mistaken. On the GIS platform, some index such as area of the report will be calculated exactly. The occur percentage and prevent percentage is also done exactly. This has an important mean for the staff of the forest farm. And of course, the system based on GIS and information management is an innovation in forest area.

The next work is to realize the connection of the PDA and the INTERNET which will transport the data on time. And also adding the RS image is a bright future for monitoring. Forecasting the diseases will be another important work of the system.

ACKNOWLEDGEMENTS

Funding for this research was provided by NCET（Programme for New Century Excellent Talents in University, Project No. NCET-06-0122）and Innovative Research Team in University (Project No. IRT0607). The first author is grateful to Zhang Xiaoli professor for helping finishing the project.

REFERENCES

Cao Yanlong, “Pithiness of the system development based on Visual Basic ”, 2005

Chen Libiao, Zhao Lianqing, “Considering of the development of HEIBEI forest resource information management”, 2003

Guo Ruijun, “Pithiness of the database development of Visual Basic ”, 2007

He Ruizhen, Zhang Yin, Zhang Jingdong, “The designing and the development of the forest resource information management based on GIS”,2005

He Zhengwei, Huang Runqiu, “The analysis of the forest information system development”, 2004

Hong Lingxia, Lu Yuanchang, Lei Xiangdong, “The information system development of the country level”, 2005

Li Xiaoli, Zhang Wei, “Examples of database system develop on Visual Basic+SQL Server”, 2003

Liang Jun, Qu zhiwei, “The information system development of Chinese forest diseases”, 2005

Xu Ruidong, “The study of forest resource information management system based on “3s” ”, 2007

Zhang Yuan, Yin Mingfang, Wang Shuhai, “The study of the application of the forest resource management system”, 2004

Zhong Qiuping, Li Diqiang, Li Jiangnan, “The development and the application of the resource management geographic system”, 2001

Zhou Guona, Gao Baojia, ”The application of GIS in the study of the plant diseases and pests”,2003

A SIMPLIFIED BAYESIAN NETWORK MODEL APPLIED IN CROP OR ANIMAL DISEASE DIAGNOSIS

Helong Yu[1,2], Guifen Chen[1,2,*], Dayou Liu[1]

[1] *College of Computer Science and Technology, Jilin University, Changchun, Jilin Province, P. R. China, 130012*

[2] *College of information Technology, Jilin Agricultural University, Changchun, Jilin Province, P. R. China 130118*

[*] *Corresponding author, Address: College of Computer Science and Technology, Jilin University, Changchun, Jilin Province, P. R. China, 130012, Tel: +86-431-84532775, Fax: +86-431-84532775, Email:guifchen@163.com*

Abstract: Bayesian network is a powerful tool to represent and deal with uncertain knowledge. There exists much uncertainty in crop or animal disease. The construction of Bayesian network need much data and knowledge. But when data is scarce, some methods should be adopted to construct an effective Bayesian network. This paper introduces a disease diagnosis model based on Bayesian network, which is two-layered and obeys noisy-or assumption. Based on the two-layered structure, the relationship between nodes is obtained by domain knowledge. Based on the noisy-model, the conditional probability table is elicited by three methods, which are parameter learning, domain expert and the existing certainty factor model. In order to implement this model, a Bayesian network tool is developed. Finally, an example about cow disease diagnosis was implemented, which proved that the model discussed in this paper is an effective tool for some simple disease diagnosis in crop or animal field.

Keywords: bayesian network, crop or animal, disease diagnosis, noisy-or, certainty factor

---

Please use the following format when citing this chapter:

Yu, H., Chen, G. and Liu, D., 2009, in IFIP International Federation for Information Processing, Volume 294, *Computer and Computing Technologies in Agriculture II, Volume 2*, eds. D. Li, Z. Chunjiang, (Boston: Springer), pp. 1001–1009.

1. INTRODUCTION

There exists a lot of uncertainty phenomenon and problem in agriculture, especially in the field of crop or animal disease. Uncertainty in disease diagnosis is more extensive and complex. So, in order to create an effective disease diagnosis system, uncertain knowledge must be dealt with.

Bayesian network is a kind of probabilistic graphical model(P.Larranaga,S. Moral.2008), namely a combination of probability theory and graph theory. By graph, Bayesian network can represent knowledge naturally and intuitively. By probability, Bayesian network can solve uncertain problem.

Bayesian network can reason in dual direction, which can be used in both prediction and diagnosis. Also, Bayesian network can use prior knowledge effectively and make the best of knowledge form domain expert.

However, the construction of the Bayesian network needs a great amount of probability. Generally, this is very complicated and difficult.

As far as this paper is concerned, a simplified Bayesian network is proposed and used in disease diagnosis system.

The simplified model is a two-layered structure and obeys noisy-or assumption.

In order to decrease the difficulty of constructing Bayesian network, domain knowledge and existing certainty factor knowledge base are used.

Then a Bayesian network tool was developed and an example about cow disease diagnosis was implemented.

2. METHODS AND TECHNOLOGY

2.1 The defination of Bayesian network

Bayesian network is a binary group, namely S=<G，P>， in which:

（1）G is a directed acyclic graph. The nodes correspond to random variable and the directed arcs represent probabilistic dependence between variables. The meaning of the arc from x to y is that x have direct influence to y.

（2）P is the set of local probability distribution , $P=P\{P(x|\pi_x)\}$ is conditional probability, which is used to measure the strength of causal dependencies and π_x is the set of parent nodes of x.

2.2 A two-layer model of disease diagnosis

There are two types of nodes in the disease Diagnosis System, which are disease nodes and symptom nodes(Fig.1).

Both disease nodes and symptom nodes are Boolean variables. Disease nodes contain states: 'happen' and 'not happen'. Symptom nodes contain states: 'find' and 'not find'(P.J.F Lucas.2005, Radim Jirousck.1997).

This BN is a two-layer network, in which the upper layer is composed of disease nodes and the lower layer is composed of symptom nodes. Obviously, the arc direction is from disease nodes to symptom nodes.

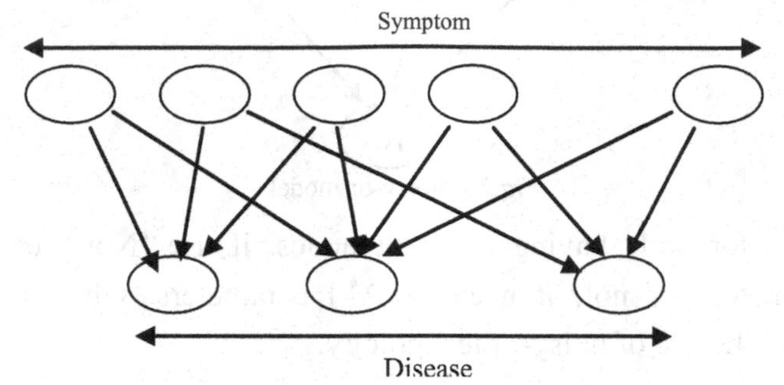

Fig. 1 two-layered structure

2.3 Noisy-or technology

One problem faced in knowledge engineering for Bayesian networks is the exponential growth of the number of parameters in their conditional probability tables (CPTs). The most common practical solution is application of the noisy-OR (or their generalization, the noisy-MAX) gates, which take advantage of independence of causal interactions and provide a logarithmic reduction of the number of parameters required to specify a CPT.

This model has three assumptions: Parents and child are Boolean variables.Inhibition of one parent is independent of the inhibitions of any other parents.All possible causes are listed. In practice this constraint is not an issue because a leak node can be added (a leak node is an additional parent of a Noisy-or node).

Now, we can have a definition of noisy-or:

(1) A child node is false only if its true parents are inhibited.

(2) The probability of such inhibition is the product of the inhibition probabilities for each parent.

(3) So the probability that the child node is true is 1 minus the product of the inhibition probabilities for the true parents.

According to Fig. 2, Given the reason nodes, suppose that $p_i = P(F = true \mid H_i = true)$, we can get the following conclusions:

If all reason nodes are false $P(F = true) = 0$;

if only one reason node is true, $P(F = true) = p_i$;

else $P(B = false) = \prod_{i=1}^{m} (1 - p_i)$,in which m is the counts of true reason nodes.

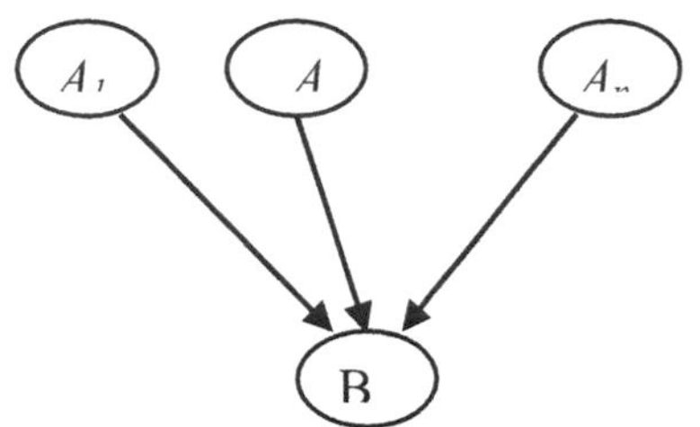

Fig.2 : Noisy-or model

Generally, for node having k parent nodes, if use 'Noisy-or', it needs $\Theta(k)$ parameters, if not, it needs $\Theta(2^k)$ parameters. Obviously, BN is simplified by the use of noisy-or technology.

3. THE CONSTRUCTION OF BN

Before being deduced, Bayesian network must be constructed. As we know, Bayesian network has two parts: structure and CPT, so the process of constructing Bayesian network is to construct structure and CPT (David J.Spiegelhalter.1993).

Construcion of a Bayesian network for a domain problem needs communication and cooperation of Bayesian network expert, domain expert and BN software tool(P.J.F Lucas.2000).

There are three methods to construct Bayesian network: manual construction, machine learning and combination of them. This article mainly introduces manual method, which constructs Bayesian network by domain expert elicitation(E.Charles,J.Kahn, etc.1997).

3.1 Elicitation of BN structure

In this process, variables and relationships between them should be determined.

First, select variable set. It is important to limit the number of variables. So, it is necessary to choose important variables which are

(1) Target variables: or Query variables, they are outputs of net and what we want to know.

(2) Observation variables: or Evidence variables, they are inputs of net and used to reason states of query variables.

In the relationship between cause and effect, there are two cases:

(1)Multi-causes, one effect.

(2)One cause, multi-effects.

According to the Bayesian formula

$$P(X_1 \ldots X_n) = P(X_n \mid X_{n-1}, \ldots, X_1)P(X_{n-1} \mid X_{n-2}, \ldots, X_1) \ldots P(X_2 \mid X_1)P(X_1) = \prod_{i=1}^{n} (X_i \mid Parents(X_i))$$

we can find that the right sequence of adding nodes is:

(1)Add symptom nodes.

(2)Add disease nodes that be influenced directly by symptom nodes.

(3)Repeat the above two steps until all nodes are added.

3.2 Elicitation of Condition Probability Table

In most of the cases, each relationship between an influenced node and its parents were estimated separately from different sources of data or—in a few cases—given by experts. The conditional probabilities were calculated from the sources in different ways (Kristian Kristensen etc.2002).

In our model, conditional probability table is obtained through expert knowledge, existing certainty factor model and parameter learning.

3.2.1 Elicitation of CPT from parameter learning

Suppose there is much disease diagnosis data, in which there exists the value of symptom node and diseases node(Table 1). S_i=f represents this symptom is found and S_i=nf represents it is not found. D_i=h represents this disease have happened and D_i =nh represents it did not happen. The amount of data record in the table is m.

Table 1. Some data record about symptom and diseases

label	S_1	S_2	…	S_p	D_1	D_2	…	D_q
1	f	nf	…	f	h	nh	…	nh
2	nf	f	…	f	nh	h	…	nh
….	…	…	…	…	…	…	…	…
m	a	a		a	b	b		b

According to the table, we can achieve the following probability.

（1）prior probability

Assume the amount of $S_i = true$ in the table is *r*, $P(S_i = true) = \frac{r}{m}$

（2）conditional probability

Assume the amount of ($S_i = true$ and $S_j = true$) is *s*, the amount of $S_j = true$ is *t*, $P(S_i = true \mid S_j = true) = \frac{s}{t}$

3.2.2 Elicitation of CPT from domain expert

In this process, the state and qualitative probability of each variable should be determined. This can be obtained by domain expert and literature.

Humans tend to think in categories ("likely", "unlikely", etc.) rather than in terms of exact probability (Chard T.1991). So the transformation from qualitative probability to quantitative probability is necessary, which was achieved by consulting domain expert repeatedly. The corresponding relationship can be represented by binary group <Qualitative Probability, Quantitative Probability >, for example, <always, 0.99>, <often, 0.78>,etc.

3.2.3 Elicitation of CPT from certainty factor model

There exists some knowledge in certainty factor model. In order to use this part of knowledge in Bayesian network, some actions should be taken(F.trai.1996, Kevin B.Korb, Ann E.Nicholson.2006).

In the certainty factor model, the knowledge given by domain expert is rule-based, and measurement for the belief is certainty factor, that is:

IF A Then $B : CF(B \mid A)$

Definition of certainty factor (CF):

$$CF(B \mid A) = \begin{cases} \dfrac{P(B \mid A) - P(B)}{1 - P(B)}, & if \quad P(B \mid A) > P(B) \\ 0, \quad if & P(B \mid A) = P(B) \\ \dfrac{P(B \mid A) - P(B)}{P(B)}, & if \quad P(B \mid A) < p(B) \end{cases}$$

However, in the Bayesian network, uncertainty is measured by probability. So, in order to construct Bayesian network, it needs to transform CF to probability(Nevin Lianwen Zhang.1996, Wang ronggui etc.2004). From above formula, we can obtain:

$$P(B|A) = \begin{cases} CF(B|A)(1 - P(B)) + P(B), if\ CF(B|A) \geq 0 \\ (CF(B|A) + 1)P(B), if\ CF(B|A) < 0 \end{cases}$$

So, in order to get probability, it needs to know *P(B)*, which is prior probability of node *B*.

P(B) can be obtained from domain expert, literature, or existing data. If not, assume *P(B)*=0.5,which represents ignorant.

If P(B)=0.5,the formula is as follows:

$$P(B|A) = \begin{cases} 0.5 \times CF(B|A) + 0.5, if\ CF(B|A) \geq 0 \\ (CF(B|A) + 1) \times 0.5, if\ CF(B|A) < 0 \end{cases}$$

The transformation from CF model to Bayesian network brings two advantages. One is that the relationship between variables can be demonstrated visually and intuitively, the other is that more information, namely the probability, can be achieved from the Bayesian net.

3.3 The developing of Bayesian network tool

In order to implement this model, a Bayesian network tool was developed. This tool includes two components. One is the Bayesian building component, the other is Bayesian reasoning component. Through the building component, the structure and CPT parameters, namely the knowledge base, can be constructed visually, which are stored in a XML file. Through the reasoning component, the posterior probability of a node can be computed.

4. AN EXAMPLE

In order to verify the disease diagnosis model, an example about cow disease diagnosis is introduced, which is from an existing certainty factor model. Table 2 shows the detail. The column of S represents symptom nodes, D representing disease nodes, CF representing certainty factor, P representing corresponding probability, IP representing inhibition probability (noisy probability).

In this model, we assume the prior probability of symptom nodes and disease nodes is 0.5

Table 2. Part of data from a cow disease knowledge base of certainty factor model

S	D	CF	P	IP
tbsz21	w13	0.25	0.625	0.375
bcs11	w13	0.15	0.575	0.425
Ydyc02	w13	0.2	0.6	0.4
sz05	w13	0.2	0.6	0.4
tbcz12	w13	0.2	0.6	0.4
sz02	w14	0.15	0.575	0.425
ydyc19	w14	0.3	0.65	0.35
bcs21	w14	0.2	0.6	0.4
ydyc02	w14	0.2	0.6	0.4
tbcz12	w14	0.15	0.575	0.425
tsjc18	w15	0.3	0.65	0.35
tbcz20	w15	0.2	0.6	0.4
tbsz15	w15	0.3	0.65	0.35
tbcz13	w15	0.1	0.55	0.45
bcs11	w15	0.1	0.55	0.45

According to the noisy-or assumption and data from Table 2, we can get conditional probability table of the Bayesian network about cow disease diagnosis.

From domain knowledge and table 3, the structure of cow diseases diagnosis can be obtained, which is constructed by the building component. See Fig.3.

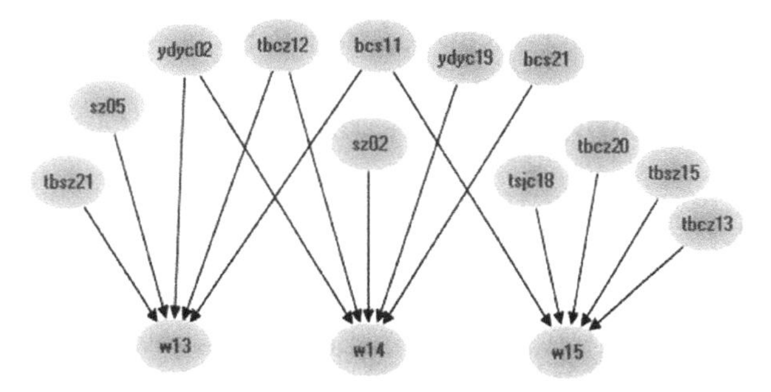

Fig.3 The Bayesian structure of cow disease diagnosis

The posterior probability of a node can be computed by reasoning component. In practice, the reasoning result is conformed with the domain knowledge and existing certainty factor model.

5. CONCLUSIONS

BN is a strong tool for representing and dealing with uncertain knowledge. There exists a lot of uncertainty knowledge in crop or animal disease diagnosis. So it is natural to use BN to build disease diagnosis system. However, in general the data for disease is scarce, so a simple and effective Bayesian network is needed.

While building the two-layered BN, Noisy-or model and transformation from CF to probability are used to decrease network scale and simplify the network structure.

In running the cow disease diagnosis system, we find that the reasoning result is conformed with the solution given by domain expert and the existing certainty factor model, which proves that it is effective to use Bayesian network to represent and deal with uncertain knowledge in disease diagnosis.

ACKNOWLEDGEMENTS

Funding for this research was provided by National "863"project "Research and application of agricultural knowledge grid" (No.2006AA10Z245-2)and China National "863"project " Research and application of precision working system for maize" (No.2006AA10A309) .

REFERENCES

Chard T.Qualitative probability versus quantitative probability in clinical diagnosis: a study using a computer simulation. Med Decis Making. 1991 Jan-Mar;11(1):38-41.

David J.Spiegelhalter. Bayesian Analysis in Expert Systems, Statistical Science, 1993.Volume 8, Issue 3: 219-247.

E.Charles,J.Kahn, etc. Construction of a Bayesian network for mammographic diagnosis of breast cancer, Comut.Biol.Med,1997:19-29.

F.trai..A Bayesian network for predicting yield response of winter wheat to fungicide programs, Computers and electronics in agriculture.1996: 111-121.

Kevin B.Korb, Ann E.Nicholson. Bayesian Artificial Intelligence, CRC Press.2006:.225-260

Kristian Kristensen etc, The use of a Bayesian network in the design of a decision support system for growing malting barley without use of pesticides, Computers and Electronics in Agriculture,2002(33):197-217

Nevin Lianwen Zhang. Exploiting causal independence in Bayesian network inference, Journal of artificial intelligence, 1996: 301-328.

P.J.F Lucas. Bayesian network modeling through qualitative patterns. Artificial Intelligence, 2005: 233-263.

P.J.F Lucas. Certainty-Factor-Like structures in Bayesian belief networks, Knowledge-based systems,2001: 327-335.

P.Larranaga, S.Moral. Probabilistic graphical models in artificial intelligence. Applied soft computing.2008:1-18.

Radim Jirousck. Constructing probabilistic models, International journal of medical informatics 1997(45): 9-18.

Wang ronggui etc, From Certainty Factor Model to Bayesian Network. computer science ,2004,31(10).

AN EARLY WARNING SYSTEM FOR FLOUNDER DISEASE

Bin Xing[1], Daoliang Li[1,*], Jianqin Wang[1], Qingling Duan[1], Jiwen Wen[2]

[1] *College of Information and Electrical Engineering, China Agricultural University, Beijing 100083, China*

[2] *Economics and Management Department, Beijing Forestry University, Beijing, P. R. China 100083, China*

[*] *Corresponding author, Address: College of Information and Electrical Engineering, China Agricultural University, Beijing 100083, China, Tel: +86-10-62736764,Fax:+86-10-62737741, Email: li_daoliang@yahoo.com*

Abstract: With the constant expansion of the scale and mismanagement in aquaculture,the diseases of flounder occur more and more frequently than before, which has brought great economic losses to fish farmers. For the sake of the problem described above, based on a great number of surveys, the early warning theory of flounder disease, the analysis of the outbreak and development of diseases and the relationship between disease and factors, the logic process of the early warning for flounder disease was confirmed. It consists of five parts: specifying the target, searching for the source, distinguishing the sign, predicting the degree and eliminating the menace. Using the expert survey method the early warning indexes which affect the normal life of the flounder and calculated the range of the water environment factors were also confirmed. Finally, an early warning system was implemented, which can reduce the damage from the flounder disease.

Keywords: flounder disease, predict, early warning, artificial neural network

1. INTRODUCTION

In recent years, flounder has become the most important aquaculture species in China's coastal areas. However, with the constant expansion of the scale and mismanagement in aquaculture, the diseases of flounder occur

Please use the following format when citing this chapter:

Xing, B., Li, D., Wang, J., Duan, Q. and Wen, J., 2009, in IFIP International Federation for Information Processing, Volume 294, *Computer and Computing Technologies in Agriculture II, Volume 2*, eds. D. Li, Z. Chunjiang, (Boston: Springer), pp. 1011–1018.

more and more frequently than before, which has brought great economic losses to fish farmers, and it has threaten the stabilization of aquiculture and sustainable development (Liu, 2006). The arbitrariness of China's aquaculture technology and its current management lead to the spread of disease, decline product quality, and even affect the food safety. In order to solve these problems many experts and scholars have done a lot of researches in the aquaculture disease diagnosis and treatment. With the guidance of knowledge-based diagnosis reasoning theory, J.W.Wen developed fish disease diagnosis expert system and proposed a theory of fish disease diagnosis knowledge conceptualization, the description of diagnosis problems, diagnosis knowledge representation, the construction and solution of diagnosis, the development of diagnosis system and so on(Wen, 2003). Wei Zhu developed an expert system for fish disease diagnosis with the method of CBR & RBR that can make accurate diagnosis (Zhu, 2006). Xiaoshuan Zhang developed an intelligent decision support system, which presents the effort to apply evolutionary prototyping model in the intelligent decision support system for fish disease and health management (Zhang, 2008). However, the majority researches in aquaculture disease are relatively for diagnosis and disease treatment, and do not pay much attention on the early prevent, which lead to fish disease management lagging behind.

So in order to reduce the crisis and loss, it is very important to develop an early warning system to monitor the water environment, examine fish disease and take treatment measures as early as possible. The aim of this paper is to design and implement an early warning system for flounder disease to reduce the damage from the disease. In this paper, based on the analysis of factors that affect flounder, eleven indexes were selected, and using the expert survey method the range of the factors were also calculated. Using expert knowledge and neural networks and genetic algorithms, status early warning and trending early warning model for water environment was established. Based on the method of case-based reasoning, we established symptom early warning model for flounder disease. Finally, an early warning system for flounder disease was implemented.

2. ANALYSIS OF EARLY WARNING SYSTEM

2.1 The logic process of the early warning system and the early warning index

The process of the implement of the early warning system is shown in Fig 1. It consists of five parts: specifying the target, searching for the source, distinguishing the sign, predicting the degree and eliminating the menace.

Each of the process is indispensable, so, analysis and design this early warning system should keep to the early warning logic process. As far as flounder are concerned, flounder diseases are the target that we should early warn. Water environments are the source which can result in flounder diseases directly, such as dissolve oxygen, water temperature and so on. Distinguishing the sign of flounder diseases is to determine the relationship between symptom and disease. After the analysis of the water environments and symptom for flounder, we can predict the degree of the flounder disease and take measures to eliminate the menace.

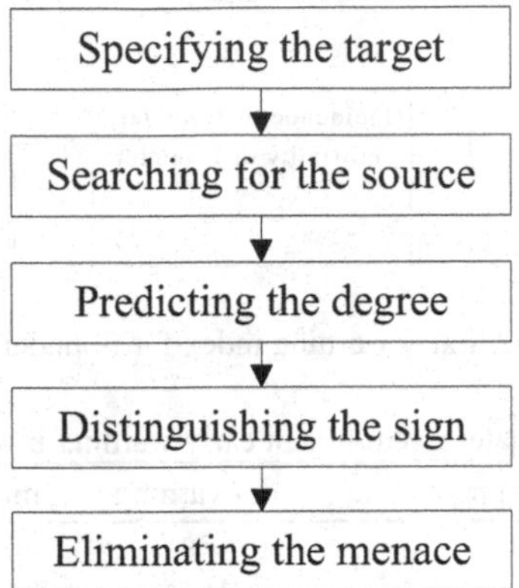

Fig1. The logic process of the early warning system and the early warning index

According to the logic process of early warning, the early warning index for flounder diseases can be divided into warning situation index, warning omen index and warning source index (Wu, 1999). After making consultation from the aquaculture disease experts and the technician of the intensive aquaculture factory, we chose the important and easy monitoring index and removed the index that is hard to monitor and not very important to flounder. The index system of flounder disease can be described as Fig 2. According to the consultation of the domain experts, the interval of no-warning, mid-warning, bad-warning for each early warning index were established. The equation confirming of the border values is

$$V = \frac{\sum E(w_i) * a_i}{N} \quad (1)$$

Where: $E(w_i)$ is the weight of the experts, and it depends on the Qualifications, age and employment time of the expert. a_i is the value set by each expert. N is the number of the domain experts. V is the border value of each water environment early warning index that the domain experts give. Based on the equation, we can calculate the border values, and some of the border values are as follows in table 1.

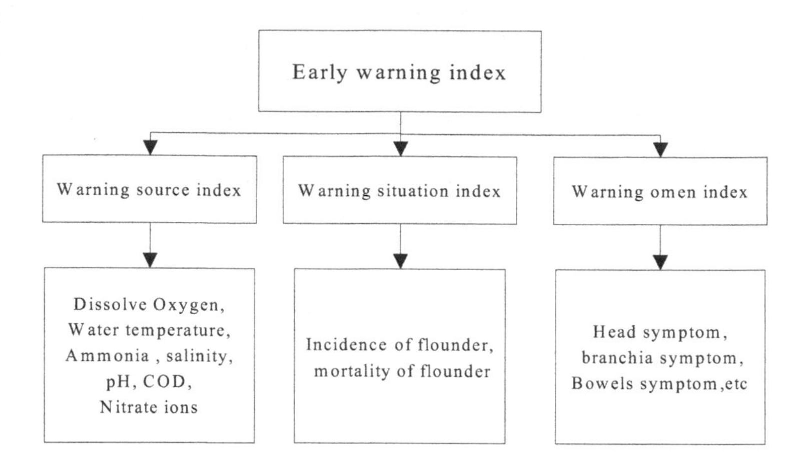

Fig2. Early warning index for flounder disease

Table 1. The interval of the water environment early warning index

Water environment early warning index	no-warning	mid-warning	bad-warning
Dissolve oxygen	>6	(3,6)	(0,3)
Water temperature	(11,24)	(24-26) or (8-11)	>26 or <8
Ammonia	<0.5	(0.5,1)	>1
Salinity	(15,32.5)	(32.5,40) or (9.5,15)	>40 or <9.5
pH	(7.0,8.8)	(6.5-7)	<6.5
Surface damage rate	(0, 1%)	(1%, 3.7%)	>3.7%
Mortality of ascites	(0, 0.21%)	(0.21%,0.48)	>0.48%
Incidence of ascites	(0, 0.75%)	(0.75%, 5.5%)	>5.5%
Mortality of white spot disease	(0, 1.9%)	(1.9%, 3.78%)	>3.78
Incidence of white spot disease	(0, 2.1%)	(2.1%, 4.85%)	>4.85
⋮	⋮	⋮	⋮

2.2 Demand analysis of the early warning system

The function demand can be described as follows: first, this system can implement the water environment status early warning and trending early warning, non water entrainment early warning and symptom early warning. The early warning results can be informed to the users through website, short message, Speaker Systems and so on. Second, the system has the diagnosis function for the flounder disease, and can gives the diagnosis results and prevent treatment to the users. Third, the function of query, update, insert and delete of all sorts of information can be implemented in the system. There are three kinds of users for this system, flounder farmers, domain experts and administrators, and each kind of users has different competence.

Flounder farmers can make early warning, enquiry the history information of early warning. As for domain experts, besides the competence that flounder farmers have, they can query and update the diseases cases database and knowledge database. Administrators can query and update any kind of information.

3. DESIGN OF EARLY WARNING SYSTEM FOR FLOUNDER DISEASE

3.1 Design of function module

According to the analysis of function demand and the consultant to the domain expert, the system can be divided into three parts: early warning module, knowledge browse module and system maintenance module. The details of each module can be shown in Fig 3. The most important module is early warning module, which contains four parts: water environment status early warning, water environment tendency early warning, non water environment early warning and symptom early warning.

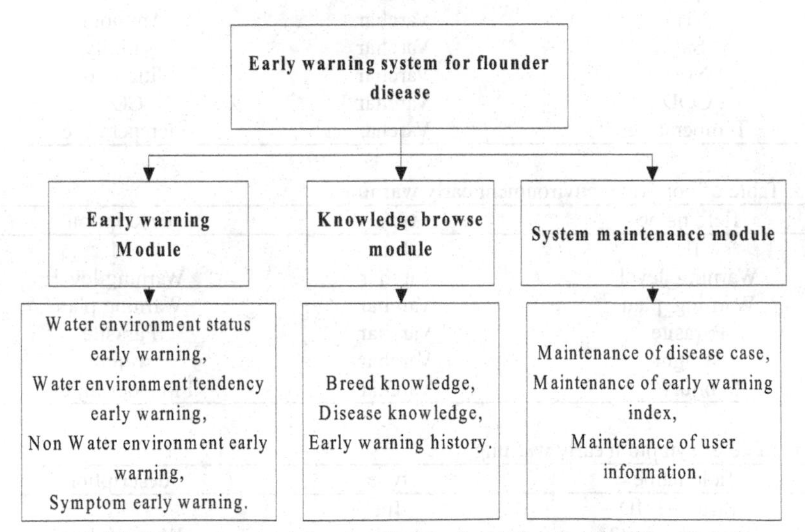

Fig3. Function module of early warning index for flounder disease

As to water environment status early warning, according to user's input and the real time water quality data from the sensors, the border values of each index from the table given above can be searched, and the single-factor early warning level could be calculated. After that, using the knowledge of domain experts, multi-factor early warning level could also be confirmed.

Combine neural networks and genetic algorithms, the next half hour index for each water environment can be predicted. And then, using the same method used in the status early warning to confirm the early warning level in the future and water environment tendency early warning can be achieved. If the Fish farmer inputs the symptoms and the mortality of the flounders, first the disease of the flounders can be diagnosed, and then according to the mortality the symptom early warning level can be calculated.

3.2 Design of databases

In this system, three tables were designed including: water environment early warning, non water environment early warning, and symptom early warning, publish of the warning level, symptom and disease cases. Some main database tables are show as follows in table 2 to 4.

Table 2. Table of Water environment early warning:

Field name	Type	Description
ID	Int	id
Warning_level	varchar	Warning level
Warning_plan	Varchar	Warning plan
DO	Varchar	Dissolve oxygen
pH	Varchar	pH
NH3	Varchar	Ammonia
Salt	Varchar	salinity
NO3	Varchar	Nitrate ions
COD	Varchar	COD
Temperature	Varchar	temperature

Table 3. Table of *n*on water environment early warning

field name	type	description
ID	Int	id
Warning_level	varchar	Warning level
Warning_plan	Varchar	Warning plan
Parasite	Varchar	Parasite
Weight	Varchar	weight
injury	Varchar	Flounder injury

Table 4. Table of symptom early warning

field name	type	description
Warning3_ID	Int	id
Warning_level3	varchar	Warning level
Warning_plan	Varchar	Warning plan
Disease_id	Varchar	Disease id
Disease_rate	Varchar	Incidence
Death_rate	Varchar	mortality

4. IMPLEMENT OF THE EARLY WARNING SYSTEM

According to the theoretical contents and the analysis and design of early warning for flounder diseases described above, the early warning system was implemented. The system is developed based on MVC, which has better reuse and can be expanded and revised more easily than before. This early warning system is developed by java language, so it can be run in any platform. Some interfaces are as follows in Fig.4. The first three figures are the input interfaces, which include water environment early warning, non water early warning and symptom early warning, and the fourth figure is the early warning results for the flounder disease.

Fig 4. Interfaces of early warning system for flounder disease

5. DISCUSSION AND CONCLUSION

Using the knowledge of domain experts, the interval of no-warning, mid-warning, bad-warning for each water environment early warning index was confirmed. Diagnosis and treatment and early warning knowledge database were established. A case database was constructed for flounder diseases. All these knowledge can provide data for the effective early warning.

The system used the expert knowledge to do early warning on water environment, non-water environment and symptoms. Trend early warning on the water environment by use of neural networks and genetic algorithms, and confirmed symptoms and diseases using case-based method, and it provided the effective decision support for a diagnosis and treatment and prevention on the flounder diseases.

The system is limited because only the influence factors that affect flounder disease in intensive aquaculture mode were considered and in other mode this system may not suitable for disease early warning. So the next step is to develop a common early warning system for flounder disease which can be adapted to any early warning for flounder disease according to the different aquaculture mode.

ACKNOWLEDGEMENTS

The research was funded by National Key Technology R&D Program (2006BAD10A02-05). We are grateful to the help from many domain experts from Aquaculture Department of college of Tianjin agricultural university, and we also would like to thank all members of our lab.

REFERENCES

Jiwen Wen. A Knowledge-based Fish Diseases Diagnosis Reasoning System. China agricultural university, 2003(in Chinese)

K. M. Saridakis, A. J. Dentsoras. Integration of fuzzy logic, genetic algorithms and neural networks in collaborative parametric design. Journal of Advanced Engineering Informatics 2006, 20:379-399

Wei Zhu. Call Center Oriented Intelligent System for Fish Disease Diagnosis. China agricultural university, 2003(in Chinese)

Xiaoshuan Zhang, Zetian Fu, Wengui Cai, Dong Tian, Jian Zhang. Applying evolutionary prototyping model in developing FIDSS: An intelligent decision support system for fish disease/health management. Journal of Expert Systems with Applications, 2008, 4:1-13

Y. X. Wu, G. M. Zhou. Early-warning indicators of regional forest resources, Journal of Zhejiang Forestry College, 1999, 16(1): 14-19, (in Chinese)

Yanqing Duan, Zetian Fu, Daoliang Li. Toward developing and using web-based tele-diagnosis in aquaculture. Journal of Expert Systems with Applications, 2003, 25: 247–254

Zhenhua Liu, Zhaojun Teng. Prevention on Ciliates Disease for Flounder. Journal of Scientific fish, 2006, 5:118-120(in Chinese)

DETERMINING PERCENTAGE OF BROKEN RICE BY USING IMAGE ANALYSIS

H. Aghayeghazvini [1,*] , A. Afzal [1] , M. Heidarisoltanabadi [1] , S. Malek [1] , L. Mollabashi [2]

[1] *Esfahan Agricultural Research Center, Engineering Department, Esfahan, IRAN*

[2] *Esfahan University of Technology, Faculty of Physics, Esfahan, IRAN*

[*] *Corresponding author, Address: Esfahan Agricultural Research Center, Engineering Department, post box 81785-199 Esfahan, IRAN, Tel: +98-311-7760061,Fax: +98-311-7757022, Email: Aghayeghazvini@yahoo.com*

Abstract: During the rice milling process, a huller system is used to remove rough rice hull, then abrasive or frictional mill product whitened rice. In such machines adjustment and control, operation is important, because breakage percentage of rice is one main factor in determination of quality. Today's image processing techniques has become an increasingly popular and cost-effective method. The estimation of rice quality by image processing as a tool to determine broken percentage is the aim of this study. The images were acquired by a digital camera (1944 x 2592 pixels) from grains spread on corrugated plate, and then analyzed by scion software. Statistical analysis showed a significant correlation (greater than 0.9) between the results obtained from the proposed method and conventional method. This work indicates that the digital processing technique can be used for estimating broken grains.

Keywords: milling process, digital image processing, whitened rice, digital camera, conventional method.

1. INTRODUCTION

Rice is the dominant staple food crop in developing countries. One of the most important aspects of rice grain is its milling process. The control of percentage of broken kernels and whiteness (degree of milling) in milled rice is required to minimize the economic loss in rice milling.

Please use the following format when citing this chapter:

Aghayeghazvini, H., Afzal, A., Heidarisoltanabadi, M., Malek, S. and Mollabashi, L., 2009, in IFIP International Federation for Information Processing, Volume 294, *Computer and Computing Technologies in Agriculture II, Volume 2*, eds. D. Li, Z. Chunjiang, (Boston: Springer), pp. 1019–1027.

Milling of rough rice (or paddy) is usually done at about 12-14% wet basis moisture content to produce white, polished edible grain, due to consumer preference. From the economic point of view, the quality of milled rice and its control are of paramount importance since the grain size and shape, whiteness, broken percentage and cleanliness are strongly correlated with the transaction price of rice (Conway et al., 1991). All these factors are closely related to the process of milling, in which rough rice is first subjected to dehusking or removal of hulls and then to the removal of brownish outer bran layer, known as whitening. Finally, polishing is carried out to remove the bran particles providing surface gloss to the edible white portion. A high percentage of broken grains in the milled product or low head rice recovery represent a direct economic loss to the millers.

The degree of milling determines the extent of removal of bran layer from the surface of milled kernels and thus it is very important to measure percentage of broken rice. Hence, in the milling process, the pressure in the milling chamber and the duration of milling must be adjusted to get the maximum output.

The extent of losses during milling depends on many factors, such as variety and condition of rough rice, degree of milling required, the kind of rice miller used, and the operators. Besides rubber roll dehuskers, two types of milling machines, namely, abrasive and frictional types are used for whitening and polishing of grain, respectively. In both types of machines, the degree of milling is also controlled by adjusting the pressure in the milling chamber by means of a spring-loaded counterweight at the discharge outlet. In practice, most control systems for rice milling equipment are essentially based on manual operation (Yadav et al., 2001). Informal contacts with several commercial mills in Esfahan province in central Iran revealed that milled rice quality is regularly monitored manually and visually at approximate time intervals about 1 h due to unavailability of continuous on-line measurement methods. Actual determination of broken percentage, grading, head rice yield (HRY) and milled degree of milling are made by laboratory measurements. The necessary adjustments made by a trained operator, based on visual inspection and the results of laboratory measurements, take effect in a few minutes to produce milled rice with a minimum amount of broken kernels and maximum degree of kernel whiteness. Usually milled rice samples obtained by milling test in the laboratory are supplied to the operator and used as reference for each grade of the degree of milling.

Despite the extensive use of image analysis in manufacturing and medical industries, its applications are almost low existent in grain-based industries. The determination of milled rice quality parameters by image processing techniques will enable regular monitoring of milling operation in an objective manner, and thus allow the operator to quickly react within a few minutes to changes in material properties.

Yadav et al., (2001) used digital image analysis to determine the (HRY), representing the proportion by weighting milled kernels with three quarters or more of their original length, and the whiteness of milled rice on ten varieties of Thai rice. They reported that in case of the whiteness of milled samples, the values provided by a commercial whiteness meter and the mean of gray level distribution determined by image analysis correlated with an R2 value of 0.99.

Fant et al. (1994) determined the gray scale intensity in the digital images of rice samples subjected to various degrees of milling and correlated the mean gray level with lipids concentration on the surface of rice kernels. Liu et al. (1998) used digital image analysis to estimate the area of the bran layer on the surface of rice kernels and correlated with the surface lipids concentration determined by chemical analysis. They reported that the degree of milling could be measured quickly and accurately in terms of the surface lipids concentration in a milled rice sample.

The main objective of the present study was to develop quick and practical techniques that could be used for estimating percentage of broken rice and rice grading based on two-dimensional imaging of milled rice kernels being sampled at regular intervals.

2. APPROACH TO THE PROBLEM

Percentage of broken rice could be expressed relative to the weight of broken rice rather than total weight of milled rice in a simple. In general, it is possible to estimate the weight of the objects having regular shapes based on their dimensional characteristics. Therefore, simple models could be developed for estimating the percentage of broken rice and grading of the milled rice samples using image analysis from the measurements of dimensional features and gray level distribution, respectively, as described in the following sections.

Based on the results of an earlier study for a single variety of rice (Yadav and Jindal, 1998), it was hypothesized that the weight of individual kernels, whether of head rice and broken fraction, is proportional to their respective dimensional features. Accordingly, the weight of a milled rice sample might be expressed as a power function of a composite characteristic dimensional feature derived from the whole and broken kernels in that sample as described in the following sections.

On the whole, the characteristic dimensional features of the kernels could be their length (L), thickness (T) and area (A) based on individual measurements. The weight of rice in the image of a milled rice sample is:

$$W= \rho.T.A \quad (1)$$

Where W is the weight of rice in a milled rice sample, ρ is the density of sample, T is the mean of grain thickness and A is the projected area of rice in an image. Generally, thickness and density are two constant parameters in a milled rice sample.

3. MATERIALS AND METHODS

3.1 Rice Samples

Milled rice samples of one variety, namely, Sazandegi were obtained from the four commercial mills in Esfahan, with moisture contents about 12% dry basis. Four samples of each mill weighting individually about 250 g were taken from each mill and kept separately in polyethylene bags. Then four 20g samples were obtained from each bag. Thus, 16 samples were used for determining the milled rice characteristics.

3.2 Imaging of Rice Grains

A diagram of the imaging system is shown in Fig. 1. It consisted of a lighting unit, a color CCD camera connected to a host Centrino 2GH Toshiba computer. The lighting unit comprised with four circular 32 W fluorescent lamps but under indoor ambient day light condition. The camera (model DSC-T7, Sony Inc., Japan) was equipped with a zoom lens (model Vario-Tessar 3x, F 3.5 and 38–114 mm zoom, Carl ziss Inc., Germany) capable of producing image output with resolution of (1944 x 2592 pixels). The operation of the system was carefully controlled for extracting reproducible features from the captured i Determining Percentage of Broken Rice by Using Image Analysis

mages of milled rice samples through various adjustments. All the captured frames were 8-bit (0–255) gray scale images. Each sample of rice kernels was spreaded manually without touching each other on a background about 100 × 70 cm areas directly under the CCD camera. The dimensional features of individual rice images were extracted using the Scion software version 4.0.3.2 developed at Scion Corporation. The imaging system was calibrated with the help of a triangular shape cardboard having dimensions 46.6-34.9-31.4 mm. The linear measurements on rice grains dimensions were subsequently converted into actual values based on the calibration results.

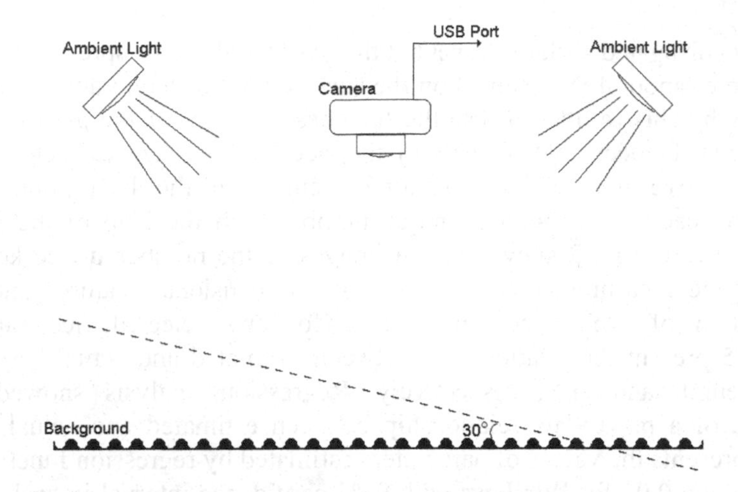

Fig. 1. Schematic diagram of the image analysis system.

3.3 Determination of Broken Rice Percentage

A laboratory rice grader (model TRG 05A, Stake Engineering Co. Ltd., Japan) was used to separate the head and broken kernels in milled rice samples through appropriate adjustment of its settings. The Broken Rice precentag was based on the kernel length equal to or more than 75% of the average length of whole whitened rice kernels, as already defined. It was found that the grader required a minimum of 12 g milled rice for obtaining reproducible values of broken rice percentage differing by less than 1%.

Therefore, representative samples of milled rice weighing 20 g were obtained with a specially fabricated sample divider and separated into head and broken fractions by the laboratory grader for determining broken rice percentage. The use of small test samples was necessary to limit the total number of rice kernels for subsequent image analysis. In view of the imprecise separation of kernels by the laboratory grader, all head and broken kernels in their respective fractions were separated manually with the assistance of imaging system and weighed to compute the actual broken rice percentage of the representative sample. Later all kernels in the head and broken rice fractions were imaged to extract their dimensional features, namely, length (L), mean thickness (D) and area (A). The values of estimated dimensions were then computed for different milled rice samples and related with their actual broken rice percentage. Finally, the broken rice percentage obtained from the laboratory grader was compared with the actual broken rice percentage estimated by image analysis.

3.4 Results and Discussion

Use of corrugated surface prevented the overlap of grains spread on it and grains were randomly positioned on the surface with different angles and by their length. This resulted to use the thickness average of the grains in the calculations. Dimensional features of the rice kernels were extracted from their respective images by separating them from the background and identifying each image with a unique number with the help of the Scion image software. Fig. 2 shows typical images of the numbered rice kernels used for the measurement of characteristic dimensional features and the computation of broken rice percentage for any selected rice sample. Figs. 3–5 present the relationships between estimated and actual based on kernel length and area, respectively. Regression analysis showed the existence of a power-law relationship between estimated and actual data. Table 2 presents the values of parameters estimated by regression function of SPSS version 9.05 for Windows with 95% confidence interval in each case. These results validated the hypothesis that the broken rice percentage of milled rice sample could be estimated with root mean square error (RMSE) of less than 2% from the dimensional features extracted from the images of the rice kernels in their natural rest position.

Fig. 2. An image of rice sample.

The area of kernel images provided the best estimation of broken rice percentage in terms of the RMSE determined from the actual and estimated values for all samples.

The separation of milled rice samples by the laboratory grader often resulted in overlapping head and broken rice fractions. Therefore, the broken rice percentage obtained from the laboratory grader was compared with the broken rice percentage estimation based on the measurement of area of kernels as shown in Fig. 1. These results confirmed the discrepancies encountered in the determination of HRY by the laboratory grader in comparison with manual inspection of kernels by image analysis. These differences in HRY were possibly due to the imprecise separation of the whole and broken kernels by the laboratory grader than the more accurate vision-based measurements.

The HRY determination by the laboratory grader was influenced by its operating settings such as the size of indentation in the rotating cylinder and the inclination angle of the receiving trough. In this study, the operating conditions of the grader were identical for all rice varieties. Also, the proportion of the overlap between the head and broken fractions depended upon the changes in the dimensional characteristics of the rice kernels during the milling operation and the differences in rice varieties.

These results further implied that the percentage of broken rice estimated and thus grading by image analysis and the laboratory grader were indeed related, and could be used for monitoring the milling operation of different rice varieties.

Table 1. Result of regression analysis for estimating percentage of broken rice from dimensional characteristics of rice grain images.

Characteristic dimension	Regression parameters	R2	RMSE (%)
Length	1.06	0.93	0. 7
Percentage of broken rice	1.07	0.92	1.75

The developments of such low-cost machine vision-based techniques that either enhance or replace currently used manual methods may pave the way for rapid assessment, and thus better control of rice milling operations in a conventional setting

It seems likely that similar procedures could be adopted for the inspection of some other crop quality, based on the techniques described in this paper.

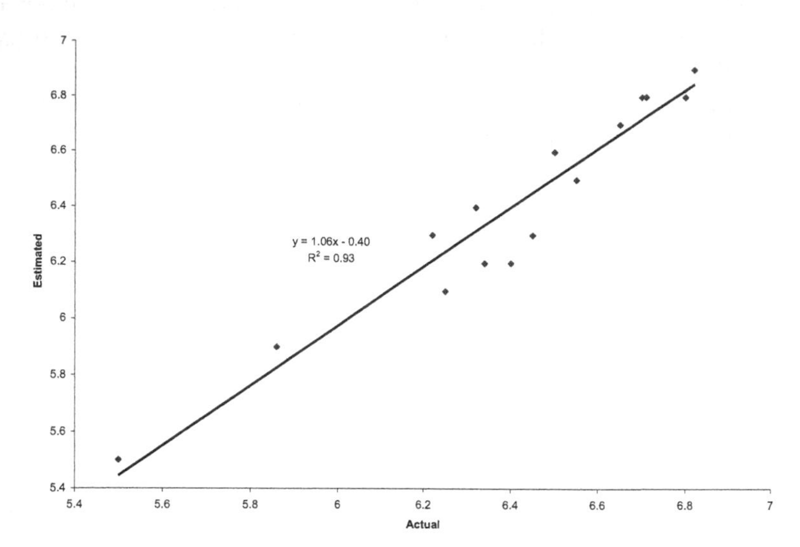

Fig. 3. Comparison of actual and estimated length of rice grains.

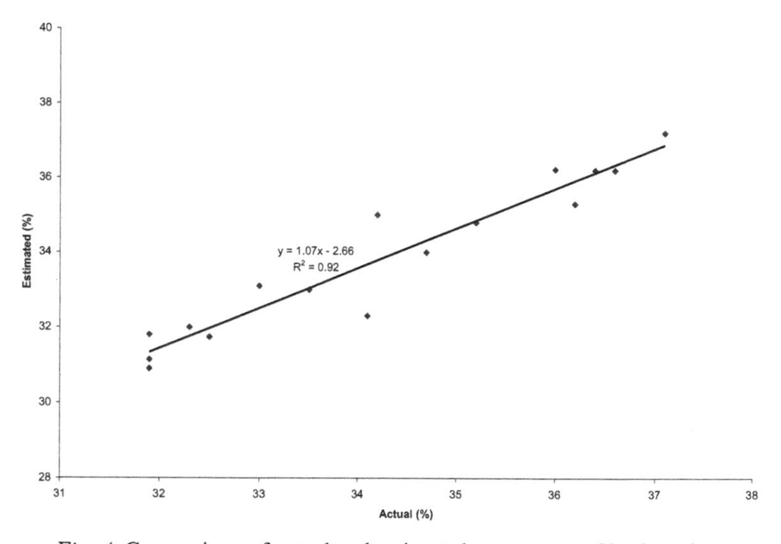

Fig. 4 .Comparison of actual and estimated percentage of broken rice.

4. CONCLUSIONS

Results showed that this project could be used practically for optimizing the controls in mills and rice buying centers, furthermore showed that broken rice percentage and grading of milled rice could be estimated from two-dimensional images of the milled rice kernels in their natural rest position. The broken rice percentage showed a distinct power-law relationship with the characteristic dimension ratio defined in terms of length and area of head and broken rice kernels. The developed empirical relationships for estimating broken rice percentage and milled rice grading could be used for regular monitoring and better control of rice milling operations. More study is required to determine the possible use of this research for other cereals.

ACKNOWLEDGEMENTS

The authors would like to thank the Mr. B.k. Yadav and V.K. Jindal .

REFERENCES

Blasco, J., Alexios, N. and Molto, E. Machine vision system for automatic quality grading of fruit. Biosystem Eng, 2003, 85(4): 415-423.

Forbe, K. Volume estimation of fruit from digital profile image. M. Sc. Thesis of Electronic Engineering. Cape Town University. Cape Town. South Africa, 2000.

Gonzalez, R.C. and Wood R.E Digital image processing. Second edition. Prentice Hall press.New Jersey. USA, 2001.

Yadav, B.K., Jindal, V.K Monitoring milled rice characteristics by image analysis. In: Salokhe, V. M., Jianxia, Z, 1998, (Eds.), Proceedings of the International Agricultural Engineering Conference, Bangkok, Thailand, 7–10 December, pp. 963–971.

Yadav, B.K., Jindal, V.K. Monitoring milling quality of rice by image analysis. Elsevier.Computers and Electronics in Agriculture, 2001, 33:19–33.

A DIGITAL IMAGE METHOD FOR ANALYSIS OF SOIL PORES

Shufang Jiang[1,2], Yaohu Kang[1,*], Zeqiang Sun[1]

[1] *Institute of Geographic Sciences and National Resources Research, Chinese Academy of Science, Beijing P. R. China 100101*

[2] *Graduate University of Chinese Academy of Science, Beijing P. R. China 100039*

[*] *Corresponding author, Address: Institute of Geographic Sciences and National Resources Research, Chinese Academy of Science, Beijing P. R. China 100101 Tel: +86-10-64856516, Fax: +86-10-64856516, Email: kangyh@igsnrr.ac.cn*

Abstract: Measurements of porosity and pore size distribution provide important data on the physical properties of soils. This paper presents a digital image method for the analysis of soil pores. In the analysis of soil thin sections, the image threshold separating pore space from the surrounding solid, as well as the capillary pore size threshold separating air pores and capillary pores, was obtained by analysis of the thin sections of calibration samples where values of total porosity and capillary porosity were pre-determined by traditional soil physics methods (IM). The total porosity and capillary porosity, as well as percentages of pores of a particular size, of all samples of similar soil type can then be determined by these image thresholds in thin section image analysis. The maximum capillary pore size in soils can also be determined. Because the thresholds for the total porosity and capillary porosity are determined based on physical soil characteristics in this method, the error associated with existing methods (caused by subjective threshold estimates) was overcome. Small variations in results proved that this method has good accuracy and is acceptable. Any personal computer and flatbed scanner, along with any commercial remote sensing software (ENVI, PCI, ERDAS, etc.) and Geographic Information System software (ArcGIS, ArcView, SuperMap, etc.) are sufficient to complete the method. In addition, the method can also be used for analysis of pore shapes and arrangements.

Keywords: digital image method, image threshold, soil pore distribution, soil porosity

Please use the following format when citing this chapter:

Jiang, S., Kang, Y. and Sun, Z., 2009, in IFIP International Federation for Information Processing, Volume 294, *Computer and Computing Technologies in Agriculture II, Volume 2*, eds. D. Li, Z. Chunjiang, (Boston: Springer), pp. 1029–1038.

1. INTRODUCTION

Measurements of porosity and pore size distribution provide important data on soil physical properties. Porosity (n) of a given volume of soil is defined as the ratio of pore volume to the total soil volume. In a traditional density measurement method (DMM), porosity is measured indirectly from the bulk density (ρ_b) and particle density (ρ_p) by the equation n = 100 [1 - (ρ_b / ρ_p)] (Hillel, 1980). The result obtained using DMM is the total porosity. It is, however, difficult to measure the percentage of pores with different sizes, such as capillary pores and air pores.

The intrusion method (IM) is another method and is the most popular method used in practice. In IM, total porosity is determined by the volume of liquid that just fills the open spaces and completely saturates a soil sample, divided by sample volume (Lawrence, 1977; Hillel, 1980). The porosity of capillary pores is determined by the field water capacity (the volumetric water content of the soil sample, determined once the saturated soil sample has been left to drain completely, under gravity, for over 8 hours) divided by the sample volume. Air porosity is calculated by the total porosity minus the capillary porosity.

With the rapid development of computational techniques, image analysis has been used to analyze the images of resin-impregnated thin sections (thin sectioning method, TSM) for determining both the total porosities and pore sizes of soils. The procedure of TSM includes: (1) obtaining the images of the resin-impregnated thin sections using optical or electron microscope (Bouma et al., 1977; Chen et al., 1980; Pagliai et al., 1983), X-ray scanning (Brandsma et al., 1999) or combinations of more than one of these techniques (Cousin et al., 2005); and (2) analysis of the images using commercial or specialized computer software to segment soil grains and pores.

The accuracy of TSM mainly depends on the image analysis. One method for image analysis is implemented by counting pixels with the same features according to a predetermined grey scale level using computer software (Murphy et al, 1977). To overcome low resolution image pixels allocated as both solid and void, Vogel and Kretzschmar (1996) implemented a program to segment the images into solid and void using the bi-level-segmentation based on the original grey scale level histogram of the images. Li et al. (2004) photographed thin sections under high intensity ultraviolet light. The photographs are scanned and image threshold operate using the maximum color intensity. The experimental errors are obtained by duplicate estimates from numerous images. Another method for image analysis has been recently developed using multi-spectral image technology. Multi-spectral images are collected from soil thin sections, always using transmitted, reflected or polarized illumination (Protz and VandenBygaart, 1998). Then supervised and unsupervised classification techniques in RS (Remote Sensing) software are used to separate voids and other features into several classes, which are then aggregated to individual features by reference to conventional micromorphology or point counting methods. For the purpose of improving multi-spectral image analysis, Adderley et al. (2002) provided a linkage, obtained by empirical observation, to traditional micromorphological description procedures.

The main principle for the pore size classification is the equivalent pore diameter (diameter of a circle having an area equal to the pore area). Different investigators have subdivided the continuous pore-size distribution of a soil into classes with different criteria based on their primary functions (White, 2006). Pagliai et al. (1983) investigated the pore size distribution of the clay loam soil using 9 classes. Pores with diameter >1.0 mm in equivalent diameter were considered to be macropores (Singh et al., 1991; Li et al., 2002). Velde et al. (1996) classified pores with less than a given width of 2.5 mm as being small. Fox et al. (2004) subdivided the pores into four size classes (50 ~ 200, 200 ~ <350, 350 ~ <500, and ⩾500 μ m). These pore size classification methods were implemented using sizes determined by empirical or given values.

This paper presents a method for analyzing soil pores by obtaining digital images of resin-impregnated thin sections of soils using a personal flatbed scanner. Image thresholds for separating the pore space and solid, and for separating capillary pores and air pores, were determined with commercial RS and GIS software on a personal computer using values of total porosity and capillary porosity of calibration samples measured by traditional soil physics methods. The total porosity and capillary porosity, as well as percentages of pores of a particular size, of all samples of similar soil type can then be determined by image thresholds. In addition, the pores can be divided into air pores and capillary pores using the capillary pore threshold. The maximum capillary pore size in this soil type can also be determined by the capillary pore threshold.

2. METHODOLOGY AND PROCEDURES

2.1 Methodology

Solid and pore spaces impregnated with resin blended with dye display different colors on a thin section made using the TSM. It was found that the color is deeper near the center of the void space while it is lighter near the center of the solid section, and changes gradually in the boundary region from void space to solid in digital images of the thin section taken using a personal flatbed scanner. A pore space may comprise several pixels. For the same pore space, pixel number depends on the scanner resolution, increasing as scanner resolution increases. According to the computer technology for image analysis in RS, each pixel has a digital number (DN), where deeper colors have a smaller DN and lighter colors have a larger DN. In this classification, pixels belonging to pore spaces have smaller DN values and those belonging to the solid have larger DN values in the thin section digital image. If the pixel number for each DN is determined, the percentage pixel number corresponding to each DN relative to the total pixel number of the digital image can be calculated. Next, the cumulative total percentage of the range from the smallest DN to any DN is calculated. If calculated starting from the smallest DN and continuing for increasing DN until the cumulative percentage is approximately equal to the value of soil porosity pre-determined by IM, the DN at this point can be taken as the threshold DN for the image, because the summation at this point is also equal to the total number of void pixels divided by the total number of pixels in the image, i.e. soil porosity. And this image threshold allows the image to be divided into two parts: pore and solid. The resulting binary image contains information specifying only whether the pixel is solid or pore.

In the binary image, each individual pixel, and each patch composed of several connected pixels, is the pores. The larger the number of pixels, the larger the pore, and the smallest possible pore size is 1 pixel. The pore size can be calculated by:

$$D = 2\sqrt{\frac{nA}{N\pi}} \tag{1}$$

where

D =equivalent diameter of a pore (mm)

n = number of pixels in the pore

N = total number of pixels in the digital image

A = area of the digital image (mm^2)

$\pi = 3.1416$

The diameter of every pore can be calculated using Eq. (1). The percentage area (P) of any size pore in the digital image can be calculated by:

$$P = \frac{C_n \times n}{N} \tag{2}$$

Where C_n = number of pores with n pixels.

The capillary porosity can therefore be calculated by the total area of capillary pores divided by the total area of the thin section in the digital image. In thin section image analysis, the cumulative total of the percentage area of pore patches, starting with the smallest patch and adding contributions from increasing patch sizes, is calculated until the summation approximately equals the capillary porosity pre-measured using IM. The cumulative percentage at this point is the total capillary pore area divided by the total area of the thin section, i.e. capillary porosity. At this point, the largest D reached corresponds to the largest capillary pore and the corresponding pixel number (n) is the threshold for capillary pores.

Using image threshold and capillary threshold, for total porosity and capillary pore size, the total porosity, capillary porosity, capillary pore size and percentage of different size pores of similar soil can be analyzed.

2.2 Procedures

According to the methodology described above, a procedure for soil pore analysis is developed as follows:

(1) Take a representative soil sample and pass it through a sieve after air-drying. Place the soil into three PVC plastic cylinders (three repetitions) with diameters of $\geqslant$ 63 mm, heights $\geqslant$ 50 mm and a filter paper at the bottom. These will be used as calibration samples.

(2) Measure soil porosity and capillary porosity of the calibration samples using IM and retain them for further analysis.

(3) Collect undisturbed soil samples and store in similar PVC cylinders to those used in Step (1). These are now the samples to be used for investigation. Avoid soil samples containing many roots or stones.

(4) Oven-dry all samples including the calibration samples from Step (2). Impregnate all samples with Epoxy resin mixed with dye, under vacuum conditions, using TSM. After the samples have solidified, make thin sections for each sample again following the TSM.

(5) Scan the thin sections of the calibration samples with a personal flatbed scanner. Considering scanning time, storage space and quality of the images, a scan resolution of 2400 dpi is recommended. Record the real scan area (A) of each thin section. Be sure to choose the same scan mode in this step, especially the color mode and zooming size. Save the images as tiff files.

(6) Input the images of the thin sections of the calibration samples into RS software. Using the statistical function in the RS software, calculate the percentage pixel numbers corresponding to each DN, relative to the total pixel number of the digital image, and calculate the cumulative percentage starting with the smallest DN and increasing. Output these results in the statistics parameter table.

(7) Find out the summation of the percentage value approximates to the corresponding samples' soil porosity measured in step (2) from the statistics parameter tables. Determine which cumulative percentage from Step (6) best approximates the corresponding samples' soil porosity from Step (2). Record their respective DN values.

(8) Considering sample variance and differences between individual samples, calculate the mean DN value from all the thin sections of the calibration samples. This number is the image threshold for total soil porosity.

(9) Convert original digital images of calibration samples to binary images according to the image threshold, using the classify function in the RS software.

(10) Input the binary images of calibration samples into GIS software, using the reclassify function to identify the patches. Pore patches are assigned value of 1 and colored black. Solid patches are assigned values of NoData and colored white.

(11) Use the region-group function to calculate the pixel number of each pore patch (n), the number of patches with n pixels (C_n), and the total pixel number of images (N) of the binary images of calibration samples. Transfer these results from an attribute table to an Excel file.

(12) Sorting the output Excel file by the field of connection pixel number, calculate the cumulative percentage area of the patches (P_c) in each digital image, starting with the smallest area and continuing for increasing areas until the cumulative percentage approximates the capillary porosity measured using IM. Record the corresponding connection pixels number in each image respectively. Calculate the mean value of these numbers (n_c).

(13) Considering sample variance and differences between individual samples, define n_c as the capillary pore threshold of this batch of samples.

(14) Substitute n in Eq. (1) with n_c, so that the maximum capillary pore size (D_C) of this batch of samples can be calculated.

(15) Input the thin section images of the samples to be investigated into RS software. Using the statistical function in the RS software, output the statistics parameter tables. Next, determine the total porosity (P_t) of the thin section of each investigation sample by using the spreadsheet column showing the cumulative percentage DN, along with the image porosity threshold from the corresponding output table.

(16) Convert original digital images of all investigation samples to binary images, using the image porosity threshold and the classify function in the RS software.

(17) Input the binary images of all the investigation samples into GIS software, record the patch areas as for step (10), and apply the region-group function to output the attribute table to an excel file as described in Step (11). Divide the pores into capillary pores and air pores according to the capillary pore threshold.

(18) Calculate the porosity of capillary pores of each investigation sample using Eq. (2).

(19) Calculate the porosity of the air pores of each investigation sample using the subtraction Pt-P.

(20) Calculate the diameter of any pore from the Excel file of each investigation sample by using Eq. (1).

(21) Calculate the porosity of any size pore from the Excel file of each investigating sample by using Eq. (2).

The binary images can also be used for further analysis of pores, such as pore shape, according to the investigation requirements.

3. VERIFICATION

Three types of soils were selected and analyzed to verify the method presented in this paper. Soil samples with different bulk densities, and consisting of sandy soil, latosolic red soil and fluvo-aquic soil, were prepared using the method in Step (1) with sample numbers of 20, 30 and 28, respectively. For the sandy soil, 2 calibration samples were selected randomly from the samples with bulk density of 1.7 g/cm^3 and 1 calibration sample was selected randomly from samples with bulk density of 1.8 g/cm^3. For the latosolic red soil, each calibration sample was selected randomly from the samples with bulk densities of 1.1, 1.3 and 1.4 g/cm^3. For fluvo-aquic soil, each calibration sample was selected randomly from the samples with bulk densities of 1.2, 1.4 and 1.6 g/cm^3. Relevant soil information is summarized in Table1.

Table 1 Parameters of soil samples

	Sandy soil		Latosolic red soil					Fluvo-aquic soil				
Sites	North China Plain		South of China					North China Plain				
Total number	20		30					28				
Density (g/m^3)	1.7	1.8	1.1	1.2	1.3	1.35	1.4	1.2	1.3	1.4	1.5	1.6
calibration samples	2	1	1	0	1	0	1	1	0	1	0	1
sample repetitions	8	9	5	6	5	6	5	5	6	5	5	4

Figs. 1(a), (b) and (c) show the binary images for one calibration sample of sandy soil, one calibration sample of latosolic red soil and one calibration sample of fluvo-aquic soil, respectively. Total porosity data for each soil, measured using IM and the method presented in this paper, are listed in Table 2.

Analysis found that relative errors associated with the three soil sample types are all < 12%. In sandy soil samples, almost all the relative errors are < 5%. For fluvo-aquic soil, 75% of samples have relative errors < 5.5%. For latosolic red soil, nearly 90% of the samples have relative errors < 5%. The mean relative error for all the soils is < 5.5% (Table 2). Statistical analysis (the paired samples T Test) shows there is no significant difference between the two methods for the three types of soils with different densities at $P < 0.01$.

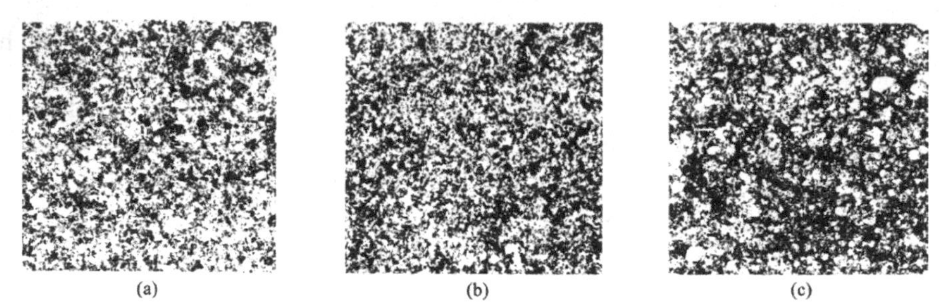

Fig. 1 Soil binary images (pore is black and solid is white) covering 1 cm ×1 cm. (a) Sandy soil; (b) Fluvo-aquic soil; (c) Latosolic red soil.

Table 2 Soil porosity (%) data using IM and the presented method

	Sandy soil		Latosolic red soil					Fluvo-aquic soil				
Density(g/m^3)	1.7	1.8	1.1	1.2	1.3	1.35	1.4	1.2	1.3	1.4	1.5	1.6
Mean_1	32.69	32.65	58.80	59.94	61.03	59.80	59.97	48.57	51.68	54.57	57.04	55.81
SD_1	1.39	1.58	1.68	1.23	1.09	1.68	1.74	1.71	2.59	3.06	2.06	2.50
Mean_2	32.54	32.20	58.82	61.04	62.68	59.79	58.25	48.61	52.17	52.63	56.83	52.96
SD_2	1.72	1.71	2.10	2.93	0.24	2.95	4.8	2.60	1.22	3.56	6.57	3.61
Mean relative error (%)	0.46	1.38	0.03	1.84	2.70	0.02	2.87	0.08	0.95	3.56	0.37	5.11

Note: Mean_1 is the mean value of soil porosity using IM; SD_1 is the standard deviation of soil porosity using IM; Mean_2 is the mean value of soil porosity using images analysis method; SD_2 is the standard deviation of soil porosity using images analysis method.

Porosities of capillary pores for each soil, measured using IM and the presented method, are listed in Table 3. Frequency analysis shows that 80% of latosolic red soils and 90% of sandy soils have relative errors $< 10\%$, while 92% of the fluvo-aquic samples have relative errors $< 15\%$. The paired samples T test shows that none of the results are significant for any types of soil between the two methods at $P < 0.01$ except for fluvo-aquic soil, which has a bulk density of 1.6 g/cm^3. The mean relative errors for all the soils are $< 4.0\%$ except for fluvo-aquic soil, which has a bulk density of 1.6 g/cm^3 (Table 3).

Table 3 Capillary porosity (%) data obtained using two methods

	Sandy soil		Latosolic red soil	fluvo-aquic soil				
Density(g/m^3)	1.7	1.8	1.35	1.2	1.3	1.4	1.5	1.6
Mean_1	19.35	20.08	50.19	31.41	33.08	37.18	39.03	41.39^a
SD_1	0.76	0.71	1.08	2.22	1.66	1.81	2.26	0.85
Mean_2	19.33	20.06	50.15	31.02	32.83	37.15	37.60	37.00^b
SD_2	0.89	1.21	3.50	6.50	4.37	3.25	3.40	0.81
Mean relative error (%)	0.10	0.10	0.08	1.24	0.76	0.08	3.66	10.60

Note: Mean_1 is the mean value of capillaries porosity using IM; SD_1 is the standard deviation of capillaries porosity using IM; Mean_2 is the mean value of capillaries porosity using images analysis method; SD_2 is the standard deviation of capillaries porosity using images analysis method; Different lowercase in same line means significant at $P < 0.01$.

Analysis also shows that the largest capillary pore sizes for the sandy soil, latosolic red soil and fluvo-aquic soil are 0.67, 0.71, and 0.74 mm, respectively. These results are consistent with their physical properties. The above results prove that the presented method is acceptable and can provide results with a good level of accuracy.

4. CONCLUSIONS AND DISCUSSION

In this paper, a digital image method was developed for the analysis of soil pores. The traditional soil physics method (IM) is used to determine the total porosity and capillary porosity, and both porosities are used to obtain image thresholds for pores and capillary pores in image analysis of thin sections of calibration samples. The total porosity and capillary porosity, as well as percentages of pores of any size, can then be determined for further soil samples. The maximum size of capillary pores in soils can also be determined. The small variation of the results proved that this method has acceptable accuracy.

Because the thresholds for the total porosity and capillary porosity are determined based on physical soil characteristics in this new method, the error associated with existing methods (caused by subjective threshold estimates) was overcome. In addition, the personal flatbed scanner is very popular, inexpensive and convenient to use with a personal computer, but also has high resolution. The commercial RS software (ENVI, PCI, ERDAS et al.) and GIS software (ArcGIS, ArcView, SuperMap et al.) are also popular and sufficient for digital image analysis of soil thin sections.

Therefore, the method presented in this paper would be convenient for use in the analysis of soil pores. It is necessary to point out that identical experimental conditions, such as the ratio of resin and dye, the scanning mode, etc., must be used for the same soil type when one uses this method, because the image thresholds are sensitive to color.

ACKNOWLEDGEMENTS

This study is part of the work of the Knowledge Innovation Project (KSCX2-YW-N-003) and the CAS Action Plan for Western China Development (KZCX2-XB2-13), and the Project for 100 Outstanding Young Scientists (Yaohu Kang) supported by the Chinese Academy of Sciences, and the Key Technologies R&D Programs (06YFGZNC00100 and 06YFGZNC06700) supported by Tianjin Municipal Science and Technology Commission.

REFERENCES

Adderley, W.P., Simpson, I.A., and Davidson, D.A. Colour description and quantification in mosaic images of soil thin sections. Geoderma. 2002. 108:181-- 195

Bouma, J., Jongerius, A., Boersma, O., Jager, A., and Schoonderbeek, D. The function of different types of macropores during saturated flow through four swelling soil horizons. Soil Sci. Soc. Am. J. 1977.41:945--950.

Brandsma, R.T., Fullen, M.A., Hocking, T.J., and Allen, J.R. An X-ray scanning technique to determine soil macroporosity by chemical mapping. Soil Tillage Res. 1999.50:95--98.

Chen, Y., Tarchitzky, J., Brouwer, J., Morin, J., and Banin, A. Scanning electron microscope observations on soil crusts and their formation. Soil Sci. 1980.130(1): 49--55

Cousin, I., Issa, O.M., and Bissonnains. Microgeometrical characterisation and percolation threshold evolution of a soil crust under rainfall. Catena. 2005.62:173--188.

Fox, D.M., Bryan, R.B., and Fox, C.A. Changes in pore characteristics with depth for structural crusts. Geoderma. 2004.120:109--120

Hillel, D. Fundamentals of Soil Physics. London: Academic Press. 1980.385pp.

Lawrence, G. P. Measurement of pore sizes in fine-textured soils: A review of existing techniques. J. Soil Sci. 1977.28:527--540.

Li, D.C., Velde, B., and Zhang, T.L. Observations of pores and aggregates during aggregation in some clay-rich agricultural soils as seen in 2D image analysis. Geoderma. 2004.118:191--207

Li, D.C., Velde, B., Delerue, F., and Zhang, T.L. Influences of experimental factors on analysis of pore structure using images of soil sections. Act Pedol. Sin. 2002.39(1):52--57

Murphy, C.P., Bullock, P., and Turner, R.H. The measurement and characterization of voids in soil thin sections by image analysis. Part I. Principles and techniques. J. Soil Sci. 1977.28:498--508.

Pagliai, M., Lamarch, M., and Lucamante, G. Micromorphometric and micromorphonlogical investigations of clay loam soil in viticulture under zero and conventional tillage. J. Soil Sci. 1983.34: 391--403.

Protz, R., and VandenBygaart, A.J.. Towards systematic image analysis in the study of soil micromorphology. Sci. Soils. 1998. 3:34--44

Singh, P., Kanwar, R.S., and Thompson, M.L. Macropore characterization for two tillage systems using resin-impregnation technique. Soil Sci. Soc. Am. J. 1991.55: 1674--1679.

Velde, B., Moreau, E., and Terribile, F. Pore network in an Italian vertisol: quantitative characterization by two dimensional image analysis. Geoderma. 1996.72:271--285.

Vogel, H.J., Kretzschmar, A. Topological characterization of pore space in soil—sample preparation and digital image-processing. Geoderma. 1996.73: 23--38

White, R.E. Principles and Practice of Soil Science: The Soil as a Natural Resource. 4th ed. UK: Blackwell Publishing. 2006.363pp.

DETECTION AND POSITION METHOD OF APPLE TREE IMAGE

Wenhua Mao [1,2*], Baoping Jia [1], Xiaochao Zhang [2], Xiaoan Hub [2]

[1] *College of Food Science & Nutritional Engineering, China Agricultural University, Beijing, P. R. China 100086*

[2] *Institute of Mechatronis Technology and Application, Chinese Academy of Agricultural Mechanization Sciences, Beijing, P. R. China 100083*

[*] *Corresponding author, Address: Institute of Mechatronis Technology and Application, Chinese Academy of Agricultural Mechanization Sciences, Beijing 100083, P. R. China, Tel: +86-10-64882667, Fax: +86-10-64882652, Email:mwh-924@163.com*

Abstract: Apples should be quickly and correctly detected from their surroundings for the apple harvesting robot. The basic color feature was extracted from FuJi apple tree images and analyzed by the statistical analysis method. Accordingly, a new apple detection method was proposed to position the centroid of picking apples. The color difference was used to segment apples from their surroundings. Then the picking apples were chosen by area parameter. After that, the conglutinated apples were segmented by bidirectional scanning line algorithm. Finally, all of picking apples were positioned by their circum-diameter matching algorithm. The experimental result showed that the correct classification rate of apple fruit achieved 90%.

Keywords: apple detection, image processing, color difference, position

1. INTRODUCTION

An apple harvesting robot must have the ability to segment an apple from its surroundings. Machine vision is an available method by which to capture images and to process color, shape and texture information. The past research on detecting fruit can be divided into two categories: local analysis (intensity and color information on the desired object) and shape analysis (fitting of circles or ellipses)(Jiménez et al.,2000, Plebe et al.,2001,

Please use the following format when citing this chapter:

Mao, W., Jia, B., Zhang, X. and Hub, X., 2009, in IFIP International Federation for Information Processing, Volume 294, *Computer and Computing Technologies in Agriculture II, Volume 2*, eds. D. Li, Z. Chunjiang, (Boston: Springer), pp. 1039–1048.

Takahashi et al., 2002, Zhao et al.,2005). The shape analysis is more robust, while the local analysis is faster.

Bulanon et al. (2001,2002) use luminance and color difference transformations of RGB color to recognize apple fruit in images. Stanjnko et al. (2004) apply thermal imaging to detect apples in the late afternoon for the purposes of calculating the fruit load. Zhao et al. (2005) used a combination of redness index ($r = 3R - (G + B)$), texture-based edge detection, and circle fitting in RGB color. Amy et al. (2006) developed a method for segmentation of apple fruit from video via background modeling. Global Mixture of Gaussians (GMOG), is based on the principles of Mixture of Gaussians (MOG), was used for motion-detection applications.

But the apple segmentation based on machine vision contains some avoidless factors. One of the main factors is variable lighting for the environment is outdoor. Another is that of occlusion, including occlusion of apples by leaves and branches of the tree, occlusion by other apples and by trellis poles and wires.

Therefore, the objectives of this research:

1.to develop a novel method for apple segmentation using the color feature to avoid the influence of variable lighting;

2.to develop a new method for position the picking apples and segmentation the occlusion of apple fruit by leaves, branches, trellis poles, and even by other apple fruit.

2. APPLES SEGMENTATION WITH COLOR FEATURE

In general, the ripe Fuji apple is red, and the surrounding of apples is composed of green leaves and weeds, taupe branches, sky and soil, as shown in Fig.1. In order to segment apples from compound background, the software SPSS 13.0 for windows was used to find the distinct color feature of apples.

Fig .1: The captured image of Fuji apples tree

2.1 Color feature analysis

50 frames color images of apple trees were taken by the use of a color digital camera (Panasonic DMC-FZ5) in outdoor natural lighting conditions during 2006/2007 in the Sino-Japanese friendship tour orchard. The captured images included various different conditions, such as the relative distance of camera and apple trees, the lighting (sunny or cloudy), the time (a.m. or p.m.).

25 frames chosen randomly from the 50 frames collected color images were used for the statistical analysis as the follow steps:

Step1. Preprocess

The original image included object (apples) and background (leaves, branches, sky, soil and weed). Their region of interesting were manually extracted from the source images to built the sub-image settings of apples, leaves, branches, sky ,soil and weed, respectively. Then, the RGB triplets of six classes were automatically acquired from the various sub-image settings. The number of captured data of each class was shown in the table 1.

Table 1. The captured data of each class in the RGB color model

Label	Label (background) 4900				
(apple)	branch	leaf	sky	soil	weed
980	980	980	980	980	980

Step2. RGB data Analysis

In the RGB color model, the box plot of each class (Fig.2) was drawn. It could be seen from the Fig.2 that:

1) The *R* index of apple was larger than the other classes'.

2) The *R* index was larger than the *G* and B indexes for apple and soil, but the *R* index of apple was larger than the soil's.

3) The *B* index of sky was larger than the other classes'.

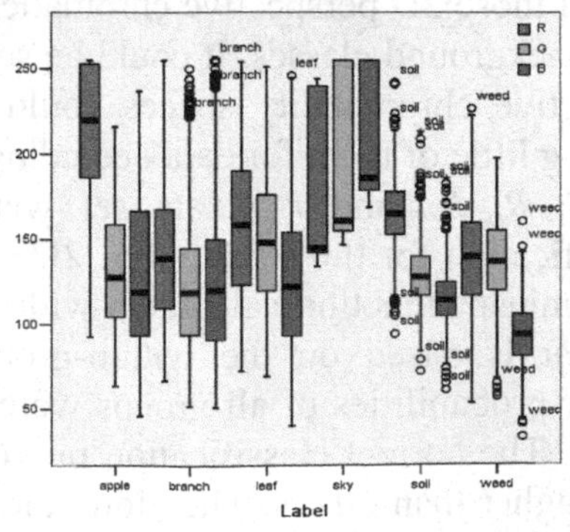

Fig .2: The box plot of six classes in the RGB triple

Step3. 2D data Analysis

The RGB colors were transformed into 2D perspective chromaticity spaces that were mainly emphasized the R index and commonly used in computer vision:

1) the difference indexing:

$$D_{rg} = R - G \tag{1}$$

$$D_{rb} = R - B \tag{2}$$

2) the HS indexing:

$$S = 1 - \frac{3}{R+G+B}[\min(R,G,B)] \tag{3}$$

$$H = \begin{cases} \frac{\pi}{2} + arctg\frac{\sqrt{3}(G-B)}{2R-G-B}, & G \geq B \\ \frac{3\pi}{2} - arctg\frac{\sqrt{3}(G-B)}{2R-G-B}, & G < B \end{cases} \tag{4}$$

3) the ratio indexing:

$$R_{rg} = \frac{R}{G} \tag{5}$$

$$R_{rb} = \frac{R}{B} \tag{6}$$

4) the normalization indexing:

$$r = \frac{R}{R+G+B} \tag{7}$$

$$g = \frac{G}{R+G+B} \tag{8}$$

The scatter plots of those 2D perspective chromaticity spaces (Fig.3) were drawn as apple and background classes. It could be seen from them that: All of those 2D perspective chromaticity spaces could segment apples from background, excluding little of them for data occluding.

The D_{rg}-D_{rb}, H-S, R_{rg}-R_{rb} and r-g data set were used to make the discrimination classification for the selection of 2D color indexing of apple detection. The discriminant function calculated with the enter independents together method, that is based on the within-group covariance, on the assumption that prior probabilities of all groups were equal. The result was shown in the table 2. The correct classification rate (abbr. *CCR*) of D_{rg}-D_{rb} color indexing was higher than others. Therefore, the D_{rg}-D_{rb} color indexing was optimal.

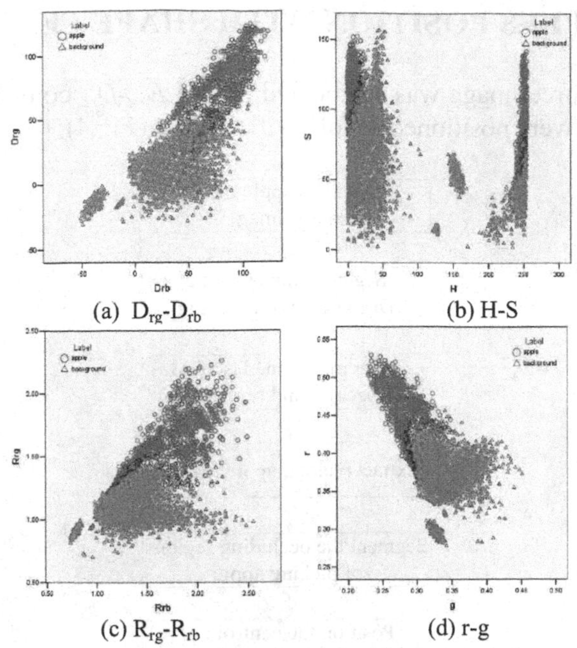

Fig .3: The scatter plots of apple and background classes

Table 2. The CCR of apple and background in 2D color indexing

Color indexing	CCR of apple	CCR of background
D_{rg}-D_{rb}	93.8%	98.2%
H-S	92.2%	86.7%
R_{rg}-R_{rb}	89.7%	97.1%
r-g	93.1%	94.2%

2.2 Segmentation method in optimal color indexing

The nearest neighbor clustering method was used to segment apples from background in D_{rg}-D_{rb} color indexing. If the distance of a pixel point $P(D_{rg}, D_{rb})$ with the apple's initial point $P(Da_{rg}, Da_{rb})$ was less than the distance with the background's initial point $P(Db_{rg}, Db_{rb})$, $P(D_{rg}, D_{rb})$ was the apple class; on the contrary, $P(D_{rg}, D_{rb})$ was the background class.

If $\left\| P(D_{rg}, D_{rb}) - P(Da_{rg}, Da_{rb}) \right\| < \left\| P(D_{rg}, D_{rb}) - P(Db_{rg}, Db_{rb}) \right\|$,

$P(D_{rg}, D_{rb}) \in Apple$; else $P(D_{rg}, D_{rb}) \in Background$.

Where, $P(Da_{rg}, Da_{rb})$ and $P(Db_{rg}, Db_{rb})$ were set the median of D_{rg}-D_{rb} color indexing of apple and background classes, respectively:

$$P(Da_{rg}, Da_{rb}) = P(82, 89)$$

$$P(Db_{rg}, Db_{rb}) = P(10, 26)$$

3. APPLES POSITION WITH SHAPE FEATURE

After the source image was segmented by the D_{rg}-D_{rb} color indexing, the picking apples were positioned as follow flowchart (Fig.4):

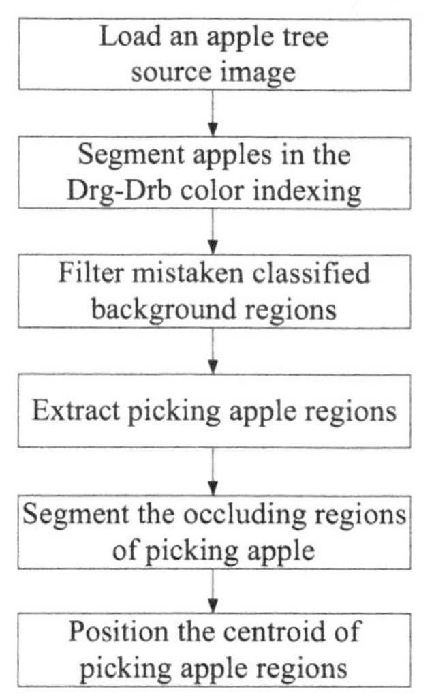

Fig.4: The flowchart of an apple tree image processing

3.1 Filter background noise

The segmentation image was composed of most apples and little background pixels. The area of apple region was much larger than the area of background region. Therefore, the mini-area erasion algorithm was used to filter the background noise.

All object regions were fast labeled and their areas were computed. The area threshold of erasing background region (T_b) was equal to the mean area of all object regions (A_m). If the area of a region was less than T_b, this region was erased from object regions, and its area data also was excluded.

3.2 Extract picking apples

If an apple had a near distance to the harvesting robot, its area was correspondingly larger and less occluding. Therefore, the picking apples were extracted as follow:

1)The edge of apple regions was smoothed by the opening and closing algorithm. The holes in apple region were filled by the hole filling algorithm.

2)The area threshold of picking apple (Ta) was equal to the mean area of all apple regions (Ama). If the area of a region was less than Ta, this region was an unpicking apple, and it was filtered also from object regions.

3.3 Segment occluding apples

An apple may be occluded by leaves, branches, trellis poles, and even by other apple fruit. And the occluding scale, orientation and shape were difference in thousand ways. But the segmentation line of occluding parts had some variation points. The pixel value of a variation point was 0, and the pixel values of left and right points connected with it were 255.

Therefore, the bidirectional scanning line algorithm was used to segment occluding apple regions, which were scanned from the horizontal and vertical direction (Fig.5). If there were three variation points in a scanning line, and all of them were not start or end pixels (*D(x,y)*), then the second variation point was a occluding point of two objects (*C(x,y)*). The segmentation line of two objects was the line connected with two *C(x,y)*.

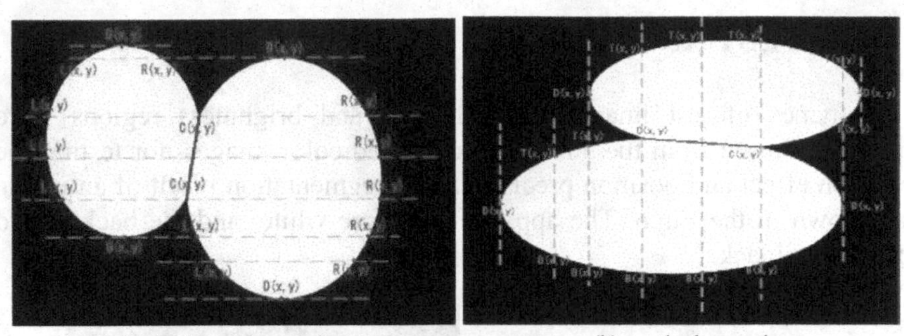

(a) horizontal scanning (b) vertical scanning

Fig.5: The sketch map of the bidirectional scanning line algorithm:

Similarly, the segmented regions of unpicking apples were filtered by the extract picking apples algorithm.

3.4 Position the centroid of picking apples

The shape feature of single apples was similar, they were approximate circle. Therefore, the centroid of picking apple was positioned by the shape feature as follows:

1) The Roundness of picking apples (R) were computed as the followed formula:

$$R = \frac{4A}{\pi L^2} \tag{9}$$

Where, A and L were the area and length of object, respectively.

The single picking apples were selected by the R: if $0.9<R<1$, then object was a rounding apple; on the contrary, object only was a missing apple.

The centroid of rounding apple(xc,yc)was positioned as the formula:

$$x_c = \frac{1}{A} \sum_{(x, y \in R)} x \tag{10}$$

$$y_c = \frac{1}{A} \sum_{(x, y \in R)} y \tag{11}$$

2) The centroid of missing apple was positioned by the circum-diameter matching algorithm. The circum-diameter (D_c)was computed as the formula:

$$D_c = 2 \times \max(\| P(x, y) - P(x_c, y_c) \|) \tag{12}$$

Where, $P(x,y)$ was an edge point of object. If the length of two edge points was furthest matching with the D_c, this line was the central axis of apple. Its' midpoint was the centroid of missing apple.

4. TEST AND RESULT

10 frames of test images with shading and brightness regions were randomly chosen from the 50 frames collected color images, for testing the detection effect and position precision. The segmentation result of an image was shown in the Fig.6. The apple pixels were white, and the background pixels were black.

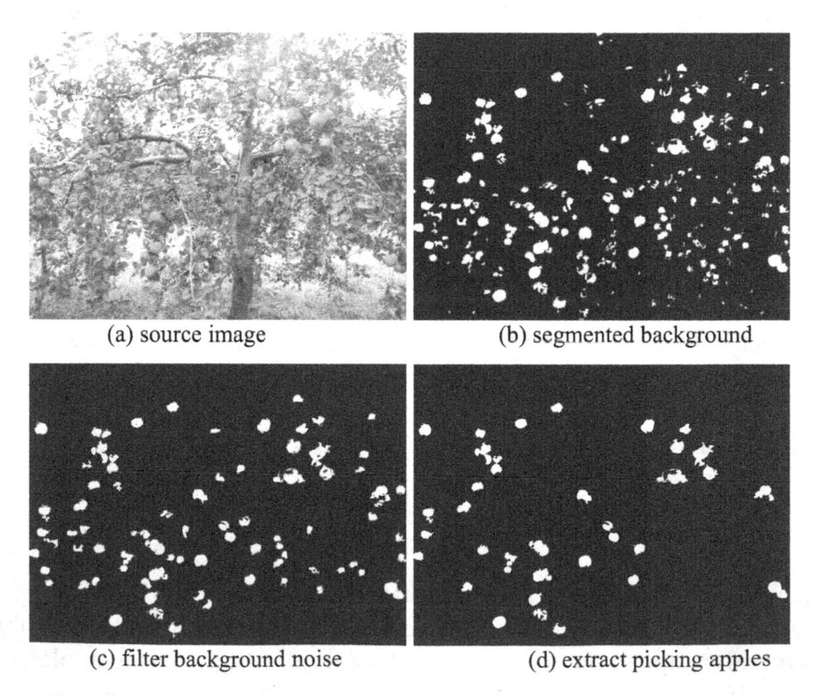

(a) source image (b) segmented background

(c) filter background noise (d) extract picking apples

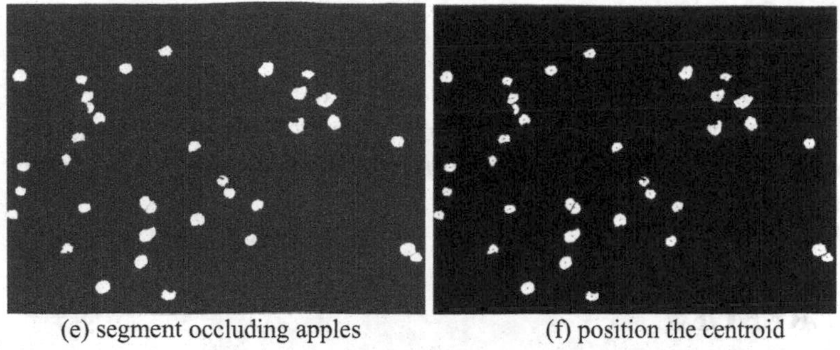

(e) segment occluding apples (f) position the centroid

Fig.6: The result images processed step by step

The result of image processing was compared with the result of eye detection (Table 3). The number of picking apple detected by image processing (abbr. IP) (*Na*) was automatically gained in the above mentioned algorithm. The number of picking apple detected by eye (*Ne*) was manually counted from the source image. The Error (*E*) was computed by the follow formula:

$$E = \frac{|N_a - N_e|}{N_e} \times 100\ \% \qquad (13)$$

Table 3. The compare of image processing and eye detection

Image	Number of picking apple detected by IP	Number of picking apple detected by eye	Error (%)
1	35	32	9.4
2	28	29	3.4
3	24	23	4.3
4	31	30	3.3
5	37	34	8.8
6	24	26	7.7
7	20	22	9.1
8	26	24	8.3
9	30	33	9.1
10	18	19	5.3

5. CONCLUSION

1. Develop an optimal D_{rg}-D_{rb} color indexing for segmentation apples from background;

2. Develop a bidirectional scanning line algorithm for segmentation the occlusion apples;

3. Develop a circum-diameter matching algorithm for position the centroid of occlusion apples.

4. Get the data of number of picking apples with a maximum error 9.4%.

ACKNOWLEDGEMENTS

Funding for this research was provided by the Postdoctors Fund in China and the Ministry of Science and Technology of the P.R.China. The first author is grateful to the China Agricultural University for providing her with pursuing a postdoctor work.

REFERENCES

A. Plebe, G. Grasso. Localization of spherical fruits for robotic harvesting, Machine Vision and Applications,2001, 13 (2): 70-79

A. R. Jiménez, R. Ceres, J. L. Pons. A survey of computer vision methods for locating fruit on trees, Trans. ASAE, 2000, 43(6): 1911-1920

D. M. Bulanon, T. Kataoka, S. Zhang, Y. Ota, T. Hiroma. Optimal thresholding for the automatic recognition of apple fruits, ASAE Paper, 2001,No. 01-3133

D. M. Bulanon, T. Kataoka, Y. Ota, and T. Hiroma. A Color Model for Recognition of Apples by a Robotic Harvesting System,Journal of the JSAM, 2002, 64(5):123-133

D. Stanjnko, M. Lakota, M. Hocevar. Estimation of number and diameter of apple fruits in an orchard during the growing season by thermal imaging, Computers and Electronics in Agr, 2004,42: 31-42

J.Zhao, J. Tow, J. Katupitiya. On-tree fruit recognition using texture properties and color data,IEEE/RSJ Int. Conf. Intell. Robots and Systems, 2005,263-268

L. T. Amy, L. P.Donald, P. Johnny. Segmentation of Apple Fruit from Video via Background Modeling, ASAE Paper, 2006, No. 063060

T. Takahashi, S. Zhang, H. Fukuchi. Measurement of 3-D Locations of Fruit by Binocular Stereo Vision for Apple Harvesting in an Orchard, ASAE Paper, 2002, No. 021102

THE RESEARCH ON THE JUDGMENT OF PADDY RICE'S NITROGEN DEFICIENCY BASED ON IMAGE

Jun Sun[1, 2,*] , Hanping Mao[1] , Yiqing Yang[1]

[1] *Key Laboratory of Modern Agricultural Equipment and Technology, inistry of Education & Jiangsu Province, Jiangsu University, Zhenjiang, Jiangsu Province, China 212013*

[2] *School of Electrical and Information Engineering ,Jiangsu university, Zhenjiang ,Jiangsu Province, China 212013*

* *Corresponding author, Address: School of Electrical and Information Engineering, Jiangsu University, Zhenjiang 212013,Jiangsu Province,China, Tel:(0)13775544650, Fax:051188780311, Email :sun2000jun@ujs.edu.cn*

Abstract: Because of the unreliability judgment of paddy rice's nitrogen deficiency depending on the traditional artificial naked eye, in this article, the way of the paddy rice's nitrogen deficiency examination based on image is put forward, to achieve the precise fast lossless detection and judgment on the paddy rice's nitrogen. Based on the sorting function of SMV, paddy rice leaf's visible images are gathered, the texture features of image are extracted, the RBF nuclear function is chosen, the penalty coefficient C and the regularity coefficient γ are set, and the SVM sorting model is constructed. The recurrence sentencing rate to the training sample achieves 100%. The examination is caught on the test sample, and the accuracy rate of examination recognition achieve 95%, which indicates that the method of paddy rice's nitrogen lossless examination judgment by image is effective and feasible to achieve the precise fast judgment on paddy rice's nitrogen.

Keywords: SVM（Support Vector Machine）, paddy rice, image, nitrogen

1. INTRODUCTION

The nitrogen is one of the most major limit factors which affects the paddy rice growing, and it is the important ingredient which constitutes the

---

Please use the following format when citing this chapter:

Sun, J., Mao, H. and Yang, Y., 2009, in IFIP International Federation for Information Processing, Volume 294, *Computer and Computing Technologies in Agriculture II, Volume 2*, eds. D. Li, Z. Chunjiang, (Boston: Springer), pp. 1049–1054.

crops organic matter. The lack of paddy rice's nitrogen nutrition directly influences the growing of paddy rice, and exerts an adverse influence to paddy rice's quality and the output. Therefore the prompt judgment of the paddy rice's nitrogen whether to lack is more and more needed in paddy rice's growth process. But the traditional naked eyes' judgment has the inevitable human factor. Therefore one intelligent examination judgment method is urgently needed. The lossless detection technology provides one new method for carrying on the rapid, accurate, non-destructive detection on the paddy rice crops nitrogen condition.

Support Vector Machine is the new generation machine learning algorithm which develops on the statistics theoretical basis. When it compares with the neural network method, this algorithm has the merits of the simply structure, the strong pan-ability and so on, and it has not the over-study problem, specially aims at the finite sample situation, in order to get the optimal solution under the existing information. Support Vector Machine transforms the actual problem to the higher dimensional feature space through the nonlinear transformation, in the higher dimensional space, the structure of linear decision function realizes nonlinear decision function of original space decision function, which can guarantee that it has the better disseminate ability(Fang,2006). This article uses the SVM method of Support Vector Machine to set up the model and judge the paddy rice whether to lack nitrogen. At present, in domestic and foreign countries, the crops nitrogen nutrition lossless monitoring and the examination research mainly concentrate on the research of the leaf spectrum index and the remote sensing spectrum index. The literature report has not been seen that uses Support Vector Machine's method to conduct the research to the paddy rice's nitrogen deficiency detection based on the image characteristic, so this article research has the significance.

2. PADDY RICE CULTIVATION EXPERIMENT

The experimental objects are the paddy rice seedlings which come from Zhenjiang suburb farmland. The paddy rice variety is new fragrant superior 80, and the growing cycle is 115 days. The testing site is the big intelligence control greenhouse vent in the Jiangsu University. On August 1, 2007, they are transplanted using the water cultivation method, and the standard nourishment liquid formulation (except the quantity of nitrogen) is compound according to the international paddy rice nourishment liquid formulation. The paddy rice is fixed in the experimental cylinder with the sponge. Row with line distance is 0.3m×0.3m, and each cylinder is a plot.

In order to gain different nitrogen-containing level of the paddy rice leaf blade, in view of nitrogen nutrition in the paddy rice nourishment liquid, it can be divided into 3 levels N1, N2, N3, that is to say, the nitrogen normal (40 mg/L), the nitrogen lacking (0.5×40mg/L), the nitrogen lacking (0.25×40 mg/L), and 4 times repetition are set. Each cylinder has 4 clumps, and at average level each clump has the paddy rice plant number of 2.5. Each cylinder installs 10L nourishment liquid, it examines one time a weak, and the nourishment liquid is replaced each time for every 30 days.

The normal nitrogen-containing sample comes from the N1 level sample, and the nitrogen deficiency sample comes from the N2 N3 levels sample.

3. GATHERING IMAGE CHARACTERISTIC OF PADDY RICE LEAF

The paddy rice leaf's images are gathered on the scene using the high speed image gathering system. To avoid the paddy rice leaf complex background effect, after being picked the paddy rice leaf in the experimental field, it is immediately placed on the experiment table's white paper in the dark awning carries on the image photography. An example image is shown in Figure 1.

The paddy rice leaf image's characteristics are mainly the texture features, color greyscale characteristic and so on. The extraction image characteristic must avoid the object's shape and size to affect the effectiveness of the extraction characteristic.

Fig. 1 paddy rice leaf's original chart

The paddy rice leaf image is processed by filtering the background having independence on leaf. The color characteristics of paddy rice leaf are red (R) , blue (B) , green (G) , chroma (H) , light intensity (I) and saturation (S) value .The other color parameter, such as R/ (G + B) , G/ (R + B) , B (R + G) , G/R , G/B and R/B, can be caught out by Excel software. Researches indicate that paddy rice leaf change obviously with the inner nitrogen change. The red light and green light increase with the nitrogen content rise. The blue light decreases with nitrogen content increasing. Color parameter R , G , G/ (R + B) , B/ (R + G) , G/B , R/B have relation with the nitrogen content, so they can be taken as the parameter on which nitrogen diagnostic based(Xue ,2002;Zhang, 2003;Lu,2000).

4. SVM PATTERN RECOGNITION METHOD

Suppose m dimensions training sample input data xi (i=1,2,3…M) belonging to the category 1 or the category 2 respectively. When the training data has the noise, the feature space cannot be separated linearly generally. In order to permit the indivisibility, it introduces nonnegative relaxation variable ξ_i:

$$y_i(W^T x_i + b) \geq 1 - \xi_i, i = 1,2,... M \tag{1}$$

The combination optimization question becomes:

$$\min \; Q(W, b, \xi) = \frac{1}{2} \| W \|^2 + C\sum_{i=1}^{M} \xi_i^P \tag{2}$$

The constraint condition is:

$$y_i(W^T x_i + b) \geq 1 - \xi_i \text{ i=1,…M} \tag{3}$$

In the formula $\xi = (\xi_1,...\xi_M)^T$, C is the marginal coefficient, P is a norm. Introducing the nonnegative Lagrange multiplier α_i with β_i , may result in

$$Q(W, b, \xi, \alpha, \beta) = \frac{1}{2} \| W \|^2 + C\sum_{i=1}^{M} \xi_i - \sum_{i=1}^{M} \alpha_i [y_i(W^T x_i + b) - 1 + \xi_i] - \sum_{i=1}^{M} \beta_i \xi_i \tag{4}$$

The above formula suppose various variables' deflection is equal to 0, and may obtain the corresponding dual question when it substitutes in the above formula.

$$\max \; Q(\alpha) = \sum_{i=1}^{M} \alpha_i - \frac{1}{2} \sum_{i,j=1}^{M} \alpha_i \alpha_j y_i y_j K(x, y) \tag{5}$$

The constraint condition is

$$\sum_{i=1}^{M} y_i \alpha_i = 0 \quad C \geq \alpha_i \geq 0 \;\; i = 1,... M \tag{6}$$

And K (x, y) is the nuclear function.
The optimal sorting decision function is:

$$D(x) = \sum_{i \in S} \alpha_i y_i x_i^T x + b \tag{7}$$

To the non-linear problem, it can transform the non-linear problem through the nuclear function as the linear question in the higher dimensional space. And then the optimal sorting surface is found in the transformation space(Zhao,2007;Ge,2007). The sorting function becomes:

$$D(x) = \sum_{i \in S} \alpha_i y_i K(x_i, x) + b \tag{8}$$

The diagnosis method is:

$$\begin{cases} \text{Normal} & when D(x) > 0 \\ \text{nitrogen deficiency} & when D(x) < 0 \end{cases}$$

5. MODEL ESTABLISHMENT AND EXPERIMENT

The establishment of recognition model must solve the question of choosing the nuclear function at first, according to the literature, the sorting question can be obtained a better result by using the radial base function. The radial base function can insinuate the nonlinear sample data to the higher dimensional feature space, process the sample data which has the nonlinear relations.

In this experiment the situation of the computer used is: Processor CPU1.40GB, 1G memory, XP operating system, Matlab7.0 edition. The number of paddy rice leaf's images which have been gathered in the research is 200, taking 150 of them as training sample, 50 of them as test sample.

The sample import value are color parameters R , G , G/ (R + B) , B/ (R + G) , G/B , R/B ,Output value is D. D value is 1 when the nitrogen content is normal and D value is 0 when the nitrogen content is lack.

After determining the nuclear function, it needs carry on the optimization to the nuclear function parameter. The radial base function needs take definite parameter as penalty coefficient C and regularity coefficient γ . In this article, the different combinations are attempted and the results obtained are shown in Table 1. At last it determines C=3, γ =15, the accuracy recurrence sentencing rate is 100%, and the accuracy forecast rate is 95%.With the traditional BP neural network contrast, the input layer node number of BP Network selected is 7, and the implicit layer node number is 10, the output layer node number is 1. The recognition results using the SVM algorithm and the BP neural network are shown in Table 2.

*Table 1.*Different parameter choice and test result

C	Gamma	Support vector number	the accuracy recurrence sentencing rate	The accuracy forecast rate
1	1	34	80%	70%
3	15	35	100%	95%
5	20	40	90%	85%
10	25	40	80%	75%
20	30	43	80%	75%

Table.2 Training times and test result reference table

Algorithm	Training time (ms)	Test sample number	Correct judgment rate
SVM	51	50	95%
BP Network	1115	50	85%

6. CONCLUSION

In the recent years, the lossless detection technology has been obtained the widespread attention in the crops nitrogen nutrition diagnosis and the nitrogen fertilizer recommendation. In this article, Support Vector Machine is applied in paddy rice whether to lack nitrogen judgment research in the first time, and the very good judgment results are achieved. The test result has indicated that the SVM algorithm has a big enhancement compared to the traditional BP neural network from the training time and the correct recognition rate, it provides one new method for the nitrogen deficiency judgment of other plants, and it has certain promoted significance.

ACKNOWLEDGEMENTS

Funding for this research was provided by China Postdoctoral Science Foundation under Grant (NO:20070420972); Jiangsu University High-grade Specialty Person Scientific Research Foundation under Grant (NO: 05JDG050); “863”Project(2008AA10Z204). The first author is grateful to the Jiangsu University for providing him with pursuing a postdoctoral degree.

REFERENCES

Fang Ruiming. Induction Machine Rotor Diagnosis using Support Vector Machines and Rough Set[J]. Lectures notes on Artificial Intelligence,2006.Vol: 631～637(in Chinese)

Ge Guangying. Algorithm of vehicle detection and pattern recognition using SVM. Computer Engineering. 2007(6):6-10(in Chinese)

Lu Renfu,Daniel E Guyer, Randolph M Beaudry. Determination of firmness and sugar content of apples using near—infrared diffuse reflectance. Journal of Texture Studies,2000,31:615- 630(in Chinese)

Xu Guili, Mao Hanping, Li Pingping. Extracting Color Features of Leaf Color Images. Transactions of the CSAE. 2002,7:150-154(in Chinese)

Xu Guili,MAO Hanping,LI Pingping. Application Algorithm to Extract Color Images Color and Textures Features. Computer Engineering. 2002,6:25-27(in Chinese)

Zhang Wei,Mao Hanping,LI Pingping, XIA Zhijun. Research on Extracting Color and Texture Features of Plant Nutrient Deficiency Leaves’ Image. Journal of Agricultural Mechanization Research. 2003,4:60-63(in Chinese)

Zhao Jiewen,Hu Huaiping,Zou Xiaobo.Application of support vector machine to apple classification with near—infrared spectroscopy. Transactions of the CSAE.2007, 23(4):149-152.(in Chinese)

DIGITAL IMAGE ANALYSIS OF REACTIVE OXYGEN SPECIES AND CA2+ IN MOUSE 3T3 FIBROBLASTS

Hongzhi Xu[1], Dongwu Liu[1,2,*], Zhiwei Chen[1,2]

[1] *Analysis and Testing Center, Shandong University of Technology, Zibo, Shandong Province, P. R. China 255049*

[2] *School of Life Sciences, Shandong University of Technology, Zibo, Shandong Province, P. R. China 255049*

[*] *Corresponding author, Address: Analysis and Testing Center, Shandong University of Technology, Zibo 255049, Shandong Province, P. R. China, Tel: +86-0533-2781987, Fax: +86-0533-2786781, Email: liudongwu@sdut.edu.cn*

Abstract: Recently, analysis of digital images with confocal microscope has become a routine technique and indispensable tool for cell biological studies and molecular investigations. Because the light emitted from the point out-of-focus is blocked by the pinhole and can not reach the detector, thus only an image of the fluorescence from the focal plane is imaged. In present studies, we use the probes 2′, 7′-dichlorof luorescein diacetate (H2DCF-DA) and Fluo-3 AM to research reactive oxygen species (ROS) and Ca2+ in mouse 3T3 fibroblasts, respectively. Our results indicate that the distribution of ROS and Ca2+ were clearly seen in mouse 3T3 fibroblasts. Moreover, we acquired and quantified the fluorescence intensity of ROS and Ca2+ with Leica Confocal Software. It was found that the quantified fluorescence intensity of ROS and Ca2+ was 123.30.26±8.99 and 125.13±12.16, respectively. Taken together, our results indicate that it is a good method to research the distribution and fluorescence intensity of ROS and Ca2+ in cultured cells with confocal microscope.

Keywords: digital image, confocal microscope, mouse 3T3 fibroblasts, reactive oxygen species, Ca2+

Please use the following format when citing this chapter:

Xu, H., Liu, D. and Chen, Z., 2009, in IFIP International Federation for Information Processing, Volume 294, *Computer and Computing Technologies in Agriculture II, Volume 2*, eds. D. Li, Z. Chunjiang, (Boston: Springer), pp. 1055–1060.

1. INTRODUCTION

Confocal microscope, which works by exciting fluorescence with a highly focused beam of laser light, is one of the most exciting advances in optical microscope. Recently, confocal microscope has become a routine technique and indispensable tool for cell biological studies and molecular investigations (Lichtman, 1994). Since there are two pinholes in confocal microscope, the light emitted from the point out-of-focus is blocked and can not reach the detector. Thus only a digital image of the fluorescence from the focal plane is observed. Moreover, the laser can scan from point to point over the sample and a single two-dimensional image of the optical section is acquired (Lichtman, 1994). Because sectioning is performed using optics rather than the physical sectioning of the sample, living cells can be analyzed with confocal microscope (Blancaflor and Gilroy, 2000).

Using confocal microscope, the cells can be analyzed in three dimensions with much more clarity than conventional microscope. There have been numerous scientific papers employing confocal microscope in cell biology, and this technology is vcry important for biology science research (Serhal et al., 2007; Wang et al., 2004; Lee et al., 2007). Confocal microscope has been used to investigate the heterogeneity of plant mitochondrial responses (Armstrong et al., 2006), Ca^{2+} level (Nichols et al., 2007), oxidative stress in cells (Kannan et al., 2006) and mitochondrial localization (Yang et al., 2006). Since confocal microscope is an appropriate and important method for quantitative and qualitative analysis of cells, we applied confocal microscope for analyzing the distribution of reactive oxygen species (ROS) and Ca^{2+} in mouse 3T3 fibroblasts in present studies.

2. MATERIALS AND METHODS

2.1 Cell culture

Mouse 3T3 fibroblasts, obtained from China Centre for Type Culture Collection (Wuhan, China), were maintained at 37°C with 5% CO2 and 95% air in Dulbecco's modified Eagle's medium (DMEM, Life Technologies, Carlsbad, CA) supplemented with 10% fetal bovine serum.

2.2 Ca^{2+} detection

To analyze Ca2+ level, cells were loaded with Fluo-3 AM (Molecular Probes, 5 μmol/L) at 37°C for 30 minutes. Excess dye was eliminated by washing the disks three times in PBS buffer.

2.3 ROS detection

Intracellular production of ROS was visualized by using 2′, 7′-dichlorof luorescein diacetate (H_2DCF-DA, Molecular Probes). This nonpolar compound is converted to the membrane-impermeant polar derivative H_2DCF by esterases when it is taken up by the cell. H_2DCF is nonfluorescent. However, it is rapidly oxidized to the highly fluorescent DCF by intracellular ROS (Genty et al., 1989). For assessment of ROS, the samples of cells were immersed in 1.5 ml of 10 μM DCFH-DA for 50 min, wiped to remove excess solution, and placed inside the growth chamber. At the end of the incubation period, the cells were rinsed briefly with PBS buffer.

2.4 Digital Images acquirement with confocal microscope

For microscopy, Cells plated on glass-bottom dishes were loaded with fluorescent probes as described above. After washing the dyes, the cells were put under an inverted microscope (Leica DM IRE 2, Germany). A laser-scanning confocal microscope (Leica TCS SP2, Germany) with an air-cooled, argon-ion laser as the excitation source at 488 nm was used to view the sites of ROS and Ca^{2+}, respectively.

Images were obtained with a 100× oil immersion objective lens. ROS and Ca2+ were respectively detected in the green and red channel. The channel settings of pinhole, detector gain, amplification offset and gain were adjusted to provide an optimal balance of fluorescent intensity of the targeted cells and background. Data were collected by a computer attached to the instrument, stored on the hard drive, processed with a Leica TCS Image Browser, and transferred to Adobe Photoshop 6.0 for preparation of figures.

2.5 Digital Images analysis with confocal software

The fluorescence intensity of ROS and Ca2+ in mouse 3T3 fibroblasts was acquired and quantified with Leica Confocal Software. Data are expressed as mean ± SD.

3. RESULTS AND DISCUSSION

Reactive oxygen species (ROS), which include superoxide ($\cdot O_2^-$), the hydroxyl radical (·OH), and hydrogen peroxide (H_2O_2), are produced by metabolic activation of molecular oxygen (Kohen and Nyska, 2002). Under normal metabolic activities such as respiration and photosynthesis, ROS are also produced. However, the production of ROS is enhanced during stresses such as nutrient limitation, exposure to xenobiotics and rehydration. Cells have evolved a variety of mechanisms to counteract the effects of reactive oxygen species, which include antioxidant enzymes and low-molecular-weight antioxidants (Scandalios, 1993; Kohen and Nyska, 2002). If not effectively and rapidly removed from cells, ROS will damage a wide range of macromolecules, possibly leading to cell death. Because ROS play an important part in a variety of biotic and abiotic stress conditions, the methods for direct in vivo identification and quantification of ROS is of special importance in stress studies.

H_2DCF-DA, which is not oxidized by superoxide, can be oxidated to fluorescent DCF by H_2O_2 and organic peroxides (Zhu et al., 1994). It has been found that H_2DCF-DA is a specific probe for intracellular H_2O_2 in a wide variety of organisms to study the mitochondrial H_2O_2 production (Pantopoulos et al., 1997; Quillet-Mary et al., 1997). Mitochondria are a major source of ROS in eukaryotic cells. In humans, aberrant mitochondrial ROS formation has been associated with conditions such as Parkinson's disease, amyotrophic lateral sclerosis, and aging (Beal, 1995). Our results indicate that ROS were produced intracellularly in mouse 3T3 fibroblasts, and was visualized clearly by laser-scanning confocal microscopy with the probe DCFH-DA (Fig.1. a).

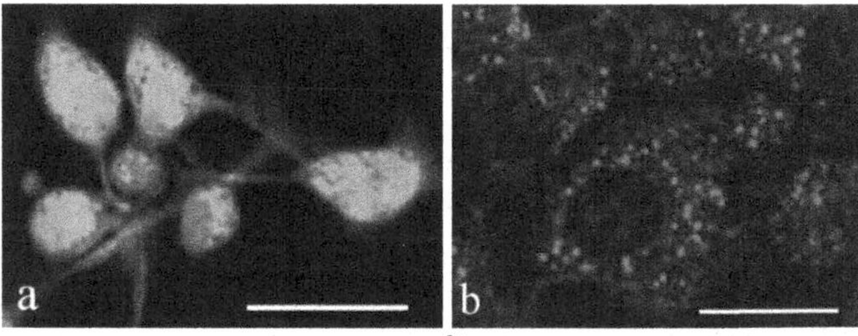

Fig.1. Intracellular localization of ROS and Ca^{2+} level. Laser-scanning confocal microscope images of ROS and Ca^{2+} level in mouse 3T3 fibroblasts. Cells were labeled with H_2DCF-DA and Fluo-3 AM, respectively. (a) Green channel, showing ROS fluorescence; Scale bars, 30 μm. (b) Red channel, showing Ca^{2+} fluorescence. Scale bars, 21.4 μm.

Since calcium is an intracellular messenger, the techniques for measuring cytosolic free Ca^{2+} concentrations have been essential. Fluo-3 fluorescence depends on concentration of free Ca^{2+} (Kao et al., 1989; Minta et al., 1989). Using Fluo-3 AM as probes, the distribution of Ca^{2+} can be clearly seen in mouse 3T3 fibroblasts (Fig.1. b). Moreover, we acquired and quantified the fluorescence intensity of ROS and Ca^{2+} in mouse 3T3 fibroblasts with Leica Confocal Software. It was found that the quantified fluorescence intensity of ROS and Ca^{2+} was 1123.30.26±8.99 and 125.13±12.16, respectively (Fig. 2).

Taken together, our results indicate that it is a good method to research the cultured cells with confocal microscope. With the fluorescence probes

H_2DCF-DA and Fluo-3 AM, not only the distribution of ROS and Ca^{2+} can be acquired, but also the fluorescence intensity of ROS and Ca^{2+} can be analyzed.

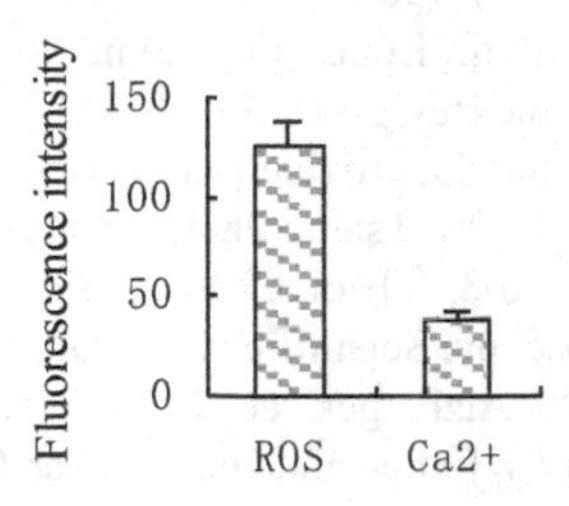

Fig.2. The fluorescence intensity of reactive oxygen species (ROS) and Ca^{2+} in mouse 3T3 fibroblasts was acquired and quantified with Leica Confocal Software. The error bars represent the SD.

4. CONCLUSION

The developments in computer technology and reagents that brought fluorescence-based methods to microscopy have become prevalent in cell biology. Our results indicate that it is a good method to research the cultured cells with confocal microscope. With the fluorescence probes H_2DCF-DA and Fluo-3 AM, not only the distribution of ROS and Ca^{2+} can be acquired, but also the fluorescence intensity of ROS and Ca^{2+} can be analyzed.

ACKNOWLEDGEMENTS

This work was supported by the Natural Science Foundation for Outstanding Young Scholars of Shandong Province, China (Grant No. 2007BS06021).

REFERENCES

A. F. Armstrong, D. C. Logan, A. K. Tobin, et al. Heterogeneity of plant mitochondrial responses underpinning respiratory acclimation to the cold in Arabidopsis thaliana leaves, Plant Cell Environ, 2006, 29: 940-949

A. Minta, J. P. Kao, R. Y. Tsien. Fluorescent indicators for cytosolic calcium based on rhodamine and fluorescein chromophores, J. Biol. Chem, 1989, 264: 8171-8178

A. Quillet-Mary, J-P. Jaffrezou, V. Mansat, et al. Implication of mitochondrial hydrogen peroxide generation in ceramide-induced apoptosis, J. Biol. Chem, 1997, 272: 21388-21395

B. Genty, J. M. Briantais, N. R. Baker. The relationship between the quantum yield of photosynthetic electron transport and quenching of chlorophyll fluorescence, Biochim. Biophys. Acta, 1989, 990: 87-92

Blancaflor EB, Gilroy S. Plant cell biology in the new millennium: new tools and new insights, Am J Bot, 2000, 87: 1547-1560

C. Wang, B. Li, H. Zhang, et al. Effect of arsenic trioxide on uveal melanoma cell proliferation in vitro, Ophthalmic Res, 2007, 39: 302-307

J. G. Scandalios. Oxygen Stress and Superoxide Dismutases, Plant Physiol, 1993, 101: 7-12

J. P. Kao, A. T. Harootunian, R. Y. Tsien. Photochemically generated cytosolic calcium pulses and their detection by fluo-3, J. Biol. Chem, 1989, 264: 8179-8184

J. W. Lichtman. Confocal microscopy, Scientific American, 1994, 271: 40-53

K. Pantopoulos, S. Mueller, A. Atzberger, et al. Differences in the Regulation of Iron Regulatory Protein-1 (IRP-1) by Extra- and Intracellular Oxidative Stress, J. Biol. Chem, 1997, 272: 9802-9808

K. Serhal, C. Baillou, N. Ghinea, et al. Characteristics of hybrid cells obtained by dendritic cell/tumour cell fusion in a T-47D breast cancer cell line model indicate their potential as anti-tumour vaccines, Int J Oncol, 2007, 31: 1357-1365

M. F. Beal. Aging energy and oxidative stress in neurodegenerative diseases, Ann. Neurol, 1995, 38:357-366

R. A. Nichols, A. F. Dengler, E. M. Nakagawa, et al. A Constitutive, Transient Receptor Potential-like Ca2+ Influx Pathway in Presynaptic Nerve Endings Independent of Voltage-gated Ca2+ Channels and Na+/Ca2+ Exchange, J Biol Chem, 2007, 282: 36102-36111

R. Kannan, N. Zhang, P. G. Sreekumar, et al. Stimulation of apical and basolateral VEGF-A and VEGF-C secretion by oxidative stress in polarized retinal pigment epithelial cells, Mol Vis, 2006, 12: 1649-1659

R. Kohen, A. Nyska. Oxidation of biological systems: oxidative stress phenomena, antioxidants, redox reactions, and methods for their quantification, Toxicol. Pathol, 2002, 30: 620-650

X. D. Yang, C. J. Dong, J. Y. Liu. A plant mitochondrial phospholipid hydroperoxide glutathione peroxidase: its precise localization and higher enzymatic activity, Plant Mol Biol, 2006, 62: 951-962

Y. Y. Lee , S. F. Yang, W. H. Ho, et al. Eugenol modulates cyclooxygenase-2 expression through the activation of nuclear factor kappa B in human osteoblasts, J Endod, 2007, 33: 1177-1182

EXPERIMENTAL STUDY FOR AUTOMATIC COLONY COUNTING SYSTEM BASED ON IMAGE PROCESSING

Junlong Fang[1], Wenzhe Li[1], Guoxin Wang[2]

[1] *Engineering College Northeast Agricultural University Harbin, Heilongjiang 150030 China swanhaha@163.com*

[2] *Hei longjiang Institute of Science and Technology*

Abstract: Colony counting in many colony experiments is detected by manual method at present, therefore it is difficult for man to execute the method quickly and accurately .A new automatic colony counting system was developed. Making use of image-processing technology, a study was made on the feasibility of distinguishing objectively white bacterial colonies from clear plates according to the RGB color theory. An optimal chromatic value was obtained based upon a lot of experiments on the distribution of the chromatic value. It has been proved that the method greatly improves the accuracy and efficiency of the colony counting and the counting result is not affected by using inoculation, shape or size of the colony. It is revealed that automatic detection of colony quantity using image-processing technology could be an effective way.

Key words: colony counting; image-processing; optimal chromatic value; automatic detection

1. INTRODUCTION

Colony counting is one of the most basic and frequent operation in colony cultivating experiments, and is both basic and important in agriculture, foods, medical analysis. At present, people usually adopt national standard detection method to count colony, which is ordinary nutrition agar pump method. When the number of sample is huge, the method becomes complex, time consuming, low efficiency, so it is necessary to ameliorate traditional

---

Please use the following format when citing this chapter:

Fang, J., Li, W. and Wang, G., 2009, in IFIP International Federation for Information Processing, Volume 294, *Computer and Computing Technologies in Agriculture II, Volume 2*, eds. D. Li, Z. Chunjiang, (Boston: Springer), pp. 1061–1066.

counting method. With the promotion of computer technology, using computer as a tool to effectively reduce the intensity of production, improve labor productivity, and realize production automation has become a development trend. In recent years, using computer image processing technology to solve the quality detection task of agricultural products has been the general concern of scholars at home and abroad, so as to colony counting. The study developed a detection system based on computer image processing which found a recognition and rapid colony counting approach.

2. HARDWARE PLATFORM OF THE SYSTEM

Hardware composition of automatic colony counting system is shown in Fig.1; the system is mainly made up of light source box, CCD camera, image acquisition card, computer, monitor and other equipments. CCD camera is the color camera of Panasonic WV-CP410. Image acquisition card is DH-CG300 which is a product of China Daheng Group, Inc., maximum resolution of image acquisition is 768×576×24bit. Configuration of computer is PENTIUM-Ⅳ 1.7G CPU, 256M memory, 80G hard disk, 64M display memory. Petri dish with cultured colony is placed under optical platform, optical platform is used to adjust amplitude and angle of optical radiation, so as to get a clear image which will be easy to process and identify. Colony image is acquired by CCD camera under preset illumination condition, then it was sent to a computer installed with image acquisition card, after that image processing software which runs in the computer will do a series of preprocessing, object segmentation and counting, then the number of colony is obtained.

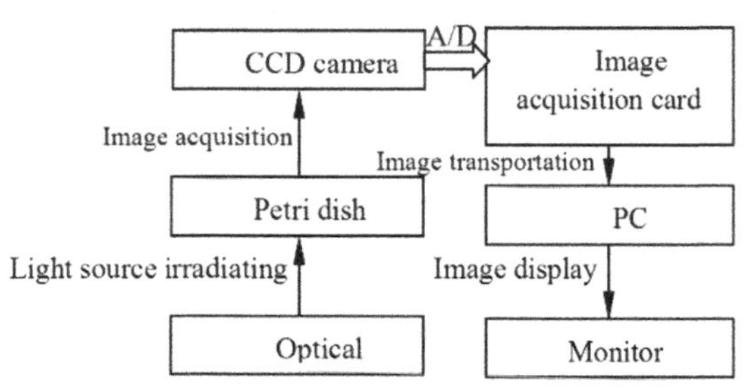

Fig.1 Hardware configuration of the system

3. IDENTIFICATION OF COLONY IMAGE

From gray image acquired from CCD camera as is shown in Fig.2, we can see that: colony image has a uniform gray value, although gray value inside and outside of Petri dish is different, both of them are uniform, what is more, gray value contrast between colony and background is large. Taking into account these characteristics, threshold segmentation method can be used to binarize the image which will separate colony from background.

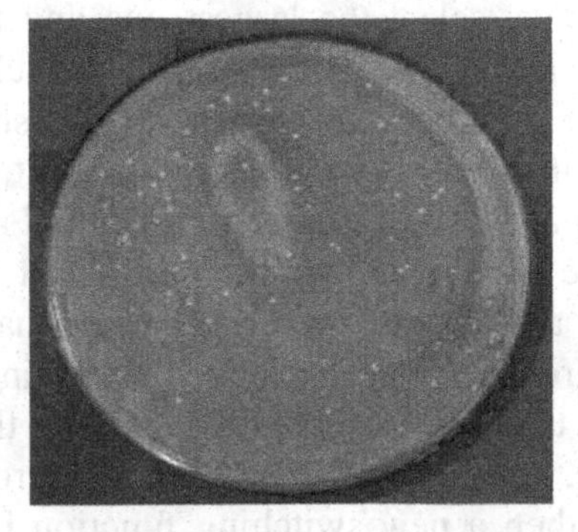

Fig.2 Gray image

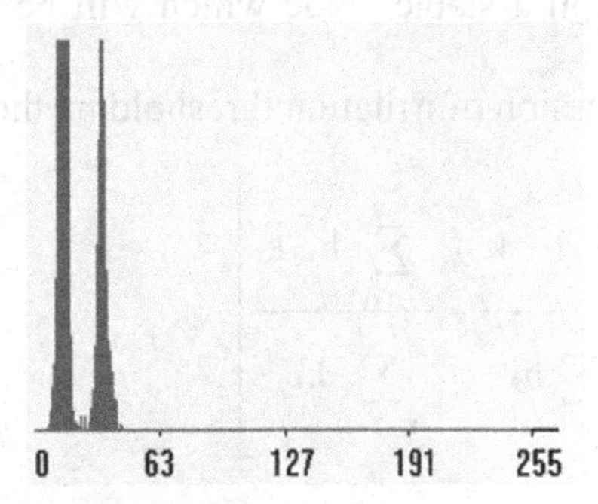

Fig.3 Histogram of gray image

3.1 Image Preprocessing

In order to enable easy target segmentation and enhance process accuracy, automatic counting system needs to preprocess acquired images. According to the system structure as is shown in Fig.1, system noise is mainly bring about from CCD circuit, threshold used to segment images is sensitive to noises, so the paper adopted grayscale, median filter, contrast enhancement and other preprocessing methods to remove noises and enhance image.

3.2 Image Extraction

From gray image as is shown in Fig.2 and corresponding histogram as is shown in Fig.3, we can see that gray value of Petri dish and colony is so close that ordinary bimodal law can not separate them but withdraw them together.

Having compared several image segmentation methods, the paper used the follow method: divide image to be detected into several small blocks first, use iterative threshold method to determine optimal threshold of each block, compare these thresholds, select the largest one and then subtract a small compensation named B (B=10 in the paper) which is used as global threshold, binarize the preprocessed colony image (shown in Fig.4). Gray value of colony in the image is higher (black represent gray value is 0, white represent gray value is 255), so this iterative threshold method can automatic find the most suitable threshold. Optimal threshold value of each block image is determined as separate pixels of the small block image into foreground and background by an initialized switching function, carry out integral computing on them separately and averaging the two number to get a new threshold value, separate the image into foreground and background using the new value, then a new switching function is generated. Iteration repeated until switch function does not change any more, that means the irritation convergence in a stable value which will be optimal threshold of the small block image.

Mathematical expression of irritation threshold method is as follows:

$$T_{i+1} = \frac{1}{2}\left[\frac{\sum_{k=0}^{T_i} h_k\, k}{\sum_{k=0}^{T_i} h_k} + \frac{\sum_{K=T_i+1}^{L-1} h_k\, k}{\sum_{K=T_i+1}^{L-1} h_{k_i}}\right] \tag{1}$$

Where: L is the number of gray level, h_k is the number of pixel with gray value of k, T_{i+1} is optimal threshold value of the small block image.

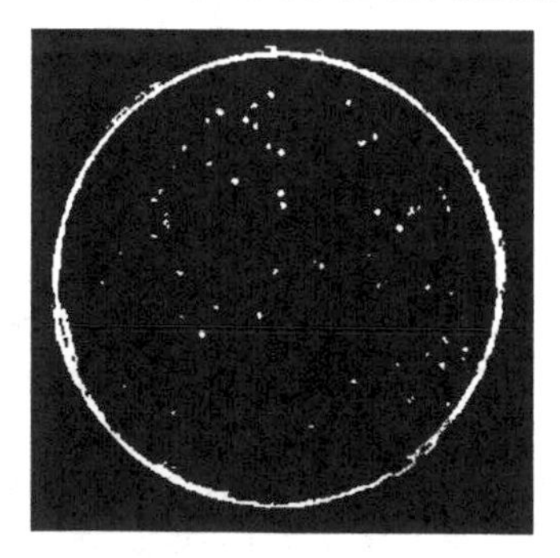

Fig.4 Binary image

Fig.5 Image removed Petri dish edge

3.3 Remove Petri Dish Edge

From binarized colony image as is shown in Fig.4, we can see that there is still obvious Petri dish edge besides colony image which will affect later counting approach. Edge image of Petri dish is always connected region with the most number of pixels, so scan the image to find the largest region and turn it into black (set gray value to 0), result is shown in Fig.5.

4. COLONY COUNTING ALGORITHM

The task to count the number of colony is to count the number of connected regions in binary image. Traditional connected region labeling algorithm is simple, its principle is: find a white point, scan its neighborhood from bottom to top, from right to left to get connected region and label it, so time complexity of the algorithm is O(n^2)(n is size of the image), when n is big, calculation speed was very slow. This paper presented a new labeling counting method, its time complexity is only O (n) (n is size of the image), thereby improved the calculate speed on the basis of accuracy. Details of the algorithm is as follow: Scan the image from left to right, from up to bottom, when come across a white pixel A, use A as a seed point, set its value with 1, find other white pixels that are eight neighborhood linked with it, set the found pixel value with 1, continue the search process with these seed point until there is no unlabeled white pixels. Now, a connected region labeling has finished. Find next unlabeled white pixel as new seed point, set its label value with 2, repeat above process until the whole image has been scanned, the largest region label is the number of colony.

5. CONCLUSION

The system used VC++ to program. Total counting time is less than 1 second; average relative error is 2.5%. Experiment shows that above image processing method is feasible for white colony counting, it does not affect by colony shape, size and inoculation method, but sensitive to illumination condition and background color. So, a closed light source system will be a good choice which may avoid the impact of changes in natural light. Besides, using black background will get a more accurate counting result.

ACKNOWLEDGEMENTS

The research work in the paper is supported by the Heilongjiang youth science and technology special funds project (QC07C39), Heilongjiang postdoctoral financing expenses project (LBH-Z06163) and Northeast Agricultural University scientific research funds.

REFERENCES

Chinese National Standards[S]. Food Hygiene Inspection Methods (part of Microbiology).Beijing: Chinese National Standards Press,1985:194-196.

Corkidi G, Diaz-uribe R, Folch-mallol J L, et al. An image analysis method that allows detection of confluent microbial colonies of various sizes for automated counting [J]. Appl Environ Microbiol, 1998, 64(4):1400-1412.

Garcia-Armesto MR, Prieto M. Modern microbiological methods for foods:colony count and dirtect count methods [J]. A review Microbiologia, 1993,9(1):1.

Lang Rui. Realization of Digital Image Processing in Visual C++ [M].Beijing: Beijing Hope Electronics Press,2003:279-284.

Mukherjee DP, Pal A, Sarma SE, et al. Bacterial colony counting using distance transform [J]. Int J Biomed Comput. 1995, 38(2): 131.

Pickett DA, Welch DF. Evaluation of the automated Bactalert system for pediatric blood culturing [J]. Am J Clin Pathol, 1995,103(3): 320.

Spadinger I, Palcic B. Cell survival measurements at low doses using an automated image cytometry device [J]. Int J Radiat, 1993, 63(2): 183.

Wang Dongsheng, Wang Xifa. BASIC Image Processing Program 150 Examples[M].Hefei: University of Science and Technology of China Press,1991:268-269.

Zhang Yujin. Image Processing and Analization [M].Beijing: Tsinghua University Press,1999:84-85.

Zhou Yingli. Automatic Colony Counting Methods and Realization Based on Image Processing. Data Acquisition and Processing. 2003,Vol.18(4):460-464.

STUDY AND REALIZATION OF IMAGE SEGMENTATION ON THE COTTON FOREIGN FIBERS

Wenxiu Zheng, Jinxing Wang[*], Shuangxi Liu, Xinhua Wei

Mechanical and Electronic Engineering College, Shandong Agricultural University, Taian, Shandong Province, P.R. China 271018

[*] *Corresponding author, Address: Mechanical and Electronic Engineering College, Shandong Agricultural University, Taian 271018, Shandon Province, P.R.China, Tel: +86-538-8246826, Fax: +86-538-8246107, Email:jinxingw@163.com*

Abstract: A method of foreign fibers image segmentation based on Mean shift、dilation and filtering algorithm is presented. For the representative gray images of hair、chicken feather and mixed foreign fibers, the Mean shift algorithm is used to carry on image segmentation; then dilation and filtering process is carried on to the divided image element. In this way the precise image segmentation of foreign fibers is realized. It's proved by experiments that the image segmentation method proposed by this article can suppress the noise well, and the segmentation results are satisfied for all kinds of foreign fibers image.

Keywords: Foreign fibers, Histogram analysis, Image segmentation, Mean shift

1. INTRODUCTION

Foreign fibers refer to the non-cotton fibers and the colored fibers which are mixed in the process of raw cotton production、processing and circulation, such as chemical fiber、hair、silk、linen、plastic film、plastic rope、dyeing lines and so on(Li Bidan et al.,2006). Although the net content is low, there is a big impact on the cotton industries. Now the general selection of manual work has low efficiency and consumes a lot of time. The professional investigation estimates that a medium cotton enterprise uses

Please use the following format when citing this chapter:

Zheng, W., Wang, J., Liu, S. and Wei, X., 2009, in IFIP International Federation for Information Processing, Volume 294, *Computer and Computing Technologies in Agriculture II, Volume 2*, eds. D. Li, Z. Chunjiang, (Boston: Springer), pp. 1067–1075.

300-400 people to select the foreign fibers everyday, and the annual cost of this item amounts to more than 200 million Yuan. Whatsoever the selection of foreign fibers is like to look for a needle in a sea which results in the eye weariness of the workers for glaring at the white cotton for such a long time and finally the foreign fibers which are not sorted still cause the problem of product quality(Kang et al.,2002; Wang Xinlong et al.,2002).So applying the machine vision technology to research rapid recognition of foreign fibers are of great significance, for it can improve the quality of cotton and promote the development of cotton textile industries. The foreign fibers detection system based on machine vision is the key technology, thus this paper mainly research the image segmentation technology of foreign fibers (Jiao Wenxing et al., 2003).

The aim of image segmentation technology is to distinguish the different special meaning areas as these areas are not intersect mutually,

image segmentation is not only one of the great important contents in the area of image processing, but also a classic problem in machine vision. For image analysis and understanding, the original image is transformed a more abstract form through separate objectives、extract parameters and survey

parameters. The common methods of image segmentation have threshold value segmentation、region growing、region splitting and merging、edge detection and boundary tracking and so on. Because of the complexity of the imaging and the information included in the image, though the researchers proposed many segmentation methods, so far there is not a general segmentation method (Zhu Zhigang et al., 2003; Zhang Xiaolu et al., 2005).

In this paper, a new method of foreign fibers image segmentation based on mean shift、dilation and filtering algorithm is presented, it can suppress the noise well. Through the histogram analysis to the foreign fibers image, the gray scale distributed situation was determined and this image segmentation method was used to realize the image segmentation. The results indicate that this method has the characteristics of high speed and accuracy (Chen Donglan et al., 2003).

2. EXPERIMENTAL MATERIALS

This experiment chooses 14 kinds of foreign fibers, such as the chicken feather、the white paper、the fluorescent polypropylene silk、non-fluorescent green-grey polypropylene silk、non-fluorescent white polypropylene silk、the white chemical fiber silk、the colored chemical fiber cloth strip、the colored cotton cloth strip、the hemp rope、the hair、the candy wrapper、the colored sheep yarn、the white plastic bag、mulching plastic and so on, and full amount of ginned cotton which doesn't contain foreign fibers. After fully-smashing by the ginned cotton machine, the uninterrupted uniform cotton layer of 80 centimeters wide and two millimeters thick were formed. In the process of smashing the above foreign fiber samples were put in gradually, also the layer of foreign fibers were preserved after the process, then the scanner would be applied to carry on the image scanning on these layers of cotton samples. As a result, 40 foreign fiber images of 24 true colors whose size is 1700×2344 were produced.

3. HISTOGRAM ANALYSIS

In order to extract the foreign fibers smoothly, the 40 samples were taken to make the histogram analysis. As the image samples are the 24 true color images, then before the gray histogram analysis, the image type was necessary to convert, 24 true color images will be converted to 256 gray images (Kapur et al., 1985). So the gray images of hair、 chicken feather and mixed foreign fibers were acquired and their histogram analysis was done. The following Fig.1-3 are the gray images and histograms:

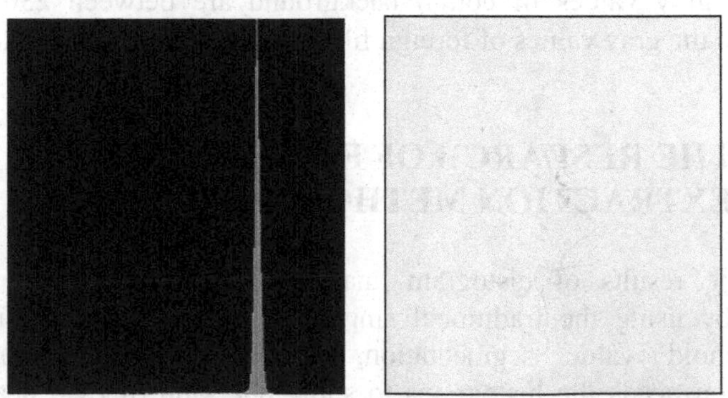

Fig.1 hair's gray image and histogram

Fig.2 chicken feather's gray image and histogram

Fig.3 mixed foreign fiber's (feather + tape + plastic sheeting) gray image and histogram

Based on the research of gray histograms from Fig.1 to Fig.3, the following conclusions are drawn:

(1) All histograms present a single peak distribution, and the aggregation is very high. These show that the gray values of cotton and foreign fibers mix together, and the cotton's background occupies the overwhelming superiority in the whole image, however the foreign fibers' has very small proportion in the whole image;

(2) The gray values of cotton background are between 230～255 in general and the gray values of foreign fibers are less than or equal to 230.

4. THE RESEARCH OF FOREIGN FIBER'S EXTRACTION METHOD

From the results of histogram analysis, the foreign fibers can't be extracted by using the traditional single-threshold value segmentation or multi-threshold value segmentation, for the single-threshold value segmentation needs the histograms to show one kind of twin peaks or the multi-peak distribution, however the multi-threshold value segmentation needs to split the image into many sub-images and in order to discriminate the objects, each sub-image was needed to be marked, what's more, after threshold value segmentation, the adjacent boundary of sub-images may have the grayscale discontinuity(Chen Donglan et al.,2003).Therefore the method of Mean-shift adaptive threshold value segmentation was decided to be chosen to complete the segmentation of the foreign fibers.

Mean shift process is a method of kernel density estimation, which is based on the Parzen window method to examine the probability density function in the pattern recognition, corresponding in the window function $\varphi(x)$ of the Parzen window method, and the definition kernel function K(x).In most cases what we care about is the symmetry of kernel function, which could be expressed for the following forms:

$K(x) = C_{k,d} k(\| x \|^2)$. And $C_{k,d}$ is constant.

Here take the Gauss kernel function as an example:

$$K(x) = (2\pi)^{-d/2} \exp\left(-\frac{1}{2} \| x \|^2 \right)$$

Obtain the convergent recurrence formula and the Mean shift vector:

$$y_{i+1} = \frac{\sum_{i=1}^{n} x_i g\left(\left\| \frac{x - xi}{h} \right\|^2 \right)}{\sum_{i=1}^{n} g\left(\left\| \frac{x - xi}{h} \right\|^2 \right)},$$

$$m_{h,G} = \frac{\sum_{i=1}^{n} x_i g\left(\left\| \frac{x - xi}{h} \right\|^2 \right)}{\sum_{i=1}^{n} g\left(\left\| \frac{x - xi}{h} \right\|^2 \right)} - x = y_{i+1} - y_i$$

The Mean shift method has many strong points. The path of the Mean shift vector movement towards convergence point is a smooth track and the angle of two consecutive vectors is usually less than 90 degrees. This advantage guarantees the stability of the convergence, and compared with other fast rise ways, the gentle path of the Mean shift can complete the image segmentation rapidly; achieving the purpose of rapid and natural portray the real lines of nature objects. Based on the above characteristics, the application of the Mean shift technology to the foreign fibers image segmentation can divide the target sector exactly and obtain the accurate foreign fibers image (Jensen et al., 2002; Jason et al., 2008).

In the course of the research, regarding the dispersion of the image information, finally 13 pixels Mean-shift adaptive threshold segmentation method was selected to carry on image segmentation through repeated experiments (Zhu Zhigang et al., 2003).The following Fig.4-6 are the processing effect:

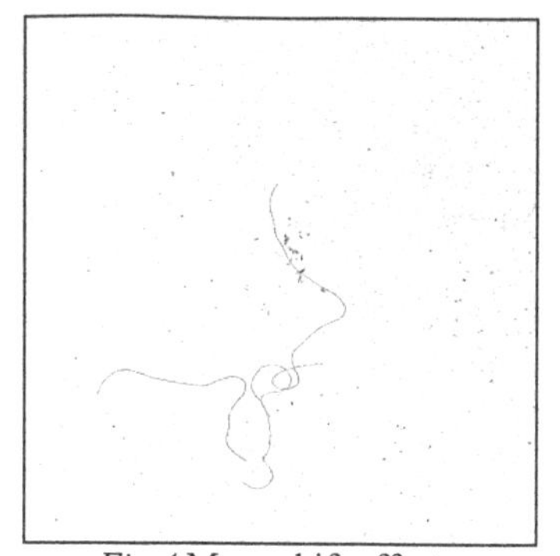

Fig.4 Mean shift effect chart of hair gray image

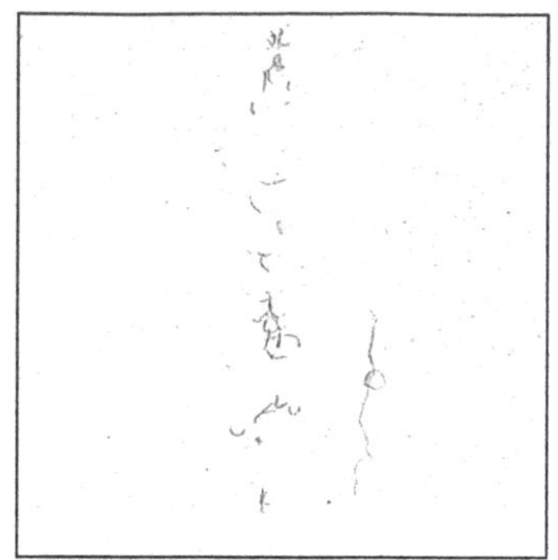

Fig.5 Mean shift effect chart of chicken feather gray image

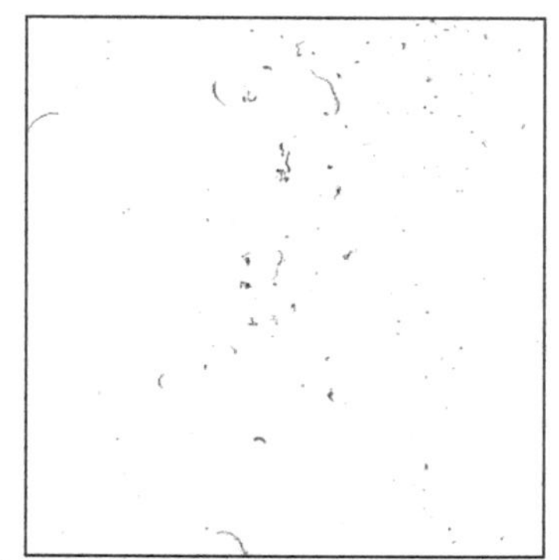

Fig.6 Mean shift effect chart of mixed foreign fiber (feather + tape + plastic sheeting) gray image

5. THE ENHANCEMENT PROCESSING OF FOREIGN FIBERS' IMAGE

Contrasting the original gray images, results could be found:
(1) Salt and pepper noise exists in the local of light uneven;
(2) There is a phenomenon of excessive processing.
In order to obtain the image which is closer to the original gray image, it is necessary to unify some processing methods to optimize or enhance the image:
(1) Median filtering technology was used to weaken the salt and pepper noise;
(2) The phenomenon of excessive processing is due to the band width of the Mean-shift algorithm is fixed and the change of two images' background.
Based on thc above two points, dilation process was carried on to the image element first, so the part phenomenon of excessive processing was revised and the region of foreign fibers was made to connect as far as

possible; then filtering process was used. After repeated testing, 3×3 rectangular structural element was decided to use to deal with a dilation process; then two 7×7 median filtering processes were used again, the experiments show that this approach can get more satisfactory results (Zhu Zhigang et al., 2003; Ying Xu et al., 1998).The effect of after dilation and filtering process is shown in Fig.7-9.

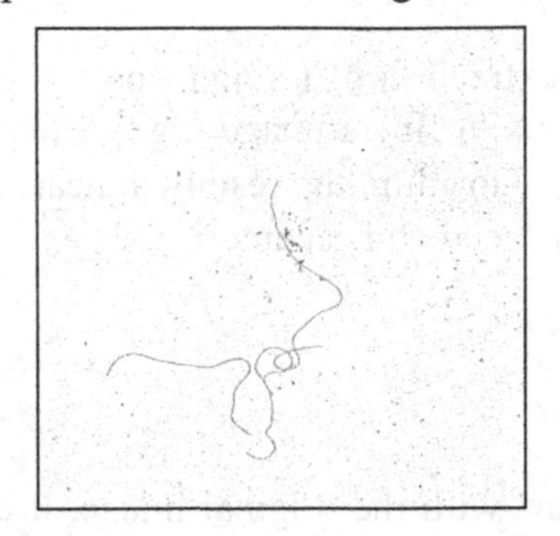
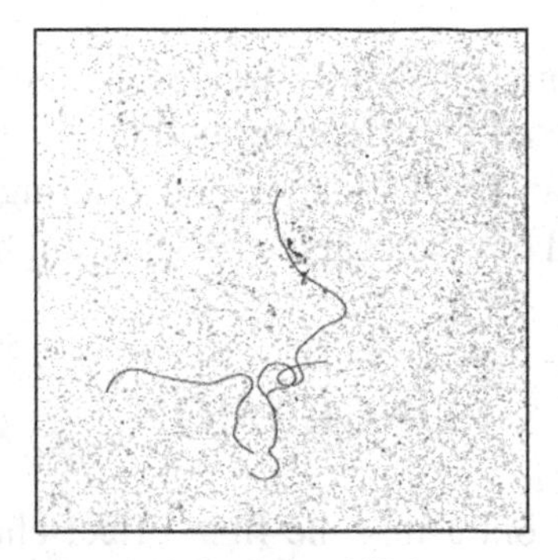

Fig.7 the dilation and filtering effect chart of hair Mean-shift image

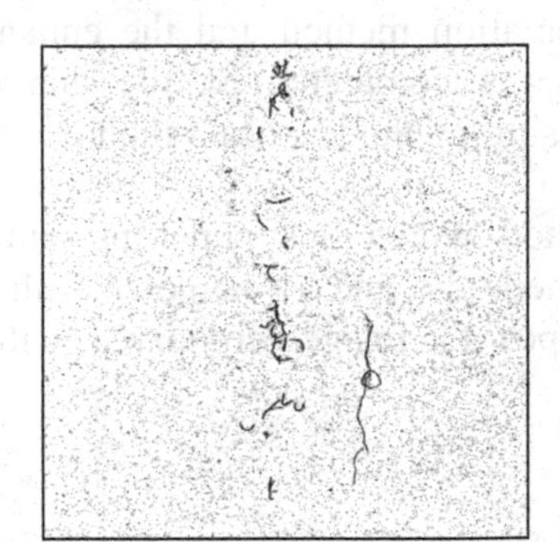
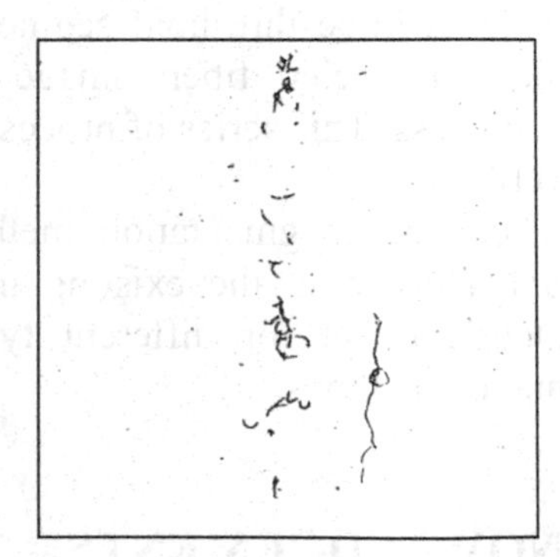

Fig.8 the dilation and filtering effect chart of chicken feather Mean-shift image

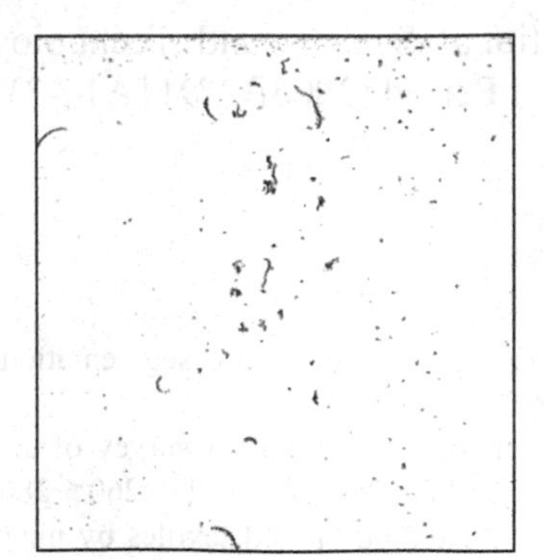
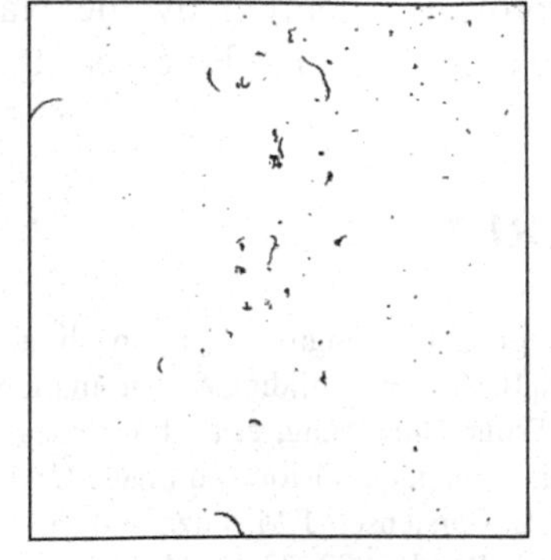

Fig.9 the dilation and filtering effect chart of mixed foreign fibers (feather + tape + plastic sheeting) Mean-shift image

The precise extraction of foreign fibers was completed after the dilation and filtering process on the foreign fibers Mean-shift image. Contrasting the original image with the final processing results, a discovery could be made:

(1) This method effectively extracts the foreign fibers;

(2) This method depends less on the environment illumination;

(3) Although effective, there are still a small portion of non foreign fibers are extracted, which could be considered to join the revision in the following work.

Through the above processes, the extraction of foreign fibers is precisely completed by using the method of foreign fibers image segmentation based on Mean shift、dilation and filtering algorithm, the results indicate that this method has the characteristics of high speed and accuracy.

6. CONCLUSION

(1) Contrasting the final effect chart with the original image, it could be concluded that the extraction of foreign fibers was effectively completed by Mean-shift adaptive threshold segmentation method and the enhancement processing of foreign fibers image was completed by the dilation and filtering process. This series of processing methods extract the foreign fibers effectively;

(2) The image segmentation method which this paper presents is an important addition to the existing methods, and it can obtain the better segmentation effect for different types of image, compared with other segmentation methods.

ACKNOWLEDGEMENTS

The project supported by the National Science and Technology Pillar Program in the Eleventh Five-year Plan Period.(2006BAD11A14-3)

REFERENCES

Chen Donglan, Liu Jingnan, Yu Ling-ling. Comparison of image segmentation threshold method[J].Machine Building & Automation, 2003, 1(1):77～80

Jason E. Fritts, Hui Zhang, et al. Image segmentation evaluation: A survey of unsupervised methods. Computer Vision and Image Understanding, 2008, 110（2）:260～280

Jensen K L, Carstensen J M. Fuzz and pills evaluated on knitted textiles by image analysis [J].Textile Res J, 2002, 72(1): 34～38

Jiao Wenxing, Pan Tianli, Li Yue. Application of computer vision technique in agricultural products quality inspection. Shanxi Journal of Agricultural Sciences, 2003, (5):29～33

Kang T J, Kim S C. Objection evaluation of the trash and color of raw cotton by image processing and neural network[J].Textile Res J, 2002, 72(9): 776～782

Kapur J N, Sahoo P K. A new method for gray-level picture threshold using the entropy of the histogram[J].Computer Vision, Graphics, and Image Processing, 1985, 29(3):273～285

Li Bidan, Ding Tianhuai, Jia Dongyao. Design of a Sophisticated Foreign Fiber Separator [J]. Agricultural Machinery Journal, 2006, 37（1）:107-110

Wang Xinlong, Li Na. Summary of the foreign fibers [J]. China Cotton Processing, 2002,（5）:29～30

Ying Xu, Victor Olman, et al. A segmentation algorithm for noisy images: design and evaluation. Pattern Recognition Letters, 1998, 19（13）:1213～1224

Zhang Xiaolu, Han Liqun. Application of Computer Vision Technology to Fiber Identification [J]. Journal of Beijing Technology and Business University, 2005, 23（2）:43-45

Zhu Zhigang, Shi Dingji, et al. Digital Image Processing[M].Publishing House of Electronics Industry, 2003

THE STUDY OF NON-DESTRUCTIVE MEASUREMENT APPLE'S FIRMNESS AND SOLUBLE SOLID CONTENT USING MULTISPECTRAL IMAGING

Muhua Liu [1,2] , Duan Wumao [2] , Huaiwei Lin [2]

[1] *Key Laboratory of Nondestructive Test in Nanchang University of Aviation of China Education Ministry, 330063,Nanchang, P.R.China 330045*

[2] *College of Engineering, Jiangxi Agricultural University, 330045 Nanchang, P.R.China 330045*

* *Corresponding author, Address: College of Engineering, Jiangxi Agricultural University, 330045 Nanchang, P.R.China 330045, Tel: +86-791-3813260, Fax: +86-797-3813260, Email: suikelmh@sohu.com*

Abstract: Firmness and soluble solid content (SSC) are two important quality attributes. This researches investigated the feasibility of using multi-spectral imaging to non-destructive measuring the apple's firmness and SSC. The spectral imaging in wavelength of 632nm、650nm、670nm、780nm、850nm and 900nm were captured. The Lorentzian distribution (LD), Gaussian distribution (GD) and Exponential distribution (ED) with three parameters were used to fit scattering profiles for all wavelengths. LD was found to be the best function for fitting gray distribution of the image. The multi-linear regression model using Lorentzian parameters for predicting apple firmness and soluble solids content were built using best single wavelength, double wavelengths, three wavelengths and four wavelengths. The best model with three wavelengths was able to predict apple soluble solid content with r=0.831, SEC=0.55 ° Brix and predict apple firmness with r=0.880, SEC=0.52 N with four wavelengths. Experimental results show that the multi-spectral scattering imaging has high potential as a nondestructive and rapid method to assess fruit internal quality.

Keywords: Apple, Spectral imaging, Soluble solid content, Firmness, Multi-linear regression

---

Please use the following format when citing this chapter:

Liu, M., Wumao, D. and Lin, H., 2009, in IFIP International Federation for Information Processing, Volume 294, *Computer and Computing Technologies in Agriculture II, Volume 2*, eds. D. Li, Z. Chunjiang, (Boston: Springer), pp. 1077–1086.

1. INTRODUCTION

Firmness and soluble solid content (SSC) are two important quality attributes used for assessing apple quality. On-line sensing using spectral imaging with several spectral bands can provide rapid method to assess fruits for their internal quality including firmness and SSC for human consumption, and thus, improve industry competitiveness and profitability.

Recently, optical techniques, especially near-infrared (NIR) spectroscopy, have been investigated as a nondestructive means for assessing quality attributes of fruits such as firmness and soluble solids content (SSC) (Lammertyn et al., 2000; Lu, 2000a). NIR spectroscopy determines the internal quality of fruit by measuring spectral reflectance/transmittance over the visible and NIR region. Although NIR spectroscopy has been successful for measuring SSC from whole fruit (Kawano et al., 1993; Lu, 2001; Slaughter, 1995).

McGlone et al. (1997) reported a moderately good correlation between scattering measurements and firmness, but with a high standard error. These reported studies have shown that light scattering is related to the condition of fruit. Diode lasers as a light source are easy to implement and low in costs, but they only provide light scattering information at one spectral band (or wavelength), which are insufficient for predicting fruit quality. Multiple wavelengths are needed in order to obtain more useful information about fruit firmness and SSC (Y Peng, 2006).

In this research, a spectral imaging technique was investigated for measuring light scattering profiles from apple fruit at selected wavelength 632nm, 650nm, 670nm, 780nm, 850nm and 900nm. Computer algorithms were developed to quantify the spectral scattering images and relate them to fruit firmness and SSC.

2. MATERIALS AND METHODS

2.1 Apple samples

One hundred fresh Red Delicious (RD) apple samples were used for experiments. They were purchased from fruit market. The apples were stored in either controlled humidity environment 56% to 58% or room temperature from 24℃ to 26℃ for at least 20 hours prior to testing. Then, the apple samples were cleaned and marked before experiments were started.

2.2 Spectral imaging system

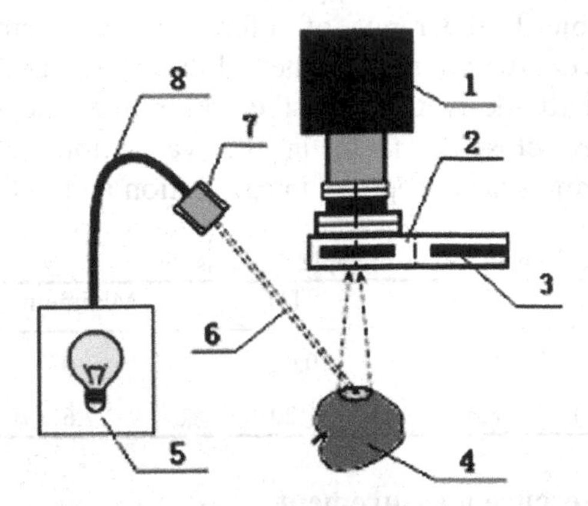

1. CCD Camera, 2.filter wheel setup, 3. filter, 4.fruit, 5.light source, 6.photon, 7.lens, 8.light fiber

Fig.1 Spectral Imaging System

Figure 1 shows a compact laboratory spectral imaging system. This system mainly consists of 250W quartz tungsten light source, a high performance air-cooled CCD camera (Model GC-755P,HoneyWell Corp., Korea) with the pixel resolution of 0.1mm per pixel, a rotating filter wheel containing six bandpass filters(632, 650, 670, 780, 850 and 900±10nm,) Matrox Meteor II /MC frame grabber (Matrox Corp., Canada)and computer with Intellicom image capture software. During image acquisition, each of the six filters was rotated in sequence to obtain six images. The exposure time was set at 1.2s for the six filters. As the light beam hit the apple surface, most of the light penetrated into the fruit and scattered in different directions, which generated scattering images at the surface of the apple sample. The Spectral imaging system captured the scattering images of different wavelength through the filters from the apple surface over a 30mm diameter area.

2.3 Reference measurements

2.3.1 Firmness reference measurement

Firmness was then measured from the same imaging location by using an 9mm probe fruit sclerometer (Model GY-1 Top Instrument Corp., China) at a loading rate of 2.0mm/s.Maximum forces recorded during 9mm penetration were used as a reference measure of apple firmness. Firmness was tested for three times from the same imaging location then averaged as apple firmness value. All of apple samples were tested, they were divided into calibration set with 75 samples and validation set with 25 samples, Statistics of firmness data for the apple samples are shown in the table 1. As seen from the table 1, the range of reference measurement value in the calibration set covers the range of in the validation set, at the same time, the mean and standard derivation of the reference measurement in the calibration set are close to them in the validation set. Therefore, the distribution of samples is appropriate in calibration and validation sets.

Tab. 1 Reference measurement results of apple firmness (N)

Apple firmness	Mean	STD	Minimum	Maximum
Calibration set (n=75)	8.91	0.92	7.0	11.1
Validation set (n=25)	8.65	1.20	7.6	10.5

2.3.2 SSC reference measurement

SSC was measured from the same imaging location by using sugar refractometer (Model WZ-103 Zhongyou Optical Instrument Corp., China).SSC was tested for two times from the same imaging location then averaged as apple SSC value. All of apple samples were tested; they were divided into calibration set with 75 samples and validation set with 25 samples, Statistics of SSC data for the apple samples are shown in the table 2. As seen from the table 1, the range of reference measurement value in the calibration set covers the range of in the validation set, at the same time, the mean and standard derivation of the reference measurement in the calibration set are close to them in the validation set. Therefore, the distribution of samples is appropriate in calibration and validation sets.

Tab. 2 Reference measurement results of apple SSC (° Brix)

Apple sugar	Mean	STD	Minimum	Maximum
Calibration set (n=75)	150	1.24	9.00	15.0
Validation set (n=25)	14.8	1.21	10.0	14.8

3. RESULTS AND DISCUSSION

3.1 Models of Scattering Profiles

All of the apple samples were placed on the carry flat roof of spectral imaging system, and then each of the six filters was rotated in sequence to obtain six images. Six wavebands were 632nm, 650nm, 670nm, 780nm,

850nm and 900nm respectively. Six hundred scattering images were acquisition in all. Figure 2 show part of spectral scattering images.

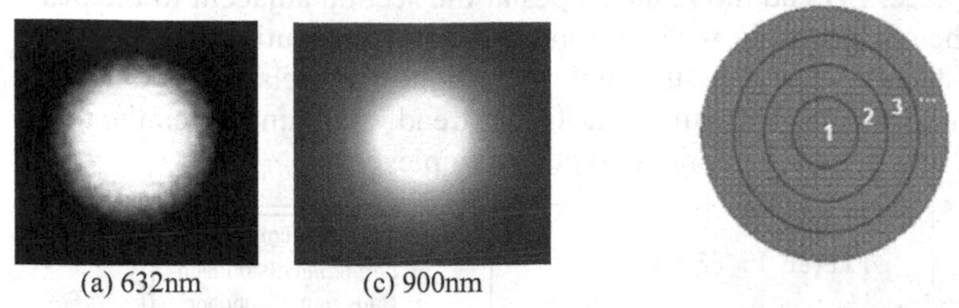

(a) 632nm (c) 900nm

Fig.2 spectral images o 2 f apple *Fig.3* quantitative scattering image with pixels

Scattering images were circular and symmetric with respect to the light incident point, and their intensity decreased rapidly as the distance from the light incident point increased. Thus, each spectral scattering imaging could be reduced to a one-dimensional scattering profile through radial distance. In this study, the scattering image is divided into N circular bands of equaled distance or pixels, as shown in Fig3. The radial intensity of the scattering profiles was calculated by averaging all pixels within each circular, and every circular was achieved at intervals of eight pixels in turn. The following three mathematical models, i.e. Lorentzian distribution (LD), Gaussian distribution (GD) and Exponential distribution (ED) functions, were proposed comparatively to fit each scattering profile respectively. The LD, GD and ED functions are as follows Eqs.1-3,

$$I_i = a_{1i} + \frac{a_{2i}}{1 + (\frac{x}{a_{3i}})^2} \tag{1}$$

$$I_i = a_{1i} + a_{2i} e^{-\frac{|x|}{a_{3i}}} \tag{2}$$

$$I_i = a_{1i} + a_{2i} e^{-0.5(\frac{x}{a_{3i}})^2} \tag{3}$$

where: I is the light intensity in the CCD count; x is the scattering distance measured from the beam incident center, in mm; i denote six different fitters, and $i = 1,2,\ldots,6$ a_{1i} , a_{2i} and a_{3i} are asymptotic values, peak values, and scattering widths, corresponding to individual filters.

Figure 4 illustrates how the three distribution functions fit the scattering profiles at the six wavelengths for an apple fruit. When the distance (x) is small, the three distribution functions have a steep descending attribute; as x increase, they have a gentler descending attribute. The ED curve was sharp at the origin, and overall it did not fit the data well. The GD curve, on the other hand, under fit the section of the profile where greatest changes in the slope took place. LD had moderate slopes at the section adjacent to the peak and fitted the entire profile well. Among the three distributions studied. LD had the best curve-fitting results with the highest correlation and lowest stand error of estimate. The same data-fitting trend was found when the three distribution functions were applied to other samples.

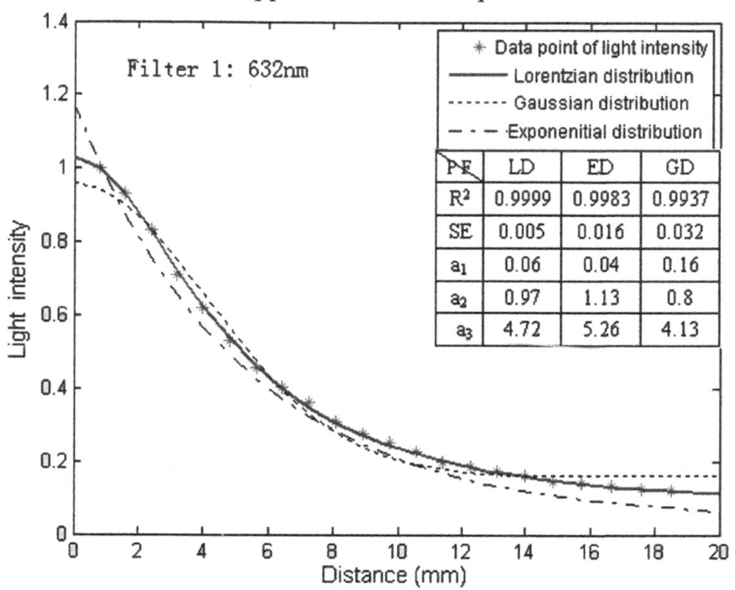

PF	LD	ED	GD
R^2	0.9999	0.9983	0.9937
SE	0.005	0.016	0.032
a_1	0.06	0.04	0.16
a_2	0.97	1.13	0.8
a_3	4.72	5.26	4.13

Fig.4 Curve fittings of functions for an apple imaging scattering profiles in 632nm

3.2 Predicting model building

According to LD parameters and reference measurement results, the firmness and SSC prediction models were built by multi-linear regression (MLR) respectively. The prediction model consists of six partitions corresponding to six wavebands filters. Each partition has three LD parameters for an individual filter. The prediction model is described by the following equation:

$$\hat{Y} = f_0 + \sum_{i=1}^{N} (f_{1i}a_{2i} + f_{2i}a_{2i} + f_{3i}a_{3i}) \qquad (4)$$

where Y is the predictive results of model; N is the total number of filters; i=1,2,3,4,5,6; a_{1i},a_{2i} and a_{3i} are asymptotic values, peak values, and scattering widths for LD, corresponding to individual filters. f_0,f_{1i},f_{2i} and f_{3i}(i=1,2,3,4,5,6) are 19 regression coefficients.

The prediction model calibrations and validations were performed. The procedures of establishing a firmness calibration model were as follows: (1) LD parameters (a_{1i},a_{2i} and a_{3i}) for Filter 1 through Filter 6 were calculated for each sample from the six spectral scattering images; (2) the 100 samples were divided into two groups randomly: 75 samples for calibration and the remaining 25 samples for validation; (3) A cross-validation method, in which one sample was left out each time, was applied to the calibration data set to remove extreme samples, as a result, about 8% of the total calibration samples were removed with 69 samples left for calibration; (4) The predictive models were built under six different wavebands according to Eq.4. The performance of the final MLR predictive models are evaluated in terms of standard error of calibration (SEC) and standard error of prediction (SEP) and the correlation coefficient (r).

3.2.1 Model of apple firmness

According to the Eq.4, the model of apple firmness was built with multi-linear regression as follows,

$$F = f_0 + \sum_{i=1}^{N} (f_{1i}a_{2i} + f_{2i}a_{2i} + f_{3i}a_{3i}) \quad (5)$$

where F is predictive results of firmness.

Multi-linear regression analysis indicated that when only one wavelength was used, the first wavelength (680nm) gave best firmness predictions among the six wavelengths, with r equal to 0.698 and SEP=0.60 N for validation samples. Firmness predictions from the other five wavelengths were not as good as those from the first one, with r being between 0.423 and 0.686.When two wavelengths were used, improved firmness predictions were observed: r=0.805 and SEP=0.51 N for the validation samples. Firmness predictions continued to improve when three or four wavelengths were used, but the improvements were much smaller in terms of r and SEP values. There results indicated that the best single wavelength was 632nm with r=0.706 and SEC=0.59; the best double wavelengths combination were 650nm and 900nm with r=0.837 and SEC=0.46; the best three wavelengths combination were 632nm, 650nm and 850nm with r=0.869 and SEC=0.52; the best four wavelengths combination were 632nm, 650nm, 670nm and 900nm with r=0.880 and SEC=0.52.More wavelength tend to improve prediction results, but the improvement is much smaller. When four wavelength combinations were used, the correlation coefficient was 0.880 for calibration and 0.869 for validation, and the SEC and SEP were 0.52 N and 0.53 N, respectively.

Through compared with best single wavelength, double wavelengths, three wavelengths and four wavelengths, the best four wavelengths combination has the best calibration and validation results for firmness prediction of apple fruit by a multi-linear regression, then the firmness prediction model was developed.

3.3 SSC Prediction Model

According to the Eq.4, the model of apple SSC was built with multi-linear regression as follows,

$$S = f_0 + \sum_{i=1}^{N} (f_{1i}a_{2i} + f_{2i}a_{2i} + f_{3i}a_{3i}) \qquad (6)$$

where S is predictive results of SSC.

Multi-linear regression analysis indicated that when only one wavelength was used, the first wavelength (680nm) gave best SSC predictions among the six wavelengths, with r equal to 0.610 and SEP=0.61N for validation samples. SSC predictions from the other five wavelengths were not as good as those from the first one, with r being between 0.336 and 0.564.When two wavelengths were used, improved SSC predictions were observed: r=0.761 and SEP=0.64 N for the validation samples. Firmness predictions continued to improve when three or four wavelengths were used, but the improvements were much smaller in terms of r and SEP values. There results indicated that the best single wavelength was 632nm with r=0.622 and SEC=0.58, best double wavelengths combination were 632nm and 650nm with r=0.776 and SEC=0.62; best three wavelengths combination were 632nm, 650nm and 780nm with r=0.831 and SEC=0.55; best four wavelengths combination were 650nm, 670nm, 780nm and 850nm with r=0.813 and SEC=0.67. More wavelengths tend to improve prediction results, but the improvement is much smaller. When four wavelength combinations were used, the correlation coefficient was 0.813 for calibration and 0.805 for validation, and the SEC and SEP were 0.67 N and 0.68 N, respectively. However the three wavelength combination with r=0.831 for calibration and r=0.819 for validation, and the SEC and SEP were 0.55 N and 0.55 N, respectively.

Through compared with best single wavelength, double wavelengths, three wavelengths and four wavelengths, the best three wavelengths combination has the best calibration and validation results for SSC prediction of apple fruit by a multi-linear regression, then the SSC prediction model was developed.

4. CONCLUSIONS

Spectral scattering images at six selected wavelengths were useful for predicting apple fruit firmness and SSC. Comparing three mathematical functions *i.e.* LD, ED, and GD, LD was the best for fitting the scattering profiles for Red Delicious apples acquired by a spectral imaging system.

Parameters of the LD were linearly related to fruit firmness and SSC. Multi-linear prediction model was established between LD parameters and fruit firmness and SSC. This research demonstrated that the best model with three wavelengths was able to predict apple SSC with r=0.831, SEC=0.55 and four wavelengths was able to predict apple firmness with r=0.880, SEC=0.52. Spectral scattering technique is a promising technique for non-destructive sensing of apple fruit firmness and SSC. Further research is being conducted to improve the spectral imaging system for real time acquisition of scattering images from apple fruit and develop more effective and efficient algorithms to classify fruit into many different grades based on their firmness and SSC.

ACKNOWLEDGEMENTS

This work has been financially supported by the National Natural Science Foundation of China (No. 30460059) and The Science Foundation of Key Laboratory of Nondestructive Test in Nanchang University of Aviation of China Education Ministry for assistance.

REFERENCES

D. C. Slaughter, 1995, Nondestructive Determination Of Internal Quality In Peaches And Nectarines. Transactions Of The Asae, 38(2): 617-623.

J. Lammertyn, B. Nicolai, K. Ooms, V. De Smedt, J. De Baerdemaeker, 1998, Nondestructive Measurement Of Acidity, Soluble Solids, And Firmness Of Jonagold Apples Using Nir-Spectroscopy. Transactions Of The Asae 41(4): 1089-1094.

R. Lu, 2001, Predicting Firmness And Sugar Content Of Sweet Cherries Using Near-Infrared Diffuse Reflectance Spectroscopy. Transactions Of The Asae, 44(5): 1265-1271.

R. Lu, D.E. Guyer, R.M. Beaudry, 2000, Determination Of Firmness And Sugar Content Of Apples Using Near-Infrared Diffuse Reflectance. J. Texture Stud., 31: 615-630.

S. Kawano, H. Watanabe, M. Iwamoto, 1992, Determination Of Sugar Content In Intact Peaches By Near Infrared Spectroscopy With Fiber Optics In Interactance Mode. J. Japan. Hort. Sci. 61(2): 445-451.

V.A. Mcglone, H. Abe, S. Kawano, 1997, Kiwifruit Firmness By Near Infrared Light Scattering. J. Nir Spectros. 5: 83-89.

Y. Peng, R. Lu, 2006, An Lctf-Based Multispectral Imaging System For Estimation Of Apple Fruit Firmness: Part I. Acquisition And Characterization Of Scattering Images. Transactions Of The Asae. 49(1): 259-267.

Y Peng, R Lu, 2006, An Lctf-Based Multispectral Imaging System For Estimation Of Apple Fruit Firmness: Part Ii. Selection Of Optimal Wavelengths And Development Of Prediction Models. Transactions Of The Asae. 49(1): 269-275.

ON-LINE DETECTING SIZE AND COLOR OF FRUIT BY FUSING INFORMATION FROM IMAGES OF THREE COLOR CAMERA SYSTEMS

Xiaobo Zou[1,*] , Jiewen Zhao[1]

[1] *School of Food and Biological Engineering, Jiangsu University , Zhenjiang 2120132*

[*] *Corresponding author, Address: School of Food and Biological Engineering, Jiangsu University , Zhenjiang 212013, Jiangsu Province, P. R. China, Tel: +86-511-88780201, Fax: +86-511-88780201, Email: zou_xiaobo@ujs.edu.cn*

Abstract: On the common systems, the fruits placed on rollers are rotating while moving, they are observed from above by one camera. In this case, the parts of the fruit near the points where the rotation axis crosses its surface (defined as rotational poles) are not observed. Most researchers did not consider how to manage several images representing the whole surface of the fruit, and each image was treated separately and that the fruit was classified according to the worse result of the set of representative images. Machine vision systems which based 3 color cameras are presented in this article regarding the online detection of size and color of fruits. Nine images covering the whole surface of an apple is got at three continuous positions by the system. Solutions of processing the sequential image's results continuously and saving them into database promptly were provided. In order to fusing information of the nine images, determination of size was properly solved by a multi-linear regression method based on nine apple images' longitudinal radius and lateral radius, and the correlation coefficient between sorting machine and manual is 0.919, 0.896 for the training set and test set. HSI (hue-saturation-intensity) of nine images was used for apple color discrimination and the hue field in 0o~80o was divided into 8 equal intervals. After counting the pixel in each interval, the total divided by 100 was treated as the apple color feature. Then 8 color features were got. PCA and ANN were used to analysis the 8 color features. There is a little overlapped in the three-dimensional space results of PCA. An ANN was used to build the relationship between 8 color characters and 4 apple classes with classification accuracy for the training/test set 88%/85.6%.

Keywords: Apple, sequential image, size, color, detection

Please use the following format when citing this chapter:

Zou, X. and Zhao, J., 2009, in IFIP International Federation for Information Processing, Volume 294, *Computer and Computing Technologies in Agriculture II, Volume 2*, eds. D. Li, Z. Chunjiang, (Boston: Springer), pp. 1087–1095.

1. INTRODUCTION

The external appearance is one of the most important factors in pricing the apples. Nowadays, several manufacturers around the world produce sorting machines capable of pre-grading fruits by size, colour and weight. Numerous studies have been conducted in order to perform non-destructive measurements of the quality parameters of fresh fruits. Characterization of apple features included the presence of defects, the size, the shape and the colour. Descriptive variables are, e.g. the roundness, the diameter, the average green colour on the apple and the colour properties of defect spots (Tao Y., et al, 1994; Nakano K., 1997; Paulus I., et al, 1997; Leemans V, et al, 1998; Blasco, J., et al 2003). On the common systems, the fruits placed on rollers are rotating while moving. They are observed from above by one camera. In this case, the parts of the fruit near the points where the rotation axis crosses its surface (defined as rotational poles) are not observed. This can be overcome by placing mirrors on each side of the fruit lines oriented to reflect the pole images to the camera, but the quality of images reflect by mirrors are blurred. Another system was presented by Guedalia. He used three cameras observing the fruit rolling freely on ropes.

As it may be seen, most researchers (except V. Leemans, Tao and Guedalia) did not consider how to manage several images representing the whole surface of the fruit. It seems that each image was treated separately and that the fruit was classified according to the worse result of the set of representative images. The objective of this paper was: first, to present a method to combine the data extracted from the different images of a fruit moving on a machine in order to dispose information related to the whole surface of the fruit. Therefore, it would be possible to build a fruit database from which grading can be operated. Second, some methods for improving rapidity and precision of apple inspection were investigated.

2. MATERIALS AND METHODS

2.1 Image acquisition

The external trigger that is composed of an emitter and an acceptor was place at the grading line. The three frame-grabbers grab images when every roller passes through the trigger. As it mentioned above that there are three apples in the view field of each camera, therefore, nine images were grabbed from an apple.

2.2 Image preprocessing

The images in the field of view are not only the apples waiting for measurement but also including the rollers in grading line and other mechanical parts above line. Image preprocessing includes background segmentation, image de-noise, child image segmentation and sequential images processing.

The background is relatively complicated. To get rid of the background, multi-thresholds method was put forward. That is, thc R value in RGB(red-green-blue) and S value in HIS (hue- intensity - saturation) were taken into account. The segmentation values are as follows

$$p(x,y) = \begin{cases} \text{background pixel}: & R < 90 \,\|\, (S < 0.20 \,\&\&R < 200) \\ \text{apple pixel}: & \text{else} \end{cases} \tag{1}$$

There may still be some noises in the image after getting rid of the background, so this paper introduces medial filter to getting rid of the noise. Fig.1 shows the image after background segmentation and de-noise.

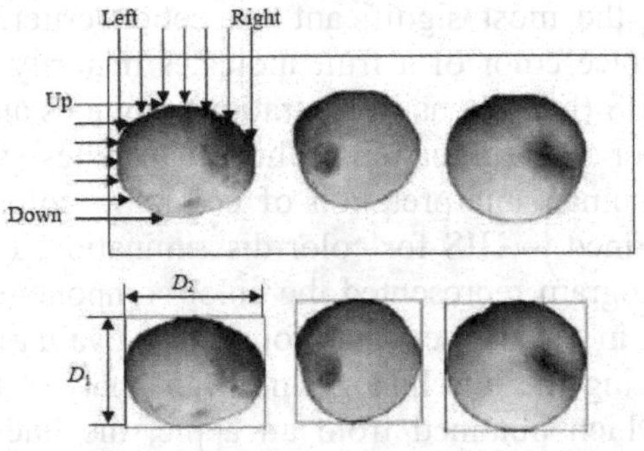

Fig.1: Single child apple images segmentation

There are three apples waiting for measurement in the field of view at most. In order to take out one's own information of the individual apple, single apple division has become inevitable operation. The minimum enclosing rectangle of each single apple was used to divide the view image to three child images as shown in fig.1.

2.3 Size grading

In the image preprocess, the minimum enclosing rectangles of each child single apple image are obtained. It is very easy to get the longitudinal radius (D_1) and lateral radius (D_2) as shown in fig.1. Therefore, 18 D_i (i=1,2, …, 18) are obtained from an apple since nine images were grabbed from am apple.

Any one of the 18 character parameters cannot represent the size of the apple as the apples are randomly oriented on the grading line.

We can imagine that there is some information of apple's size in the 18 characters. We sort the 18 characters in ascending first. Then, a multi-linear regression model was build between apple's size and the 18characters which got from nine apple images of an apple as following:

$$Size = a_1 D_1 + a_2 D_2 + \cdots + a_{18} D_{18} + C \quad (2)$$

where, $a_i (i = 1,2,\cdots 18)$, C are constants. D_i (i=1,2, ..., 18) are characters in sort ascending.

2.4 Color grading

2.4.1 Color feature parameters extraction

Color is one of the most significant inspection criteria related to fruit quality, in that surface color of a fruit indicates maturity or defects. Color representation in HIS (hue-intensity-saturation) provides an efficient scheme for statistical color discrimination. These attributes were the closest approximation to human interpretation of color. So color RGB signals of apple were transformed to HIS for color discrimination. For digitized color image, the hue histogram represented the color components and the amount of area of that hue in the image. Therefore, color evaluation of apples was achieved by analyzing the hue histogram. After analysis the hue values of the nine images which obtained from an apple, the hue values of "Fuji" apple images are mainly between 0°-100°. The hue field in 0°-80° can be divided into 8 equal intervals. The number of pixels in each interval divided by 100 was treated as apple's color feature. Then 8 color features were got. The hue curve of the different category apples is presented in Fig.2. The maximum feature appeared in 0°~20° for Extra "Fuji" apples, 20°~40° for first degree, 40°~60° for substandard degree. There is no maximum feature for second degree.

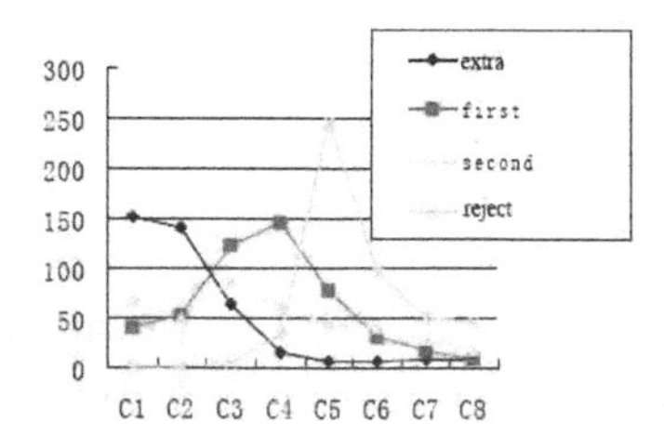

Fig.2: Hue curves of different 'class' Fuji apples

2.4.2 Color pattern recognitions

The 8 color parameters data obtained by the process mentioned above was subjected to PCA and ANN for pattern recognition.

PCA (principal component analysis) is a projection method that allows an easy visualization of all the information contained in a dataset (Buratti et al., 2004; Falasconi et al., 2005). In addition, PCA helps to find out in what respect a sample is different from others and which variables contribute most to this difference.

ANNs (artificial neural networks) are one of the promises for the future in computing. They offer an ability to perform tasks outside the scope of traditional processors. They can recognize patterns within vast datasets and then generalize those patterns into recommended courses of action. A major area where neural networks are being built into pattern recognition systems is the processors for sensors. Sensors can provide so much data that a few meaningful pieces of information can be lost. These neural network systems have been shown successfully in recognizing targets.

While determining the suitable network topology, the network processes the inputs and compares its resulting outputs against the desired outputs. Errors are then propagated back through the system, causing the system to adjust the weights that control the network. This process occurs over and over as the weights are continually tweaked. During the training of a network the same set of data is processed many times as the connection weights are ever refined.

2.5 Fruits

318 apples used in this study were sent directly to our laboratory from a farmer. The size of an apple was measured by manual calipers according to the China grade standards. The 318 "Fuji" apples were divided into 2 sets. An initial experiment was conducted with 200 fruits ("Training set"). The samples were inspected by the machine vision system. Reference measurement for color was then taken. An independent set of 118 samples ("Test set") was fed into the robotic device to assess the efficiency of the on-line machine vision procedure and to test the precision of the on-line machine vision process. The apples in "Training set" and "Test set" were classified into four classes: the three categories Extra, I, II and the reject, as Table 1 shows.

Table 1 318 apples in training set and test set detection by manual

Size and color detection		Samples	
		Training set (200 apples)	Test set (118 apples)
Size measurement	Max (mm)	96.3	89.3
	Min(mm)	56.1	62.6
	Mean (mm)	74.8	75.5
	r	0.919	0.896
Color classification	Extra (fruits)	50	20
	Category I (fruits)	50	41
	Category II (fruits)	50	40
	Reject (fruits)	50	17
	p	88%	85.6%

* *r: Correlation coefficient of size regression model; p:classification accuracy of ANN model*

3. RESULTS AND DISCUSSION

3.1 Apples size determination

200 apples with size between 56 and 96mm were randomly selected. The size (maximal diameter) of each apple was measured twice by the experts using a caliper. Both measurements were compared and the precision was calculated by averaging the differences. Some details of apple quality parameters used for the training and test sets are summarized in Table 1. Then, the 200 apples put to the grading line. 18 D_i (i=1,2, …, 18) characters in sort ascending were got for each apple, and they were used as independent value in the regression model. Regression models were obtained by stepwise algorithm. Finally the size model was determined as follows:

$$Size = 0.508 \times D_{14} + 0.330 \times D_{18} + 12.898 \quad (3)$$

The performance of the size model for training and test sets was shown in table 1, with correlation coefficient of training set/ test set 0.919/0.896.

As form (3) shows that it is not all characters got from the nine images were useful, only D_{14} and D_{18} have high relationship with Size. Apples are randomly oriented on the grading line, therefore, D_{14}, and D_{18} may come from any one of the nine images. There should be mentioned again that the 18 D_i (i=1, 2, …, 18) characters are in sort ascending.

3.2 Apples colour grading

The hue histogram of the nine images was obtained by statistical evaluation of the nine images. Then, the 8 apple color character parameters were calculated and chosen as pattern recognition parameters. Fig.3 exhibits the results of principal component analysis (PCA) for the four different color classes apples. PCA is a simple method to project data from several feature parameters to a three-dimensional space. The values of 85.26% of PCA1, 4.21% of PCA2 and 1.62% of PCA3(Fig.3) indicate contribution rate to pattern separation. It shows that the pattern separation is not sharp.

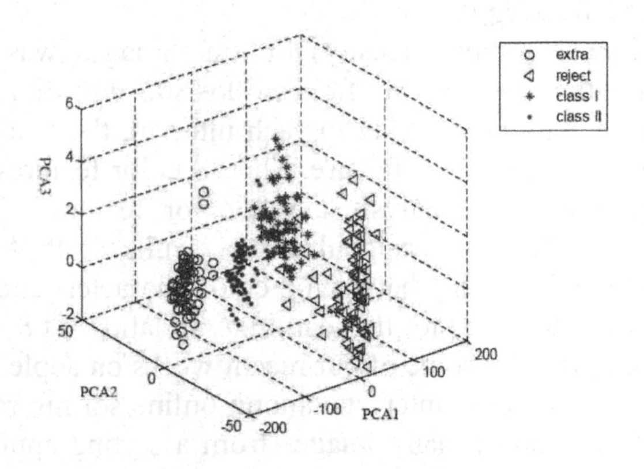

Fig.3: Results of the PCA of the 8 color features for the different apple grades.

In this study, the ANN with a standard back-propagation algorithm was applied. The 8 apple color character parameters were served as the input values for the neural network. The apple's four color grades were coded to serve as the output layer of the neural network: extra (1,0,0,0); class I (0,1,0,0); class II (0,0,1,0); and reject (0,0,0,1). There are 9 hidden nodes in the neural network. Other parameters of the BP-ANN were: Activation: Logistic; Learning Rate: 0.02; Momentum: 0.9. The artificial neural network was trained with the 200 training samples in training set 20,000 times. It was then used to classify the test set, which consisted of 118 'Fuji' apples with different color grades. The classification accuracy for the training set and the classification accuracy for the test set were 88% and 85.6% respectively as shown in table 1.

4. CONCLUSION

The grading of apples into quality classes is a complex task involving different stages. The main conclusions of this study are as follows:

(1) Nine images of an apple were grabbed by three CCD cameras during the motion of the fruit on the grading line.

(2) The apple is segmented from the black background by multi-thresholds *method* with R < 90 || (S < 0.20 & &R < 200) for background pixels, allowed fruits to be precisely distinguished from the background.

(3) 18 D_i (i=1,2, …, 18) characters in sort ascending were got from nine images of each apple and were used as size and shape character parameters. The *width* and *height* regression models were got by stepwise with correlation coefficient of learning set/validation set 0.949/0.936 for *width*, and 0.886/0.853 for *height*.

(4) HSI (hue-saturation-intensity) of nine images was used for apple color discrimination and the hue field in 0o~80o was divided into 8 equal intervals. After counting the pixel in each interval, the total divided by 100 was treated as the apple color feature. Then 8 color features were got. PCA and ANN were used to analysis the 8 color features. There is a little overlapped in the three-dimensional space results of PCA. An ANN was used to build the relationship between 8 color characters and 4 apple classes with classification accuracy for the learning/validation set 88%/85.6%.

This research provides some of our recent works on apple grading projects, and we hope to raise some interests among online sorting researchers about the fusing information of many images from a sorting apple. Enhancement of the grading process should come from every stage, and particularly, from the image acquisition stage.

ACKNOWLEDGEMENTS

This work is supported by "Natural Science Foundation of China" and "863 high-tech fund of China" Fund. We also wish to thank many of our colleagues for many stimulating discussions in this field.

REFERENCES

Blasco, J., Aleixos, N., Molt, E.(2003) Machine Vision System for Automatic Quality Grading of Fruit, Biosystem Engineering, 85(4), 415–423.

Davenel, A., Guizard, C., Labarre, T., Sevila, F., (1988). Automatic detection of surface defects on fruit using a vision system. J. Agric. Eng. Res. 41, 1-9.

Leemans, V., Magein, H., & Destain, M.-F. (2002). On-line apple grading according to European standards using machine vision. Biosystem Engineering, 83(4), 397–404.

Li Q. Z., Wang M. H. (1999) Study on High-Speed Apple Surface Defect Segment Algorithm Based on Computer Vision. Proceedings of 99 International Conference on Agricultural Engineering, Beijing, China, December.

Miller W M (1995). Optical defect analysis of Florida citrus. Applied Engineering in Agriculture, ASAE, 11(6), 855–860

Nakano, K. (1997). Application of neural networks to the color grading of apples. Computers and Electronics in Agriculture, 18, 105–116.

Paulus I., Busscher R. De, Schrevens E.. (1997) Use of image analysis to investigate human quality classification of apples. J. agric. Engng Res. 68,341-353

Penman DW(2002). Determination of stem and calyx location on apples using automatic visual inspection. Computers and Electronics in Agriculture, 33(2002), 7–18

Tao, Y., & Wen, Z. (1999). An adaptive spherical image transform for high-speed fruit defect detection. Transactions of the ASAE, 42(1), 241–246.

Wen, Z., & Tao, Y. (1998). Brightness-invariant image segmentation for on-line fruit defect detection. Optical Engineering, 37(11), 2948–2952.

Yang Q (1993). Finding stalk and calyx of apples using structured lighting. Computers and Electronics in Agriculture, 8, 31–42

A METHOD OF TOMATO IMAGE SEGMENTATION BASED ON MUTUAL INFORMATION AND THRESHOLD ITERATION

Hongxia Wu, Mingxi Li*

Huangshi Institute of Technology,Huangshi, Hubei Province, P.R.China 435003

* *Corresponding author, Address: Editorial Department of Journal of Huangshi Institute of Technology ,Huangshi 430003, Hubei Province, P. R. China, Tel: 15972372916, Email: limx10920@yahoo.com.cn*

Abstract: Threshold Segmentation is a kind of important image segmentation method and one of the important preconditioning steps of image detection and recognition, and it has very broad application during the research scopes of the computer vision. According to the internal relation between segment image and original image, a tomato image automatic optimization segmentation method (MI-OPT) which mutual information associate with optimum threshold iteration was presented. Simulation results show that this method has a better image segmentation effect on the tomato images of mature period and little background color difference or different color.

Keywords: Image segmentation, mutual information, threshold optimization, tomato image

1. INTRODUCTION

Traditional threshold method has only considered gray levels but neglect spatial distribution of gray scale and internal relation between segment and original of the image , such as Otsu algorithm, fuzzy C means algorithm(FCM),adaptive algorithm(Rigau et al.,2004;Zhou Xiaozhou et al.,2007;Lv Qingwen et al.,2006).When lighting is not uniform and noise out-burst or larger background gray scale changed, massive information will lost and the area obtained can't represent the shape of original subject after

---

Please use the following format when citing this chapter:

Wu, H. and Li, M., 2009, in IFIP International Federation for Information Processing, Volume 294, *Computer and Computing Technologies in Agriculture II, Volume 2*, eds. D. Li, Z. Chunjiang, (Boston: Springer), pp. 1097–1104.

segmentation, besides FCM algorithm has the problems such as depended on initial value too much, converge on local maxima and classification number should be pre-determined, and people still can't solve these after a lot of research and improvement. So according to the internal relation between segment image and original image, a tomato image automatic optimization segmentation method which mutual information associate with optimum threshold iteration was presented.

2. OPTIMUM THRESHOLD SEGMENTATION ALGORITHM BASED ON MUTUAL INFORMATION

2.1 Image mutual information measures

Mutual information of image A and B can be defined:

$$I(A,B) = H(A) + H(B) - H(A,B) \tag{1}$$

H(A) and H(B) is the average information quantity of image A and B, and H(A, B) is their correlated average information quantity.

The average information quantity and correlated average information quantity of A and B can be calculated as follows:

$$\begin{cases} H(A) = \sum_{a} -p_A(a)\log p_A(a) \\ H(B) = \sum_{b} -p_B(b)\log p_B(b) \\ H(A,B) = \sum_{a,b} -p_{A,B}(a,b)\log p_{A,B}(a,b) \end{cases} \tag{2}$$

$P_A(a)$ and $P_B(b)$ is the probability density function with gray levels a of image A and gray levels b of image B, $P_{A,B}(a,b)$ is the joint probability density function of image A and B. They can be calculated as follows:

$$\begin{cases} p_{A,B}(a,b) = \dfrac{h(a,b)}{\sum_{a,b} h(a,b)} \\ p_A(a) = \sum_{a} p_{A,B}(a,b) \\ p_B(b) = \sum_{b} p_{A,B}(a,b) \end{cases} \tag{3}$$

$h(a,b)$ is the joint-histogram of image A and B, it means the number of the correlated point pairs with gray levels a of image A and gray levels b of image B.

Mutual information doesn't balance consistency of different image directly dependent on gray levels but the probability of each image and the joint probability when two images combined. So it's not sensitive to gray change or one-to-one gray transformation, and can process the relationship between the positive and the negative image gray simultaneously.

2.2 Optimum threshold iteration algorithm

This algorithm chooses one approximate threshold as the initial value of the estimate, generates sub-images after image segmentation, and chooses new threshold according to the characteristic of sub-images, then segment again by using new threshold. After several circulation, incorrect pixels will minimized. The steps are as follows:

$$T_0 = \frac{Z_{\min} + Z_{\max}}{2} \quad (4)$$

(1) Select the initial threshold: choose the median of the image gray levels as the initial threshold.

$Z_{\min}$, $Z_{\max}$ means the minimum and the maximum of the image gray levels.

(2) Segment the image into two groups by using threshold T_k, and

$$R_1 = \{f(x,y) \mid f(x,y) \geq T_k\} \quad (5)$$

$$R_2 = \{f(x,y) \mid 0 < f(x,y) < T_k\} \quad (6)$$

(3) Calculate the gray means Z_1 and Z_2 of area R_1 and R_2, and

$$Z_1 = \frac{\sum_{f(i,j)<T_k} f(i,j) \times N(i,j)}{\sum_{f(i,j)<T_k} N(i,j)} \quad (7)$$

$$Z_2 = \frac{\sum_{f(i,j)\geq T_k} f(i,j) \times N(i,j)}{\sum_{f(i,j)\geq T_k} N(i,j)} \quad (8)$$

$f(i,j)$ is the gray level of (i,j) point, $N(i,j)$ is the weight coefficients of (i,j) ,usually $N(i,j) = 1.0$.

(4) Calculate new threshold T_{k+1}:

$$T_{k+1} = \frac{Z_1 + Z_2}{2} \quad (9)$$

(5) If $T_k = T_{k+1}$, then circulation ends, otherwise $k = k + 1$, turns to step 2. T_{k+1} is the optimum segment threshold after circulation ends.

2.3 Optimization segmentation algorithm

Mutual information Registration is the best creation of the retrospective registration method of registration accuracy and robustness, segmentation can be regarded as the image degeneration, and segment can be treated as a kind of special form, mutual information reaches maximum when spatial location of segment and original image is consistent and calculated area matches original subject's shape, and optimization segmentation results can be considered contains the most original image information quantity. So, image segmentation algorithm based on mutual information and optimal threshold iteration can be described: assume I, I_T is original and image segmented choose initial value T which is determined by using optimal threshold iteration algorithm, maximum mutual information quantity is optimal goal, searching the optimal threshold $T_{optimal}$ in the range $[T-\delta, T+\delta]$:

$$T_{optimal} = \arg\max(MI(I, I_T)) \tag{10}$$

Its algorithm flow shows as Fig. 1

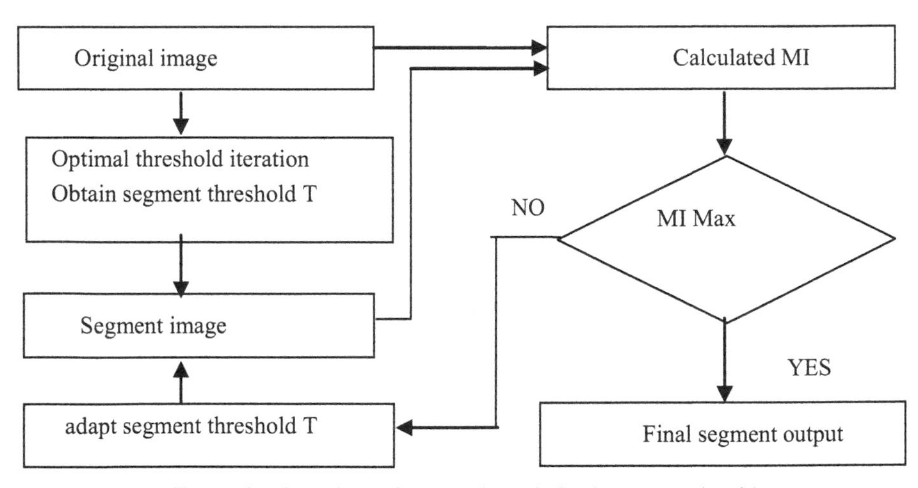

Fig. 1. the flow chart of automatic optimized segment algorithm

3. TOMATO IMAGE SEGMENTATION COMPARISON EXPERIMENT

According to the segmentation principle above, fig 2～6 gives three different tomato segment results of different growing status and color by using Otsu algorithm, EN-2D algorithm(Zhang Honglei et al., 2007; Vincent,

1993; Gao Hai et al.,2006)and MI-OPT algorithm. For better visual effect, each object contour of binary processed images stack s on original images in every picture (black line border of left images).

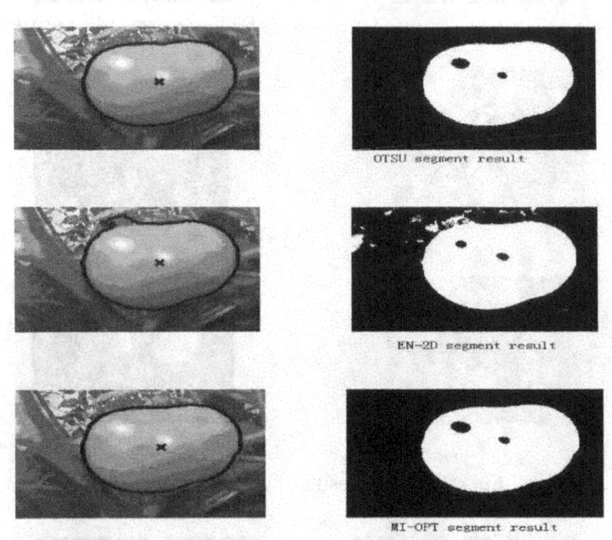

Fig. 2. simple fruit color consistent segment result

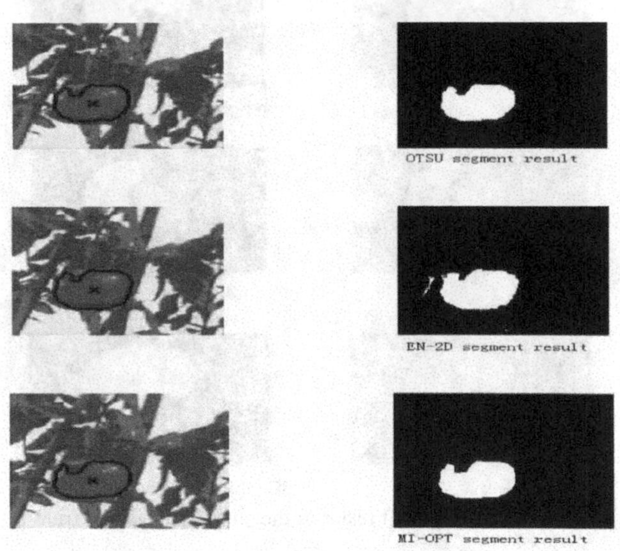

Fig. 3. slight mask segment result

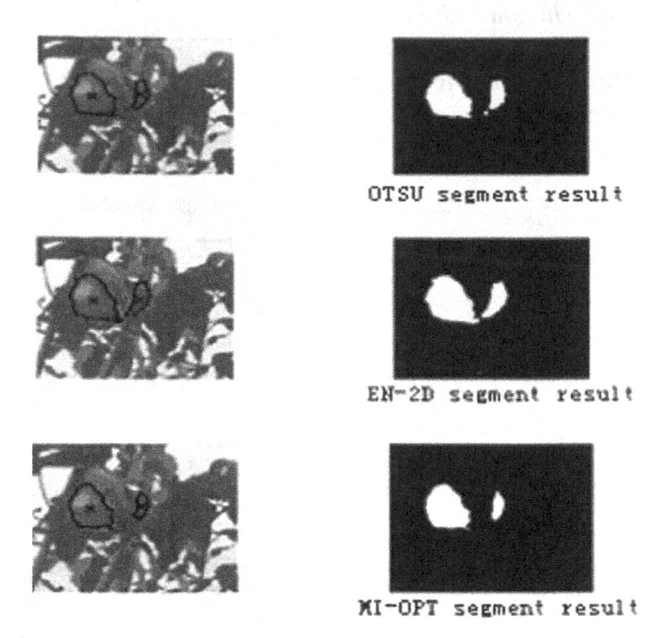

Fig.4. segment result of the serious mask fruit

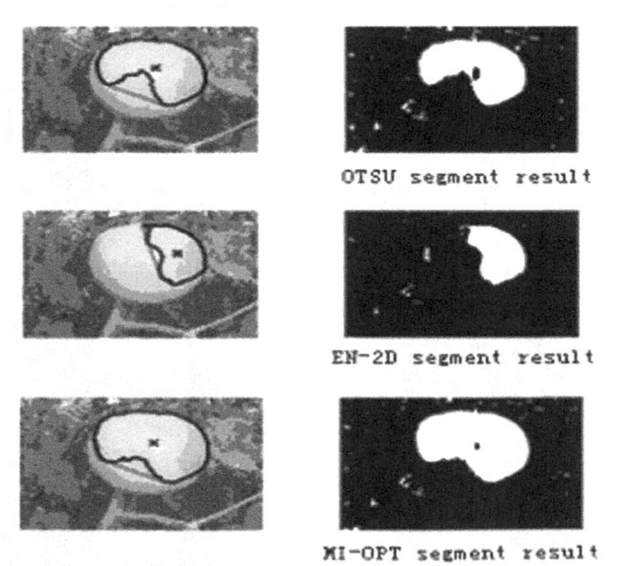

Fig.5. segment result of the color inconsistent fruit

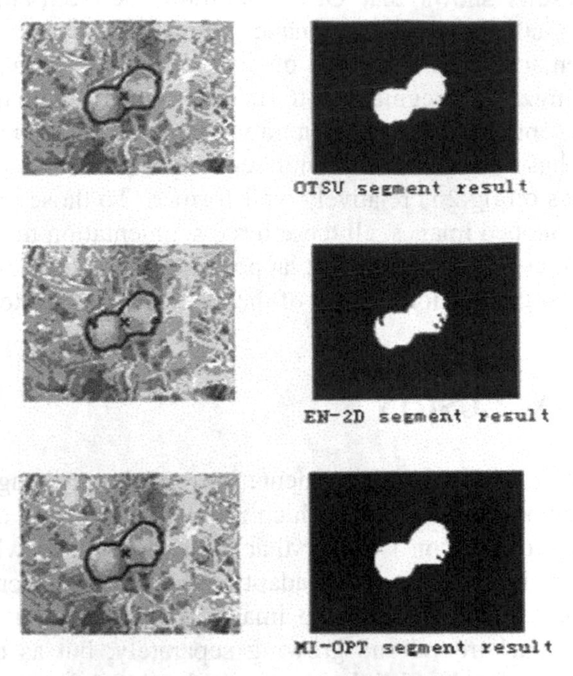

Fig.6. segment result of connected fruits

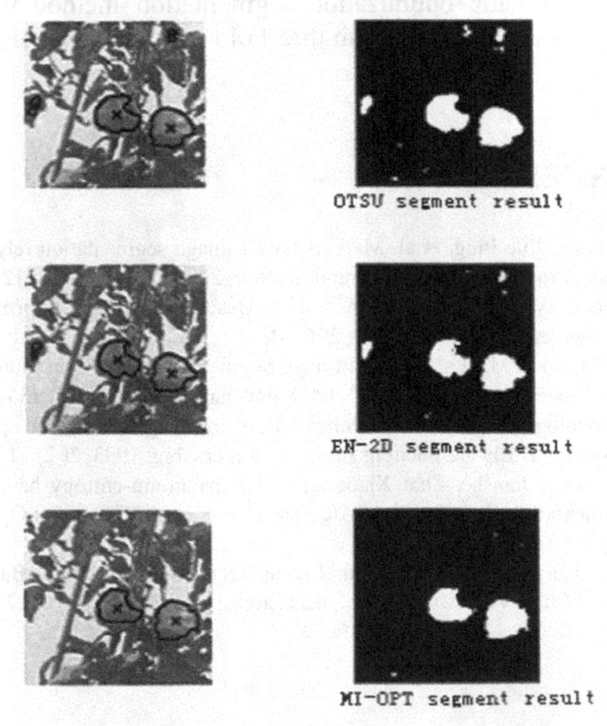

Fig. 7. segment result of color inconsistent fruits

All these results shows that Otsu maximum between-cluster variance segmentation method based on image color difference, 2D-maximum entropy segmentation method based on image color difference and image automatic optimization segmentation method which mutual information associate with optimum threshold iteration can all segment ripe tomato images which has consistent color and separated, object contour of binary processed shows on fig 2 is relatively well-formed. To those different colors or occlusion or lapped images, all these three segmentation method can't get satisfied results, especially multi-fruit lapped and serious occlusion shows on fig 4, and other segmentation method of these images needed to be explored.

4. CONCLUSION

In terms of these three segmentation method, image automatic optimization segmentation method which mutual information associate with optimum threshold iteration is better than other two, like what shows on fig3 and 5.All in all , these three self-adaptive threshold segmentation have a stable and good effect to the tomato images of larger color difference of tomato with its background and growing separately, but as to the tomato images of mature period and little background color difference or different color, image automatic optimization segmentation method which mutual information associate with optimum threshold iteration is better(shows on fig 5).

REFERENCES

Gao Hai, Lin Weisi, Xue Ping, et al. Marker- based image segmentation relying on disjoint set union,Signal Processing: Image Communication, 2006, 21(2) : 100- 112

Lv Qingwen,Chen Wufan. Image Segmentation Based on Mutual Information, Chinese Journal of Computers,2006，29（2）：296-301

Rigau J, Feixas M, Sbert M, et al. Medical Image Segmentation Based on Mutual Information Maximization,Proceedings of MICCAI’04, Saint-malo, France, 2004: 135-142

Vincent L.Morphological grayscale reconstruction in image analysis: applications and efficient algorithms,IEEE Transactions on Image Processing, 1993, 2(2) , 176- 201

Zhang Honglei, Song Jianshe, Zhai Xiaoying.A 2D maximum-entropy based self-adaptive threshold segmentation algorithm for SAR image processing,Electronics Optics & Control, 2007，14（4）：63-65

Zhou Xiaozhou,Zhang Jiawan,Sun Jizhou. Image Segmentation Method Based on Mutual Information and Chan-Vese Model ,Computer Engineering,2007，33（22）：220-222

KEY OF PACKAGED GRAIN QUANTITY RECOGNITION——RESEARCH ON PROCESSING AND DESCRIBING OF “FISH SCALE BODY”

Ying Lin[1,2], Xinglin Fang[1], Yueheng Sun[2,*], Yanhong Sun[1]

[1] *College of management, Chongqing Jiao Tong University, Chongqing, P.R China,400074;*
[2] *School of Computer Science and Technology, Tianjin University, Tianjin, China, 300072;*
[*] *Corresponding author, Address: School of Computer Science and Technology, Tianjin University, Tianjin, China, 300072, Tel: +86-22-27401016, Email: yhs@tju.edu.cn*

Abstract: The key to identifying the packaged grain is the shape of package, and the key to identifying shape is processing and describing the boundary of package. Based on a lot of analysis and experiment, this article select the canny operator and chain code to process and describe the boundary of package. Aiming at the boundary is not absolute connectivity, the closure operation of Mathematical Morphology is introduced to do pretreatment on binary image of packaged grain. Finally the boundary is absolute connectivity. Experiments show that the proposed method enhances the anti-jamming and robustness of edge detection.

Keywords: edge detection; Mathematical Morphology; Chain code

1. INTRODUCTION

The technique core of grain reserves automatically supervision and audit system which based on video is to identify the grain deport scene video, accurately gets the sum of grain quantity to replace the mode that pass by manual's supervision, and eradicate completely empty grain deport and make a false report to get more interest phenomenon. Thus, administration section is satisfied with requirement of grain reserves automatically supervision and audit (Lin et at., 2007).

Please use the following format when citing this chapter:

Lin, Y., Fang, X., Sun, Y. and Sun, Y., 2009, in IFIP International Federation for Information Processing, Volume 294, *Computer and Computing Technologies in Agriculture II, Volume 2*, eds. D. Li, Z. Chunjiang, (Boston: Springer), pp. 1105–1114.

The grain packages in the grain deport three-dimensional pile up by pursueing layer toward the vertex layer from the first floor, forming a big cube type structure. In the image catched by camera device, the big cube contains three sides, is vertex face and two on the sides respectively. Because of illumination and the mode of pileing up, in the video image, the gray level in different place of the grain package surface will have very big difference, the center region of grain package would be brighter, but the peripheral region is relative ash dark, so on the vision, the grain package edge presents an approximate ellipse and all grain packages ellipse edge connectivity together seems to be fish scale, so we call "fish scale body" to the ellipse that the grain packages forming. Each grain package can see three different sides in the video image, so the grain package will form three classes different ellipse "fish scale body". If we accurately identify the quantity of the "fish scale body" on the three sides of the big cube, so the quantity of grain package can identify. Because the weight of each packaged grain is fixed and have already knew, we will obtain the weight of the whole cube grain pile, therefore, the packaged grain quantity recognition is to the recognition of the ellipse "fish scale body" quantity. So the key to identifying the quantity of packaged grain of grain deport is to find a fast, accurate and valid method to extract and describe the shape of ellipse "fish scale body" to assured accurately identify each ellipse "fish scale body".

The image processing of ellipse "fish scale body" mainly includes ellipse "fish scale body" edge location and description. This paper final select Canny operator to detect the edge and edge direction chain code to describe the edge of grain package based on analyzing, researching and doing experiment on the existing edge detection operator. Aiming at the detected ellipse "fish scale body" edge has a little incomplete occlusive character, this paper adopts the close operate of mathematical morphology to repair and link the edge nick. Finally the experiments show that canny operator and edge direction chain code detect and describe ellipse "fish scale body" have a higher robustness, can acquire better effect..

2. "FISH SCALE BODY" EDGE DETECTION RESEARCH

Edge means that the difference of gray value of adjacent pixel in the image contrasts sharply, it is extensively exist between target and other target, target and background (Yao et at., 2006). The image edge is one of the most important features of image, which is an important basis of image analysises, such as image segmentation, texture feature and shape feature. Specific to the packaged grain image, the edge described the shape information of ellipse "fish scale body", and the shape information is the basis to exactly identify and analyze the ellipse "fish scale body", so the chief operation to the packaged grain image is the edge detection of ellipse "fish scale body".

So far people have put forward various different edge detection algorithms, such as Roberts operator, Sobel operator, Prewitt operator, Log operator, Canny operator an so on(Gong et at., 2006). In the digital image, edge and noise pixel point both belong to the gray value abrupt change point, so in the edge detection, edge detection capability is in contradiction with noise suppression capability. Some algorithms have a better capability of edge detection, others have a better antinoise ability. Each operator has its own advantage.

2.1 Operators

The Roberts operator is also called the operator of gradients cross, it makes use of a local difference operator to look for edge, and provides a simple approximate method for calculating the gradient amplitude:

$$G(i,j) = |f(i,j) - f(i+1,j+1)| + |f(i+1,j) - f(i,j+1)| \quad (1)$$

The Roberts operator has powerful ability in edge detection, and would be highly effective to steep low noise image. But it easily to lose partial edges, is better to have a steep low noise image substitution.

The Sobel operator adopts neighborhood means, so it can avoid calculating gradient between the pixels inside points. The Sobel operator puts a focal point in the pixel which is close to template center, and the order of operate is weighted average, then differential, lastly calculated gradient. The Sobel operator is preferable in processing the image which its gray value gradual changed and has low noise, but it isn't isotropic, so edge detection is not complete connectivity, and the edge detected easily appears many pixel widths.

The fundamental principle of the Prewitt operator is the same with the Sobel operator, but the two convolution templates is not the same with the Sobel operator.

The first derivative will get extremum at the position where the pixel gray value tremendous change, in contrast, the first derivative value is zero at others position. The extremum point of the first derivative definitely lead to the second derivative arise passing zero positions, which apropos correspond to edge of original image. So we can make use of passing zero positions of the second derivative to detect the image edge.

In the two-dimensional space, a kind of common second derivative operator is the Laplace operator, Laplace expression at a point of a continuous function is:

$$\nabla^2 f(x,y) = \frac{\partial^2 f(x,y)}{\partial x^2} + \frac{\partial^2 f(x,y)}{\partial y^2} \tag{2}$$

When it comes to the digital image, we can also resort to some templates to calculate the Laplace value. But, the Laplace operator has two defects, one is to lose a edge direction information and the other is to dobule the influence of noise because the Laplace operator is a second difference operator.

Marr and Hildreth combine Gaussian filter and Laplace edge detection together, forminged a LOG operator. First, we use a Gaussian function first to smooth the image, and then use the Laplace operator to calculate gradient value. So this operation smoothes image as well as reduces noise and the isolated noise pixel and smaller structure organization will be filtered.

While applying a LOG operator, it is very important to select the variance parameter σ of the Gaussian function, which plays very great impact to the image edge detection precision. The Gaussian filter is a low pass filter, so the σ is bigger, and can repress of higher frequency noise, avoiding detecting the deceitful edge, but the image edge of pixel signal also smoothing and result in some edges point to lose. Whereas, the σ is smaller, can detect the detail of the high frequency of image, but the capability of repress noise drooped, and easily appear a deceitful edge.

The Canny operator is a optimization approach operator of three criterion such as signal to noise ratio, positioning precision, single pixel edge (Canny et at., 1986).

The Canny operator is according to optimization method, it adopts a Gaussian filter to carry on smoothing the image, so it has stronger capability of noise suppression control and can detect precise weak edge, as well the Canny operator will also lead to smooth some high frequency edge, result in losing edge information. The Canny operator adopted two thresholds algorithm to detect and link edge, it adopts many scale detection and directive search, so it better than the LOG operator, but it also increased complications of the algorithm in the meantime.

2.2 Experiment results

The edge detection effect of ellipse "fish scale body" by every edge detection operator is shown in Fig.1.

After Carrying out gray scale transformation, brightness correction and filter, the packaged grain image as figure 1(a) show. Then separately use Roberts operator, Sobel operator, Prewitt operator, LOG operator, Canny operator to detect edge of figure 1(a), results shown as figure 1(b)~(f). Obviously, the edge detection by the Roberts operator is coarse, handles the detail does not sufficient and omits a great deal of edge information. The edge detection effect of Sobel operator and Prewitt operator is similar, and handle the detail isn't very good. By contrast, LOG operator and Canny operator handle the detail is better than other operators, but in the aspects of detected edge smoothing and connectivity, the Canny operator is obviously better than LOG operator.

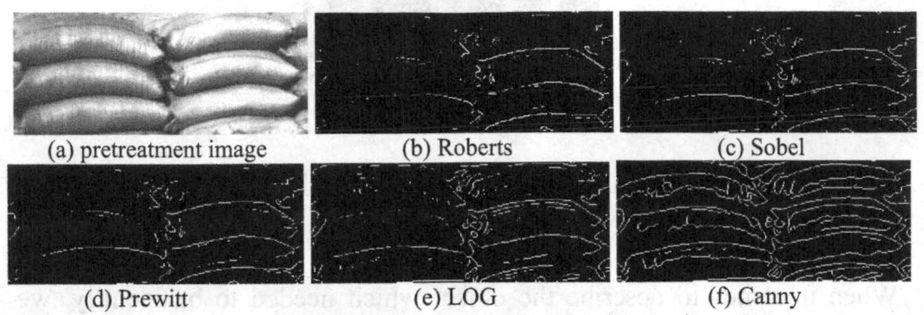
(a) pretreatment image (b) Roberts (c) Sobel
(d) Prewitt (e) LOG (f) Canny

Fig. 1: Each operator edge detection result

In order to effectively detect the edge of the ellipse "fish scale body" of packaged grain image, the edge detection algorithm choose by us should be precision positioning edge as well as make the edge curve is as far as possible close, and easy to descript edge and fetch feature. Finally we select the Canny operator by experiments results, finally we choose the Canny operator to detect the bag food image. Though the Canny operator has better detection effect, but we see from the figure 1(f) that the ellipse "fish scale body" edge still isn't complete close, and has great quantities of deceitful edge, which make ellipse "fish scale body" of the edge description will can't carry on. The reasons for phenomenon are that the contrast degree of the gray image is lower; On the other hand, the image is complicated and has many target body. Therefore, if we only depend on an edge detection operator, the edge curve is very rare close.

Matheron proposed the Mathematical Morphology in 1964, then Meyer, Serra, Sternberg and other scholars spent a lot of energy on researching on the Mathematical Morphology, which has already become a kind of important means in the non - linear image processing (Fan et at., 2007).The Mathematical Morphology implements target feature extraction through selecting suitable structure element (probe), and its fundamental operations include dilatation and decay operation. The open and close operation is a compound operation combinated by dilatation and decay operation. The close operation can clean the eyelet and fill up narrow split and long thin gutters and so on in the image edge region, which is suitable for getting rid of split of "fish scale body" edge to make edge is close. Because the Mathematical Morphology is according to the binary image, we must add image binaryzation operation to the image pretreatment. Aiming at the ellipse "fish scale body" shape character, we select a elliptic structure element to carry on the close operation of the Mathematical Morphology in figure 1(a) which is binarization processed, then carry on a Canny edge detection, finally, we get the edge profile of "fish scale body" shown in Fig.2.

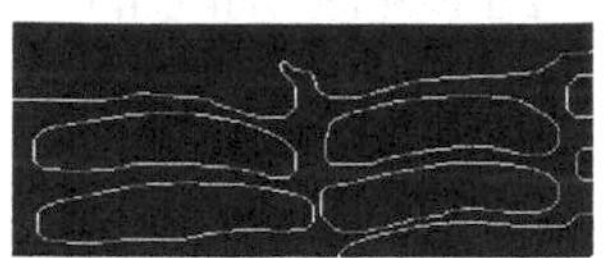

Fig. 2: Edge detection after close operation

3. "FISH SCALE BODY" EDGE DESCRIPTION RESEARCH

When it comes to describe the object which needed to be identify, we wish that we can a method which can provide more abundant detail information than a single parameter instead of image to describe the object. The image description can according to its internal feature, can also according to its exterior feature, so we can divide the description of image into the edge description (chain code, edge segmentation etc.) and region description (quad tree, skeleton etc.). Usually, edge description concerns about the shape feature of region in the image, but the region description is inclined to features, such as gray value, color and texture...etc. In the packaged grain automatic recognition, what we concern is the shape of the ellipse "fish scale body", therefore, we select edge description method to describe ellipse "fish scale body".

In the operation of identifying the ellipse "fish scale body", we request that strictly to carry on the statistics various characteristic parameters value according to the actual shape of ellipse "fish scale body", therefore, the accurate description edge of ellipse "fish scale body" is very important. The domestic and international scholars have already researched how to describe two-dimensional profile curve from different angles, and proposed various edge description methods.

Approximative polygon matching profile curve (Kong et at., 2001). The sixth reference proposed a new method which use approximative polygon of profile curve to express edge. There is many common methods such as based on contractive minimum circumference polygon method, based on polymeric minimum mean square error line segment approximation method, and based on split minimum mean square error line segment approximation method. In the digital image, many pixel points constituted edge curve. If each adjacent pixel points constitute a line, so we can use many line to accurately express edge. Obviously, this kind of method have excessive amount of computation and the real-time is worse. If we use discretization edge pixel points according to the certain algorithm to express edge, computation amount decreases, but edge description precision consumedly descends, so this method is not suitable for the description of "fish scale body" edge.

Based on B Splines edge curve representation method(Cohen et at., 1995). The seventh reference proposed a new method which used B Splines to express edge. The advantage of this method is in consideration geometrical characteristic of profile curve to some extent. But sampling and fitting to the profile curve is a very complicated, so this method is not suitable for the description of "fish scale body" edge.

Edge chain code method. In 1977, Freeman firstly proposed 4 chain code and 8 chain code, so the chain code is also called Freeman code (Freeman H et at., 1977). The chain code method uses coordinates of edge starting point and point direction code of edge curve to describe a edge curve, and it is often used to express curve and edge region in the domain, such as image processing, computer graphics and pattern recognition etc. The method makes use of a series of particular length and connect with each other of direction line to express the edge of target. Because of length stationary and direction number finite of each line segment, only the starting point needs coordinates, other points can use direction to express an offset, so we can use a direction number to replace two coordinates numbers for saving bit. Obviously, the chain code can consumedly reduce the data quantity which edge representation needs. The chain code has some advantage such as simple, saving storage space, easy to compute, translation invariant and so on, so it is easy to the statistics characteristic parameter of close region.

There are many "fish scale body" in the packaged grain image, so we demand to carry on description and recognition for each "fish scale body", and need to adopt a calculate easily, small memory space, real-time high algorithm. The edge of "fish scale body" is a close curve, therefore, this paper uses chain code method to describe "fish scale body" edge.

The sampling of digital image is according to stationary spacing of the mesh, so the rule of the most simple chain code is to track edge and assign a direction value to each two adjacent pixels. The four directions and eight directions chain code is common (Gong et at., 2006), and the eight directions chain code increases four directions than four directions chain code. In the digital image, each pixel point has eight adjacent points, and the eight directions chain code corresponds with the practical situation, so it can accurately describe center pixel point and its adjacent pixel points information. In eight directions chain code, chain code along edge curve pixel points moves and codes by eight adjacency mode. two close together The offset of two adjacent pixels uses digit 0~7 representation. The code value add 1, the direction revolve 45° by inverse hour.

A edge curve can be only definite by the starting point and chain code of the edge curve. If curve S shows as Fig.3, and its chain code expresses by array Lianma[].

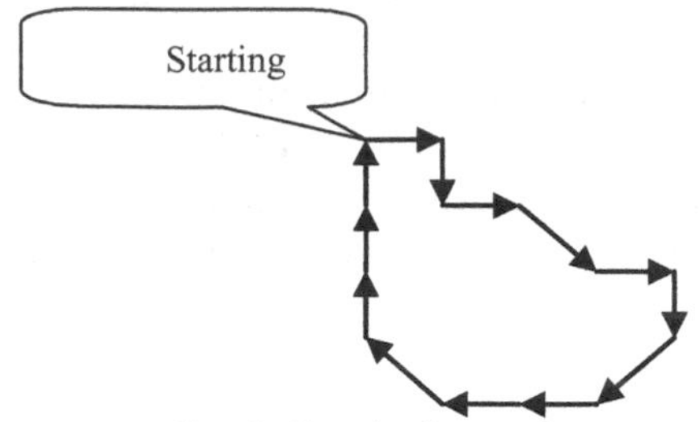

Fig. 3: Sketch of curve S

We take eight directions chain code to describe the curve S, then we get an array Lianma[]={0, 6, 0, 7, 0, 6, 5, 4, 4, 3, 2, 2, 2}. Obviously, as long as any point coordinates and others chain code value are known, we can accurately describe the edge of region. For getting a integrated edge chain code of, and easy to follow-up operation, such as storing the chain code, feature extraction, reconstruction...etc. we can store the starting point coordinates, chain code and chain code length of curve in a table(Li et at., 2008). Chain code table type such as Fig.4shows.

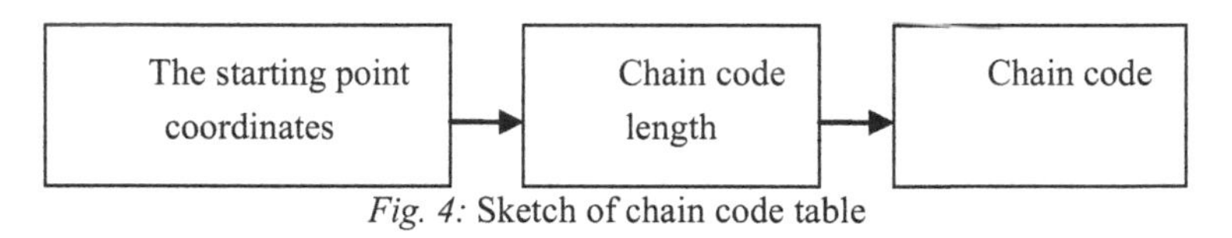

Fig. 4: Sketch of chain code table

The edge of an object can express by the coordinates of starting point and a sequence of direction code. Aiming at binary packaged grain image which carried on edge detection, we take the most top left corner of "fish scale body" edge profile which got by line scan method for the trail starting point, adopting eight directions chain code to traversal "fish scale body" edge. The algorithm steps of chain code tracks the edge of packaged grain image as Fig.5 shown:

We take Edge of image tracking operation for extracting the feature of object body in the image, such as the endpoint of edge curve, point of intersection, corner - point, centroid of close curve etc., in order to identify object body. After edge detecting the edge curve of ellipse "fish scale body" in the packaged grain image, we find that it looks like an ellipse. According to mathematical knowledge we can pass judgments as follows: (1) Establish an ellipse variance threshold, judge whether edge curve is a close ellipse shape or not. (2) If the edge curve is an approximate ellipse, we can take linear combination by a series of ellipse mathematics characteristic for the membership function to identify packaged grain image. The mathematics characteristic of ellipse has circumference, centroid, area, eccentricity etc., and these parameters very easy to get by edge chain code computation (Li et at., 2008). We can adopt linear combination of characteristic to identify the ellipse edge of packaged grain image, to distinguish "fish scale body" on the different face.

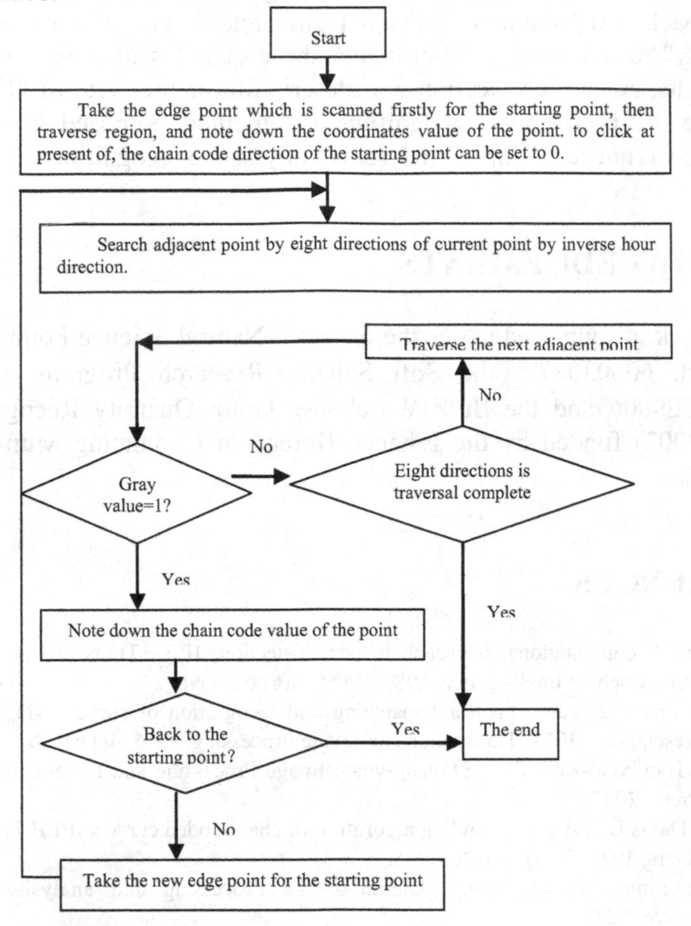

Fig. 5: Sketch of chain code method flowchart

4. CONCLUSION

In contrast, Canny operator is a well-defined operator, its definition strict, has high signal to noise ratio and detection precision, and the edge detection is more smoother, continuous, legible, so it is fit for edge detection of ellipse "fish scale body". Aiming at the detected ellipse "fish scale body" edge has a little incomplete occlusive character, this paper adopts the close operate of mathematical morphology to repair and link the edge nick. Finally, we get a complete close edge. Compared with other methods, Chain code method is simple, can reduce storage data and raise image processing velocity and match efficiency, and have a translation invariant characteristic, so it is easy to statistics the characteristic parameter of the closed curve region.

This text combines Canny operator and edge direction chain code together used for packaged grain image edge detection and description of ellipse "fish scale body" in the image. Experiments show that this method can acquire good results. Accurate detection and description of the edge of "fish scale body" are the premise of recognition, the method proposed by this text satisfied the request of ellipse "fish scale body" edge recognition.

ACKNOWLEDGEMENTS

This work is supported under the National Natural Science Foundation of China No. 60603027 , the Soft Science Research Program of CSTC N0.2007CE9006 and the Bulk Warehouse Grain Quantity Recognition, a project (2007) funded by the Finance Bureau of Chongqing Municipality, P.R. China.

REFERENCES

Canny, John. A computational approach to edge detection, IEEE Transactions on Pattern Analysis and Machine Intelli-gence, 1986, PAMI-8(6):679-698

Cohen F S, Zhang Z Yang, Invariant matching and recognition of curves using B-splines curve representation, IEEE Transactions on Image Processing, 1995, 4(1):1-10

Fan Li-nan, Han Xiao-wei,, Zhang Guang-yuan. Image Processing and Pattern Recognition, Science Press, 2007

Freeman H, Davis L S. A corner finding algorithm for chain-coded curves, IEEE Transactions on Computers, 1977, 26(5):297-303

Gong Sheng-rong, Liu Chun-ping. Digital Image Processing and analysis, Tsinghua University Press, 2006

Hari A, Suri S, Paruikar G. Detecting and resolving packet filter conflicts, Proc of Policy 2001 workshop, Tel Aviv: IEEE Press, 2000:1203-1212

Kong Weixi, Kimia B B. Onsolving 2D and 3D Puzzles Using Curve Matching, The IEEE Conference on Computer Vision and Pattern Recognition, Hawaii, 2001

Li Fu-yu, Li Yan-jun, Zhang Ke. The use of chain-code technique in extracting feature point in scene image, Journal of Image and Graphics, 2008,13(1):114-118

Lin Ying, Fu Yang. Key of bulk warehouse grain quantity recognition—rectangular benchmark image recognition, Journal of Zhejiang University (Engineering Science), 2007,41(10):1643-1646

Yao Min. Digital Image Processing, China Machine Press, 2006

QUANTITY INTELLIGENT RECKONING FOR PACKAGED GRANARY GRAIN BASED ON IMAGE PROCESSING

Ying Lin [1,2], Yong Liu [1], Yueheng Sun [2,*], Yanhong Sun [1]

[1] *College of management, Chongqing Jiao Tong University, Chongqing, China,400074;*

[2] *School of Computer Science and Technology, Tianjin University, Tianjin, China, 300072;*

[*] *Corresponding author, Address: School of Computer Science and Technology, Tianjin University, Tianjin, China, 300072, Tel: +86-22-27401016, Email: yhs@tju.edu.cn*

Abstract: This paper presents a quantity intelligent reckoning approach for packaged granary grain based on image processing. The actual scene video was taken as the analysis object, and the dual-threshold Canny operator and the morphology processing method are used to extract the object grain bags' characteristic outline-- the boundary of the counter-band of light. Then, a counting algorithm which integrates mode theory and variance analysis technology is presented for the quantity second-judgment. Experimental results show that by accurately extracting the characteristic outline and counting the number of the characteristic outline, the algorithm presents an effective method for grain quantity detection with high recognition precision and efficiency.

Keywords: quantity intelligent reckoning, packaged granary grain, image processing, counting algorithm

1. INTRODUCTION

From the nearly 20 years' practice of storage grain regulatory in China, the quantity of grain reserves supervision and auditing is still manual regulation. Because of the geographical dispersion and the lack of supervision of reserve granary, it is difficult for management departments to carry on the effective supervision and investigation, leading to the virtual, false subsidies, theft, and other widespread phenomena, bringing the huge

---

Please use the following format when citing this chapter:

Lin, Y., Liu, Y., Sun, Y. and Sun, Y., 2009, in IFIP International Federation for Information Processing, Volume 294, *Computer and Computing Technologies in Agriculture II, Volume 2*, eds. D. Li, Z. Chunjiang, (Boston: Springer), pp. 1115–1123.

economic losses to the country, and what is more serious is that it affects the national macro-manipulation and price stability. So it is necessary to seek convenient and effective tools for reserve granary grain quantity reckoning (Lin Ying et al., 2007) . And the conveniences of digital images acquisition, transmission and identifiability cause the grain quantity examination tool which is based on image pattern recognition to become an important technological measure that solves the problems of long-range automatic monitoring and auditing (M.D. Kelly., 1973). In this paper a computerized intelligent recognition technology to achieve the precious quantity of grain reserves was adopted.

The key of video-based grain reserves automatic monitoring and auditing technology is based on the result of actual scenes video image recognition to calculate the real-time quantity of grain reserves. The grain pile is cubic, and each bag of grain's weight is fixed. If the number of grain bags were distinguished, the total quantity of grain reserves was able to be calculated. Therefore, the key of packaged granary grain quantity intelligent reckoning is various superficial grain bags recognition and quantity reckoning (Pavlidis, T., 1982). Based on the analysis of grain reserves characteristic, an effective and convenient method to recognize the object grain bag' characteristic outline -- counter-band of light boundary was proposed (Chen Xiaochun et al., 2006; Luan Xin et al., 1999). In this paper we introduced a smart method of video-based packaged granary grain quantity intelligent reckoning to extract the characteristic outline and count the number of the characteristic outline (Chen-Chau Chu et al., 1993).

2. INTELLIGENT DETECTION METHOD

The key of image recognition is whether the computer accurately understood the image information and effectively extracted the characteristics of region of interests (ROI). By comparing the effects of various local edge detection operators dealing with the actual scene images, the dual-threshold Canny operator as a local edge operator was chosen, and a counting algorithm based on statistical theory was involved (Kakarala, R. et al., 1992).

This method includes following four major steps:

1. Format conversion. Fully considered the complexity of the image information and the efficiency of image processing, the actual acquisition of the RGB images were converted into gray image.

2. Noise elimination. Regarding the influence of noise, the value filtering was used to carry on smooth processing to gray images and improved the quality of the gray images.

3. Characteristic outline extraction. Based on the characteristic outline of the counter-band of light boundary category of rectangle, the dual-threshold Canny operator and the morphology processing method were effectively extracted ROI region.

4. Outline statistical counting. On the basis of determining ROI, a simple pixel statistic counting algorithm was presented, to some extent reduce the computational time and the complexity of the intelligent reckoning.

2.1 Space conversion

In practical application, the gray image processing is quite convenient, and the processing efficiency is high. According to images throughout the different gray level, the value of two-dimensional function $f(x, y)$ can be denoted as the gray value of the coordinates (x, y) (Rosenfeld, A. et al., 1979). The monitor collects images for the red, green, blue (RGB) form, and the grain bags cross-section characteristic information is relatively simple, so first of all the gray conversion formula $f(x, y) = 0.3R + 0.59G + 0.11B$ was used to convert from the RGB space to gray space, and the result was shown in Fig.1 and Fig.2.

Fig.1: Original image

Fig 2: Gray image

2.2 Image denoising

The primitive packaged granary scene video images have a certain degree of noise pollution, making images become fuzzy, even blurred the image characteristic information. In order to improve the quality of the images, we often need to reduce or eliminate noise pollution for the accuracy of image recognition (Moulin, P. et al., 1999). Common methods of image enhancement, such as the neighbor average filtering, the median filtering, Wiener filtering, homomorphic filtering, and so on. Considered the complexity of the system software and the efficiency of image processing, the median filtering was adopted to carry on smooth processing with the scene gray images, and to eliminate noise pollution. By way of eliminating noise and preserving the information of characteristic outline effectively, a small 3×3 filter mask was used.

2.3 Characteristic outline extraction

The edge which is the partial characteristic discontinuous expression of the image depicts the characteristic outline. The main purpose of characteristic outline extraction is more accurately to identify the boundary of ROI in the scene image, on the basis of image pretreatment result acquired, in order to effectively improve the accuracy of characteristic outline extraction. According to the specialty of packaged-grain bags stacking in the reserve granary, the rectangular counter-band of light which was caused by the pressure among bags formed characteristic outline of each grain bag (John F. Haddon et al., 1990).

Characteristic outline extraction carried on the source gray image, using the rate of change of intensity and direction of changes in the suitable boundary segmentation local edge detection operator method checks each pixel point for the neighborhood, and the completion of the pixels in a neighborhood of gray rate of changes, which is in the direction of quantifying the identification, making the same rate of difference of gray pixels constitute closure and connectivity region. The main consideration of the fuzzy images of high level noise is how to choose the edge detection operator. For two-dimensional image $f(x, y)$, x, y respectively, on behalf of the pixels in a two-dimensional pixel-point benchmark of the abscissa, longitudinal coordinates, the position $f(x, y)$ of the gradient can be expressed as a vector, using G_x and G_y. Specific formula as follows: Gradient vector can be expressed as the following

$$\nabla \mathrm{f} = \left[G_x, G_y \right]^T = \left[\frac{\partial f}{\partial x} \frac{\partial f}{\partial y} \right]^T \quad (1)$$

Set $\alpha(x, y)$ represent gradient direction

$$\alpha(x, y) = \arctan\left(\frac{G_x}{G_y} \right) \quad (2)$$

In the direction of $\alpha(x, y)$ the rate of change velocity will be

$$\nabla f = mag(\nabla \mathrm{f}) = \left[G_x^2 + G_y^2 \right]^{\frac{1}{2}} \quad (3)$$

In terms of practical application of the margin calculation, gradient operator is equivalent to the following calculated norm

$$\nabla f = |G_x| + |G_y| = \left|\frac{\partial f}{\partial x}\right| + \left|\frac{\partial f}{\partial y}\right| \quad (4)$$

Based on the characteristics of digital image processing and the analysis of actual scene images of the reserve granary, the form of a difference instead of the above calculation of the differential operator was used.

In the ideal circumstance, suppose the bending of each grain bag is the same, and the gathered counter-band of light image has the uniform width band of light. However, there are a lot of random factors in the actual packaged-grain bags stacking, such as the stacking-fault, changes in location of the bags, and so on, leading to the acquisition by showing a count-band of inhomogeneous width and local banding. Taking the complexity of the scene image into account, the first derivative may be unable to find the boundary and the information of the second derivative will be adopted. Five kinds of the edge detection algorithm were selected to analyze the scene image, and the results were shown in Fig.3. From the analysis of results, the Roberts algorithm was bad, lost many edges, and presented many isolated points. The Prewitt algorithm and the Sobel algorithm's effect were similar, and eliminated the noise well. But the local boundary presented the partial break. The LOG algorithm examined the characteristic outline of the counter-band of light well, but still presented the partial boundary break, and was relatively quite sensitive to the noise. The Canny algorithm obtained continual outline boundary, can distinguish characteristic of each grain bag's counter-band from the results, and effectively enhanced the anti-jamming capability (Marr, D. et al., 1980). In this paper we adopted the Canny edge detection operator to achieve the extraction of the characteristic outline and binaryzation.

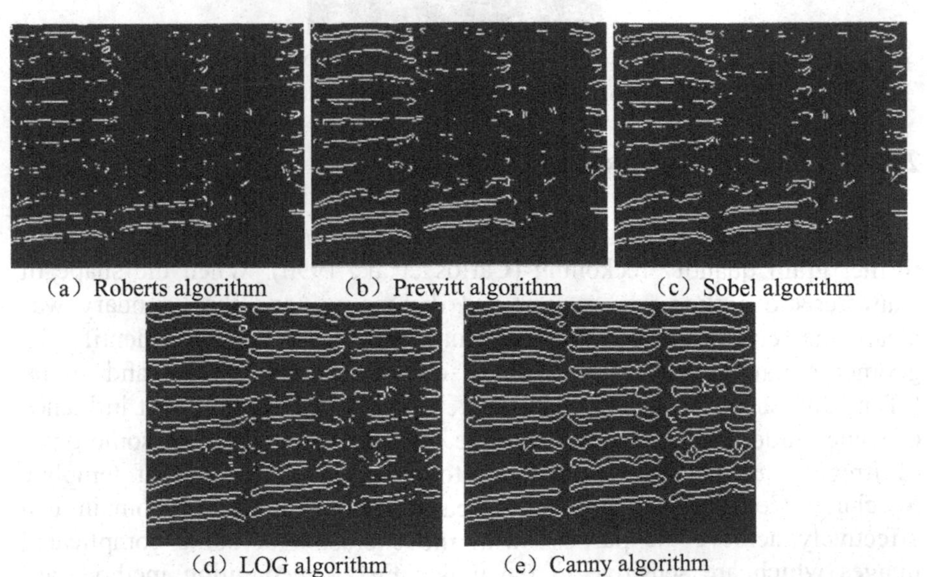

(a) Roberts algorithm (b) Prewitt algorithm (c) Sobel algorithm

(d) LOG algorithm (e) Canny algorithm

Fig. 3: Comparison of counter-band of light boundary image edge detection algorithms

In the actual process of reckoning, as the result of the limit of the scene light condition, the brightness and the contrast gradient of the images are non-uniform, and the boundary of the counter-band of light is not clear. There are two measures to improve the quality of the images: first, improve the scene illumination condition of the environment; second, take some effective measure to strengthen the ROI. In order to heighten the accuracy of reckoning, the full consideration in the design process is given. After the above image processing, we have already identified outline boundary of the counter-band of light of each grain bag, namely ROI region. But it still exist some noise spot which was misjudging as the pixels of ROI. The purpose of the outline enhancement is to wipe off the pixels of misjudgment and further identify the boundary of the counter-band of light. Based on morphological analysis of image processing, the specific structural element *B* was introduced to refine ROI region A:

$$A \otimes \{B\} = ((\cdots((A \otimes B^1) \otimes B^2)\cdots) \otimes B^n) \quad (5)$$

When the processing result of the region A no longer changed, the refinement finished. The result was shown in Fig.4.

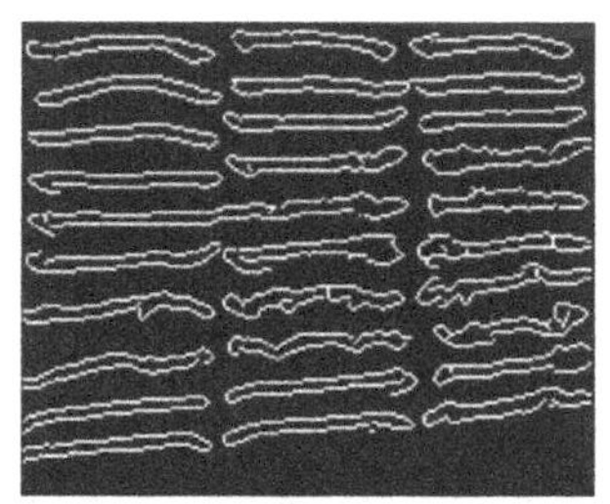

Fig 4: The extraction of characteristic outline boundary based on the Canny operator and the morphology processing method

2.4 Outline statistical counting

Pattern recognition of the identified ROI edge region was prepared for further grain quantity reckoning (Carlos et al., 1990). When the shape of characteristic outline was regular and disjunctive, and the boundary was clear, the corresponding template matching was used to identify its geometric parameters easily, such as central location, radius and so on (Tang Jinkuan et al., 2007) . However, as the scene environment influence of some random factors and the outline of actual detection with some types of irregular rectangle, it is difficult to adopt the algorithm of template matching. Counting the number of each outline connected domain can effectively achieve the purpose with the exclusion of some complicated images which are sensitive to the noise. Fuzzy recognition method was restricted in the practical application because of the great computation load

（Chen Songcan et al., 2000）. Full consideration was given to the effectiveness of the recognition and the outline characteristic of the counter-band of light, and we introduced a method based on a simple pixel statistic counting algorithm to achieve reckoning.

Assume $f(i, j)$ is the binary image of characteristic outline, whose size is M× N, and $f(i, j)$ will range from 0 to 1, where 0 denote the white pixels, and 1 the black pixels; $i \in \{0,1,2,\cdots,M-1\}$, $j \in \{0,1,2,\cdots,N-1\}$; Let $[n]$ denote the maximal integer which is less than n (Carlos et al., 1990). Check each pixel of the binary image followed by a vertical line from the top to bottom in turn, when the pixel value jump from 0 to 1 or 1 to 0, accumulator plus one automatically. It is convenient to reckon the quantity of grain bags by accumulated value. Consider the timeliness of the algorithm, the measure of accumulation equipartition is adopted to select the position of vertical direction line segment, to avoid counting the vertical direction line segment of all. To improve the accuracy of reckoning, we selected many vertical directions to measure, introduced the analysis of combining the mode theory and the variance theory, set the corresponding threshold, and computed the final result. The algorithm is as follows:

1. Select vertical directions which will be

$$x = \left[\frac{1}{n}M\right], \left[\frac{2}{n}M\right], \left[\frac{3}{n}M\right], \cdots, \left[\frac{n-1}{n}M\right], n \geq 9 \tag{6}$$

check each pixel of the binary image followed by a vertical line from the top to bottom in turn, and write down various accumulation results, namely $k_1, k_2, k_3, \cdots, k_{n-1}$;

2. Count the number of the results of Step 1, let k denote the mode of this group of data and n_0 the corresponding number, and we have frequency

$$p = \frac{n_0}{n-1} \tag{7}$$

3. If $p \geq 0.8$, then the variance will be

$$Var(k) = E\{[k_i - k]^2\} \tag{8}$$

Where: $i = 1,2,\cdots,n-1$, Or let $n = n+1$, and return to Step 1;

4. If $Var(k) < T$, then stop computing and give the final result; Or let $n = n+1$, and return to Step 1, until the process finish.

3. THE EXPERIMENTAL ANALYSIS AND CONCLUSIONS

In the actual quantity reckoning, according to the three neighboring surface images of the three-dimensional warehouse scene, the key technology to extracting various characteristic outlines and reckoning them was needed. Taking the timeliness and accuracy of the system into account, it is low efficiency to select so many initial vertical directions, whereas to select few directions will not achieve the high standard precision. Using the design methods in the actual recognition, all of the 20 samples of the characteristic outline can be correctly recognized, and the correct detection rate is 99%. The experimental analysis indicates that the method used to achieve grain quantity intelligent reckoning is feasible.

From the above analysis, we can educe the conclusion of this intelligent reckoning, as follows:

1. Using the space conversion in pretreatment process, the image information can be suitably compressed, and the efficiency of intelligent reckoning is improved.
2. In view of the grain bags counter-band of light's charactcristic boundary, using the concept of ROI, and integrating the dual-threshold Canny operator and the morphology processing method is adopted to achieve high-quality recognition.
3. It is the innovation which is based on a simple pixel statistic counting algorithm that is presented to meet management department's demand, and the advanced technology has certain value in practical application of the regulatory process of the reserve granary.

In view of the problems based on the reserve granary automatic video supervision and auditing, we presented a smart method that is through the pattern recognition of real-time scenes image to acquire the actual grain reserves. According to the characteristic of the counter-band of light, the dual-threshold Canny operator and the morphology processing method are integrated to achieve the extraction of ROI boundary. Then we presented a simple algorithm based on the pixel statistic counting to carry out the reliable quantity reckoning. However, as a result of the actual scene illumination limit and various grain bags characteristic's randomness, some boundary of processed images still has a certain shortage will influence high-accuracy of the quantity reckoning, therefore ,the refinement of the characteristic outline boundary still need further study.

ACKNOWLEDGEMENTS

This work is supported under the Bulk Warehouse Grain Quantity Recognition, a project funded by National Natural Science Foundation of China (Grant No.60603027) and Science & Technology Commission of Chongqing Municipality, P.R. China (Grant No.2007 CE9006).

REFERENCES

Carlos A. Cabrelli and Ursula M. Molter. Automatic Representation of Binary Images, IEEE Transactions on Pattern Analysis and Machine Intelligence, 1990, 12(12):1190-1196

Chen Songcan, et al. Studies and Implementation of Fuzzy Recognition Methods for Image, ACTA ELECTRONICA SINICA, 2000, 28(11):50-54.

Chen Xiaochun, et al. A method of shape recognition, Pattern Recognition and Artificial Intelligence,2006,19(6):758-763. (in Chinese)

Chen-Chau Chu, Aggarwal, J.K. The integration of image segmentation maps using region and edge information. IEEE Transactions on Pattern Analysis and Machine Intelligence, 1993, 15(12):1241 - 1252

John F. Haddon, James F. Boyce. Image segmentation by unifying region and boundary information, IEEE Transactions on Pattern Analysis and Machine Intelligence, 1990, 12(10):929 – 948.

Kakarala, R., Hero, A.O. On achievable accuracy in edge localization, IEEE Transactions on Pattern Analysis and Machine Intelligence, 1992,14(7):777 - 781

Lin Ying, Fu Yang. The key of bulk warehouse grain quantity recognition ---- Rectangular benchmark image recognition, Journal of Zhejiang University(Engineering Science), 2007, 41(40): 1643-1646 (in Chinese)

Luan Xin, et al. The New Recognition Algorithm for Irregular Quasi circular Object. Journal of Image and Graphics, 1999, 4(3):202-206 (in Chinese)

M.D. Kelly, Edge Detection in Pictures by Computer Planning, Machine Intelligence, Vol. 6 (American Elsevier, New York, 1973), pp. 397-409.

Marr, D. & Hildreth, E. Theory of Edge Detection, Proceedings of the Royal Society London, 1980, B207: 187-217

Moulin P, Liu J. Analysis of Multiresolution Image Denoising Schemes Using Generalized-Gaussian and Complexity Priors, IEEE Trans Information Theory, Special Issue on Multiscale Analysis, 1999, 45(3):909-919.

Pavlidis, T. (1982). Algorithms for Graphics and Image Processing, Computer Science Press, Maryland, USA.

Rosenfeld, A., Davis, L.S. Image segmentation and image models, Proceedings of the IEEE, 1979, 67(5):764 – 772

Tang Jinkuan, et al. A Novel Head-location Algorithm Based on Combined Mask, Journal of Image and Graphics, 2007, 12(8):1389-1394. (in Chinese)

AN ADAPTIVE ERROR-CHECK SOLUTION

Ying Lin[1,2], Liang Ge[2], Yueheng Sun[1,*]

[1] *School of Computer Science and Technology, Tianjin University, Tianjin, China, 300072;*

[2] *College of management, Chongqing Jiao Tong University, Chongqing, China,400074;*

[*] *Corresponding author, Address: School of Computer Science and Technology, Tianjin University, Tianjin, China, 300072, Tel: +86-22-27401016, Email: yhs@tju.edu.cn*

Abstract: Because of the traditional methods of video error checking are unfitted to the network environment in which the packet loss rate frequently changes, this paper proposes an adaptive error checking solution on the basis of estimation of the change trend of packet-loss rate. By setting the threshold value of packet-loss rate, the solution takes advantage of improved auto repeat request supported by RDP to realize real-time transportation in a low packet-loss rate, and uses improved forward error correction, a method of dynamically decollating images at the basis of varying sending rate, to solve the problem of error checking in a low transmitting rate.

Keywords: auto repeat request, forward error correction, packet-loss rate, transmitting rate

1. INTRODUCTION

Unstable Network seriously influences the real-time transportation of Video streams. When there exist great jams in the communicating channel, the quality of transportation decreases fiercely. If the rate of packet loss is more than 15 percents, streams are too broken to be rectified totally (Zhang et al., 2001). So in order to keep video complete and clear, the receiver always uses many methods checking errors to amend and restore video damaged partly. ARQ and FEC, which are used usually, can make a perfect effort in large scopes. However, as they are insensitive to the frequently changing situation of transporting channel, so they perform terribly in such a condition and worse bandwidth obviously (Shan et al., 2003). In a word, in the circumstance of dynamically varying bandwidth, it doesn't exist a ideal

Please use the following format when citing this chapter:

Lin, Y., Ge, L. and Sun, Y., 2009, in IFIP International Federation for Information Processing, Volume 294, *Computer and Computing Technologies in Agriculture II, Volume 2*, eds. D. Li, Z. Chunjiang, (Boston: Springer), pp. 1125–1134.

solution to the real-time transportation of video streams with high quality (Fan et al., 2007). Considering, this paper proposals an adaptive method to solve the problem of checking error for streams in the condition of the rate of packet-loss infirmly varying.

2. THE COMPARISON OF CURRENT ERROR-CHECK METHODS

2.1 Auto Repeat Request

The principle of Auto Repeat Request (ARQ) method is that after checking error, the receiver sends the sender data message about the list of false data if there are some mistakes producing in the transportation (Zhang et al., 2004). Usually, the third method is more common to be used (Wang et al., 2007). After having found out mistakes, the receiver sends the data message containing serial number of false video frames and transmits correct frames instead of all frames.

ARQ method can reduce quantity of information and consummation of resource (Moore et al., 2004). Although ARQ is an efficient way, it isn't fit to the situation of narrow bandwidth. When channel is obstructed, it's easy to lose packets and frames and consequently increase the quantity of rollback data packets; the condition of network becomes worse and worse. Besides, since the rate of packet-loss increases, the receiver has to send back repeatedly the same data, and it will cause serious time-delay. Finally, ARQ can not assure the real-time transportation.

2.2 Forward Error Correction

Forward Error Correction (FEC) method uses simplex communicating channel style, the receiver operates the total tasks of checking error and restoring video (Wang et al., 2007). In the process, the sender creates error correct codes to support the receiver to rectify mistakes, such as RS code or BCH code. In the process of checking, the receiver doesn't send reports of checking result and the sender keep sending streams all the time.

Advantages of FEC are that, because receiver checks mistakes by itself and sender takes no operations to adjust the transmitting rate and send correct data, it will reduce the times of communication between double sides and decrease the redundant data to enhance utilization rate; Even though some redundant codes lose in thc transportation, it effects little to the quality of video.

Disadvantages of FEC are that, firstly, if the bandwidth changes suddenly, the sender can't percept the change of channel and keeps the stable speed to transmit, which leads to lose many packets (Liu et al., 2004); secondly, it will affect terribly checking error. And because it doesn't send correct data, video will be hard to restore (Gu et al., 2002). thirdly, error correct codes contain lots of information of motion compensation, there will be bad obstruction as large volume of correct codes.

3. DESIGN OF ADAPTIVE ERROR-CORRECTION SOLUTION

3.1 Error-Retransmit solution based on RDP protocol

In order to ensure complete transmission in the unstable circumstance, the receiver needs to check error for video data, as soon as finding out mistake, the receiver should send report containing the result of rectification to request the sender to send the correct data back. It needs the support of transmitting protocol to achieve feeding back.

It's known that, traditional ARQ method designed on the basis of TCP, which sets up mechanisms that provide safe-guarantee to the peer-to-peer transportation（Lin et al., 2002）. But because TCP uses complex process of interaction and tedious retransmitting mechanism leading to heavy burden for resource and serious time-delay, so TCP is not available to the real-time transportation of huge video streams. With the development of web technologies, transmitting protocols have been improved greatly. Nowadays, there have been many protocols that are suitable to support the transportation of mass streams, and RDP (Reliable data protocol) gets the best comprehensive effect（Hu et al., 2004）.

RDP provides each transmitting layer with duplex communicating channels, it tries to reliably send all user information and show mistakes if there are some failures in the process. Its extended IP data service ensure reliable transmission, and it takes advantages of serial number and correcting code packaged in the head of data packet to detect and remove the false data. Compare to TCP, RDP is convenient to user, because it supply simple functions for controlling and buffering and managing （Ma et al., 2007）. For the transportation of stream, RDP provide powerful index-searching function containing synchronization code and serial number to load frame timing and spacing order. It is available for the receiver to finish the sorting and complex statistic of losing frames. The structure of RDP packet is shown in Fig.1. In view of perfect functions, this paper sets up SARQ method on the base that uses RDP data packets to record the list of video streams.

SYN	ACK	EAK	RST	NUL	0	Ver No	Header Length
SourcePort							
DestinationPort							
Data Length							
Sequence Number							
Acknowledgement Number							
Checksum							
Variable header area							

Fig.1: The structure of RDP packet

Video transportation is commonly divided into two styles: video-frame model and video-packet model, they use individually particular ways to transmit streams （Xu et al., 2005）. In order to be compatible with two modes, this paper improves original formation of RDP data packets deleting all useless features and adding some user-defined characters. In case that one packet may encapsulate several video streams, the receiver has to verify that all packets have been gotten completely. So, this paper proposes adding data sequence feature to RDP data packets, which is used to record list of video frames. As removing some useless data items, the volume of packets become a bit small so that it can avoid obstructing the channel because of large volume. Improved structure of RDP data packet is shown in Fig.2.

SYN	ACK	EAK	RST
SourcePort			
DestinationPort			
Data Length			
Data Sequence			
Checksum			

Fig.2: Improved structure of RDP data packet

As sending video frames and packets, the receiver will add with RDP data packet and record detailed information. After getting video packets, the receiver firstly examines additional information in RDP data packet, and check error. If the object transmitted is real-time stream, the receiver tries to find whether streams are in the right order or lost by the SYN and ACK item. Taking advantage of data sequence, it is clear whether stream frames encapsulated in video packet were lost. Finishing checking error, the receiver will record the list of serial number of frames or packets lost, and send feed-back report to inform the sender sends the correct data again.

For instance, in order to adjust the volume of data packets, the receiver will package one key frame and several referenced frames in one packet, as shown in Fig.3, data sequence item will record their serial number: 1-0 and 1-1 and so on. Having attained packets, the receiver will examine whether the current order of frames is matching the original order, if some frames were lost, the receiver will request the sender to transmit data again according to the data sequence.

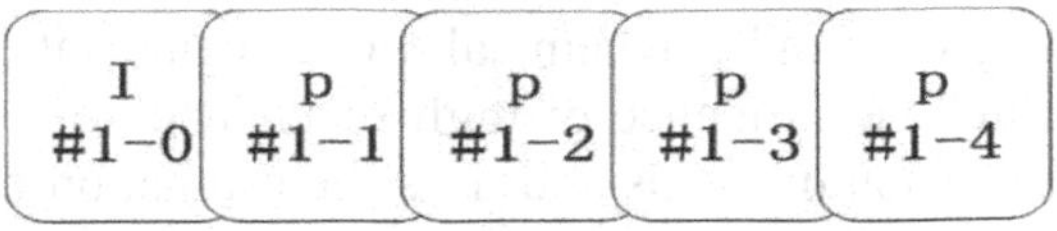

I #1−0	p #1−1	p #1−2	p #1−3	p #1−4

Fig.3: The structure of encapsulated frame packet

3.2 The solution of cutting apart dynamically image based on varying code-rate

Traditional FEC method is a way that with the constant speed, the receiver takes charge of checking error and correcting image by itself (Mei et al., 2004). This method lacks elemental communication between two sides that are responsible to the transportation, even though there are critical mistakes, which are hard to rectify, the receiver is difficult to modify the transmitting speed, because it can't get any report of the situation of network (Lin et al., 2007). The final result is that there are more and more false streams and the transportation is tough to continue. So, this paper proposes a new solution – improved FEC based on predicting current sending rate. The receiver will take advantage of RDP protocol to send periodically RDP feed-back reports, which are used to trace the changing trend of channel. In accordance to reports, the receiver calculates the rational speed to ensue the normal transmission. This paper uses TFRC (TCP—Friendly Rate Control) algorithm to evaluate ideal speed of transmission. The formula of TFRC is depicted as followed.

$$T = \frac{S}{RTT\sqrt{\frac{2bP}{3}} + t\_RTO\sqrt{\frac{3bP}{8}}(1+32P^2)} \qquad (1)$$

T is upper limit of sending speed, RTT is round trip time, t_RTO is time of overtime retransferring, P is pocket loss event ratio, S is size of data packet. If the receiver can get data in a RTT interval, it will feed back to the sender with packet-loss rate, and ensue that the sender can obtain continuingly this parameter.

Traditional FEC uses the motion estimation of all pixels of an image and formats RS correcting code containing estimation of the time and space region. In the process of checking error, the receiver examines whether the motion of pixels accord with original estimation, when there are some false motion, it can be corrected by the RS code (Gao et al., 2006). This approach is more available to use in the situation that the bandwidth is quite good, but, because high redundancy of RS code will heavy the transferring burden of channel , if the bandwidth becomes very narrow, images will not be reduced resulting from loss of lots of RS code. Moreover, FEC will increase complexity of reducing image, particularly as the receiver could not get whole RS code. So, in the premise of speed estimation, this paper proposes a checking solution by dynamical image segmentation: SFEC.

In order to reduce the influence of high redundancy of correcting code, SFEC expends the region of pixels and uses the estimation of image blocks instead of pixels. Through dynamically cut apart the image, it will decrease the complexity of RS code. This paper adopts even-square dynamic algorithm to divide an image into several fragments, such as 2×2、4×4、8×8 and so on.Fig.4 shows segments of the image.

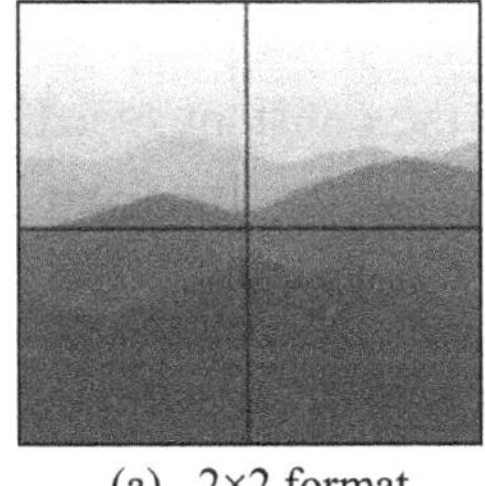

(a) 2×2 format

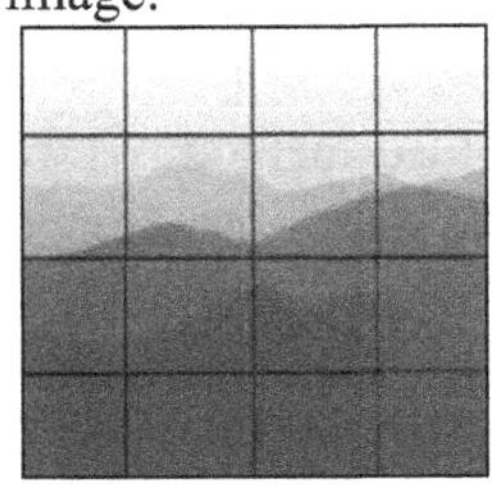

(b) 4×4 format

Fig.4: Image segments operated by SFEC

The first step is, Calculate the expected transmitting speed. At first set k regions of changing blocks, and through several experiments measure the average value of each changing block, and set the weight according to the frequency of appearance of each region, finally calculate the theoretic speed value.

The second step is, during the transmission, firstly estimate value of p_t that stands for current transmitting speed, and calculates the segmenting coefficient depicted with h.

$$h = \frac{\sum_{i=1}^{k} m_i p_c}{p_t} \tag{2}$$

Where:

$$0 < m_i < 1 \ , \ \sum_{i=1}^{k} m_i = 1 \tag{3}$$

The third step is, If h<=1, it certificates the current speed is more than expected speed, and this situation is fit for real-time transmission, it is unnecessary to slit the image; If h >1, it stands for that the current communicating channel is terrible, and h is larger as packet-loss rate is increasing, the possibility of losing correcting code is larger.

Calculate the quantity of video blocks:

$$N = \left(\frac{1}{l_a \times h}\right)^2 \quad (4)$$

Where:

$$0 < la < 1, \ 1/(la \times h) \text{ is even} \quad (5)$$

l_a is relative coefficient of segments, which is related to the size and the style of image.

N is the quantity of blocks, inverse with the h and is proportional with p_t, it stands that the quantity of blocks becomes less as the current speed is lower, and decreases redundancy of RS by extending the region of pixels and reducing the resolution of image.

3.3 Realization of adaptive solution

This paper proposes an adaptive error-check solution based on estimating the change of packet loss rate. Before starting transmitting, it should set a rational threshold value accepted by both the receiver and the sender, considering the request of the quality of video and the real-time transportation. During the transmission, supported by RDP protocol to create the network connection, the sender captures the feed-back data packet containing the list of video packets obtained by the receiver at last transmission, and uses formula 3.4 to calculate the current packet-loss rate and transmitting speed by TFRC. Finally, according to packet-loss rate, adjust the transmitting approach.

Calculate the packet-loss rate.

$$r_{loss} = \frac{n_t - n_r}{n_t} \times 100\% \quad (6)$$

r_{loss} is packet-loss rate, n_t is the number of total packets transmitted, n_r is number of all packets received.

If packet-loss rate is less than threshold value, it's better to use SARQ. The sender calculates the sending speed at first and the sender checks error and rectifies mistake. Finishing upper work, the sender sends video frames and correcting codes that have been operated by the algorithm of image-segment with the predicting speed. The receiver takes responsibility of checking error and reduce image, and send the feed-back report to the sender. The detailed process is shown as Fig.5.

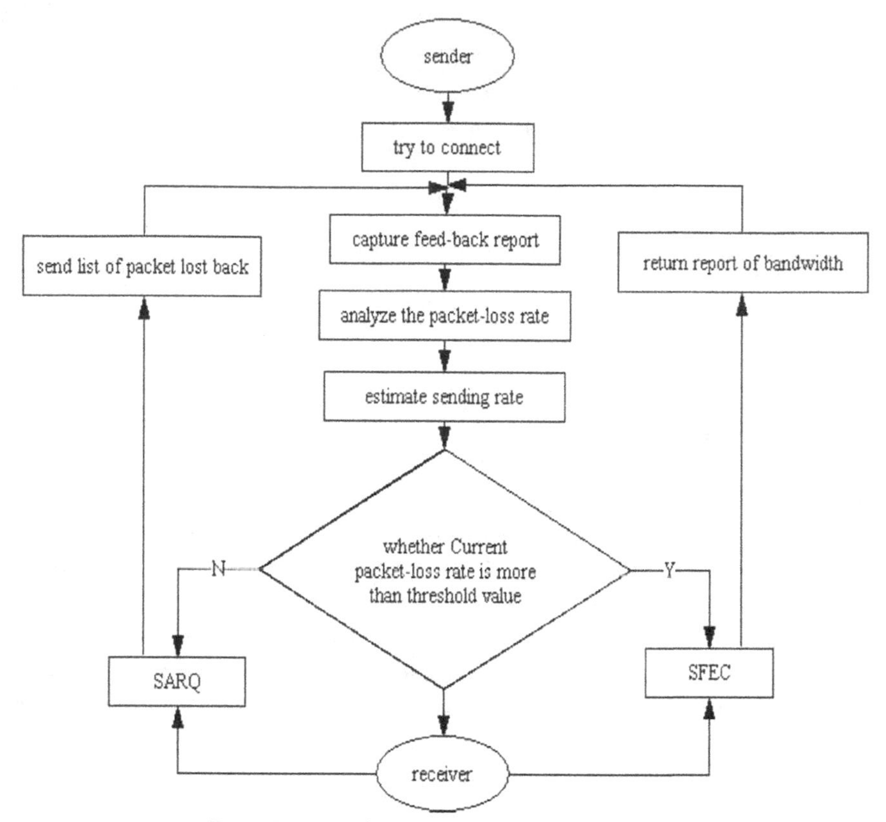

Fig.5: Process of the realization of adaptive solution

4. THE SIMULATION AND ANALYSIS

This paper does tests with video segments of "Foreman", whose format is QCIF and the style of YUV is 4:2:0 and the coding frame number is 20 and post-code frame frequency is 10FPS. The switching time is subjected to exponential distributions, and the value is 400ms. The original speed is set by 512kbps. The threshold value of packet-loss rate is 0.1 and this value is controlled in the region from 0 percent to 30 percent. The result is shown as Table1. and Table2. and Fig.6.

*Table 1.*The comparison of the result of two methods as the packet-loss rate keeps from 0 to 10%.

	SARQ	ARQ
PSNR（dB）	37.14	36.88
Bit rate（Kbps）	29.45	25.13
Coding time（s）	37.92	45.17

*Table 2.*The comparison of the result of two methods as the packet-loss rate keeps from 10% to 30%.

	SFEC	FEC
PSNR（dB）	35.11	28.46
Bit rate（Kbps）	26.29	20.36
Coding time（s）	43.73	54.82

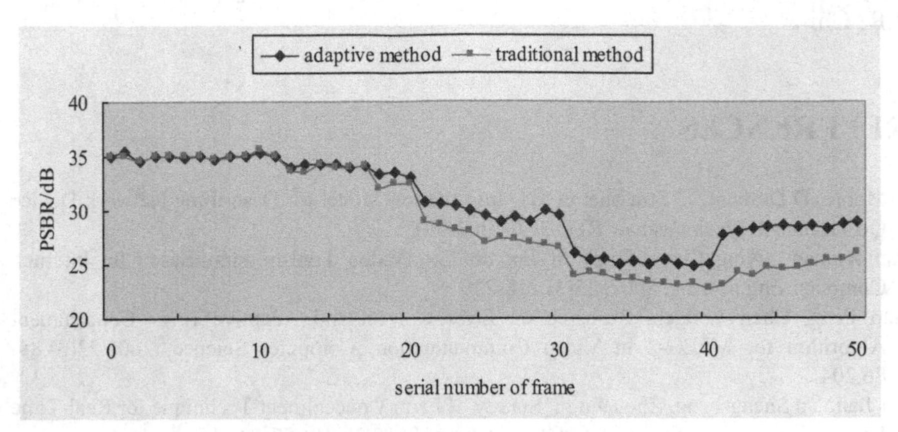

Fig.6: Comparison of signal-to-noise ratio as the packet-loss rate keeps from 10% to 30%

Under the condition that packet-loss rate keeps from 0 to 10 percent, we test with SARQ method. The receiver sends lists of false video streams with the improved format of RDP data packet. Compare to the traditional method, the result of experiment is not obvious, because the packet-loss rate is a bit low and the quality of channel is relatively good, but we also find the effect of SARQ is better. If the packet-loss rate is from 10 percent to 30 percent, we use SFEC to realize transportation and checking error. After estimating the trend of changing packet-loss rate, we slit image with two formats, such as 8×8 and 16×16. Compared with traditional FEC, the result certificates, not only at the aspect of optimization of sending speed but video quality, the adaptive solution can get better effort in the varying circumstance.

5. CONCLUSION

Unstable bandwidth is a main reason that affects the real-time transportation and the quality of video images. In the condition of low packet-loss rate using SARQ and the condition of high packet-loss rate using SFEC, to solve the problem, this paper proposes an adaptive error-check solution. In practice, this method can bring very good effects to the real-time transmission and the quality of video.

ACKNOWLEDGEMENTS

This work is supported under the National Natural Science Foundation of China No. 60603027, the Soft Science Research Program of CSTC N0.2007CE9006 and the Bulk Warehouse Grain Quantity Recognition, a project (2007) funded by the Finance Bureau of Chongqing Municipality, P.R. China.

REFERENCES

B Moore, D Durham. J Strassner et al. Information Model for Describing Network Device QoS Datapath Mechanisms. RFC 3670, 2004-01

Fan Xinnan. Xing Chao. Study 0f End to End Video Transmission Based 0n Internet, Computer Engineering, 2007, 33(3):218-220

Gao Peng, Chen Yongen. Research on Error Correct and Adaptive Error Concealment Algorithm for MPEG-2 in Video Communication, Computer Science, 2006,33(3):84-86,204

Gu Jian, Yu Shengsheng, Zhou Jingli. Survey of Error Concealment Technique for Real-Time Video, Computer Engineering and Applications, 2002,38(10):57-59,118

Hu Weijun, Li Kefei. Error Resilient Video Coding for Video Streams Transmission, CHINA CABLE VISION, 2004, 2:25-28

Lin Ying, Ge Liang. Study on Real-time Video Transportation for National Grain Depot, CCTA 2007: 533-541

Liu Jieping, Yu Yinglin. A Scheme for Video Transmission Based on Multiple Description, Computer Engineering and Applications, 2004, 40(35):119-121

Ma Xin, YANG Xiaokang, Song Li. An Adaptive Temporal Error Concealment Technique, Journal of Image and Graphic, 2007, 12(10):1782-1784

Mei zheng, Li Jintao, An Adaptive Forward Error Correction Algorithm for Streaming Video, Journal of Software, 2004,15(9):1405-1412

Q Zhang, W W Zhu, Y Q Zhang. Channel-Adaptive Resource Allocation for Scalable Video Transmission Over 3G Wireless Network[J]. IEEE Transactions on Circuits and Systems for Video Technology.2004,14(8): 104 9-1063

Shan Y, Kalyanaraman S. Hybrid Video Downloading / Streaming over Peer-to-Peer Networks, Proc. of International Conference on Multimedia and Expo, 2003: 665-668

Wang Feng, Zhu Guangxi, Zhang Zhenming. Multi-product Code for FGS Video Transmission over IP-Based Wireless Networks, Computer Engineering and Applications, 2006, 29:143-146

Wang Ke, Liu Zhiqin. Strategy of Stream Media High Quality Transmission Based on P2P Overlay Network, Computer Engineering, 2007, 33(15):137-139

Xu Fan, Zeng Zhiyuan,MPEG-2 Internet Video Using TCP-friendly Rate Control and Forward Error Correction, Microcomputer Applications, 2005, 26(5): 556-559

Zhang Qian, Zhu Wenwu, Zhang Yaqin. Resource Allocation for Multimedia Streaming over the Internet Multimedia, I EEE Transactions, 2001, 3(3): 339-355

FORTIFY METHOD OF MOVING OBJECT DETECTION BASED ON COLOR AND EDGE GEOMETRICAL FEATURES

Ying Lin[1,2], Yang Fu[1], Yueheng Sun[2,*], Yanghong Sun[1]

[1] *School of Management, Chongqing Jiao Tong University, Chongqing, P. R. China 400074*

[2] *School of Computer Science and Technology, Tianjin University, Tianjin, P. R. China 300072*

[*] *Corresponding author, Address: School of Computer Science and Technology, Tianjin University, Tianjin, China, 300072, Tel: +86-22-27401016, Email: yhs@tju.edu.cn*

Abstract: A Fortify method of moving object detection is proposed in this paper. Based on color and edge geometric features, the new method can automatically organize a supervised area. First, by the color features in a real supervised scene, the method extracts several regions of interest (ROI) with noise, then matches them with geometric shape by Fourier descriptors in the database and sequentially achieves an automatic-organizing supervised area. Experimental results show that this method has low operation cost, high efficiency, strong anti-jamming, high accuracy and robustness.

Keywords: Fortify method, moving object detection, geometric features, color features

1. INTRODUCTION

Along with the computer science and digital image processing technology developing, intelligent video surveillance system has been in intelligent transport, social security, intelligent buildings, grain monitoring and other digital video surveillance areas be widely implied. The most signified point of intelligent video surveillance system is the motion detection and alarm technology. At present the main concentrated of motion detection focus on splitting off the object from the background in the designated area. There are

---

Please use the following format when citing this chapter:

Lin, Y., Fu, Y., Sun, Y. and Sun, Y., 2009, in IFIP International Federation for Information Processing, Volume 294, *Computer and Computing Technologies in Agriculture II, Volume 2*, eds. D. Li, Z. Chunjiang, (Boston: Springer), pp. 1135–1143.

three methods in the general realization of motion detection, as follows: background subtraction; temporal difference; optical flow.

Above are three traditional motion detection methods. In the selection of detection region, the traditional approach can not automatically select the characteristic object as the detection region. In the practical applications, usually taking above three traditional motion detection in the surveillance system for the detection motion by the manual setting, it can't meet the need of automatically select of detected region, and the target objects automatic deployment of intelligent request. Also the manual methods can not meet requirements for providing specific measure changes in the certain practical application, it only provides the judgment of whether or not occur.

Especially, when the monitoring scene is complex, manual setting of the detection region maybe wrongly select the non-motion detection region, thereby increasing the additional computing complexity of the motion detection system. To address the above motion detection setting problem, this paper proposed the intelligent motion region setting method by the HSI (Hue, Saturation, Intensity) color space feature model and use the Fourier descriptors to describe the geometrical shape by (Zahn et at., 1972), and combined with the edge match. The methods applied the sensitivity of color space, and it has the nature of rotation, translation and normalization unchanged character of the normalized Fourier descriptor (Persoon et at., 1986) to descript accurately. Experimental results prove that this method has low operation cost, high efficiency, accuracy and strong anti-jamming, robustness.

2. INTELLIGENT MOTION DETECTED REGIONAL SETTING

The computer automatically detects the regional segmentation mainly based on the color characteristics of detection region and the geometrical shape characteristics in the setting of detection region. For this purpose, the extraction of image processing as a key step for the image, which including a large number of various features information. Usually in order to carry out the further image analysis, identification, coding in the main part of pretreatment, the accuracy of segmentation is directly impact on the effectiveness of the following process. However, the application of segmentation technology is constrained by the technical conditions, in the field of image recognition applications is mainly implied by the gray image of segmentation. This paper proposed the initial decomposition of the image by color characteristics in the HSI color space, and use the results of decomposition match with the geometrical shape database that described by the Fourier descriptors to achieve the motion detection. The method mainly constitutes by the following two parts:

1、Determination of color characteristics region (HSI space) (Wang Shuan et at., 1999)

Firstly, converting the key frame of real image to the hue sub-space and extracting the color feature corresponding the scope of the H component in the HSI color space, initially identify the regions of ROI in the color image.

2、The match of geometrical shape characteristic region

The Identified preliminary ROI regions to binarization and using Canny operator to do the edge segmentation, Fourier descriptors to describe the broader of the segmentation, then matching with the geometrical shape database described by Fourier descriptors.

2.1 Determination of color characteristics region (HSI space)

Based on the specific color of rectangular benchmark, firstly extract the color from the hue、saturation and intensity model (HSI) color space, initially identifying ROI (region of interests). In the actual image acquisition process, from the acquisition of Monitor RGB format for images, since the first need to achieve a conversion to RGB space HSI space Fig.1 (Zhu Minghan et at., 2005).

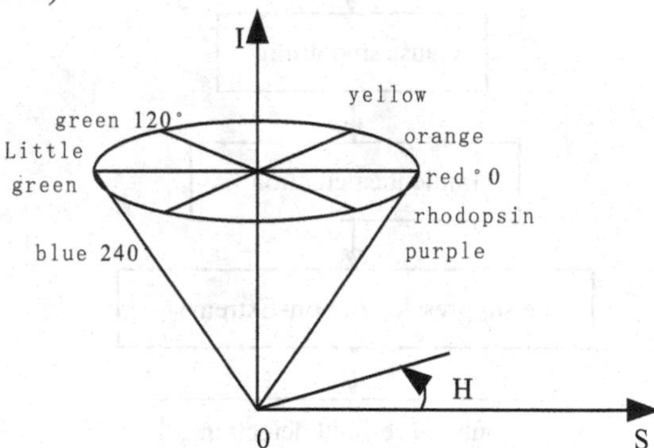

Fig.1. Sketch picture of HIS color model

Firstly will be a normalized: (Lin Ying et at., 2008)

$$r = \frac{R}{255}, b = \frac{B}{255}, g = \frac{G}{255} \quad (1)$$

Then proceed into:

$$H = 90 - \arctan\left(\frac{F}{\sqrt{3}}\right) \times \frac{180}{\pi} + \{0, g > b; 180, g < b\} \quad (2)$$

$$I = \frac{r + b + g}{3} \tag{3}$$

Where:

$$S = I - \frac{\min(r, g, b)}{I} \tag{4}$$

$$F = \frac{2r - g - b}{g - b} \tag{5}$$

HIS color model from the diagram, it indicted that we extract any color from the region based on the color model as the actual needs, thus needed to get the color model of regional for the automatic division to a number of the color characteristics of ROI region to binarization and tag.

Then double-edge detection threshold Canny operator (Shridhar et at., 1984)to detect the marginal binarization of ROI region, dual-threshold Canny edge detection operator using different detection threshold can guarantee that the edge of connectivity, traditional threshold method is more vulnerable and the detection edge is not closed, the smaller threshold of detection to repair the edge result of the larger detection threshold, so as to ensure the detected result connectivity and easy identification of detection edge, the specific implementation of the following as shown in Fig.2 :

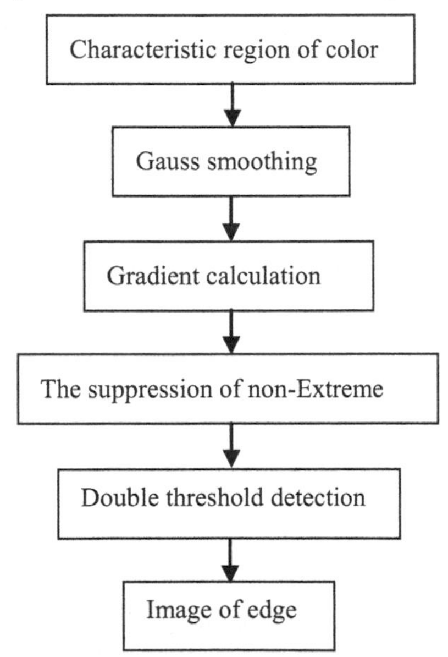

Fig.2: Extract the edge of characteristic region

2.2 The match of geometry Regional characteristics

The process of geometrical shape match including the edge extraction, the analysis of the relationship of several segment regions in the spatial distribution, with regard to the knowledge of known geometrical shape to analyze and understand the regional segmentation, and these regional characteristics of the region shape extraction through is the basis of the analysis.

First of all, it needs to address the problem how to descript the extracted geometrical shape. This method should meet the description of the same kind of shape has the most similarity, but the different kind of shape has greatest diversity, also should meet the translation, rotation and zoom invariance. In this paper, we take the normalization of shape Fourier descriptors without any relationship to the starting point and the direction to describe the edge. (Rasehom et al.,)

Firstly, we take the extracted region as the complex plane, and the vertical coordinates as the virtual axis, the horizontal coordinates as the real axis. The region on the border point (x, y) equivalent to the point on the plane (x + iy), the border point sequence can be written in the complex are as follows:

$$z_j = x_j + iy_j (j = 0,1,2 \cdots N-1) \tag{6}$$

The definition of Fourier descriptors (FDs) as follows:

$$A(k) = \frac{1}{N}\sum_{j=0}^{N-1} Z_j \exp(-\frac{i2\pi kj}{N}) \quad (k = -\frac{N}{2},\cdots,-1,0,1,\cdots,\frac{N}{2}-1) \tag{7}$$

Inversion as:

$$Z_j = \sum_{j=0}^{N-1} A(k)\exp(\frac{i2\pi kj}{N}) \quad (j = 0,1,2,\cdots,N-1) \tag{8}$$

Fast Fourier Transform calculation A (k), followed by the normalization of Fourier descriptors:

$$NFD(k) = \begin{cases} 0, k = 1; \\ A(k)/A(1), k = 1,2,\cdots,N/2; \\ A(k+N)/A(1), k = -1,-2,\cdots,-N/2+1 \cdot \end{cases} \tag{9}$$

A (0) in the tube center of mass, the rest of the Fourier coefficients are complex number. If set the A (0) as 0, Fourier descriptors has non-relationship with the location. On the other hand, with the exception of A (0), the value of the other Fourier coefficient A (1) or A (-1) are largest. Therefore, all coefficients divided by $\|A(1)\|$ or $\|A(-1)\|$ will normalized Fourier coefficient between 0 and 1. If all coefficient divided by the A (1) or

A (-1), the Fourier descriptors also has non-relationship with the direction and the starting location.

The definition of following Fourier description to do the geometry match:

$$FF = \frac{\sum_{k=-N/2+1}^{N/2} \|NFD(k)\| / |k|}{\sum_{k=-N/2+1}^{N/2} \|NFD(k)\|} \qquad (10)$$

Based on the description of the geometrical shape of the Fourier, matching the shape in the geometrical shape database the "shape of feature extraction and the description" of modules, normalized by the Fourier descriptors, the normalization Fourier descriptors to organize the database of geometric shape features by the index.

Through the submission of modules of "the extraction of shape feature and the description", acquired the normalized Fourier descriptors FFi. Do the similarity calculation between FFi and geometrical shape characteristics of the description of FFq, the results of calculation are reflected by the distance between FFi and FFq, the value of the numerical size reflects the degree of similarity between the image to be matched with and the geometric shape database, if those are not matched, then updated the geometric shape database and finished match computing. Fig.3 shows specific matching process (Zhang Kai et al., 2006, Tou et al., 1981):

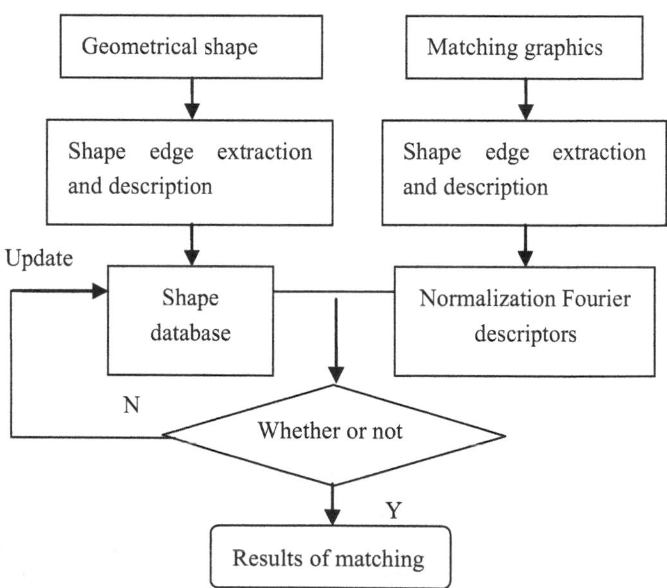

Fig.3: Structure of Shape Feature Match

Through the normalization Fourier descriptors NFD(k), we can calculate the degree of similarity between any two shapes, and identify the object shape of rotation, translation and scale invariant. Because of the frequency components of Fourier transform each other Orthogonal, Euclidean distance to the calculation of a Fourier description of the shape difference, that is:

$$dist = \sqrt{\sum_{k=2}^{M} \left\| FF_i(k) - FF_q(k) \right\|^2} \tag{11}$$

Because the energy of shape is mostly concentrated in the part of low-frequency, Fourier transform the high frequency components generally smaller and vulnerable to high-frequency noise interference, usually it only uses low-frequency components of normalized Fourier description to calculate the objects' similarity difference.

3. THE EXPERIMENTAL RESULTS AND ANALYSIS

Because of the pixel-value of uncompressed 256 bmp image is consistent with the real digital image, so it's suitable for the digital image processing. Therefore, in this paper we take the uncompressed 256 bmp image file format as the experimental image file format.

Applying the designed method to complete the regional selection automatically, based on color and geometrical shape in the real images to select the right region, the correct detection rate is 100 percent, and no misjudged situation occur. The use of color and geometrical shape is the basic characteristic which to complete automatically fortify, from Fig.4 to Fig.5 are the experimental pictures:

Fig.4. Monitoring Scene

Fig.5. Extract of Color Feature

Fig.6. Image of Binarization

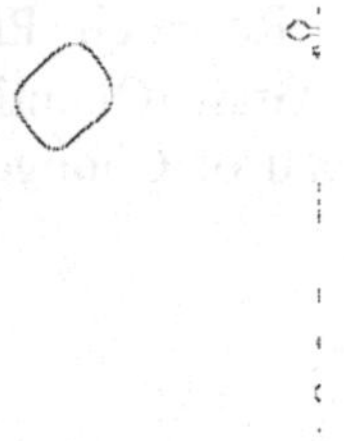

Fig.7. Edge Detection of Canny Operator  *Fig.8.* Match Result of Fourier Descriptor

Through the above experiments, we can acquire the following conclusions, with regarding to the fortify method which based on the color and geometrical shape as basic characteristics:

(1) This method used the conversion of HIS color space, and the application of hue in the sub-space made it has good robustness in different lighting conditions, adopting the double threshold Canny edge detection operators to complete ROI edge extraction, the results prove that this method meets the real application requirement.

(2)This method used the concept of ROI, it can adapt to the requirements of real-time monitoring in the supervisory process, and quickly determining the target area, providing the basis for the latter part of the motion detection.

(3)Adopting nature which the same kind of shape has larger similarity, but the different kinds of shape have greater difference. We take Fourier descriptors to describe the geometrical shape, to a certain extent strengthen the system application scope.

4. CONCLUSION

From above experiments, we can see that this method adopts the color and the geometrical shape characteristics descript by the Fourier descriptor as the basic conditions to determine the detection region, with characters of high efficiency, low computation, strong anti-jamming and robustness. It provides a new thinking for the intelligent motion detection study, and makes up the additional computing complexity weakness of traditional manual setting method. The experiments show that this method is feasible, also effective, but it can not work for those situations which the objects can't simply distinguished by the colors and geometrical shapes, usually has misjudge phenomenon, so the description of abstract target still needs further study and exploration.

ACKNOWLEDGEMENTS

This work is supported under the National Natural Science Foundation of China No. 60603027 , the Soft Science Research Program of CSTC N0.2007CE9006 and the Bulk Warehouse Grain Quantity Recognition, a project (2007) funded by the Finance Bureau of Chongqing Municipality, P.R. China.

REFERENCES

Barron J , Fleet D and Beauchemin S . Performance of optical flow techniques[J]. International Journal of Computer -Vision, 1994, 12(1): 43-77

C T Zahn, R Z Roskies. Fourier descriptors for plane closed curves [J]. IEEE Transactions on Pattern Analysis and Machine Intelligence, 1972, C-21(3): 269-281

Cheng Xiao-chun, a method of shape recognition[J], Pattern Recognition and Artificial Intelligence,2006,6, 126-132(in Chinese)

E, Persoon and K.S. Fu. Shape discrimination using Fourier descriptors[J]. IEEE Trans. on PAMI, 1986, 8(3): 388-397

Gao, J., Zhou, M. & Wang, H. A Threshold and Region Growing Combined Method for Filament Disapearance Area Detection in Solar Images[C], In Proceedings of The Conference on Information Sciences and Systems. 2001,The John Hopkins University: 243-245

Lin Ying, Fu Yang. The key of bulk warehouse grain quantity recognition ---- Rectangular benchmark image recognition[J].Journal of Zhejiang University(Engineering Science), 2007,41(40): 1643-1646 (in Chinese)

M.D. Kelly. Edge Detection in Pictures by Computer Planning[J], Machine Intelligence, 1973,Vol. 6 (American Elsevier, New York), pp. 397-409

Marr, D. & Hildreth, E.. Theory of Edge Detection[J], Proceedings of the Royal Society London 1980, B207: 187-217

Shridhar M, Badreldin A. High accuracy character recognition algorithm using Fourier and topological descriptors[J]. Pattern Recognition, 1984, 17(5):515～524

T. Poggio, H. Voorhees and A. Yuille(1985).A Regularized Solution to Edge Detection[J], A. I. Memo 883, M.I.T, 24(3): 23-26

Tou, J.T. & R.C. Gonzalez. Pattern Recognition Principles[M], Addison-Wesley Publishing, Reading, 1981, MA, USA

Wang Shuan,AI Hai-zhou, HE Ke-zhong. Difference image based multiple motion targets detection and tracking[J]. Journal of Image and Graphics, 1999, 4 (A):471-475(in Chinese)

Zhang Kai, the application of Visual C++in image processing[J], Internal Science and Technology, 2006,2, 245-252(in Chinese)

Zhang, T.Y. & C.Y. Suen. A fast parallel algorithm for thinning digital patterns[J], CommACM 1984, 27(3): 236-239

Zhu Minghan, Luo Dayong. Moving Objects Detection Algorithm Based on Two Consecutive Frames Subtraction and Background Subtraction[J]. Computer Measurement & Control, 2005, 3(6): 215-217(in Chinese)

ENCRYPTION OF DIGITAL IMAGE BASED ON CHAOS SYSTEM

Jingtao Jian[1], Yan Shi[2], Caiqi Hu[1], Qin Ma[3,*], Junlong Li[4]
[1] *College of Mechanical and Electrical Engineering, Qingdao Agricultrual University, Qingdao, China, 266109; Email: jjtao_2518@163.com*
[2] *Foodstuff Science and Engineering college, Qingdao Agricultural University, Qingdao, China, 266109;*
[3] *College of Information and Electrical Engineering, China Agricultural University, Beijing, China, 100083*
[4] *Office of Agricultural Machine,Laixi, China,266600*
[*] *Corresponding author, Address: College of Information and Electrical Engineering, China Agricultural University, 17 Tsinghua East Road, Beijing, 100083, P. R. China, Tel:+86-10-62736973, Email: mei6668@163.com*

Abstract: In this paper, four kinds of chaos mapping equations such as Logistic, Henon, Quadratic and MacKeyGlass were discussed, and the numerical characteristics of those chaos mapping equations were analyzed and compared by histogram and correlation coefficient. Then the better chaos encryption system was selected according to analyzing result, and the encryption method of poor chaos encryption systems were modified. So the good performance of encryption was obtained, the degree of image scrambling transformation was improved greatly, and good effect of image encryption was gained.

Key words: Chaos encryption system; Digital image; Histogram; Correlation coefficient

1. INTRODUCTION

In recent years, with rapid development of the internet and the computer communications technique, the information transmission security becomes a hot subject of research currently subject of research. The chaos map equations have the property that a tiny fluctuation of the initial value can change the corresponding chaos code greatly, and so it is very difficult to

---

Please use the following format when citing this chapter:

Jian, J., Shi, Y., Hu, C., Ma, Q. and Li, J., 2009, in IFIP International Federation for Information Processing, Volume 294, *Computer and Computing Technologies in Agriculture II, Volume 2*, eds. D. Li, Z. Chunjiang, (Boston: Springer), pp. 1145–1151.

decrypt the chao coding file. Therefore, the chaos code are often used to encript the sound and image information. (Mei shuli et al., 2006; Yang wei et al., 2005; Yin xiandong et al., 2005) Different chaos codes can derived from different iterative equations such as Logistic, Henon, Quadratic, Mackeyglass and so on. But most researchers pay their attention on the Logistic chaos coding, in fact, each chaos code serial has its different value range and distribution from others. These two parameters impact the encrypt effect directly, and they are correlative to other parameters an the initial values in the chaos map equation. The object of this paper is to analyze the numerical properties of each chaos signal and find the optimal value range of each parameter in image encryption.

2. COMPARISON OF FOUR COMMON CHAOS CODE SERIES

2.1 Fundamental theory of imagc chaos encryption

Consider four common chaos code series such as the Logistic, the Henon, the Quadratic and the MackeyGlass, the corresponding map equations are

$$X_i = a \times X_{i-1} \times \left(1 - X_{i-1}\right), \quad \text{where } a = 4, \quad X_0 = 0.39 \tag{1}$$

$$\begin{cases} X_i = 1 - a \times (X_{i-1})^2 + Y_{i-1} \\ Y_i = b \times X_{i-1} \end{cases}, \quad \text{where} \begin{cases} X_0 = 0.1, a = 1.4 \\ Y_0 = 0.1, b = 0.3 \end{cases} \tag{2}$$

$$X_i = a - (X_{i-1})^2, \quad \text{where } X_0 = 0.1, a = 1.95 \tag{3}$$

$$X_i = X_{i-1} + \frac{a \times X_{i-s}}{1 + (X_{i-s})^{10}} - b \times X_{i-1}, \tag{4}$$

where $X_0 = 0.1, a = 0.2, b = 0.1, s = 17$

The secret communication technique based on the chaos theory has four methods: the chaos spread spectrum, the chaos shift keying, the chaotic parameter modulation and chaotic masking. In these methods, the chaotic parameter modulation is applied widely being simple and mature. The chaos encryption principle based on the modulation technique and the chaos map method is shown in figure 1 (Mei shuli et al., 2006)., where s(n) is the wavelet transform series of the original signal, e(n) is the encryption signal which will be transmit to the receiving end, and $\hat{s}$ (n) is thc final response series, that is the decryption signal series. The corresponding chaos encryption result is shown in Fig.2.

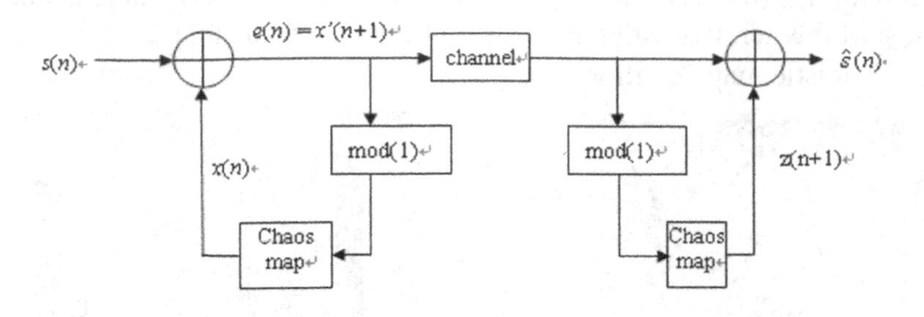

Fig.1. Chaos encryption theory

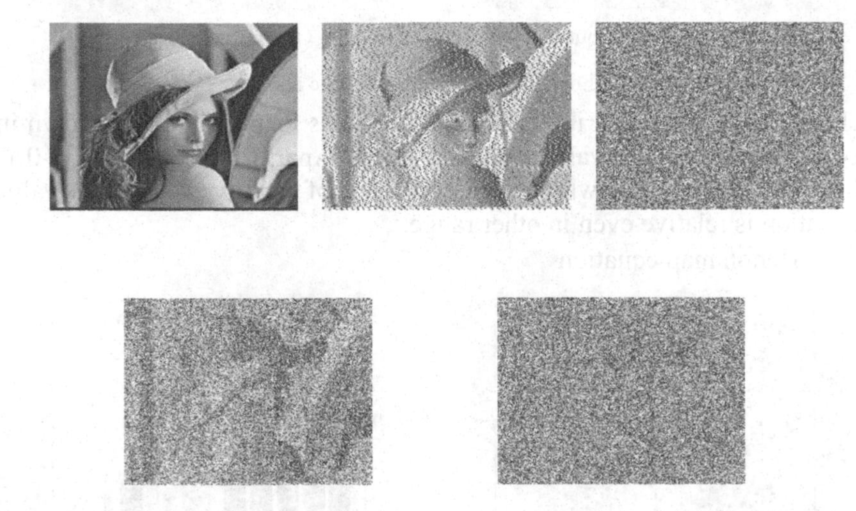

Fig.2. Encryption image based on different chaos series

(1) MacKeyGlass map equation

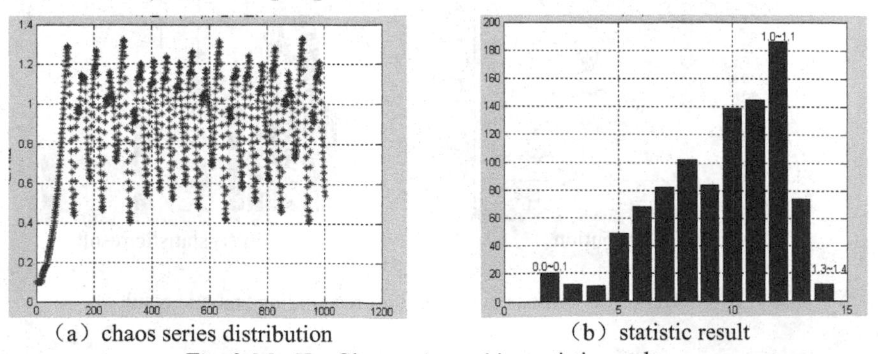

(a) chaos series distribution (b) statistic result

Fig. 3. MacKeyGlass series and its statistic result

The iterative value distribution of the MackeyGlass map equation is shown in figure 3. It is easy to see that the value range of the MacKeyGlass map equation is (0 ~ 1.4), and the distribution of the iterative value is uneven, the most of the iterative value is focus on the range (1.0~1.1).

(2) Logistic map equation

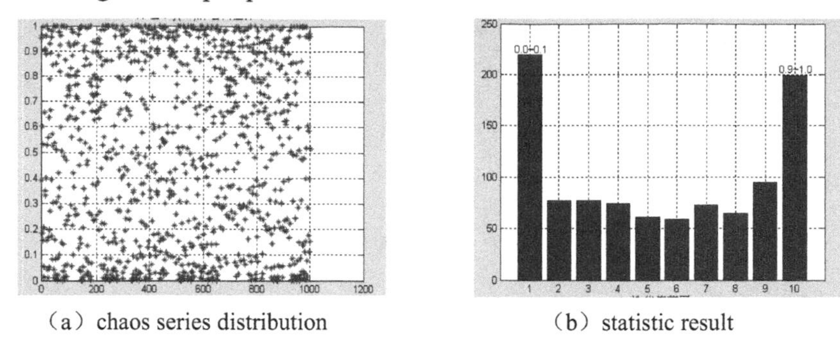

(a) chaos series distribution (b) statistic result

Fig.4. Logistic series and its statistic result

The iterative value distribution of the Logistics map equation is shown in Fig.4, the corresponding value range is (0~1.0). Apart from thc range (0~0.1) and the range (0.9~1.0) where focus on a lot of iterative value, the value distribution is relative even in other range.

(3) Henon map equation

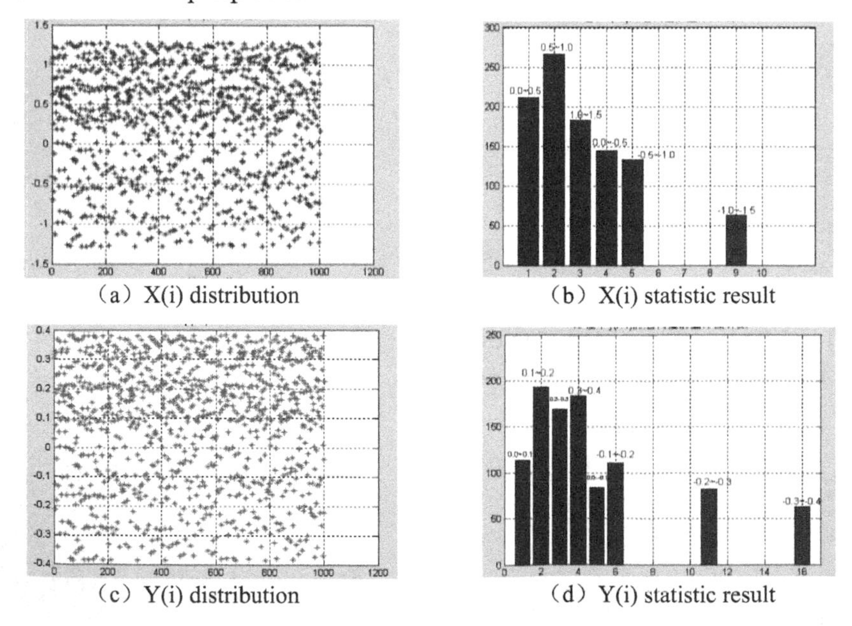

(a) X(i) distribution (b) X(i) statistic result

(c) Y(i) distribution (d) Y(i) statistic result

Fig.5 Henon series distribution and corresponding statistic result

The solution x[i] of the Henon map equation is shown in Fig.(a) and (b), from which we know that the value range of x[i] is （-1.5 ~ 1.5）, and the most of the values fall in the range （-1.5 ~ 0.25） and the range （0.25 ~ 1.5）.

The soulution y[i] of the Henon map equation is shown in Fig.5(c) and (d), it is clear that the value range of y[i] is （-0.4 ~ 0.4）, and the most of the values fall in the range （0.1~0.4） and the range (-0.4~0.1).

（4） Quadratic map equation

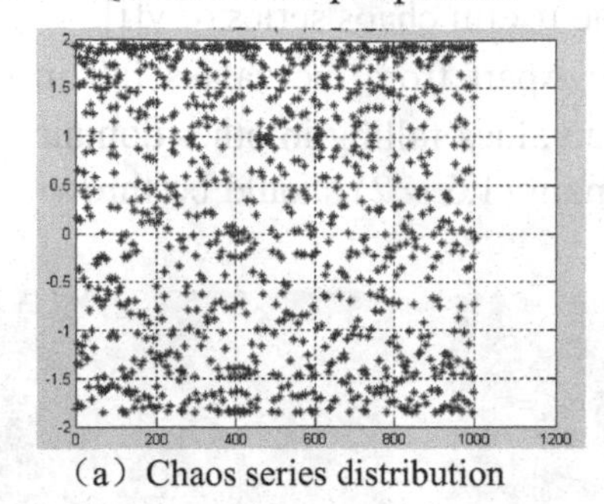

（a） Chaos series distribution

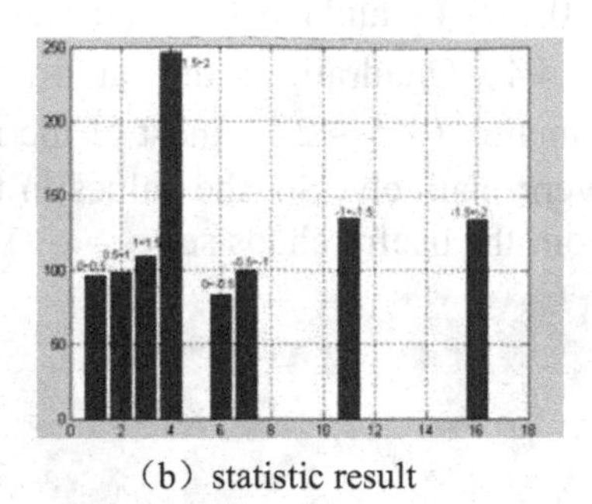

（b） statistic result

Fig.6 Quadratic series distribution and the corresponding statistic result

The iterative value distribution is shown in Fig. 6. It is easy to see that the value range of this map equation is （-2~2）, and most of the values are distributed in the domains (1.5~2) and (-2~2.5). It should be noted that the value distribution in the domain (-2~1.5) is even in some degree.

It is easy to see that the distribution of the Logistic chaos series is even relatively, and the corresponding encryption effect is better. In fact, it could be concluded that the encryption effect is correlative with the evenness of the chaos series. And so we can improve the encryption algorithm by building a map function which can turn the uneven chaos series into the relative even ones.

3. OPTIMIZATION OF THE CHAOS SERIES

（1） Most of the iterative values obtained by the MacKeyGlass map equation are focus on the range （1.0~1.1）， the rest iterative values distributed in other range are not even in some extent.

Leaving all the iterative value falled in the range （1.0~1.1） and throwing off other ones, that is, only the values falled in the domain （1.0~1.1） are taken as the useful chaos series.

（2） Apart from the domains （0~0.1） and （0.9~1.0）, the iterative value derived from the Logistics map equation distributed in the domain

(0.1~0.9) evenly. So, we can abandon the values falled in the domains (0~0.1) and (0.9~1.0) and remain the ones falled in other domains.

（3） The iterative values derived from the Henone equation can be devided into two parts: x[i] and y[i] . Most values of the x[i] fall in the domains（-1.5~0.25）and（0.25~1.5）. Most values of the y[i] fall in the domains（0.1~0.4）and (-0.4~0.1).

Obviously, the values fall in the domains (-1.5~ 0.25) and (0.25 ~ 1.5) are taken as the useful chaos series of x[i], and the values fall in the domains（0.1~0.4）and (-0.4~0.1) are taken as the useful chaos series of y[i].

（4）Quadratic is the same as above, Apart from the values fall in the domain （1.5 ~ 2）, most of the iterative values falling in other domain are even relatively. So, the values in the domain (1.5 ~ 2) should be thrown off from the useful chaos series.

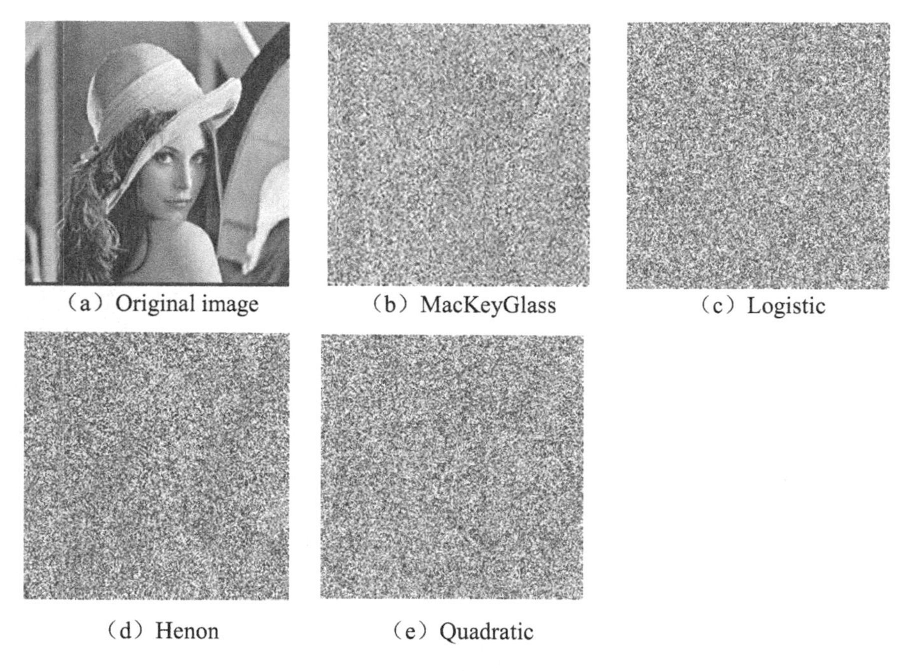

（a）Original image （b）MacKeyGlass （c）Logistic

（d）Henon （e）Quadratic

Fig.7 Improved chaos series image encryption

4. CONCLUSION

The image encryption properties based on different chaos series were discussed in this paper. And then, an new improved method was proposed. In this new method, the encryption properties was improved by optimizing the distribution of the chaos series. In fact, the amount of the chaos map equation is more than 10, analyzing their properties and screening out the best chaos series for the digital image encryption is very useful in net communications.

ACKNOWLEDGEMENTS

The project is supported by the national natural science fundation of China(No.60772038)

REFERENCES

Mei shuli, Gao wanlin. Image Chaos Encryption Technology Based on Interval Wavelet Coding [J]. Computer engineering, 2006, 32(16): 203-204,234

Wang Ying, Zheng Deling, Ju Lei. Digital Image Encryption Algorithm Based on Three-Dimension Lorenz Chaos System [J]. Journal of University of Science and Technology Beijing, 2004,26 (6)

Xu Quansheng,Li Zhen,Du Xuqiang. An Image Encryption Algorithm Based on Chaotic Sequences [J]. Journal of Chinese Computer Systems, 2006, 27 (9)

Yang wei, Chen Xiyou. Image Encryption Algorithm Based on Chaotic Mapping [J]. Techniques of Automation and Applications, 2005, 24(7)

Yi Kaixiang, Sun Xin, Shi Jiaoying. An Image Encryption Algorithm Based on Chaotic Sequences [J]. Journal of Chinese Computer Systems,2002,12(9)

Yin xiandong, Yao jun,Tang dan. Image Encryption Technology Based on DWT Domain[J]. Techniques of Automation and Applications, 2005, 3(1): 1-5

AN IMPROVED EDGE DETECTION METHOD FOR IMAGE CORRUPTED BY GAUSSIAN NOISE

Xiao Wang[1], Hui Xue[1,*]

[1] *College of Information and Electrical Engineering, China Agricultural UniversityBeijing, P.R. China, 100083*

[*] *Corresponding author, Address: College of Information and Electrical Engineering, China Agricultural University Beijing, P.R. China, 100083, Tel:86-13520508835, Email: xue_huicn@yahoo.com.cn*

Abstract: Due to the difficulty with extracting edge points and eliminating noise points from images, an improved maximizing objective function algorithm was proposed. More directions were added to relocate the edge points, at the same time, edge and noise characteristics were analyzed to separate and the noise points were eliminated by a proper threshold T. The comparison based on principle of the improved method, classical methods and the references methods is done, the simulation results indicated that the performance of the improved edge detection method was better than that of other compared algorithms.

Keywords: edge detection, Gaussian noise, directions, threshold

1. INTRODUCTION

Pattern recognition and machine vision were applied to extract the objects from the background in the images, as well in the agriculture field, contours of the objects can be used to recognize the plumpness of the plants. In order to extract the configuration parts from an image, the complete information of the image edge was necessary. Meanwhile, since the experimental images were always corrupted by Gaussian noise, the results of the edge detection were turned to be blurry or with some speckles.

Please use the following format when citing this chapter:

Wang, X. and Xue, H., 2009, in IFIP International Federation for Information Processing, Volume 294, *Computer and Computing Technologies in Agriculture II, Volume 2*, eds. D. Li, Z. Chunjiang, (Boston: Springer), pp. 1153–1159.

In classical edge detection methods, such as Sobel Operator, Laplacian-Gaussian Operator, Roberts Operator, and Prewitt Operator etc, the maximums of the gradients were utilized to find image edge points (Duan et al., 2005). Because being sensitive to the noise, classical edge detection operators may not be practically used in the actual image processing. In order to discriminate edges with exact location, high signal-to-noise ratio and thin contours, lots of new edge detections algorithms are studied.

Zhao Chunjiang et al.(Zhao et al., 2005) proposed a new method based on the image gray-level characteristic, it made the process faster, but the location of the edge points in complex images was still a problem and some of the parameters needed to be settled by using plenty of experiments. Different masks were used to search the edge points: half neighborhood algorithm (Xu et al., 2006), more directions (Yao et al., 2007) etc., which could get better locations but noise always be preserved as edge points. Chung-Chia Kang and Wen-June Wang (Kang et al., 2007) proposed a method based on the maximizing objective function, which got the edges much better than the classical methods, but as well it was very sensitive to the noise. The proposed new edge detection method can be reasonably considered Gaussian noise reduction and accurate location of edge by adding more masks and separating edge and noise respectively.

2. THE IMPROVED ALGORITHM

The algorithm proposed in Kang et al., 2007 can get the edge points clearly because of its good discrimination capability, but when it combined with Gaussian noise, the image came to be blurry. Fig. 1 showed the 4 directions used in the original method.

Each direction mask was divided into two parts: S_0 and S_1, it was known that if the distances between S_0 and S_1 were large and the distances inside S_0 and S_1 were small, then the edge intensity may be large. There was an objective function describing these distances were given as follows (Kang et al., 2007):

$$f(i) = (L - 1)\frac{N_f}{D_f} \tag{1}$$

Where,

$$N_f = \min\left(1, \frac{|m_0 - m_1|}{\omega_1}\right) \tag{2}$$

$$D_f = 1 + \frac{1}{15} \sum_{\substack{p_m, p_n \in S_1 \\ m > n, m \neq n}} \min\left(1, \frac{|p_m - p_n|}{\omega_2}\right) + \frac{1}{3} \sum_{\substack{p_m, p_n \in S_1 \\ m > n, m \neq n}} \min\left(1, \frac{|p_m - p_n|}{\omega_2}\right) \tag{3}$$

And ω_1=90, ω_2=40, and $m_0 = (1/6)\sum_{p_i \in S_0} p_i, m_1 = (1/3)\sum_{p_i \in S_1} p_i$ (Kang et al., 2007), N_f describes the interset between S_0 and S_1, when the distance is larger thanω_1, the edge intensity is large enough. D_f describes the intraset distance, if the edge intensity is small D_f shows to be large. For an 8-bit gray level image, L is equal to 256. $f(i)$ shows the relationship of interset and the intraset distances, and provides values $f(i)$ for estimating .

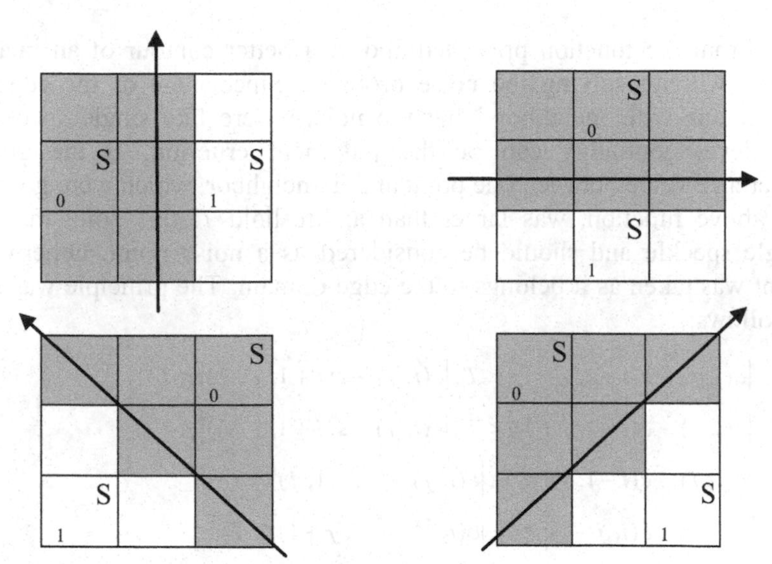

Fig. 1. Original 4 directions

Therefore, the edge points were located to differentiate the edge points and the noise point on the first step. For getting the exact location of the edge points, another 4 new added directions proposed by the improved method were showed in Fig.2. Hence, edge points and noise points were separated better, and the edge points can be located more accurately.

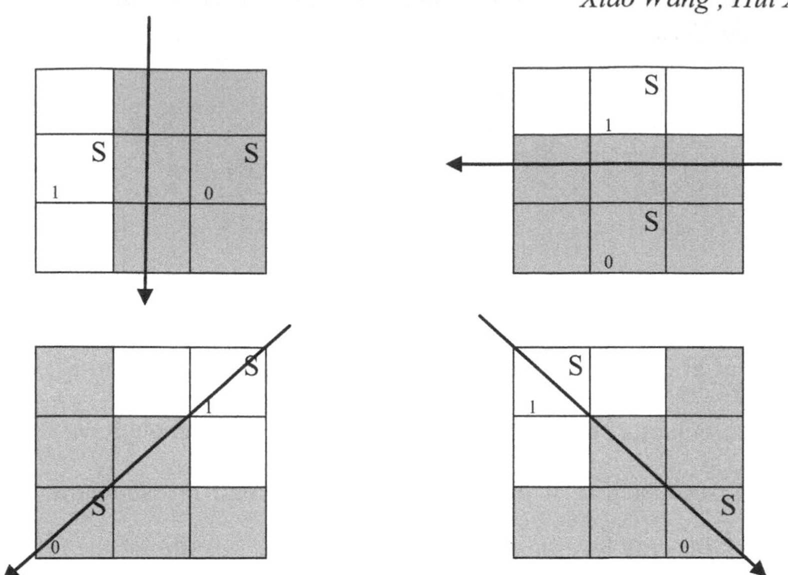

Fig. 2. The new 4 directions

From the function presented above, a better contour of an image was gotten without solving the noise problem. Since most of the edges were continuous with 'neighbors' but the noises were like single speckles, the gray-level continuity can be the judgment criterion. If the gray-level difference value between one point and its neighbor, which were got through the above function, was larger than a threshold *T*, this point mostly is a single speckle and should be considered as a noise point. Otherwise, the point was taken as it belongs to the edge domain. The principle was showed as follows:

$$\begin{aligned}
&|e(i,j)-e(i-1,j-1)|<T, |e(i,j)-e(i+1,j+1)|<T \\
&|e(i,j)-e(i-1,j+1)|<T, |e(i,j)-e(i+1,j-1)|<T \\
&|e(i,j)-e(i-1,j)|<T, |e(i,j)-e(i+1,j)|<T \\
&|e(i,j)-e(i,j-1)|<T, |e(i,j)-e(i,j+1)|<T
\end{aligned} \tag{4}$$

Where $e(i, j)$ was a pixel in a picture, $e(i-1, j-1), e(i+1, j+1)$ etc. were the neighbors of $e(i, j)$, and T was the threshold.

If a pixel $e(i, j)$ had one of the characteristic showed in (4), the proposed method took it as an edge point, otherwise it should be taken as a noise point and eliminate it. The threshold *T* was settled as 1.8 times of the gray-level average value, which was selected from plenty of experiments. Then, the job of edge detection and the noise problem both were completed.

3. RESULT AND DISCUSSION

A sample picture was used to compare the classical edge detection operators, the original algorithm and the references methods with the proposed method in Fig.3.

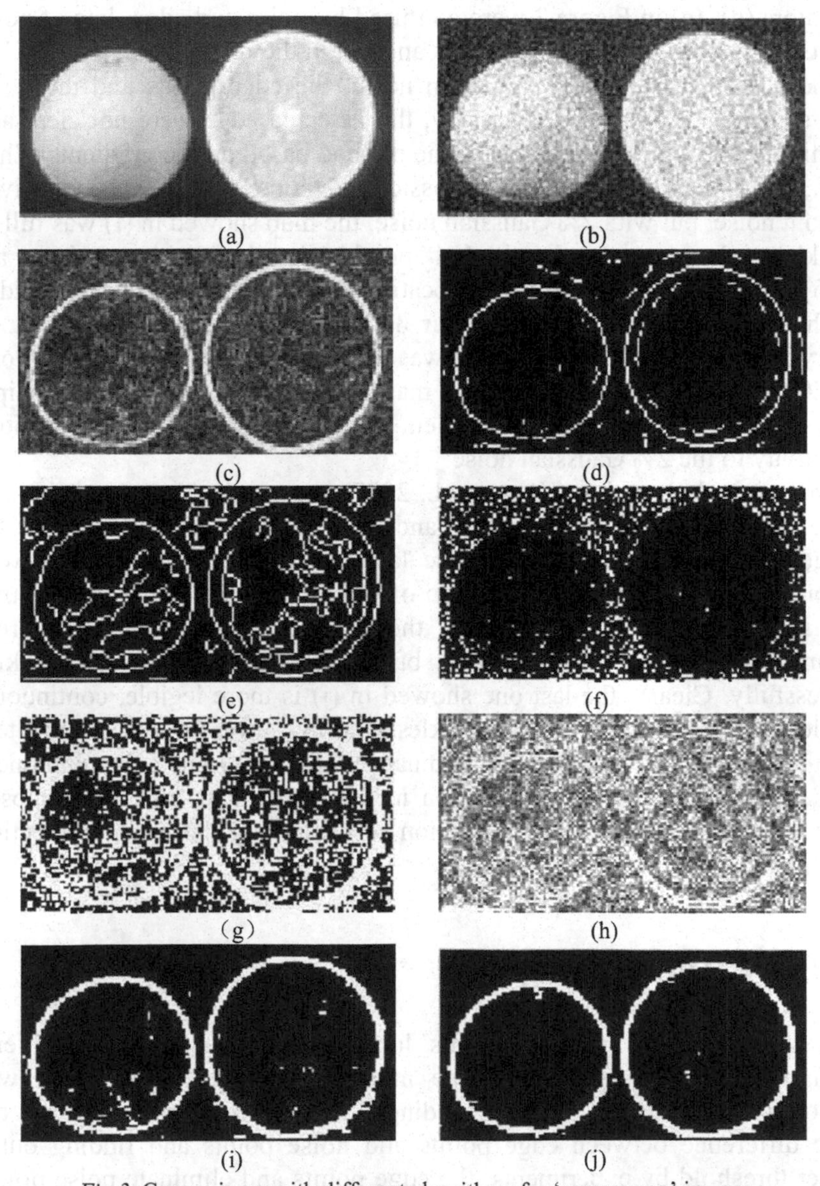

Fig.3. Comparisons with different algorithms for 'cameraman' image:

(a) the original 'cameraman' image; (b) original image added 2% gaussian noise; (c) the edge map by using original algorithm (Kang et al., 2007); (d) the result of LOG operator; (e) the result of Canny operator; (f) the result of the method base on gradient(Zhao et al., 2005) (g) the result of half neighborhood algorithm (Xu et al., 2006) (h) the result of the improved Sobel operator (Yao et al., 2007) (i) the result of original method with the denoising process;(j) the edge map by using the proposed improved method with 8 directions and the denoising process.

According to the images in Figure 3, the results of classical edge detection operators (d), (e) in Figure 3 were confused by noise and edge, lots of noise points were considered as edge points and left in the results, they were full of speckles caused by the 2% Gaussian noise; the edge points and the noise points cannot be located respectively, the detected edge were not clear and definitude enough. The principle of the method based on the gradients (Zhao et al., 2005) was similar with the classical methods, better result was given without noise, but with 2% Gaussian noise, the map showed in (f) was full of speckles. (g) was gotten by the half neighborhood algorithm (Xu et al., 2006), 8 masks were used and the location of the edge points were decided by the pivotal threshold, the contour hardly can be seen in the result. A blurry image was seen in map (h), it was the improved Sobel method (Yao et al., 2007)which was added another 2 masks, because of the similar principle with classical methods and the gradient methods, the result showed serious sensitivity to the 2% Gaussian noise.

The original algorithm (Kang et al., 2007) showed in (c) was sensitive to noise, the detected edge was blurred and the whole map was a little dark for image analyzing. In comparison, the last two images (i) and (j) showed better edge responses, in the situation of 4 directions in (i), the bottom part still had some speckles caused by the Gaussian noise, it had apparent advantages than the prior methods, but it cannot eliminate the speckles successfully. Clearly the last one showed in (j) is more legible, continuous, and less infected by noise. The speckles seen in (i) were mostly eliminated, and the edge was maintained well and hardly affected. Hence, the plumpness of the fruit can be more clearly seen in the result map, and the proposed method obtains the best edge detection result among those edge detection algorithms.

4. CONCLUSION

In order to improve the edge points' location and solve the noise problem, an improved method based on the maximizing objecting function was proposed in this paper. Through adding 4 more directions, analyzing the basic difference between edge points and noise points and finding out a proper threshold by experiments, the edge points and eliminate noise points were relocated from the edge image. The simulation results had shown that the proposed improved method provides a better edge detection results with exact location and responses to edge points, it clearly showed the shape and the plumpness of the main objects in the picture. And the capacity to eliminate the noise makes the proposed method can be used in agricultural recognitions and some more practical image edge detection situations.

REFERENCES

Chung-Chia Kang, Wen-June Wang, A novel edge detection method based on the maximizing objective function, Pattern Recognition 40(2007), 609-618

DUAN Rui-ling, LI Qing-xiang, LI Yu-he, Summary of image edge detection, Optical Technique, Vol.3 No.3, May 2005, 415-419

XU Wei-dian, XU LI-chun, HU Yue-li, Adaptive edge detection using a half neighborhood algorithm, Journal of Shanghai University (Natural Science), Vol.12 No.2, April. 2006, 146-149

YAO Xing-zhong, HU Han-ping, LU Tong-wei, An improved Sobel operator method based on cat visual cortex, Computer Engineering and Applications, 2007,43(31), 64-67,70

ZHAO Chun-jiang, SHI Wen-kang, DENG Yong, Novel edge detection method based on gradient, Opto-Electronic Engineering, Vol.32 No.4, April. 2005, 86-88

COMPARSION OF MULTISPECTRAL REFLECTANCE WITH DIGITAL COLOR IMAGE IN ASSESSING THE WINTER WHEAT NITROGEN STATUS

Liangliang Jia [1,2], Xinping Chen [1,*], Minzan Li [3], Zhenling Cui [1], Fusuo Zhang [1]

[1] *College of Resources and Environmental Sciences, China Agricultural University, Beijing, 100094, China*

[2] *Institute of Agricultural Resources & Environment, Hebei Academy of Agriculture and Forestry Sciences, 050051, Shijiazhuang, China*

[3] *Key lab of Precision Agriculture, China Agricultural University, Beijing, 100083, China*

[*] *Corresponding author, Address: College of Resources and Environmental Sciences, China Agricultural University, Beijing 100094, Beijing, P.R. China, Tel: +86-10-62733454, Fax: +86-10-62731016, Email: chenxp@cau.edu.cn*

Abstract: Previous researches have shown that the digital image color intensity could reflect the crops N status, but there is little information about the comparision of spectrum reflectance in the visible bands with the digital imagery color intensities. A field experiment was conducted to compare the wheat canopy reflectance at visible bands (400-700 nm) at shooting stage with near ground digital image to detect N deficiencies. Single color bands of R, G, B and ratio indices of G/R, G/B, R/B, R/(R+G+B), G/(R+G+B) and B/(R+G+B), which derived from digital image and spectral measurments, were regressed with wheat N status. The R, G, G/B, R/B, R/(R+G+B) and G/(R+G+B) all had negative correlations, while the G/R and B/(R+G+B) indices had positive correlations, with plant N status. For the B band, the digital image analysis data got positive correlations while the spectral measurements got negative correlations. With higher correlation coefficient than other indices, the R/(R+G+B) was the best index in this research. Considering the easiness of getting digital images and the accurate prediction of crops N status, the digital image analysis method seems to be a better way for in field plant N status evaluation.

Keywords: Spectrum reflectance; Image analysis; Color intensity; Winter wheat

---

Please use the following format when citing this chapter:

Jia, L., Chen, X., Li, M., Cui, Z. and Zhang, F., 2009, in IFIP International Federation for Information Processing, Volume 294, *Computer and Computing Technologies in Agriculture II, Volume 2*, eds. D. Li, Z. Chunjiang, (Boston: Springer), pp. 1161–1170.

1. INTRODUCTION

Nitrogen is one of the most important plant nutrients for winter wheat production and application of N fertilizer can help farmers obtaining higher yields. But the N over fertilization had become a common problem on the North China Plain and resulted low N use efficiency and increased the possibility for groundwater nitrate contamination (Zhao et al., 2006; Chen et al., 2000).

Traditional methods of estimating in-season optimum N requirements for wheat are based on soil Nmin testing (Wehrmann et al., 1988), tissue N concentrations (Vaughan et al., 1990, Tyner and Webb, 1946), and stem sap nitrate concentration (Zhen and Leigh, 1990). But all these methods require multiple and destructive sampling during the crop growth, can be expensive and time consuming (Blackmer and Schepers, 1995). Chlorophyll meter provides a rapid and inexpensive way of assessing N status of crops without destruction, but it could only test a small number of leaves and lead to uncertainty in characterizing variability within large fields (Fox et al., 1994; Smeal and Zhang, 1994).

Remote sensing provides a cheap and quick solution to assess the growing conditions or crops than traditional analysis methods. Spectrum reflection of crops canopy have been shown its advantage in assessing N status of crops (Graeff et al., 2001; Filla et al., 1995; Gitelson and Merzlyzk, 1994). Some indices such as NDVI (Normalized Difference Vegetation Index, Rouse et al., 1973), RVI (Ratio Vegetation Index, Jordan, 1969), and GNDVI (Green Normalized Vegetation Index, Gitelson and Merzlyzk, 1997) have been developed to analyze the reflection of crop canopy. These vegetation indices have been widely used in estimating the crops growth status and nutrient status. On the other hand, remote sensing via aerial color or CIR photography has been used in detecting the N status or characterizing N requirements of crops (Sripada et al., 2005, 2006; Flowers et al., 2003; Sharpf and Lory, 2002; Blackmer and Schepers, 1996) by utilizing thc ditital number of canopy image for R (red), G (green) and B (blue). With the quick development of digital camera image acquisition technology, some researches have used digital camera to analyze the crops nutrients disorder (Jia et al., 2004a; Graeff et al., 2001), vegetation coverage (Lukina et al., 1999), tiller density (Adamsen et al., 1999) and N fertilizer recommendation (Jia et al., 2004b) The calculated G/R (Adamsen et al., 1999) or R/(R+G+B) (Jia et al., 2004a) ratios showed good correlations with plant total shoot N concentration, SPAD chlorophyll meter readings, or the predicted N supply rates.

The digital image covers the spectrum reflection in the visible bands ranging from 400 nm to 700 nm. It is still no comparsion of the correlations for R, G, B and their combinations G/R, G/B, R/B or G/(R+G+B),

R/(R+G+B) and B/(R+G+B) with the spectrum reflection of crops canopy, and which one could be used in crops N status detection. In this context, the objective of this study was to compare the methods of spectral reflectance in the visible part of the spectrum with the digital image analysis to make plant N status assessment.

2. MATERIALS AND METHODS

2.1 Experiment background

A field experiment comprised a winter wheat/summer maize double cropping system, which involved planting and harvesting of one wheat and one maize crop in one year, was conducted in Dongbeiwang Experiment Site, near Beijing, China since October 1999. The experiment was a 3 x 2 x 3 factorial split-split block experiment with four replications, which contained (i) three methods of irrigation, sub-optimal irrigation as a control, conventional irrigation as farmers' practice and optimized irrigation based on soil water measurement, as main plots of 60 x 30 m; (ii) two straw levels, with straw (straw recycling) and without straw (without straw recycling), as sub-plots of 60 x 15 m, and (iii) three levels of N application as sub-sub-plots of 20 x 15 m (control, farmer's N fertilization and 'optimized' N fertilization). Farmer's N fertilization consisted of 150 kg N ha-1 applied as NH_4HCO_3 before sowing and an additional 150 kg N ha-1 as urea at booting, which is typically practiced by local farmers. The optimized N fertilization treatment took the soil mineral N (Nmin) in the soil profile before sowing, at regreening and at booting, and yield goal into consideration to determine target N value (split for three growth stages).

The first crop in this experiment was summer maize, which was sown at June 1999 and harvested at Sept. 1999. From Oct. 1999 to Sep. 2004, five winter wheat/ summer maize rotations followed. In this study, all results were obtained in the 2003/2004 winter wheat growing season, totally 3 treatments with different N treatments but all with optimized irrigation and without straw recycling were selected in each of the 4 replicates (Table 1).

2.2 Data collection and processing

On shootingting stage of Apr. 25th, digital pictures of the winter wheat canopy were obtained with an Olympus E-20P digital damera. To work at a comparable solar angle and light intensity, all images were taken between 12 to 13 hour on cloudless day at 1.2 m above the ground and at an angle of 60°.

The digital images of 2560×1960 pixels of 8 bit for red, green and blue were transferred in JPEG format to a computer and processed with Adobe Photoshop® to extract color information of the digital number of Red, Green and Blue. Ratio indices of G/R, G/B, R/B, G/(R+G+B) and R/(R+G+B) were calculated (Lukina et al., 1999; Scharpf and Lory, 2002; Jia et al., 2004).

Table 1 Soil mineral nitrogen (Nmin in kg/ha) at different wheat growth stage and N application rates (kg/ha) in the selected treatments at Dongbeiwang, Beijing, China

Treatments†	No N	Conventional N	Optimized N
Nmin before sowing (0-90 cm)	19	398	47
Basal mineral N fertilization	0	150	0
Nmin at regreening (0-60 cm)	26	291	40
Topdressing at regreening	0	0	57
Nmin at shooting (0-90 cm)	13	554	42
Total N applied before shooting	0	150	57

†: Treatments fertilization background. No N treatments, 1 crop growth season no N fertilization; Conventional N, continuously 10 crop seasons with conventional high N fertilization of 300 kg/ha per crop season; Optimized N, continuously 10 crop seasons with optimized N fertilization based on soil-plant analysis.

The wheat canopy hyper spectral reflectance measurements were conducted at the same day with digital image acquisition. A handheld ASD FieldSpec spectral radiometer with a range of 325-1075 nm and 1 nm resolution was used to measure the spectral reflectance of wheat canopy. Measurements were made 5 sites over each plot, looking straight down from 20 cm above the canopy. With a field view of 28°, the sensor viewed an area about 20 cm in diameter. Radiometer calibration was conducted before each measurement with a white panel. Data were obtained between 11 and 14 hour in order to have a stable solar angle and light intensity. 460 nm, 560 nm and 690 nm were selected as compared with Blue, Green and Red bands of digital image. Ratios of G/R, G/B, R/B, G/(R+G+B) and R/(R+G+B) were calculated.

Plant SPAD readings were taken with a Minolta SPAD®--502 chlorophyll meter on 21 April as an average of the first fully expanded leaves from 30 randomly selected plants per plot. The stem sap nitrate concentration of wheat was tested with a Reflect Meter (Merck Co., Darmstadt, Germany) at the same day. Above ground plant biomass was harvested in a 1 m2 sampling area per plot, dried to constant weight at 70℃, and analyzed for total N using the Kjeldahl method. And at maturity three separate sub-samples (each of size 3 m2) were harvested to determine grain yield for each plot. For analysis of soil Nmin at planting, regreening (156 days after sowing, DAS) and shooting (180 DAS), five cores were collected from each plot and pooled at 0 to 30 cm, 30 to 60 cm, and 60 to 90 cm depth intervals. All samples were dried and sieved, extracted with 0.01 mol L-1 CaCl2 and analyzed for NH4+ and NO3- by continues flow analyzer TRAACS 2000 (Bran+Luebbe, 1996). Soil Nmin in 0-90 cm soil profile was tested at before sowing and shooting stage, but only 0-60 cm soil layer depth was tested at regreening stage.

2.3 Data analysis

Nitrogen effects were analyzed quantitatively by comparing the means of agronomic parameters and reflectance spectra of each treatment through Duncan's multiple range tests at a probability of 0.05. Data sets of single band of R (red), G (green) and B (blue) and ratio indices of G/R, G/B and R/B, and normalized indices of G/(R+G+B), R/(R+G+B) and B/(R+G+B), which derived from digital image (near ground and aerial photography) or from the ASD FieldSpec spectral measurements, were regressed with plant N status parameters of chlohrophyll meter SPAD readings, upland biomass, sap nitrate concentration and total N content by using the SAS software (SAS Inst., 1990). Linear or nonlinear models were tested to fit the plot patterns and best-fit R2 values for the relationship.

3. RESULTS

3.1 Nitrogen effects on N status and spectral reflectance of wheat at shooting stage

Wheat growth was significantly affected by soil Nmin and N fertilizer application rate before shooting (Table 2). The plots with no N treatment were characterized by lower SPAD readings, above ground biomass, total N content and sap nitrate concentration. Optimized N treatments significantly lower than traditional N treatments in above ground biomass, total N concentration and plant sap nitrate concentration, but no difference was found in SPAD readings.

Table 2 Average SPAD readings, above ground biomass, total plant N content and plant sap nitrate concentration of winter wheat at shooting stage and grain yield at harvest

Treatments	SPAD reading	Biomass (kg/ha)	Total N Cont. (%)	Stem sap nitrate Cont. (mg/kg)
No N	33.4b*	1040c	2.12c	38c
Traditional N	43.6a	2149a	3.58a	3875a
Optimized N	40.3ab	1743b	3.26b	2176b

**Values followed by different letters are significantly different at P<0.05.*

The N management effects were manifested in canopy reflectance. Spectral reflectance of wheat canopy under different N treatments measured at shooting stage (21 April, 2004) is shown in Figure 1. Reflectance decreased in the visible wavebands region with increasing N supply. Two reflectance peaks were found in 460 nm and 560 nm and an absorption valley was found in 690 nm. The no N treatments had the higher reflectance percentage than those of optimized and traditional N treatments throughout the visible spectrum from 400 nm to 700 nm. Reflectance from no-N treatment was significantly higher than those of optimized and traditional N treatments in 560 and 690 nm. There were no differences of spectral reflectance between optimized and traditional N treatments in visible wavebands of 460, 560 and 690 nm, but the spectral reflectance of optimized N treatment was higher than traditional N-treatment (Table 3).

Table 3 Average reflectance (%) at 460, 560 and 690 nm of wheat with different N treatments at shooting stage

Treatments	460 nm	560 nm	690 nm
No N	4.8a	18.5a	6.9a
Traditional N	4.5a	13.9b	4.3b
Optimized N	4.6a	14.9b	4.6b

**Values followed by different letters are significantly different at $P<0.05$*

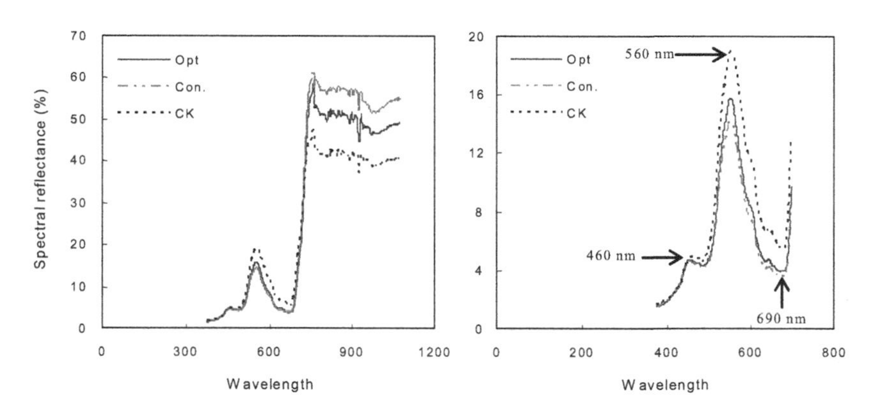

Figure 1 Wheat canopy reflectance with different N treatments at shooting stage

3.2 Correlations of color intensity and ratio indices with with wheat N status

Linear correlations were found to be the best correlation model between the color intensities and ratio indices with wheat N status at shooting stage. The correlation coefficients were listed in table 4. Significant linear correlations were found for single band of R, G and B with wheat N status at shooting stage. The R and G bands all had significant negative correlations while the B band had positive correlations with N status parameters for digital image. And the B band had negative correlations with plant N status for spectral measurement.

For the ratio indices, G/B and G/R had sigificant positive correlations with wheat N status while the R/B had negative correlations both for digital image analysis and spectral measurement. The normalized indices of R/(R+G+B) and B/(R+G+B) had significant correlations with plant N status for both digital image and spectral measurements. But significant correlations for G/(R+G+B) were only found for digital image.

Table 4 Correlation coefficients of wheat N status parameters at shooting stage with digital image color intensity and various ration indices for near ground and aerial photography and spectral reflectance

r	R	G	B	G/R	G/B	R/B	R/(R+G+B)	G/(R+G+B)	B/(R+G+B)
Digital image analysis									
SPAD	-0.431*	-0.370	0.842**	0.508*	-0.807**	-0.690*	-0.809**	-0.794**	0.825**
Nitrate	-0.762**	-0.741**	0.752**	0.723**	-0.780**	-0.850**	-0.892**	-0.767**	0.855**
Biomass	-0.738**	-0.712**	0.725**	0.718**	-0.784**	-0.837**	-0.869**	-0.736**	0.828**
N%	-0.734**	-0.704**	0.846**	0.712**	-0.880**	-0.880**	-0.946**	-0.842**	0.922**
Spectral measurements by ASD FieldSpec									
SPAD	-0.820**	-0.721**	-0.480*	0.793**	-0.679*	-0.679*	-0.824**	0.203	0.669*
Nitrate	-0.805**	-0.795**	-0.360*	0.625*	-0.899**	-0.899**	-0.733**	-0.241	0.859**
Biomass	-0.849**	-0.846**	-0.435*	0.637*	-0.828**	-0.828**	-0.730**	-0.234	0.867**
N%	-0.892**	-0.787**	-0.283	0.835**	-0.894**	-0.894**	-0.909**	-0.066	0.937**

4. CONCLUSION

In this study, single bands of R, G and B with ration indices G/R, G/B, R/B and normalize indices of R/(R+G+B), G/(R+G+B) and B/(R+G+B), which derived from data sets of digital image and spectral measurement, had been tested the correlations with wheat N status. And in this research, not only G/R ratio (Adamasen et al., 1999), but also the R/B, R/(R+G+B) and B/(R+G+B) had good correlations with N status of wheat at shooting stage. The high correlations for digital image analysis indices with N status showed its potential be used in crops N status evaluation.

The digital image acquisition had the advantage of integrating a large sampling area and would involve less labor than the use of spectrum reflectance measurement for N management. And the correlations inconsistency also happened in the B band, which got high negative correlations with spectra testing, but positive correlations with near ground digital image. This was different with the previous research of nutrients deficiency plants reflect more light over the whold visible spectral bands

(Al-Abbas et al., 1974). More researches should be conducted before definitive conclusion can be draw. And the higher correlation coefficient values of R/(R+G+B) with wheat N status for the digital camera and spectral reflectance suggested it maybe the best index in this research.

A shortcoming for digital image was not able to use the infrared spectrum reflectance from crops, more sensitive spectrum band to crop growth status, while lots of spectrum reflection indices have shown very useful in assessing N status of crops. The NIR spectral reflectance and the relative vegetation indices in this research had showed better correlations with N status than digital image (Table 5). But with the quick development of technology, the CCD digital camera now can directly record the NIR band. Considering the easiness and quickness of getting digital images and the accurate prediction of crops N status, the digital image seems to be a better way for in field plant N status assessment.

Table 5 Correlation coefficients between wheat N status parameters at shooting stage with NIR, NDVI, GNDVI, RVI derived from spectral reflectance of wheat canopy

r	NIR	NDVI	GNDVI	RVI
SPAD	0.757**	0.828**	0.811**	0.850**
Nitrate	0.849**	0.863**	0.910**	0.881**
Biomass	0.743**	0.853**	0.884**	0.846**
N%	0.907**	0.958**	0.950**	0.929**

**Significant at P<0.05; **Significant at P<0.01*

ACKNOWLEDGEMENTS

This research was financially supported by the Key Project of Eleventh Five-year National Plan (2006BAD02A15), Natural Science Foundation of China (Project number: 30571080), the Program for Changjiang Scholars and Innovative Research Team in University of China (IRT 0511).

REFERENCES

A. A. Gitelson, M.N. Merzlyak. Remote estimation of chlorophyll content in higher plant leaves. International Journal of Remote Sensing, 1997, 18: 291-298.

A. A. Gitelson, M.N. Merzlyak. Spectral reflectance changes associated with autumn senescence of Aesculus hippocastanum L. and Acer platanoides L. leaves. Spectral features and relation to chloropyll estimation. Journal of Plant Physiology, 1994, 143: 286-292.

A. H. Al-Abbas, R. Barr, J. D. Hall, F. L. Crane, M. F. Baumgardner. Spectra of normal and nutrient-deficient maize leaves, Agronmy Journal, 1974, 66: 16-20

B. Vaughan, K. A. Barbarick, D. G. Westfall, P. L. Chapman. Tissue nitrogen levels for dryland hard red winter wheat, Agronomy Journal, 1990, 82(3): 561-565.

Bran and Luebbe. Bran+Luebbe Traacs 2000 continuous flow analyzer operation manual. MT9, GB-352-87A and GB-352-87E. Publication No. MT7-50EN-01. Bran+Luebbe GmbH, Norderstedt, Germany. 1996.

C. F. Jordan. Derivation of leaf area index from quality of light on the forest floor, Ecology, 1969, 50: 663-666

D. Smeal, H. Zhang. Chlorophyll meter evaluation for nitrogen management in corn, Communications in Soil Science and Plant Analysis, 1994, 25(9&10): 1495-1503.

E. H. Tyner, J. W. Webb. The relation of corn yields to nutrient balance as revealed by leaf analysis, Journal of American Society of Agronmy, 1946, 38: 173-185.

E. Lukina, M. Stone, W. Raun. Estimating vegetation coverage in wheat using digital images, Journal of Plant Nutrition, 1999, 22(2): 341-350

F. J. Adamsen, J. Paul, J. Pinter, E. M. Barnes, R. L. LaMorte, G. W. Wall, S. W. Leavitt, B. A. 1999. Kimball. Measuring wheat senescence with a digital camera, Crop Science, 1999, 39(7): 719-724

I. Filla, L. Serrano, J. Serra, J. Peñuelas. Evaluating wheat nitrogen status with canopy reflectance indices and discriminant analysis, Crop Science, 1995, 35: 1400-1405

J. H. Wehrmann, C. Scharpf. M. Boehmer, J. Wollring. Determination of nitrogen fertilizer requirements by nitrate analysis of the soil and plant, In: Plant Nutrition 9th International Colloquium on Plant Nutrition, 1982, 202-208

J. W. Rouse, R. H. Has, J. A. Schell, D. W. Deering. Monitoring vegetation systems in the great plains with ERTS. Third ERTS Symposium, 1973, NASA SP-351, Vol. 1: 309-317. NASA, Washington, DC

L. L. Jia, X. P. Chen, F. S. Zhang, A. Buerkert, V. Römheld. Low altitude aerial photography for optimum N fertilization of winter wheat on the North China Plain, Field Crop Research, 2004a, 89: 389-395

L. L. Jia, X. P. Chen, F. S. Zhang, A. Buerkert, V. Römheld. Use of digital camera to assess the nitrogen status of winter wheat in the North China Plain, Journal of Plant Nutrtion, 2004b, 27(3): 441-450

M. Flowers, W. Randall, H. Ronnie. Quantitative approaches for using color infrared photography for assessing in-season nitrogen status in winter wheat, Agronmy Journal, 2003, 95: 1189-1200.

P. C. Scharf, J. A. Lory. Calibrating corn color from aerial photographs to predict sidedress nitrogen need, Agronomy Journal, 2002, 94: 397-404

R. F. Zhao, X. P. Chen, F. S. Zhang, H. L. Zhang, J. Schroder, and V. Roemheld. Fertilization and nitrogen balance in a wheat-maize rotation system in North China, Agronomy Journal, 2006, 98: 938-945

R. G. Zhen, R.A. Leigh. Nitrate accumulation by wheat (Triticum aestivum) in relation to growth and tissue N concentrations, Plant and Soil, 1990, 124: 157-160

R. H. Fox, W. P. Piekielek, K. M. Macneal. Using a chlorophyll meter to predict nitrogen fertilizer needs of winter wheat, Communications in Soil Science and Plant Analysis, 1994, 25(3&4): 171-181.

R. P. Sripada, R. W. Heiniger, J. G. White, A. D. Meijer. Aerial color infrared photography for determining early In-season nitrogen requirements in corn, Agronomy Journal, 2006, 98: 968-977

R. P. Sripada, R. W. Heiniger, J. G. White, and R. Weisz. Aerial color infrared photography for determining late-season nitrogen requirements in corn, Agronomy Journal, 2005, 97: 1443-1451

S. Graeff, D. Steffens, S. Schubert. Use of reflectance measurements for the early detection of N, P, Mg, and Fe deficiencies in Zea mays L, Journal of Plant Nutrition and Soil Science, 2001, 164: 445-450

SAS Institute. SAS/STAT user's guide, Version 8.1, SAS Inst., Cary, NC, 1998.

T. M. Blackmer, J. S. Schepers. Use of a chlorophyll meter to monitor nitrogen status and schedule fertigation for corn, Journal of Production Agriculture, 1995, 8(1): 56-60

T. M. Blackmer, J.S. Schepers. Aerial photography to detect nitrogen stress in corn, Journal of Crop Physiology, 1996, 148: 440-444

Xinping Chen, Hongjie Ji, Fusuo Zhang. The integrated evaluation on effect of excess fertilizer application on nitrate concentration of vegetable in Beijing. p. 270-277. In X. L. Li et al. (ed.) Fertilizing for sustainable production of high quality vegetables, Chinese Agriculture Publishing, Beijing, 2000 (In Chinese)

THE STUDY ON SCALE AND ROTATION INVARIANT FEATURES OF THE LACUNARITY OF IMAGES

Lidi Wang[1 *], Xiangfeng Liu[2], Li Min[3]
[1] *College of Information & Electrical Engineering, Shenyang Agricultural University.*
[2] *College of Sciences. Shenyang Agricultural University.*
[3] *Department of Science, Shenyang Jianzhu University*
[*] *Corresponding author, Address: College of Information & Electrical Engineering, Shenyang Agricultural University, 120 Dongling Road, Shenyang, 100161, P. R. China, Tel:+86-24-88487130, Email:wanglidi@gmail.com*

Abstract: It has been shown that the fractal dimension has a strong correlation with human judgment of surface roughness. Besides the fractal dimension, which is the most important fractal feature, lacunarity describes the characteristics of fractals. This feature is used in some fields and has good performances. In the field of image processing and recognition, it is important to study the scale and rotation invariant features. In this paper, the` scale and rotation invariant features of the Lacunarity are studied and the rule of varies is proposed.

Key words: Lacunarity, Invariant Feature, fractal, image processing

1. INTRODUCTION

It is well known that the fractal geometry is suitable to describe the feature of irregular natural image owing to the fact that fractal theory has been studied in the field of image processing and has achieved success in many areas. It has been shown that the fractal dimension has a strong correlation with human judgment of surface roughness. Besides the fractal dimension, which is the most important fractal feature, lacunarity describes the characteristics of fractals that have the same fractal dimension but different appearances(Marie-Pierre Dubuisson et al.,1994; Yueh-Min Huang et al.,

Please use the following format when citing this chapter:

Wang, L., Liu, X. and Min, L., 2009, in IFIP International Federation for Information Processing, Volume 294, *Computer and Computing Technologies in Agriculture II, Volume 2*, eds. D. Li, Z. Chunjiang, (Boston: Springer), pp. 1171–1174.

2005). This feature is used in some fields and has good performances such as partial discharge defect identification(Candela, R. et al., 2000), locating address blocks in postal envelopes(Jacques Facon1 et al., 2005), etc. In the field of image processing and recognition, it is important to study the scale and rotation invariant features. In this paper, the scale and rotation invariant features of the Lacunarity are studied and the rule of varies is proposed.

The remainder of this paper is organized as follows. The Lacunarity Feature Extraction is explained in section II. Section III includes the Scale and Rotation Invariant features of Lacunarity. Finally, the Conclusions and discusses are given in section Ⅳ.

2. LACUNARITY FEATURE EXTRACTION

Lacunarity was initially introduced by Mandelbrot to describe different textures which has the same fractal dimension. To compute the Lacunarity feature, there are several methods such as box counting, fuzzy C-means clustering(C. R. Tolle et al., 2003), etc.

Traditionally, Lacunarities are derived from the box counting algorithm by computing the following moment, in addition to M(L):

$$M^2(L) = \sum_{m=1}^{N} m^2 \cdot p(m, L) \tag{1}$$

the lacunarity for a box of size L is defined as:

$$\Lambda(L) = \frac{M^2(L) - (M(L))^2}{(M(L))^2} \tag{2}$$

where M(L) is the mass of the fractal set, $p(m,L)$ is the probability that m points fall within a box of side length L centered at a box of side length L centered at a point.

3. SCALE AND ROTATION INVARIANT FEATURES OF LACUNARITY

Scale and Rotation Invariant features are very important to pattern recognition. Including Lacunarity, fractal features usually have the property of scale and rotation invariant in theory, especially for the ideal fractal model. But in practice, there are differences between the signal and the fractal model. In order to study the invariant feature of Lacunarity, five kinds of different texture images from Brodatz texture image databases are selected shown in Fig.1. The five kinds of images are sand, water, brick, grass and straw, the size of the image is 512×512.

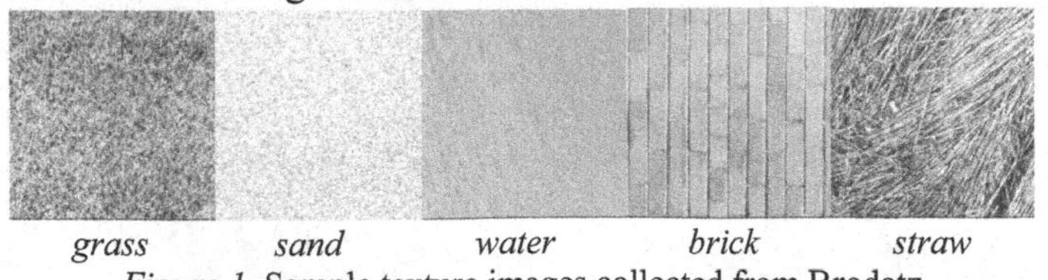

Figure 1 Sample texture images collected from Brodatz

In order to improve calculation efficiency and reduce calculation chanciness, when calculating the feature of images, 10 blocks of pixels in the image are selected randomly and the mean of their calculated Lacunarity feature are used as the Lacunarity feature of the whole image. In the progress of computation, let size L=7. One of the calculated results of the scale variant for the sand image is shown in figure 2.

From experiments, it is clearly shown that the Lacunarity's value become higher when the image's scale become bigger.

On the other hand, one of the calculated results of the rotate angle variant for the sand image is shown in figure 3.

From experiments, it is clearly shown that the Lacunarity's value hold invariant when the image's rotated.

4. CONCLUSIONS AND DISCUSSES

The experiment result shows the scale and rotate feature of the Lacunarity for images. In order to e, other conditions must be considered such as image interpolation method, window's size used in the computation of Lacunarity, etc. and these are the next work need to do in the further study.

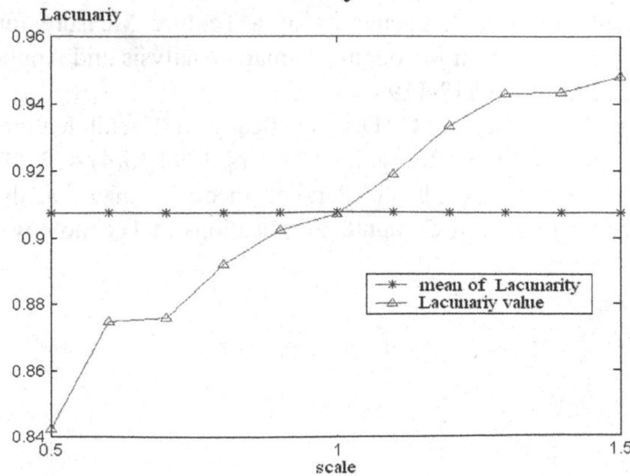

Figure 2 Zoom influence of Lacunarity feature (sand)

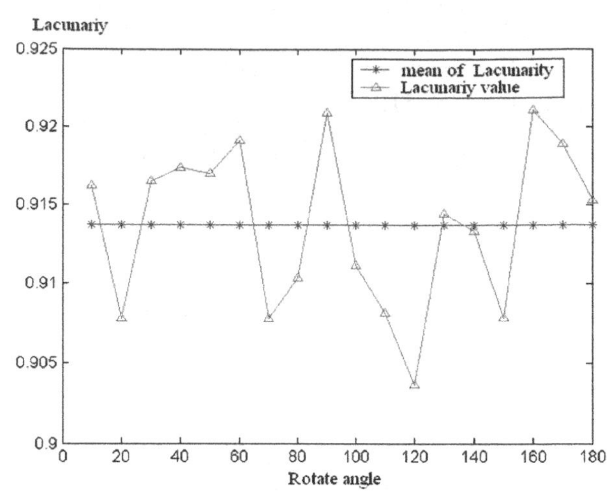

Figure 3 Rotation influence of Lacunarity feature (sand)

ACKNOWLEDGEMENTS

The research is supported by: Scientific Research Foundation for Youth Supervisors of Shenyang Agricultural University.

REFERENCES

C. R. Tolle, T. R. McJunkin, D. T. Rohrbaugh. Lacunarity Definition for Ramified Data Sets Based on Optimal Cover. Physica D: Nonlinear Phenomena. 2003,179(3-4): 129-152

Candela, R.; Mirelli, G.; Schifani, R. PD recognition by means of statistical and fractal parameters and aneural network. Dielectrics and Electrical Insulation, IEEE Transactions on. 2000, 7(1):87 – 94.

Jacques Facon1, David Menoti1 el. Lacunarity as a Texture Measure for Address Block Segmentation. Progress in Pattern Recognition, Image Analysis and Applications. Springer Berlin / Heidelberg. 2005,3773:112-119.

Marie-Pierre Dubuisson and Richard C. Dubes.Efficacy of fractal features in segmenting images of natural textures. Pattern Recognition Letters, 1994,15(4):419-431

Yueh-Min Huang and Shu-Chen Cheng. Parallel medical image analysis for diabetic diagnosis. International Journal of Computer Applications in Technology. 2005, 22(1):34 - 41

STUDYED ON THE EXTERNALDEFECTS SEGMENTATION BASED ON THE COLOR CHARACTER OF POTATOES

Min Hao[1], Shuoshi Ma[1,*]

[1] *College of Electrical and Mechanical Engineering, Inner Mogolia Agriculture University, Huhot,010018,China*

[*] *Corresponding author, Address: College of Electrical and Mechanical Engineering, Inner Mogolia Agriculture University,Huhot,010018, Inner Mongolia Autonomous Region China Tel: 0471-4209285, Email:mashuoshi@imau.edu.cn*

Abstract: Potato quality detection in China remains at the stage of dependent on human sense organ to identify and judge. According to the characteristics and requests of potatoes' detection, The original image was disposed fast and smoothly by the median filtering, and based on the threshold segmentation by setting up the values of B(blue), the background was effectively wiped off. By analyzing the circumscription of color characters between the normal and external defect potatoes, the external defect segmentation was realized. This way is simple and feasible.

Keywords: potatoes external defect; segmentation; color character

1. INTRODUCTION

Global awareness of the potato's key contribution to agriculture, the economy and world food security. The United Nations has declared 2008 as "International Year of the Potato", and potatoes was known as the "hidden treasures." Potatoes is a nutrient-rich agricultural products and is the world's number-four food crop after rice, wheat and maize, with annual production of more than 300 million tonnes in more than 100 countries. In China, potatoes is a major dominant crop in the Inner Mongolia Autonomous Region, potato yield is high. With the constant development and growing of

Please use the following format when citing this chapter:

Hao, M. and Ma, S., 2009, in IFIP International Federation for Information Processing, Volume 294, *Computer and Computing Technologies in Agriculture II, Volume 2*, eds. D. Li, Z. Chunjiang, (Boston: Springer), pp. 1175–1180.

the potato processing enterprises, potato quality inspection and grading is a necessary topic that needs to be solved.

The way of agricultural products' quality detection, includes automatic and semi-automatic and nondestructive detection methods. When the growth potato affected by man-made and natural growth conditions and other complex factors, its shape, size, color, and other feature are different. While the existing potato classifier based on the classifying principle of only size detection, can not meet the requirements for the composite indicator detection. So in China, detections of potatoes are mainly depended on the human's sense organs, which lead to lowness of efficiency and bigness of error. In addition, such a subjective assessment depend on the individual ability, color resolution, fatigue and other conditions, can not meet the requirements of a high standard classification and go against automation.

Color is important in evaluating quality and maturity level of many agricultural products. Color grading is an essential step in the processing and inventory control of fruits and vegetables that directly affects profitability(Dah-Jye Lee et al.,2008). The researches were more in quality testing of seeds, rice, fruit, eggs, vegetables.etc(Ying Yibiin et al.,2005). But so far there has been little research on potatoes.

2. TEST DEVICE AND RESEARCH OBJECT

2.1 Research Object

Hohhot Wuchuan country is known as the "land of the Chinese Potato".Famous brand agricultural products "Wuchuan potato" had been selected special for the 2008 Beijing Olympic Games by Beijing Olympic Organizing Committee. Took image of variety KeXin 1st potato from Hohhot Wuchuan country as the specific study object, make detection according to the standard of NY/T1066-2006 “Grades and specifications of potatoes”.

2.2 Hardware System

The main function of the hardware system was data acquisition, transmission and processing. The hardware system included Basler A101fc face CCD camera made in Germany with high sensitivity and SNR (Signal-to-noise Ratio);IEEE1394a image acquisition card; computer with AMD 2800 +, RAM 512M, hard drives 120G; 60W annularfluorescent lamp fixed on top of the light box(Fig 1).The image, an analog signal, which was exported from the camera and transformed into the digital signal by image acquisition card. The digital signal was storied in RAM for next processing .

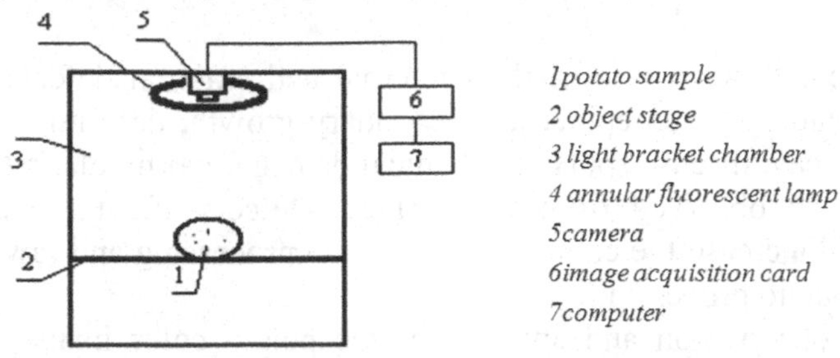

Fig.1: Hardware system

2.3 Software Technology

Potato original image was 24 bit true color image, the image resolution was 1030 * 1300.The processing was realized by using MATLAB 7.0.

3. IMAGE PREPROCESSING

The original image was disposed fast and smoothly by the linear low pass filtering. For the major color of the potato is yellow and the color yellow is compound of red and green, based on the threshold segmentation by setting up the values of B(blue), the background was effectively wiped off. Speckle noise was eliminated by the median filtering of 3*3. The effect was obviously and was shown as Figure 2.

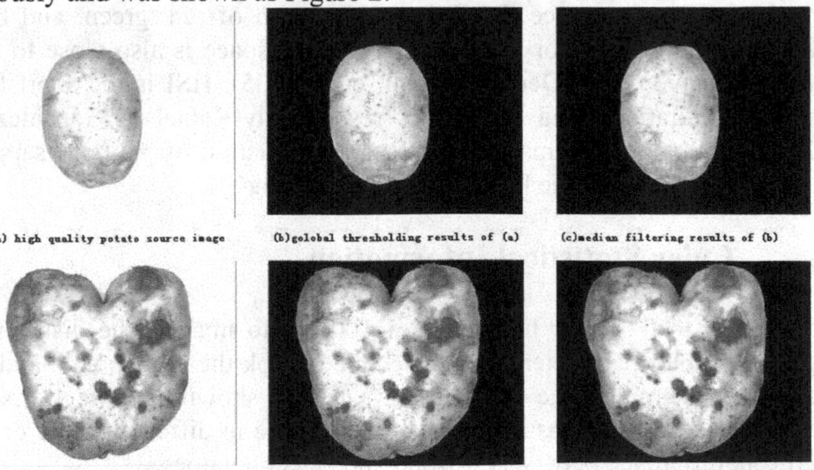

(a) high quality potato source image (b)golobal thresholding results of (a) (c)median filtering results of (b)

(d) external defects image (e)golobal thresholding results of (d) (f)median filter results of (e)

Fig.2: Image Preprocessing

4. THE EXTERNAL DEFECT SEGMENTATION BASED ON THE COLOR CHARACTER OF POTATO

Potato defects were classified as internal and external defects. External defects, including green epidermis, secondary growth, deformities, crackle, dry rot or decay, disease spots, mechanical damage, wormhole, rat bites and so on, can be observed from the surface. Defects affect the quality of products and increase the complexity of potato processing and raw materials loss in the potato processing.

Through observation and analysis of the potato color image, the color character between the high-quality potato and the detect one is obvious different. Good potato was light yellow, while dry rot and decay was grey or black grey, green epidermis was green or celadon. So it was suggested that these color characteristics could be considered as the recognition features of defects. In the paper, defects region color features which were extracted from color image according to the different color information of the different components in HSI and RGB color model, acted as the color features of the object Segmentation.

4.1 Color Model

The color is as to geometirc characteristic of the picture, have certain stability, to scale, translation, rotate, have a strong one stupid and getting wonderful quite(Wu Funing et al.,2005). Because the color information is the inherent character of the object, color characteristics is a simple and effective feature. Commonly, the format of images taken by a digital camera is in the RGB color space. RGB is the acronym of red, green, and blue. Apart from the RGB color space, the HSI color space is also close to how humans perceive colors(Deng-Fong Lin et al.,2005). HSI is the short form for hue, saturation, and intensity, respectively(Rafael C.Gonzalez et al.,2005). In this study, image color feature extraction of potatoes samples are investigated both in the RGB and HSI color space.

4.2 Color Statistical Information

According to the color histogram of 100 potato images, the distribution range of six color parameters were calculated. Took the good potato and dry rot ones for examples, the distribution state was shown as table 1. Except saturations had no obvious difference, others were in different range or had significant difference between the good and detect potatoes.

Table 1 Color Statistical Information of potatoes

	Red	Green	Blue	hue	saturation	intensity
Good potatoes	150~255	150~255	90~255	0~180	0~75	0.5~1
Dry rot potatoes	50~150	40~150	45~160	0~360	0~102	0.2~0.5

4.3 Defects Segmentation

For a detects potato image, scanning from left to right and from top to bottom, meanwhile calculating each pixel of the R, G, B, H, S, I value. When the parameters values were between the scope of defect range, set the RGB value of the pixel to RGB(255,0,0), otherwise value remained constant. As Figure 3 shows that after segmentation, background was black, defects were red, and the color of good parts did not change.

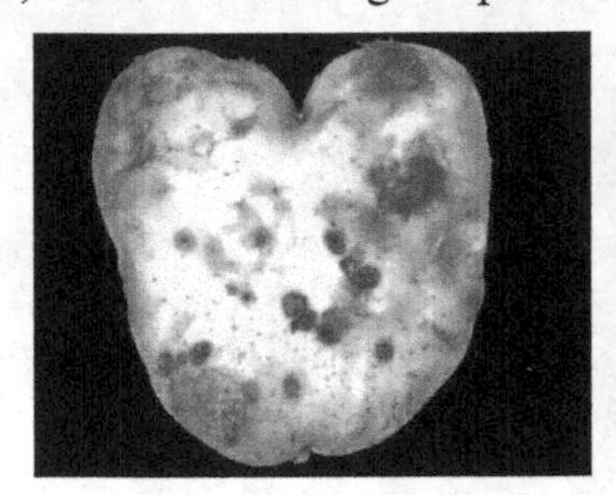
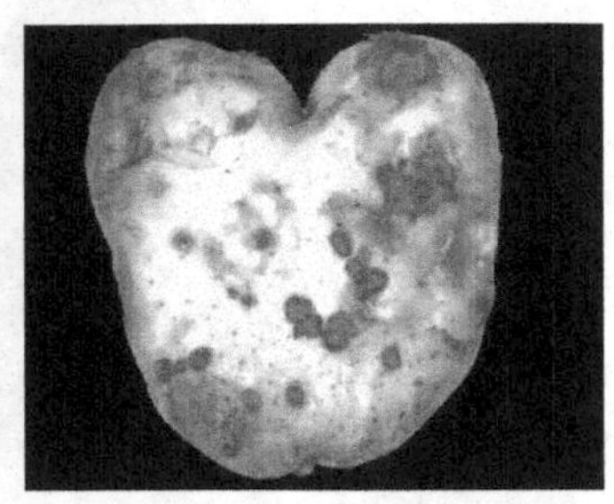

Fig3 External defects segmentation

5. CONCLUSION

According to the color characteristics of potato image, the background was effectively wiped off and the external detects segmentation was realized. This method was simple and effective. To find the appropriate color characteristics was the key.

ACKNOWLEDGEMENTS

The first author is grateful to the College of Electrical and Mechanical Engineering for providing her with pursuing a PhD degree at the Inner Mogolia Agriculture University. (Hu Zhiquan et al.,2005)

REFERENCES

Dah-Jye Lee, James K. Archibald, Yu-Chou Chang and Christopher R. Greco, Robust color space conversion and color distribution analysis techniques for date maturity evaluation. Journal of Food Engineering, Volume 88, Issue 3, October 2008, Pages 364-372

Deng-Fong Lin, Huan-Lin Luo, Fading and color changes in colored asphalt quantified by the image analysis method.Construction and Building Materials, Volume 18, Issue 4, May 2004, Pages 255-261

Rafael C.Gonzalez,Richard E.Woods, Digital Image Processing Second Edition. Electronic Industry Press of China, 2003

Wu Funing,Yang Zibiao,Zhu Hong,etc.Applied researchers of classifying and identifying crop image.Review of China agricultural science and technology, Volume 5,2003,76~80

Ying Yibiin,Rao Xiuqin,Zhao Yun,Jiang Yiyuan.Application of machine vision technique to quality automatic identification of agricultural products.Transactions of the CSAE, January 2000, 103~108

THE RESEARCH OF PADDY RICE MOISTURE LOSSLESS DETECTION BASED ON L-M BP NEURAL NETWORK

Jun Sun[1,2*], Hanping Mao[1], Jinjuan Liu[2], Bin Zhang[3]

[1] *Key Laboratory of Modern Agricultural Equipment and Technology, inistry of Education & Jiangsu Province, Jiangsu University, Zhenjiang, Jiangsu Province, China 212013*

[2] *School of Electrical and Information Engineering ,Jiangsu university, Zhenjiang ,Jiangsu Province, China 212013*

[3] *ChangZhou Institute of Technology, Changzhou,Jiangsu Province,China 213002*

* *Corresponding author, Address: School of Electrical and Information Engineering, Jiangsu University, Zhenjiang,212013,Jiangsu,Province,China,Tel:(0)13775544650,fax:0511887803 11, Email :sun2000jun@ujs.edu.cn*

Abstract: The method of the quantitative analysis on the paddy rice moisture condition is studied, which is based on the spectral reflectivity of the leaf crest layer. Several subsections are carried on the entire spectrum curve by the equidistance, The sensitive characteristic wave-length is selected based on the table of molecular spectrum sensitive wave band, obtains the characteristic spectral reflection index value to take as the characteristic value. The convergence rate of the BP neural network is slow, so the L-M algorithm is introduced to carry on the renewal of the neural network weights. The paddy rice water moisture quantitative analysis forecast model is established by making use of the fast study function of the L-M algorithm neural network. The forecasting results indicate that the highest prediction error of the paddy rice water content is 6.72% and the average error rate is 4.23%. The prediction effect is better than the traditional BP network arithmetic, and it can be used in the lossless inspection of paddy rice moisture.

Keyword: Paddy rice; Characteristic wave-length; Spectrum reflection index ;L-M BP Network

Please use the following format when citing this chapter:

Sun, J., Mao, H., Liu, J. and Zhang, B., 2009, in IFIP International Federation for Information Processing, Volume 294, *Computer and Computing Technologies in Agriculture II, Volume 2*, eds. D. Li, Z. Chunjiang, (Boston: Springer), pp. 1181–1188.

1. INTRODUCTION

The paddy rice is one of the most important grain crops in the world, and its all cultivated area and the whole output are only inferior to the wheat. Because the 80% of populations in the Asian and the 1/3 of the populations in the African and Latin America take the paddy rice as the principal food, the paddy rice production is important outstandingly to the developing countries. In our country, the average rice production of every year occupies the first one in the world in the nearly 10 years, and the paddy rice sown area occupies the 22.8% of the world total sown area. The average per acre output of the paddy rice approximately is 380 kilograms, which is to be among the most ones in main production paddy rice's country . Because the water content is the main component of paddy rice, the loss of the water content is one of the most universal factors which limits paddy rice primitive forces. The water content influences immediately paddy rice's physiological biochemistry process and shape structure, thus it affects the final paddy rice output and the quality, therefore the water content management is one of the most important measures in the crops production. The prompt and accurate monitoring on or diagnosis on paddy rice water content is very important to raising the paddy rice water content management level and the instruction of saving water of agricultural production.

Domestic and foreign countries have researched paddy rice's water content examination. Michio and so on monitors the water content scarcity condition of paddy rice using the 960nm place derivative spectrum(Michio S, 1989). Tian Yongchao and so on have studied the relativity of the paddy rice's crown level near-infrared spectral reflectivity and the water condition of plant under the different holard nitrogen condition(Tian, 2005). The reports about the research using the L-M neural network algorithm to the paddy rice's water content condition quantitative analysis have not been seen in domestic and foreign countries at present.

2. EXPERIMENTAL DESIGH AND EXPERIMENT

2.1 Experimental design

The experiment of growing paddy rice is carried on in big shed in the agricultural engineering research institute of Jiangsu University, and the selection of paddy rice species is new fragrant superior 80, whose vegetative cycle is 115 days. Paddy rice is planted in the environment of the greenhouse sand bonsai, keeping the line distance of rows as 0.3m×0.3m. The paddy rice's nourishment liquid configuration is according to international paddy rice nourishment liquid formulation.

The paddy rice seeds are sow on July 1, 2007. They are divided into 3 groups according to 3 water content level on August 1. The first group guarantees the sufficient water content supply in entire experiment period. The second group keeps the water immersing the sand just in experiment period. The third group keep the sand on the water scarcity condition in the experiment period.

2.2 Measure instrument and method

The instrument used in this experiment is the FieldSpec 3 hand-hold portable spectrum instrument which is produced in American ASD Corporation. This instrument's spectrum measuring range is 350-2500nm. The sampling interval is 1.4nm in the 350-1000nm spectrum area, and the resolution is 3nm. In the 1000-2500nm spectrum area the sampling interval is 2nm, the resolution is 10nm, the angle of view is 15 degrees. In the spectral region it has 512 array element light PDA array detectors, the concave achromatism diffraction grating, the even field image formation, the easy wave length demarcation, and the sensitivity high characteristics.

The measurement experiment is carried on the sunny and non-clouds weather, the measuring time is 11:00-13:00. The contributory data are collected at the same time, such as the measuring environment parameter, the measuring time, gathering location and so on. Pokes head is vertical downward, goes against 0.8m to the adult plant crown level, the data are gathered in the identical place each time. The standard operation procedure is to be optimized, to collect dark current, to measure the white tabula, measuring the terrain feature. Each time the data measured the white tabula are taken as the relative reflectivity (standard tabula rasa reflectivity is 1). The system program must be optimized one time each 10 minutes in order to guarantee the data quality.

The measurement is repeated 3 times in every measuring point, and the average data of 3 spectrum data are taken as the characteristic spectral reflectivity data.

In the visible light and short-infrared regions, with the water content rising, the spectral reflectivity drops. The paddy rice's crown level spectral reflectivity is shown as Fig.1. In visible light region, the wavelengths region of 450-470nm, 540-570nm, 700-730nm have the obvious reflection peak, and have high correlativity. In the short-infrared region, near 930-970nm, 1430nm-1460nm and 1630nm-1670nm spectral reflectivity also has very high relevance with the leaf water content. The average data of the spectral reflectivity in the above six regions are taken as the characteristic value.

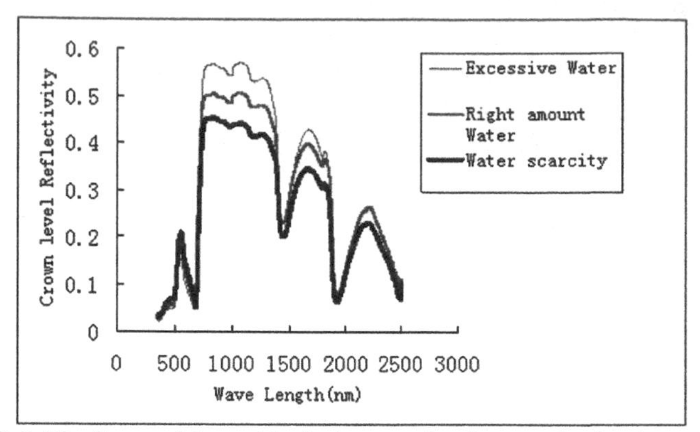

Fig1. The figure of the paddy rice's crown level spectral reflection rate

2.3 The leaf water contents

After spectrum measures, the paddy rice leaves are selected randomly and maintained in the keeping freshness bags. Then them are brought back to the laboratory to gain fresh weight and are dried for 1 hour with the constant temperature 100℃, afterward they are carried on 12 hour being dried with the constant temperature 60 ℃ ,so the leaves' dry weight are measured separately. The calculating formula of leaf water content is shown as formula 1. The part experiment data of paddy rice leaf water content data are shown as Fig.1.

$$z = \frac{y - x}{y} \quad (1)$$

Z denotes the leaf's water content rate,y denotes the leaf fresh weight, and x denotes the leaf dry weight.

Table 1. The partial experiment data of paddy rice leaf water content

Water volume	Date			
	07.9.27	07.10.4	07.10.9	07.10.16
Water scarcity	0.7546	0.7292	0.7089	0.686
Right amount water	0.7974	0.7701	0.7533	0.7357
Excessive water	0.7856	0.7584	0.744	0.7159

3. L-M ALGORITHM BP NETWORK

The BP neural networks is one kind of multi-layer-forward neural networks which is used most widely. The BP algorithm uses a gradient descent searching method, whose parameter moves along with the opposite direction to error gradient. The error between the practical output and the expected output of neural network through computing is computed. Backward each networks weight value is regulated until the error catch the smallest. Its shortcoming lies in its complexity of computation, and its speed is very slow. It possibly falls into the local minimum point. The training time is long, its value stability is bad. Otherwise, the initial weight value, learning rate and momentum coefficient parameters and so on are adjusted difficultly.

The L-M algorithm is one kind of the fast algorithms using standard value optimization techniques. It combines the gradient descent method and Gauss Newton's method, may also take it as the improvement form of a Gauss Newton's method, it has both local convergence of a Gauss Newton's method and the global characteristic of the gradient descent method. Because the L-M algorithm has used the approximate two-rank differential coefficient information, it is much quicker than the gradient method.

The search direction used in the Levenberg -Marquardt's method is the following group of linear equality solution:

$$(J(x_k)^T J(x_k) + \lambda_k I) d_k = -J(x_k) F(x_k) \quad (2)$$

Among them: the scalar λ_k decides the search direction and amplitude size, when λ_k =0, the search direction is same as the GaussNewton method's; When λ_k tends to infinity, d_k tends to zero vector, thus finds the most quickly descent direction, so long as λ_k is big enough, it may guarantee $F(x_k + d_k) < F(x_k)$, therefore λ_k can guarantee the function value drop in iteration.

In order to control the size question of λ_k in each iteration, we linearly predict sum of squares $F_p(x_k)$ and cubic interpolated value of the smallest $F_k(x^*)$ to estimate relative non-linearity of $F(x)$.The linear prediction sum of squares carries on the computation according to the equation below:

$$F_p(x_k) = J(x_{k-1})^T d_{k-1} + f(x) \quad (3)$$

$$f_p(x_k) = F_p(x_k)^T F_p(x_k) \quad (4)$$

Among them: $f_k(x^*)$ and the length of stride parameter α^* may through cubic interpolated value $F(x_k)$ and $F(x_{k-1})$, α^* is the minimum estimated length of stride. If $f_p(x_k)$ is bigger than $f_k(x^*)$, λ_k will be reduced to $\frac{\lambda_k}{1+\alpha^*}$, otherwise λ_k will be increased to $\lambda_k + \frac{f_k(x^*) - f_p(x_k)}{\alpha^*}$.

With the renewal of λ_k , the solution of formula (1) constitutes a search direction d_k, then d_k increases a unit iteration length of stride to carry on the next search step, so analogizes, this linear search process guarantees $f(x_{k+1}) < f(x_k)$ in iteration each time. This method has better robustness than the GaussNewton's method(Yuan,2007 ; Zhang 2004).

4. EXPERIMENTAL RESULTS

Because the big shed's air temperature, humidity as well as six spectral interzone's average spectral reflectivity value are taken as the characteristic parameters, therefore neural networks' input layer node number is 8, the implicit layer node number takes 15, the output layer node number takes 1.The LM algorithm is applied in the neural networks' weight value renewing, and the network is set up and trained.

The 200 samples are selected to carry on the experiment, among them, 120 samples are taken as the training samples of LM algorithm neural network, and the residual 80 samples are taken as the test samples. Real water content and forecast water content contrast figure based on LM BP network is shown as Fig.2, and the real water content and forecast water content contrast chart based on BP network is shown as Fig.2. From Fig.1, we can see that the points corresponding the real water content and the forecast water content approach a line. It also illuminates that the forecast water content is very near to the real water content. The forecasting results indicate that the highest prediction error of the paddy rice water content is 6.72% and the average error rate is 4.23%.

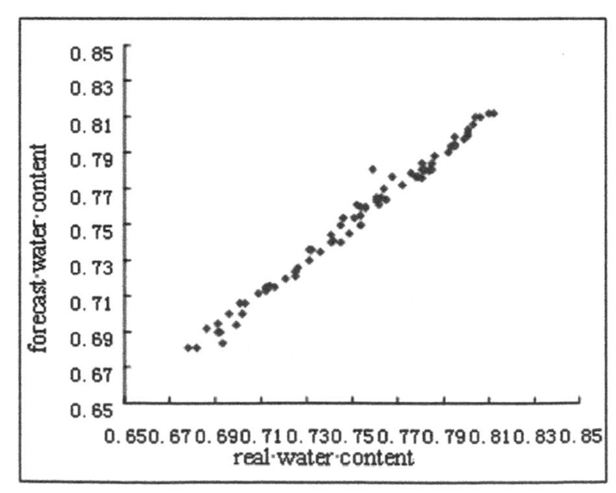

Fig. 2 real water content and forecast water content contrast figure based on LM BP network

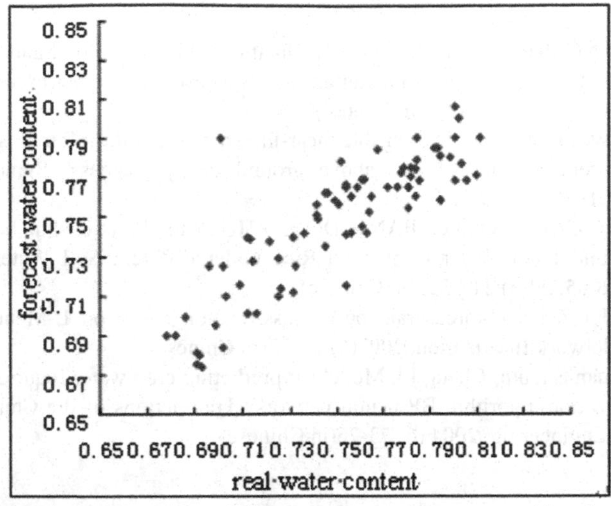

Fig. 3 The real water content and forecast water contentcontrast chart based on BP network

5. CONCLUSION

In this paper, the L-M algorithm is imported into BP network to carry on the renewal of the neural network weights. And put the improved BP network into the prediction of the paddy rice water content. The prediction effect is better than the traditional BP network arithmetic. It can provide the reliable water-loss basis for paddy rice planters and it can be used in the lossless inspection of paddy rice moisture. It also has the very significant practical value.

ACKNOWLEDGEMENTS

Funding for this research was provided by China Postdoctoral Science Foundation under Grant (NO:20070420972); Jiangsu University High-grade Specialty Person Scientific Research Foundation under Grant(NO: 05JDG050); "863"Project(2008AA10Z204); Changzhou Young Scientific and Technological Talent Training Plan (NO:CQ2008009); Jiangsu University College Student Scientific Research Project(NO: 07A087).

REFERENCES

Ji Hai-yan, WANG Peng-xin, YAN Tai-lai. Estimations of Chlorophyll and Water Contents in Live Leaf of Winter Wheat with Reflectance Spectroscopy. Spectroscopy and Spectral Analysis, 2007,27(3):514-516.(in Chinese)

Michio S,T suyoshi A. Seasonal visible near-infrared and mid-infrared spectral of rice canopies in relation to LAI and above ground dry phytomass. Remote Sensing of Environment,1989,27:119-127

Tian Yongchao, CAO Weixing, JIANG Dong, ZHU Yan. Relationship between Canopy Reflectance and Plant Water Content in Rice under different Soil Water and Nitrogen Conditions. 2005, 29(2):318-323(in Chinese)

Yuan Jin-li,GUO Zhi-tao.Stored-grain pests classification based on L-M neural networks. Agricultral Network Information. 2007(6):29-32(in Chinese)

Zhang Bing, Yuan Shouqi, Cheng Li. Model for predicting crop water requirements by using L- M optimization algorithm BP neural network. Transactions of the Chinese Society of Agricultural Engineering. 2004 (6):73-76(in Chinese)

EVALUATION OF REGIONAL PEDO-TRANSFER FUNCTIONS BASED ON THE BP NEURAL NETWORKS

Zhongyi Qu[1,*], Guanhua Huang[2], Jingyu Yang[3]

[1] *College of Hydraulic and Civil Engineering, Inner Mongolia Agricultural University, Huhhot, China, 010018;*

[2] *Chinese-Israeli International Center for Research and Training in Agriculture, China Agricultural University, Beijing 100083, P. R. China*

[3] *Institute of City Planning and Design, China Architecture Institute, Beijing, China, 100007*

[*] *Corresponding author, Address:, College of Hydraulic and Civil Engineering, Inner Mongolia Agricultural University y, 306 Zhaowuda Road, Huhhot, 010018, P. R. China, Tel:+86-471-4300181, Fax:+86-471-4300249, Email:qzy682000@yaohoo.com.cn*

Abstract: The unsaturated soil hydraulic properties, including soil water retention curve and hydraulic conductivity, are the crucial input parameters for simulating soil water and solute transport through the unsaturated zone at regional scales, and are expensive to measure. These properties are frequently predicted with pedo-transfer functions (PTFs) using the routinely measured soil properties. 110 soil samples at 22 soil profiles from Jiefangzha Irrigation Scheme in the Hetao Irrigation District of Inner Mongolia, China were collected for the analysis of soil properties i.e. soil bulk density, soil texture, particle size distribution, organic content, and soil water retention curve (SWRC). The Brooks-Corey (BC) model and van Genuchten (VG) model were used to fit the measured SWRC data for each soil sample by using the RETC software. Pedo-transfer functions (PTFS), which describes relationship between the basic soil properties and the parameters of the BC and VG models, were then established with the artificial neural networks (ANN) model. It is found that the ANN model has better effect on the clay loam, loamy clay, loam soil and silty clay to simulate BC model. However, it has better effect on the loam soil, loamy clay and sandy clay to simulate VG model. So, we can draw the conclusion that the ANN model can conveniently establish PTFS between soil basic feature parameters and SWRC model and has reasonable precision. This will be a good method to estimate soil water characteristic curve model and soil hydraulic parameter in the regional soil water and salt movement simulation and water resources evaluation.

Please use the following format when citing this chapter:

Qu, Z., Huang, G. and Yang, J., 2009, in IFIP International Federation for Information Processing, Volume 294, *Computer and Computing Technologies in Agriculture II, Volume 2*, eds. D. Li, Z. Chunjiang, (Boston: Springer), pp. 1189–1199.

Keywords: Hetao Irrigation District, BP model, pedo-transfer functions, soil water retention curve, soil basic property

1. INTRODUCTION

Soil hydraulic properties, including soil water retention curve (SWRC) and hydraulic conductivity, are the crucial parameters for simulating soil water and solute transport through saturated and unsaturated zone. Many empirical models were developed to describe the soil hydraulic properties, among them both the van Genuchten (VG) model (van Genuchten, 1980) and Brooks-Corey (BC) (Brooks and Corey, 1974) model are widely used in the simulation of water and solute transport. However, for large-scale problems, when the temporal and spatial variability of the region is considered, the required measurements of soil hydraulic properties are tremendous, time-consuming, and very expensive. The pedo-transfer functions (PTFs) are the promising tool to estimate the soil hydraulic properties with more easily measured or basic soil properties in the attribute database of a digital soil survey map, in which soil hydraulic properties are not always available. The PTFs for soil hydraulic properties are those functions, with which the parameters in VG and /or BC model are expressed as the linear or nonlinear empirical functions of different land characteristics and soil properties, i.e. soil texture (including sand, silt and clay contents), bulk density and organic mater content (Pachepsky et al., 1996; Tamari et al., 1996; Schaap and Leij, 1998; Minasny et al., 1999; Schaap et al., 2001).

The PTFs were firstly described by Bouma and van Lanen (1987), .a recent approach for fitting PTFs is to use artificial neural networks (ANN) (Pachepsky et al., 1996; Schaap et al., 1998). Tamari and Wosten (1999) gave a review on the ANN and its application on predicting soil hydraulic properties. Most researchers have found that the ANN performs better than multi regression (Schaap et al., 1998, Koekkoek and Booltink, 1999). More recently, based on the ANN, Budiman et al. (2002) and Nemes et al. (2003) developed computer codes for the PTFs. An advantage of using the ANN based approach is that there is needed to assume a prior relationship. Minasny et al.(1999) found that the ANN based approach performs as well as the extended nonliner regression. Because the PTFs are site specific, the PTFs developed in one region are not always applicable in other regions with acceptable accuracy (Tietje and Tapkenhinrichs, 1993; Kern, 1995; Tietje and Hennings, 1996; Cornelis et al., 2001; Wagner et al., 2001; Nemes et al., 2003). Therefore, it is necessary to develop the PTFs for any specific study areas.

The objective of this study is to develop the PTFs for estimating the parameters of the VG and BC models with the use of ANN, based on the datasets collected in Jiefangzha Irrigation Scheme (see the Fig.1) of Hetao Irrigation District in Inner Mongolia, China. And the PTFs will be used to develop the digital soil map for the spatial distribution of SWRC in the study area.

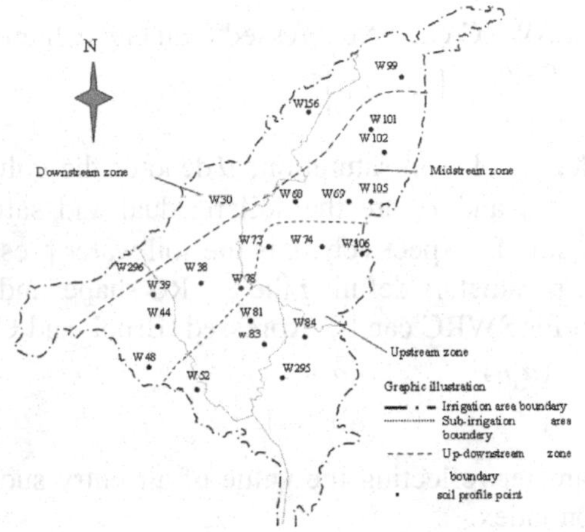

Fig.1: The sampling sites of 22 soil profiles in Jiefangzha Irrigation Scheme.

2. MATERIALS AND METHODS

2.1 Soil sampling

The soil samples were collected from Jiefangzha Irrigation Scheme in the Hetao Irrigation District of Inner Mongolia, northwest of China(see the Fig.1). The total area of Jiefangzha Irrigation Scheme is 2.16×10^5 hm^2. 22 points were selected for monitoring soil water and salt content in this area. Undisturbed soil samples were collected by using two kinds of soil samplers from soil layers 0-10 cm, 10-20 cm, 20-40 cm, 40-70 cm, and 70-100 cm along soil profiles at each monitoring point. The first kind of soil sampler has diameter 50.46mm and height 35mm, while the second kind of soil sampler has diameter50.46mm and height50mm . Soil samples with the first kind of soil sampler are collected for each layer, and they were used to determine the bulk density and porosity. While soil samples with the second kind of soil sampler are collected for each layer, and they were used to determine the SWRC by using a pressure apparatus (Soilmositure, USA-SEC) at the different suction pressures 0, 100, 200, 300, 400, 500, 600, 900, 1100, 2200, 3300, 4400, 5500, 7700, 9900, 12400, 14800cm.. At the same time, 500g disturbed soil samples each at the same layer as that of the undisturbed sample for all profiles. The disturbed soil samples were used to determine soil particle size distribution by using the sieve method and gravimeter. The organic matter content was determined by the Walkey-Black method.

2.2 VG and BC models

The VG model for SWRC can be expressed (van Genuchten, 1980):

$$S_e = \frac{\theta - \theta_r}{\theta_s - \theta_r} = \left[1 + (\alpha h)^n\right]^{-m} \quad (1)$$

where S_e is the degree of soil saturation; θ denotes the volumetric soil water content ($cm^3 .cm^{-3}$), θ_r and θ_s are the soil residual and saturated volumetric water contents ($cm^3 cm^{-3}$), respectively, h is the soil water pressure head (cm); α in cm^{-1}, n and m are parameters defining the SWRC shape, and m=1-1/n.

The BC model for SWRC can be expressed (Brook and Corey, 1974):

$$S_e = \begin{cases} (\alpha_1 h)^{-\lambda} & (\alpha_1 h > 1) \\ 1 & (\alpha_1 h \le 1) \end{cases} \quad (2)$$

Where α_1 is parameter reflecting the value of air entry suction; λ is the soil pore-size distribution index.

The model parameters θ_r, θ_s, α, α_1, n, m and λ were obtained by fitting the two functions to the measured soil water retention data with the nonlinear least-squares optimization program RETC (van Genuchten et al., 1991).

2.3 Pedo-transfer functions of neural networks

Neural network model is quite powerful and according to Gershenfeld (1999) with on hidden layer that has enough hidden units, it can describe any continuous function. Conventionally, parametric PTFs train the network to fit the estimated van Genuchten parameters. But there are some problems (Budiman Minasny and McBratney, 2002). Based on the ANN theory (Gershenfeld, 1999), we proposed a new objective function for neural network training, which can predict the VG and BC model parameters with minimizing the difference between the measured water contents and the predicted values. The detail procedure can be seen in Fig.2. The steps are as follows:

1). fit the individual water-retention curve to van Gentuchten function and Brooks-Corey function and estimate the parameters θ_r, θ_s, α, α_1, n and λ.

2).Train the neural network to predict the parameter vector $\mathbf{p}$ = [θ_r, θ_s, α, n] from basic soil properties by minimizing objective function.

The above steps are usually used for parametric PTFs. It is a neural network with an objective function that matches the parameters. The proposed method continues with fine-tuning steps described below:

3). Use the trained weights as an initial guess for the second training, which fine tunes the estimates.

4). For each soil sample, predict the hydraulic parameters with the trained weights, and calculate the water content using the van Genuchten equation and Brooks-Corey equation at each of the measured potentials.

5). Adjust the weights and error **U**, to minimize the difference between the predicted and measured water content with the optimization routine. A neural network with an objective function can match the measured or observed values.

The neural network based PTFs can predict the model parameters θ_r, θ_s, α, α_1, n, m and λ with basic soil properties as their inputs. Four kinds of inputs were considered: (1) when particle size distribution is only available, the inputs are sand (>0.05mm), silt (0.05-0.002 mm) and clay (<0.002 mm) contents;(2) the inputs are particle size distribution and soil bulk density; (3) the inputs are particle size distribution and soil organic matter content; (4) the inputs are particle size distribution, soil bulk density and soil organic matter content. The data was randomly divided into a calibration set with 80 samples and a validation set with 30 samples. Because α and n are log-normally distributed, the outputs of prediction are θ_s、θ_r、$\ln(\alpha)$、$\ln(n)$. The network consists of one hidden layer with sigmoid action function in the hidden layer and linear function in the output. The parameters of ANN model were determined by trial and error method. The BP network was performed with the Neural Network Toolbox in MATLAB program ver.7.0 .

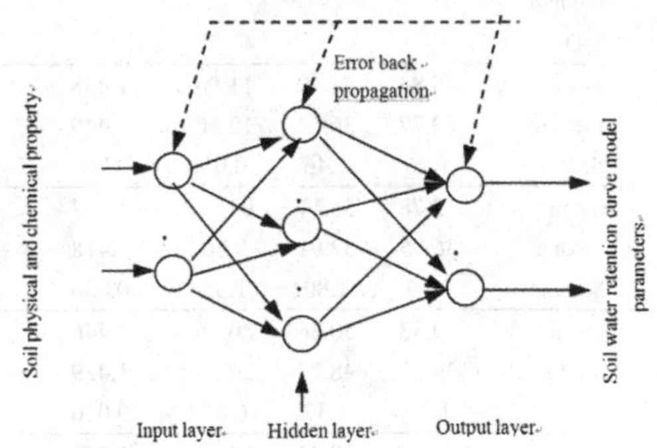

Fig.2 structure of a neural network predicting the Van Genuchten and Brooks-Corey model parameters

3. RESULTS AND DISCUSSION

3.1 Physical and chemical properties

The physical and chemical properties of 110 soil samples at 22 profiles were shown in Table1. Loamy soils account for 68.2% of the total soil samples, clay soils accounts for 30% of the total soil samples, where sandy soil only accounts for less than 3% of the total soil samples. The spatial distributions of soil texture in the study area are sand clay loam, loam and clay loam in the upstream zone (southern part). Most soil textures are clay loam in the midstream zone (middle part), while loamy clay is the main soil texture in the northern part. In the soil horizon of 0~40 cm, sandy clay loam accounts for 70% of the soil textures in the upstream zone (southern part), clay loam accounts for 65% of the soil textures in the midstream zone (middle part), while loamy clay land clay loam are the major soil textures in the downstream zone (northern part). The statistics results of total 110 soil samples see the table 1. It is founded that the organic matter content of sandy loam, loam and clay loam are relative high. This can indicate that the soil texture has closely relation with physical and chemical property of soil.

Table 1 Physical and chemical properties and their statistical results of soil samples

Soil texture	Sample number	Statistical properties	Sand (%)	Silt (%)	Clay (%)	Bulk density (g/cm^3)	Organic matter content (g/kg)
Loamy sand	3	mean	88.96	6.68	4.36	1.436	9.06
		median	86.95	8.26	4.16	1.444	2.091
		S.D	2.12	2.02	0.22	0.003	1.014
Sandy loam	13	mean	68.64	23.36	8.00	1.446	7.488
		median	69.21	23.18	7.01	1.441	4.927
		S.D	2.98	2.86	0.86	0.007	0.687
Sand clay loam	1	mean	60.46	21.03	18.51	1.474	14.420
		median	-	-	-	-	-
		S.D	-	-	-	-	-
Loam	10	mean	49.84	38.20	11.96	1.438	8.948
		median	50.77	36.54	12.56	1.449	10.126
		S.D	1.26	1.46	0.61	0.009	0.661
Silt loam	7	mean	35.78	55.17	9.042	1.437	4.654
		median	36.95	52.91	9.50	1.418	0.794
		S.D	2.19	1.80	1.32	0.006	4.301
Silt clay loam	18	mean	29.38	50.36	20.26	1.446	8.671
		median	30.23	48.77	20.24	1.429	4.370
		S.D	1.25	1.37	0.67	0.010	0.512
Clay loam	26	mean	42.88	37.43	19.70	1.493	9.047
		median	41.99	36.93	19.54	1.486	4.327
		S.D	1.14	1.02	0.51	0.012	0.441
Loam clay	19	mean	27.00	41.10	31.90	1.405	12.773
		median	27.74	41.45	31.56	1.401	9.468
		S.D	1.278	0.78	0.95	0.008	0.564
Silt clay	13	mean	15.95	51.62	32.43	1.454	8.144
		median	14.14	51.22	32.35	1.432	3.819
		S.D	1.61	1.26	1.17	0.021	0.616

S.D : Standard deviation

3.2 The fit and establishment of soil water retention curve (SWRC) model

By fitting the VG and BC model to the measured data for each soil sample, we obtained the model parameters as shown in Table 2 and Table 3. According to the fitted results we found that the model parameters of VG model and BC model soil property have significant influence on parameter of model. For example, the *n* of VG model will decrease with the decrease of organic matter content (see the table 1). At the same time, the soil clay particle content has good relations with *n*, *α* of VG model.

According to the fitting results, it is found that the two kinds of models have good effects. The model parameter of 9 kinds soil texture was calculated their average (see the table 2 and table 3). The maximum error of measured value and fitted value is less than 5%. Furthermore, the Van Genuchten is better than Brooks-Corey model.

Table 2 The fitted parameters for van Genuchten model by using the RETC program

Soil texture	Sample number	Statistical properties	θ_r ($cm^3 cm^{-3}$)	θ_s ($cm^3 cm^{-3}$)	α (cm^{-1})	n	m=1-1/n	Determination coefficient (R^2)
Loamy sand	3	Mean	0	0.512	0.009	1.536	0.345	0.98
		Maximum	0	0.530	0.012	1.690	0.408	0.99
		Minimum	0	0.489	0.008	1.408	0.290	0.97
		S.D.	0	0.021	0.002	0.142	0.059	
Sandy loam	13	Mean	0.111	0.482	0.013	1.373	0.251	0.98
		Maximum	0.175	0.543	0.046	2.062	0.515	0.99
		Minimum	0.171	0.442	0.002	1.096	0.088	0.98
		S.D.	0.037	0.030	0.014	0.263	0.122	
Sand clay loam	1	Mean	0.135	0.450	0.034	1.150	0.131	0.97
		Maximum						
		Minimum						
		S.D.						
loam	10	Mean	0.088	0.500	0.028	1.287	0.194	0.98
		Maximum	0.139	0.619	0.053	2.060	0.514	0.99
		Minimum	0.038	0.430	0.002	1.056	0.053	0.98
		S.D.	0.071	0.056	0.016	0.304	0.140	
Silt loam	7	Mean	0.03	0.53	0.01	1.33	0.23	0.99
		Maximum	0.126	0.56	0.02	1.78	0.44	0.99
		Minimum	0	0.47	0	1.10	0.09	0.98
		S.D.	0.05	0.04	0.01	0.24	0.12	
Silt clay loam	18	Mean	0.149	0.508	0.048	1.151	0.128	0.98
		Maximum	0.226	0.56	0.165	1.352	0.260	0.99
		Minimum	0.069	0.443	0.001	1.07	0.065	0.97
		S.D.	0.065	0.037	0.046	0.079	0.055	
Clay loam	26	Mean	0.095	0.485	0.025	1.230	0.177	0.98
		Maximum	0.137	0.565	0.083	1.674	0.403	0.99
		Minimum	0.026	0.427	0.002	1.079	0.074	0.97
		S.D.	0.060	0.033	0.021	0.149	0.087	
Loam clay	19	Mean	0.167	0.501	0.098	1.179	0.139	0.98
		Maximum	0.222	0.617	1.098	1.671	0.401	0.99
		Minimum	0.050	0.429	0.003	1.059	0.056	0.97
		S.D.	0.073	0.044	0.247	0.159	0.095	
Silt clay	13	Mean	0.202	0.542	0.072	1.129	0.112	0.98
		Maximum	0.294	0.604	0.179	1.258	0.205	0.99
		Minimum	0.118	0.454	0.003	1.056	0.053	0.97
		S.D.	0.071	0.049	0.066	0.064	0.048	

Table 3 The fitted parameters for Brook and Corey model by using the RETC program

Soil texture	Sample number	Statistical properties	θ_r ($cm^3\ cm^{-3}$)	θ_s ($cm^3\ cm^{-3}$)	α_1 (cm^{-1})	λ	Determination Coefficient(R^2)
Loamy sand	3	Mean	0	0.464	0.023	0.268	0.97
		Maximum	0	0.480	0.048	0.433	0.99
		Minimum	0	0.455	0.010	0.013	0.96
		S.D.	0	0.031	0.021	0.22	
Sandy loam	13	Mean	0.072	0.469	0.018	0.234	0.97
		Maximum	0.120	0.510	0.065	0.409	0.99
		Minimum	0.011	0.433	0.004	0.080	0.96
		S.D.	0.047	0.025	0.017	0.094	
Sand clay loam	1	Mean	0.095	0.450	0.045	0.119	0.98
		Maximum					
		Minimum					
		S.D.					
loam	10	Mean	0.087	0.486	0.033	0.236	0.97
		Maximum	0.117	0.621	0.058	0.622	0.99
		Minimum	0.020	0.413	0.004	0.055	0.96
		S.D.	0.045	0.057	0.019	0.164	
Silt loam	7	Mean	0.02	0.51	0.01	0.24	0.97
		Maximum	0.07	0.55	0.03	0.55	0.97
		Minimum	0	0.46	0.00	0.09	0.96
		S.D.	0.03	0.04	0.01	0.16	
Silt clay loam	18	Mean	0.145	0.501	0.048	0.141	0.96
		Maximum	0.209	0.56	0.153	0.254	0.97
		Minimum	0.110	0.43	0.004	0.066	0.94
		S.D.	0.039	0.039	0.038	0.059	
Clay loam	26	Mean	0.114	0.466	0.022	0.239	0.97
		Maximum	0.185	0.540	0.087	0.570	0.98
		Minimum	0.023	0.430	0.004	0.073	0.95
		S.D.	0.047	0.028	0.021	0.125	
Loam clay	19	Mean	0.160	0.495	0.044	0.158	0.98
		Maximum	0.260	0.610	0.171	0.577	0.99
		Minimum	0.019	0.430	0.004	0.061	0.96
		S.D.	0.085	0.043	0.044	0.127	
Silt clay	13	Mean	0.178	0.537	0.090	0.108	0.96
		Maximum	0.248	0.600	0.200	0.211	0.99
		Minimum	0.063	0.453	0.006	0.055	0.94
		S.D.	0.081	0.047	0.073	0.051	

3.3 Pedo-transfer functions based on ANN model

The training of BP model was performed by using the randomly selected 80 soil samples, and the parameters of the two models were estimated according to the basic soil property data. And then the established ANN PTFs were tested with another 30 sample data. The tested results were shown in Tables 4 and 5. It can be found that the average relative error (ARE) decreases gradually with inputs factors increase. Therefore, it is better to estimate the parameters of SWRC by using soil bulk density, particle size distribution and soil organic matter contents.

From the comparison of two kinds of models, VG model is better than BC model. From the table 4 and table 5 we can see that, the ANN model has good prediction ability.

As shown in table 4 and table 5, it can be found that the predicted values and approach the measured values well with the inputs of particle size distribution + soil organic matter content + soil bulk density. The four parameters R2 are greater than 0.5. Through the comparison of the soil water retention curve (SWRC) model fitted by BC and VG with that fitted by basic soil properties and measured SWRC, it can be found that the different SWRC model has different effect for the different soil texture. The BC model has better effect for the clay loam, loamy clay, loam and silty clay. However, the VG model has better effect for the loam, loamy clay and sandy clay.

Table 4 Error analysis of predicted results for BC model parameters

Inputs	Output	Average absolute error	Average relative error(%)
Particle size distribution	θ_r	0.0465	26.22
	θ_s	0.0111	6.27
	$\mathrm{Ln}(\alpha_1)$	0.0474	26.72
	$\mathrm{Ln}(\lambda)$	0.3671	6.99
Particle size distribution+soil organic matter content	θ_r	-0.0031	0.62
	θ_s	-0.0111	2.21
	$\mathrm{Ln}(\alpha_1)$	0.0022	0.44
	$\mathrm{Ln}(\lambda)$	-0.0144	2.84
Particle size distribution+soil bulk density	θ_r	-0.1738	4.33
	θ_s	-0.1679	4.19
	$\mathrm{Ln}(\alpha_1)$	-0.0233	4.58
	$\mathrm{Ln}(\lambda)$	-0.3962	9.88
Particle size distribution+soil organic matter content+ soil bulk density	θ_r	-0.0020	0.22
	θ_s	-0.1578	4.37
	$\mathrm{Ln}(\alpha_1)$	-0.0480	3.29
	$\mathrm{Ln}(\lambda)$	-0.0230	2.54

Table 5 Error analysis of predicted results for VG model parameters

Inputs	Output	Average absolute error	Average relative error (%)
particle size distribution	θ_r	0.0420	23.70
	θ_s	0.0211	11.89
	$\mathrm{Ln}(\alpha)$	0.0388	21.85
	$\mathrm{Ln}(n)$	0.0150	8.44
particle size distribution+ soil organic matter content	θ_r	-0.0050	0.98
	θ_s	0.0039	0.76
	$\mathrm{Ln}(\alpha)$	-0.0077	1.52
	$\mathrm{Ln}(n)$	-0.0097	1.92
particle size distribution+soil bulk density	θ_r	-0.3217	7.10
	θ_s	-0.1848	4.08
	$\mathrm{Ln}(\alpha)$	0.0253	0.55
	$\mathrm{Ln}(n)$	-0.3788	8.36
particle size distribution+soil organic matter content+ soil bulk density	θ_r	0.0247	6.09
	θ_s	0.0029	0.71
	$\mathrm{Ln}(\alpha)$	0.0081	1.99
	$\mathrm{Ln}(n)$	0.0139	3.41

4. CONCLUSION

The van Genuchten model and Brooks-Corey model was selected as the optimal equation to describe the soil water retention characteristic of Jiefangzha irrigation area soils. PTFs based on the ANN model for estimating soil hydraulic characteristics were derived from basic soil properties (particle-size distribution, soil organic matter, and bulk density). Among the four parameters of Eq. (1) and Eq. (2), the saturated water content (θ s) was best predicted through the entire soil data set, while prediction of the value of n and residual water content (θ r) was the poorest, The developed ANN models for estimatingθ s, θ r, ln(α) ln(α 1)/ and n/λ were tested for their stability and predictability by the double cross-validation method. It was found that the signs of the regression coefficients and the determination coefficients were stable. The PTFs obtained from this study appear superior in predicting the soil hydraulic parameters, compared to multi-regression PTFs. The PTFs derived in this study were used to estimate soil water retention curve and has better effects. So, we can determine the spatial distribution of regional soil parameters through this method and PTFS which based on the ANN model.

ACKNOWLEDGEMENTS

This study was partially supported by the National Natural Science Foundation of China (Project No. 50669005).

REFERENCES

Arya, L.M., Paris, J.F., A physic empirical model to predict the soil moisture characteristic from particle-size distribution and bulk density data. Soil Science Society of America Journal, 1981, 45:1023-1030.

Baumer, O.M., Predicting unsaturated hydraulic parameters. In: van Genuchten, M.Th., et al. (Ed.), Proceedings of the International Workshop on Indirect Methods for Estimating the Hydraulic Properties of Unsaturated Soils. Riverside, CA, 11 - 13 Oct. University of California, Riverside, CA, 1992, : 341 - 354.

Cornelis, V. M., Ronsyn, J., van Meirvenne, M., Hartmann, R., Evaluation of pedotransfer functions for prediction the soil moisture retention curve. Soil Science Society of America Journal, 2001, 65 (3):638 - 648.

Goncalves, M. C., Pereira, L. S., Leij, F. J., Pedo-transfer functions for estimating unsaturated hydraulic properties of Portuguese soils. European Journal of Soil Science, 1997, 48: 387 - 400.

Huang G.H, Zhang R.D, Evaluation of soil water retention curve with the pore–solid fractal model, Geoderma, 2005,127: 52– 61

Kern, J. S., Evaluation of soil water retention models based on basic soil physical properties. Soil Science Society of America Journal, 1995, 59:1134 - 1141.

Minasny, B., Mcbratney, A. B., The neuro-m for fitting neural network parametric pedotransfer functions. Soil Sci.Am.J. 2002, 66:352-361.

Rawls, W. J, Gish, T. J., Brakensiek, D. L. Estimating soil water retention from soil physical properties and characteristics. Adv. Soil. Sci. Soc. Am. J. 1991, 16:213-234.

Tyler, S.W., Wheatcraft, S.W., 1990. Fractal processes in soil water relation. Water Resour. Res. 26: 1047–1054.

Tyler, S.W., Wheatcraft, S.W., 1992. Fractal scaling of soil particle size distributions: analysis and limitations. Soil Sci. Soc. Am. J.56:362– 369.

van Genuchten, M.Th., 1980. A closed-form equation for predicting the hydraulic conductivity of unsaturated soils. Soil Sci. Soc. Am. J. 44: 892– 898.

Vereeken, H., Diels, J., Van Orshoven, J., et al.(with all author names), Functional evaluation of pedo-transfer function for the estimation of soil hydraulic properties. Soil. Sci. Soc. Am. J. 1992, 56:1371-1378.

Wosten, J.H.M., Pachepsky, Y.A., Rawls, W.J., 2001. Pedo-transfer functions: bridging the gap between available basic soil data and missing soil hydraulic characteristics. J. Hydrol. 251:123– 150.

Yang, J. Y., Qu, Z. Y.,The determination and evaluation on soil water retention curve model in Hetao irrigation district. Journal of Arid Land Resources and Environment, 2008, 04:56-61 (in Chinese).

STRAWBERRY MATURITY NEURAL NETWORK DETECTNG SYSTEM BASED ON GENETIC ALGORITHM

Liming Xu [1,*] , Yanchao Zhao [1]
[1] *College of Engineering, China Agricultural University, Beijing, P. R. China 100083*
* *Corresponding author,address: College of Engineering, China Agricultural University, Beijing, 100083,P.R China,Tel:86-10-62737291 Email: xlmoffice@126.com*

Abstract: The quick and non-detective detection of agriculture product is one of the measures to increase the precision and productivity of harvesting and grading. Having analyzed H frequency of different maturities in different light intensities, the results show that H frequency for the same maturity has little influence in different light intensities; Under the same light intensity, three strawberry maturities are changing in order. After having confirmed the H frequency section to distinguish the different strawberry maturity, the triple-layer feed-forward neural network system to detect strawberry maturity was designed by using genetic algorithm. The test results show that the detecting precision ratio is 91.7%, it takes 160ms to distinguish one strawberry. Therefore, the online non-detective detecting the strawberry maturity could be realized.

Key words: genetic algorithm; neural network; maturity; strawberry

1. INTRODUCTION

High-quality products not only need the feasible planting management, but also are harvested in the optimum mature period. There are many researches about fruit maturity test by using image processing.The machine vision system (Miller et al., 1989) has been carried to check the fresh peach and confirm the peach maturity by comparing the peach natural colour to the standard hue of different maturity. The broccoli has been classified into no-

Please use the following format when citing this chapter:

Xu, L. and Zhao, Y., 2009, in IFIP International Federation for Information Processing, Volume 294, *Computer and Computing Technologies in Agriculture II, Volume 2*, eds. D. Li, Z. Chunjiang, (Boston: Springer), pp. 1201–1208.

ripe, ripe and over-ripe according to the average of special frequency respond(Qiu et al.,1992)。The colour image processing system has been constructed (Yoshitaka et al., 1997), the test results show that the influence of the quantity of light and the camera angle have no influence on Hue (H) and Saturation (S), H could be used to distinguished the different maturity banana and tomato. The RGB model has been transformed into HIS model (Choi et al.,1995), the cumulating H distributing map of fresh tomato was divided into 6 grades, and tomato maturity Index was constructed. The watermelon maturity distinguishing system has been built based on computer vision(Masaru et al.,1997). The green skin stripe image of samples were collected, the RGB model was transformed into the HIS model, the watermelon was divided into three grades: on-ripe, ripe and over-ripe. The maturity, sugar and surface damage of the strawberry has been determinanted by mean of a* value in L a*b* model and sugar meter(Masateru et al.,1997).

At present, the Chinese scholars have begun to study the fruit maturity through image procession. The external colour feature from the hue histogram by using computer vision has been extracted to established a three-layer feed-forward neural network system by using genetic algorithm, through which could realize on-line automated detection of apple maturity(Yang et al.,1997). The peach maturity according to the peach feature frequency of impedance has been determined(Ye et al., 1999). the watermelon maturity by its librating frequency respond has been judged (Wang et al.,1999). The external feature by using H has been distilled to detected automatically the tomato maturity by mean of multilayer feed forward neural network with GA.(Zhang et al.,2001). The change of tomato maturity under different light intensity, presented the maturity detecting foundation has been simulated (Cao et al., 2001). The triple-layer feed-forward network to extract colour feature from citrus image has been constructed to show the ratio of total soluble solid (TSS) to titratable acid (TA), then estimate the citrus maturity (Xu et al., 2001) .The colour and vein of a lot of lack-nutrition leaves image planted by using genetic algorithm has been optimized and chosen to recognize the lack of nutrition (Mao et al.,2003). The non-destructive detecting electron nose according to the apple smell has been built to detect the samples by using main-element analysis method and genetic algorithm network, the results show that the correcting ratio of genetic neural network is higher(Zhao et al.,2004).

For the traditional strawberry, the red fruit surface indicates ripeness. The “Tong-Zi-Yi-Hao” strawberry is different to the traditional strawberry, while the surface is whole red, the fruit is not mature. The colour is changing from light-red, fresh-red to black-red, the black-red strawberry is the optimum maturity. Based on the closed light simulation box, this paper analyzed the changing rule of H frequency in different light intensity, constructed the strawberry neural network detecting system based on genetic algorithm to distinguish its maturity on-line by the fruit image.

2. MATERIALS AND METHODS

2.1 Design of the image information collecting system

The hardware and software of the image collecting system are same with Xu et al (2007).

2.2 Chosen of strawberry maturity distinguishing value

1) the colour feature value of different strawberry maturities

For every strawberry, the pixel (or frequency) is different, so it could reflect the colour feature. The strawberry colour is mainly red, the colour depth is not different, this paper mainly analyzes the H hisgram.

In HIS model, the value of H is from 0° to 360°. The H hisgram of different mature strawberries in different light intensities (Fig. 1) were got.

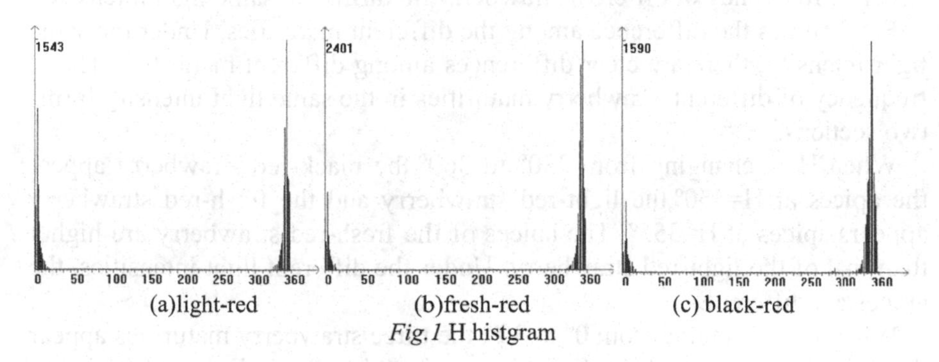

(a)light-red (b)fresh-red (c) black-red

Fig.1 H histgram

In Fig.1, the H value in different mature strawberry is distributed in two regions, 0°~30°and 330°~360°. The H hisgram comparison of different maturity strawberries in different light intensity is difficult to distinguish the different maturity strawberry. So the statistic of H hisgram is gained, the distance is 5°, the formula of H frequency is following:

Fi=(the pixel of a* per i) /the total pixels in the image (1)

(1) H frequency of the same strawberry maturity in different light intensities

The standard light intensities were chosen, they were 875lx, 2510lx, 3390lx, 4660lx. In these light intensities, the H frequency of every strawberry was gained (Fig.2).

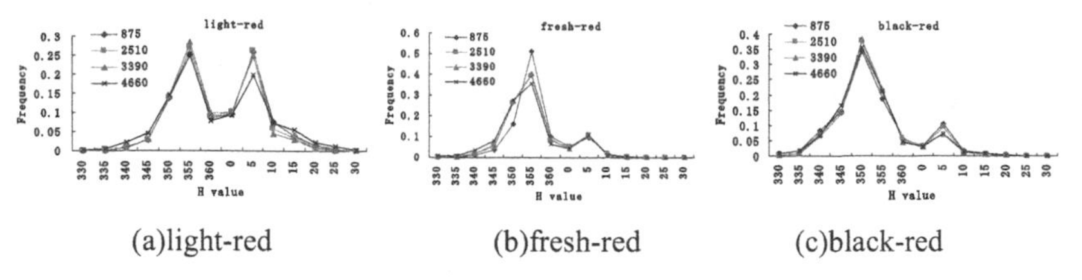

(a)light-red (b)fresh-red (c)black-red

Fig.2 H frequency of same maturity strawberry in different light

Under the different light intensities, H frequency of the same maturity strawberry has little change. Therefore, H frequency is not affected by the light intensity.

For the light-red strawberry, H frequency has "two-apices" when H is 355° and 5°respectively, the tops are 0.26 and 0.24 respectively.

The H frequency of the fresh-red strawberry and black-red strawberry is very similitude, there are also two apices, but one is higher, the other is lower. The fresh-red strawberry gets 0.41 higher apices when H is 355°. The black-red strawberry reaches the 0.37 higher apices when H is 350° .The two strawberries go down to the lower at 0.1 and 0.09 respectively.

(2) H frequency of different strawberry maturities in same light intensity

Fig.3 shows the difference among the different maturities. Under the same light intensity, there are clear differences among different maturities. The H frequency of different strawberry maturities in the same light intensity forms two sections.

When H is changing from 330° to 360°,the black-red strawberry appear the apices at H=350°,the light-red strawberry and the fresh-red strawberry appears apices at H=355°. The apices of the fresh-red strawberry are higher than that of the light-red strawberry. Under the different light intensities, the apices are different.

When H is changing from 0° to 30°, the three strawberry maturities appear the apices at H=5°, and the H frequency is going down in turn. Therefore, this section is able to be used to distinguish the strawberry maturity (Fig.4).

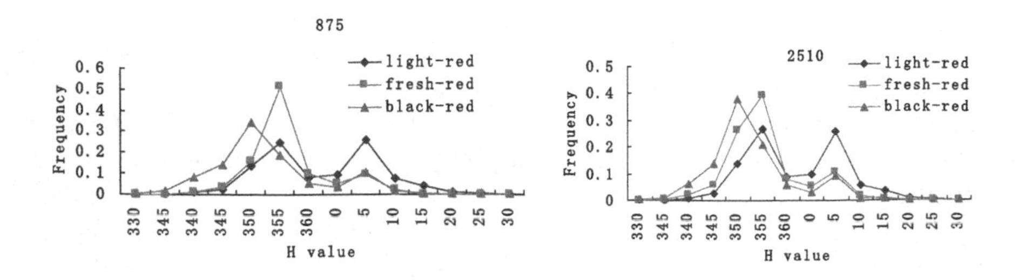

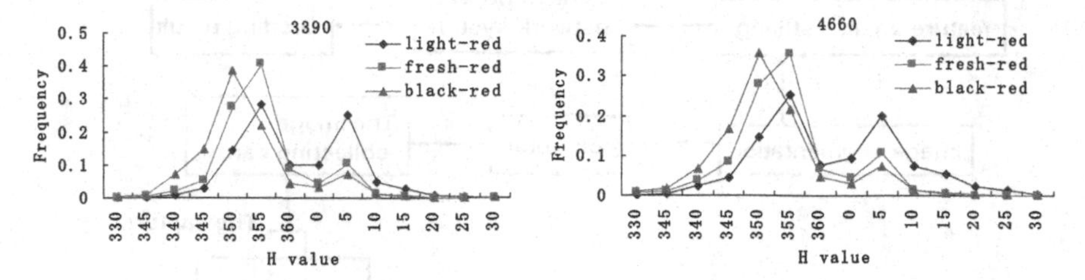

Fig.3 H frequency of different strawberry maturities in same light intensity

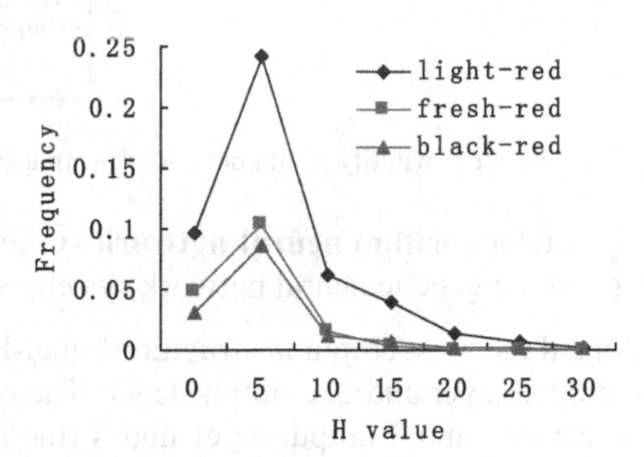

Fig.4 H frequency average at H= 0~30°of different maturity strawberry

2) Chosen of colour feature value of strawberry maturity

According to the analysis of H frequency of different strawberry maturities, H frequency is confirmed to test non-destructively strawberry maturity。H is 0, 5, 10, 15 and 20 respectively, the change from high to low is light-red, fresh-red and black-red.

2.3 Strawberry maturity neural network detecting system based on genetic algorithm

The neural network is designed by using genetic algorithm, an improved genetic algorithm(Wang et al 1996) is used to optimize the structure and gain the optimum network structure.

1) Structure of genetic algorithm neural network system

To detect the strawberry maturity non-destructively, this paper constructed the strawberry maturity genetic neural network detecting system by mean of the image collecting system, the image processing system and neural network system (Fig.5).

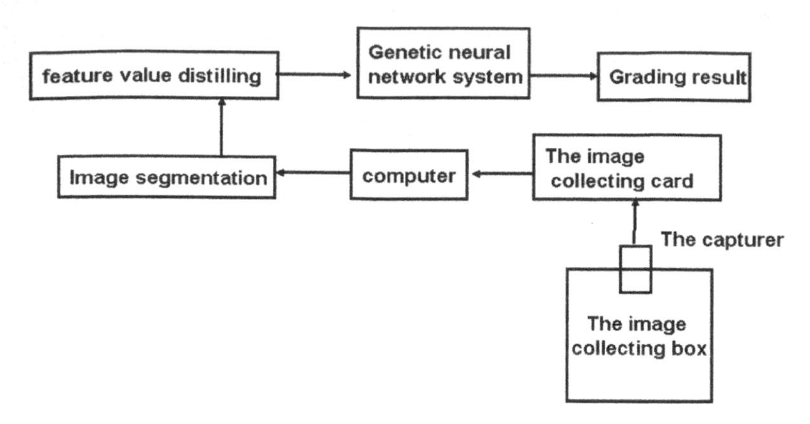

Fig. 5 The strawberry maturity neural network detecting system frame

2) Design of genetic algorithm neural network system

The design of the genetic neural network system is following:

This paper adopted the most common structure: triple-layer structures, one input layer, one hidden layer and one output layer. The network had 5 input layer nodes(5 H feature), and 3 output layer nodes (the maturity: black-red, fresh-red and light-red), the output layer nodes were described by a binary system (1 1 1) , (0 1 1) and (0 0 1) (Fig.6).

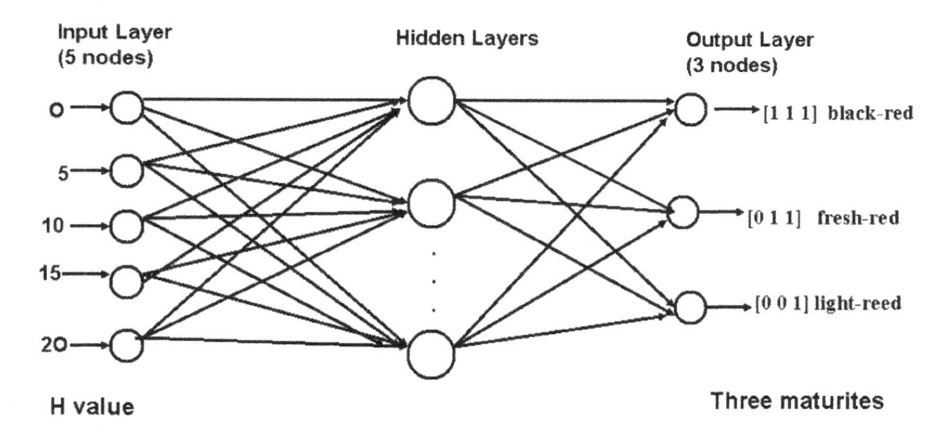

Fig. 6 The triple-layer strawberry maturity neural network detecting system

The colony scale was 20.The original number of hidden-layer node was 3 , the largest number of hidden-layer node was 13. The crossing probability P_c was 0.05, the variance probability P_m was 0.001, the terminating condition $f \leq 0.05$.

3. RESULTS AND DISCUSSION

The different mature strawberries were picked from Beijing Xiaotangshan Maizhuang Strawberry Production Base, and classified into three maturities by manual according to the surface colour, twenty exemplars in every maturity were used to be trained. After the strawberry maturity genetic neural network detecting system was studied and designed, the optimum number of the hidden layer node was 7. Therefore, the structure of neural network was 5-7-3, the training numbers were 1198, the training time was 154s. The test to detect strawberry maturity by using the trained neural network was completed, the training exemplars were 60 strawberries, every maturity had 20 strawberries, the test results were showed in Table 1.

The total detecting precision ratio is 91.7%, the time to classify the every strawberry is 120ms. The collecting image time is 40ms, so it takes 160ms to detect a strawberry. Therefore, the online detecting the strawberry maturity has been realized.

The major reason not to detect precision was that the colour of the fruit surface is not uniform.

Table 1 The result of strawberry maturity trial by genetic neural network distinguishing system

grade by manual	Grade by mechanical system			detecting precision ratio (%)
	Black-red	Fresh-red	Light-red	
Black-red	18	2		90
Fresh-red		18	2	90
Light-red		1	19	95

4. CONCLUSION

The quick and non-detective detection of agriculture product is one of the measuremeant to increase the precision and productivity of harvesting and grading. The strawberry maturity neural network detecting system based on genetic algorithm was completed. Specifically:

1) Under the different light intensities, the change of H frequency of different strawberry maturities was analyzed. The H frequency in same strawberry maturity is not affected by the light intensities. Under the same light intensity, the change of different strawberry maturities is in order, Therefore, the section of H frequency is confirmed to distinguish the strawberry maturity.

2) According to the H frequency, the strawberry maturity neural network detecting system was conducted. In order to decrease the training time, the genetic algorithm was adopted, the number of hidden layer node of triple-

layer feed-forward neural network was optimized, so the training time is shorted. The results show that the total detecting precision ratio of the system is 91.7%, it takes 160ms to detect a strawberry. Therefore, the online detecting the strawberry maturity has been realized.

REFERENCES

B K Miller, Michael J.Delwiche. A Colour Vision System for Peach Grading. Transactions of the ASAE , 1989,32(4):1484-1490

Cao Qixin, Liu Chengliang, Yin Yuehong, et al.Colour Image Processing Based on Quality Feature Extraction of Tomato.Robt, 2001,23(7):652-656

Choi K, Lee G, Han Y J, et al. Tomato maturity evaluation using colour image analysis.Transaction of the ASAE, 1995,38(1):171-176

Mao Hanping,XU Gui li,LI Ping ping. Study on Application of Genetic Algorithm to Feature Selection of Leaves Image for Diagnosing Vegetable Disease of Nutrient Deficiency. Journal of Jiangsu University of Science and Technology, 2003,24(2):1-5

Masaru TOKUDA,Tsuneo KAWAMURA.Deveolpment of Visual System for Watermelon Harvesting Robot(Part2). Journal of Japanese Society of Agricultural Machinery, 1997,59(4):47-52

Masateru NAGATA. Study on Product Quality Estimation based on image process. Study report,March,2000

Qiu W,Shearer S A. Maturity assessment of broccoli using the discrete Fourier Transform. Transactions of the ASAE, 1992,35(6):2057-2062

Wang Qiang, Shao Huih. Genetic Evoloved Neural Network and its Application in Formaldehyde Process Modeling and Optimization. Journal of Shanghai Jiaotong University, 1996,30(4):143-150

Wang Shumao,JiaoQunying,Ji Junjie.An Impulse Response Method of Nondestructive Inspection of the Ripeness of Watermelon. Transaction of the CSAE , 1999,15(3):241-245

Xu Liming,Zhang Tiezhong.Influence of light intensity on extracted colour feature of different mature strawberry.New Zealand Journal of Agricultural Research. 2007,50:559-565

Xu Zhenggang.Investigation of Non-destructive Citrus Maturity Determining Method Based on Image Information.Unpublished Master thesis.Zhe Jiang University,June,2001

Yang Xiukun, Chen Xiaoguang, Ma Chenglin, et al. Study on Automated Colour Inspection of Apples Using Genetic Neural Network. Transaction of the CSAE, 1997,40:173-176

Ye Qizheng,Yao Honglin,Li LI, et al.a Method for Measure Maturity According to the Feature Frequency of Resistance of Post-Harvest Fruit.Plant Physiology Communications. 1999,35(4):304-307

Yoshitaka MOTONAGA, Takaharu KAMEOKA,Atsushi HASHIMOTO.Constructing Colour Image Processing System for Managing the Surface Colour of Agricultural Products.Journal of Japanese Society of Agricultural Machinery, 1997,59(3):13-21

Zhang Changli,Fang Junlong,Fan Wei.Automated Identification of Tomato Maturation Using Multilayer Feedforward Nural Network with Genetic Alorithms(GA).Transaction of the CSAE , 2001,17(3):153-156

Zhao Jie-wen; Zou Xiao-bo; Pan Yin-fei; Liu Shao-peng. Research on method of apples odorant recognition based on GA-neural network. Journal of Jiangsu University of Science and Technology, 2004,25(1):1-4

ONE PREDICTION MODEL BASED ON BP NEURAL NETWORK FOR NEWCASTLE DISEASE

Hongbin Wang*, Duqiang Gong, Jianhua Xiao, Ru Zhang, Lin Li
College of Veterinary Medicine, NorthEast Agricultural University, Harbin,Heilongjiang Province, P. R. China, 150030
* *Corresponding author, Address: College of Veterinary Medicine, NorthEast Agricultural University, Harbin,150030, Heilongjiang Province, P. R. China, Tel: +86-451-55191940, Fax: +86-451-55190470, Email: neau1940@yahoo.com.cn*

Abstract: The purpose of this paper is to investigate the correlation between meteorological factors and Newcastle disease incidence, and to determine the key factors that affect Newcastle disease. Having built BP neural network forecasting model by Matlab 7.0 software, we tested the performance of the model according to the coefficient of determination (R2) and absolute values of the difference between predictive value and practical incidence. The result showed that 6 kinds of meteorological factors determined, and the model's coefficient of determination is 0.760, and the performance of the model is very good. Finally, we build Newcastle disease forecasting model, and apply BP neural network theory in animal disease forecasting research firstly.

Key words: Newcastle disease; forecasting model; BP neural network ; meteorological factors

1. MATERIALS AND METHODS

1.1 Data

One province in southwest of china was selected as object of this experiment. The meteorological factors data from 1999 to 2005 obtained

Please use the following format when citing this chapter:

Wang, H., Gong, D., Xiao, J., Zhang, R. and Li, L., 2009, in IFIP International Federation for Information Processing, Volume 294, *Computer and Computing Technologies in Agriculture II, Volume 2*, eds. D. Li, Z. Chunjiang, (Boston: Springer), pp. 1209–1216.

from China Meteorological Bureau. Epidemic situation data of ND in the same period obtained from Veterinary bulletin published by China Ministry of Agriculture. The cultivation scale of domestic fowl obtained from China Stock Farming Statistics Yearbook. Because the data of cultivation scale in China Stock Farming statistic by year, to unify the calculation method, the amount of domestic fowl for one year was treated as cultivation quantity for each month. The incidence rate of ND calculates as the follow formula:

Disease Incidence for each month of ND=morbility amount of ND / amount of domestic fowl by end year

1.2 Principle for BP Neural Network

BP neural network contain one input layer (Greenough G, 2001), some imply layer and one output layer, the treatment unit for data in every layer be called treatment unit. The transmission procedure of information is unidirectional transmission to input layer. The inferent information be treat in input layer, imply layer, and be transmit to output layer. The status of every layer can only be affected by next layer. If none anticipant outcome be generate in output layer, then change to back-propagation, the error between outcome and expected value be return along origin path. By modify weigh of neuron in every layer, the error be reduce gradually. This circulation would not stop until all error in net be restrained to a defined value. To analyze data by neural network, a great mount of train must be done for adjust the weigh value among neurons. The study procedure was completed by continual adjustment of weight (Li Jing, 2006).

1.3 Methods

1.3.1 Associativity analysis between meteorological factors and incidence of ND

To set meteorological data of this month as independent variable and incidence rate of ND for next month as dependent variable, associativity analysis was done by SPSS11.0, so as to provide evidence for the building of prediction model.

1.3.2 The building of prediction model of ND based on BP neural network

The data for Output layer was incidence rate of ND for each month in experiment ragion.The sample data make its value range in [−1,1]. The formula is $X'=(X\text{-}X\min)/(X\max\text{-}X\min)\times 2\text{-}1$; The data for output layer was incidence of ND, and its formula is $Y' = (Y - Y\min)/(Y\max - Y\min) \times 0.8 + 0.1$, output value range in [0.1, 0.9]. the result must be treated after train and fitting for model, so as to get actual data. The data that including temperature, humidity, wind speed, cloudage, time for sunshine and amount of precipitation obtained from October 1999 to may 2004 became input layer of network. And the incidence rate for ND of this region obtained from November 1999 to June 2004 became anticipant output layer of network. The accurate of this network can be validating by check data block. One implied layer was used in this neural network. The quantity scope of implied layer units determined by $h=\sqrt{N+M}+a$ was referred in this research. Among the total, M was set as the count of output units, N was the count of input units, a was constant range from 1 to 10[9]. After the count of input layer units-N(meteorological factors) and count of output layer units – N have been determined, the BP neural network was train and adjusted by alter the count of implied layer units h which change by the value of constant a. The train function was trainlm, the activation function was sigmoid function: $F(x)=1 / (1+e^{(-x)})$. The structures of BP neural network see fig.1.

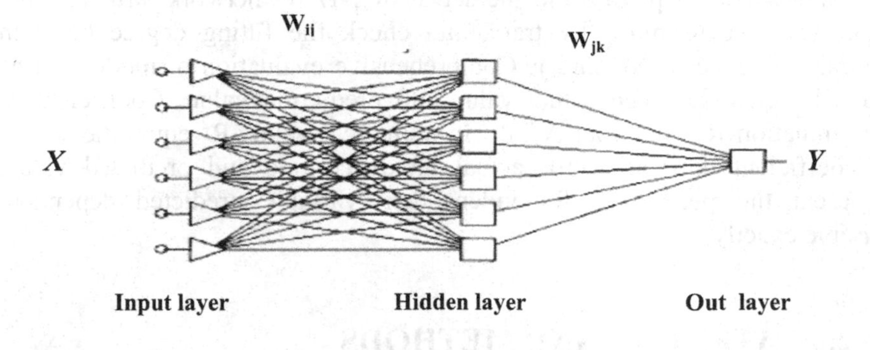

Fig.1 Structure of BP neual network Forecasting model

X means input unit(meteorological factor), Y means output unit(disease incidence); W_{ij} means the weight which connects the unit$_i$ in the Input layer with the unit$_j$ in the Hidden layer, W_{jk} means the weight which connects the unit$_j$ in the hidden layer with the unit $_k$, in the output layer, and all the connective weights can be adjusted according to Sigmoid training function.

1.3.3 The implement of BP neural network based on MATLAB7.0

1.3.3.1 The initalize of BP neural network

net=newff(minmax(*p*),[*m*,*n*],{'logsig','logsig'},'trainlm');minmax(*p*) was the scope of input value *p*. before applied, the meteorological data *p* should be normalized, and change into [–1,1]. 3-layer network was adopted in this experiment, *m* was the count of implied units, *n* was the count of output units. Because output variable *n* was incidence of ND, *n*=1,logsig was activation function among layers in BP neural network.

1.3.3.2 The train of BP neural network

net=train(net,P,T);P was input vector, T was target vector. Base on the reverse transmission algorithm of error for network study, new weight of network and threshold value was obtained from last train. The min error of network was set as 1×10^{-5}, the max frequency of train was 5 000, other parameters was acquiesced by system. The end purpose of train was make the error between incidence rate simulated by network and expected output value T.

1.3.3.3 The prediction of neural network and evaluation of effects

y=sim(net,P);To predict the incidence of ND by network structure and input vector determined by train, and check the fitting degree between actural incidence of ND and *y*. Comprehensive evaluation of model can be made by error between actual value and prediction value. Coefficient of determination R^2=(corrcoef(A,T))^2, In numerical value, R^2 equal the square of coefficient correlation for actual incidence rate and predicted value, represent the percentage for independent variable predicted dependent variable exactly.

2. ATERIALS AND METHODS

2.1 The associatively between meteorological factors and incidence rate

By correlation analysis between incidence rate of ND each month and meteorological factors(tab.1),The incidence rate of ND positive correlated with medial humidity each month and cloudage each month sigificantly, and negative correlated with average wind speed each month significantly.

Tab.1 Correlation matrix for every meteorological factors and ND incidence

Coefficient correlation	Air Temperature (X1)	humidity (X2)	Cloudage (X_3)	Wind speed (X_4)	Precipitation (X5)	Time for sunshine (X6)	Incidence rate (Y)
air temperature (X1)	1	0.499**	0.870**	-0.173	0.822**	-0.605**	0.244
humidity (X2)	0.499**	1	0.773**	-0.849**	0.709**	-0.861**	0.431**
cloudage (X3)	0.870**	0.773**	1	-0.431**	0.922**	-0.884**	0.299*
wind speed (X4)	-0.173	-0.849**	-0.431**	1	-0.414**	0.609**	-0.485**
precipitation (X5)	0.822**	0.709**	0.922**	-0.414**	1	-0.778**	0.263
time for sunshine (X6)	-0.605**	-0.861**	-0.884**	0.609**	-0.778**	1	-0.253
incidence of a disease (Y)	0.244	0.431**	0.299*	-0.485**	0.263	-0.253	1

** *means very significant difference* ($P < 0.01$ *at two tailed level*) , * *means significant difference* ($P < 0.05$ *at two tailed level*)

2.2 The determination of BP neural network and the prediction result

By many times debugging, it was determined that 6 was the optimum count of implied layer in network. 6 units in input layer was 6 kinds of different meteorological factors, there are 6 units in implied layer, the output layer unit was incidence rate of ND, therefore, the network structure be determined as 6—6—1. The error reaches smallest: 1×10^{-5} after 68 times trains (see fig.2), the weigh between input layer and implied layers; the weigh value and threshold value between implied layers and output layers listed in table 2, the fitting for BP neural network training value and practical incidence see fig.3.

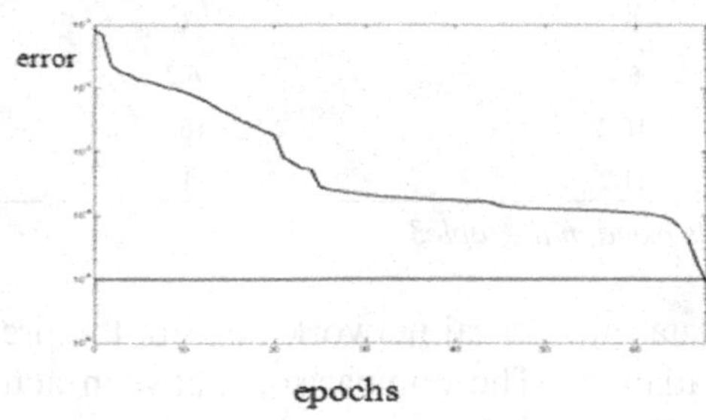

Fig.2 Training curve of BP neural network

Tab.2 Threshold and connective weight between each unit

Neuron layer	weight (Wij)	Shreshold
Input layer-imply layer	4.4712 -2.5107 -25.9255 31.0318 -12.6838 -24.1211	-7.3887
	-8.1328 4.2805 30.0656 12.2717 -5.3577 14.5849	-3.9163
	-1.8181 3.5385 13.1725 18.762 -16.603 2.02	6.8882
	8.1653 -16.1253 -12.2964 -9.2546 -9.9482 -34.8834	2.3161
	14.2829 16.0193 -10.8602 10.7641 1.8061 -14.7833	-25.2292
	-2.1933 3.6 -5.1103 -5.4201 5.1412 -3.4249	-6.1514
Imply layer-output layer	-1.1287 -1.2789 -10.4971 -1.2455 10.3956 -13.0465	10.7271

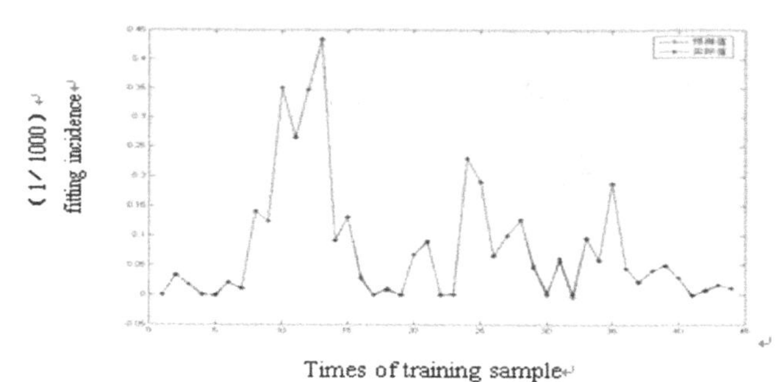

Fig. 3 Fitting for BP neural network training value and practical incidece

After train of neural network by inlet input vector and expected output vector, the weigh value and threshold value among every layers can be obtained, then the value can be save and fixed, the result for train and fitting see tab. 3.

Tab.3 Value of BP neual network training value and practical incidence

Time (year.month)	Fitting value for train of BP artificial neuron network (1/1000,000)	Practical incidence rate of disease (1/1000,000)	Absolute value of error(1×10^{-6})
2003.09	187.5	187.8	0.3
2003.10	44.6	44.1	0.5
2003.11	21.6	20.7	0.9
2003.12	40.1	40.4	0.3
2004.01	49.0	49.1	0.1
2004.02	27.3	28.1	0.8
2004.03	1.1	0	1.1
2004.04	6.1	6.2	0.1
2004.05	16.2	16.1	0.1
2004.06	11.2	11.1	0.1

* *Part of training data is listed in the table3*

By input check data into neural network trained, the prediction result can be get after run Matlab7.0. The comparison between actual incidence and prediction was list in tab. 4.

Fig.4 Value of Absolute error and forecast and incidence

Time (year.month)	Predictive value for BP neuron artificial network (1/1000,000)	Practical incidence rate of disease (1/1000,000)	Absolute value of error (1×10^{-6})
2004.07	6.2	5.8	0.4
2004.08	5.3	4.4	0.9
2004.09	-2.4	0	2.4
2004.10	0.1	0	0.1
2004.11	3.8	4.6	0.8
2004.12	3.9	1.4	2.5

2.3 Evaluation for effects of prediction model

By train, the predicted incidence rate of ND is Correspond with actual value, the result that coefficient of determination is $R^2=1.000$ show that the effects of fitting is good, and the check can be done later.

After input the check data into well-build prediction model, the prediction results were obtained and listed in table 4. combined with the prediction result and actual incidence, the coefficient of determination of prediction model $R^2=0.760$, this result illustrate that the prediction performance are good.

3. CONCLUSION

In human medical, BP neural network was used not only in prediction of infectious disease but also in diagnosis, prognosis, etiological factor analysis, risk analysis etc(Deng wei, 2002). The mathematic model applied in those domains generally requests a high foundation. Because of complex situation in practice, the building of prediction model is very difficult. Many scholar applied the artificial neural network in epidemic of tumor。The reason for meteorological factors became main object for this study was meteorological data easy-collection, affected ND greatly. By study the non-linear relationship between meteorological and incidence was build. The structure of network has been build and can be used to predict incidence rate of ND. In this study, the main work concentrated in building model, the model need to be train and check repeatedly, and the count of implied units need be adjusted endlessly. Data in Input layer need be normalized, different activation function and algorithm be selected. Different model need be screened and check many times, the model will not be selected until the result of checking has been reach the requirement. Because the selection of coefficient for network is random, the weight obtained in every train is different from others, therefore the train need be repeated endlessly. Otherwise, the quantity of data must be very great. Consequently, to investigate more data is the one importent work in future.

REFERENCES

Deng wei, Jin Peihuan. The application of artificial neural network in prophylactic medicine. Chinese Public health, 2002, 18(10):1265~1267

G Greenough, M McGeehin, S M Bernard. The potential impacts of climate variability and change on health impacts of extreme weather events in the United States. Environ Health Perspect,2001, 109(Suppl 2):191~198.

Li Jing, Wang Jingfei, Wu Chunyan, et al. the development of evaluation framework for risk of high pathogenicity bird flu. Chinese agricultural science, 2006, 39(10):2114~2117

CLASSIFICATION OF WEED SPECIES USING ARTIFICIAL NEURAL NETWORKS BASED ON COLOR LEAF TEXTURE FEATURE

Zhichen Li, Qiu An, Changying Ji*

College of Engineering, Nanjing Agricultural University, Nanjing, Jiangsu, 210031, China

** Corresponding author, Address: College of Engineering, Nanjing Agricultural University, Nanjing, Jiangsu, 210031, China, 025-58606571 Email: chyji@sohu.com*

Abstract: The potential impact of herbicide utilization compel people to use new method of weed control. Selective herbicide application is optimal method to reduce herbicide usage while maintain weed control. The key of selective herbicide is how to discriminate weed exactly. The HIS color co-occurrence method (CCM) texture analysis techniques was used to extract four texture parameters: Angular second moment (ASM), Entropy(E), Inertia quadrature (IQ), and Inverse difference moment or local homogeneity (IDM).The weed species selected for studying were Arthraxon hispidus, Digitaria sanguinalis, Petunia, Cyperus, Alternanthera Philoxeroides and Corchoropsis psilocarpa. The software of neuroshell2 was used for designing the structure of the neural network, training and test the data. It was found that the 8-40-1 artificial neural network provided the best classification performance and was capable of classification accuracies of 78%.

Keywords: weed, texture feature, artificial neural network, neuroshell2

1. INTRODUCTION

The application of herbicides in agricultural crops has been practiced for a long time. The utilization of herbicides result in significant increasement in crop production. However the environmental pressure grows day by day with the enhancement of herbicides utilization. How to reduce the herbicides and maintain the crop production has stimulated many researchers to study.

Please use the following format when citing this chapter:

Li, Z., An, Q. and Ji, C., 2009, in IFIP International Federation for Information Processing, Volume 294, *Computer and Computing Technologies in Agriculture II, Volume 2*, eds. D. Li, Z. Chunjiang, (Boston: Springer), pp. 1217–1225.

Thompson et al. (1991) suggested that selective herbicide application (variable-rate or intermittent application technology) to weed-infested area could result in decrease of herbicide use rather than the entire field. Since herbicides are applied only to weed patches, large areas of the field remain untreated. For the technology to be effective, weeds species and spatial location must be identificated from crop field. The application of machine for weed identification has increased rapidly for the recent two decades.

Burkes (2000) suggested three steps for weed discrimination using machine vision. First digital images of weed and crop must be acquired. Second the characteristic feature must be generated for discrimination between several weed species. Finally, a sufficient classification algorithm must be implemented using the feature as input and the weed class as output.

The use of color features in classical gray image texture analysis techniques was first reported by Shearer (1986). Shearer and Holmes (1990) reported accuracies of 91% for classifying different types of nursery stock by the color co-occurrence matrix (CCM) method. The traditional gray image texture features were expanded by Shearer(1986) to utilize HIS color co-occurrencc mcthod consisted of three co-occurrence matrices, one each for the hue, saturation, and intensity color features.

Meyer et al. (1998) used red, green, blue (RGB) true color to produce an excess green color feature for discriminating between four different species of weeds and soil regions with a 99% accuracy. They used the traditional gray scale co-occurrence matrix to generate four texture statistics: angular second moment, inertia, entropy, and local homogeneity. They observed classification accuracies of 93% for grasses and 85% for broadleaf categories when using the four texture feature. Tang et al. (1999) used a Gabor wavelets-based feature extraction method and neural net works to classify images into broadleaf and grass categories. They achieved 100% classification accuracies when testing 20 sample images from each of the two categories.

The resulting 11 texture feature equations are defined by Burks . The Color Co-occurrence Method (CCM) was also utilized by Burks to discriminate between six different classes of groundcover. The result showed that the CCM texture statistic procedure was able to classify five species of weed and soil with accurate of 93%.

The CCM method was utilized by Burks to evaluate three different neural network classifiers for real-time weed control systems. The result showed that the BP neural-network classifier provided the best classification performance with accuracies of 97%.

Detecting inter-row weed was researched by Mao Wenhua, the location feature of crop within field was used for discriminating weed with accuracies of 86%. The texture and position feature were uscd for studying by Cao Jingjing, the result of the research showed that the correct classification of weed was 93%.

Artificial Neural Network (ANN) is a data processing system based on the structure of biological neural system. Prediction with ANN is not like modelling and simulation, but by learning from the data generated experimentally or using validated models, ANN differ from conventional programs in their ability to learn about the system to be modelled without priori knowledge of the process variables relationship.

A discriminating method was developed based on image features of the peanut kernels and artificial neural network by Chen Hong. The image characteristics parameters such as the color parameters HIS, and veins characteristics parameters RW,GW,BW were used as the input to the neural network set up by MATLAB. The results of the experiment show that the accuracies of the identification of the method are 95% for normal peanut kernels, 90% for slightly moldy peanut kernels and 100% for severely moldy peanut kernels, respectively.

BP neural network model was developed to map the relationship of process variables and distributing quality coefficient of Profile Modeling Spray of the Fruit Trees by Lin Huiqiang. The result shows that the correlation coefficient R between simulating outputs of the BP network and the results of experiments is 0.99, which has wide adaptability. And the BP network can be used conveniently to carryout various quantificational calculations.

The improved LVQ neural network algorithm was applied in the process to identify the grade of apples by Bao Xiaoan, The research result was that the correct identification rate of 88.9% with good stability.

A three layer feed forward neural network was established by Chen JiaJuan. The purpose of the research was to identify the corn and the background of field. The experiment showed that corn could be recognized correctly by using this automated measurement system of corn leaf color value, the judging accuracy could attain 91.6%, and corn leaf color value could also be calculated correctly.

The objectives of this paper were to develop an ANN to classify species of six weed using four texture features. The classification process using the texture feature data is very simplicity, convenience, fast and accurate. This will also assist in identifying weed from field crop real-time based on machine vision. The work of designing the structure of neural network and train or test the neural network is all done by the software neuroshell2.

2. COLOR TEXTURE FEATURE EXTRACTION METHOD

The image texture feature analysis technique selected for this study includes three processes based on the Color Co-occurrence Method (CCM). One, the image is transformed from RGB to HIS using the following equation. The purpose of this process is to reduce the storage and computing time. It can be classify the observation with the accuracy of 93% only using hue and saturation texture feature by the research of Burks[7] . Only hue and saturation texture feature was used in this study.

$$I = (R + G + B)/3 \quad (1)$$

$$S = 1 - \frac{3 * Min(R, G, B)}{R + G + B} \quad (2)$$

$$H = ARCCOS \left\{ \frac{(R - G) + (R - B)}{2[(R - G)^2 + (R - B)(G - B)]^{1/2}} \right\} \quad (3)$$

Two, one CCM matrix is generated from the HIS image. The matrix for intensity was represented by the function P(i,j,d,θ). The CCM measures the probability that a pixel at one particular gray-level will occur at a distinct distance and orientation from any pixel given that pixel has a second particular gray-level. In the function P(i,j,d,θ) i represents the gray-level of location(x,y), and j represents the gray level of the pixel at a distance d and an orientation ofθfrom location(x,y). 1 was selected for d and 0 was selected for θin this study. For example: the image and CCM matrix are showed in figure1. The process may be understood by looking for the number of occurrence of zero adjacent to three in $I(x, y)$ in figure1 (a),we find a total of two occurrences of zero adjacent to three.This corresponds to the two value at CCM matrix location (0,3).

$$I(x,y) = \begin{bmatrix} 0 & 0 & 3 & 1 \\ 2 & 1 & 0 & 2 \\ 3 & 2 & 0 & 3 \\ 1 & 2 & 1 & 3 \end{bmatrix}$$

(a)

$$P(i,j,1,0^\circ) = \begin{bmatrix} 1 & 1 & 2 & 2 \\ 1 & 0 & 3 & 2 \\ 2 & 3 & 0 & 1 \\ 2 & 2 & 1 & 0 \end{bmatrix}$$

(b)

Fig.1. the image and CCM matrix example: (a) gray-level image, (b) CCM matrix

Three, two CCM matrix of hue and saturation is used to generate eight texture features using the following equation.

Angular second moment (ASM):

$$H_{ASM} = \sum_{i=0}^{N_g-1}\sum_{j=0}^{N_g-1}[p(i,j)]^2 \tag{4}$$

Entropy(E):

$$H_E = \sum_{i=0}^{N_g-1}\sum_{j=0}^{N_g-1} p(i,j)\ln(p(i,j)) \tag{5}$$

Inertia quadrature (IQ):

$$H_{IQ} = \sum_{i=0}^{N_g-1}\sum_{j=0}^{N_g-1} p(i,j)(i-j)^2 \tag{6}$$

Inverse difference moment (IDM):

$$H_{IDM} = \sum_{i=0}^{N_g-1}\sum_{j=0}^{N_g-1} \frac{p(i,j)}{1+(i-j)^2} \tag{7}$$

Where:

$$p(i,j) = \frac{P(i,j,1,0)}{\sum_{i=0}^{N_g-1}\sum_{j=0}^{N_g-1} P(i,j,1,0)} \tag{8}$$

And N_g = the total number of attribute levels. The calculate method for the saturation is the same as the hue equations.

3. MATERIALAND METHOD

3.1 Weed Species

Six weed species common to south china row crops were selected for this study. The weed species were Arthraxon hispidus, Digitaria sanguinalis, Petunia, Cyperus, Alternanthera Philoxeroides and Corchoropsis psilocarpa. Every of the six species were grown from one experimental field of Nanjing Agricultural University. The weed species have been grown under normal ambient conditions until they reached an appropriate maturity level. The number of different size leaves for studying were 14,11， 8， 9， 8， 10 respectively for the above species.

3.2 Image Acquisition and processing System

A three CCD camera of Olympus N438 was used for collecting the digital RGB images. The weed leaves were put on a white paper immediately after getting from the field. The image was taken outdoors under natural sunlight.

The image was acquired and digitized into 24-bit (RGB) images with a resolution 640*480. All of the leaves images were recorded on an U-disc of the camera and then downloaded into the personal computer of PentiumIII 700 MHz CPU .

Six species of weed were captured in 640*480 pixel digital images which included both white paper and weed leaf. The images for every species of weed are presented in figure 2. A computer software program using VC++ was developed to generate RGB image files, create a HSI color model of the image, generate color co-occurrence matrices, and calculate CCM statistics for the specified matrices.

3.3 NeuroShell 2

NeuroShell 2 is a software program by Ward System Group.Inc. The program can mimic the human brain's ability to classify patterns or to make predictions or decisions based upon past experience.

NeuroShell 2 combines powerful neural network architectures, a Microsoft® Windows icon driven user interface, sophisticated utilities, and popular options to give users the ultimate neural network experimental environment. It is recommended for academic users only, or those users who are concerned with classic neural network paradigms like backpropagation. Users interested in solving real problems should consider the NeuroShell Predictor, NeuroShell Classifier, or the NeuroShell Trader. The central interface of NeuroShell 2 is as example of figure 3.

NeuroShell 2 enables you to build sophisticated custom problem solving applications without programming. You tell the network what you are trying to predict or classify, and NeuroShell 2 will be able to "learn" patterns from training data and be able to make its own classifications, predictions, or decisions when presented with new data.

Arthraxon hispidus

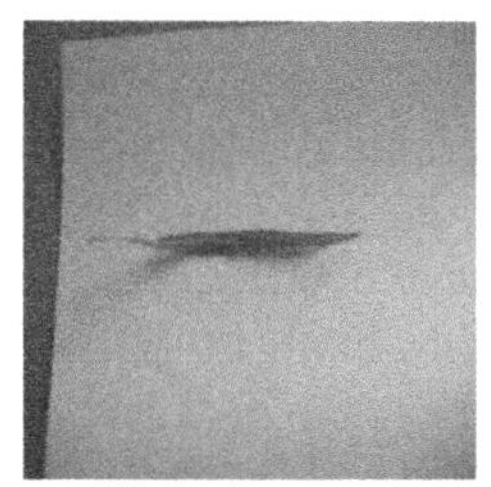
Digitaria sanguinalis

Petunia

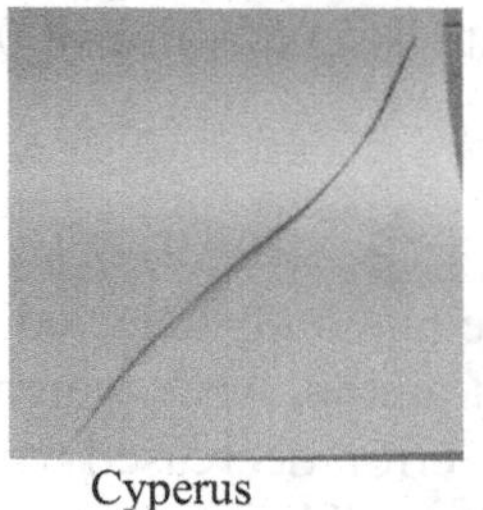
Cyperus

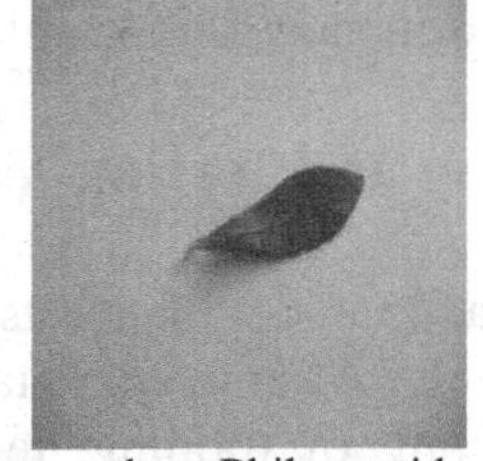
Alternanthera Philoxeroides

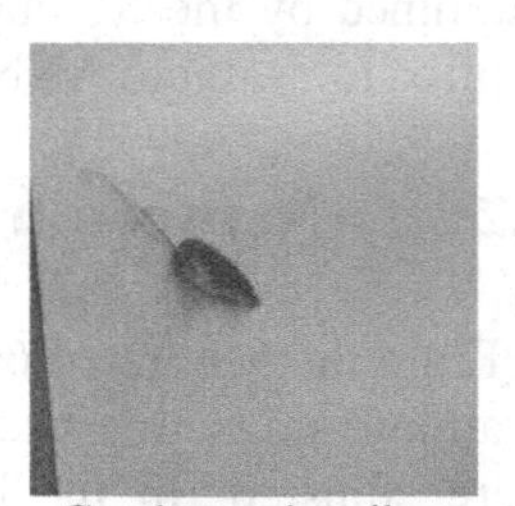
Corchoropsis psilocarpa

Fig.2 Six species of weed

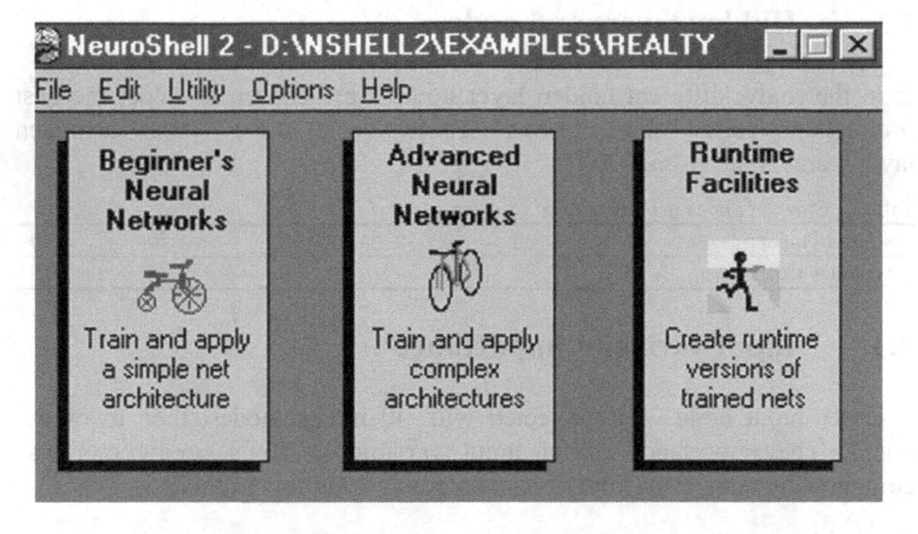

Fig.3 The central interface of NeuroShell 2

4. RESULT AND DISCUSSION

4.1 Neural network model

Three layered ANN was developed to classify leaf species of six weed. Eight in the input layer represented eight input parameters and one nodes in output layer to classify Arthraxon hispidus, Digitaria sanguinalis, Petunia, Cyperus, Alternanthera Philoxeroides and Corchoropsis psilocarpa. The number of hidden layer nodes was chosen according to ANN performance.

The backpropagation algorithm was used for ANN training. The best testing error was taken as criterion to stop training. The errors were examined by the testing sets, which were not used for ANN training, were applied to evaluate ANN performance. The learning rate is set to value of 0.1.

4.2 Network training and testing

From the generated 60 data, 45 and 15data sets were randomly selected as Training and tseting respectively. After every data set training, ANN weights were adjusted. In the beginning of training, testing error decreased with training process. Training was continued until testing error did not decrease.

4.3 Hidden layers and nodes

In the study, different hidden layer nodes were chosen to select the best production results (Table1). Learning rate was set at 0.1. Nodes in hidden layer were varied from20 to 60.

Table 1. Effect of nodes in hidden layer

Nodes of hidden layer	20	30	40	50	60
Accuracy (%)	72	76	78	78	78

4.4 Input variables importance

Every input node was connected with 40 hidden nodes, then to output nodes. The importance of an input variable to ANN predictions was compared by comparing sum of its connecting weights (Table.2).

Table.2. Attribution of the input invariable

Variable	ASM	E	IQ	IDM
Weight	0.312	0.308	0.25	0.13

5. CONCLUSION

Classification species of weed by ANN using simulation data is a simple and convenient method. Classification accuracy could be increased by careful selection of hidden nodes and appropriate of learning rate. In this study, the appreciate structure of artificial neural network is 8-40-1 with optimal learning rate is 0.1. The Angular second moment and entropy of weed texture feature were important to all of the input variables. The inverse difference moment of weed texture feature is the minimum affective to the outputs.

ACKNOWLEDGEMENTS

Funding for this research was provided by Hubei Provincial Department of Education (P. R. China). The first author is grateful to the Wuhan Polytechnic University for providing him with pursuing a PhD degree at the Wuhan University of Technology.

REFERENCES

Bao Xiaoan, Zhang Ruilin , Zhong Lehai. Apple grade identification method based on artificial neural network and image processing. Transaction of the CSAE 2004,20(3),109-111

Cao Jngjing, Wang Yiming, Mao Wenhua et al. Weed detection method in wheat field based on texture and. Position features. Transaction of the CSAM 2007.38(4),107-110.

Chen Jiajuan, Ji Shouwen, Ma Chenglin. Investigation on automated color measurement of corn leaves based on genetic neural network. Transaction of the CSAE 2000,16(3),115-117

Cheng Hong, xiong Lirong, Hu Xiaobo, et al. Identification method for moldy peanut kernels based on neural network and image processing. Transaction of the CSAE 2007,23(4),158-161.

G.E.Meyer, T.Mehta, M.F. Kocher,etl.1998. Textural image and discriminant analysis for distinguishing weeds for spot spraying.Transactions of ASAE,41(4),1189-1197

Lin Huiqiang, Xiao Lei, Liu Caixing, et al. Neural network model for profile modeling spray of chemical to fruit trees and its applications. Transaction of the CSAE 2005,21(10),95-99

Mao Wenhua ,Wang Yiming, Zhang Xiaochao et al. Machine vision system used for real-time detection inter-row weed. Transaction of the CSAE 2003,19(5),114-117.

Ramesh, M. N., Kumar, M. A. & Rao, P. N. S. (1996). Application of artificial neural networks to investigate the drying of cooked rice. Journal of Food Process Engineering, 19, 321-329.

Shearer,S.A., and R.G.Holmes.1990. Plant identification using color co-occurrence matrices. Transactions of the ASAE 33(6):2037-2044.

Shearer.S.A.1986. Plant identificationusing color co-occurrence matrices derived from digitized images. PH.D. thesis. Ohio university, Agricultural Engineering.

T.F.Burks, S.A.Shearer, F.A.Payne.2000. Classification of weed species using color texture features and disciminant anylysis. Transactions of ASAE,43(2),441-448

T.F.Burks, S.A.Shearer, J.R.Heath.etl.2005. Evaluation of neural-network classifiers for weed species discrimination. Biosystems Engineering(2005) 91(3),293-304

T.F.Burks, S.A.Shearer, R.S. Gates, K.D. Dono.2000. Backpropogation neural network design and evaluation for classiffing weed species using color image texture feature.Transactions of ASAE,43(4),1029-1037

Tang L.， L.F. Tian, B.L.Steward, et al. Texture-based weed classification using Gabor wavelets and neural network for real-time selective herbicide applications. ASAE ,1999 ,Paper No.993036

Thompson,J.F.,J.V.Stafford, and P.C.Miller. 1991. Potential for automatic weed detection and selective herbicide application. Crop Prod. 10(4):254-259.

SPATIAL ESTIMATION OF SOIL MOISTURE AND SALINITY WITH NEURAL KRIGING

Zhong Zheng [1,2] , Fengrong Zhang [1,*] , Xurong Chai [1] , Zhanqiang Zhu [1] , Fuyu Ma [2]

[1] *College of Resources and Environment, China Agricultural University, Beijing, P. R. China 100094*

[2] *College of Agriculture, Shihezi University, Shihezi, Xinjiang Province, P. R. China 832003*

[*] *Corresponding author, Address: College of Resources and Environment, China Agricultural University, Beijing, P. R. China 100094, Tel: +86-10-62732643, Fax: +86-10-62732643, Email: zhangfr@cau.edu.cn, zhenglxx@gmail.com*

Abstract: The study was carried out with 107 measurements of volumetric soil water content (SWC) and electrical conductivity (EC) for soil profile (0-30 cm) and the estimating accuracy of ordinary kriging (OK) and back-propagation neural network (BPNN) was compared. The results showed that BPNN method predicted a slightly better accurate SWC than that of OK, but differences between both methods were not significant based on the analysis of covariance (ANOVA) test ($P > 0.05$). In addition, BPNN performed much better in EC prediction with higher model efficiency factor (E) and ratio of prediction to deviation (RPD) (E=0.8044 and RPD=3.54) than that of OK (E=0.7793 and RPD=0.39). Moreover, a novel neural kriging (NK) resulting from the integration of neural network (NN) and ordinary kriging (OK) techniques was developed through a geographic information system (GIS) environment for obtaining trend maps of SWC and EC. There was no significance between results of NK and OK through trend maps. Comparing with OK, NK gives better spatial estimations for its great advantage of establishing spatial nonlinear relationships through training directly on the data without building any complicated mathematical models and making assumptions on spatial variations.

Keywords: soil moisture, soil salinity, spatial estimation, ordinary kriging, artificial neural networks

---

Please use the following format when citing this chapter:

Zheng, Z., Zhang, F., Chai, X., Zhu, Z. and Ma, F., 2009, in IFIP International Federation for Information Processing, Volume 294, *Computer and Computing Technologies in Agriculture II, Volume 2*, eds. D. Li, Z. Chunjiang, (Boston: Springer), pp. 1227–1237.

1. INTRODUCTION

In recent years, agricultural development on the oasis plain of Maigaiti, Northwest China, has being threatened by soil secondary salinization irrigation-induced due to excessive and inefficient water use. Salt accumulation and excessive salt concentration in farmlands has led to land degradation, crop yield decrease, abandoned lands increase, water quality deterioration and environmental degradation (Kitamura et al., 2006). Therefore, consistent and early stage identification of soil salinity as well as assessment of soil moisture is vital for crop production, especially in arid areas where harsh climatic conditions together with rapidly increasing population densities (Farifteh et al., 2007).

In the farmland, it is impractical to sample all the points with the desired temporal frequency in order to research the variability of soil water and salinity content. Optimizing spatial sampling scheme to reduce sampling density and estimation of unsampling values can save time and money (Ferreyra et al., 2002; Li et al., 2007). However, its effectiveness relies on the accuracy of the spatial interpolation used to define the spatial variability. Multivariate techniques such as geostatistics and artificial neural network (ANN) have been widely used as estimation tools. Geostatistics provides descriptive tools such as kriging to directly implements the prediction of an attribute at an unsampled location according to known data points within a local neighborhood surrounding (Emery & Ortiz, 2007). ANN has the ability to model extremely non-linear and complicated relationships between a set of inputs, and are operated by using the available input and output responses without considering inherent system parameters (Sarangi et al., 2006). It can be used as an alternative to predict regionalized variables (RV) which are functions on geographic locations (Huang & Foo, 2002; Farifteh et al., 2007). The performances of different interpolation methods such as ordinary kriging (OK), inverse distance weighting(IDW), splines and so on, have been analyzed in several studies, whereas there have been many conflicting reports concerning the performances of different interpolation methods (Gotway et al., 1996; Patel et al., 2002; Brocca et al., 2007). In addition, many comparisons of various interpolation techniques have been made in respect to different data sets used, different mathematical procedures and different input parameters (Boken et al., 2004; Robinson & Metternicht, 2006). Moreover, very few studies compare the performance of OK and ANN methods simultaneously (Rizzo & Dougherty, 1994).

Hence, the aim of this paper is to: (i) identify the performance of OK and ANN for estimation of soil moisture and salt content in a given area within a threshold of error. (ii) investigate the applicability of neural kriging (NK) resulting from the integration of neural network (NN) and ordinary kriging (OK) techniques; (iii) illustrate trend maps of soil moisture and salinity distribution for study area.

2. METHODS

2.1 Site description

The study site is located at the west margin of Taklamakan desert in Northwest China. Field research was conducted on a cotton farmland (30°14'29"-39°14'57"N, 78°06'21"-78°07'00" E), which is located in northeast region of Maigaiti County of Xinjiang (Fig. 1). The research field covers 0.54 km^2 (900m x 600m) with a 2-4% slope northwest to southeast. The area is 442 m above sea level and experiences an arid climate with mean annual temperature, precipitation, evaporation and frost-free period of 12.4 °C, 46.5 mm, 2526 mm and 212 d over a 20-year period, respectively. The soil texture is dominant sand soil and its distribution and variability in the topsoil are influenced by Taklamakan desert in arid regions.

The farmland for the present study was reclaimed in 2000 and cotton was planted from 2003. Since the groundwater in the study site has higher mineral degree, soil secondary salinization and drought are the main limiting factors for crop production in arid regions.

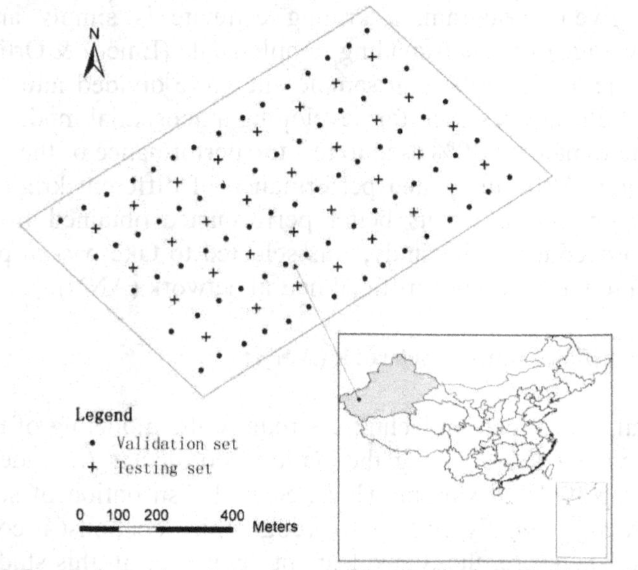

Fig.1: Research area location and map of the distribution of soil samples.

2.2 Data acquisition

A grid sampling scheme (40-60 m sampling space) was imposed on the field with 107 sample measurements of volumetric soil water content (SWC) and electrical conductivity (EC). SWC was measured using a portable Time Domain Reflectometry (Soil Moisture Equipment Corp., TRASE®TDR) and

EC using a portable WET sensor (Delta-T Devices Ltd., Cambridge, UK). On each sampling point, we inserted vertically a triple wire TDR probe to monitor soil moisture and a wire WET sensor probe for soil salinity in the soil profile (0-30 cm). Each EC and SWC measurement was geo-referenced using a Differential Global Positioning System (DGPS). At each sampling grid point, five EC or SWC measurements were made within a 1-m diameter circle. The average reading for each grid point was computed as EC or SWC datum point. Sample measurements were implemented before cotton cultivation at March 12, 2007. Among this set of 107 measurements, a set of 75 data were selected to constitute validation samples, remaining set of 32 for testing samples (Fig. 1).

2.3 Spatial estimation methods

2.3.1 Kriging

Kriging estimate relies on a weighting scheme where closer sample locations have greater impact on final prediction. At an unsampled location and for a given variogram, a kriging estimate is simply an optimally weighted average of the surrounding sampled data (Emery & Ortiz, 2007). In this study, the total numbers in sample site were divided into training sets with 70% of all samples used for developing a geospatial model, and testing sets with the remaining 30% used to test the performance of the models (Fig. 1). According to the integrated performance of different kriging methods, ordinary kriging(OK), for its better performance obtained from a cross-validation procedure in the study, was selected to take part in performance comparison to the following artificial neural network (ANN).

2.3.2 Artificial neural network (ANN)

In general, all estimation techniques require the modeling of the function Z=f(*X, Y*), where (*X, Y*) being the station coordinates (latitude–longitude) and Z the regionalized variable (RV). Spatial distribution of soil moisture and salinity is generally related to geographic locations(*X* coordinate, *Y* coordinate). Therefore, the network input layer used in this study relates to geographic *X* coordinate and *Y* coordinate, while the network output layer relates to soil moisture and soil salinity. The number of neurons in the hidden layer is of great importance, as too many neurons may cause over-fitting problems, and it can be defined using a formula recommended or using a trial-and-error approach (Huang & Foo, 2002). Thus, a 3-layer feed-forward back propagation neural network (BPNN) (topology structure: 2 x 5 x 2) was established (Fig. 2) and used within Neural Network Toolbox of

Matlab 7.0 (The MathWorks Inc. Natick, MA). Tan-sigmoid transfer functions and log-sigmoid transfer functions (non-linear) were selected for the hidden and output layers, respectively. The Levenberg-Marquardt algorithm, which provides a fast optimization, was used for network training. Total of 107 soil samples were divided into two groups as 75 for the development (training and validation) and 32 for the test (Fig.1). Simulated error is 3% and 0.02 mS/cm for SWC and EC in training and testing process, respectively.

However, ANN only allows the RV estimation, but not the predictor variance, which is possible with kriging (Rizzo & Dougherty, 1994; Koike et al., 2001). In some applications ANN is coupled with kriging estimation, which was called neural kriging (NK) by Rizzo and Dougherty (1994). NK is divided into two steps: the first is neural and uses neural network, and the second uses OK. The final estimates are produced as a sum of NN estimates and OK estimates. Hence NK is an integrated interpolation technique. In spatial estimation its utilization is justified by the fact that it extracts its knowledge only from data, which contain information about the spatial distribution of the RV. The present paper follows the NK approach to estimate soil moisture and salinity and compare the obtained estimates with OK technique, and then illustrate trend maps of SWC and EC.

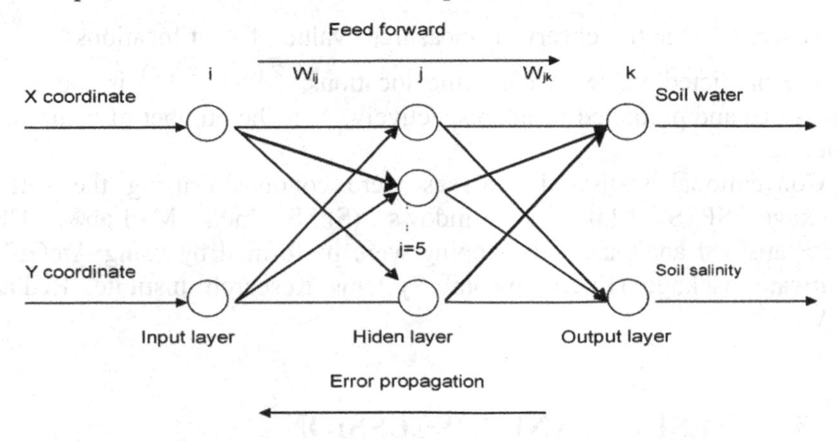

Fig.2: Three-layer feed forward back propagation neural network structure.

2.4 Evaluation criteria

We used cross-validation to validate the accuracy of interpolation algorithms and examine the difference between the measured values and the predicted values using mean absolute errors (MAE) or relative MAE (MAE%) (Eq. (1)), root-mean-square error (RMSE) or relative RMSE (RMSE%) (Eq. (2)), ratio of prediction to deviation (RPD) (Eq. (3)) and model efficiency factor (E)(Eq. (4)). The RPD indicates strength of statistical correlation between measured and predicted values. MAE, RMSE and E values indicate degree of agreement between measured and predicted values. Detailed descriptions and definitions of these model performance parameters are given by Robinson & Metternicht (2006) and Farifteh et al. (2007).

$$MAE = \frac{1}{N}\sum_{i=1}^{N}\left|Z^{*}(x_i) - Z(x_i)\right| \qquad MAE\% = 100\frac{MAE}{\bar{Z}(x_i)} \tag{1}$$

$$RMSE = \sqrt{\frac{1}{N}\sum_{i=1}^{N}\left(Z^{*}(x_i) - Z(x_i)\right)^{2}} \qquad RMSE\% = 100\frac{RMSE}{\bar{Z}(x_i)} \tag{2}$$

$$RPD = \frac{\sqrt{\sum_{i=1}^{N}\left(Z^{*}(x_i)\right)^{2} - \left[\left([\sum_{i}^{N} Z^{*}(x_i)]^{2}/N\right)/(N-1)\right]}}{\sqrt{\sum_{i=1}^{N}\left(Z^{*}(x_i) - Z(x_i)\right)^{2} - \left[\left([\sum_{i}^{N}(Z^{*}(x_i) - Z(x_i)]^{2}/N\right)/(N-1)\right]}} \tag{3}$$

$$E = 1 - \frac{\sum_{i=1}^{N}\left(Z^{*}(x_i) - Z(x_i)\right)^{2}}{\sum_{i=1}^{N}\left(\left|Z^{*}(x_i) - \bar{Z}(x_i)\right| + \left|Z(x_i) - \bar{Z}(x_i)\right|\right)^{2}} \tag{4}$$

Where $Z(x_i)$ is the observed (measured) value of Z at locations x_i, $Z^{*}(x_i)$ is the predicted value at the same locations, $\bar{Z}(x_i)$, $\bar{Z}^{*}(x_i)$ is the average measured and predicted value, respectively. N is the number of values in the dataset.

Conventional statistical analyses were conducted using the software package SPSS 12.0 for Windows (SPSS Inc., MatLab®, USA). Geostatistical analyses and mapping were performed by using ArcGIS 9.0 software package (Environmental Systems Research Institute, Redlands, CA).

3. RESULTS AND DISCUSSION

3.1 Descriptive statistics

Based on above estimation methods as discussed, descriptive statistics for SWC and EC used in the development and validation of BPNN and OK are summarized in Table 1. It is observed than both variables are approximately normal distribution avoiding the need for data transformation according to the coefficient of skewness. The soil of study area has wide ranges of SWC and EC with 27.3–48.3% and 0.07–1.90 mS/cm, respectively. The predicted values of SWC and EC with BPNN method, ranging between 31.4–43.2% and 0.24–1.12 mS/cm respectively, have smaller range and distributions closer to the average measured values than that of OK, which rang between 28.4-47.3% and 0.18-1.47 mS/cm respectively. Besides, there is less standard deviation of BPNN than that of OK. Hence, we primarily conclude that BPNN method could perform somewhat better than OK from the descriptive statistics obtained from cross-validation.

Table 1 Summary statistics of SWC (%) and EC (mS/cm)

Soil property	Mean	Maximum	Minimum	Std.Dev.	Skewness	Kurtosis
SWC (%)						
Measured	36.7	48.3	27.3	5.2098	0.2290	2.1629
OK method	37.3	47.3	28.4	4.4309	-0.0659	2.2501
BPNN method	37.2	43.2	31.4	3.4303	-0.2273	1.7673
EC (mS/cm)						
Measured	0.72	1.90	0.07	0.4208	0.5471	2.6463
OK method	0.66	1.47	0.18	0.3081	-0.0465	1.8989
BPNN method	0.70	1.12	0.24	0.2544	-0.0903	1.8407

3.2 Spatial estimation and its performance

SWC and EC can all be estimated by both methods while the former variable yields very good results (Fig.3a) and the latter shows a bad generalization (Fig.3b). Besides, Fig.4a and Fig.4b also shows good performance of SWC by both methods, respectively. The worse performance of EC can be seen in Fig.4c and Fig.4d, where measured and predicted values are more scattered. Comparing to coefficient of determination (R^2) (Fig.4), R^2 between predictions and observations ranges from 0.5704 to 0.5763 for SWC and from 0.4697 to 0.4948 for EC by both methods. This result reveals that SWC prediction value has better performance than EC. Furthermore, spatial variability of EC is being under-predicted, which is in good agreement with the field finding similar to the study reported by Li et al. (2007). It has to be pointed out that the quality and quantity of the data used for the study were not adequate to the complexity of spatial variability of soil properties. Boken et al. (2004) considered that the accuracy estimate can be improved by enhancing the representation of sampling sites as well as by limiting the estimations to irrigated areas within counties.

Based on the statistical parameters as discussed, both OK and BPNN model are validated using the model efficiency factor E and RPD of predicted values. The E and RPD between reference measurements, i.e. accurate or good prediction if RPD and E values are higher than 2.5 and 0.80 respectively by Farifteh et al. (2007), suggest an accurate to good prediction. It is observed from Table 2 that BP method (with E=0.8578 and RPD=15.06) predicts an accurate SWC nearly similar to that of OK (with E=0.8294 and RPD=16.76), whereas different performances occurred in predicting EC.

BPNN method with much higher E =0.8044 and RPD=3.54 performs well while OK with E=0.7793 and RPD=0.39 performs poorly in EC prediction. In addition to MAE% and RMSE%, it is considered that OK performs slightly better, but differences between BPNN and OK methods in predictions are not significant for both variables based on the analysis of covariance (ANOVA) test (P >0.05).

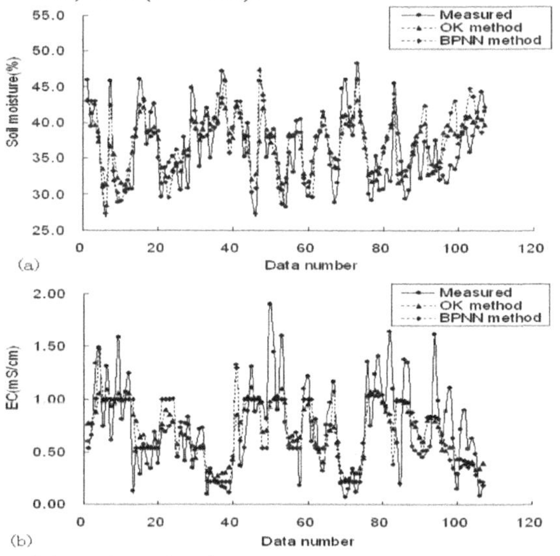

Fig.3: Comparison of the estimated and measured values with both methods for (a) SWC and (b) EC.

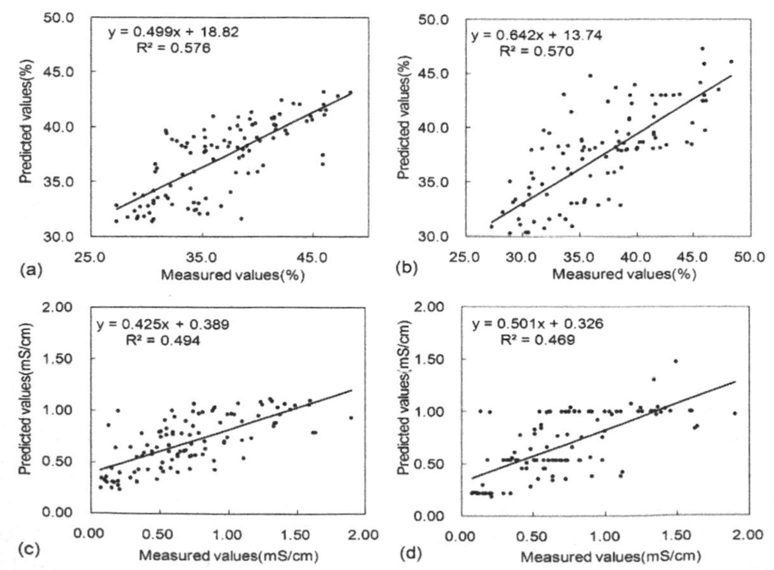

Fig.4: Plots of the measured versus predicted values and the fitted regression line and equation for (a) SWC with OK, (b) SWC with BPNN, (c) EC with OK and (d) EC with BPNN method.

Table 2 Performance of predicted values using OK and BPNN methods

Soil property	MAE	MAE%	RMSE	RMSE%	E	RPD
SWC (%)						
OK method	2.88	7.84	3.45	9.39	0.8294	16.76
BPNN method	2.79	7.63	3.49	9.51	0.8578	15.06
EC (mS/cm)						
OK method	0.24	32.67	0.30	41.9	0.7793	0.39
BPNN method	0.23	32.32	0.31	42.7	0.8044	3.54

3.3 Trend maps of SWC and EC by NK and OK

As explained in section 2.3, the data nonlinear trend can be estimated by neural kriging (NK) better than the ordinary kriging (OK) estimator if neural networks are coupled with kriging. The NK approach presented here for spatial estimation is the result of the integrated of two different techniques: BPNN evaluation and ordinary kriging (OK).

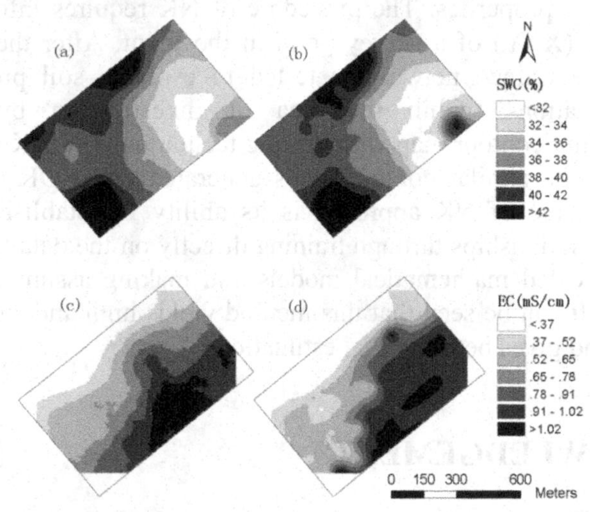

Fig.5: Trend maps obtained from estimated values of (a) SWC by OK, (b) SWC by NK, (c) EC by OK and (d) EC by NK method.

Trend maps of soil moisture and salinity produced by OK and NK are illustrated in Fig.5. Positive values show areas where SWC or EC predictions are higher and negative values represent areas where prediction values are lower. By looking at the maps of Fig.5, differences between results of OK and NK are not significant. In addition, the smoothed contour maps in Fig.5a and Fig. 5b display quite similar patterns with low soil moisture in the eastern section and high in the western and southern parts of the study area, whereas contour maps in Fig.5c and Fig. 5d with high salinity in the eastern section and low in the western sections including northwestern and southwestern parts. Because of the research field with a 2-4% slope northwest to southeast and soil texture with high sand content and permeability, salt leaching with irrigation and upward transport with evaporation are frequent. This may be result in rapid salt leaching and accumulation in the topsoil in southeastern section of this field. Therefore, water flooding for irrigation should be reduced to a great extent and water-saving irrigation should be promoted in irrigation areas.

4. CONCLUSIONS

The key aim of the work is to contribute to the problem of spatial estimation of soil properties with a novel solution, through the combined utilization of statistical, geostatistical and artificial neural network (ANN) techniques. The approach of neural kriging(NK), coupled neural network (NN) with ordinary kriging (OK), has a great potential for predicting and mapping soil properties. The procedure of NK requires information on the coordinates (X, Y) of a survey point in the input. After the completion of training, the trained network is tested to estimate soil properties for all sample locations within the area of investigation by producing a corresponding contour map with kriging technique. The NK results compare very well with similar contour maps generated using OK techniques. The main advantage of NK approach is its ability in establishing patterns or nonlinear relationships through training directly on the data without building any complicated mathematical models and making assumptions on spatial variations. It can be seen that this method yields high and significant spatial relations and gives better spatial estimations.

ACKNOWLEDGEMENTS

This work is supported in part by grants from the National Natural Science Foundation of China (No.70673104) and from the Xinjiang Bingtuan Science & Technology Research Program of China (No.2007YD24、2006YD43 and 2006GJS13). The authors also wish to thank the key oasis eco-agriculture laboratory of Xinjiang Bintuan for offering research workstation.

REFERENCES

Boken, V. K., Hoogenboom G., Hook, J.E., Thomas, D.L., Guerra, L.C., Harrison, K.A.. Agricultural water use estimation using geospatial modeling and a geographic information system, Agric. Water Manage, 2004, 67: 85–199.

Brocca, L., Morbidelli, R., Melone, F., Moramarco, T.. Soil moisture spatial variability in experimental areas of central Italy, Journal of Hydrology, 2007, 333: 356–373.

Emery, X. and Ortiz J. M.. Weighted sample variograms as a tool to better assess the spatial variability of soil properties. Geoderma, 2007, 140: 81–89.

Farifteh, J., Van der Meer, F., Atzberger, C., Carranza, E.J.M.. Quantitative analysis of salt-affected soil reflectance spectra: A comparison of two adaptive methods (PLSR and ANN). Remote Sensing of Environment, 2007, 110: 59–78.

Ferreyra, R.A., Apezteguia, H.P., Sereno, R, Jones, J.W.. Reduction of soil water spatial sampling density using scaled semivariograms and simulated annealing. Geodenna, 2002, 110: 265–289.

Gotway, C.A., Ferguson, R.B., Hergert,G.W., Peterson, T.A.. Comparison of kriging and inverse-distance methods for mapping soil parameters. Am. J. Soil Sci., 1996, 60: 1237–1247.

Huang, W.R. and Foo, S.. Neural network modeling of salinity variation in Apalachicola River. Water Research, 2002, 36: 356−362.

Kitamura, Y., Yano, T., Honna T., Yamamoto S. and Inosako, K.. Causes of farmland salinization and remedial measures in the Aral Sea basin—Research on water management to prevent secondary salinization in rice-based cropping system in arid land. Agricultural Water Management, 2006, 85: 1–14.

Koike, K., Matsuda, S., Gu, B.. Evaluation of interpolation accuracy of neural kriging with application to temperature-distribution analysis. Mathematical Geology, 2001, 33: 421-448.

Li, Y., Shi,. Z., Wu, C.F., Li, H.X., Li, F.. Improved prediction and reduction of sampling density for soil salinity by different geostatistical methods. Agricultural Sciences in China, 2007, 6: 832–841.

Patel, R.M., Prasher, S.O., Goel, P.K., Bassi, R.. Soil salinity prediction using artificial neural networks. J. Am. Water Resour. Assoc, 2002, 38: 91–100.

Rizzo, D.M., Dougherty, D.E.. Characterization of aquifer properties using artificial neural networks: neural kriging. Water Resources Research, 1994, 30: 483–497.

Robinson, T.P. and Metternicht, G.. Testing the performance of spatial interpolation techniques for mapping soil properties. Computers and Electronics in Agriculture, 2006, 50: 97–108.

Sarangi, A., Singh M., Bhattacharya, A.K., Singh, A.K.. Subsurface drainage performance study using SALTMOD and ANN models. Agricultural Water Management, 2006, 84: 240–248.

ARTIFICIAL NEURAL NETWORK ANALYSIS OF IMMOBILIZED LIPASE CATALYZED SYNTHESIS OF BIODIESEL FROM RAPESEED SOAPSTOCK

Yanjie Ying [1] ,Ping Shao [2] , Shaotong Jiang [3] , Peilong Sun [2,*]

1 College of Enformation Engineering, Zhejiang University of Technology, Hangzhou,China, 310014.

2 College of Biological and Environmental Engineering, Zhejiang University of Technology, Hangzhou,China, 310014.

3 School of Biotechnology and Food Engineering, Hefei University of Technology,Hefei, China,230009.

** Corresponding author, Address: College of Biological and Environmental Engineering, Zhejiang University of Technology, Hangzhou,China, 310014. P. R. China, Tel: +86-571-88320604, Fax: +86-571-88320345. Email: pingshao325@yahoo.com.cn*

Abstract: Refined vegetable oils are the predominant feedstocks for the production of biodiesel. However, their relatively high costs render the resulting fuels unable to compete with petroleum-derived fuel. Artificial neural network (ANN) analysis of immobilized Candida rugosa lipase (CRL) on chitosan catalyzed preparation of biodiesel from rapeseed soapstock with methanol was carried out. Methanol substrate molar ratio, enzyme amount, water content and reaction temperature were four important parameters employed. Back-Propagation algorithm with momentous factor was adopted to train the neural network. The momentous factor and learning rate were selected as 0.95 and 0.8. ANN analysis showed good correspondence between experimental and predicted values. The coefficient of determination (R2) between experimental and predicted values was 99.20%. Biodiesel conversion of 75.4% was obtained when optimum conditions of immobilized lipase catalysed for biodiesel production were methanol substrate molar ratio of 4.4:1, enzyme amount of 11.6%, water content of 4% and reaction temperature of 45℃. Methyl ester content was above 95% after short path distillation process. Biodiesel conversion was increased markedly by neural network analysis.

Keywords: rapeseed soapstock, artificial neural network, biodiesel, immobilized lipase

---

Please use the following format when citing this chapter:

Ying, Y., Shao, P., Jiang, S. and Sun, P., 2009, in IFIP International Federation for Information Processing, Volume 294, *Computer and Computing Technologies in Agriculture II, Volume 2*, eds. D. Li, Z. Chunjiang, (Boston: Springer), pp. 1239–1249.

1. INTRODUCTION

Biodiesel, namely fatty acid alkyl esters, has become a new kind of clean burning fuel that can be used as a mineral diesel substitute for engines produced from renewable sources such as vegetable oils and fats which are mainly constituted by triglycerides(Marchetti et al., 2007; Michael et al., 2005). The major problem for production of biodiesel is the price of raw material, which accounts for about 70% of the total costs. Therefore, researchers are always looking for the suitable materials to produce biodiesel on a larger scale. Soapstock(SS), a byproduct of the refining of vegetable oils, is an important biodiesel feedstock. SS is generated at a rate of about 5% of the volume of crude oil refined, which consists of a heavy alkaline aqueous emulsion of lipid(Jon et al,. 2005; Ma et al,. 1999; Haas et al,. 2000) There are several uses for this waste, such as the production of soaps or acid oil, and more recently the production of biodiesel. Many attempts have been made to develop an enzymatic process using lipase as catalyst(Haas et al,. 2003; Shashikant et al,. 2005; Shao et al,. 2006).

However the components of SS were so complicated that it was difficult to obtain better results when lipase used in production of biodiesel. The optimization processes involved labor-intensive and low conversion of fatty acid methyl ester. Artificial neural network (ANN) was a mathematical algorithm which had the capability of relating the input and output parameters without requiring a prior knowledge of the relationships of the process parameters(Balaraman et al,.2005). This meaned a short computing time and a high potential of robustness and adaptive performance. The ANN was able to model chemical processes based on linear or non-linear dynamics.

ANN was now the most popular artificial learning tool in biotechnology, with applications ranging from pattern recognition in chromatographic spectra and expression profiles, to functional analyses of genomic and proteomic sequences. The use of advanced non linear data analysis tools such as ANN was often used in food science (Shao et al,. 2007; Montague et al,. 1994). Lou et al. used an artificial neural network (ANN) method, computer model system, which match the functionality of the brain in a fundamental manner, represented the nonlinearities in a much better way (Lou et al,. 2001). Few studies on the optimization of predicting biodiesel conversion from SS have appeared in the literature. There appeared to be a need for the optimization of lipase synthesis for production of biodiesel by ANN analysis.

In this study, it involved first saponification of the rapeseed soapstock followed by acidification to produce free fatty acids. An artificial neural network analysis of immobilized lipase on chitosan catalyzed preparation of biodiesel from rapeseed soapstock was carried out. This would be very interesting for further applications of biochemical processes on a larger scale, which do not require any kind of mechanistic premises but only input and output variables.

2. MATERIALS AND METHOD

2.1 Materials and reagents

Candida rugosa lipase was purchased from Sigma Chemical Co. (USA). The activity was measured by titrating FA liberated from olive oil with 50mM KOH as described previously. One unit was defined as the amount of enzyme that liberated 1 μmol FFA per minute.

Chitosan powder was obtained from Golden Shell Biochemical Co.Ltd. The material was obtained from prawn's shells with a degree of deacetylation of 85%. Glutaraldehyde was purchased from Fluka.

All other chemicals and reagents used were of analytical grade.

Rapeseed soapstock was supplied by Hangzhou Oil and Fat Co. Ltd. (Hangzhou, China).

Pope2# wiped-film molecular still (Pope Scientific, Inc. USA).

2.2 Immobilized of lipase to chitosan beads

Chitosan powder of 3% (w/v) was completely dissolved in 1% (v/v) acetic acid. This solution was poured into a coagulant bath of 1N sodium hydroxide solution containing 26% (v/v) ethanol under stirring to form spherical gels and allowed to stand for 3h. chitosan was obtained by filtration and rinsed with distilled water until neutrality(Foresti et al,. 2007). One gram of chitosan beads were mixed with 3mL of 0.01% (v/v) glutaraldehyde was added to the beads. After 20min, the supernatant was removed and 3ml of 0.5% (v/v) lipase in deionized water was added to the beads and allowed to react for 45min. Finally, the beads were washed thrice in deionized water. The beads were resuspended in deionized water and stored at 4℃.

2.3 Saponification and acidification

Saponification of soapstock was carried out with the use of 0.5 M sodium hydroxide followed by acidification. Acidification of SS was essential for the reaction, which destroy the emulsion of the complicated mixtures. SS was acidified to acid oil in a 500ml round bottom flask, which required the addition of sulfuric acid. The pH of the mixture was adjusted to 2-3 so as to fully convert the soap to fatty acid. In addition, the acidification was carried out at high temperature of 90℃ for 1.5h. The mixture was settled in separation funnel to remove the bottom fraction, then top was washed to neutrality and dried. Its FFA content was determined by a standard titrimetry method. This acid oil had an initial acid value of 186 mg KOH/g corresponding to a FFA level of 96%.

2.4 Immobilized lipase catalyzed reaction

Experiments were conducted in a laboratory-scale setup which consisted of 100ml glass flasks with condensation tube. The flasks were kept in a water bath maintained at some temperature. Immobilized lipase was added the mixture of oil and methanol. The molar amount of the oil was calculated from its saponification value. The contents were stirred after tightly closing the tube. The reaction conditions were optimized by ANN analysis. At end of reaction, the enzyme was separated out by filtration and filtrate was washed with distilled water after transferring it to a separating funnel. The ester phase was then dried using anhydrous sodium sulfate and the solvent was removed under reduced pressure.

2.5 Distillation of crude fatty acid methyl esters

The product of the above mentioned reaction was washed and dried for the sake of the distillation using short path distillation(Jiang et al,. 2006). The conditions were determined as follows: evaporating temperature 110℃ at 5.32 Pa, rolling speed of 150 min-1 and feed temperature of 80℃. The light component was the fatty acid methyl esters we required. GC-MS was implemented to characterize the biodiesel.

2.6 Determination of FAME and saponification value

Saponification value was determined by standard method.
FAME composition was determined by the procedure(Shao, et al,. 2007).

2.7 Artificial neural network model

There were various types of ANNs. The type chosen for use in this study was the back propagation learning algorithm that was very powerful in function optimization modeling. In these networks, signals were propagated from the input layer through the hidden layers to the output layer. A node thus received signals via connections from other nodes, or the outside world in the case of the input layer. To train an ANN model, a set of data containing input nodes and output nodes are fed. Once the training was over, ANN was capable of predicting the output when any input similar to the pattern that was has learned was fed. The ANN was tested for the remaining set of experimental data (Balaraman et al,. 2005).

Artificial neural network (ANN) was applied here comprises of three layers to provide a nonlinear mapping, in the BP network used here, the input consists of methanol substrate molar ratio, enzyme amount, water content and reaction temperature. The central composite design was showed in Table 1. All experiments were carried out in a randomised order to minimize the effect of unexpected variability in the observed response due to extraneous factors.

Table 1 Independent variables and their levels for central composite design.

Independent variables	codes	variable levels		
		-1	0	+1
methanol molar ratio	X_1	2:1	4:1	6:1
enzyme amount(%)	X_2	4	8	12
water content(%)	X_3	2	6	10
reaction temperature(℃)	X_4	35	45	55

The node number of network's input layer (iN) was set at 4 and the node number of output layer was oN = 1 corresponding to biodiesel conversion. Each layer had a bias except output layer.

According to the dimension of samples, the training precision should be improved to avoid the training progress becoming "under-fitting". Large values of training epochs, hidden layer's node number (hN) and learning rate (lr) were imported to reach this aim. They were set as epochs = 10000, hN = 6 and lr = 0.8. So the number of network's weights that need be identified was decided(Haykins 1994):

$$W = (iN + 1)\, hN + (hN + 1)\, oN = (4 + 1)\, 8 + (8 + 1)\, 1 = 49 \qquad (1)$$

The number of sample subset for training was assumed as Ntrain and the number of sample subset for test was assumed as Ntest .We could calculate the ratio between Ntrain and Ntest (ropt) just using the equation:

$$r_{opt} = \frac{N_{train}}{N_{test}} = 1 - \frac{\sqrt{2W - 1} - 1}{2W - 2} \qquad (2)$$

Now the total number of samples is known :

$$Ntrain + Ntest = 27 \qquad (3)$$

Computing Equation (1), (2) and (3), it was easily gotten: Ntrain = 24.23. Because the number should always be an plus integer, the value was changed as Ntrain = 25, which means we should choose 25 units of samples as the training samples and the other 2 units as testing samples.

At the same time, to avoid the training progress becoming “over-fitting”, a method called as estimation-followed-by-validation was introduced. The principle of this method was that after each epoch of network's training, samples for testing were imported to validate which the training progress was over-fitting or not.

Back-Propagation algorithm with momentous factor was adopted to train the neural network. Here, the momentous factor was selected as mom = 0.95. Random value in the range [-1, 1] was used as the initial weight of network. Mean Square Error (MSE) was chosen as the performance function. In ANN computation, we used the toolbox Matlab 6.5.

ANN tends to implicitly match the input vector to the output vector. ANN had been applied for the purpose of simulation on the same experimental data used for ANN analysis (Table 2).

Table 2 Central composite design and experiment data.

Run	X_1	X_2	X_3	X_4	True model (%)	ANN model (%)	Error (%)
1	-1	-1	0	0	40.30	42.251	4.6176
2	-1	1	0	0	42.60	42.298	0.7099
3	1	-1	0	0	38.10	38.038	0.1619
4	1	1	0	0	55.80	53.225	4.6148
5	0	0	-1	-1	43.20	44.653	3.3630
6	0	0	-1	1	47.40	45.711	3.5625
7	0	0	1	-1	54.80	55.039	0.4362
8	0	0	1	1	53.70	53.174	0.9789
9	-1	0	0	-1	45.60	45.217	0.8391
10	-1	0	0	1	47.90	48.131	0.4830
11	1	0	0	-1	50.40	49.766	1.2584
12	1	0	0	1	53.50	55.636	3.9930
13	0	-1	-1	0	35.90	35.288	1.7054
14	0	-1	1	0	42.70	41.892	1.8934
15	0	1	-1	0	34.20	33.151	3.0671
16	0	1	1	0	49.70	49.857	0.3168
17	-1	0	-1	0	38.20	39.307	2.8971
18	-1	0	1	0	46.10	46.143	0.0931
19	1	0	-1	0	36.40	37.279	2.4146
20	1	0	1	0	48.70	48.591	0.2231
21	0	-1	0	-1	42.20	42.843	1.5243
22	0	-1	0	1	42.40	43.146	1.7589
23	0	1	0	-1	44.10	44.295	0.4423
24	0	1	0	1	46.30	47.513	2.6189
25	0	0	0	0	62.90	63.504	0.9602
26	0	0	0	0	64.40	63.504	1.3914
27	0	0	0	0	63.60	63.504	0.1511

3. RESULTS AND DISCUSSION

3.1 Establishment of model

Artificial neural network analysis was employed to solve a wide variety of problems in science and engineering(Montague et al,. 1994). Unlike other modeling techniques such as simultaneous heat and mass transfer, kinetic models and regression models, an ANN could accommodate more than two variables to predict two or more output parameters. ANN differed from conventional programs in their ability to learn about the system to be modeled without a need of any prior knowledge on the relationships of the process variables.

After neural networks were trained successfully, all domain knowledge extracted out from the existing samples was stored as digital forms in weights associated with each connection between neurons. The relationship of the model was expressed as $Y = purelin\ (W21 Tansig\ (W11 p_1 + b1) + b2)$. Fig.1 illustrated the scheme of three layers neural network. Comparison of experimental data with simulated data was shown in Fig.3. The coefficient of determination (R2) between experimental and predicted values was 99.20%. The connection weights value (W11) and bias value (b1) between input layer and hidden layer, the connection weights value (W12) and bias value (b2) between hidden layer and output layer were as follows, respectively.

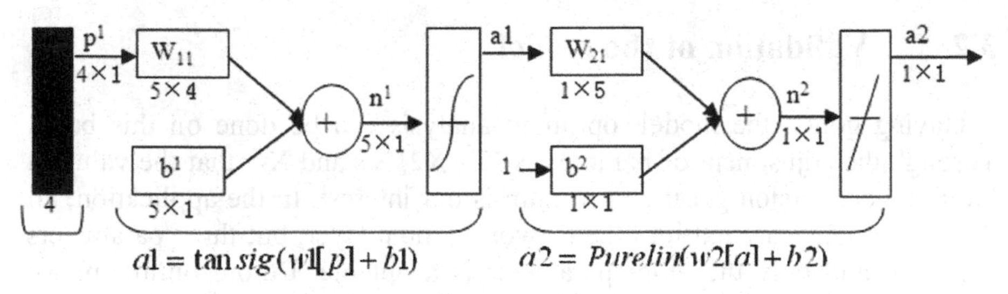

Fig.1. The scheme of three layers neural network.

W11=[0.6791 -0.1797 -1.1696 -0.0430; 0.5828 -1.8350 1.4063 0.0115; 1.5787 0.9839 -0.5267 -0.9566; -1.3268 1.2835 -0.4042 0.9861; 1.7453 -2.6483 0.6485 -0.0039]T

W21=[-0.2203 -0.1000 0.0438 0.0543 0.1562]

b1=[-1.2544 -0.7808 0.3163 -0.1443 1.2315]T

b2=[0.2372]

The effects of methanol concentration on extent of predicted conversion was shown in Figure.2 (A). It was indicated that the methyl ester yield was sensitive to the methanol molar ratio. An increase in methyl ester yield was observed with the increasing of methanol molar ratio at first. But the trend was reversed when the ratio reached a certain value. It could be interpreted that, under a certain amount, the methanol was used to improve the solubility of water in oil and reaction mixture would become well mixed. So with the increasing of methanol amount, methyl ester first increased and then decreased as a result of the decrease of enzyme activity caused by excessive methanol. The effect of enzyme amount on extent of predicted conversion was shown in Figure.2 (B). The biodiesel conversion increased with the increased enzyme amount because of improvement of lipase activity. The optimum methanol molar ratio and enzyme amount for the maximum methyl ester conversion was around 4.4:1 and 11.6%, respectively.

Influence of water content on production of methyl ester showed significant variation both above and below the optimum values (Fig.2 (C)). With the increase of water, we observed an increase at first and decrease subsequently in the amount of methyl ester; It was likely that Candida rugosa lipase was inactivated by methanol due to insufficient water in the reaction mixture and methyl esterification reaction was inhibited when much water content above 4% (Manohar et al,. 2003). Water content about 4% was therefore suggested. When lipase powder was used as the catalyst, the reaction rate was much lower. In this experiment, the amount of lipase powder used was used with some amount of water.

3.2 Validation of the model

Having gotten the model, optimum analysis can be done on this basis. Through the adjustment of parameters X1, X2, X3 and X4, that the value of biodiesel conversion get thc maximum is our interest. In the application, all these parameters are put into the network's input layer, but three parameters are fixed and only the other parameter is adaptable. Before optimum, we choose an experiment data as initial data. The simulation shows that when X1 = 0.2, X2 = 0.9, X3 = -0.5 and X4 = 0, biodiesel conversion has the maximum value 0.7720.

According to ANN result, an experiment with an methanol substrate molar ratio of 4.4:1, enzyme amount of 11.6%, water content of 4% and reaction temperature of 45℃ was conducted in order to investigate the effect of ANN. The experiment was carried out at the optimized conditions. Methyl ester conversion of 75.4% was obtained and was in good agreement with the predicted one. The accuracy of the model was validated with triplicate experiments under the aforementioned optimal reaction conditions. As a result, the model was considered to be accurate and reliable for predicting the conversion of methyl ester.

According to previous result, an experiment with molecular distillation was conducted in order to obtain high content of methyl ester. The content and recovery of biodiesel was above 95%, 89.5% after short path distillation respectively.

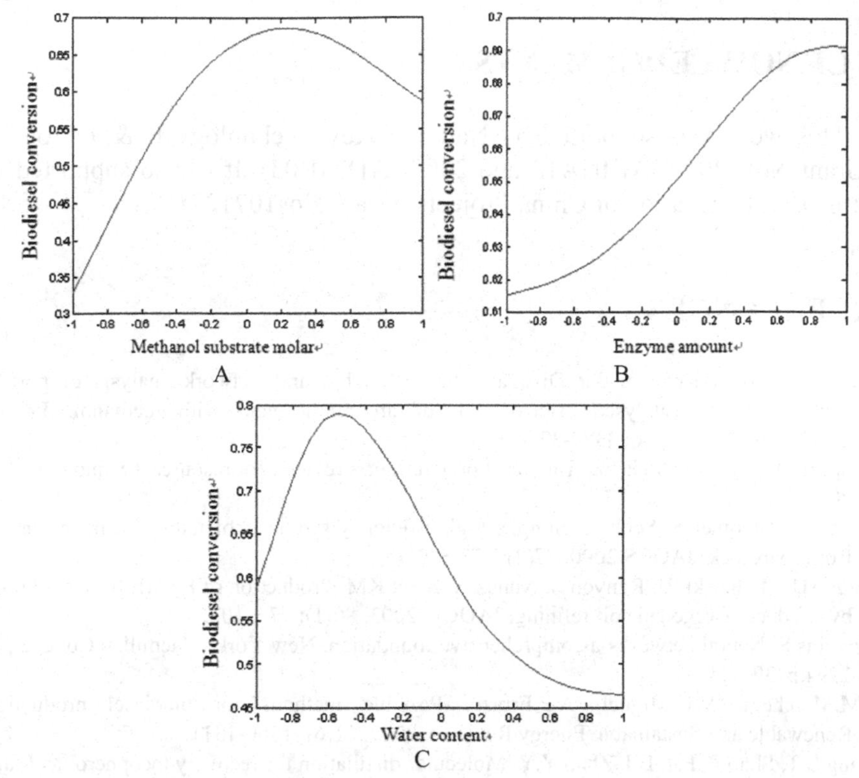

Fig.2. The effects of methanol substrate molar (A), enzyme amount (B) and water content (C) on extent of predicted conversion.

4. CONCLUSIONS AND FUTURE WORKS

From the above analysis, the main outcomes can be outlined as follows:

(1) Artificial neural network was used to optimize the production of biodiesel. The coefficient of determination (R2) for the model is 99.20%. Biodiesel conversion of 75.4% was obtained when optimum conditions of immobilized lipase catalyzed for biodiesel production were methanol substrate molar ratio of 4.4:1, enzyme amount of 11.6%, water content of 4% and reaction temperature of 45℃.

(2) Methyl ester content was above 95% after short path distillation. Validation experiments verified the availability and the accuracy of the model. The predicted value was in agreement with the experimental value.
(3) Future work is using other lipases or combined with Candida rugosa lipase to develop an integrated reaction system for production of biodiesel with low cost.

ACKNOWLEDGEMENTS

This work was supported by National Key Technology R &D Program (Grant Nos. 2006BAD05A12 and 2007BAD34B03). It is also supported by Ministry of education of China Projects (Grant Nos107127).

REFERENCES

Balaraman Manohar, Soundar Divakar. An artificial neural network analysis of porcine pancreas lipase catalysed esterification of anthranilic acid with methanol. Process Biochemistry. 2005, 40: 3372-3376.

Fangrui Ma, Milford A.Hanna. Biodiesel production: a review. Bioresource Technology.1999, 70: 1-15.

Haas MJ, Bloomer S, Scott K. Simple, high-efficiency synthesis of fatty acid methyl esters from soapstock. JAOCS 2000, 77(4): 373 - 379.

Haas MJ, Michalski PJ, Runyon S, Nunez A, Scott KM. Production of FAME from acid oil, a by-product of vegetable oil refining. JAOCS, 2003. 80(1): 97 - 102.

Haykins S. Neural networks-a comprehensive foundation. New York: Macmillan College pub, 1994.p. 397-443.

J.M.Marchetti, V.U.Miguel, A.F.Errazu. Possible methods for biodiesel production. Renewable and Sustainable Energy Reviews. 2007, 11(6): 1300-1311.

Jiang S.T, Shao P, Pan L.J, Zhao Y.Y. Molecular distillation for recovery tocopherol and fatty acid methyl esters from rapeseed oil deodoriser distillate. Biosystems Engineering, 2006, 93(4): 383-391.

Jon Van Gerpen. Biodiesel processing and production. Fuel Processing Technology. 2005, 86(10): 1097-1197.

Lou W. G., Nakai S. Application of artificial neural networks for predicting the thermal inactivation of bacteria: a combined effect of temperature, pH and water activity. Food Research International. 2001. 34: 573-579.

Manohar B, Divakar S. Application of surface plots and statistical designs to selected lipase catalysed esterification reactions. Process Biochemistry. 2003, 39(7): 847-854.

Michael J.Haas. Improving the economics of biodiesel production through the use of low value lipids as feedstocks: vegetable oil soapstock. Fuel Processing Technology. 2005, 86: 1087-1096.

M.L.Foresti, M.L. Ferreira. Chitosan-immobilized lipases for the catalysis of fatty acid esterifications. Enzyme and Microbial Technology. 2007, 40: 769-777.

Montague G, Morris JN. Neural network contributions in biotechnology. Trends Biotech. 1994, 12: 312-324.

P.Shao, S-T Jiang, Y Li. Molecular distillation and enzymatic reaction for concentration of vitamin E from rapeseed oil deodorizer distillate. Scientia Agricultura Sinica. 2006, 39(12): 2570-2576.

Shao P, Jiang S-T, Ying Y-J. Optimization of molecular distillation for recovery of tocopherol from rapeseed oil deodorizer distillate using response surface and artificial neural network models. Food and Bioproducts Processing. 2007, 85: 1-8.

Shashikant Vilas Ghadge, Hifjur Raheman. Biodiesel production from mahua (Madhuca indica) oil having high free fatty acids. Biomass and Bioenergy. 2005, 28:601 - 605.

A HYBRID APPROACH OF NEURAL NETWORK WITH PARTICLE SWARM OPTIMIZATION FOR TOBACCO PESTS PREDICTION

Jiake Lv[1,2] ,Xuan Wang[1,2] , Deti Xie[1,3] , Chaofu Wei[1,3*]

[1] *Chongqing Key laboratory of digital agriculture,Chongqing,400716,P.R.China*

[2] *College of Computer and Information Science, Southwest University, Chongqing, 400716,P.R.China*

[3]*College of Resources and Environment, Southwest University, Chongqing, 400716,P.R.China*

[*] *Corresponding author, Address: College of Resources and Environment, Southwest University, Chongqing, 400716, P.R.China, Tel: +86-23-13983377663 ,Email: lk@swu.edu.cn*

Abstract: Forecasting pests emergence levels plays a significant role in regional crop planting and management. The accuracy, which is derived from the accuracy of the forecasting approach used, will determine the economics of the operation of the pests prediction. Conventional methods including time series, regression analysis or ARMA model entail exogenous input together with a number of assumptions. The use of neural networks has been shown to be a cost-effective technique. But their training, usually with back-propagation algorithm or other gradient algorithms, is featured with some drawbacks such as very slow convergence and easy entrapment in a local minimum. This paper presents a hybrid approach of neural network with particle swarm optimization for developing the accuracy of predictions. The approach is applied to forecast Alternaria alternate Keissl emergence level of the WuLong Country, one of the most important tobacco planting areas in Chongqing. Traditional ARMA model and BP neural network are investigated as comparison basis. The experimental results show that the proposed approach can achieve better prediction performance.

Keywords: tobacco pests prediction, particle swarm optimization, neural network, learning algorithm

---

Please use the following format when citing this chapter:

Lv, J., Wang, X., Xie, D. and Wei, C., 2009, in IFIP International Federation for Information Processing, Volume 294, *Computer and Computing Technologies in Agriculture II, Volume 2*, eds. D. Li, Z. Chunjiang, (Boston: Springer), pp. 1251–1260.

1. INTRODUCTION

Alternaria alternate Keissl and *Ascochyta gossipii* are major pests in southwest china tobacco planting areas, causing damage to the tobacco by larval stages feeding on flower buds and berries (Wang, 2004). Accurate forests of the pests emergence in advance can help determining a timely treatment schedule. In general, there are some *complex and non-linear* relationships between emergence level and several environmental factors like rainfalls, solar radiation, temperature, air humid etc (Reichert and Omlin, 1997). Modeling such relationships with conventional techniques such as time series, regression analysis, ARMA model etc, has been attempted before with variable success (Shaffer and Gold, 1985; Roditakis and Karandinons, 2001; Satake and Ohgushi et al., 2006). However, these models are based on statistical methods and show some deficiency in the presence of an abrupt change in environmental or sociological variables which are believed to affect pests emergence. In addition, the employed techniques for those models use a large number of complex relationships, require a long computational time, and may result in numerical instabilities, thus not achieving the desired accuracy.

During the past decade, neural networks and in particular, feed forward backward propagation perceptions, were widely applied in different fields (Hippert and Pedreira et al., 2001). The multi-layer perceptions could be trained with non-linear transfer to approximate and accurately generalize virtually any smooth, measurable function while taking no prior assumptions of the data distribution (Russell and Norvig, 2003). Several characteristics, including built-in dynamism in forecasting, data-error tolerance and lack of requirements of any exogenous input, make neural networks attractive for use in pests prediction. Park et.al used artificial neural network to predict Korean pine trees insect pests hazard rating (Park and Chung, 2006). Tourenq et.al adopted neural network to forecast rice crop damage of Camargue,France and the results show better performance than the traditional regression model(Tourenq and Aulaginer et al., 1999). A.Drake developed and used a multi-layer perception ANN to model *Alternaria sp* emergence in Queensland, Australia(Drake, 2001). Although the back propagation (BP) algorithm is commonly used in recent years to perform the training task, some drawbacks are often encountered in the use of this gradient-based method. They include: the training convergence speed is very slow and easy entrapment in a local minimum(Carpinteiro and Otavio et al., 2004).

Particle swarm optimization(PSO), with capability to optimize complex numerical functions, is initially developed as a tool for modeling social behavior(Kennedy, 1997).Unlike other evolutionary algorithms, PSO relies on *cooperation* rather than *competition* and can provide promising solution due to the ability of performing global search for best forecast model(Clerc,

1999; Shi and Eberhart, 1999).Moreover, it is ideally suited for solving discrete and/or combinatorial type optimization problems(Russell and Norvig, 2003).In this paper, a hybrid approach of neural network with particle swarm optimization is developed by adopting PSO to train multi-layer perceptions. To forecast *Alternaria alternate Keissldaily* emergence level of the WuLong country, one of the most important tobacco planting areas in Chongqing. Comprise wih Traditional ARMA model and BP neural network, the results show that the proposed approach can achieve better prediction performance.

This paper is organized as follows. Following the introduction, some preliminaries and notations are described in Section 2.Hybrid approach and detailed procedure used in this study are depicted in Section 3.In Section 4, experimental results and discussion are illustrated. Conclusions are finally made in Section 5.

2. PRELIMINARIES AND NOTATIONS

2.1 Particle swarm optimization

Particle swarm optimization (PSO) is a kind of algorithm to search for the best solution by simulating the movement and flocking of birds (Eberhart and Shi, 2000). The algorithm works by initializing a flock of birds randomly over the searching space, where every bird is called as a "particle" (Shi and Eberhart, 1999). These "particle" fly with a certain velocity vector, based on its momentum and the influence of its best position (P_b) as well as the best position of its neighbors (p_g), and then compute a new position that the "particle" is to fly to. Supporting the dimension for a searching space is D , the total number of particles is n , the position of the i th particle can be expressed as vector $X_i = (x_{i1}, x_{i2},..., x_{iD})$; the best position of the i th particle being searching until now is denoted as $P_{ib} = (p_{i1}, p_{i2},..., p_{iD})$; and the best position of the total particle swarm being searching until now is denoted as vector $V_i = (v_{i1}, v_{i2},... v_{iD})$ .Then the PSO algorithm is described as :

$$v_{id}(t+1) = v_{id}(t) + c_1 * rand(\ ) * [p_{id}(t) - x_{id}(t)] + c_2 * rand(\ ) * [p_{gd}(t) - x_{id}(t)] \quad (1)$$

$$x_{id}(t+1) = x_{id}(t) + v_{id}(t+1) \quad 1 \le i \le n \ 1 \le d \le D \quad (2)$$

Where c_1, c_2 are the acceleration constants with positive values; $rand(\ )$ is a random number between 0 and 1; w is the inertia weight. In additional to the parameters c_1, c_2 , the implementation of the original algorithm also requires placing a limit on the *velocity* (v_{max}).

In the literature (Shi and Eberhart, 1998; Shi and Eberhart, 1999), several selection strategies of inertial weight w have been given. Generally, in the beginning stages of algorithm, the inertial weight w should be reduced rapidly, when around optimum, the inertial weight w should be reduced slowly. So in this paper, we adopted the following selection strategy:

$$w = \frac{w_s - w_e}{Max_{iter} - iter} + w_e \quad (3)$$

Where w_s the starting is weight, w_e is the finial inertia, Max_{iter} is the maximum number of iterations and *iter* is current iteration. After adjusting the parameters w and V_{max} , the PSO can achieve the best search ability.

2.2 Neural networks

A neural network (NN) of arbitrary topology can be represented by a directed graph $N = (V, E, w)$ (Ahn and Cho et al., 2000). V is the set of nodes, which is divided into the subset V_1 of input nodes, the subset V_H of hidden nodes and the subset V_O of output nodes. $E \subseteq V \times V$ is the set of connections. Each connection $(j,i) \in E$ is associated with a weight $w_{ij} \in \Re$ .For each unit $i \in V$, let us define its "projective" field $P_i = \{j \in V | (i,j) \in E\}$ and its "receptive" filed $R_i = \{j \in V | (j,i) \in E\}$.We denote by p_i and r_i the cardinality of P_i and R_i respectively.

Every non-input node $i \in V_H \cup V_O$ receives from its receptive field R_i a net input given by

$$u_i = \sum_{j \in R_i} w_{ij} y_j \quad (4)$$

Where y_j represents the output value of node j , and sends to its prospective field P_i an output equal to

$$y_i = f(u_i) \quad (5)$$

Where f is an arbitrary activation function

Input nodes do no computation: they just transmit an n -dimensional input pattern $x = (x_1, ... x_n)$ .Thus, the output of input node $h \in V_I$ is the h attribute x_h of the input pattern.

3. HYBRID PSO-NEURAL APPROACH FOR PESTS PREDICTION

3.1 The structure of model

In this paper , the principle of hybrid approach is to utilize particle swarm optimization learning algorithm to train the weights of feed forward neural network, which not only can make use of neural network ability to handle the *non-linear* relationships between emergence level and the factors affecting it directly from historical data but also take advantage of strong global searching ability of PSO. Therefore, a multi-layer perception (MLP) neural network is firstly used. As a three-layer connected feed-forward network has been proved to approach any continuous function, in this study, we use a three-layer feed-forward network for pests' emergence level forecast. It is built as shown in Fig.1.

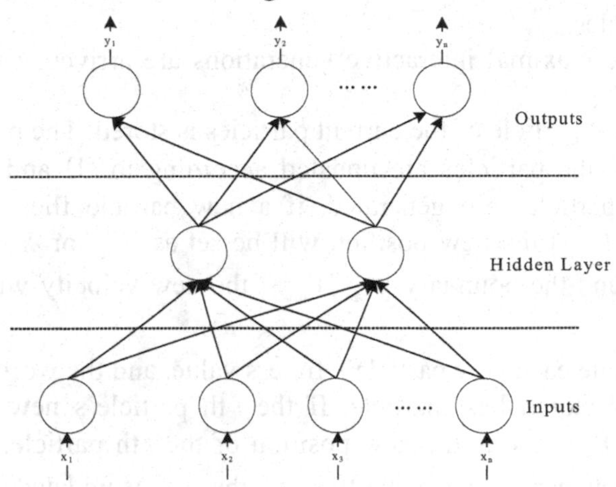

Fig.1. The structure of neural network for pests prediction

In this type of network the first layer is composed of the input variables such as rainfalls, solar radiation, temperature, air humid, and the last layer is composed of the output variables like emergence level of pests. The second layer consists of hidden nodes whose size plays a critical role for neural network *non-linear* mapping. In order to avoid over-fitting and can truly reflect the relationship between emergence level and the factors affecting, we use function in (WANG, 1999) to determine the number of hidden nodes. It is depicted as follows:

$$j = \sqrt{n+m} + a \tag{6}$$

Where j is a hidden node numbers, and n , m represents the number of input nodes and output nodes respectively, $a \in [1,10]$.

3.2 The procedure of hybrid approach

Once the network topology is specified, corresponding weights of each connecting nodes need to be trained. In the study, the total weights of the network will be thought of as the position vector of the particle in the PSO and the weights of each particle vector are initialized randomly and are evolved using the PSO algorithm.

The procedure for the hybrid approach can be summarized as follows:

Step 1: Assign the input variables, output variables and the number of hidden nodes for pests emergence level forecast model

Step 2: Initialize the positions and velocities of a group of particles randomly in the range of $[0,1]$

Step 3: Evaluate each initialized particle's fitness, and P_b is set as the position of the current particles, while p_g is set as the best position of the initialized particles.

Step 4: If the maximal interactive generations are arrived, go to *step 9*, else, go to *step 5*

*Step 5:*The best particle of the current particles is stored. The positions and velocities of all the particles are updated according to (1) and (2), then a group of new particles are generated, if a new particle flies beyond the boundary $[V_{min}, V_{max}]$,the new position will be set as X_{min} or X_{max} ;if a new velocity is beyond the boundary $[V_{min}, V_{max}]$,the new velocity will be set as V_{min} or V_{max}

Step 6: Evaluate each new particle's fitness value, and the worst particle is replaced by the stored best particle. If the i th particle's new position is better than P_{ib} , P_{ib} is set as the new position of the i th particle. If the best position of all new particles is better than P_g ,then P_g is updated.

Step 7: Reduce the inertia weights w according to the selection strategy described in section 2.1 of section 2.

Step 8: If the current p_g is unchanged for ten generations, then go to *step 9*;else, go to *step 4*

Step 9: Output the output variables results.

3.3 Performance measure indicators

In order to evaluate the performance of solutions, root mean squared error (RMSE) and the absolute maximum error(MAXIMAL) measurements, in this paper are used as performance measure indicators.

The root mean squared error(RMSE) is expressed as

$$RMSE = \sqrt{\frac{1}{N}\sum_{i=1}^{N}\left(Z_i - \hat{Z}_i\right)^2} \tag{7}$$

The absolute maximum error (MAXIMAL)is expressed as

$$MAXIMAL = \max_{i=1,\ldots,N}\left|Z_i - \hat{Z}_i\right| \tag{8}$$

Where z_i is the actual electrical load and $\hat{z}_i$ is the forecasted value, N is the number of samples.

4. EXPERIMENTAL RESULTS AND DISCUSSION

Alternaria alternate Keissl is the major insect pests in tobacco planting of southwest china region and emergence levels is greatly influenced by local climate factors. We choose data from WuLong Country Tobacco Corporation in Chongqing, China from May to October of 1996 to that of 2004 as training and testing. All environmental data are normalized into the range between -1 and 1 by using the maximum and minimum values of the variable, and the emergence is classified by 4 level according to china plant pests prediction assessment standard. The whole data set is shown as table 1.

Ten thousand training epochs are adopted as the stopping criteria. The sigmoid function is adopted at the hidden and output nodes. In the PSO training, the number of population is set to 100 while the maximum and minimum velocity value is 0.25 and -0.25 respectively. These values are obtained by trial and error. The w_s and w_e are set 1.0 and 0.4 respectively. By trial and error, it was found that the best values for c_1 and c_2 are 3.0 and 5.0.

For evaluating and comparing the performance of our proposed hybrid PSO-NN approach, traditional BP neural network and statistical autoregressive moving average method for the ARMA model are also utilized as comparison basis. For BP neural network, the learning rate is 0.09 and activation function is sigmoid. Ten thousand training epochs are also adopted as the stopping criteria. All procedures were implemented by using a specific MATLAB toolbox called netlab(Nabney, 2002).ARMA model is built by means of SPSS 14.0.

Fig.2 lists the prediction result using our hybrid PSO-NN approach with true data. BP-Neural network and ARMA model are also compared for the same data.

Table 1. Standardized data set

Year	Month	C_1	C_2	C_3	C_4	C_5	D_1
1996	5	0.12	0.11	-0.16	0.02	-0.19	0001
	6	-0.09	-0.09	-0.00	0.04	0.21	0001
	7	1.00	-0.53	0.36	0.25	0.10	0001
	8	0.43	-0.18	0.80	0.70	0.12	1000
	9	-0.86	0.96	0.90	0.93	-0.17	1000
	10	-0.67	0.28	0.32	0.32	-0.25	0010
1997	5	0.14	-.06	-.17	0.24	0.26	0001
	6	0.10	0.03	-0.32	-0.21	0.20	0001
	7	-0.41	-0.34	0.50	0.48	0.12	0001
	8	-0.07	0.12	0.87	0.90	0.15	0010
	9	-0.86	0.68	0.88	0.91	0.16	0010
	10	-0.67	-0.03	0.04	0.11	-0.23	0001
…	…	…	…	…	…	…	…
2004	5	-0.29	-0.25	-0.14	-0.09	0.17	0001
	6	-0.60	0.81	-0.0	-0.09	0.18	0010
	7	0.07	0.68	1.00	0.95	0.21	0100
	8	-0.39	0.28	0.92	0.86	-0.09	0100
	9	-0.46	0.15	0.01	0.09	0.13	0010
	10	-0.42	-0.56	-1.00	-0.98	0.16	0001

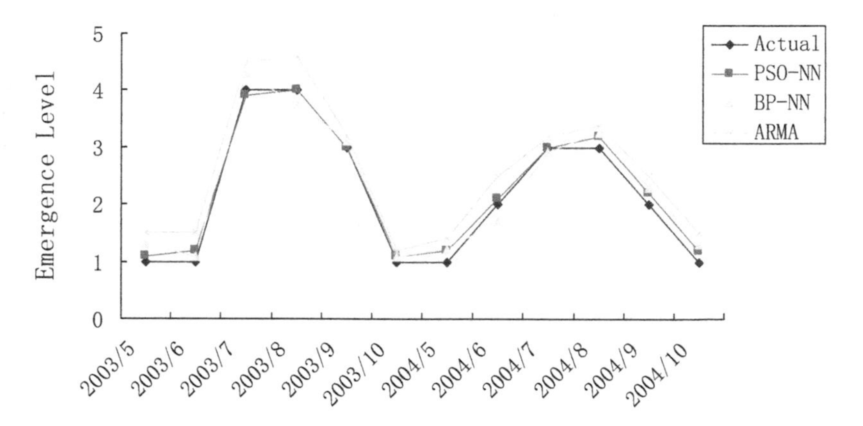

Fig.2 The actual and predictions about the *Alternaria alternate Keissl* emergence level

From Fig.2, it can be worked out that there is only one point at which BP performs a little better than PSO-NN on 2003/6 and none at which ARMA outperforms than PSO-NN. With reference to all individuals, the precision of predictions from PSO-NN is much better when we take the tendency as a whole.

Table 2 shows performance measure indicators values when we apply three different approaches. It can be observed that for both RMSE and MAXIMAL, PSO-NN exhibits better performance in the training process as well as better prediction ability in the validation process than those by BP-NN and ARMA.

Table 2 Performance measure indicators comparison for different forecasting approach

	Training set			Test set		
	PSO-NN	BP-NN	ARMA	PSO-NN	BP-NN	ARMA
RMSE	0.18	0.26	0.41	0.17	0.25	0.42
MAXIMAL	87.41	220.37	237.2	99.76	224.01	267.15

The above results indicate that the hybrid PSO-NN outperforms in pests emergence level forecasting and the prediction is more concise than that of BP-NN and ARMA. It is because PSO can effectively perform global search for better forecast model when it is utilized to train network learning, and avoid easy entrapment in a local minimum.

5. CONCLUSIONS

In this paper, we present a hybrid approach of neural network with practical swarm optimization learning algorithm for pests prediction. The approach is applied to tobacco Alternaria alternate Keissl emergence level and two performance measure indicators including RMSE and MAXIMAL are presented and compared. The results show that prediction is more accurate when compared with the traditionally used BP-based perception and ARMA model. Considering the factual circumstance of pests prediction, there are also some issues that need to be further discussed and improved. For example, some parameters like the maximum and minimum velocity values in PSO training initialization are now determined by trial and error, how to design a learning algorithm for the automated parameters selection will be an important issue in future research.

ACKNOWLEDGEMENTS

This paper was supported by the grants from National Science &Technology Pillar program in the Eleventh Five-year Plan Period (No.2006BAD10A01-02) and Chongqing Tobacco Corporation Development Foundation (NO.2006016).

REFERENCES

Ahn, B. S. and S. S. Cho, et al. (2000). "The integrated methodology of rough set theory and artificial neural network for business failure prediction." Expert Systems with Applications 18 (2): 65-74.

Carpinteiro and A. S. Otavio, et al. (2004). "A hierarchical neural model in short-term load forecasting." Applied soft computing 4 (4): 405-412.

Clerc, M. (1999). The swarm and the queen: towards a deterministic and adaptive particle swarm optimization. Proceeding Congress on Evolutionary Computation, Washiongton DC.

Drake, A. (2001). "Use of remote sensing and ANN in prediction of pests in Queensland." Remote sensing of environment 12 (4): 32-35.

Eberhart, R. C. and Y. Shi (2000). Comparing Inertia Weights and Constriction Factors in Particle swarm optimization. 2000 congress on Evolutionary Computing.

Hippert, H. and C. Pedreira, et al. (2001). "Neural networks for short-term load forecasting: a review and evaluation." IEEE Trans Power System 16 (1): 45-55.

Kennedy, J. (1997). The particle swarm: Social adaptation of knowledge. Proceedings of the 1997 International Conference on Evolutionary Computation, Indianapolis.

Nabney, I. T. (2002). NETLAB:Algorithms for pattern recognition. London, Springer.

Park, Y. and Y. Chung (2006). "Hazard rating of pine trees from a forest insect pest using artificial neural networks." Forest ecology and management 222 : 222-233.

Reichert, P. and M. Omlin (1997). "On the usefulness of over parameterized ecological models." Ecology modeling 95 : 289-299.

Roditakis, N. E. and M. G. Karandinons (2001). "Effects of photoperiod and temperature on pupil diapause induction of grape berry moth lobelia botrana." Physiol Entomol 26 : 329-340.

Russell, S. and P. Norvig (2003). Artificial intelligence: A modern approach, Prentice-Hall International Inc.

Satake, A. and T. Ohgushi, et al. (2006). "Modeling population dynamics of a tea pest with temperature-dependent development: predicting emergence timing and potential damage." Ecology research 21 : 107-116.

Shaffer, P. L. and H. J. Gold (1985). "A simulation model of population dynamics of the coding moth cydia pomonella." Ecology modeling 30 : 247-274.

Shi, Y. and R. C. Eberhart (1998). A modified particle sarm optimizer. IEEE World Conf on Computation Intelligence.

Shi, Y. and R. C. Eberhart (1999). Empirical study of Particle Swarm Optimization. IEEE World Conference on Evolutionary Computation.

Tourenq, C. and S. Aulaginer, et al. (1999). "Use of artificial neural networks for predicting rice crop damage by greater flamingos in the Camargue, France." 120 (2-3): 349-358.

WANG, N. (1999). "The research of hybrid optimization strategy in neural networks." The Journal of Tsinghua University: 66-70.

Wang, G. (2004). "Advances and outlook for forecast work of tobacco disease and insect pests in China." Journal of China tobacco Science 1 : 44-46.

RESEARCH ON OPTIMIZATION OF ENCODING ALGORITHM OF PDF417 BARCODES

Ming Sun[1,*] ,Longsheng Fu[1] ,Shuqing Han[1]
[1] *College of Information and Electrical Engineering, China Agricultural University, Beijing, China, 100083*
[*] *Corresponding author, Address: P. O. Box 63, 17 Tsinghua East Road, Haidian District, Beijing, 100083, P. R. China, Tel:+86-10-62737591, Email: drmingsun@163.com*

Abstract: The purpose of this research is to develop software to optimize the data compression of a PDF417 barcode using VC++6.0. According to the different compression mode and the particularities of Chinese, the relevant approaches which optimize the encoding algorithm of data compression such as spillage and the Chinese characters encoding are proposed, a simple approach to compute complex polynomial is introduced. After the whole data compression is finished, the number of the codeword is reduced and then the encoding algorithm is optimized. The developed encoding system of PDF 417 barcodes will be applied in the logistics management of fruits, therefore also will promote the fast development of the two-dimensional bar codes.

Keywords: Two-dimensional barcode, PDF417, encoding algorithm, visual C++6.0

1. INTRODUCTION

In recent years, the consumption conceptions of the agricultural products such as fruits have changed significantly. People's focus is transferring from the amount to the quality. However, it is difficult to distinguish a high-quality product from a counterfeit or an inferior product from its appearance alone. The immediate concerns are to ensure the quality of fruit products from the market to the consumers and to guarantee consumer benefits; simultaneously, to protect manufactures' economic benefits and to encourage manufacturers to plant healthy fruit products without harmful effects. Currently, however, inferior fruits pass for genuine ones on the market;

Please use the following format when citing this chapter:

Sun, M., Fu, L. and Han, S., 2009, in IFIP International Federation for Information Processing, Volume 294, *Computer and Computing Technologies in Agriculture II, Volume 2*, eds. D. Li, Z. Chunjiang, (Boston: Springer), pp. 1261–1270.

moreover, the difficulty of identifying the manufacturer responsible is compounded by the lack of product marking.

The two-dimensional (2D) barcode, considered a new kind of practical technique in the field of automatic recognition and information carrier, is understood by more and more people, and can be used for marking fruit products. The primary advantage of 2D code is the ability to encode a large quantity of information in a small space. Since PDF417 code is a type of wide application 2D code, we store the basic information of fruits and their manufacturer in a PDF417 code, which is glued or printed on the fruit box. The consumer can request salespeople to identify PDF417 code with the aid of a decoding system and print out the basic information of fruits, their manufacturer, and also the web address to an online database containing further information such as: the time of harvest, fertilizer information, national national attestation with a prize certificate, etc. Alternatively, the consumer can also make use of a digital camera or a web camera to take a PDF417 code picture and read its information using a decoding software. Since 2D barcodes permit faster and more accurate recording of information, work in process can move quickly and be tracked precisely. Quite a bit of time can be spent tracking down the location or the status of projects, folders, instruments, materials, or anything else that moves within an organization. A 2D barcode can help you keep better track of them so you can save time and respond more quickly to inquiries and changes. Thus, 2D barcodes can improve operational efficiency and management level of manufacturers, enhance market order of agricultural products and foods including the fruits, and realize the modern management of agricultural products and foods in China.

2. BRIEF INTRODUCTION OF PDF417

PDF417 is a 2D barcode which can store up to about 1,800 printable ASCII characters or 1,100 binary characters per symbol, as shown in Figure 1. The symbol is rectangular; the shape of the symbol can be adjusted to some extent by setting the width and allowing the height to grow with the data. It is also possible to break large amounts of data into several PDF417 symbols which are logically linked. There is no theoretical limit on the amount of data that can be stored in a group of PDF417 symbols.

Each PDF417 2D symbol is composed of 4 bars and 4 spaces, as shown in Figure 2 (GB/T 17172-1997, 1997). Each bar or space includes 1 to 6 mold piece and for a total mold number of 17, their structure accounts for their name “417 barcode”.

Fig.1: Structure of PDF417

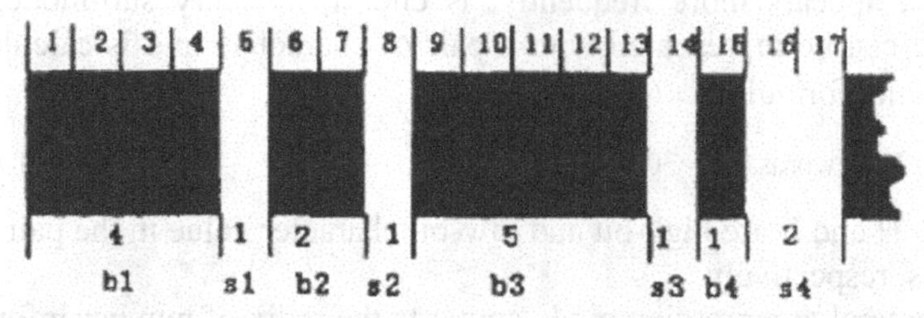

Fig.2: Module distribution of PDF417

3. ENCODING OF PDF417

PDF417 2D barcodes are first put forward by the USA, so application outside China has already been very widespread. A barcode Symbology defines the technical details of the barcode and the ISO standard defines many concrete encoding rules, but the particularities of Chinese characters have not been taken into account in the encoding of PDF417 barcode (Qi, 2003). Though China has already established a national standard of PDF417, the encoding rules suitable for Chinese characters still have not been put forward. This paper focuses the encoding of Chinese characters, and effectively realizes the encoding of PDF417 barcode of Chinese characters in the computer.

In fact, the computer encoding process corresponds to conversion of input information into codewords. The concrete process is as follows:

Firstly, encode the input information with compression according to the certain mode structure, and obtain data codewords. Secondly, calculate error-correction codewords according to the data codewords. Then, according to the number of the data codewords and the error-correction codewords, make the arrangement of codewords matrix, and obtain the finder of both sides of each row. Finally, search the code character list number, i.e. character set, obtain the sequence consisting of bars and spaces corresponding to the codewords, and then draw a PDF417 image.

3.1 Data encoding

The PDF417 has three kinds of data compression mode structure: text compression mode, byte compression mode, numeral compression mode. These modes can be conversed each other (Chen and Liu, 2006).

The text compression mode includes four sub-modes: Alpha, Lower Case, Mixed, and Punctuation.

Sub-modes can make representation of data more effective. The character set, which appears more frequently, is chosen in every sub-mode. Every codeword can be represented with a pair of characters and is calculated by the following formula:

$$\text{Codewords} = 30 * H + L \quad (1)$$

Where: H and L are high-bit and low-bit character value in the pair of characters, respectively.

The numeral compression mode converts the radix of number information according to the numeral total. By the conversion of the radix from 10 to 900, the numeral sequencc is converted to a codeword sequence. We can represent three numeral bits with one codeword in numeral compression mode. The numeral sequence is divided into a group every 44 bits. A pre-bit '1' is added in the front of every numeral sequence group. Then by the conversion of the radix from 10 to 900, 15 codewords are obtained.

The byte compression mode converts the radix of the number information according to byte total. By the conversion of the radix from 256 to 900, the byte sequence is converted to codeword sequence. The byte compression mode contains two locked modes (901/924). When byte total is multiples of six, the 924 mode is taken. By the conversion of radix from 256 to 900, every 6 bytes from left to right in the sequence can be represented by 5 codewords. When byte total is not times of 6, the 901 mode is taken. The conversion of first 6 bytes is the same as 901 mode. Every one in the left bytes corresponds to a codeword, and can be represented directly by codeword. 913 mode is used for temporary conversion from text compression mode (TC) to byte compression mode (BC), which is represented directly by its codeword. The conversion only works on the first codeword, the succeeding codewords return to the current sub-mode of text compression mode (TC).

When encoding is done in this mode, we need to judge whether the number of data streams is times of 6 at first, and choose the right mode to perform the radix conversion. If it is times of 6, 924 mode is chosen; if not, 901 mode is chosen.

3.2 Optimization of byte compression encoding algorithm

Inasmuch as this system aims at the application of marking fruits, its input information contains fruit name, habitat, internal quality, appearance and other details such as: the address, telephones, and faxes of manufacturer and clients, as well as the production date. The system is mainly used in China, so only Chinese character encoding is employed in the system. A Chinese character is stored as machine code in two bytes. Machine code is used to compression encoding for Chinese characters. Byte compression mode is adopted for other numbers and alpha. Since English characters and numbers are represented in one byte in the computer, to conform to Chinese characters, they are converted to representation in two bytes. Then we can implement byte compression encoding for all the input information. The concrete process is as follows.

(1)Judgment of the input information.
(2)Conversion from the acquired GB Code to QW code.
(3)Compression encoding of radix conversion for QW code.

3.3 Error-correction codewords

Since PDF417 is oriented to users, we can choose an error-correction rank for it as required and obtain the related error-correction codeword Ci. When error-correction rank S is chosen, the number of error-correction codewords is k = 2^(S+1). The error-correction rank can be determined automatically by the system as shown in Table 1, or designated by users. The rules, by which the system determines an error-correction rank, are showed in table 1. However, users designate an error-correction rank by themselves, which cannot be lower than the recommended rank. Otherwise, there will appear an error prompt. For a given codeword set, an error-correction codeword is calculated by the Reed–Solomon error-correction algorithm. However, the VC++ program can not run in the equation with the unknown. Therefore, Ci has to be calculated indirectly as follows:

Table 1. Recommended error-correction rank

Number of codewords	Error-correction ranks
1 ~40	2
41 ~160	3
160~320	4
321~ 863	5

(1) Establishment of a polynomial.
Polynomial is established as follows:

$$d(x) = d(n-1)\,x\hat{}\,(n-1) + d(n-2)\,x\hat{}\,(n-2) + \ldots + d(1)\,x + d(0) \quad (2)$$

Where: coefficients in the polynomial are composed of data codewords. The first codeword is the coefficient of the highest-order term. The last one corresponds to the lowest-order term.

(2) Establishment of generated polynomials of error-correction codewords.

K generated polynomials of error-correction codewords are shown as follows:

$$g(x) = (x-3)(x\ \ 3\hat{}2)\ldots(x\ \ 3\hat{}\,k) = x\hat{}\,k + g(k-1)\,x\hat{}\,(k-1) + \ldots + g(1)\,x + g(0) \quad (3)$$

(3) Calculation of error-correction codewords.

To a set of given codewords and a chosen error-correction rank, an error-correction codeword is the complement of the coefficient of the remainder obtained by polynomial d (x) multiplied by x^k and divided by generated polynomial g(x). The coefficient of the highest-order term in the remainder is the first error-correction codeword. The coefficient of the highest-order term corresponds to the last codeword, which is also the last valid codeword in the module. If c(i)>-929, the minus number in GF(929) is equal to the complement of itself. If c (i)<=-929, the minus number is equal to the complement of the remainder (c(i) /929).

(4) Optimization of error-correction algorithm

He and Kang (2002) gave a method, which several coefficients in generated polynomials of error-correction codewords were determined by computer. In this method, every level of powers of three was calculated many times directly. For high levels of powers, overflow would occur because of large data and error would happen. It also was difficult to perform the storage (Dai,2004; Dai and Wu, 2006). Therefore, we adopted to calculate the mod of 929 for several coefficients and store them in a two-dimensional array, called coefficient[9][512]. The value of the two-dimensional array can directly be employed in calculation. The concrete procedure is as follows.

```
for(j=0;j<=8;j++)     // error-correction rank
{
int n=(int)pow(2, (j+1));// the number of error-correction codeword
const int mask=929;
int coefficient[9][512];// store the coefficient of g(x) after divided by 929
int i, k, p;
coefficient[j][0]=1;p=1;// initialization
```

```
for(i=1;i<=n;i++)
{
p=p*3%mask;// 3^i(mod929)
   coefficient[j][i]=0;// error-correction rank is j, the coefficient of g(x) x^i
for(k=i;k>=1;k--)// begin iteration
   {
         coefficient[j][k]= (coefficient[j][k-1]-p* coefficient[j][k])%mask
   }
coefficient[j][0]=-coefficient[j][0]*p%mask
}
for (i=0;i<=n;i++)
   coefficient[j][i]= (coefficient[j][i]+mask)%mask
}
```

3.4 Optimization of the arrangement of codeword matrix

The arrangement of codeword matrix should comply with the following rules:

3 <= the number of rows <= 90

1 <= the number of column <= 30

Filling-code number (Fillnum) should be as small as possible

Filling-code number (Finalnum) should be smaller than 2700

In the application of PDF417 bar codes, the reasonable utilization of space should be taken into sufficient consideration. The barcode matrix should economize valid space and also be user-friendly, so it is necessary to arrange the barcode matrix in optimal form (Sun, 2005; Zheng and Liu, 2006). The ratio of row to column can be user-defined, but in order to save space, the filling-codeword number should be ensured to be the smallest. The number of column (column_temp) can be determined temporarily by the row-column ratio (lwratio) and total number of codewords (totalnum), which includes data codeword and the error-correction codeword. The rule is as follows:

column_temp=(int)sqrt(totalnum/lwratio)+1.

When totalnum/lwratio is the square of some number r or the difference between totalnum/lwratio and r is not significant, we can use r as the column number of the barcode data codewords. Then, we add filling-codeword to the column as the preliminary column number. Because the image to be produced is saved in bmp format, it is requested that the byte number of each row must be times of 4. The rule is as follows:

```
if(column_temp%2==1)
{
```

```
  column=column_temp;
}
else
{
  column=column_temp+1;
}
```

By the row-to-column ratio, we can obtain the number of rows as follows:

$$row = lwratio * column \quad (4)$$

If neither the number of row nor column comply with the rule, in the other word, it is not in the range which the rule defines, a new row-to-column ratio is needed to be set unless the rule is complied with. Data codewords, filling codewords and error-correction codewords are put together in the matrix. According to the following formulas, the value of finders of left and right row can be obtained:

When imod3=0, Li=30Xi+Y, Ri=30Xi+V
When imod3=1, Li=30Xi+Z, Ri=30Xi+Y
When imod3=2, Li=30Xi+V, Ri=30Xi+Z
Where Xi=INT (Layers No./3), i=0, 1, ……89
Y=INT[（Layer No.-1）/3]
Z=error-correction rank*3+[(Layer No.-1) mod3]
V=number of columns every layer -1

Thus, the arrangement of the codeword matrix is accomplished.

3.5 Drawing according to the character set

The PDF417 barcode character set is composed of three clusters. Every cluster contains all the 929 PDF417 codewords represented in different bar-space form. Every symbol character corresponds to only one codeword in every cluster. The three cluster character sets are stored in two-dimensional array symbolset[3][929]. The row 0, 1 and 2 correspond to 0, 3 and 6 cluster in the character set, and the column 0 to 928 correspond to the sequence of bar-space corresponding to codewords 0 to 928, respectively. The sequence of bar-space can be obtained from codewords by searching the two-dimensional array symbolset[3][929], when a PDF417 image is drawn. It can be realized by the following program:

```
// code_matrix[i][j]  matrix of codewords bar-space sequence
// [code_temp[i][j]]  codewords matrix ////
for(j=1;j<=column+2;j++)
{
for(i=0;i<row;i++)
{
  if(i%3==0) // barcode symbol adopts the first cluster of character set
```

```
    {
        code_matrix[i][j] = symbolset[0][code_temp[i][j]];
    }
    if(i%3==1)// barcode symbol adopts the third cluster of character set
    {
        code_matrix[i][j] = symbolset[1][code_temp[i][j]];
    }
    if(i%3==2)// barcode symbol adopts the sixth cluster of character set
    {
        code_matrix[i][j] = symbolset[2][code_temp[i][j]];
    }
  }
}
```

Thus, the sequence of bar-space is obtained; thereupon, drawing can be done directly according to the rule.

4. CONCLUSION

The proposed improved algorithm of codeword byte compression was proven to be an effective algorithm for making PDF417 barcodes used for the Chinese characters, which realized encoding optimization for the same data information with less codewords, and improved the encoding efficiency of Chinese characters greatly. The coefficients of generated polynomial of the error-correction codewords were obtained by iterative calculation, and saved in a two-dimensional array. Since the value can be read directly from the array, the encoding process is greatly accelerated. Because the arrangement of barcode matrix was optimized, the barcode space was economized on condition that user's requirement was satisfied. Under the VC++6.0 programming environment, the encoding algorithm of PDF417 was realized by programming. For example, Figure 3 is a PDF417 barcode image produced by the developed software.

Fig. 3: A example of PDF417

REFERENCES

D. Chen, H. Liu. Two-dimensional Barcode Technology & Application, Chemic Industry Press, 2006

D. Zheng, E. Liu: Optimization of PDF417 Bar Code's generation. Packaging Engineering, 2006, 27(1): 84-86

GB/T 17172-1997. National Standard of the People's Republic of China: 417 Bar Code, China State Bureau of Quality and Technical Supervision, 1997

J Qi. Research on the Generation and Recognition of Two-dimensional Barcode, Graduate School of Harbin Engineering University, 2003

J. He, J. Kang. Computer Coding and Decoding of Bar Code, Computer Measurement & Control, 2002, 10(4): 263-266

J. Sun. Coding technology of PDF417 two-dimensional bar code and its implementation in Visual Basic. Journal of Xi'an Shiyou University(Natural Sciences Edition), 2005, 20(1): 77-80

S. Dai, X. Wu. PDF417 Error Correcting Code and its Implementation, Journal of PLA University of Science and Technology, 2006, 7(2): 137-140

Y. Dai. Research on the Application and Principle of Two-dimensional barcode'encoding and decoding , Nanjing University of Aeronautics and Astronautics, 2004

RESEARCH ON PERSONALIZED INFORMATION FILTERING OF SEARCH ENGINE

Shu Zhang[1], Xinrong Chen[1,*], Changshou Luo[2,*]

[1] Department of Computer Science, College of Information and Electrical Engineering, China Agricultural University, Beijing, P. R. China 100083，China

[2] Beijing Academy of Agriculture and Forestry Sciences, Beijing 100089, China

[*] Corresponding author, Address: Xinrong Cheng, Department of Department of Computer Science, College of Information and Electrical Engineering, China Agricultural University, Beijing 100083, P.R.China, Tel: +86-13521369058, Email: hh0188@sina. com

[*] Corresponding author, Address: Changshou Luo, Beijing Academy of Agriculture and Forestry Sciences, Beijing 100089, P.R. China,Tel:+86-13683248103, Email: luochangshou@163.com

Abstract: Since network has been created and developing rapidly in recent years, the age of information exploding is coming. The Search Engine becomes more and more important for people, but the traditional search engine retrieves and provides information just according to the keywords that users input. How to recommend the right information to users has become the hot point. The technology of personalized information filtering brings people hope .The paper I present analyzed the achievements of those filtering technologies ,and adopted user-system complex-operating modeling to build User-activity-collecting module, User-interest-updating module and User-searching module, in order to meet theme-oriented searching's needs. Experiments showed that the user-interest model can provide personalized service and enhance search engine's precision.

Keywords: information filtering, user interest-model, personalized searching, vector space model

---

Please use the following format when citing this chapter:

Zhang, S., Chen, X. and Luo, C., 2009, in IFIP International Federation for Information Processing, Volume 294, *Computer and Computing Technologies in Agriculture II, Volume 2*, eds. D. Li, Z. Chunjiang, (Boston: Springer), pp. 1271–1280.

1. INTRODUCTION

With the rapid development of Internet information, billions of webpages have been piling up, it makes the most valuable information searching more and more difficult .Search engine (Li Xiaoming et al., 2004) becomes an important retrieval tool for people. Traditional search engines can retrieve and provide information according to the keywords users input, then the system will recommend the same results for the same keywords input by different users. In fact, it depends on what users need. Subject search engine (Liu Huilin et al., 2007) and personalized service (Zeng Chun et al., 2002) emerge as the times require. Subject search engine can filter the unrelated field information during the crawling to reduce the searching resources. The search engine based on technology of personalized service will provide users information which they require, and prevent users from confusing in the information sea.

2. INFORMATION FILTERING AND PERSONALIZATION

Information filtering (Liu Baisong et al., 2003) (ab. IF), contains two significations: the first one which is applicable to rubbish E-mail filtering is how to delete the unrelated from the massive inordinate information; the other used in information recommending in favor of user interests. is to extract the related information which is demanded by users from dynamic information streams. Compared with huge dynamic information, user interests will not change in general. Information filtering on search engine means it can recommend users how to get the required information from information repository.

Search engine's personalized service will select the different information for different users from huge information repository according to their different interesting. Technology of information filtering is the implemental technology of search engine personalization, which can recognize the users' need from dynamic web resources.

Information filtering (IF) technologies (He Jun et al., 2001) can be classified into three, such as:

1) IF technology based on rules, by means of designed rules library to filter information.

2) IF technology based on collaboration, using the comparability of the different users to filter information.

3) IF technology based on content, taking advantage of the comparability between resource and user interest to filter information.

In terms of the research on three personalization technologies' application in theme-oriented search engine, we find that: 1) based on rules. It is difficult to design the rules. This technology has high uncertainty, which should not be adopted. 2) based on collaboration. The key point of this technology is how to cluster the users, but theme-oriented search engine has its own field-orientability. That is said, theme-oriented search engine has locked the users from the certain field. It is unnecessary to cluster users, which is time-consuming and will gain little significant effect. 3) based on content. This technology will lead us to the result information based on the user-interest model, which makes the tiny change on the results which can be gotten from general search engines.

It can also provide the users more related information, and be applicable to the personalized service of theme-oriented search engines. This research is based on theme-oriented search engines which are applied to a certain field and have certain users, equaling to achieving user clustering in general search engines. Therefore, this research adopts the information filtering technology based on content to implement the personalization of theme-oriented search engines.

Information filtering system contains 4 foundation models (Pang Yali et al., 2007):

1) Web resource analyzing model, analyzing and describing the webpages crawled by spider.

2) User-interest model, obtaining and describing user information by obvious or hidden meanings.

3) User-interest updating model, tracking and analyzing the users' behaviors to get the users' current interest.

4) Filtering model, matching the web information description with user interest information in terms of given rules, the filtering model will afford users the web information required in descending order.

From above 4 foundation models, we can see how to obtain the user interests and create user interest models, which will influence the effect of information filtering. The creating of user interest model is one of the key technologies to implement personalized information filtering. Furthermore, filtering model's design also is the key point during the creating process of search engine personalization.

3. USER-INTEREST MODEL

3.1 User-interest Description

User interest description is the key technology in information filtering, which is related to the filtering effect directly. At present, there are 3 descriptions (Zhang Meixiang et al., 2005):

1) Keyword Description. According to certain rules, each interest's weight has not only been valued, but the user interest vector has been set up, so that each interest keyword can possess the branch vector and each keyword's weight can get the value of the branch vector.

2) Fixed Document Set Description (ab. FDS). This description selects the most representative FDS which can reflect all kinds of user information of a certain field sufficiently. The FDS description is used to solve some problems that are hard to be described with exact keywords.

3) Paragraph Description. In lengthy web text information, what an user is interested maybe just has several paragraphs. Paragraph is the minimum unit of articles, so the meaning of paragraph description analyzes is to find out some which the users are interesting in.

The theme-oriented search engine in this paper is the search engine facing the certain field. It possesses its own thematic words library, which can express the user interest exactly. It is hard to use FDS and Paragraph description which are applicable to general search engine. Therefore, the study adopts Keyword Description to express user interest.

The Keyword Description is implemented in Vector Space Model (Zeng Chun et al., 2002).

Given a user-interest vector I, I= （I_0, I_1, I_2…I_i, I_{i+1} …I_n）, i=0, 1, 2…n， Where, I_i is the branch vectors of user-interest， which are interest keyword; the value of I_i is the weight W_i of each keyword .

3.2 Selection of User-interest Modeling

At present, in the field of personalized information service, there are 3 main methods (Ji Meijun et al., 2006) to build user-interest models.

1) Manual customization modeling. It is a modeling method through users' self-input or selection. But it counts on user interests totally and can not track the changes of an user's interests timely;

2) Demonstration modeling. It is up to users to provide the demonstrations of relevant interests. However, the method requires users to mark webpages in order to obtain corresponding demonstrations. Therefore, users' normal browsing behaviors will be disturbed;

3) Automatic modeling. It is constructed automatically according to users' browsing behavior, which would not interfere with users.

Considering the disadvantage of the front two modeling methods above, the study adopts the complex user-interest modeling methods, as illustrated in Fig.1, which is the cooperation of users' inputting and collecting of users' behaviors by the Log-collector. Adopting complex modeling method can avoid two things, which are negative information out of date and dynamic IP false work in manual customization modeling of users.

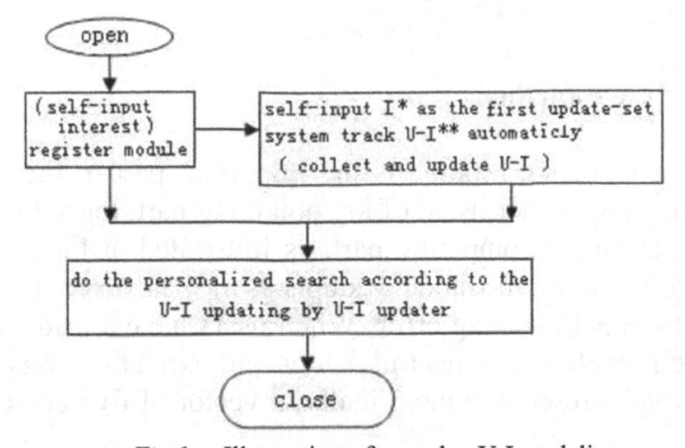

Fig.1: Illustration of complex U-I modeling
* I is the abbreviation of interest; ** U-I is the abbreviation of user-interest

4. INFORMATION FILTERING

During the implementation process of IF, the system adopts user-behavior collector to obtain user behaviors, then update user-interests and analyzes them, finally uses search-recommender to achieve information's filter and recommendation, as illustrated in Fig.2.

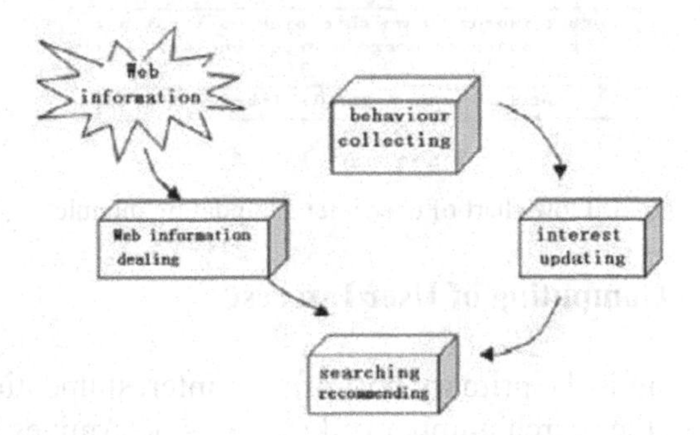

Fig.2: Framework of personalized information filtering

4.1 User-behavior Collecting Module

Users' historic searching behavior can reflect the trend of users' interests, which is collected by the user-behavior collector. The behavior contains search time, search content, search number and so on.

4.2 User-interest Updating Module

4.2.1 Module Function

User-interest Updating Module is the important part of the personalized search system, which is consisted of log obtaining part, log splitting part, log stating part and weight computing part, as illustrated in Fig.3. When users have no search behavior, the module adopts users' registered interests as first updating set to avoid updating error. When users have already some search behaviors, the module will collect user logs and extract the logs considering the factor of logs' preserved time. Finally, a vector of five branch vectors is figured out.

Fig.3: Flow chart of user-interest updating module

4.2.2 Weight Computing of User Interest

Weight computing is the primary part of user-interest updating. During the weight computing, the search number of keywords determines how much the user is interested in . So we regard it as the basic factor of computing weight. Otherwise, we can also consider the search date as the important factor to hope this module will do its best to obtain users' closest interests.

User interests may be changed, so the date as the keyword is searched determines whether the user's interest about this keyword is reduced or enhanced. What would be happened, when some keywords were paid more attention previously but little attention now. The time function can solve this problem. It is a degressive function, which would depress the former keywords' weights and increase the more lately searched keywords' weights.

Below is the formula to compute weight by this system.

$$W=\sum f_i t_i,$$

Where, i =1,2,3....T (T is the period of user interest updating);

f_i is the number of the keyword's searched times at day i in period T;

$$t_i=1-(T+1-i)/T$$

this function is time-factor function, which is aiming at decreasing the former keywords' weights.

4.2.3 About Keywords long time unvisited

The capacity of hardware is so limited that it can not preserve all the users' search logs and information processed by log processor endlessly. Thus, we regard the keywords long time unvisited as the users' uninterested and discard them.

The formula below is to process the keywords,

$$C_{i,j}=\begin{cases} C_{i,j}, & D_{now}-D_{visited} \leq 5T \\ 0, & else \end{cases}$$

Where, T is the period of User-Interest Updating;

$C_{i,j}$ is the search frequency of keyword. When becomes zero, it will be deleted from the log library;

D_{now} is the current date of system;

$D_{visited}$ is the final visited date of $C_{i,j}$;

D_{now} -$D_{visited}$ is the days between current and final search of $C_{i,j}$.

If a keyword can not be searched by user more than 5 times, its search frequency will become zero and it would be canceled from user logs library.

4.3 Implementation of Information filtering

The implementation of information filtering includes four steps. Firstly, the search module with User-interest Model can get the user interest vector from User-interest Updating Module; secondly, extract the keywords and their weights from webpages model; then compute the relevance degree between user interest vector and webpage model vector; the last step is to sort the webpages obtained by general keyword searching according to interest-webpage relevance degree. The flow chart is illustrated below, Fig.4.

During the computing of relevance degrees, the study adopts typical Vector Space Model relevance degree method (Zeng Chun et al., 2002),

$$r=\cos<\alpha,\beta>=(\alpha,\beta)/|\alpha||\beta|=\sum_{i=1}^{n}(F(i)x_i w_i^2)/\sum_{i=1}^{n}(F(i)x_i w_i)^2$$

Where, α equals the user interest vector I; β equals the vector F extracted by the webpage model;

the value of their branch vector is the weight of the branch one.

This module adjusts the webpages tinily according to the magnitude of the relevance degree, and then attains the search result.

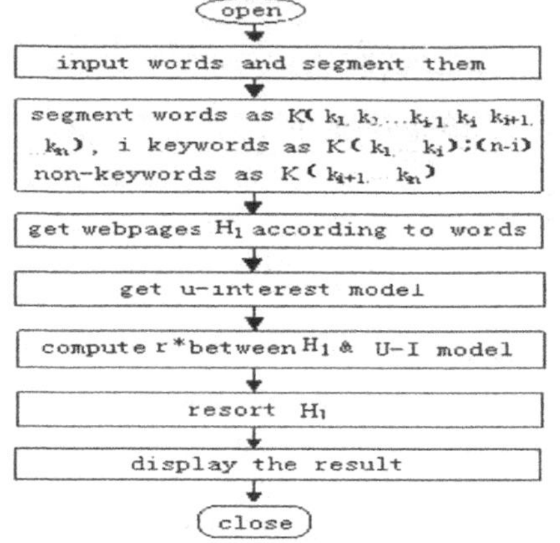

Fig.4. Implementation of information filtering

5. EXPERIMENTS AND ANALYSIS

The experimental system adopts Windows platform, PHP and Visual c++ programming language, and Apache server, as well as SQL Server database. In the experiments, the sample set came from the flower-relevant webpages crawled by thematic spider; the user behavior was preserved by logs; the user interest updating period was 5 days; the quantity of webpages was 8000.

5.1 Micro-comparison Experiment

According to above works, we designed 3 micro-comparison experiments:

1) Same user at two states, both personalized search and general search, inputting the same keyword.

2) Same user at two states, both personalized search and general search, inputting non-flower keyword, which is experimental by non-keyword "technology".

3) Different user at the state of personalized search and by inputting the same keywords.

From above experiments, it can be seen that:

1) The results show that:

Non-personalized search returns the same results for different people;

Search with registered information, which is static personalized search, turned out that the results were not changed as the user interest changed. The logs show that the user was interested in "yulan" then.

Search with user-interest updater, that is dynamic personalized search, can return the result which could mostly reflect the information needed by users.

2) The results gained by inputting general word "technology" in two states show that:

Search without the personalized technology obtains mess result;

Search with the personalized technology acquires user-interest relevant information.

3) The results we gain by inputting the same keyword from different user shows that:

The system leads to different results by inputting the same keyword for different users according to each user's different interest updated by themselves.

All above analyses prove that the system could be used to implement personalization.

5.2 Macro-comparison Experiment

This experiment involved user Xia Yan, who has used the system for 7 days. His search behavior was analyzed by user-interest updating module to: Plum blossom,1.1666669999999999;Peony,1.;Clove,1.; Rose,0.83333299999999999; Jasmine,0.33333299999999999 (sorted by each flower's weight and the data was the lately updated). Keywords were plum, blossom and peony.

The comparison experiment was performed among Baidu, Google, Thematic search engine and Personalized thematic search engine. We searched keywords relevant website both in baidu and google, then got 31 seed websites by picking out the same, finally crawled 61358 webpages as the experimental webpage sample.

Meanwhile, the keywords "peony" and "plum blossom" were searched at above 4 systems. We got the relation between webpages quantity and average relevance degree, as illustrated in Fig.5.

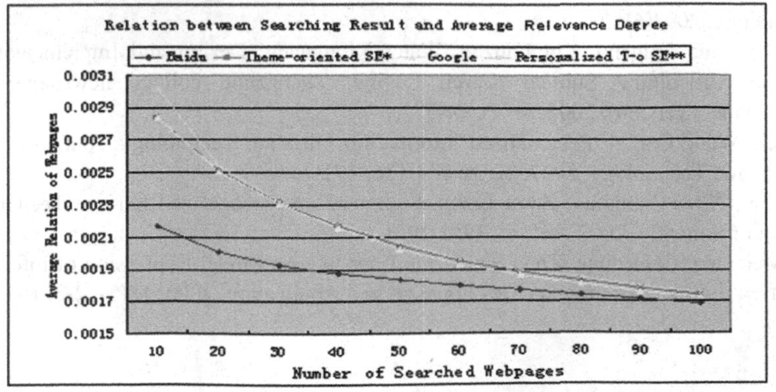

* SE is the abbreviation of search engine.

** T-o SE is the abbreviation of theme-oriented search engine.

Fig.5: Relation between Webpages Quantity and Average Relevance Degree

6. CONCLUSION

The study took advantage of manual customization modeling and automatic modeling to construct a complex user-interest model, and proposed the algorithm of building and updating user personal interests based on the model. Experimental results prove that the model and algorithm can enhance the searching precision. The development of search engine can easily integrate the user profile, semantic and syntax technology to service the users. The next step is to join ontology technology effectively in the preprocessing and service part of the search engine, consequently achieve the true semantic search.

ACKNOWLEDGEMENTS

Fund for this research was provided by Beijing Natural Science Foundation Committee (P. R. China). The first author very appreciates what the College of Information and Electrical Engineering has contributed at the moment she pursues a master degree at the China Agricultural University.

REFERENCES

He Jun, Zhou Mingtian. Information Filtration Technology in Information Network. Journal of Systems Engineering and Electronic Technology, 2001, 23(11): 76-79.

Ji Meijun. Research on Related Question of Personalized User Modelling. Journal of Information, 2006, 25(3): 77-79.

Li Xiaoming, Yan Hongfei, Wang Jimin. Search Engine— Theory, Technology and System, Science and Technology publishing house, 2004, 29-54.

Liu Baisong. Research on Information Filtration. Journal of Modern Books and Information Technology, 2003(6): 23-26.

Liu Huilin, Guo Laigang, Liu Lanzhe, Wang Xingguang. Design and Implementation of Chinese Agricultural Subject Search Engine. Zhengzhou college newspaper (Neo-Confucianism version), 2007, 39(2): 74-77.

Pang Yali, Wang Caifen. Personalized Information Filtration Technology. Journal of Gansu Science and Technology, 2007, 23(3): 124-126, 171.

Zeng Chun, Xing Chunxiao, Zhou Lizhu. Summary on Personalized Service Technology . Journal of Software, 2002, 13(10): 1952-1961.

Zhang Meixiang, Chen JunjieZhao Shuanzhu. User Interest Model Expression of Information Filtration. Journal of Computer Development and Application, 2005, 18(5): 2-3, 14.

AGENT-BASED WHEAT SIMULATION MODEL COOPERATION RESEARCH

Shengping Liu[1], Yeping Zhu[1,*]

[1] *Agricultural Information Institute, Chinese Academy of Agricultural Sciences, Beijing, China, 100081;*

[*] *Corresponding author, Address: Library 311, Agricultural Information Institute, Chinese Academy of Agricultural Sciences, No.12 Zhongguancun South Street, Beijing, 100081, P. R. China, Tel:+86-10-68919652-2342, Fax:+86-10-68919886-2339, Email: zhuyp@mail.caas.net.cn*

Abstract: Cooperative multi-agent systems (MAS) are ones in which several agents attempt, through their interaction, to jointly solve tasks or to maximize utility. Due to the interactions among the agents, multi-agent problem complexity can rise rapidly with the number of agents or their behavioral sophistication. This paper propose a kind of agent-based cooperation design thought and the realization for wheat simulation model, hangs together the growth model and knowledge model of wheat, realize organic coupling and integration between the function of forecast and decision-making.

Keywords: Multi-agent System, Wheat Simulation Model, Cooperation

1. INTRODUCTION

Agriculture system is complicated & open, since each element within system is highly alternant and cooperative. Applying modern information technology to agriculture production has facilitated the realization of agriculture modernization. As an important aspect, Crop Simulation Model technology is applied to crop production management. There have been lots of outcomes through over 40 years' development, such as CERES、GROPGRO. Meanwhile, in order to provide much better technology for agriculture production and decision-making, many researchers are trying to

Please use the following format when citing this chapter:

Liu, S. and Zhu, Y., 2009, in IFIP International Federation for Information Processing, Volume 294, *Computer and Computing Technologies in Agriculture II, Volume 2*, eds. D. Li, Z. Chunjiang, (Boston: Springer), pp. 1281–1289.

combine Crop Simulation Model technology with other new technology as well, such as Expert System, Visualization technology, Artificial Intelligence technology, Network technology.

As the same time as information technology successfully applied to agriculture, Agent Theory and Technology, as a part of distributed artificial intelligence, has developed rapidly since it arose from 1970s, it is a popular direction of artificial intelligence now (CHEN Ying-chun, 2003). Multi-Agent Technology currently provided new methods and tools for solving complicated problems decision, and for setting up distributed, intelligent, integrated, and man-machine harmony decision making supporting system(Wooldridge, et al. 2000). MAS (Multi-Agent System) plays significant roles in plenty of application fields, such as Industry (Process Control, Remote Telecommunication, Air Traffic Control and Transportation System), Commerce (Information filtration, Information Collection, E-business, Business Process Management), Entertainment (Games, Interactive theater and cinema), Medical Treatment (Remote Medical Treatment and Health Care) and Education (ZHU Ye-ping, et al. 2005).

Current agriculture information system, such as Crop simulation model management system, Expert System, Database, Knowledge Model System, they are independent and unilateral since they aim at solving limited problems. Self-restraint, openness, Interactivity and collaboration are key characters of Agent System (DeLoach, 2004). If we apply Multi-Agent technology to agriculture decision system development, we can reuse and share existed resource, exerting advantages of original system, as well as achieving functional organic coupling and integrated coordination, then improving accuracy and overall of management decision (F. Bousquet and et al. 2002). Building crop management cooperation decision system based on Agent technology, implementing crop information share, software share, cooperation work and group decision support under distributed environment, which will enhance the usage of information, fully exerting efficiency of network, achieving several computers cooperative running (DING Wei-long, 2005).

Meanwhile, it shall react, consult and solve the emergency during crop production, and offer new methods and technology for agriculture production management and environment protection harmony development.

By analyzing characters of Agent, taking advantage of its cooperative work and high intelligent, and combining wheat growth simulation model and wheat growth management knowledge model, then developing wheat production management cooperation decision system based on Agent, in order to realize forecast and decision function organic coupling and integration.

2. AGENT-BASED WHEAT SIMULATION MODEL COLLABORATIVE DECISION SYSTEM ANALYSIS AND DESIGN

Multi-Agent System Analysis and Design is the key of agent technology application, which is not widely used in agent-oriented software engineering field. Now the main modeling methods include AAII modeling methodology, Gaia way, Agent UML methods and Z language method, and so on (N.HARNOS. 2006). Object-oriented technology and agent-oriented is inextricably linked to each other, at present object-oriented technology has relatively mature development and a lot of practical methods and tools. Visualization tool UML is a excellent object-oriented approach and UML-based multi-agent modeling technology is more and more recognized (WENG Wen-yong, and et al. 2004).

Based UML visual modeling technology, form the point of view of solving agriculture system problems this paper realizes agent-based wheat model collaborative decision system analysis and design.

2.1 Main system function design

Main system function includes cultivation design, real-time growth simulation, dynamic management control, expert knowledge advisory, data query maintenance, system manage.

Cultivation design function is mainly based on the weather, soil, species and other data of the decision-making point and operates wheat knowledge manage model and wheat growth model so that they cooperate with each other. Finally we build a seemly cultivation program which includes Variety choice, seedtime, seed density and quantity, fertilizer operation and water operation.

Growth simulation forecasting system is based on data and cultivation program. Through operating growth simulation model it can forecast wheat growth dynamic feature which includes wheat daily organ dry weight, root density index, leaf area index, water use and stress, nitrogen use and stress.

Dynamic management control function uses suitable dynamic growth targets (such as leaf age, leaf area index, dry matter accumulation and source library target, etc.) as standard expert curve. When the growth shape which is forecasted by growth simulation model significantly deviates from the shape of the standard expert curve, system analysis the cause and recommend an appropriate control measures (such as irrigation, fertilization, etc.) and the regulation period, at the same time, system revise a suitable growth index and let wheat growth in the light of standard expert curve. Finally system output specific control measure, forecasting wheat growth dynamics and yield components.

Data manage function mainly includes nationwide weather, soil and variety data of decision point. It could not only query and extract required data in database but also add and modify data according to the need.

Expert knowledge advisory includes various kinds of knowledge of wheat growth and development process, which from variety choice to cultivation techniques, irrigation and fertilization, diagnosis and extermination of plant disease and insect pest.

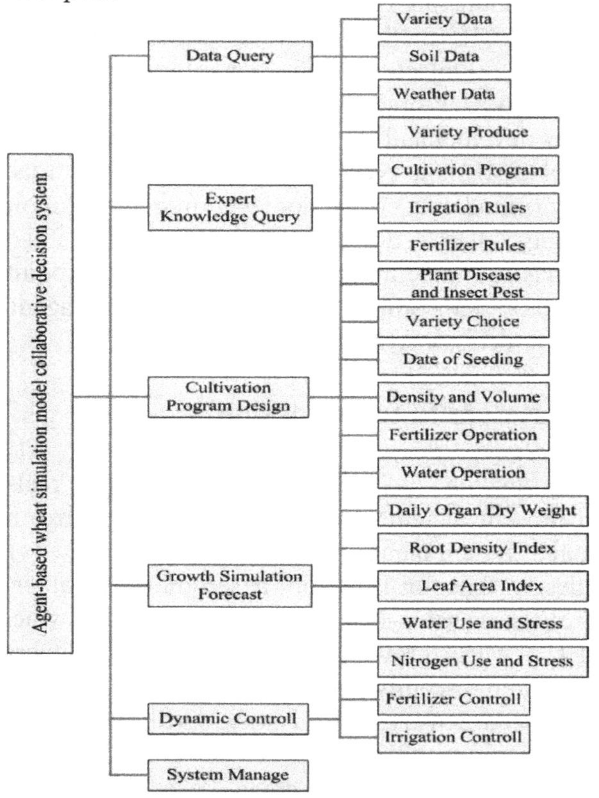

Fig.1: system function diagram

2.2 System Use Case Analysis

In UML language use case describe system function from the system outside user point of view. When we analyze multi-agent system first of all we establish use case model according to their functional requirements (Gu Shaoyuan, and et al. 2001). Traditional use case diagram has two main objects: use case and actor. Multi-agent system is made of many agents with autonomy, sociality, reaction and initiative features, which are collaboration with each other and interact on users. In agent-oriented analysis and design we make some expansion in traditional UML and add agent in model as a new activities. Agent-based wheat simulation model collaborative decision system specific includes data manage agent, knowledge manage agent, knowledge model agent, growth model agent and dynamic control agent. Not only as part of the system, but also as independent users and administrators, they achieve interactive collaboration and cooperative work, help realizing system function and completing customer needs.

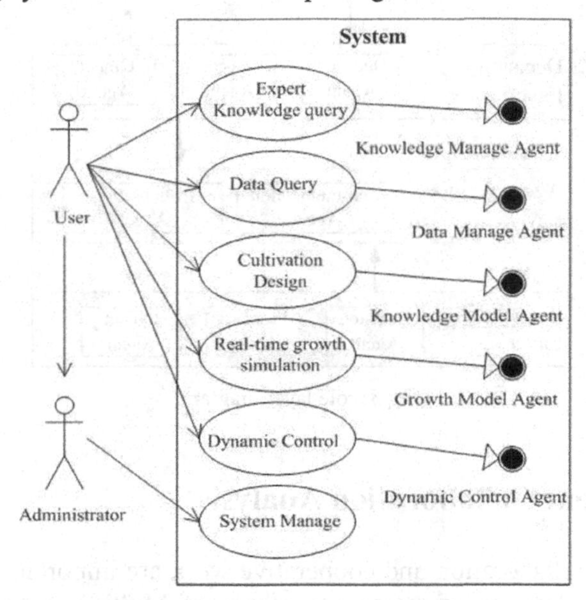

Fig.2: system use case diagram

2.3 System Role Analysis

In current agent-oriented method research, some researchers introduce role concept and regard role as basic unit of multi-agent system modeling. Compared with specific agent, role abstracts a common character of some agent. In multi-agent system, system object is assumed by every role agent (Ma Jun, Yan qi and et al. 2004). Making role as basic modeling unlit reflects goal-driven thinking of system modeling process, role have very important effect from analysis to the entire development process.

Agent-based wheat simulation model collaborative decision system designs 9 kind of role agent at four layers. At user layer we design Interface Agent. At resource layer we design Data Manage Agent, Knowledge Manage Agent and Model agent. Among Model Agent role agent registers as Knowledge Model Agent and Growth Model Agent, which belong to same role and complete different function. At decision layer role agent includes Decision Agent, Forecast Agent and Object Agent. At agent service layer role agent includes Communication Agent and Collaboration Agent.

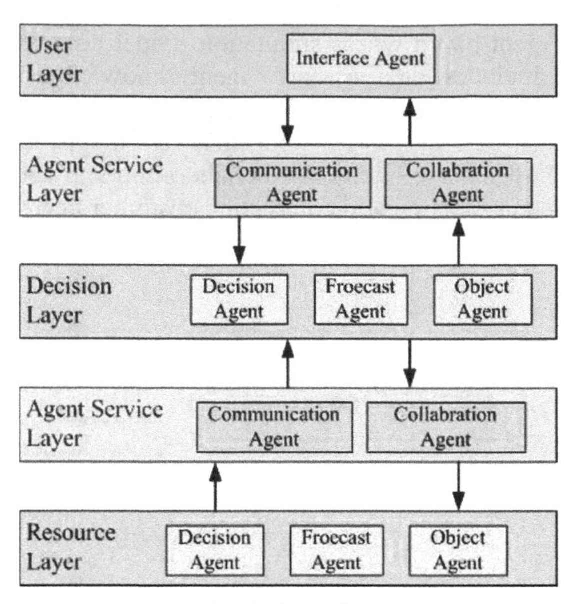

Fig.3: role layer diagram

2.4 Agent Collaboration Analysis

Interactive collaboration and cooperative work are important reflection of sociality, and also one of major advantages of Multi-Agent system. They express the interactive and dynamic cooperative relationship between Agents by cooperation diagram. Take achieving cultivation programs designing function in agent-based wheat simulation model collaborative decision system on the basis of MAS for an example, explain the interactive and cooperative relationship between Agents in the system.

When users need cultivation programs, firstly, put forward the task requests to Interface Agent which will send the task to Object Agent by calling Communication Agent. After Object Agent receives information and analyzes it, it will send messages to Data Manage Agent and pick up relevant data. Data manage agent will look up the perennial data on the decision-making point such as weather, soil and breeds etc. according to object agent's command, and send to Knowledge Model Agent which can run relevant models to produce a set of cultivation programs including breeds-choosing, sowing time, density, sowing quantity, fertilizer planning and moisture planning programs etc..

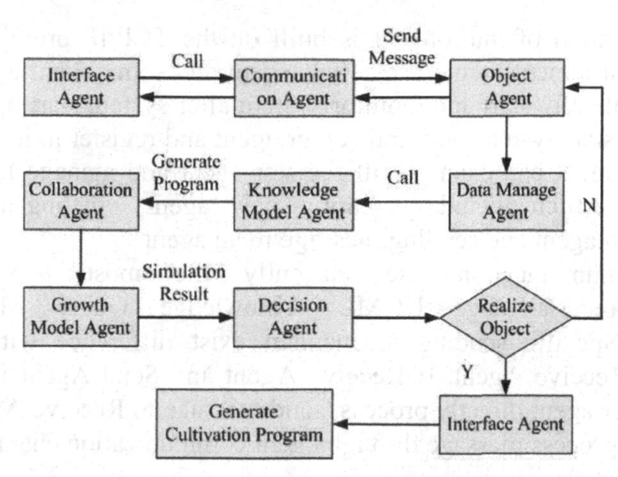

Fig.4: Collaboration diagram

And then Knowledge Model Agent will call Collaboration Agent, by which Knowledge Model Agent will transfer cultivation programs and the perennial data on the decision-making point such as weather, soil and breeds etc. produced by itself to Growth Model Agent. Growth Model Agent will simulate growth according to the information, and predict the yield and quality indexes gained by cultivation programs produced by carrying into Knowledge Model Agent, and send the prediction result to Decision Agent. Decision Agent will analyze the prediction result. If the yield and quality indexes predicted by Growth Model agent achieve the aim which users request, the result will be send back to Interface Agent, and export a set of cultivation programs. If not, the result will be send back to Object Agent and Interface Agent. After amending the aim program, simulate and predict, recycle as this until create a set of cultivation programs satisfying the demand.

3. PROGRAM INTEGRATION AND DEVELOPMENT

System development and operation choose Windows platform. To assure cross-platform commonality of multi-agent system and easy portable we choose JAVA language. We construct a number of different functions role agent through agent development and software integration. The wheat growth simulation, knowledge management, and other functional modules can fully use existing software programs for encapsulation and call so that we can construct the corresponding functional agent, save development workload and realize reusing and sharing of existing resources.

Communication of multi-agent is built on the TCP/IP protocol. System platform environment includes a server agent and some client agent, server agent automatically start and monitor system after system start up. When all client agents start system first find server agent and register in it, at the same time, server agent can examine all register agent and manage life cycle of client agent which includes creating new agent, deleting an agent , suspending an agent and sending message to an agent.

Agent communication use currently the most popular agent communication language—KQML (Knowledge Query Manipulation Language). Specific sending mechanism exist difference with different location of Receive Agent. If Receive Agent and Send Agent in the same system, server agent directly process, send message to Receive Agent, if not, system need process message through agent communication channel in agent platform.

4. CONCLUSION

This research aims to improve the unilateralism of traditional crop growth simulation model system. It discusses integration mechanism for crop growth simulation model and knowledge model under Multi-Agent environment, pointing out crop management cooperation decision system based on Agent UML, analyzing and designing through several angle like function usage and cooperation decision and argument system Integration and development. Multi-Agent is intelligent and cooperative through analyzing the practice, which offers new technology means and methods for solving complicated problems of agriculture system.

Setting up Multi-Agent system platform is a new idea to integrated utilize current unilateral agriculture information system, meanwhile, it realizes resource share and cooperation decision, as well as improves information usage and management decision accuracy.

ACKNOWLEDGEMENTS

This research was supported by Digital Agriculture Program of State High-tech Research and Development Project of China (No. 2006AA10Z220, 2007AA10Z237), Special Fund of Basic Scientific Research and Operation Foundation for Commonweal Scientific Research Institutes (2008J-1-10, 2008J-1-02) and by National Scientific and Technical Supporting Programs Funded by Ministry of Science and Technology of China (2006BAD10A12).

REFERENCES

CHEN Ying-chun, 2003. Intelligent decision support system based on Multi-Agent. Journal of Hefei University of Technologic (Social Sciences), Vol. 17 No.6

DeLoach, S.A. 2006. Engineering Organization-based Multiagent Systems. LNCS Vol. 3914, Springer, 109-125

DING Wei-long, 2005. Research of the agricultural expert system based on artificial plant growth model. Journal of Zhejiang University of Technology, Vol. 27 , Supp. 2

F. Bousquet and et al. 2002. Multi-agent systems and role games: collective learning processes for ecosystem management. In Complexity and ecosystem management: The theory and practice of multi-agent systems, pages 248{285. Edward Elgar.

Gu Shaoyuan, Zhu Chenchen, Shi Hongbao,2001. A Agent—Based Method for Requirement Analysis and Modeling. Computer Engineering and Applications, Vol.5

M. Wooldridge, N. Jennings, and D. Kinny. 2000. The Gaia Methodology for Agent-Oriented Analysis and Design. Journal of Autonomous Agents and Multi-Agent Systems, 3(3).

Ma Jun, Yan qi and et al. 2004, Role-Based Software Design Method for Multi-Agent System. Computer Engineering and Applications, Vol.6

N.HARNOS. 2006. Applicability of the AFRCWHEAT2 wheat growth simulation model in Hungary. APPLIED ECOLOGY AND ENVIRONMENTAL RESEARCH 4(2): 55-61.

WENG Wen-yong, WANG Ze-bing, FENG Yan, 2004. Research on applying UML to analyze agent-oriented system. Computer Engineering and Design, Vol.25, No.7

ZHU Ye-ping, FENG Zhong-ke, 2005. Application of agent in agricultural & forestry economy decision support system. Journal of Beijing Forestry University, Vol. 27 , Supp. 2

A REASONING COMPONENT'S CONSTRUCTION FOR PLANNING REGIONAL AGRICULTURAL ADVANTAGEOUS INDUSTRY DEVELOPMENT

Yue Fan[1] , Yeping Zhu[1,*]

[1] *Agricultural Information Institute, Chinese Academy of Agricultural Sciences, Beijing, P. R. China, 100081*

[*] *Corresponding author, Address: Library Room 310 Agricultural Information Institute, Chinese Academy of Agricultural Sciences, Beijing 100081, P. R. China, Tel: +86-10-82105172, Fax: +86-10-82105172, Email: zhyp@mail.caas.net.cn*

Abstract: The regional characteristic of China's agriculture and the present research conditions on regional agriculture have been studied first. After that, a new method is proposed that developing a system for planning regional agricultural advantageous industry development to help leaders and decision-making departments at all levels to make decisions. In order to achieve the construction of a reasoning component of such a system, the key technologies planned to be adopted are clarified in details, and the structure graph and functions graph of it have also been designed.

Key words: regional agricultural advantageous industry, reasoning component, forward chaining, knowledge presentation

1. INTRODUCTION

Compared with other industries, agriculture in our country bears obvious regional characteristic. China has a vast land area, throughout which land conditions vary quite differently, and different crops, different varieties and different planting management methods have to be altered from area to area. Thus an industry in a region shows an industrial advantage compared to other industries and regions. In order to reach the goal of high-quality, high-

Please use the following format when citing this chapter:

Fan, Y. and Zhu, Y., 2009, in IFIP International Federation for Information Processing, Volume 294, *Computer and Computing Technologies in Agriculture II, Volume 2*, eds. D. Li, Z. Chunjiang, (Boston: Springer), pp. 1291–1297.

yield and high-efficiency in our country's agriculture development, detailed analysis of every industry in every region is indispensable. After knowing the advantageous industry in every region, its unique advantageous industry should merely be developed for one certain region, to ultimately fulfill the purpose of carrying out agricultural production by utilizing resources unique to their region.

At present, a great many scholars have been studying on the fundamental theories of regional agriculture, and a great deal economic analysis have been done, such as researches on comparative advantages of regional agricultural products, on regional agricultural competitiveness analysis, on regional agricultural industrial structure optimization, and on regional agricultural resources analysis. During these works, only a few information systems have been developed to assist analysis by using modern information technologies. These developed systems focus mostly on analysis of one industry or one kind of resource in some regions or on comparison of every industry in one region rather than on economic analysis and development planning of regional agriculture from the whole view and hardly on analysis of every kind of resources in one region. Thus it is necessary to develop a system for planning regional agricultural advantageous industry development to help leaders and decision-making departments at all levels to quickly, precisely, and adequately analyze developing conditions of regional agriculture, to understand regional agricultural production and development trends, and to make decisions. Such a system can comprehensively compare and analyze every industry and every kind of resources in every region from the nation's view, and can implement reasoning processes based on previous analysis results to get some reasoning results, through which to assist leaders make decisions.

This paper tries to study on the key technologies of constructing a reasoning component for planning regional agricultural advantageous industry development, clarifies some key technologies to be adopted, has done demand analysis, and completes the design of system structure graph and functions graph.

2. KEY TECHNOLOGIES

The research aims to develop a reasoning component to assist leaders make decisions on regional agricultural advantageous industry development planning. Thus component technology is firstly studied. In order to reason, knowledge on expert system is also studied, such as knowledge representation and storage, design of inference engine and inference processes. After that, such key technologies as COM and .NET component technology, building knowledge base based on database technology, and forward chaining are chosen to be adopted in the component's construction, and are detailed discussed as follows.

2.1 COM and .NET component technology

According to S. Kalaimagal and R. Srinivasan (2008), a component can be defined as "Any piece of independently executable binary code written to a specification, which can only be accessed via a set of well published interfaces and which can be integrated into any kind of software application irrespective of language/platform. A component always offers a set of services via its interfaces and may be encapsulated inside a container depending on the kind of middleware technology used to develop the component."

COM (Component Object Model) is a Microsoft centric interface standard for component-based software development introduced by Microsoft in 1993. The essence of COM is a language-neutral way of implementing objects such that they can be used in environments different from the one they were created in, even across machine boundaries. For well-authored components, COM allows reuse of objects with no knowledge of their internal implementation, as it forces component implementers to provide well-defined interfaces that are separate from the implementation. As the development of technology, COM has been one of the basic technologies integrated in Microsoft .NET, and the COM platform has also largely been superseded by the Microsoft .NET initiative. Users can easily develop COM components by using their familiar programming language in a .NET environment.

COM is very similar to other component software interface standards, such as CORBA and Java Beans, although each has its own strengths and weaknesses. It is likely that the characteristics of COM make it most suitable for the development and deployment of desktop applications, for which it was originally designed. Hence Microsoft .NET is adopted in this developing processes.

2.2 Building knowledge base based on database

The database technology bears such characteristics as structured data, low redundancy, high independence, easy expandability, and the non-procedural usage of data and shared data. As the development of database technology, especially the increasingly powerful functions and calculation speed of relational database, its application is more and more extensive.

Knowledge text files are usually adopted in traditional knowledge base management. A file in a system serves specially for a certain application, and data's logical structure is the best for it. This leads to the bad condition when database alters that application program and the definition of file structure must be adapted to data's logical structural alterations. Thus it lacks independence between database and application program in this way. This method also cannot meet the demands of sharing data between several users or programs and has difficulty in database maintaining.

Both differences and similarities exist between a database and a knowledge base, while the former manages and stores data, and the latter focuses on the management and storage of knowledge. A database manages information data, while a knowledge base processes intelligent information. A database tries to implement effective storage and search from large amount of data, while a knowledge base to fulfill effective knowledge representation and reasoning mechanism. The similarities between them lie in their realization that rules and facts in one knowledge base could be built in form of database tables to be easily added, altered, deleted or queried. Hence the combination of database and knowledge base, which combines factual data with abstract knowledge, and processing capability of data with that based on knowledge, could contribute a lot to the construction of a more effective knowledge base system.

In this realization, knowledge is represented in the general form of production rules: ♋$_1$✈♋$_2$✈♋$_3$→♌, that is: when precondition ♋$_1$, ♋$_2$, and ♋$_3$ are met at the same time, action ♌ is triggered. Rules are stored in the form of rule tables in the database. Each table is composed of elements of such fields as Rule No., Precondition 1, Precondition 2, Precondition 3, Action, Interpretation Rule No., as follows:

Table 2. A production rules' table

Rule No.	Precondition 1	Precondition 2	Precondition 3	Action	Interpretation Rule No.
1001	Yield of wheat > 10 t	Yield of maize < 1 t		Better to plant wheat	001
...					...

2.3 Forward chaining

There are two main methods of reasoning when using inference rules: backward chaining and forward chaining.

Forward chaining starts with the available data and uses inference rules to extract more data (from an end user for example) until an optimal goal is reached. An inference engine using forward chaining searches the inference rules until it finds one where the antecedent (If clause) is known to be true. When found it can conclude, or infer, the consequent (Then clause), resulting in the addition of new information to its data. Inference engines will often cycle through this process until an optimal goal is reached.

Because the data determines which rules are selected and used, this method is called data-driven, in contrast to goal-driven backward chaining inference.

One of the advantages of forward-chaining over backward-chaining is that the reception of new data can trigger new inferences, which makes the engine better suited to dynamic situations in which conditions are likely to change.

3. SYSTEM DESIGN

3.1 System structure graph

Based on forward chaining discussed before, the system structure graph of this reasoning component has been designed. It is designed into three levels: Graphical user interface level、Inference engine realization level、and Data maintenance level.

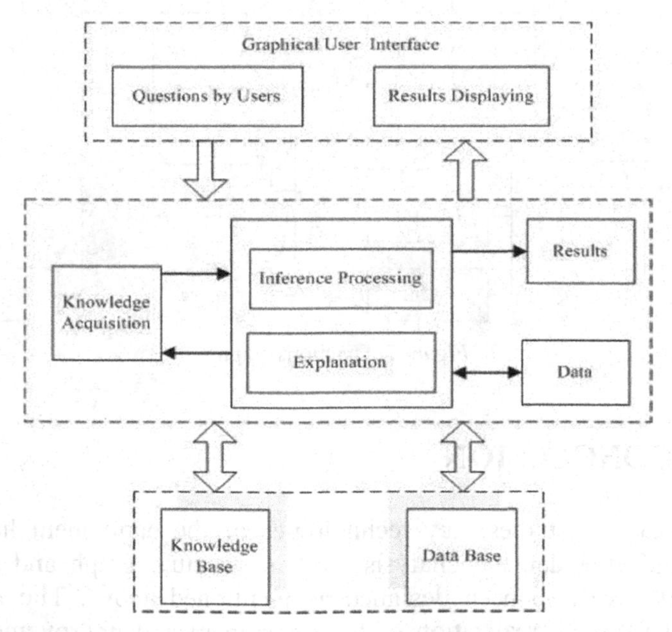

Figure 1. System structure graph

By adopting visual programming language, graphical user interface level fulfills the interaction between program and users, such as the questions raised by users, the program required data input of users to initiate inference, the inference results displaying, and etc.

The middle level is critical to the component, which accomplishes the whole inference engine based on forward chaining. It achieves the realization of the whole inference processes, including the implements of explanation mechanism, production rules' selection and inference, the feedback data's input, and so on.

The data level manages and maintains data base and knowledge base. Knowledge's representation and storage, design of data tables and knowledge tables, data's alteration and edit are all implemented here.

3.2 Functions graph

The component's functions graph is also designed here. It is divided into four parts: Data alteration、Inference and planning、System management and Help. Details are depicted in the following figure:

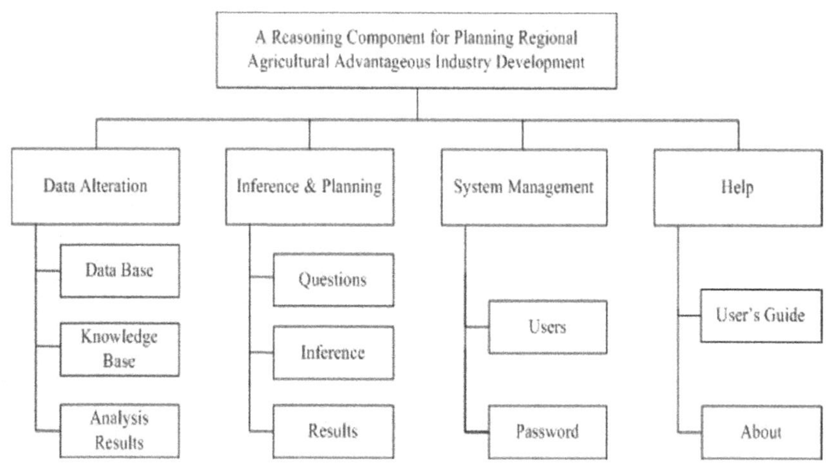

Figure 2. Functions graph

4. CONCLUSION

After continual studies, key technologies of the component have been chosen. And after demand analysis, system structure graph and functions graph of it have also been designed as mentioned above. The following works will focus on realization of the component and improvement of its functions, the obstacles confronted will be overcome after continuing studying and researches.

ACKNOWLEDGEMENTS

This research was supported by National Scientific and Technical Supporting Programs Funded by State High-tech Research and Development Project of China (2007AA10Z237 , 2006BAD10A06), and Beijing Municipal Natural Science Foundation (4042026).

REFERENCES

http://en.wikipedia.org/wiki/Main_Page

S. Kalaimagal, R. Srinivasan. A TRUSTAD Component Nomenclature, Journal of Object Technology, Vol.7 No.4, Issue May-June 2008, pp 159-173

Wang Wei, He Jianjun. Establishing of Expert System for Pneumatic Drying Process Based on Database Technology, Automation & Instrumentation, 2007(3): 9-11, 72(in Chinese)

Yang Changbao, Wu Xiuyuan, Ma Shengzhong. The Study of Intelligent GIS and Its Application in Info-Agriculture, Journal of Jilin Agricultural University, 2004, 26(1): 111-115(in Chinese)

THE ACQUISITION OF CLASS DEFINITIONS IN THE COMMODITY ONTOLOGY OF AGRICULTURAL MEANS OF PRODUCTION

Lu Zhang[1] , Li Kang[1,*] , Xinrong Cheng[1] , Guowu Jiang[1] , Zhijia Niu[2]

[1] *College of Information and Electrical Engineering, China Agricultural University, Beijing, P.R.. China 100083*

[2] *Agricultural bank of China, Beijing, P. R..China 100073*

* *Corresponding author, Address: College of Information and Electrical Engineering, China Agricultural University, Beijing 100083, P. R.. China, Tel: +86-13691086472, Email: kangli.cau@gmail.com*

Abstract: The agricultural means of production is also called the means of agricultural production. The previous work focused on constructing the means of agricultural production commodities ontology taxonomy. After finishing constructing the ontology taxonomy, adding the detail information to the ontology are the following work. The detail information includes classes' definitions, properties, relations, instances and axioms. The classes' definitions are the concrete domain knowledge manifestation. Therefore they can be used to learn class's properties and relations. This paper focuses on obtaining these definitions. To this end, the most important work is to determine where to obtain class's definitions. This paper discusses the selective process. According to authority, completeness, accuracy, practicability and computerization, compares three kinds of knowledge sources, and then selects online "Encyclopedia of China" as the knowledge source. Analyzes the encyclopedia website and its entries, and then proposes an automatic method to get class definitions. The experiment shows that nearly 70% of classes can get their definitions. Using Jena API to add the definitions to the ontology model represented in OWL format.

Keywords: means of agricultural production commodities, ontology, class definitions

Please use the following format when citing this chapter:

Zhang, L., Kang, L., Cheng, X., Jiang, G. and Niu, Z., 2009, in IFIP International Federation for Information Processing, Volume 294, *Computer and Computing Technologies in Agriculture II, Volume 2*, eds. D. Li, Z. Chunjiang, (Boston: Springer), pp. 1299–1308.

1. THE METHODOLOGY FOR BUILDING THE COMMODITY ONTOLOGY OF AGRICULTURAL MEANS OF PRODUCTION

Choose the seven-step method (Noy et at., 2001) to construct the ontology. The previous work has completed the ontology taxonomy. Under the seven- step method, the next step is for the ontology to add attributes, relationships, instances and axioms. The implementation of this step must have knowledge sources, which contain the knowledge of the means of agricultural production commodities. In fact, knowledge source is equivalent to class definitions in the ontology. Therefore getting class definitions is the prerequisite for the future work. In order to obtain definitions, where to find the definitions is the first issue to be considered. Because the total number of classes in the ontology was over 700, it is not suitable for manual input definitions. So the next step is to identify suitable knowlcdgc sources from which we can get definitions automatically. Section II discusses the knowledge source used by the process of getting definitions and its selective process.

2. THE CHOICE OF KNOWLEDGE SOURCES

2.1 The characteristics of knowledge sources in this paper

The commodity ontology of agricultural means of production will be applied for intelligent agricultural information systems used by peasants. So the knowledge described by the ontology should be recoganized generally to be authoritative and practicable within the subject and circulation domain of the means of agricultural production. Besides, the completeness and accuracy of the knowledge contained in the ontology are other factors that decide qualities of services provided by the agricultural information systems. So related knowledge sources from which we get the class definitions should have these characteristics above. For the purpose of getting definitions automatically, it is necessary to consider whether the knowledge source could be processed by computers conveniently. In a word, the characteristics of the knowledge source in this paper are authoritative, complete, practicable, accurate and digitized.

2.2 The selective process of the knowledge sources

If we see sci-tech documents on this domain (the means of agricultural iaotangshan Station. production commodities) as knowledge sources (Wang Qian, 2004), the domain experts must find out suitable ones among lots of documents. It is time-consuming. Besides, although a large number of sci-tech documents stored in database are easy to access, domain knowledge is widely distributed and thus all of them requires amount of time to be find out.

Professional books include textbooks and monographs. Their content not only could reflect author's academic level, but also is a more systematical explanation to the domain knowledge than sci-tech documents'. However, there are many different levels books in a field, so ontology engineers have to spend a lot of time on finding out the books with authoritativeness and accuracy. Besides, electronic versions of some books are hard to access, and thus ontology engineers will have to enter a large amount of text when they find class definitions in these books.

Encyclopedias and dictionaries both cover and contain everything. They are related to all human subject and knowledge (Jin Changzheng, 2007). Therefore encyclopedias could give a comprehensive explanation to all domain knowledge. Furthermore persons usually say that encyclopedias are "a university without walls" (Jin Changzheng, 2007), so encyclopedia is also an educational book for persons and the knowledge explained by it could be authoritative, accurate and practicable. At present some online encyclopedias could be looked up easily and it is probable that a program based on the structure of online encyclopedias can get every entry's explanation automatically.

"Encyclopedia of China" is the first large and comprehensive encyclopedia (Encyclopedia of China Publishing House, 1980), and it is also a large-scale encyclopedia in the world. From 1978 to 1993, the chief editor committee for "Encyclopedia of China" and the Encyclopedia of China Publishing House organized more 20,000 experts and scholars to complete this book. "Encyclopedia of China" is composed of different volumes explaining different subject knowledge. This paper chose the online "Encyclopedia of China" as the knowledge source to get class definitions.

By these consideration above, we have laid solid foundation for the future work of acquiring class definitions.

3. THE ACQUISITION OF CLASS DEFINITIONS

In some volumes of online "Encyclopedia of China", the names of some entries are the same as the classes of ontology and thus these entries' explanation can be seen as the definitions of corresponding classes. The aim of this paper is to get these entries' explanations and add them to rdfs:comment element of corresponding classes in the ontology model represented in OWL format.

The development platform used is Eclipse, and the program language is Java.

Table 1. The characteristics of three knowledge sources

The characteristics of knowledge sources	Knowledge sources		
	Encyclopedias	Sci-tech documents	Professional books (textbooks and monographs)
Authoritativeness	Educational books and writtcn by domain experts, knowledge contained is authoritative	Some are written by domain experts, high authoritativeness; it will take ontology engineers some time to find them	Some are written by domain experts, high authoritativeness; it will take ontology engineers some time to find them
Completeness	Contain all domain knowledge about any subject	A paper usually focuses on a small part of a domain, and it will take ontology engineers a lot of time to find all	Better than sci-tech documents, not as good as encyclopedias, and it will take ontology engineers some time to find all
Accuracy	Educational books and written by domain experts, knowledge contained is accurate	Some are written by domain experts, high accuracy; it will take ontology engineers some time to find them	Some are written by domain experts, high accuracy; it will take ontology engineers some time to find them
Practicability	Educational books for persons, knowledge contained is practicable	It will take ontology engineers some time to find the documents which are related to practical application	It will take ontology engineers some time to find the books which are related to practical application
Computerization	Online versions could be processed by computers	Electronic versions are easy to processed by computers	Difficulty to be processed in the absence of electronic versions

Process of getting definitions is divided into three phases based on the type of work that is being done. In the first phase, write a program allowing obtaining explanations of entries in the form of HTML from online "Encyclopedia of China". In the second phase, HTML tags can be removed from HTML files with the benefit of HTML parser (Oswald et at., 2006) to get pure text content of explanations. In the third phase, with the help of Jena API (Battle et at., 2001), the explanations will be processed and added to rdfs:comment element of corresponding classes which don't have class definitions in ontology.

Get explanations of entries in the form of HTML

Online "Encyclopedia of China" is organized by different volumes. Type http://202.112.118.40:918/search?ChannelID=2 in browser and the browser will show a list of entries in the form of hyperlink. Every hyperlink pointed to a file whose content is an explanation of current entry. There are 78203 entries and 3911 pages on online "Encyclopedia of China". These entries are showed according to the sequence of volumes. It is shown in Fig.1.

Click 'view source' in browser to view the HTML source code of the web page from http://202.112.118.40:918/search?ChannelID=2. They are shown in Fig.2-4.

In Fig.3, the FORM element whose name is "OutlineForm" represents the choice of pages from page 1 to page 3911. In this form, the INPUT element whose name is "pagetext" is an input box where write page number (In Fig.1, it is 1). A JavaScript function "Outline_onsubmit( )" in Fig.3 save the number in "pagetext" to the INPUT called "page" and then submit this form. When submitting this form, according to the action attribute of the "OutlineForm" form, the browser opens another URL representing another page which contains a few other entries. The action attribute in Fig.3 is "/outline?ChannelID=2&randno=3960" and it is also called "page base URL" because every new page URL is composed of the "page base URL" and a page number. The new URL's format is "host + page base URL + &page=pagenumber". For example if when I wanted to open the URL of page 12, the wanted URL is

http://202.112.118.40:918/outline?ChannelID=2&randno=3960&page=12

In Fig.4, "DetailForm" form shows every entry's explanation. Every entry is represented in the form of an "<a>" element. A JavaScript function "javascript:gotorec('1','32037')" in Fig.2 submit the "DetailForm" form with the first parameter whose value is 1 in Fig.4. In Fig.2, the function of "gotorec" saves the number of clicked entry to an INPUT tag called "record". The same as opening another page URL above, the URL for opening an entry's explanation is http://202.112.118.40:918/detail?ChannelID=2&randno=19265&&record=1

The string of "/detail?ChannelID=2&randno=19265" is used to indicate the constant part of URL for opening an entry explanation, which is also called "detailInfoFormBaseURI" in program written for the first phase. The first phase's flowchart is shown in Fig.5.

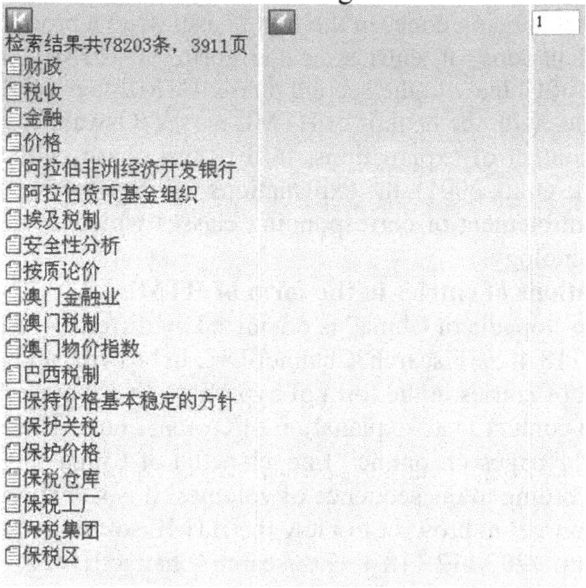

检索结果共78203条，3911页
财政
税收
金融
价格
阿拉伯非洲经济开发银行
阿拉伯货币基金组织
埃及税制
安全性分析
按质论价
澳门金融业
澳门税制
澳门物价指数
巴西税制
保持价格基本稳定的方针
保护关税
保护价格
保税仓库
保税工厂
保税集团
保税区

Fig.1: Display a list of entries in a browser

```
function gotorec(currec, rand)
{
        document.DetailForm.record.value=currec;
        document.DetailForm.submit();
}

function gotopage(type)
{
        var curpage=1;
        var pagenum=3911;

        if(type=="head"){
                curpage=1;
        }
        if(type=="tail"){
                curpage=pagenum;
        }
        if(type=="prev"){
                curpage--;
                if(curpage<1)return;
        }
        if(type=="next"){
                curpage++;
                if(curpage>pagenum)return;
        }
        document.OutlineForm.page.value=curpage;

        document.OutlineForm.submit();
}
```

Fig.2: Part of source code from http://202.112.118.40:918/search?ChannelID=2

```
<form method="post" name="OutlineForm" action="/outline?ChannelID=2&randno=3960"
target=_self>
<input type="hidden" name="presearchword" value="">
<input type="hidden" name="presortfield" value="">
<input type="hidden" name="preextension" value="">
<input type="hidden" name="page" >
       <td nowrap><a href="javascript:gotopage('head')"><img src="\web\gotohead.bmp"
border=0 alt="首页"></a></td>
       <td nowrap><a href="javascript:gotopage('prev')"><img src="\web\gotoup.bmp"
border=0 alt="上一页"></a></td>
       <td nowrap><input type="text" name="pagetext" size="4" value="1"
onkeydown="Outline_onsubmit()"></td>
       <td nowrap><a href="javascript:gotopage('next')"><img src="\web\gotodown.bmp"
border=0 alt="下一页"></a></td>
       <td nowrap><a href="javascript:gotopage('tail')"><img src="\web\gototail.bmp"
border=0 alt="尾页"></a></td>
</form>
```

Fig.3: Part of source code from http://202.112.118.40:918/search?ChannelID=2

```
<table cellpadding="1" cellspacing="0" >
<form method="post" name="DetailForm" action="/detail?ChannelID=2&randno=19265"
target="main">
<input type="hidden" name="presearchword" value="">
<input type="hidden" name="presortfield" value="">
<input type="hidden" name="preextension" value="">
<input type="hidden" name="record" >

  <tr align=left>
    <td valign=top width=1 align=left>
       <a href="javascript:gotorec('1','27980')" style="text-decoration: none; font-size:
9.0pt;"><img src="\web\page.gif" border=0></a>
    </td>
    <td valign=top align=left>
       <a href="javascript:gotorec('1','32037')" style="text-decoration: none; font-size:
9.0pt;"><font size=3 style="font-family: ZYSongDbk">财政</font></a>
    </td>
  </tr>
```

Fig.4: Part of source code from http://202.112.118.40:918/search?ChannelID=2

We try to get entries' explanation from five volumes including agriculture, modern medicine, chemical engineering, mechanical engineering and Chinese traditional medicine. In online "Encyclopedia of China", the start page number and total page number of volume on Agriculture are 1563 and 119, on Modern Medicine are 2947 and 88, on Chemical Engineering are 565 and 67.

Finally, we get files whose file name is entries' name and the file content is corresponding entry's explanation. The volumes and their total number of files are shown in Table 2.

2 Extract pure text content from HTML-format text files

The second phase removes some extra information from files generated by the first phase, such as the icons of print, and then extracts pure text content from HTML-format text files with the help of HTML Parser

API(Oswald et at., 2006). The obtained pure text content replaces the HTML-format content generated by the first phase.

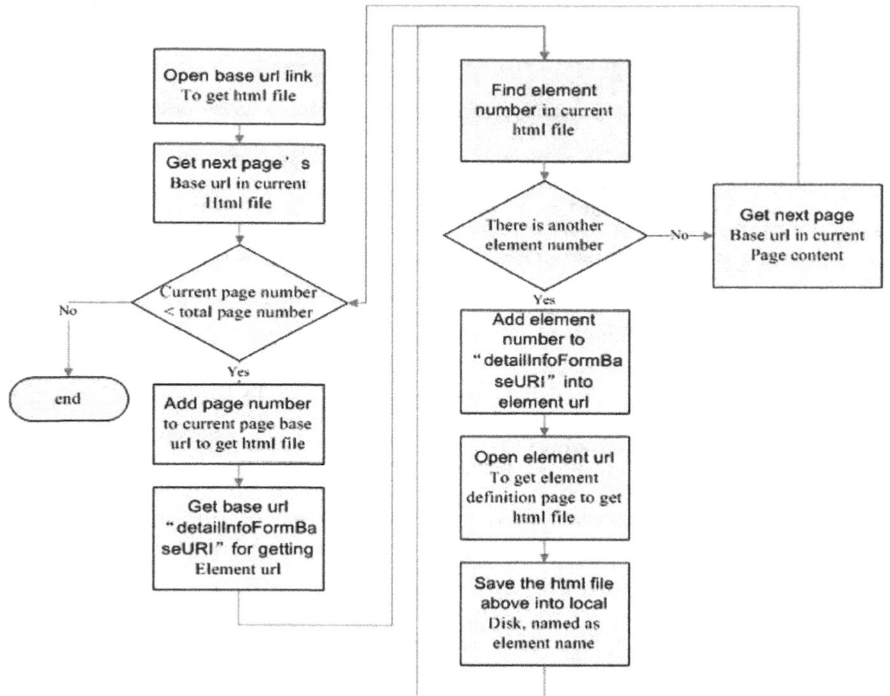

Fig.5: The flowchart of the first phase

Table 2. The volumes and their total number of files

Volume name	The total number of files
Agriculture	2391
Modern medicine	1759
Chemical engineering	287
Mechanical engineering	1438
Chinese traditional medicine	306

3 Process the content of files and add them to ontology

The third phase starts to extract class definitions from files generated by the second phase. Three methods are used to finish this task. If a class's name of ontology is equal to a file's name then the file's content must be the class's definition, which is the first method. If a class of ontology doesn't find any definition with the first method, the second method will be applied to check whether a class name and a file name have intersection. For example, if a class's name is ABC and a file's name is BC, then the file's content may be the class's definition. The third method will work under the situation when the second method doesn't find any definition. This method is designed to regard a paragraph of the file as the class's definition when the class name appears in that paragraph.

Besides, definitions acquisition programs could get subclass definitions in the class definition with the help of regular expressions based on language rules.

水溶性磷肥(water soluble phosphorous fertilizer)即能溶解于水的磷肥。主要品种有：↵

①过磷酸钙(ordinary super phosphate)，由磷矿粉经硫酸处理而成，为磷酸一钙、石膏及少量游离酸的混合物,有效磷(P2O5)含量 12～20%,易吸湿、结块。② ...↵

The content above is a definition of water soluble phosphorous fertilizer. Ordinary super phosphate is a subclass of water soluble phosphorous fertilizer. Some other class definitions also have this pattern. Corresponding regular expression is "[\u2460-\u2473]([\u4e00-\u9fa5]+), ([[[\u4e00-\u9fa5\u002d-\u0039\uff08-\uff09]+ 、]* ，]*[\u4e00-\u9fa5\u002d-\u0039\uff08-\uff09]+)". Before using this regular expression to extract subclass definitions, definitions require a preprocessing to replace some punctuation and then only "，" and "、" can appear in subclass definition.

4. RESULTS AND DISCUSSIONS

Table 3. The results of experiments

First level class	All	Has def	First level class	All	Has def
veterinary drug	109	63	fertilizer and manure	19	16
pesticide	46	28	small and middle farm implements	78	72
feedstuff	16	7	seed	13	11
land	13	10	agricultural machinery	414	287
Breeder & breeder bird	20	18	agricultural film	27	4

In Table 3, first level classes include 10 classes under the root of the ontology. For a first level class, the total number of its subclasses is recorded under the column of "all" and the total number of its subclasses which have class definition is recorded under the column of "has def". Both statistics count the first level classes themselves. The total number under the column "all" is 755, under the column "has def" is 516. Nearly 70 percent classes of ontology have their definitions.

5. CONCLUSIONS

The work of this paper focuses on the acquisition of class definitions for the constructing of the commodity ontology of agricultural means of production. First of all, various knowledge sources are comprehensively discussed according to the characteristics of authoritativeness, accuracy, completeness, practicability and computerization. Finally online "Encyclopedia of China" is chose as the knowledge source for getting class definitions automatically. Experiment shows that the definitions of most classes (nearly 70%) in the ontology can be obtained automatically. It will save ontology engineers a lot of time and labor and provide basic data to support following extracting work.

The future work is to add class definitions for classes whose definitions don't exist in online "Encyclopedia of China". Then different from traditional methods that rely on domain experts to manually build the ontology, a highly efficient and intelligent method will be used to mine properties, relations, instances and axioms from class definitions.

ACKNOWLEDGEMENTS

Funding for this research was provided by the sub-topics of the National Science and Technology Support Plan of China (Grant name: The Study on Constructing Technology of the Commodity Ontology of Agricultural Means of Production. Grant Number: 2006BAD10A050103.)

REFERENCES

Derrick oswald, somik raha, ian macfarlane, david walters.

Encyclopedia of china publishing house. Encyclopedia of china,

Html parser,http://htmlparser.sourceforge.net/ ,2006

Http://protege.stanford.edu/publications/ontology_development/ontology101.pdf, 2001

Introduction: http://202.112.118.40:918/web/bzxx.htm#bkqs, 1980(in chinese)

Jin changzheng. Encyclopedics and a science of compiling encyclopedia. Editors monthly,2001,(5):24-25(in chinese)

Natalya f. Noy and deborah l. Mcguinness. Ontology development 101: a guide to creating your first ontology. 2001.

Online version: http://202.204.214.134:918/web/index.htm, 1980(in chinese)

Steve battle et at. .jena, http://jena.sourceforge.net, 2001

Wang qian. Research on approach to construct dynamic ontology based on text mining[doctor's degree paper,jun,2007]. Beijing: china agricultural university, 2007(in chinese)

DEVELOPMENT OF THE FARM MACHINERY STATISTICAL MANAGEMENT SYSTEM BASED ON WEB

Xindan Qi[1], Hua Li[2], Wenqing Yin[2,*]

[1] *College of Mechanical and Power Engineering, Nanjing University of Technology, Nanjing, Jiangsu Provice,P.R.China 210009*

[2] *College of Engineering, Nanjing Agriculture University, Nanjing, Jiangsu Provice, P.R. China 210031*

* *Corresponding author,Address:College of Engineering Nanjing Agricultural University, Nanjing 210031 Jiangsu province,P.R.china.Tel:+86-25-58606607,Fax:+86-25-58606585, E-mail:lihua@njau.edu.cn*

Abstract: The farm machinery statistics is an important task of the farm machinery management, the statistics of farm machinery in Jiangsu Province are very wide. Improving the working of statistical analyses by advanced computer technology, enhancing the efficiency, is a topic which the farm machinery superintendent and the scientific and technical worker explore continuously. The present situation of the farm machinery statistics management informationization was analyzed .Mainly, the status of the farm machinery statistics management in Jiangsu Province was investigated. Subsequently, the farm machinery statistics management system based on the Web database was developed and the introduction on main function of the system was made.

Key words: farm machinery, statistics, management information system, B/S mode

1. INTRODUCTION

Agricultural machinery statistics is one of the important tasks of agricultural machinery management. The data is the main basis for the governmental decision-making. With the development of the computer hardware and software, the application of advanced technology have made it

Please use the following format when citing this chapter:

Qi, X., Li, H. and Yin, W., 2009, in IFIP International Federation for Information Processing, Volume 294, *Computer and Computing Technologies in Agriculture II, Volume 2*, eds. D. Li, Z. Chunjiang, (Boston: Springer), pp. 1309–1317.

possible that we can develop an efficient and convenient application which has drawn the attention of the domestic scholars (Guosheng, 1998) . Yao Zhong, from Beijing Agro-business Management University ,came up with the thought of distributed agricultural statistical information, and has explored it theoretically (Zhong, 1995) ; The application , named agricultural mechanization computer management and application of experts consulting system, designed by Hebei Provincial farm machinery management station, and Lanfang civic Agriculture and Forestry Science College and China University of Agriculture ,has passed the evaluation of the experts organized by the Department of the Agriculture in 1995.And most functions of the application has been used by the Mechanization Office of Department of Agriculture and other national concerned branches of farm machinery management of with a satisfying effect[1]. He Yong, from Zhejiang University ,developed a system of farm mechanization database management on the microcomputer according to the design principle of management information system and applying the method of systems engineering and mathematics, which is to meet the need of agricultural mechanization data processing (Yong et al., 1995) . He Ruiyin, from Nanjing Agricultural University, together with his colleagues, developed an farm machinery statistics system, which is a convenient system based on Windows interface, and suitable to the agricultural cultivating system in Jiangsu province. The system can meet the need of the informatization management of the provincial farm machinery statistics (Ruiyin, 1998) . As the main participator, I developed a system of agricultural machinery statistics management based on Windows platform, which target at the provincial agricultural cultivating system (Xindan et al., 2000) . The Agricultural Mechanization Office of the Department of Agriculture put an agricultural mechanization statistical system into use, which is based on PC and developed by PB, which can be used in data input, transportation and storage. To the upper layer of the managerial object, just import the statistical data disc reported by the various areas. Thus the system brings convenience to the statistical job for the manager. But during the data transporting process, it is still necessary to mail the floppy disk or transport the file and then import the data to the upper system by certain person.

All those application of these systems have greatly increased the efficiency of the agricultural machinery management, and saved numerous manpower and material. But the reported examples are all single-user systems which can not meet the need of the digitalized and informationized society. There are disadvantages such as low efficient, lagged information and errors during the transporting and disposing the data. Therefore developing such a management system based on Browser/Server to realize the distributed management through network is of rich practical value. (Singh et al., 1994)

2. SYSTEM ANALYSES

There are 13 prefecture-level cities in the province of Jiangsu at present. Report forms of agricultural machinery from various regions are reported to the upper layer in the forms of paper and floppy disk. The statistical working conference is held each year. The agricultural machinery bureaus from various cities are asked to attend. According to the investigation and analysis about the system, we can get the flow chart of operational information.

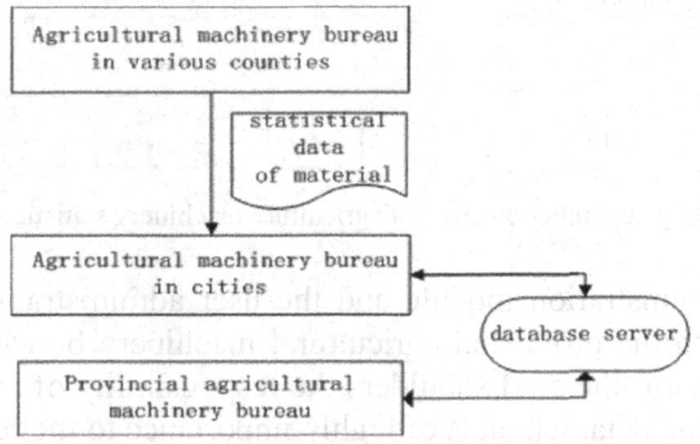

Fig.1. The flow chart of operational information

To those statistical staff who work in the county-country , they can only receive the information by traditional ways because of the grass-root units, comparative low-level application of computer, and inaccessibility of network. Thus the paper material is the basis of the overall data system.

To those staffs who work in the cities, some application on PC such as Excel is available to the statistic. Then after signed by the chief of the bureau, they can fill and submit the data on-line and mail the signed material to the provincial agricultural machinery bureau.

To those staffs who work in the provincial agricultural bureau, they have to complete the collection of the all the statistical material after the new statistical material is generated by the system automatically and provide the result of statistic and analysis to the chief leader to make decision.

3. SYSTEM DESIGN

3.1 Function design

The overall plan was made. The subsystem was divided into smaller subsystems while designing the system. To get a satisfying system with a better structure, we have to optimize the system according to some principles such as coupling rule and cohesion rule among modules.

Figure 2 is the physical function chart of farm machinery statistic system.

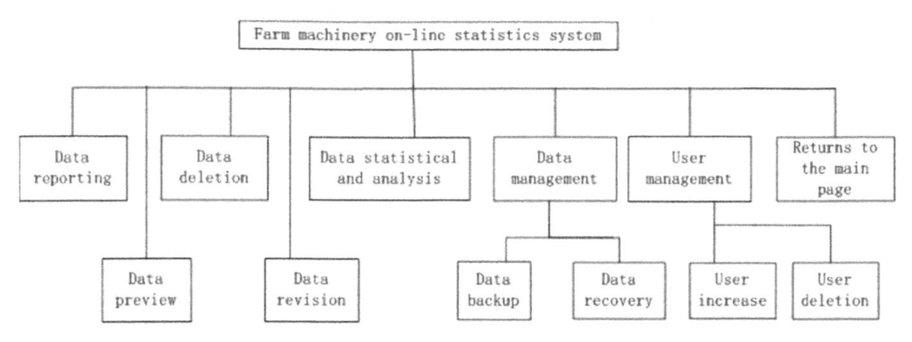

Fig .2. Physical function chart of agriculture machinery statistic system

The data administration module and the user administration module are only available to the provincial agricultural machinery bureau staff. As the super administrator, the staff shoulders the responsibility of maintaining and administrating the data, which is of highly importance to the reliability of the system. This setting will prevent the data from destruction on occasion of incorrect manipulation or some interference. At the same time, the staff can add or delete any user and give them varies of management authority.

3.2 System development platform

Considering that the ASP technology developed by Microsoft Company is comparatively early with a mature operation and wide spread, ASP.net is chosen to be the compiling environment of the system. Thus the platform will be Windows 2000 Server + IIS 5.0 + ASP.net.

ASP is the short form of Active Server Page. It is a script language based on Web. Seen as the integration of HTML, Script and CGI, it provides with a much more flexible programming than HTML, a higher security than Script and a higher efficient than CGI (Yuying et al., 2002) . Figure 3 shows the working principle of ASP

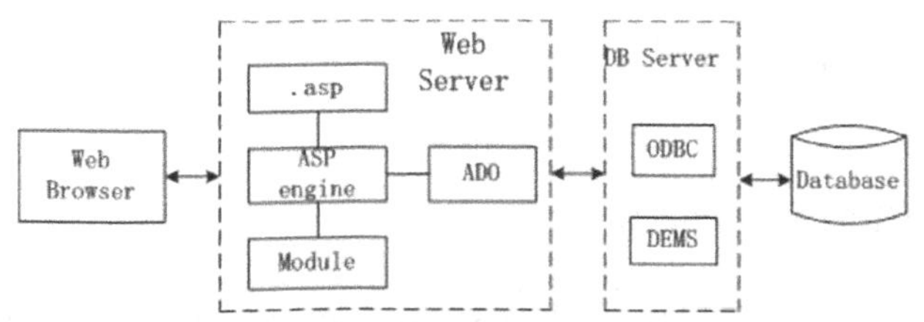

Fig.3. The chart of the work principle of ASP

4. THE REALIZATION OF THE MAIN FUNCTION

4.1 The realization of the data statistical and analytical function of the civic bureau

After the administrator of the agricultural machinery of the civic bureau enters the system, he can manage the data in the interface. Statistical analysis is one of the important functions. Because of the large quantity of the statistical data, we divide the database into 17 data tables according to the requirement of the Agricultural Machinery Office of the Department of Agriculture. Thus the system can easily do the statistical and analytical job of any table.

When the civic administrator selects the statistics & analysis, he will be asked to select the table to do. Then he will enter the main interface. If the chart analysis is chosen, he will get the yearly statistical chart of the table. There is the list box with several fieldnames under the chart for the administrator to choose, and then the yearly column diagram in accordance with the fixed fieldname is on show automatically.

Figure 4 is a column chart of an agricultural machinery overall power analysis of 3 years exemplified by the Agricultural Machinery Bureau of Nanjing.

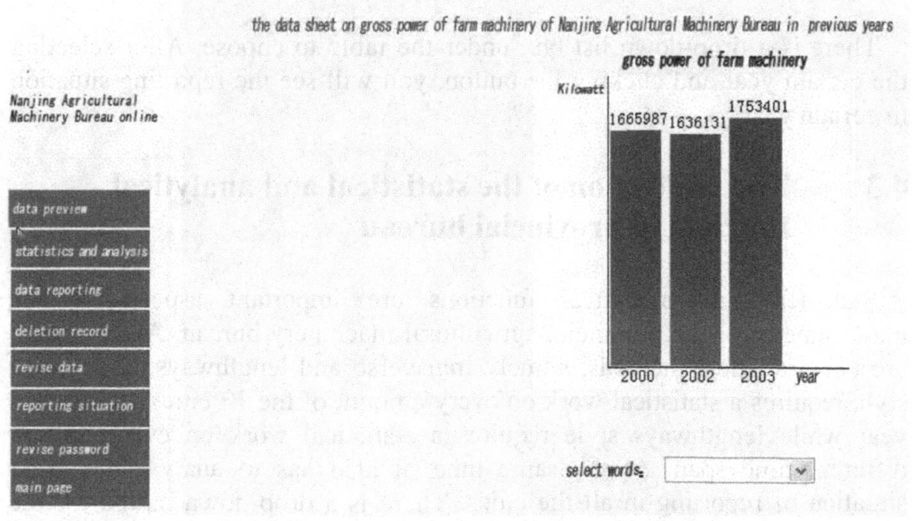

Fig .4. The example chart of statistics and analysis

4.2 The realization of the data reporting function of the civic bureau

With so many data tables to be reported, the data reporting on-line will cost around a month. The administrator must know the report status of all of the tables and then give directions to the users of the system. The function of reporting aims mainly at the convenience for the administrator to examine the overall situation of reporting such as the exact time, the concrete items of each table. While designing the system, the 17 tables will be searched, and then demonstrated the reporting in the form of table. Figure 5 is the realization:

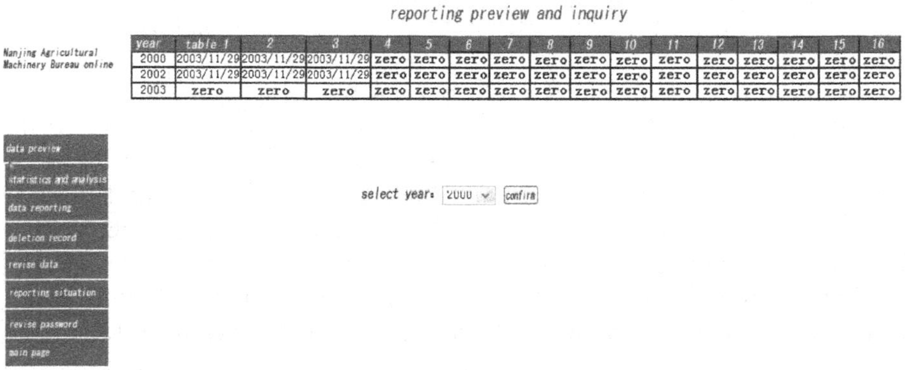

reporting preview and inquiry

Nanjing Agricultural Machinery Bureau online

year	table 1	2	3	4	5	6	7	8	9	10	11	12	13	14	15	16
2000	2003/11/29	2003/11/29	2003/11/29	zero	zero	zero	zero	zero	zero	zero	zero	zero	zero	zero	zero	zero
2002	2003/11/29	2003/11/29	2003/11/29	zero	zero	zero	zero	zero	zero	zero	zero	zero	zero	zero	zero	zero
2003	zero	zero	zero	zero	zero	zero	zero	zero	zero	zero	zero	zero	zero	zero	zero	zero

data preview
statistics and analysis
data reporting
deletion record
revise data
reporting situation
revise password
main page

select year: 2000 confirm

Fig.5. The example chart of reporting data

There is a drop-down list box under the table to choose. After selecting the certain year and clicking the button, you will see the reporting situation in certain year.

4.3 The realization of the statistical and analytical function in provincial bureau

Statistical and analytical functions are important aspects of the management for the provincial agricultural machinery bureau. Always there are two statistical methods, namely transverse and lengthways. Transverse style requires a statistical work on every amount of the 13 cities in a certain year while lengthways style requires a statistical work on every unit in different time span. At the same time, it also has to analyse the exact situation of reporting in all the cities. There is a drop-down button for the administrator to select a table to continue, then will enter the corresponding interface after the clicking. There is a chart statistical button at the bottom of the interface. The administrator will be requested to select the abscissa value after clicking the button and then he can choose a certain region or year in the corresponding drop-down list box. After clicking the button the statistical work will be done by the chosen abscissa. Supposing the agricultural machinery total power table is chosen, with the interface as the result in Figure 6:

Manager online

data preview
statistics and analysis
reporting situation
increases the new user
deprives user's qualification
revise password
data backup
data recovery
main page

gross power of farm machinary

area	year	the power of diesel oil engine (kilowatt)	the power of gasoline engine (kilowatt)	the power of electric motor (kilowatt)	gross power of farm machinaer (kilowatt)
Changzhou	2002	768629	69346	812509	1650484
Nanjing	2002	990541	41140	604450	1636131
Wuxi	2002	728155	152161	728713	1609029
Xuzhou	2002	3147172	39069	835223	4021464
Suzhou	2002	1245568	111961	1040450	2397979
Nantong	2002	1381720	144835	1071020	2597575
Huaian	2002	2009991	23432	388225	2421648
Yancheng	2002	2606413	223329	983048	3812790
Yangzhou	2002	1081417	39202	619697	1740316
Zhenjiang	2002	592856	69720	540586	1203162
Lianyungang	2002	1612637	37621	514197	2164455
Taizhou	2002	1315404	12870	652363	1980637
Suqian	2002	2165739	18504	401342	2585585

graph analysis

remark:

no submitted data this year:

Fig .6. The main screen of statistics and analysis

The analytical button is stimulated in the chart to choose the chart statistical function. And the cities which have not reported the data will be listed in the memo automatically by the system. The development of various regions can be easily seen on the transverse chart (Hua et al., 2005) . Figure 7 is the chart of different cities comparing.

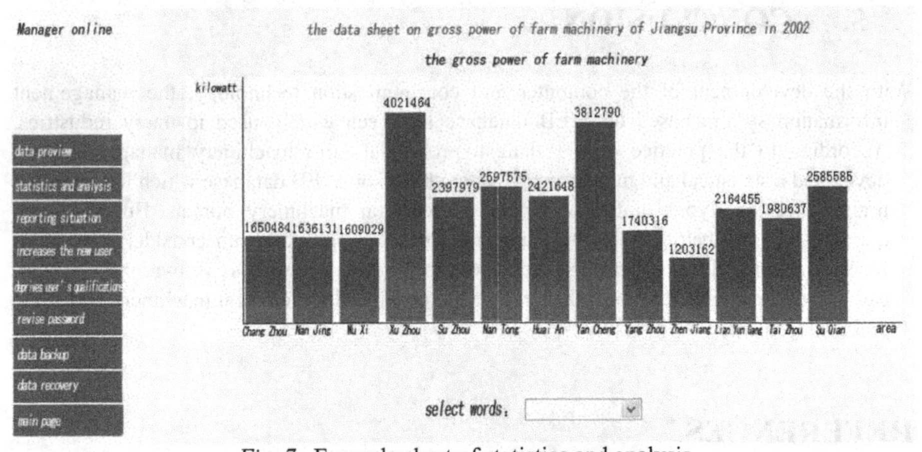

Fig .7. Example chart of statistics and analysis

The administrator can select different fields in the drop-down list box. Then the chart will be given automatically in the form of column diagram to show the differences between cities.

The difference between various years in the certain region can be easily showed by the column diagram. There are two choices in the list box, namely region and year. When the region is chosen, the statistical chart in time span in accordance with the year will be shown as well as the reporting status in the year. The statistical button is available and it can provide with a statistical chart. Figure 8 is the example of the lengthways style.

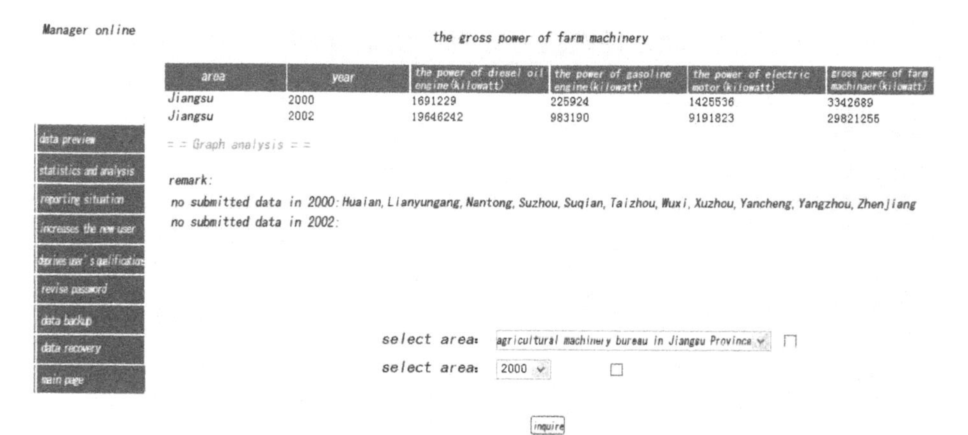

Fig.8. Example chart of the lengthways style

There is a drop-down list box including region and year choices in the analytical chart to make a multifactor output.

5. CONCLUSION

With the development of the computer and communication technology, the management information system based on WEB database has been widely used in many industries. According to the practice of the Jiangsu provincial farm machinery management, we developed a statistical and management system based on WEB database which has received praise after the tryout in the provincial agricultural machinery bureau. But the large quantity of users belonged to the country have not been taken into consideration. With further economy development and the access to hardware as well as software, the system can be further extended to meet the need of the provincial agricultural machinery statistical and management.

REFERENCES

Guosheng Yan . The application of computer in the management of agricultural mechanization[M]. BeiJing, Qinghua University Press,1998:

He Yong Bao Yidan. The investigation on agricultural mechanization database management system[J]. Science and technology Journal. 1995,(3): 147-151.

Hua Li,Wenqing Yin. The realization of the agricultural machinery statistical chart based on the WEB environment[J]. Journal of Anhui Agricultural Sciences. 2005,33 (8): 1395-1396.
Ruiyin He. The development of agricultural machinery management system in large and medium farms[J]. The research on agricultural mechanization. 1998,(4): 84-87.
Singh Gajendra,Pathak B. K. A decision support system for mechanical harvesting and transportation of sugarcane in Thailand[J]. Computers and Electronics in Agriculture. 1994,(11): 173-182.
Xindan Qi,Hua Li. The development of the software of statistical management system[J]. Nanjing Chemical and Industrial University Transaction. 2000,(6): 10-12.
Yuying Wang, Yangyu Ou,Chuanjiu Han. The realization of dynamic statistical chart in WEB[J]. Guilin Engineering College Transaction. 2002,22 (2): 201-203.
Zhong Yao. Distributed agricultural statistical information management system[J]. Computer and Agriculture. 1995,(3): 20-22.

DESIGN AND IMPLEMENTATION OF ACULOPS LYCOPERSICI POPULATION DYNAMIC MODEL PROTOTYPE BASED ON CELLULAR AUTOMATA

Shuai Zhang*, Dongsheng Wang, Linyi Li, Yongda Yuan

Center of Information Technology in Agriculture, Shanghai Academy of Agriculture Sciences, Shanghai, P. R. China 201106

** Corresponding author, Address: Center of Information Technology in Agriculture, Shanghai Academy of Agriculture, Shanghai 201106, P. R. China, Tel: +86-21-62200281, Fax: +86-21-62204989, Email: zhangshuai@lreis.ac.cn*

Abstract: Faced upon the research status of Aculops lycopersici, the importance of population dynamic has been put forward. The feasibility and superiority of cellular automata applied in the simulation of Aculops lycopersici has been discussed. This paper has put forward an Aculops lycopersici population dynamic model prototype based on cellular automata, the result showed that this model can be used to simulate the dynamic population of Aculops lycopersici. When it is applied, the improvement of parameter should be considered, at the same time, this model could provide reference for the simulation of other species of insect.

Key words: Aculops lycopersici, cellular automata, population dynamic

1. INTRODUCTION

Aculops lycopersici is classified as Arthropoda, Arachnida (Acari: Eriophyidae). It has been first discovered in Australia (Tryon, 1937), and until 1986 there ware 47 countries which has reported the occurrence of Aculops lycopersici (Nemoto, 2000). Now Aculops lycopersici has became a world wide serious pest with tomato as main host (Jeppson et al., 1975).

Please use the following format when citing this chapter:

Zhang, S., Wang, D., Li, L. and Yuan, Y., 2009, in IFIP International Federation for Information Processing, Volume 294, *Computer and Computing Technologies in Agriculture II, Volume 2*, eds. D. Li, Z. Chunjiang, (Boston: Springer), pp. 1319–1328.

Aculops lycopersici has been first reported in China by Kuang of Nanjing Agriculture University in 1983(Kuang et al., 1983). In 1999 Chen made an introduction of harm on the protected tomato by Aculops lycopersici in Kunming (Chen, 2000). As the harm to the protected vegetable by Aculops lycopersici has became more and more serious, some scholars has do researches on the physiological characteristics, ecological characteristics and control techniques of Aculops lycopersici. The impact of temperature and relative humidity on the growth and propagation of Aculops lycopersici has been researched (Xu et al., 2006). Wu has done the research of the impact to the tomato plants by Aculops lycopersici (Wu et al., 2006). Zhang has made clear the distribution of Aculops lycopersici in China (Zhang et al., 2007). And the work of biological control and chemical control has been well developed (Xu, 2006).

Many works have been done about Aculops lycopersici, but the rule of population dynamic of Aculops lycopersici is still unknown to us. Without the knowledge of population dynamic, the control work of Aculops lycopersici would stay the passive prevention stage. As there are many works about Aculops lycopersici which provide a good basis to research of population dynamic, if the research of population dynamic about Aculops lycopersici has been done, the result will be quite important to Aculops lycopersici, and the control work of Aculops lycopersici would go up to initiative prevention stage.

Now there are many scholars has do the population dynamic research about other kinds of insect pests to predict the occurrence of them. The methods which have been used are as follows:

1) Mathematics model

Logistic model: Some scholars predicted the number and area of insect pests by use logistic equation (Shen et al., 1985; Cui et al., 1998), the relationship between insect pests and environment factors could be generalization by analysis of insect pests occurrence system data, and the insect appearance area, damaged extend, population density can be predicted (Paul et al., 2004; Baumgartner, et al., 1998).

Matrix model: Matrix model can used to simulate the respond to environment of insect pests in different life stages (Leslie, 1945). Based on the Leslie matrix model, the variable dimension has been developed (Pang et al., 1980; Xu et al., 1981).

Mathematical equation: Differential equation and difference equation are availability tools to simulate the population dynamic of insect pests. Zhang has derived the nonlinear partial differential equation to describe the population dynamic of insect based on the principles of density dependence (zhang et al., 2001). Ruesink has simulated the population dynamic of alfalfa weevil by use difference equation, after that other scholars has improved this model to make the result more realistic (Wu et al., 1990; Zhang et al., 1994).

This method has higher accuracy because of advanced mathematic method, occurrence period, occurrence area and damage degree can be predicted by use this method, but it only can predict the change of quantity, the spatial difference couldn't be expressed.

2) GIS Technology

Spatial analysis of GIS (Geographic Information System): It is availability method to research the spatial distribution of insect pests by analysis the relationship between geographic condition, climate resources and occurrence of insect pests (Shepherd et al., 1988; Schell et al., 1997).

Geo-statistics Combined with GIS: Some scholars also use GIS combined with geo-statistics to simulate the occurrence area of insect pests. The spatial missing data could be supplemented, and the result is more exact.

This method makes the distribution of insect pests simple and clear, but it is suit for researches in large scale, the relationship between environment factors and population dynamic of insect pests is not quantitative, though geo-statistics can make the damage degree quantitative, the result is at current time, the state in future couldn't be predicted.

3) Complex system modeling

Agriculture system is a complex system. The population dynamic of agricultural insect pests is related with many factors. As the development of information technology, the application of complex system modeling has became more and more extensive.

Artificial Neural Networks: Artificial Neural Networks has a good effect in predicting the trend of insect occurrence (Chon et al., 2000). Drake has applied BP Neural Network in the prediction of Plague Locust in Australia (Drake, 2001). Wei has put forward a predict method of agricultural insect pests base on fuzzy neural network, the result showed that this method is easy to use, the predict result is accuracy and it has extensive value in use (Wei et al., 2007).

Intelligent agent: Agent is a focus of artificial intelligence. It has been applied in the simulation of insect pests. Parry established a model based on model to simulate the number of aphid in cropland, each aphid was considered as an agent, the propagation, growth, death, and movement of aphid was simulated. Some domestic scholars also use agent to establish the population dynamic model of agricultural insect (Sa et al., 2005).

Cellular automata: Cellular automata have extensive applications in other field, such as the simulation of traffic flow (Zhou et al., 2005) and the change of land use (Fang et al., 2005), and it also has been used in simulation of plant invasion and forest insect pests. Cole V has established a model by using GIS and cellular automata to simulate the invasion of plant, it is a random parameters model and the species is not definite, so this model has a universal adaptability (Cole et al., 1999). Zhou G has simulate the break out of gypsy moth by cellular automata, four different ways was used to implement the cellular automata model, and the result shows that the effect of inverse distance weighted model is more accurate (Zhou et al., 1995).

The mathematic model can express the quantitative change quite well, but the spatial change of insect pests couldn't be expressed by it. While this problem can be solved by use GIS technology, GIS is good at space expression, but the quantitative change couldn't be described well, combined with geo-statistic the quantitative change could be described while the change is spatial, temporal change is hard to be expressed by this method. Complex system modeling can simulate the complex system accurately, and it is easy to use. But the Artificial Neural Networks is also couldn't express spatial change, and Intelligent agent put more attention on the behavior of the agent, the relationships between agent and experiment, agent and agent are not expressed well.

Cellular automata is a kind of complex system model It can express the spatial change and temporal change quite well. There is two state of a cell: live and die. The state of cell would be influenced by neighbor cell and the environment. The mode is simile to the occurrence of insect pests, especially Aculops lycopersici is a kind of micro insects which couldn't be observed by eyes, it is a problem to use other method, but it can be solved by cellular automata (the concrete method is introduced in next paragraph). So, cellular automata can be used to simulate the population dynamic of Aculops lycopersici.

2. DESIGN OF ACULOPS LYCOPERSICI POPULATION DYNAMIC MODEL PROTOTYPE

The model prototype is made up of three main parts: tomato plants, Aculops lycopersici and environment factors. Tomato plants can be ingested by Aculops lycopersici, and environment factors impact the growth of tomato plants and Aculops lycopersici.

This model described the dynamic change of Aculops lycopersici in a 2-dimension space.

Aculops lycopersici is a kind of micro insect pests, so the body of Aculops lycopersici couldn't be considered as the basic cell, the grid with the size of 10 centimeters * 10 centimeters was considered as the basic cell in the model. In the range of a cell, only the growth state and number of Aculops lycopersici were considered, the position of Aculops lycopersici was not be considered, so the spatial change of Aculops lycopersici only could happened between each cells. Day is the basic time unit of the model.

2.1 Object

2.1.1 Tomato plants

Nutrition is abstracted from tomato plants, and the change of nutrition will represent the growth process of Aculops lycopersici.

2.1.2 Aculops lycopersici

The duration of Aculops lycopersici has contain four states: egg, larva, nymph and adult.

Egg has a certain survival rate, survival eggs will grow to larva after egg stage, the

Larva has a certain survival rate, survival larva will grow to nymph after larva state, and larva can absorb the nutrition.

Nymph has a certain survival state, survival nymph will grow to adult after nymph stage, and nymph can absorb the nutrition of tomato plants. Nymph also has the ability of diffusion which egg and larva don't have.

Adult has a certain survival rate, and the survival adult has ability of spawning. And the same to the nymph, adult can absorb the nutrition, and has the ability of diffusion.

2.1.3 Environment factors

The environment factors of this model are mainly the temperature and relative humidity, which has an important impact on the process of tomato plants growth and Aculops lycopersici growth, diffusion.

2.2 Rules

According to the growth, diffusion process of Aculops lycopersici and the growth process of tomato plants, the rules of cellular automata have been established.

The rules of the model prototype based on cellular automata are composed by development rule, propagation rule, extinction rule, diffusion rule, tomato growth rule. The concrete contents of these rules would be different to Aculops lycopersici in different stage.

Development rule: The developmental threshold temperature (DTT) of the egg is 10.51℃, the relationship between egg stage and temperature is Y (egg stage) =0.0228T (temperature) – 0.2343, the relationship between egg stage and relative humidity is Y (egg stage) =0.4387 – 0.1397RH (relative humidity); The DTT of the larva is 9.02℃, the relationship between larva stage and temperature is Y (larva stage) =0.0148T (temperature) – 0.0935, the relationship between larva stage and relative humidity is Y (egg stage) =0.4449 – 0.1310RH (relative humidity);The DTT of the nymph is 9.02℃, the relationship between nymph stage and temperature is Y (nymph stage) =0.0148T (temperature) – 0.0935, the relationship between nymph stage and relative humidity is Y (nymph stage) =0.4449 – 0.1310RH (relative humidity).

Propagation rule: Only adult has the ability of oviposition. The relationship between oviposition period and temperature is Y (oviposition period) = -1.0274T (temperature) + 47.508, the relationship between oviposition period and relative humidity is Y (oviposition period) =0.098 RH (relative humidity) +12.536; the relationship between fecundity and temperature is Y (fecundity) = 2.3367T (temperature) - 17.313 ($X < 26$), Y (fecundity) = -3.007T (temperature) – 122.9 ($X > 26$).

Extinction rule: Egg has a certain survival rate, only survival eggs will grow to larva after egg stage, and the survival rate of egg is 95%, the survival rate of larva is 85%, the survival rate of nymph is 85%, the survival rate of adult is 80%. The adult will die after oviposition.

Diffusion rule: if the total number of Aculops lycopersici in a cell has reached the threshold, the nymph and adult will diffuse to the cells in which there is enough nutrition around the center cell, and when the nutrition in a cell in not enough for the Aculops lycopersici in it, the egg and larva will die, and the nymph and adult will diffuse to the cells in which there is enough nutrition around the center cell.

Tomato growth rule: set the initial nutrition of tomato plants is 100 thousand unit of nutrition. And the relationship between tomato nutrition and temperature is $Y = 50000 + (X - 20) * 2500$. The nutrition growth would stop when it reached

Nutrition ingested rule: each larva ingests 3 units of nutrition a day; each nymph ingests 4 units of nutrition a day; each adult ingests 5 units of nutrition a day.

2.3 Steps

Step 1: the growth of tomato. According to the tomato growth rule and the value of tomato nutrition in each cell at last time, the new value of the tomato nutrition in each cell has been determined, and the initial distribution map of tomato nutrition has been formed.

Step 2: the growth of Aculops lycopersici. According to the development rule and propagation rule and the number of Aculops lycopersici in each cell at last time, the number of Aculops lycopersici at different growth periods in each cell has been determined, and the initial distribution map of Aculops lycopersici has been formed.

Step 3: the diffusion of Aculops lycopersici. According to the nutrition ingested rule and the initial distribution map of Aculops lycopersici, the quantity of nutrition that would be ingested by Aculops lycopersici can be determined, and it will be overlapped with the initial distribution map of tomato nutrition, the final distribution map of tomato nutrition has been formed after that, and the final distribution map of Aculops lycopersici also can be formed according to the diffusion rule.

3. IMPLEMENTATION OF ACULOPS LYCOPERSICI POPULATION DYNAMIC MODEL PROTOTYPE

The model prototype is implemented using ArcGIS Engine 9.1, the integrated development environment is Visio Studio.Net 2005.

Set the space is 25 meters * 30 meters, and the tomato plant in the center is set as the initial position of Aculops lycopersici. And the number of Aculops lycopersici is 40000, the temperature is 26℃, the relative humidity is 53%.

According to the rules and steps above, the state of the 40000 Aculops lycopersici in 60 days has be simulated, the result of the model is showed in Fig.1 and Fig.2 as follows.

In the 60 days, there are three peaks of Aculops lycopersici population. About 55 days later, the number of Aculops lycopersici has reached maximum. Correspond in Fig.2, the distribution of Aculops lycopersici has get a maximum area in the 50th days, after that many tomato plants died because of Aculops lycopersici, and has became more centralized as showed in the 60th day.

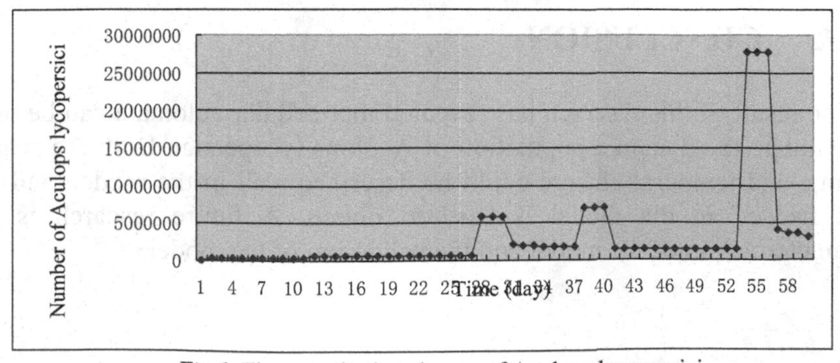

Fig.1. The quantitative change of Aculops lycopersici

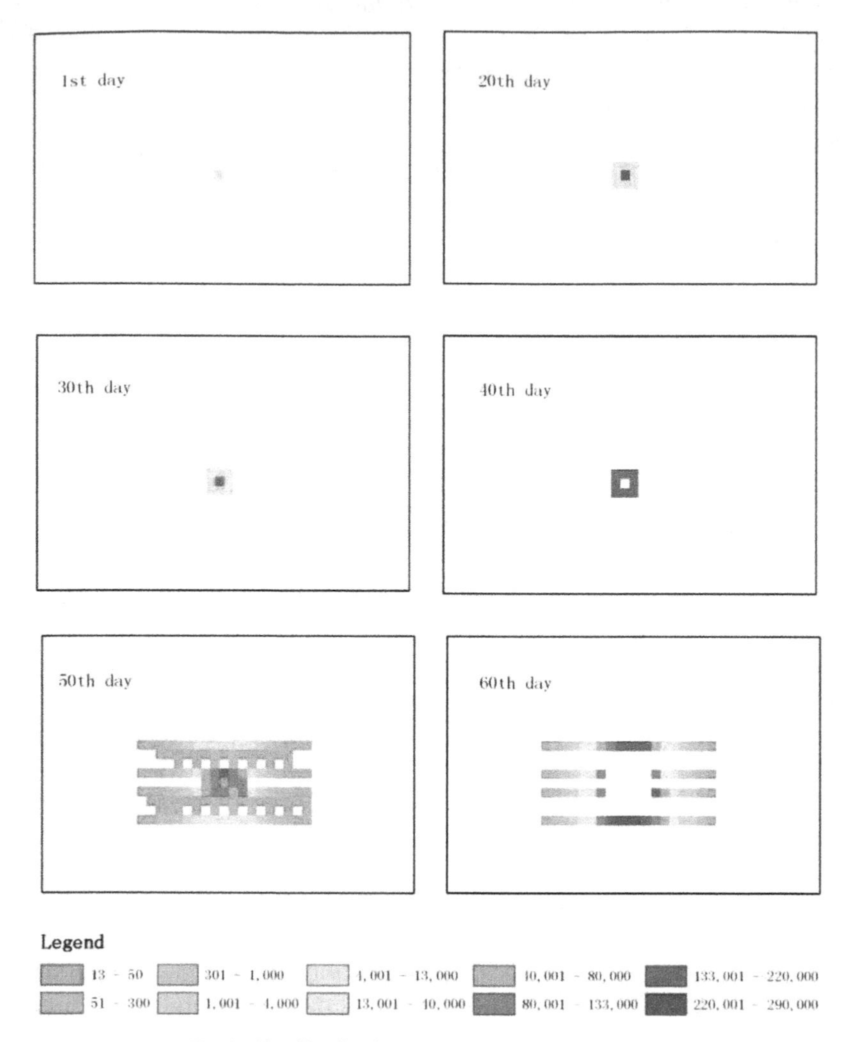

Fig.2. The distribution change of Aculops lycopersici

4. CONCLUSION

The result of this research has revealed that cellular automata can be used to simulate the dynamic population of Aculops lycopersici. Both the spatial change and temporal change could be described well in the model, and the data needed in the model is easy to obtain. A future research is the parameterization of the model and the validation of the model.

ACKNOWLEDGEMENTS

We are grateful to colleagues who did the experiment of Aculops lycopersici and provide the data of ecological characteristics about Aculops lycopersici – the plant protection research group in Eco – Environment and Plant Protection Institute, Shanghai Academy of Agricultural Sciences. This work is supported by Extension and Application of vegetable insect and diseases diagnoses, prevention and cure expert system Project (Contract Number: 075119N85) of National Transformation of Agricultural Science and Technology Achievement Fund Program.

REFERENCES

Baumgartner, et al. Quantitative analysis of gypsy moth spread in the central appalachians. Population and Community Ecology for Insert Management and Conservation, 1998:99-110.

Chen Bin, Luo Youzhen, Yin Suigong. A new insect pests of tomato - Aculops lycopersici. Yunnan Agricultural Science and Technology, 2000, 2:32-33. (in Chinese)

Chon TaeSoo, Kim JaMyung, Kim JaMyung. Use of an artificial neural network to predict population dynamics of the forest pest pine needle gall midge (Diptera:Cecidomyiida). Environmental Entomology. 2000, 29(6): 1208–1215.

Cole V, et al.. Modelling the spread of invasive species paramter estimation suing cellular automata in GIS, 1999. Department of Geography University of Auckland, New Zealand.

Drake. Use of remote sensing and ANN in prediction of pets in Queensland. Remote Sensing of Environment, 2001, 12(4):32-35.

Fang S.F., Gertnera G.Z., et al.. The impact of interactions in spatial simulation of the dynamics of urban sprawl, 2005. Landscape and Urban Planning.73: 294-306.

Jingan Cui, Lansun Chen. The effect of diffusion on the time varying logistic population growth. Computers & Mathematics with Applications, 1998, 36(3):1-9.

Kuang haiyuan. Two new species and a new record species of gall mites from China. Acta Zootaxonomic Sinica, 1983, 8(4): 389-391. (in Chinese)

L.R.H. Jeppson, H.H. Keifer, E.B. Baker. Mites Injurious to Economic Plants, University of California Press, Berkeley, 1975, 614.

Leslie P H. On the use of matrices in certain population mathematics. Biometrika, 1945, 33: 183-211.

Pang Xiongfei. On The Use of Population Matrix Models For The Studies of Insect Ecology. Journal of South China Agricultural University, 1980，1（3）：27-37. (in Chinese)

Paul W F, et al. Simulation model of Rhyzopertha dominica population dynamics in concrete grain bins. Journal of stored products research, 2004, 40:39-45.

Sa Li, Xiong Fanlun, Ding Jing, Zuo Honghao. An Agent-based Cellular Automata Models for Agroecosystems, Jouranl of University of Science and Technology of China, 2005,35(2):270-276.

Schell S P, J A Lockwood. Spatial analysis of ecological factors related to rangeland grasshopper (Orthoptera: Acrididae) outbreaks in Wyoming. Enbiron Entomol, 1997, 26(6):1343-1353.

Shen Zuorui. The Modification of Logistic Fomula and Its application in Vegetable Aphides. Journal of Beijing Agricultral University, 1985, 11(3): 297-304. (in Chinese)

Shepherd R F. Proc Lymantriidae. A comparison of features of new and old world tussock moths. Washington DC: USDA, 1988.

Wei Mingshe, Guo Yong. Prediction of Crops Pests Level Based on Fuzzy Neural Network. Journal of Taiyuan University of Science and Technology, 2007，28（6）：442-445. (in Chinese)

Wu Juan, Li Linyi, Xu Xiang, Yang Yizhong, Wang Dongsheng. Physiological Variation of Damaged Leaves of Tomato by Aculops lycopersici. Acta Horticulturae Sinica, 2006, 33(6): 1215-1218. (in Chinese)

Wu Zhongfu. Research on Simulating The Dynamic of Nilaparvata lugens Population. Journal of Fujian College of Agriculture, 1990, 19(2):115-122. (in Chinese)

Xu Rumei. Application of A Dimension-Changeable Matrix Model on The Simuluation of The Population Dynamics of Greenhouse Whiteflies. Acta Ecologica Sinica, 1981，1（2）：147-158. (in Chinese)

Xu Xiang, Li Linyi, Wang Dongsheng, Hong Xiaoyue, Wu Juan, Yuan Yongda, Xie Xianchuan. Effect of temperature and relative humidity on development and reproduction of the tomato russet mite, Aculops lycopersici (Massee)(Acarina, Eriophyidae). Acta Entomologica Sinica, 2006, 49(5): 816-821. (in Chinese)

Xu Xiang. Ecology and Control of the Tomato Russet Mite Aculops lycopersici(Tryon). Nanjing Agricultural University, Master Degree Thesis. (in Chinese)

Zhang Shuai, Li Linyi, Wang Dongsheng, Zhao Jingyin. Research on Adaptive Distribution of Aculops Lycopersici in China，Proceedings of the 4th International Symposium on Intelligent Information Technology in Agriculture(ISIITA)，2007,491-494.

Zhang Wenjun, Gu Dexiang. A Non-Linear Partial Differential Equation to Describe Spatial and Temporal Changes of Insect Population. Ecological Science, 2001, 20(4): 1-7. (in Chinese)

Zhang wenqing, Gu Dexiang, Pu Zhelong. Improvement In The Method For Simulating The Dynamic of Insect Population: A Study on The Dynamic Population Simulation Model of Paddy Stem Borer (Tryporyza Incertulas Walker). Acta Ecologica Sinica, 1994,14(3):281-289. (in Chinese)

Zhou G, et al, Forecasting the spatial dynamics of gypsy moth outbreaks using cellular transition models. Landscape Ecology, 1995, 10 (3): 177-189.

Zhou Zili, Wang Xinwei, Wang Yanna. Simulation System of City Traffic Flow Based on Cellular Automata. Computer Engineering, 2005, 31(13): 183-185. (in Chinese)

THE DESIGN OF EMBEDDED VIDEO SUPERVISION SYSTEM OF VEGETABLE SHED

Jun Sun[1, 2,*], Mingxing Liang[2], Weijun Chen[2], Bin Zhang[3]

[1] *Key Laboratory of Modern Agricultural Equipment and Technology, inistry of Education & Jiangsu Province, Jiangsu University, Zhenjiang, Jiangsu Province, China 212013*

[2] *School of Electrical and Information Engineering ,Jiangsu university, Zhenjiang ,Jiangsu Province, China 212013*

[3] *ChangZhou Institute of Technology, Changzhou,Jiangsu Province,China 213002*

[*] *Corresponding author,Address:School of Electrical and Information Engineering, Jiangsu University, Zhenjiang 212013,Jiangsu Province,China, Tel:(0)13775544650, Fax:0511887 80311, Email :sun2000jun@ujs.edu.cn*

Abstract: In order to reinforce the measure of vegetable shed's safety, the S3C44B0X is taken as the main processor chip. The embedded hardware platform is built with a few outer-ring chips, and the network server is structured under the Linux embedded environment, and MPEG4 compression and real time transmission are carried on. The experiment indicates that the video monitoring system can guarantee good effect, which can be applied to the safety of vegetable sheds.

Keywords: embedded system, safety monitoring, MPEG4

1. INTRODUCTION

Along with the function of the vegetable sheds construction being perfected day by day, it turns to be an important topic for the technical personnel to strengthen the safe management of the vegetable sheds. The former small scaled vegetable sheds sends a person on duty by turn inside the shed. This kind of way can attain the purpose of defending, but it also needs a large amount of labor and wastes lots of human resources. At present, some vegetable sheds adopt surveillance nearby the sheds. This method carries out the supervision of all important equipments of the shed, but it is

Please use the following format when citing this chapter:

Sun, J., Liang, M., Chen, W. and Zhang, B., 2009, in IFIP International Federation for Information Processing, Volume 294, *Computer and Computing Technologies in Agriculture II, Volume 2*, eds. D. Li, Z. Chunjiang, (Boston: Springer), pp. 1329–1335.

limited to only a short distance(Kallesoe,2004;Behzad,2004;Zhang,2006). With the fast development of the communication technique of the calculator network and the compress technique in multi-media graphs, a new way for the video supervision in a long range can be provided, in which governor can control the actual condition in the shed much more easily.

Traditional supervision system uses a video frequency collection card to set up a network video frequency in the PC. However, the performance of real-time and opening is not good, price is high, and the extent of integration and the stability of system is bad. At present, in the digital monitoring system, embedded technology and MPEG-4 compress coding technology are doubly subjected to concern. Development of embedded MPEG-4 digital video surveillance system has higher technology and good market prospect. This literature is mainly about the research of the supervision and regulation of the vegetable sheds based on embedded system and network technology.

2. SYSTEM MODEL

We set cameras in key positions both in and out of the vegetable sheds house and use analog shoots to carry on the real-time analog graphs of the shed scene. Because of the special environment of the vegetable sheds, the lighting in night and even the rainy and windy weather should be considered. So in order to collect the video frequency of high quality, generally we choose low shine shoots. The numerical video frequency signal conversion mold piece can change the analog graphs into digital pictures, while the compress mold can reduce the saving space of video frequency pictures. With a network video frequency server built up, users only need to put the IP address of the server into the IE browser address column to browse the picture.

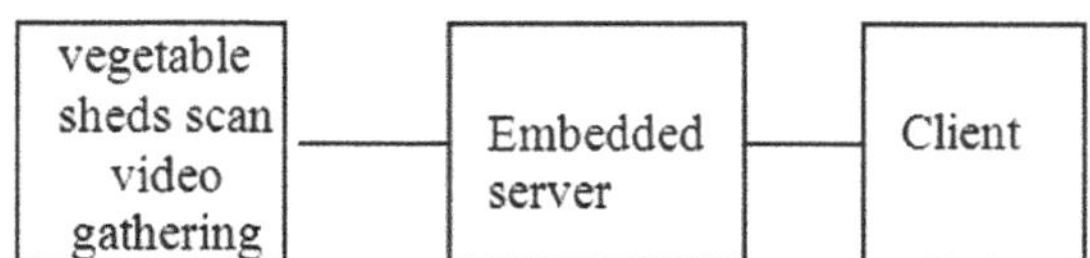

Fig.1 The whole frame image of video monitoring system

3. HARDWARE DESIGN OF EMBEDDED SERVER

The hardware structure includes the video frequency A/D chip SAA7111, core processors ARM7 embedded chip S3C44BOX, MPEG compress chip IME6400, fast Ethernet physical layer chip RTL8019AS. The system hardware frame shows in figure 2.

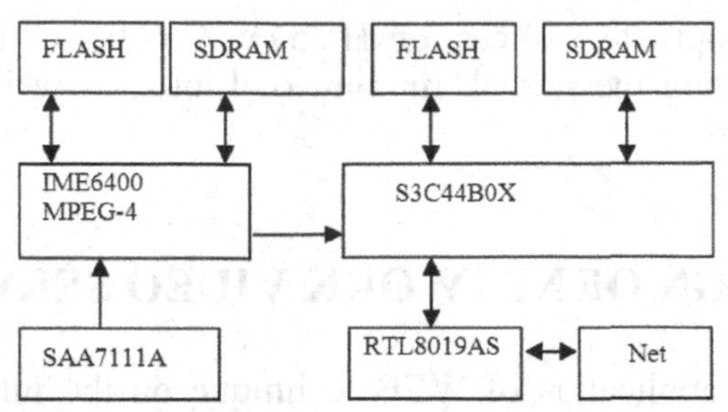

*Fig.2:*The hardware frame image of embedded server

SAA7111A is a kind of enhanced video input processor chip made by PHILIPS, which gathers A/D and decoding at the whole body. Both of the PAL and the NTSC TV standard are fit. The SAA7111A contains I2C interface internal through which work conditions can be set. Signals of SAA7111A such as VREF, HREF, RESO and LLC2 are all lead directly from the pins and have good reliability. Though A/D converter, the video frequency collection mold piece change the original analog video signals into digital signals. In order to deliver the video frequency signal on the net which is limited by its bandwidth, the numerical signal needs chip IME6400 to carry on MPEG4 compressions and then be sent to embedded processor S3C44BOX. In the end, the compression data is sent to Internet through the interface RTL8019AS and RJ45. Users can access the video data which is sent by the network interface to the network by WEB server(Xue., 2006) .

The IME6400 delivers flow for network application or program flow for saving application. The IME6400 is a kind of ideal chip made by INTIME which carries out MPEG-4 compression. Compared with MPEG-1 and MPEG-2, MPEG-4 standard has a higher compression ratio, saving storage space and better image quality. Specially, under the low bandwidth, it can adjust the bit rate of the compressed data from 128 Kbps to 6Mbps, while keeping the quality of the pictures to adapt to various demands of users(Jia.,2005) .

RTL8019AS is a kind of highly integrated Ethernet chip, compatible with NE2000 and supporting IEEE802.3 standard. It can install into 8 or 16 lines with total data. With 16KB SRAM memory inside, the chip can be used as a receiving cache for sending data, supporting full-duplex mode. There are three kinds of the interface modes: jumper mode, non-jumper mode and plug-and-play mode, with the automatic examination equipments to check whether the plug-and-play jumper has been connected or not.

The FLASH chooses the NAND type K9F1208 of 32 Mbytes to deposit the boot codes, the kernel codes and root file systems. In the management of FLASH, root document system adopts the latest system called YAFFS which can be both read and written. We can record the users connection logs into the root file system so that it's easy for the server to call the logs.

The RAM adopts two slices of HY57V561620, total 64 Mbytes of SDARM, facilitating the smooth running of Linux, as well as the network applications.

4. DESIGN OF NETWORK VIDEO SERVER

The extensive application of WEB technique on the Internet causes the popularity of B/S mode, which can be also called “thin customer’s machine”. With the main application procedures existing at the servers, users only demand the browser environment, then it is easy to download application procedure from the severs to accomplish the match tasks according to the needing.

The embedded network video frequency sever adopts embedded integrated structure(He.,2003; Zhang.,2003). With the hardware platform for real-time processing, both of the video frequency compression and network function can be concentrated to a small individual. After collection, compression and compounding, the video frequency signals then turn to the IP packet, and with suitable network protocol, the video compressed data stream can carry on real-time transmission. Users only need a browser to watch instead of installing any hardware equipments.

The software of the network video frequency server contains 5 main modules: system initialization, acquisition and compression, as well as the packing and transmission of the data, responding to users’ requests and management of users.

System initialization module is mainly responsible for reading parameters which are important for the enactment of the server port and the network bandwidth, as well as the picture manifestation. Besides, it can also start the other mold pieces of the sever; build a set of connections with the HTTP service provided according to the port parameters; build up a conjunction, respond to the WEB interview of customers’ machine and handle the claim from users.

In the module of the acquisition and compression of the video frequency data, the main mission is to collect and digitize the video analog signals. As well as, making the full use of the mature compression technique to realize high compression of the data with satisfying the technique index is also very important. This system adopts the MPEG-4 chip to carry on the compression and coding of the original data, then put into the buffer area. It turns easy to carry on the compression and coding of the video with the use of the member function. After the setting parameters are delivered into a type, the member function can work such as showing in figure 3.

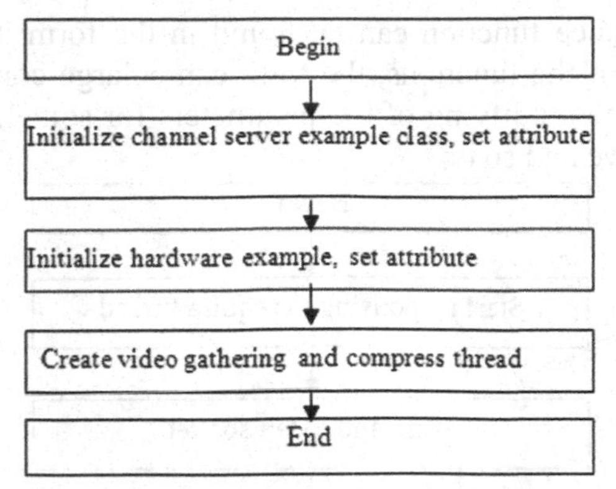

Fig.3: The flow image of video's capturing and compression

The module of packing and transmission of the data compress the data collected into the MPEG-4 format, and then carry on packing according to the RTP protocol. Finally, it's sent out to the particular port of the users who ask for help through the network. The member function of type RTPSeesion reads video frequency data from the buffer area, and then begins to pack. Before the packing, the data needs to be incised into some segments, and then plus the RTP protocol as a heading file to each part. The sending course of the video frequency flow shows in figure 4.

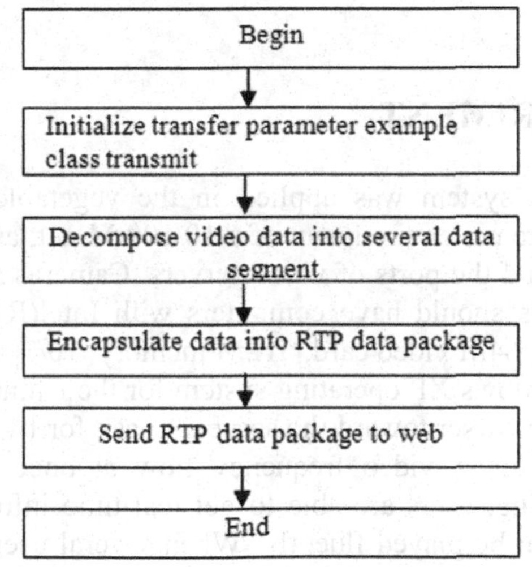

Fig.4: The flow image of sending video

The module of responding to users' requests responds to the instructions sent by users through the video server. The users send out a format instruction toward the particular port of the servers. After obtaining the instruction, according to analysis, the mold piece carries out a processing function to complete a concrete control. The module, in advance, keeps a form in which the orders and functions are match. After analyzing the order, then the performance function can be found in the form in response. With the development of the functions, the form can enlarge continuously, which can accomplish the modifying of the parameters for some servers and drive the camera to move and so on.

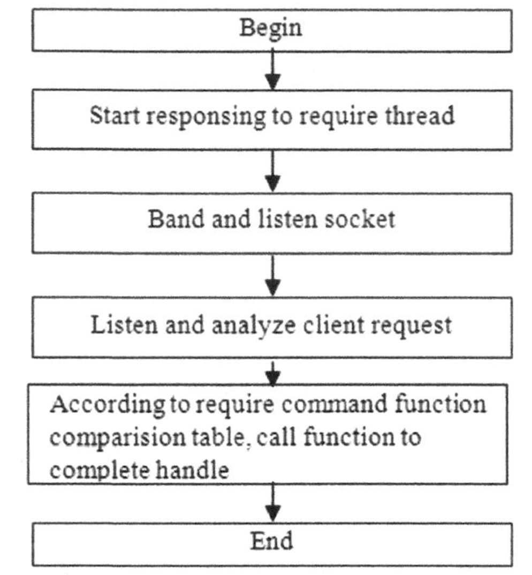

Fig.5: The flow image of answering the client's request

5. EXPERIMENT

The monitoring system was applied in the vegetable shed in Jiangsu University, with the net environment of a 10/100 M net, and hubs were used to connect users and the ports of video servers. Cameras are used to collect spot images. Users should have computers with Intel(R) Pentium(R) four CPU @ 2.66 GHz, 64M video card, 512M memory, 10/100 M gateway, 80G hard disk, and windows XP operating system for the Chinese version. When users carry an IE browser toward the servers to ask for live video streaming, servers can send out a video frequency flow at once. Without obvious wobble and delaying, users are able to get real-time information and clear pictures, which can be played fluently. When several users send out claims in the meantime toward the server, it can carry out multicast, and can satisfy the segmentation switch of daily pictures. It can also playback live video and signal scene. Especially, in the bad environment outside, it can also carry on the real-time monitoring normally.

6. CONCLUSION

Aiming at the practical application needs of the video surveillance of the pump station, we design a set of build-in network video frequency supervision system with several new technologies such as coding image acquisition, build-in system and network technique. Based on the ARM7 chips of S3C44BOX and build-in Linux operating system, the system adopts cameras to catch the video frequency, and after compressing codes by MPEG-4 chip, it can connect the network directly. With standard network browser and the medium player procedure, users can observe the long range video frequency images inside the big shed. It has certain expansion meaning and can satisfy the need of the safe supervision management for the pump station at present.

ACKNOWLEDGEMENTS

Funding for this research was provided by China Postdoctoral Science Foundation under Grant (NO:20070420972);Jiangsu University High-grade Specialty Person Scientific Research Foundation under Grant(NO:05JDG050); Changzhou Young Scientific and Technological Talent Training Plan (NO:CQ2008009).Jiangsu University College Student Scientific Research Project(NO:06A110).

REFERENCES

Behzad, M. Fault diagnosis of a centrifugal pump by vibration analysis, Proceedings of the 7th Biennial Conference on Engineering Systems Design and Analysis, 2004,3:221-226

He Zhi-xia, YUAN Jian-ping, LI De-tao, ZHANG Li-qun. Performance Prediction for Pumping Station System Based on Neural Network Technique. Journal of Jiangsu University(Natural Science Edition), 2003,24(4):45-48(in Chinese)

Jia Guo-fang. Research of the online Intelligent Monitoring Technology of the Water Pump. Drainage and Irrigation Machinery.2005(6):28-30(in Chinese)

Kallesoe,Carsten Skovmose. Model based fault diagnosis in a centrifugal pump application using structural analysis. Proceedings of the IEEE International Conference on Control Applications, 2004, 2:1229-1235

Xue Yun-hui. Image inspect system in Jianbi pump station. Drainage and Irrigation Machinery. 2006(6):27-30(in Chinese)

Zhang Feng-chen, ZHANG Xiao-jing, FENG Qin. Remote Monitoring System on the Pump Station of TM1300. Journal of Heilongjiang Hydraulic Engineering College. 2006,33(2):90-92(in Chinese)

Zhang Haiying, Hu Bin, ZHAO Xingtian, WU Shengyan. Design of Analog Video Conversion Interface . Electronic Engineer,2003(01):53-54(in Chinese)

Zhang Shaobin, LIN Li. Design of Numerical Remote Concentration Surveillance System Based on MPEG-4. Computer Engineering. 2007,33(1):265~267(in Chinese)

REALIZATION OF WORKFLOW SERVICE INVOCATION INTERFACE FOR INTEGRATION OF AGRICULTURAL NETWORK RESOURCES

Dong Wang[1], Ruizhi Sun[1,*]

[1] *College of Information and Electrical Engineering, China Agricultural University, No.17, Qinghua Dong Lu, Haidian, Beijing 100083, China*

[*] *Corresponding author, Address: College of Information and Electrical Engineering, China Agricultural University, No. 17, Qinghua Dong Lu, Haidian, Beijing 100083, China, Tel: 01062737945, Email:sunrz_cn@sina.com*

Abstract: Service Invocation Interface is the interface which workflow engine uses to communicate with external application procedures. Analyzed the difficulties caused by diversity of external network resources to realize service invocation interface, took the description, search and evaluation of these resources into consideration, put forward an integrated service invocation strategy based on SOA structure. This strategy combined the invocation method for resources that could and couldn't be packaged into web services, offered the system a uniform interface to invocate various external agricultural information resources dynamically.

Keywords: Service Invocation Interface, workflow engine, Web Service, SOA Structure, resource evaluation

1. INTRODUCTION

Workflow technology separates application logic from business process logic, which gives it the ability to change system function by modifying only the business process model without involving the realization of specific module, thus makes the collaboration, integration and reorganization of the actual business to be realized swiftly. Using workflow technology in network information service area, integrate info-service resources of a

Please use the following format when citing this chapter:

Wang, D. and Sun, R., 2009, in IFIP International Federation for Information Processing, Volume 294, *Computer and Computing Technologies in Agriculture II, Volume 2*, eds. D. Li, Z. Chunjiang, (Boston: Springer), pp. 1337–1346.

specific industry according to its business process, so that we can accomplish the collaboration and reorganization of info-service resources without changing the specific realization. Take current agricultural information service for example, the build-up of agricultural network information system now is lagged behind those developed countries to some extent, information services provided by agricultural network platforms are not in good cooperation, the problem of so called "information island" is not accidental. Therefore, to provide an integrative platform for agricultural network resources becomes a necessity, and in order to achieve this goal, we need to integrate recourses by building up a collaborative platform for heterogeneous network resources via workflow technology.

In the Workflow Reference Model raised by WFMC, a Workflow Management System (WFMS) has five unified interfaces to exchange data with other resources. These interfaces can also be called the Workflow Application Programming Interface (WAPI), among which the Service Invocation Interface identifies the invocation mechanism between WFMS and external application programs. There is a little specialty of Service Invocation Interface from other interfaces, that is it's not a necessary function of the workflow system, but once a workflow system was ineffective to support Service Invocation Interface, it would be lack of usage in practice to some extent (Chen Shi et al., 2005).

Based on SOA structure, this passage faced the integration of agricultural resources, designed an agricultural information resources collaborative platform, and would introduce in detail the design and realization of its Service Invocation Interface.

2. KEY PROBLEMS OF THE REALIZATION OF SERVICE INVOCATION INTERFACE

Workflow technology could integrate agricultural information resources of different network resource platforms, thus enabled users to check various agricultural information, gain information services of every step in the business process without knowing where exactly each of the resources came from. Therefore, how to make full use of the resources provided by these agricultural info-service platforms effectively, which was also the issue of how to design the system's service invocation interface, became the key point of the design and realization of the system.

It is not an easy job to design a uniform invocation interface for all the service resources of different kinds from various external information service platforms, and the reasons go like follows:

(1) The system needs to collaborate and integrate various agricultural information resources provided by different platforms, which use different

corresponding protocol, and have different access mechanism. So the service invocate interface has to have the ability to correspond with external resources of different kinds.

(2) The critical function of workflow management system is the routing of activities as well as the distribution of tasks. But for the real business function which should be accomplished by each task, the system could not make a prediction, only when it came to the point that the very activity got motivated, the system could decide whether external services should be called and which service was the right choice. This unpredictability of business logic results in the unpredictability of service resource interface. So that the system has to have a resource identification mechanism to identify the interface of the resource we want to invocate.

(3) Resources that need to be integrated come from various agricultural information resource platforms, the differences of these platforms in aspects of programming language and execution environment request the invocation service interface the ability to invocate cross-platform and language-independent network services.

3. REALIZATION STRATEGY OF THE SERVICE INVOCATION INTERFACE

SOA (Service-Oriented Architecture) is a component model which can connect different functional units of applied procedures (also called services) through well defined interfaces and constraints between these units. Interfaces are identified in a neutral way, apart from the hardware platforms, the operating systems and the programming languages used to realize the services. This makes the services from different platforms can interactive in a uniform and general way. It's the best choice to solve problems such as information island and legacy procedures(Liang Xiaodong et al., 2007). he realization of service invocation interface soa is only a design principle. Its structure is shown in Fig.1, which is composed of service provider, service petitioner and service register center.

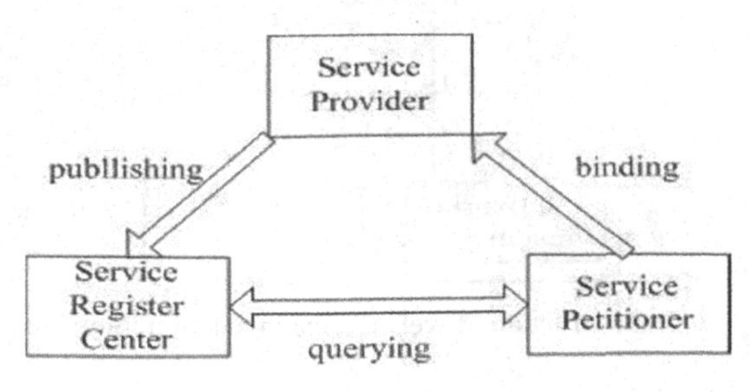

Fig.1: SOA structure

The service invocation interface web service is the best way now to support SOA structure, through its real language, platform independency and unified interface definition. And resources of different languages from different platforms can be invocated uniformly when being packaged into web services. However, there still remain quite a lot of legacy agricultural resource platforms in our country, which are inconvenient or even impossible to be packaged into web services. So, resources from these legacy resource platforms need one other strategy to be described and invocated by the way of showing their unified interfaces to service callers as independent services.

3.1 Realization of web services' invocation

Web service has inherent advantages in solving the integration of distributed applications(Ding Zhaoqing et al., 2007). The basic protocol stack of Web service is composed of SOAP, WSDL and UDDI. Simple Object Access Protocol (SOAP) provides a standard packaging structure to bind XML document to transfer protocol. Web Service Description Language (WSDL) provides a standard method to describe service interface. Uniform Description, Discovery and integration (UDDI) defines the method to discovery and publish.

The existence of SOAP makes the relation of service provider and service petitioner extremely simple: XML document exchange. For the resources which have been packaged into web services, service invocation interface only has to package related data into SOAP messages according to their WSDL documents, and then send the messages to corresponding addresses to invocate them. The structure of web service invocation interface is shown in Fig.2:

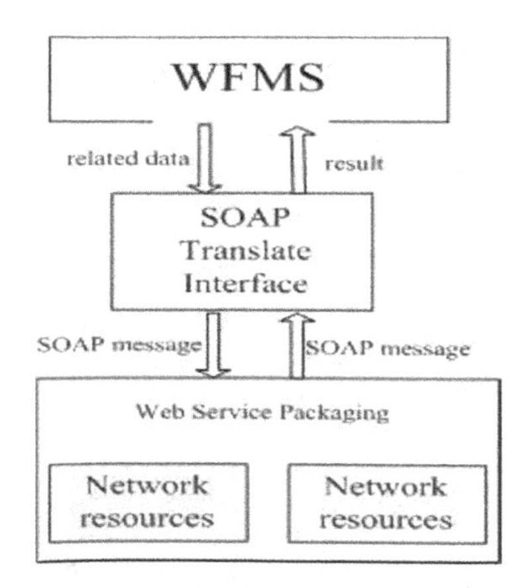

Fig.2: the structure of web service invocation interface

3.2 Realization of legacy network resources' invocation

As for legacy network information resources which cannot be packaged into web services, we divided them into independent and self-contained function units, which would be the basic units of resources integration. Although these resources can't be invocated by using SOAP message exchange, each of them would choose a way to communicate with services petitioners. And to service petitioners this difference of communication method will reflect in difference of transfer protocols.

We divided these resources into different types according to the transfer protocols they have chosen, including HTTP, FTP, RTSP etc. Although resources of the same type may be different in programming languages and platforms, they use the same transfer protocol, which means they have the same way to be corresponded with. So we would develop service invocation template for each kind of the resources. The system could accomplish the invocation of resources by using the instances of these templates.

Besides, the uncertainty of parameters when invocated these services make it necessary for the system to maintain service descriptions of the resources, including resource type, function description, uniform resource location, name of each parameter and its data type. The legacy network information resource will be described in XML form as follows:

```
- <service>
    <stype>HTTP</stype>
    <url>resource address</url>
    <description>service function description</description>
  - <param>
      <name>param1</name>
      <ptype>String</ptype>
    </param>
  - <param>
      <name>param2</name>
      <ptype>Int</ptype>
    </param>
    <rtype>String</rtype>
  </service>
```

Legacy resource invocation interface was composed by service analysis module and communication module. Fig.3 is the structure of legacy network resource invocation interface based on SOA, the concrete steps to invocate a legacy info-service resource is:

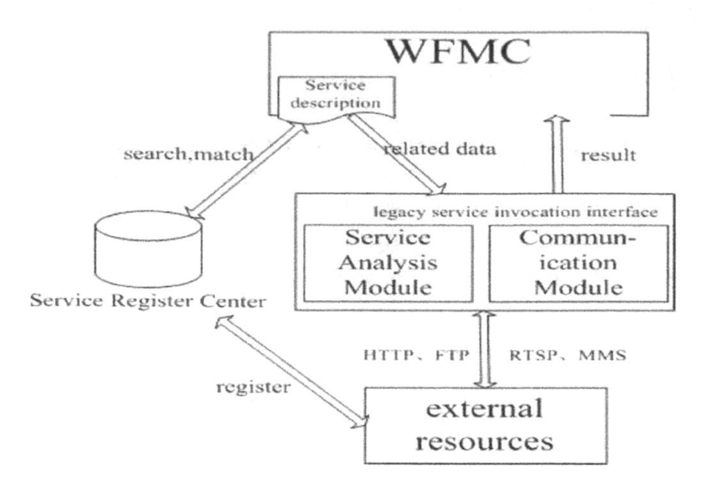

Fig.3: the structure of legacy network resource invocation interface

(1)Service analysis module analyzed the service description document sent by workflow execution logic, identified the resource type, access address, and the information of parameters etc.

(2)Communication module packaged related data sent by workflow execution logic into message of specific format, sent it to corresponding resource according to its address, and then got results sent back.

After the steps above, resources of legacy network information platforms would be seemed by workflow business process logic as independent services which approximately met the features of SOA services.

3.3 Dynamic maintenance, discovery and invocation of network resources

In order to maintain the description of legacy network resources uniformly, and further raise the reliability of web services, the system conserves a private service register center, which maintains the description of legacy network services and web services that might be invocated. Register information of each service includes the sign of service's category, the description of its function, information of each parameters and its address. And to web services, we obey the International Standard to describe them by WSDL. Compared with the web services registered in UDDI center, services registered in private service register center are more reliable, with less information redundancy, and easier to be obtained.

Services which can be invocated by the system must be the services that have been registered in UDDI or private service register center. When needed to invocate a service, the system would firstly consult the private service register center according to the service function, if a related service description was fond, then visited the resource according to the service's category and other related data via corresponding communication method; otherwise, searched the UDDI center, looked for corresponding web service and invocated it. Besides, to the same service function, when more than one related resources were discovered, we would need a resource evaluation strategy to select the best one, we would use the strategy of alternative priority based on service category and resource responding time to decide the priority of each resource. Concretely, the method of evaluation goes like this:

(1)Every service would get a primary priority when it got registered in the private service register center. This priority was different according to different service categories. The primary priority of web service was higher than that of other network information services because of its advantages in reliability, regular and analytic of the information structure.

(2)When a service was invocated, the system would find all the resources that have similar function description with it and invocated the one had the highest priority. If failed, then the second highest would be invocated.

(3)When a resource failed in invocation, its priority would be reduced relatively.

(4)When the invocation finished, the system would remand the priority of the invocated resource according to its responding time "t". The revised value would be calculated by the formula "w=wb+K（Tr-t）",in which "wb" was the priority of the resource before invocation, "Tr" was the reference resource responding time the system set originally, amended parameter "K" was a constant used to modify the weight of resource responding time. When one resource's responding time was less than the reference responding time, its revised priority would increase after being invocated successfully, otherwise it would fall.

(5)The system had set the lowest priority values of web service resources and legacy network resources, when one resource's priority became lower than the lowest priority of its category's, it would be abandoned by the system and thus got deleted from the private service register center.

When a service registered in UDDI center was invocated successfully during the execution of a process ,which means it hadn't been registered in the private service register center, the system would add it to the private service register center automatically and gave it an initial priority; when the priority of one resource was lower than the lowest priority of its type, the system would delete it from private service register center, through this way to maintained dynamically the service descriptions in private service register center. Besides, the system maintenance staffs have privilege to add, revise and delete service descriptions in private service register center directly.

3.4 Overall realization structure of service invocation interface

Based on SOA structure, the system regarded external agricultural network resources as functional independent services. The private service register center (PSCR) maintained all the legacy network services and web services that might be in use. The system was only able to invocate the services registered in PSRC or UDDI center, and it provided a strategy to evaluate the priority of services related to the same function, thus realized the dynamic discovery, match and invocation of services. The overall structure of the realization of service invocation interface is shown in Fig.4.

We combined the two types of invocation interface above to be an integrated one when put in practice. When a service was needed, the system would check the private or the public service register center for a related service description, analyzed the description and decided which kind of service invocation type would be chosen, then sent the relative parameters packaged in certain form to corresponding address to invocate the service, and finally analyzed the result sent back from the resource and sent it back to the system. In this way, we mapped unpredictable agricultural information resources into some certain types, and integrated the invocation methods of these types of resources into one uniform interface, so that we could shield the heterogeneous of external agricultural resources for the system's business process logic.

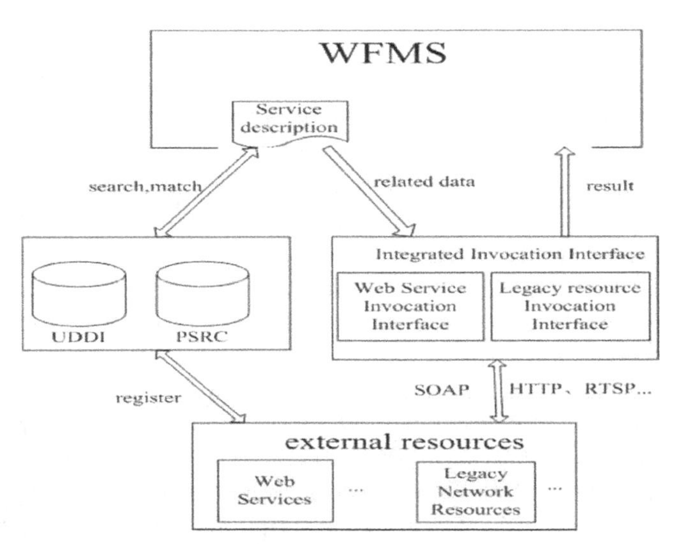

Fig.4: Structure of the overall realization of Service Invocation Interface

After isolated by the service invocation interface, all the external agricultural network resources became language and platform independent services. And these services could be invocated through an unified way, so that the workflow business logic could effectively realize the collaboration of external agricultural network resources of different platforms.

4. CONCLUSION

Service invocation interface is the part of workflow management system which is closest to practical use. The raise of SOA structure provides a basic framework for the design and realization of workflow system's cross-platform service invocation interface. Based on SOA structure, this passage analyzed the external network info-resources' similarities and differences, raised a strategy of design and realization of the service invocation interface, and realized the collaboration of part of agricultural resources.

ACKNOWLEDGEMENTS

The research presented in this paper was partly carried out in the Institute of Geographic Sciences and Natural Resources Research, CAS and Supported by National Key Technology R&D Program No.2006BAD10A05 and China National Advanced Science & Technology (863) Plan under Contract No.2006AA10Z237. The authors are grateful to the experts of Institute of Geographic Sciences and Natural Resources Research, CAS for useful comments during the review process.

REFERENCES

Aiello M. and Dustdar S. Service Oriented Computing: Service Foundations. Proc. of the Dagstuhl Seminar 2006. Service Oriented Computing, vol. 05462. Germany, July 2006

Chen Shi, OuYang Song. Research on implement of interface 3 in workflow reference mode. Computer Engineering and Design, 2005,26(7):1862-1864(in Chinese)

Cherbakov, L., Galambos, G., Harishankar, R., Kalyana, S. andRackham, G. Impact of service orientation at the businesslevel. IBM System Journal 2005 44, 653–668

Ding Zhaoqing, Dong Chuanliang. Research on Distributed Application Integration Based on SOA. Computer Engineering, 2007,33(10):246-248(in Chinese)

Edvardson, B. and Olsson, J. Key concepts for new service development. The Service Industries Journal 1996 16(2): 140–164

Erl, T. Service-Oriented Architecture: Concepts, Technology, and Design. Prentice Hall, Upper Saddle River, NJ 2005

Ling Xiaodong. A review of SOA. Computer Applications and Software 2007,24(10):122-124(in Chinese)

Shi Meilin, Yang Guangxin, Xiang Yong, Wu Shangguang. WFMS:Workflow Management System. Chinese Journal of Computers, 1999 (3) : 326 - 328.

Workflow Management Coalition (WFMC). Workflow Process Definition Interface —XML Process Definition Language. No. WFMC - TC - 1025 , Version1. 0 , 2002

Workflow Management Coalition. Workflow Management Coalition The Workflow Reference Model. WFMC T000 -1003 ,1995

Zheng Xiao, Wan nong, Lin Guoxiang, et al. Research on microstructure of cold pressed cakes from decorticated rapeseed based on porosity, China Oils and Fats, 2004, 29(12):14-17(in Chinese)

OPTIMAL MODEL ON CANAL WATER DISTRIBUTION BASED ON DYNAMIC PENALTY FUNCTION AND GENETIC ALGORITHM

Wenju Zhao, Xiaoyi Ma* ,Yinhong Kang, Hongyi Ren, Baofeng Su

Key Laboratory of Agricultural Soil and Water Engineering in Arid and Semiarid Areas, Ministry of Education, Northwest Agriculture and Forest University, Yangling, Shaanxi 712100, China

* *Corresponding author, Address: Key Laboratory of Agricultural Soil and Water Engineering in Arid and Semiarid Areas, Ministry of Education, Northwest Agriculture and Forest University, Yangling, 712100 Shaanxi, Province, P. R. China. Tel: +86-29-87082860, E-mail address: xiaoyimasl@yahoo.com.cn*

Abstract: The present optimal water delivery scheduling models are based on the assumed equal design discharges of lateral canals, which are not in accordance with practical water delivery scheduling demand in most irrigation systems. In order to solve this problem, a model of lateral canals with unequal discharges and a solution method were proposed; At present, traditional fixed penalty factor have some problem, such as it is difficulty to use unified dimension and to get a higher searching precision, besides, it prematurely converge to local optimal solution. Therefore, the thought of simulated annealing was referred to design a dynamic penalty function. In the progress of genetic operation, the SGA (Simple Genetic Algorithm) adopted adaptive crossover mutation method, and compared distinct solutions of model which based on the method in this paper, Adaptive genetic algorithm (AGA) and traditional methods used in irrigation district widely respectively. Comparing with water delivery plan compiled using traditional methods, the results illustrate that using this method can get much more reasonable lateral canals water delivery time and homogeneous discharges of upper canal. AGA can adjust the genetic controlling parameters automatically on the basis of values of individual fitness and degree of population dispersion, and get a high precision solution. So it has a higher practical value in irrigation system management.

Keywords: delivery model; penalty function; AGA; irrigation canal system; optimization delivery

---

Please use the following format when citing this chapter:

Zhao, W., Ma, X., Kang, Y., Ren, H. and Su, B., 2009, in IFIP International Federation for Information Processing, Volume 294, *Computer and Computing Technologies in Agriculture II, Volume 2*, eds. D. Li, Z. Chunjiang, (Boston: Springer), pp. 1347–1357.

1. INTRODUCTION

Optimum canal water distribution based on scientific decision-making can reduce the seepage loss and invalid disposable water in the course of water transfer, and improve the utilization of irrigation water. Optimization of water distribution in irrigation canal system means that certain method and technology are used to optimize the rotation irrigation combination of distribution canal, under the condition that the capacity of water transfer in distribution canals and lateral canals is limited to meet the particular irrigation needs of crops in irrigation district (Kang et al., 1996). There are ample of literature on the method which can optimize the decision-making for canal water distribution. Under the presupposition that the upper canals are made up of stream tubes with same water discharge and the amount of water discharge in stream tubes is equal to the amount of the water discharge in lateral canals. Suryavanshi et al.(1986) established linear programming water distribution model to fix the numbers of stream tubes and its algorithm with the purpose of reducing the investment in canal project . Wang et al. (1995) applied it into his research on optimal distribution of canal water. Having supposed the lateral canals had the same water discharge, he put forward the optimal canal water distribution model based on 0-1 programming method. However, the earlier models were expanded by Reddy et al. (1999) to allow for different users to request different discharges. Reddy et al. (1999) also introduced the concept of a time window, a window within which a user must be scheduled to receive water. Anwar and Clarke (2001) took this concept further to suggest that every user could have a requested start time and developed an integer programming model that developed a schedule where the difference between scheduled and target start time was minimized. Lv et al. (2000) improved the objective function in document (Wang et al. 1995), and pointed out the method to equally process water distribution time for all the lateral canals, and made the upper canal closed at the same time in order to decrease the times to regulate the gate. Wardlaw and Bhaktikul (2004) discussed a GA developed for solution of two lateral canal scheduling problems, results that the GA approach has been demonstrated to be robust and efficient in application to lateral canals scheduling problems. Song et al. (2004) analyzed the GA used in document (Lv et al.,2000) and did some contribution to significantly increasing the speed of getting solutions to the model and solving the problem that it difficult to get the results when there are many lateral canals. However, application of this model is limited to irrigation systems where the distribution outlets along the canal have the same discharge capacity and such systems are hypothetical. On the other hand, these researches fall short of the ability to solve the problem that majority of the lateral canals in irrigation system have unequal discharge amount of water distribution; Meanwhile, traditional fixed penalty factor have some problems, such as it is difficulty to use unified dimension, can not get a higher searching precision, prematurely converge to local optimal solution. Therefore, this paper study the canal water distribution model based on dynamic penalty function and Genetic Algorithm, it is firmly believed that the model and method will provide powerful technical support for making decision on optimal water distribution of irrigation canal system.

2. DESIGN AND APPLICATION IN THE GA OF PENALTY FUNCTION

At present, there are three commonly used methods in dealing with constraints in intelligent algorithms abandon infeasible solution (Wu et al.,1998), repair infeasible solution and penalty function method. Penalty function method, the most widely-used one, imposes certain penalty through infeasible solution and makes the resolution gradually close to feasible extreme point after iterations . The key point is to select proper penalty function for infeasible solutions: if the penalty value is too large, it is likely that the algorithm converge to non-extreme points; if the penalty value is too small, it's very likely that the convergence performance will be rather poor. In terms of constrained optimization question, the design of GA fitness function has some relations with the processing methods of constrained conditions. In the course of processing constrained conditions, having been inspired by the idea of simulated annealing, this paper takes dynamic penalty function to process constrained conditions. Annealing penalty factors σ_k is used to construct penalty function $P(x,\sigma_k)$. By using penalty function method, the equation (1) with the constrained optimization is changed into the unconstrained one shown as follows:

$$\begin{cases} \min & f(x) \\ s.t. & g_i(x) \geq 0, \quad i=1,2,m \\ & h_i(x)=0, \quad i=1,2,n \\ & l_i(x) \leq 0, \quad i=1,2,p \end{cases}$$

$$F(x,\sigma_k)=1/[f(x)+P(x,\sigma_k)]=1/\{f(x)+\sigma_k[\sum_{i=1}^{m}\left|\min(0,\sigma_1 g_i(x))\right|+\sum_{i=1}^{n}\left|\sigma_2 h_i(x)\right| \quad (1)$$

$$+\sum_{i=1}^{p}\max(0,\sigma_3 l_i(x))]\}$$

In this equation, supposing σ_k is penalty factor, σ_1、σ_2、σ_3 is penalty factor of different constrained conditions, according to actual engineering dimension and the value of feasible solution. $\sigma_k = 1/T_p$, $T_{p+1} = \alpha T_p$, $\alpha \in [0,1]$ are fixed, then x is decision variable after decoding, F is fitness function, $f(x)$ is objective function, and m、n、p are numbers of constraints, standing for "≥、>", "=","≤、<" respectively.

3. OPTIMAL CANAL WATER DISTRIBUTION MODEL FOR THE UNEVEN WATER DISCHARGE IN LATERAL CANALS AND DESIGN OF ALGORITHM PARAMETERS

3.1 Foundation of the model

In the process of canal water distribution, the lateral canals are required to meet the following requirements: (1) the practical distributed water ought to be adequate to irrigate the farmlands; (2) the time for water distribution ought to change in the scheduled irrigation duration ("T" in short); (3) the total distributed water discharge of lateral canals should be equal to the one in upper canals at any time, and the distributed water discharge in lateral canals and upper canals should be limited to 0.6~1.0 times of the already designed discharge so that the requirement of the distributed water level can be satisfied and the canal burst resulted from over input can be avoided. When the distributed water discharge is same, the greater water discharge and the shorter time for water distribution, the smaller the lost water will be. Under the condition that the requirements for the decision scheme of optimal canal water distribution are satisfied, the distributed water discharge in upper and lateral canals at any time can be close to the designed discharge of each canal, the even discharge in upper canals can be realized and the times for regulating the gate can either be reduced.

3.2 The Establishment of Optimal Model

Supposing the upper canals' designed discharge is Q_u;and there are also several lateral canals("j" in short), their designed discharge and distributed water discharge are q_{dj} and q_j(j=1, 2, …, n) respectively. Aiming at the minimal canal seepage loss of all the upper and lateral canals in all the distribution time during irrigation duration, the model is established as follows.

$$\text{Min } W = W_u + W_d \tag{2}$$

$$W_u = \sum_{i=1}^{T} f(A_u, m_u, q_u, l_u, t_u) \tag{3}$$

$$W_d = \sum_{i=1}^{T} \sum_{j=1}^{N} f(A_j, m_j, q_j, l_j, t_j) \tag{4}$$

Where, W_u and W_d is the total amount of water seepage loss while transferring in upper and lateral canals respectively during irrigation duration (m^3) ; q_u and q_j are distributed water dischageof upper and lateral canals respectively(m^3/s); A_u 、 A_j are permeability coefficients of upper and lateral canal beds, and m_u 、 m_j their permeable indexes respectively; l_u 、 l_j are the length of water transport in upper and lateral respectively(m), and t_u 、 t_j their corresponding distributed time(d).Therefore, the formula to measure the amount of canal seepage losses goes like this:

$$W = f(A, m, q, l, t) = [A \cdot l \cdot q^{(1-m)} \cdot t] / 100 \tag{5}$$

(parameters are same with the above-mentioned)

And the model mainly has the flowing constrained conditions:

Constrained conditions for distributed water discharge in lateral canals:

Any lateral canal's distributed water discharge "q_j"is α_j times of its designed discharge.

$$q_j = \alpha q_{dj} \, (j=1,\ 2,\ \ldots,\ N) \tag{6}$$

$$0.6 \le \alpha_j \le 1.0 \tag{7}$$

Constrained conditions for irrigation duration:

The distributed time is t_j, the starting time is "t_j'" , and ending time is "t_j''" . The t_j' and t_j'' should be in the irrigation duration "T ".

$$0 \le t_j' \tag{8}$$

$$t_j' = t_j'' - t_j \tag{9}$$

$$t_j'' \le T \tag{10}$$

Constrained conditions for water volume:

The product of any lateral canals' distributed water discharge and delivery time "t_j " should be equal to the required distributed water discharge "W_j " in the area.

$$W_j = q_j \times t_j \quad (11)$$

Constrained conditions for water balance:

The practical distributed water discharge of upper canals at any time ("q_u" in short) should be equal to the total amount of the discharge of lateral canals at the same period of time.

$$q_u = \sum_{j=1}^{n} q_j \times x_{ij} \qquad (i=1,\ 2,\ \ldots,\ T) \quad (12)$$

$$\text{if } t_j' \le i \le t_j'',\ x_{ij} = 1 \text{ ;otherwise } x_{ij} = 0 \quad (13)$$

Constrained conditions for distributed water discharge in upper canals:

The practical distributed water discharge of upper canals at any time ("q_u" in short) gets close to designed discharge and smaller than the maximally allowed discharge.

$$q_u \approx Q_u \quad (14)$$

$$0.6Q_u \le q_u \le 1.2Q_u \quad (15)$$

3.3 Genetic Operations and the Setting of Algorithm Parameters

It's extremely difficult to get the solutions by the above-mentioned model. Using the method in this paper, it is easy get solutions to the global optimal values of this mathematical model, because the ordinary irrigation canal system has lots of lateral canals and irrigation duration lasts 10 to 20 days, there are lots of decision-making periods based on the criterion that a period lasts 12 hours. The constrained conditions which can satisfy the requirements of formulas (6)～(13) and the balanced distributed water discharge in upper and lateral canals are accounted to several hundreds. The obvious problems of genetic algorithm are that it's hard to get feasible solutions to the questions at the beginning of evolution. Therefore, properly cope with the problems of constrained conditions, scientific selection of decision factors, coding design and determination of genetic operators are all included in this paper.

3.3.1 Coding design

After the construction of irrigation district, the length of water delivery (l), permeability parameter (A and m) and design discharge (q_{dj}), the maximum and minimum of the allowed discharge through canals are fixed. The decision of canal water distribution in irrigation district is optimize the decision variables like the starting time, the duration, and the ending time of the distributed water and the distributed water discharge under the condition that the water distribution quantity is fixed in all the canals. Having considered the variables are closely-linked and thoroughly analyzed the model's characteristics, this paper regards the ratio of distributed water discharge in lateral canals and their designed discharge ("α_j" in short)and the ending time of water distribution (t_j'' in short) as decision variables to encode . Therefore, the variable α_j can make the algorithm automatically satisfy the constrained conditions in formulas (6)~(7); the ratio of declared water quantity and distributed water discharge can satisfy the constrained conditions in formula(11); the encoding of ending time of the distributed water t_j'' can satisfy the constrained conditions in formula(10); the starting time, which is got from the margin between the ending time and the irrigation duration time, can satisfy the constrained conditions in formula(9). This method of dealing with constrained conditions remarkably facilitates the algorithm to get feasible solutions.

3.3.2 Selection of operators

Based on the evaluation of population's individual fitness degree, some individuals are selected according to certain proportion and regarded as parents for propagation. Roulette bet method is adopted in this paper to select the individuals.

3.3.3 Crossover operator

After the match and single crossover of the parental individuals, the offspring individuals are produced. In this process, the adaptive crossover rate is adopted in order to reduce the possibility of damaging some individuals who have high fitness(Srinivas et al.,1994; Wang et al.,2002). In AGA , P_c according to formulas (16).

$$P_c = \begin{cases} P_{c1} - \dfrac{(P_{c1} - P_{c2})(f' - f_{avg})}{f_{\max} - f_{avg}}, f' \geq f_{avg} \\ P_{c1}, f < f_{avg} \end{cases} \tag{16}$$

3.3.4 Mutation operators

The individuals and the location of their genetic codes are randomly selected, and the individual can freely mutated in the gene's allowed range. The adaptive mutation probability is also used for mutation probability. In AGA , P_m according to formulas (17).

$$P_m = \begin{cases} P_{m1} - \dfrac{(P_{m1} - P_{m2})(f - f_{avg})}{f_{\max} - f_{avg}}, f \geq f_{avg} \\ p_{m1}, f < f_{avg} \end{cases} \quad (17)$$

In the formulas, $f_{\max}$ is the maximal individual fitness value of population; f_{avg} is the average individual fitness value of population; f' is the larger fitness value of the two in the crossover; f is the fitness value of the individual that is involved in variation. $P_{c1} = 0.9$, $P_{c2} = 0.6$, $P_{m1} = 0.1$, $P_{m2} = 0.001$.

4. APPLICATION EXAMPLE

11 branch canals have 24 lateral canals in northern main branch, Feng Jiashan irrigation district, Shaanxi Province are chosen as application example. During the practical process of water distribution in certain irrigation duration in the spring of 2005. The designed discharge of the upper canals Q_u is 1.2m^3 /s, the designed discharge of lateral canals varies among 0.03～0.18m^3/s, the soil in canal bed is medium loam and the permeability parameters of soil "A" and "m" are 1.9 and 0.4 individually(Wang ,2000). Besides the present experience, the AGA and the method based on dynamic penalty function and AGA were adopted to make contrast verification for the marshalling scheme of canal water distribution model, the results are shown as follows in Fig.1 、 Fig.2 and Fig.3.

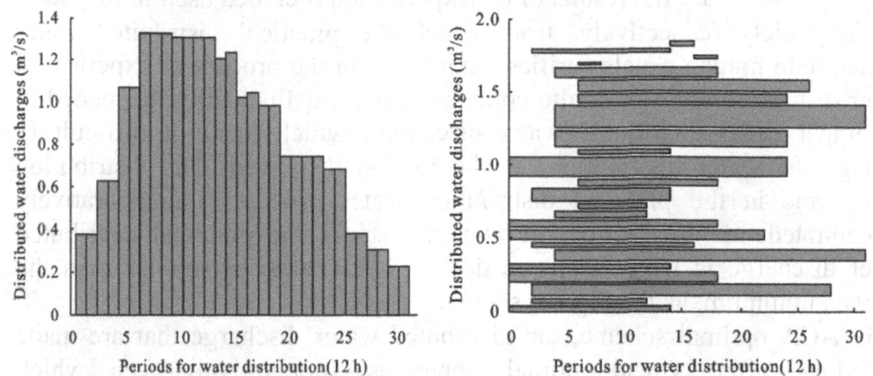

Fig. 1 Procedures of water distribution of a) upper and b) lateral canals by experiential scheme

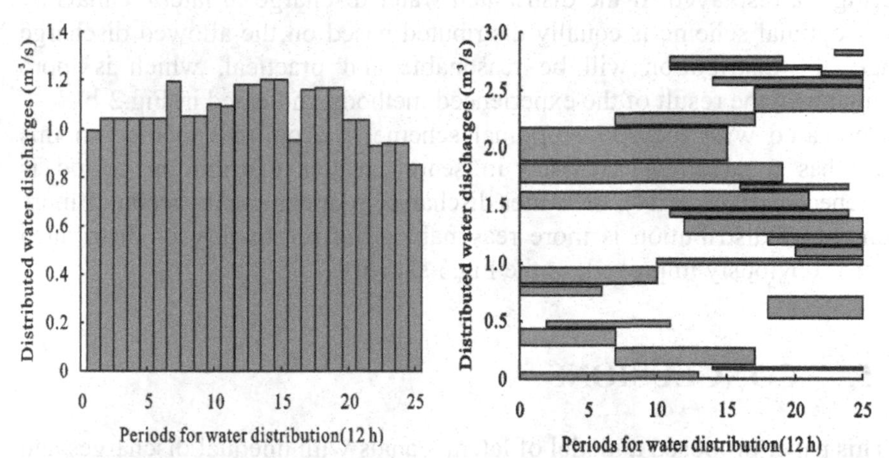

Fig.2 Procedures of water distribution of a) upper and b) lateral canals by AGA optimal scheme

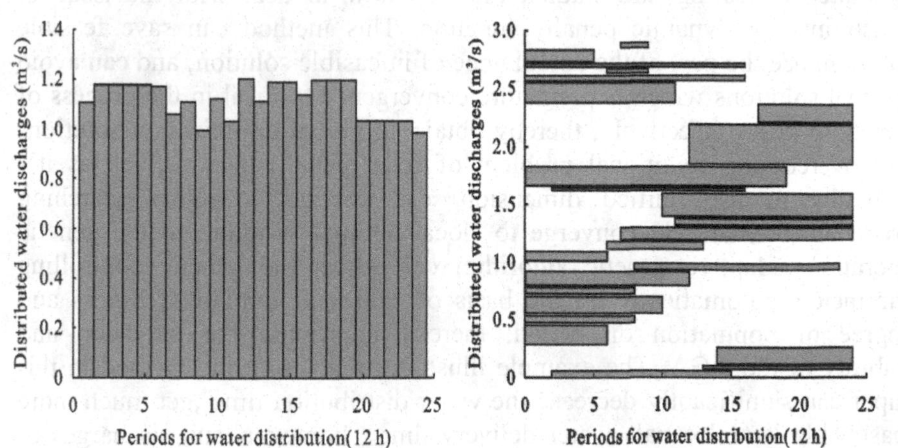

Fig.3. Procedures of water distribution of a) upper and b) lateral canals by this paper optimal scheme

In Fig.1, a、b are the results of the experienced methods used in irrigation district widely respectively, from which the practical distributed water discharge in branch canals verifies remarkably in the process of experiential water distribution and it's quite common that great fluctuation happened in the upper canal distributed water discharge, which make it difficult to manage the water distribution, as the Fig.1-a displayed. The distribution water time in the practical distribution water process is comparatively concentrated, the total discharge is rather small, the practical distributed water discharge is larger than the designed discharge or smaller than the allowed minimum, as the Fig.1-b shown.

By AGA optimal scheme, the distributed water discharge that are made according to the typical optimal scheme is comparatively even, which facilitates the water distribution and shortens 6 periods (from 31 to 25), as the Fig.2-a displayed. If the distributed water discharge in lateral canals by AGA optimal scheme is equally distributed based on the allowed discharge range, the distribution will be reasonable and practical, which is more optimal than the result of the experienced methods, indicated in Fig.2-b.

Compared with the AGA optimal scheme, the optimal method in this paper has a excellent accuracy in search, water distribution period is shortened(from 25 to 24), the water discharge in upper canals becomes more even, water distribution is more reasonable, and the quality of distributed water is obviously improved, as the Fig.3 shown.

5. CONCLUSIONS

This paper proposed a model of lateral canals with unequal discharges and a solution method, and cope with constraint condition adopted the thought of simulated annealing and studied the way how to deal with the issue of constraints by dynamic penalty function. This method can save feasible solution, use the part of the useful gene of infeasible solution, and can avoid optimal solutions which is premature convergence to local in the process of genetic process effectively, thereby obtaining global optimization solutions and overcoming traditional problem of fixed penalty factor ,such as it's difficulty to use unified dimension, can not get a higher searching precision ,prematurely converge to local optimal solution. In the genetic operation, adaptive genetic algorithm can adjust the genetic controlling parameters automatically on the basis of values of individual fitness and degree of population dispersion, thereby improving the precision and stability of the AGA. The example illustrates that using the method in this paper can significantly decrease the water distribution time ,get much more reasonable lateral canals water delivery time , homogeneous discharges of upper canal and reduce the water seepage loss while transferring. So this theory can provide powerful technical support for optimal water allocation of irrigation canal system.

ACKNOWLEDGEMENTS

Thanks to the research grants from Chinese National Natural Science Fund (50479052); Project Supported by National Science and Technology (2006BAD11B04).

REFERENCES

Anwar, A.A., Clarke, D. Irrigation scheduling using mixed-integer linear programming. J. Irrigat. Drain.Eng. , 2001,127 (2), 63–69.

Kang, S.Z., Cai, H.J.. Agricultural Water Management Science. China Agriculture Press, Beijing, 1996. (in Chinese).

Lv, H.X., Xiong, Y.Z., Wang, Z.N. Optimal Model of Rotation Irrigation Distribution Channel and Branch Canal and Delivery Time. Trans. CSAE.,2000, 16(6),43-46. (in Chinese).

Reddy, J.M., Wilamowski, B., Cassel-Sharmasarkar, F. Optimal scheduling of irrigation for lateral canals. ICID J., 1999,48 (3), 1–12.

Song, S.B., Lv, H.X. Optimization Model of Rotation Irrigation Channel Distribution and Solution with Genetic Algorithm. Trans. CSAE. , 2004, 20(2):40 - 44. (in Chinese).

Srinivas, M., Patnaik, L.M. Adaptive Probabilities of Crossover and Mutation in Genetic Algorithms. IEEE Transon Systems Man and Cybernetics. , 1994, 24(4), 656 - 667.

Suryavanshi, A.R., Reddy, J.M. Optimal operation schedule of irrigation distribution systems. Agric. Water Manage, 1986,11(1), 23–30.

Wang, X.P., Cai, L.M. Genetic Algorithm-Theory, Application and Software Realization. Xi'an Jiaotong University Press. Xi'an, 2002. (in Chinese).

Wang, Z., Reddy, J.M., Feyen, J. Improved 0–1 programming model for optimal flow scheduling in irrigation canals. Irrigat. Drain. Syst, 1995,14(9), 105–116.

Wang, Z.N. Irrigation and Drainage Engineering. China Agriculture Press. Beijing, 2000. (in Chinese).

Wardlaw R., Bhaktikul K. Comparison of Genetic Algorithm and Linear Programming Approaches for Lateral Canal Scheduling. J. Irrig. and Drain. Engrg.,2004, 130(4), 311-317.

Wu, Z.Y., Shao, H.H, Wu, X.Y. Annealing Accuracy Penalty Function Based Nonlinear Constrained Optimization Method with Genetic Algorithms. Control and Decision. , 1998, 13(2):136-140. (in Chinese).

APPLICATION OF OPERATION RESEARCH AND SYSTEM SCIENCE APPROACH TO FISHERIES MANAGEMENT

Lin Sun[1], Shouju Li[2,*], Hongjun Xiao[1], Dequan Yang[1]
[1] *School of management of Dalian University of Technology, Dalian, P. R. China 116024*
[2] *Department of Engineering Mechanics, Dalian University of Technology, Dalian, P. R. China 116024*
[*] *Corresponding author, Address: Department of Engineering Mechanics, Dalian University of Technology, Dalian, P. R. China 116024, Email:lishouju@dlut.edu.cn*

Abstract: Marine fisheries are highly complex and stochastic. A simulation model, therefore, is required. Simulation-based optimization utilizes the simulation model in obtaining the objective function values of a particular fishing schedule. The decision support system for fishery management will assist the government agencies and the fishing industry to use sound data and management science techniques in making policy decisions for fishing activities. Transferable rights to fish have proved a reliable and effective means of creating incentives to conserve marine resources. By strengthening individual fishing rights under flexible quota management systems make a significant contribution to conserving fish stocks, to reducing excess capacity and to raising the profitability of the fisheries industry.

Key words: fishery management, optimization approach, decision support system, sustainable fisheries management

1. INTRODUCTION

The ocean provides a substantial resource that could yield considerable benefits through fisheries management. The major concern in managing the fisheries issue is to take advantage of this huge resource without damaging the supply in the long run. As a result the government sets certain rules and regulations on fishing vessels in terms of the amount and the species of the

Please use the following format when citing this chapter:

Sun, L., Li, S., Xiao, H. and Yang, D., 2009, in IFIP International Federation for Information Processing, Volume 294, *Computer and Computing Technologies in Agriculture II, Volume 2*, eds. D. Li, Z. Chunjiang, (Boston: Springer), pp. 1359–1368.

fish they could catch in any given day and in any given area. Laying out these decisions and implementing them forces the fishery management organization to make some decisions that they may not be sure themselves if they could result desirable solutions. In practice, these decisions force fishery vessels to fish a specified amount of each species at any given time. Such decisions are difficult due to the complexities associated with multiple management objectives and multiple alternatives under considerations. To aid the decision making process, managers need tools to formalize these complexities into a common framework consisting of relationships among management measures, sources of uncertainty, and possible outcomes of actions (Azadivar et al, 2008).

Unregulated fisheries tend to be characterised by overexploitation of the stocks. In relation to this, there has been an extensive search for management measures to reduce fishing effort. A number of neo-classical economists have argued that two measures will reduce fishing effort in an efficient way: taxes and individual transferable quotas (ITQs). Other measures, such as nontransferable individual quotas, a total allowable catch with temporary closure of the fishery, and effort regulations are not Pareto-efficient. Institutional economists have criticised neo-classical theory for neglecting political problems, enforcement problems, and transaction costs. They argue that because of such problems, the superiority of taxes and ITQs for fisheries management is questionable.

Even though the main objective of fisheries management is to move towards sustainable fisheries, most efforts have failed. Management of some fish stocks is facing dysfunctional national and international cooperation, and available resources are decreasing. Academic disciplines such as biology, economics, and social sciences provide valuable information to fisheries management. Still, the solutions they offer only solve parts of the complex problems the management is facing, as most of the solutions are limited to the discipline within which they are proposed. Individual transferable quotas (ITQ) may, for example, be considered as economic efficient by economists, marine protected areas (MPA) as ecosystem-friendly by biologists, and community based management (CBM) as a goal to social anthropologists (Ingrid et al, 2006).

The Decision Support System (DSS) is a computer program that transfers information from research surveys, and commercial fishing reports into advices for policy makers on decisions of when, where and how much fishing effort should be allocated (Truong et al, 2005). The core of the DSS is the component that intensively applies simulation modeling and operations research techniques. The Decision Support System for Fishery Management will assist the government agencies and the fishing industry to use sound data and management science techniques in making policy decisions for fishing activities.

Technical and social systems increase in their complexity and vulnerability, and human-made systems, such as the fisheries, are not in conformance with natural systems. Industrial ecology is a framework for organizing production and designing consumption systems in ways that resemble natural ecosystems to achieve a more efficient industrialization, adjusted to the tolerances of natural systems (Fet, 1997). Technology should work with natural systems, not against them. In a natural system there is a causal relation between causes and their effects. The natural cause and effect relationships are the designer's advantage when human-made systems are being designed (Asbjornsen, 1998).

2. SYSTEM SCIENCE APPROACH TO FISHERIES MANAGEMENT

Management science, the scientific study of problem-solving, has developed a body of literature and methodologies over the past four decades on decision theory methods, evaluation of management and systems performance, and analysis of systems under uncertainty. These methods have been influential and applied successfully to many industrial decision-making and strategic-planning settings. Fisheries management has not yet embraced these innovations, which involve the "scientific method of problem-solving" and which offer considerable opportunity for improved fisheries decision-making. Fisheries science and management can be integrated with management science in what we term "fisheries-management science". The broader framework allows management in the face of uncertainty and, at the spatial and temporal scales, is appropriate for the complexity of fisheries systems. This paper presents an implementation plan for fisheries-management science in a commercial fishery. The implementation plan is presented as a possible remedy for issues that have plagued fisheries management to date. It is argued that decision theory methodologies are needed to analyse the management problem context, including strategic planning and objectives setting, appropriate spatial and temporal scale definition, interdisciplinary systems modelling methods, the assessment and management of risk, and ongoing in-season decision-monitoring. The fisheries-management science problem-solving context also provides a basis for reshaping the central agency responsible for fisheries into a more action-orientated organization consisting of multidisciplinary teams acting in support of participative, real-time decision-making through enhanced industry and stakeholders' responsibility for resource sustainability. Fisheries institutions are often hindered by rigid, disciplinary organizational structures and decision-making processes that are unable to account, in a timely fashion, for the multiple and conflicting objectives and the inherent variability that characterize a fishery system. The fisheries-management science approach addresses these issues by focusing on the problem-orientated nature of fisheries management, by managing according to objectives, and by supporting a holistic view for stock conservation and resource sustainability.

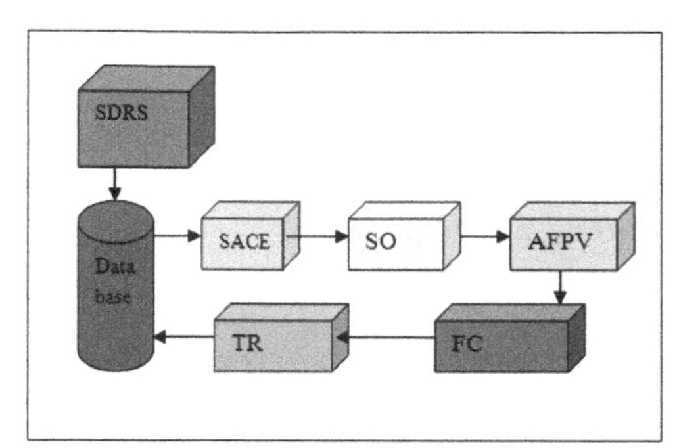

Fig.1: Decision support system for fisheries management

Decision support system for fisheries, as shown in Fig.1, consists of seven components: (1) sampling design and research survey (SDRS), (2)database and data management, (3) statistical analysis on catch or effort(SACE), (4) simulation optimization(SO), (5) allocation of fishing permits to vessels, (6) fishing activities, and (7) trip reports. A systems engineering approach to the fisheries involves connecting and describing objects and events. A life-cycle orientation that addresses the various system phases, i.e., design and development, production, distribution, operation, maintenance, retirement and disposal, is emphasized. If a system is to be analyzed in a holistic manner, it is necessary to consider all system phases, and identify the most important of these. When considering the fisheries, there are several life cycles, e.g., the life cycles of the fleets, of the individual vessels, of the equipment onboard each vessel, and of the fish. The fish is influenced by the fisheries during its entire life cycle, by ecosystem disturbances, and by emissions from production and disposal of technologies. The main interaction between the fish and the fishing fleet is in the fishing fleets' operational phase. It is common to divide systems into four main categories; hardware, software, bioware (or personnel), and economy. The hardware category is related to all physical parts of the system, i.e., parts that can be manufactured by means of technology, such as the fishing vessels. The software category is related to computer programs, instructions or general procedures, laws and regulations, and may be associated with information science and science of law (Ingrid et al, 2006).

The bioware category is related to human elements; human interaction with the system, connected to social science. Publications on Actor-Network Theory in fisheries are examples of research on human interaction within the fisheries. The economy category is related to monetary aspects, i.e., financial science. A system may be viewed as a combination of some or all of these categories. The four system categories can be related to the economic, ecological, and social dimensions of sustainable fisheries. Inputs from biologists are important to the fisheries management in order to optimize and design the best management system. The way the systems engineering principles are related to fisheries management in this article, means that fisheries management is the system to be analyzed, not the fish itself and the marine ecosystems. The overexploitation of fish is the primary driver for the need for sustainable fisheries and a well-functioning management system. In this context, fish may be considered as part of the natural system interacting with the human-made system of the fisheries and fisheries management.

3. OPTIMIZATION PROCEDURE TO FISHERIES MANAGEMENT

In case of multi-species fishery this problem is even more complex due to the fact that the spatial factors must be explicitly incorporated into the decision making process. There are several important management policy issues that cannot be evaluated without spatial analysis and modeling. Examples of these issues include determination of the location and size of marine protected areas, and judgment on whether the fishing effort displaced to protect the reserves may end up doing more harm elsewhere (Azadivar et al, 2008). The core component of the decision support system is the engine for application of simulation modeling and optimization techniques. The simulation function evaluates the performance of an area management plan based on a set of criteria. The optimization algorithm determines the optimal fishing effort allocation in terms of time, location and amount of catch. For example, fishing efforts in each sub-area and time period are controlled to maximize the value of landings, while bycatch is minimized and other management goals are achieved.

Multispecies fishery is characterized by the geographical overlap of healthy and overfished stocks. The total allowable catch (TAC) for each stock is often calculated using single species models such as maximum sustainable yield or yield per recruit model. By using single species models, technical interactions or bycatch cannot be taken into account. As a result, target catch for all stocks cannot be reached at the same time. In this section, we present a multispecies age-structured model that can be used to calculate the optimal inter-annual management strategy. The optimal inter-annual fishing trajectory is determined by solving the following problem with respect to decision variable

$$Max \qquad J = \sum_{i=1}^{I} V_i \tag{1}$$

$$ST. \qquad B_{sI} \geq B_s^* \tag{2}$$

Where J is objective function. B_s^* is the biomass target for stock s at the end of the planning horizon. V_i is the value of the total landings in year i, which can be calculated by

$$V_i = \sum_{s=1}^{S} y_{si} \times e_s \tag{3}$$

Where e_s is the price per ton of species s. y_{si} is the annual landing in weight of stock s in year i is calculated as [1]

$$y_{si} = \sum_{a=1}^{A} C_{asi} \times w_{as} \tag{4}$$

Where w_{as} is the average weight of a species s fish at age a. C_{asi} is the number of fish in age a cohort of stock s caught during year i. This optimization problem is non-linear, and may be stochastic if uncertainty factors in recruitment and catch are taken into account. In order to solve this problem, we use the simulation-based-optimization approach. A Genetic Algorithm is employed as the optimization engine for controlling the optima searching process. The value of the objective function, the total worth of the catch, is estimated for each scenario by simulating the corresponding system.

4. SUSTAINABLE FISHERIES MANAGEMENT

The world's fisheries are facing an impending crisis. This crisis will not only affect the natural biological diversity and ecology of the planet's oceans and aquatic ecosystems, but also the social and economic well-being of the individuals and communities dependent on fisheries resources for their livelihoods. With the declines in global fish stocks and the uncertainty in the ability of some stocks to rebound from extremely low population levels, there is now mounting evidence that the world's commercial fisheries may face complete collapse by the middle of the 21st century. It is imperative now, more than ever, for fisheries to become truly sustainable and for people who exploit fisheries resources to shift towards more sustainable practices in the world's fisheries. What's more, the current crisis facing the world's commercial fisheries seriously brings into question the ability of centralized government agencies to act as the primary managers of these common pool resources. It is becoming ever more apparent that such 'top-down' fisheries management techniques are not effective in achieving either ecological or social sustainability. Despite the best efforts of government agencies to manage fisheries through market-driven individual transferable quotas (ITQs) based on scientific models for maximum sustainable yield (MSY) and total allowable catch (TAC), fisheries continue to decline and the fisheries communities that depend on them are slowly decaying. In recognizing the need to decentralize the management of coastal resources, many scholars have advocated for local communities and/ or resource users to play a greater role in managing the fisheries resources on which they depend (Thompson, 2008).

Government-based attempts have often failed to manage marine resources in a sustainable way. Co-management and the recognition of fishers' knowledge is hence increasingly considered a remedy for overfishing in coastal fisheries. Co-management can be defined as the collaborative and participatory process of regulatory decision-making among representatives of user groups, government agencies and research institutions. Despite the recognition of the potential benefits of fisheries co-management and its close linkages to ecosystem-based approaches to fisheries, difficulties are frequently encountered during its implementation (Jentoft et al, 2008). A marine biological study confirms this observation by depicting a reduction in the average sizes and lower reproduction rates. Several fishing grounds are considered either fully exploited or overexploited, which is threatening the local fishers' main source of income and the region's most important economic activity (Pollack et al, 2008).

In recognition of the limitations to traditional conservation awareness campaigns, participants and environmental educators involved in CBCM fisheries should be seeking new techniques to more effectively bring about sustainable fishing activities. Marketing (CBSM) is an emerging tool that has been shown to be extremely effective at fostering sustainable behavior. CBSM combines social psychology and social marketing in a systematic way in order to maximize the success of implementing sustainable development projects. CBSM has been successfully utilized in many resource management sectors such as energy and water conservation initiatives, waste recycling, and reduction programs, and watershed management programs. Despite its successes in these areas of resource conservation and management, CBSM has rarely been used to foster sustainable behaviors related to fisheries conservation and management. In particular, this system has never been applied within the context of CBCM fisheries, despite the fact that CBSM and CBCM fisheries both have a strong focus on community level education and direct participation in nurturing more sustainable behaviors for the conservation of resources. Community-based co-management (CBCM) is a peoplecentered, community oriented, resources-based partnership approach to fisheries management in which government agencies, the community of local resource users, nongovernment organizations, and other stakeholders share the responsibility and authority for the management of a fishery. It aims to devolve power from central agencies, while empowering and building capacity within communities to manage local fisheries resources in a more sustainable manner (Thompson, 2008).

Sustainable management regimes fail if stakeholder participation is merely conceived to create 'a sense of ownership', to enrich available knowledge and to bargain compromises between a-priori conflicting interests. Preferences and attitudes of actors are subject to change during dialogue in a joint effort: Their views on other actors, their understanding of other actors' perspectives and their modes of relating with the natural environment evolve in response to new insights and positive experiences in a deliberative participatory arena. This requires skilled facilitation and an enabling environment for such an arena—often unmet preconditions. But our point here is that non-cognitive aspects play an important role, which cannot be captured by a restrictive understanding of ''trustbuilding'' in terms of ''acceptance of a given structure''. The procedural dimension of jointly developing a management structure is maybe as relevant as the regulations that come out of it. The quality of societal interaction, the identification with an area and its appreciation, the experienced plurality of positions, the empowerment of actually influencing a decision, but also the responsibilities coming along with it, and finally the experienced limitations imposed by legislation and the existing power relations—these aspects can develop during a continued and meaningful stakeholder process and are highly relevant for the sustainability of a management regime (Escobar, 1984).

Despite its popular application by local governments and municipalities, utility providers, and some environmental groups, CBSM has rarely been used to deal with fostering sustainable behaviors in common pool resource sectors such as fisheries. Perhaps the most well documented use of CBSM that provides the best rationale for its useful application in fisheries resource management comes from the agricultural sector and watershed protection campaigns. Like fishers, farmers tend to have a fairly high level of community cohesion, and are directly involved in the use of the environment on which they depend for their livelihoods. Thus, farmers are likely to have similar concerns when it comes to managing their resources. Recognizing the importance of water and soil conservation, reducing pollution and pesticides, and preserving local watersheds from excessive agricultural runoff has prompted community-based social marketers to design campaigns to foster sustainable behavior amongst local farmers. For example, Lynne et al. demonstrated the successful use CBSM techniques to get Florida strawberry farmers to adopt state enforced water saving irrigation technologies. Using surveys, the authors were able to identify the barriers to implementing such technologies, and found that farmers needed to feel that they had greater control over adopting water saving devices. As in fisheries management, empowerment has been identified as an important element of local fisheries management. Such empowerment and control is more easily secured when resource users are in a position to participate and give input on management decisions to ultimately achieve greater self-control (Jentof, 1984).

The technological development within the fisheries fleet leads to environmental problems, i.e., increased engine size leads to increased emissions of greenhouse gases. It may be more difficult to measure the exact amount of greenhouse gas emissions than increased average breadth of the fishing vessels; at least such figures cannot be calculated without introducing uncertainty. Since reduction of overcapacity in the fisheries fleet is in accordance with management objectives of achieving sustainable fisheries, and since overcapacity is related to technical parameters of fishing vessels, then frequent measurement of performance indicators based on technical parameters, would be a valuable contribution to the achievement of sustainable fisheries (Standal, 2005).

5. CONCLUSION

The optimization approach is applied to a harvesting and scheduling problem for a fishing fleet. The comparison between the simulation of status quo and the result of the optimization program shows that the current fishing activities are far from the optimum and the net profit can be improved significantly by adjusting the time, location and targeting species. Implementation of industrial ecology into fisheries management brings system thinking in ecology together with systems engineering and economics. The systems engineering method mainly uses the top-down approach, based on the point of view that the whole is more than just the sum of its parts. One of the most important attributes of systems engineering is the continuous evaluation process. As the system is analyzed, it is also evaluated at the different life-cycle stages in order to improve the system through the whole process. A potential benefit from implementing systems engineering into fisheries management is more transparency and a reduction of the risks associated with the decision-making process. Increased visibility is provided through the perspective on the system from a long-term and life-cycle perspective.

REFERENCES

A. Escobar. Discourse and power in development: Michel Foucault and the relevance of his work to the Third World. Alternatives, 1984; 10: 377–400

A. M. Fet. Systems engineering methods and environmental life-cycle performance within ship industry. PhD thesis, Norwegian University of Science and Technology, Trondheim, Norway; 1997.

B. U. Ingrid. Systems engineering principles in fisheries management, Marine Policy, 2006, 30: 624–634

D. Standal. Nuts and bolts in fisheries management—a technological approach to sustainable fisheries? Marine Policy 2005, 29: 255–63

F. Azadivar, T. Truong, Y. Jiao. A decision support system for fisheries management using operations research and systems science approach, Expert Systems with Applications 2008, doi:10.1016/j.eswa.2008.01.080

G. Pollack, A. Berghofer, U. Berghofer, Fishing for social realities—Challenges to sustainable fisheries management in the Cape Horn Biosphere Reserve, Marine Policy 2008,32: 233–242

H. Myles, Thompson. Fostering sustainable behaviours in community-based co-managed fisheries, Marine Policy 2008,32: 413–420

N.A. Carrick, B. Ostendorf. Development of a spatial Decision Support System (DSS) for the Spencer Gulf penaeid prawn fishery, South Australia, Environmental Modelling & Software 2007,22: 137-148

O. A. Asbjornsen. Systems engineering and industrial ecology. Maryland, USA: Skarpodd; 1998

S. Jentoft. Fisheries co-management as empowerment. Marine Policy, 2005, 29:1–7

S. Jentoft, B. J. McCay, D. C. Wilson. Social theory and fisheries comanagement. Marine Policy 1998; 22(4–5):423–36

T. H. Truong, J. B. Rothschild, Decision support system for fisheries management, Proceedings of the 2005 Winter Simulation Conference, 2005, 2107-2111

MACHINE VISION ON-LINE DETECTION QUALITY OF SOFT CAPSULES BASE ON SVM

Fucai Ge[1], Jiyong Shi[2], Youyi Xu[1], Xiaobo Zou[2,*], Jiewen Zhao[2]

[1] *Electro-mechinery General Works Co.Ltd, Jiangsu University, Zhenjiang, Jiangsu 212013, China*

[2] *Agricultural Product Processing and Storage Lab, Jiangsu University, Zhenjiang, Jiangsu 212013, China*

[*] *Corresponding author, Address: Agricultural Product Processing and Storage Lab, Jiangsu University, 301 Xuefu Road, Zhenjiang, Jiangsu 212013, China, Tel:+86-0511-88780201, Fax:+86-0511-88780201, Email: Zou_xiaobo@ujs.edu.cn*

Abstract: Nowadays the quality inspect of soft capsules is mainly by manual. Despite the intensive of this work, the accuracy of inspection by manual is very low. This paper proposed soft capsules online sorting system based on machine vision. The inspection process are following: (1) soft capsules were placed on rollers are rotating while moving. The image of each soft capsule was grabbed. (2) automatic threshold based on ostu was used to segmentation capsule image from background, and morphological filter was used to eliminate noise and regional markings. (3) 4 features were extracted which were perimeter, area, girth, altitude diameter and latitude diameter. Support Vector Machine (SVM) and was used to analyze these features. 15460 soft capsules were tested by the online sorting system. The overall grading accuracy was up to 94.1%. Furthermore, the grading speed of the sorting line resches10 capsules per second.

Keywords: Soft Capsules; Machine Vision; On-line Grading

1. INTRODUCTION

Soft Capsules is a new kind of capsules in which oil functional material, liquor, suspension mash or even powder is sealed. Soft capsules industry is developing very fast and more than 600 hundred million soft capsules are

Please use the following format when citing this chapter:

Ge, F., Shi, J., Xu, Y., Zou, X. and Zhao, J., 2009, in IFIP International Federation for Information Processing, Volume 294, *Computer and Computing Technologies in Agriculture II, Volume 2*, eds. D. Li, Z. Chunjiang, (Boston: Springer), pp. 1369–1378.

produced every year over the world which cost 4 hundred million dollars. There are 3 hundred million soft capsules in our nation each year ,the product is export to Japan, South East Asia, USA, Western Europe, Singapore etc.

As most of contents of Soft Capsules have viscosity, a fraction of content was adhere to injector and filling pump while it flow into wedge injector and was pushed into two pieces of colloidal film by the filling pump of automatic rotating rolling capsules machine(G. Reich et al., 2004).This process caused measurement error and fluctuation of Soft Capsules's weight which has closely correlation to it's efficacy. Therefore Soft Capsules are graded in order to keep their weight uniform. Nowadays, many companies use workers who were trained to judge Soft Capsules weight according their figures subjectively. But the grading accuracy and repeatability was low in that the grading process was based on workers personal experience, also because of the large labor intensity, the expensive cost and the low efficiency, this grading method can not apply to modern produce.

At present, there is not a on-line grading quality equipment for Soft Capsules, so we proposed to use machine vision to resolve the situation and designed a on-line grading system that could classify soft capsules according weight standard.

2. SYSTEM SETUP OVERVIEW

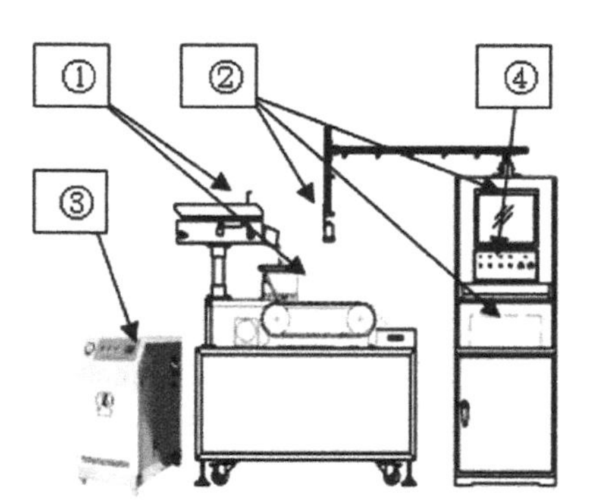

Fig. 1 Soft Capsule online sorting device entire map

A system capable of grading soft capsule at on-line throughput was developed. The setup of the system is show in Fig. 1. It consisted of feeding unit, machine vision system, grading unit and electric control unit. The basic feeding conveyor transported the soft capsule to the uniform spacing conveyor. Then, the capsule were fed to the machine vision system for the defect inspection. Finally, the automatic sorting unit accomplished the soft capsule grading operation.

The machine vision system included a lighting chamber for the desired spectrum and light distribution for soft capsule illumination, a CCD camera and an image grabbing card with four input channels which provided by Euresys company inserted in a microcomputer (processor speed: 1.66GHz).

3. IMAGE PROCESS

3.1 Image background removal

There are many ways to remove background of a image (Milan Sonka et al., 2003), according to the histogram, gray distribution of the soft capsule images is double peak. As this distribution we chose ostu maximum variance between clusters(OSTU method) to remove the background. Fig2 (a) is source image and fig 2 (b) is the result image processed by OSTU method, from the image we can get that the soft capsule was segmented completely.

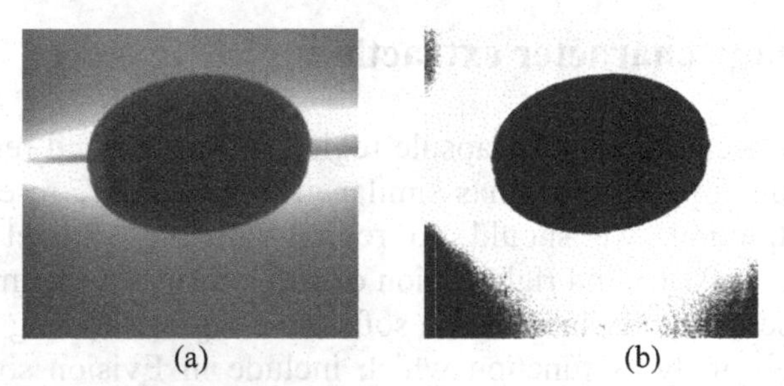

(a) (b)

Fig. 2 Before and after removing the background image Capsule

3.2 Noise removal

Although background was removed from source image, the result image still existed some noise which will influence with future processing. There are many methods to remove noise from a image such as mean smoothing, Low-pass Filter and median filter. In the research we choose 3×3 mean smoothing filter,

$\frac{1}{16}\begin{bmatrix}1 & 2 & 1\\ 2 & 4 & 2\\ 1 & 2 & 1\end{bmatrix}$ Low-pass Filter and median filter to remove noise(Rafael C.Gonzelez Richard et al., 2003) .The results of these method are as follow images in fig 3.Compared their effects ,we found fig 3 (d) was the best image for future processing.

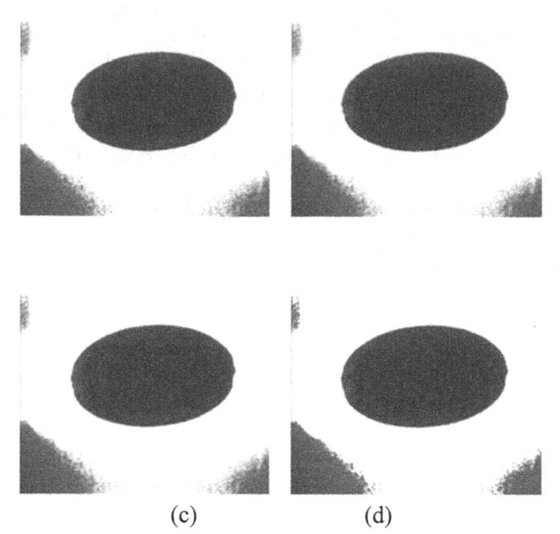

(c) (d)

Fig3 effect of mean smoothing、Low-pass Filter and median filter
(a)source image (b)processed by mean smoothing
(c)processed by Low-pass Filter (d)processed by median filter

3.3 Image character extraction

In order to keep whole soft capsule region in background removing step, some background pixel which has similar gray value were reserved. Before character extraction, we should do region labeling (Rafael C.Gonzelez Richard et al., 2003)to find right region of soft capsule in the image. On this paper, because many machine vision softs have region labeling algorithm so we chose blob analyses function which include in Evision soft to do this work. The result is show in fig4. In the image, we define that a region whose pixels is more than 50000 is soft capsule region.

After the region of soft capsule was found in image, we should extract characters for grading judgement. In this research, we used area, girth, altitude diameter and latitude diameter to represent soft capsule character. Their definition are shown in fig 5.

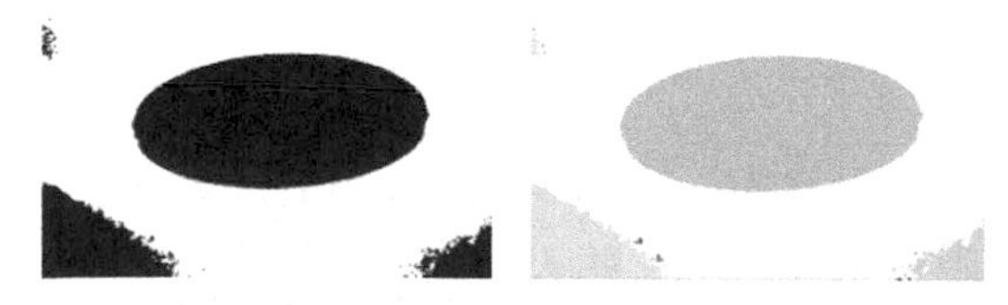

Fig 4 region labeling

(1) area S as shown in fig5, the number of pixels whose gray value is 0.
(2) girth L as shown in fig5, the number of the edge of soft capsule region,

(3) altitude diameter H the distance between the most left and right pixel.

(4) latitude diameter W the distance between the most top and bottom pixel

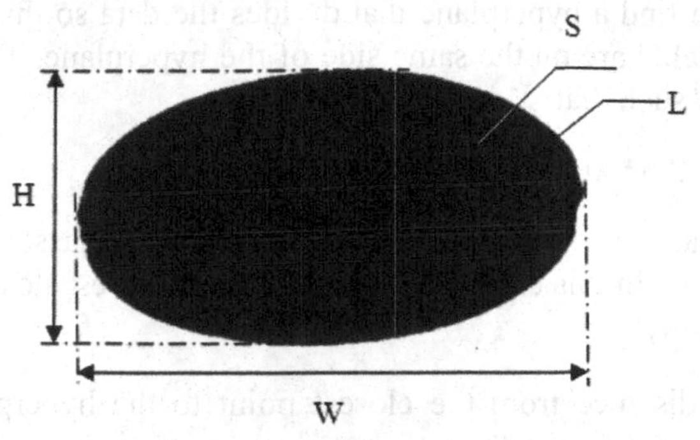

Fig 5 characters of soft capsule

4. DETECTION QUALITY BY SVM MODEL

4.1 Support Vector Machine (SVM)

Support Vector Machine(SVM) procedure is based on statistical learning theory as proposed by Vapnik and Chervonenkis (V.N. Vapnik, A.Y et al., 1995),which is discussed in detail by Vapnik (V.N. Vapnik, A.Y et al., 1995). The SVM can be seen as a method of training polynomial, radial basis function, or multilayer perception classifiers, in which the weights of the network are found by solving a quadratic programming (QP) problem with linear inequality and equality constraints. The SVM uses structural risk minimization, rather than a non-convex, unconstrained minimization problem, as in standard neural network training technique using empirical risk minimization. Empirical risk minimizes the misclassification error on the training set, whereas structural risk minimizes the probability of misclassifying a previously unseen data point drawn randomly from a fixed but unknown probability distribution(S.R. Gunn et al., 1995).

Assume that the training data with *k* number of samples is represented by $\{x_i,y_i\}=1, 2, . . . k$, where $x \in R^n$is an n dimensional vector and $y \in \{-1,1\}$ is the class label. These training patterns are said to be linearly separable if a vector ω and a scalar β can be defined so that inequalities (1) and (2) are satisfied:

$$\omega \bullet x_i + \beta \geq +1, \text{ for all } y = +1 \quad (1)$$

$$\omega \bullet xi + \beta \leq -1, \text{ for all } y = -1 \quad (2)$$

The aim is to find a hyperplane that divides the data so that all the points with the same label are on the same side of the hyperplane. This amounts to finding ω and β such that:

$$yi(\omega \bullet xi + \beta) > 0 \quad (3)$$

If a hyperplane exists that satisfies (3), the two classes is said to be linearly separable. In this case, it is always possible o rescale ω and β so that $\min_{1 \leq i \leq k} yi(\omega \bullet xi + \beta) \geq 1$.

That is, the distance from the closest point to the hyperplane is $1/\|\omega\|$. Then (3) can be written as

$$yi(\omega \bullet xi + \beta) \geq 1 \quad (4)$$

The hyperplane for which the distance to the closest point is maximal is called the optimal separating hyperplane (OSH) (Fig. 6). As the distance to the closest point is $1/\|\omega\|$, the OSH can be found by minimizing $\|\omega\|^2$ under constraint (4). The minimization procedure uses Lagrange multipliers and

quadratic programming (QP) optimization methods. If $\alpha i \geq 0$, $i=1,\ldots,k$ are the non-negative Lagrange multipliers associated with constraint (4), the optimization problem becomes one of maximizing:

$$L(\alpha) = \sum_i \alpha_i - \frac{1}{2} \sum_{i,j} \alpha_i \alpha_j y_i y_j (x_i x_j) \quad (5)$$

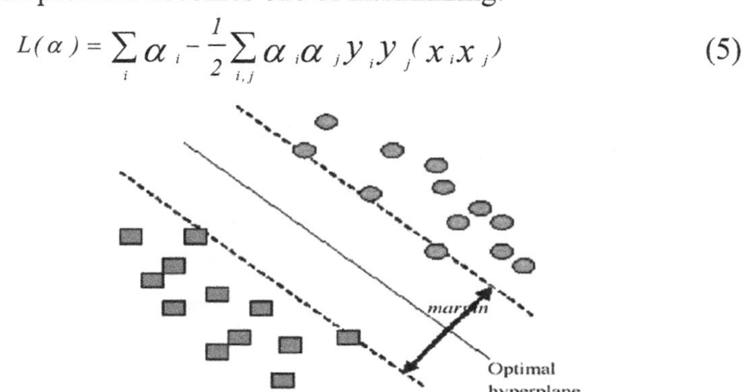

Fig. 6. Hyperplanes for linearly separable data. Dashed line passes through the support vectors

Under constrains: $\alpha i \geq 0$, $i=1,\ldots,k$. If $\alpha m = (\alpha m1 \ \ldots, \ \alpha mk)$ is an optimal solution of the maximization problem (5) then the optimal separating hyperplane can be expressed as

$$\omega^m = \sum_i y_i \alpha_i^m x_i \quad (6)$$

The support vectors are the points for which *ami*> 0 when the equality in (4) holds. If the data are not linearly separable then a slack variable $\xi i,i$=1,. . .,*k* can be introduced with ξi = 0 such that (10) can be written as Eq. (7), and the solution to find a generalized OSH, also called a soft margin hyperplane, can be obtained using the conditions: Eqs. (8)–(10):

$$y_i(\omega \bullet x_i + \beta) - 1 + \xi_i \geq 0 \tag{7}$$

$$\min_{\omega,\beta,\xi_i,\ldots,\xi_k} \left[\frac{1}{2} \|\omega\|^2 + c \sum_{i=1}^{k} \xi_i \right] \tag{8}$$

$$y_i(\omega \bullet x_i + \beta) - 1 + \xi_i \geq 0 \tag{9}$$

$$\xi_i \geq 0, i = 1, \ldots, k \tag{10}$$

The first term in (8) is same as in the linearly separable case to control the learning capacity, while the second term controls the number of misclassified points, while parameter *C* is chosen by the user. Larger value of *C* means assigning a higher penalty to errors. In situations in which it is not possible to have a hyperplane defined by linear equations on the training data, the techniques discussed for linearly separable data can be extended to allow for non-linear decision surfaces. A technique introduced by

Vapnik (V.N. Vapnik, A.Y et al., 1995) maps input data into a high dimensional feature space through some non-linear mapping. The transformation to a higher dimensional space spreads the data out in a way that facilitates the finding of linear hyperplane (Fig. 1). After replacing **x** by its mapping in the feature space $\Phi(x)$, Eq. (5) can be written as

$$L(\alpha) = \sum_i \alpha_i - \frac{1}{2} \sum_{i,j} \alpha_i \alpha_j y_i y_j (\phi(x_i) \phi(x_j)) \tag{11}$$

It is convenient to introduce the concept of the *kernel function K*, in order to make the computation easier in feature space, such as Eq. (12):

$$K(x_i, x_j) = \phi(x_i) \phi(x_j) \tag{12}$$

It is the kernel function that performs the non-linear mapping. Popular kernel functions are the following Eqs. (13)–(15):

RBF kernel function : RBF kernel function:

$$K(x_i, x_j) = \exp\left(-\frac{\left\| x_i - x_j \right\|^2}{2\theta^2}\right) \tag{13}$$

polynomial kernel function :

$$K(xi, xj) = (1 + xixj)^{\sigma} \tag{14}$$

sigmoid kernel function :

$$(xi, xj) = \tanh(\sigma(xixj) + \upsilon) \tag{15}$$

Thus, to solve Eq. (5), only the kernel function is computed rather than $\Phi(x)$, which could be computationally expensive. Eq.(16) can be used to classification function:

$$Y = sign\{\sum_{i,j} \alpha_i \alpha_j y_i y_j K(x_i, x_j) + \beta\} \tag{16}$$

In brief, SVM firstly maps the data which are not linearly separable into a high dimensional feature space, and then classifies the data by the maximal margin hyperplanes (Fig. 7).

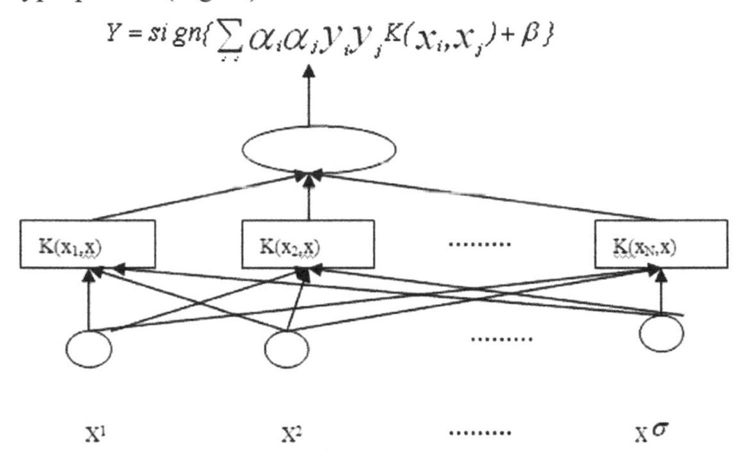

Fig 7 The construction of SVM.

4.2 Results

Radial Basis Function was choose to build SVM model to recognize unqualified soft capsules. In order to get strong Generalization Ability of SVM model, The influences of the error penalty parameter C and the Gaussian kernel parameter σ were studied. 540 soft capsules(180 unqualified and 360 qualified) were chose to extract area, girth, altitude diameter and latitude diameter to build SVM. 100 unqualified and 200 qualified in training set, the rest of soft capsules in testing set. According to the accurate rates of training samples and testing samples, the error penalty parameter C and the Gaussian kernel parameter σ can be determined. Results is shown in table 1. The accurate rate of testing set reached highest while C=32 σ=2 and the SVM model get strongest Generalization Ability.

Table 1 Selection of parameter and result of the experiment

Parameter	C	σ	C	σ	C	σ	C	σ	C	σ	C	σ	C	σ	C	σ	C	σ
value	2^{-3}	2^{-7}	2^{-1}	2^{-5}	2^{0}	2^{-3}	2^{1}	2^{-1}	2^{3}	2^{0}	2^{5}	2^{1}	2^{7}	2^{3}	2^{9}	2^{5}	2^{11}	2^{7}
Accuracy of training set （%）		97.2		97.2		96.3		95.4		95.4		94.4		94.4		93.5		92.6
Accuracy of testing set （%）		61.1		87.5		90.3		93.1		94.4		91.7		90.3		86.5		93.1

15460 soft capsules produced by Hengshun company were tested by the on-line grading system based on SVM. The accurate rate of grading is show in table 2. The solft capsules were first detected by manual using electronic scale (FA1604), and sorted into two classes: accepted and rejected.

Table.2 The detection accuracy rate of Capsule by SVM

Total number of samples	Number of Accepted	Number of Rejected	Accuracy of detection by SVM (%)
15460	14547	913	94.1%

The detection results of SVM is 94.1% as shown in table 2. Compared with the manual detection by huaman eyes (the accurate rate of detection is 84.9%), the machine detection with SVM is much higher.

5. CONCLUSION

The quality of soft capsules online detection system based on machine vision was developed in this paper. SVM is discussed and used to distinguish the rejected capsule from accepted ones. The image of each soft capsule was grabbed from the detection line when they are rotating while moving. Automatic threshold based on ostu was used to segmentation capsule image from background, and morphological filter was used to eliminate noise and regional markings. 4 features were extracted which were perimeter, area, girth, altitude diameter and latitude diameter. Support Vector Machine (SVM) and was used to analyze these features. 15460 soft capsules were tested by the online sorting system. The overall grading accuracy was up to 94.1%. Furthermore, the detection speed of the sorting line resches10 capsules per second.

The results can be used in soft capsule Production Line to inspect the quality of soft capsules accurately and rapidly. It fills the bank of soft capsule on-line grading technology in the industry. Furthermore, this system can be used to grade other productions.

ACKNOWLEDGEMENTS

It is gratefully acknowledged that this work is supported by the National High Technology Research and Development Program of China and National Natural Fund of China.

REFERENCES

G. Reich, Formulation and physical properties of soft capsules, in: F. Podczek, B.E. Jones (Eds.), Pharmaceutical Capsules, Pharmaceutical Press, London, 2004, pp. 201–212.v

Milan Sonka. Image Processing Analysis and Machine Vision[M]. Beijing. Posts & Telecom Press. 2003

Rafael C.Gonzelez Richard, Digital Image Processing [M].Beinjing :Publishing house of electronics industry,2003

S.R. Gunn, Support Vector Machines for Classification and Regression. Technical Report: Image Speech and Intelligent Systems Research Group. Paper available on http://www.isis.ecs.soton.ac.uk/ resources/svminfo/.

V.N. Vapnik, A.Y. Chervonenkis, Theory Prob. Appl. 17 (1971) 264.[18] V.N. Vapnik, The Nature of Statistical Learning Theory, Springer-Verlag, New York, 1995.

V.N. Vapnik, The Nature of Statistical Learning Theory, Springer-Verlag,New York, 1995.

PARTICLE SWARM LEARNING ALGORITHM BASED ON ADJUSTMENT OF PARAMETER AND ITS APPLICATIONS ASSESSMENT OF AGRICULTURAL PROJECTS

Shanlin Yang [1], Weidong Zhu [1], Li Chen [1,2,*]

[1] *School of Management, Hefei University of Technology, Hefei ,Anhui Province, P. R. China 230009*

[2] Department of Management Engineering, Anhui Institute of Architectural and Industry, *Hefei, Anhui Province, P. R. China 230022*

[*] *Corresponding author, Address: Department of The Management Engineering, Anhui Institute of Architectural and Industry, Hefei 230022, Anhui Province, P. R. China, Tel: 13339296584, Fax: 0551-3828127, Email: chinalichina@163.com*

Abstract: The particle swarm, which optimizes neural networks, has overcome its disadvantage of slow convergent speed and shortcoming of local optimum. The parameter that the particle swarm optimization relates to is not much. But it has strongly sensitivity to the parameter. In this paper, we applied PSO-BP to evaluate the environmental effect of an agricultural project, and researched application and Particle Swarm learning algorithm based on adjustment of parameter. This paper, we use MATLAB language .The particle number is 5, 30, 50, 90, and the inertia weight is 0.4, 0.6, and 0.8 separately. Calculate 10 times under each same parameter, and analyze the influence under the same parameter. Result is indicated that the number of particles is in 25 ~ 30 and the inertia weight is in 0.6 ~0.7, and the result of optimization is satisfied.

Keywords: parameter, the particle swarm optimization, agricultural projects measurement

1. INTRODUCTION

Artificial neural network (ANN) is a rising borderline science. Compared to the mathematical statistics, Artificial neural network doesn't need exact

Please use the following format when citing this chapter:

Yang, S., Zhu, W. and Chen, L., 2009, in IFIP International Federation for Information Processing, Volume 294, *Computer and Computing Technologies in Agriculture II, Volume 2*, eds. D. Li, Z. Chunjiang, (Boston: Springer), pp. 1379–1388.

mathematical model and it can solve some problems that traditional statistical methods failed to resolve. Up to now, various types of ANN have been developed in order to address different problems, such as classification, optimization, pattern recognition, data reduction, control and prediction. About 80%-90% percent of ANN is BP network or its change form. BP neural network is one of the most widely used neural networks, and it is a kind of multi-layer back propagation neuron network. It contains three layers: the input layer, the output layer, and the hidden layers. Its output is continuous variable from 0 to 1 and it can realize any nonlinear mapping. The BP network application in agriculture project includes: (1) Modeling and simulation of post-evaluation based on rough set-neural network. (2) Application of rough neural network in agricultural engineering project evaluation. (3) Appraise a model of agricultural high sci-tech agriculture projects base on BP (Chen Li et al.,2006; Chen Li, ZHU Wei-dong,2006; Chui W F et al.,2006).

There are a lot of neural network train methods, but there are few defects, such as it is slow to disappear, easy to converge to the local extreme point. BP is difficult for the function to get out when it gets into a local extreme point. Too many nodes in the hidden layer will lead to the long time of network learning, even the failure of convergence. An overfit phenomenon exists in the BP network (Li Xiaoqing,2006).

In order to overcome this shortcoming of BP, in this paper, we improve BP and adopt the method of Particle Swarm learning algorithm based on adjustment of parameter and Its Applications assessment of agricultural projects.

The PSO, one kind of swarm intelligence, is proposed by Kennedy and Eberhart in 1995. The underlying motivation for the development of PSO algorithm was animal social behavior such as fish schooling, birds flocking, schooling, and swarming. Some of the attractive features of the PSO include the ease of implementation and the fact that no gradient information is required (LIU Yi-jian, et al.,2005).

PSO is the simulation process of birds looking for food. During the birds' looking for food, they adjust the flight direction and speed based on external information at any time. Each bird will be regarded as a non-cubage particle (Cui Guang-zhao, et al.,2007).

Initially, a population of *n* particles is randomly generated in the particle swarm optimization algorithm and searches for optima by updating generations. Particle swarms have two primary operators: velocity update and position update. The swarm direction of a particle is defined by the set of particles neighboring the particle and its history experience. During flight, each particle adjusts its position according to its own experience, and the experience of neighboring particles, making use of the best position encountered by itself and its neighbors (PAN Hong-xia, et al.,2006).

PSO is a generic heuristic optimization algorithm based on the concept of swarm intelligence. It requires less computation times and less memory. Till now PSO has fewer parameters to adjust, and it has been used to solve many engineering and economic problems. PSO as the data analysis tool of a kind of complicated non-linear course, has already extensively applied in pattern-recognition, knowledge engineering, trend analysis, ect. However, it has not been employed so far in an environmental assessment of agricultural project. Here, we apply the adapted PSO approach to search for good environmental assessment of agricultural project sequences with fitness function (Zhao Bo, et al.,2004).

2. PARTICLE SWARM OPTIMIZATION ALGORITHM

2.1 Mathematical models of PSO algorithm

Particle swarm optimization is a member of the wide category of swarm intelligence methods that traces its evolution to the emergent motion of a flock of birds searching for food. It uses a number of particles that constitute a swarm. Each particle traverses the search space, looking for the global minimum (or maximum). During each generation, each particle moves in the search space with a velocity according to its own previous best solution and its group's previous best solution.

The basic particle swarm model consists of an initial swarm of random particles. At the beginning, the PSO algorithm randomly initializes the population (called swarm) of individuals. If there are m particles which conclude a group in D-dimensional space, the ith particle is $X_i=(x_{i1}, x_{i2},\ldots, x_{iD})$, where $(i=1, 2,\ldots, m)$, that is, the position of the ith particle in dimensional space is X_i. While searching optimal solution in search space , each particle remembers two variables : one is the best position found by its own so far , denoted by *pbest* and another is the best position among all particles in the swarm , denoted by *gbest* . The ith particle's flying speed can be noted as $V_i=(v_{i1}, v_{i2},\ldots,v_{iD})$. (LI Qiang et al.,2007).

Each particle updates its own velocity and position according to formula Equ.(1) and Equ.(2)

$$v_{id}^{n+1} = wv_{id}^{n} + c_1 r_1^n (p_{id}^n - x_{id}^n) + c_2 r_2^n (p_{gd}^n - x_{id}^n) \quad (1)$$

$$x_{id}^{n+1} = x_{id}^{n} + v_{id}^{n+1} \quad (2)$$

where x_{id}^{n+1} is the current position of Particle i ; p_{id} is the searched optimal position of Particle i ; p_{gd} is the searched optimal position of the whole particle swarm . In the above equations, c_1 is called self-confidence range; c_2 is called swarm range, and they pull each particle towards *pbest* and *gbest* positions; r_1 and r_2 are randomly generated value between 0 and 1. They determine the affection of p_{id} and p_{gd},the relative influences of the social and the cognition components; w is the inertial weight factor, which is able to adjust the abilities of overall and local search; v_{id}^{n+1} the velocity of the *i*th particle, must lie in the range [v_{dmin}, v_{dmax}]; v_{dmax} represents for the fast iterative speed substitution calculation; it is a constant set by the user. Large values of v_{id}^{n+1} can result in particles moving past good solutions, while small values can result in insufficient exploration of the search space (YANG Hua-chao et al.,2007) .

2.2 PSO algorithm procedures

The whole process of the proposed algorithm can be described as follows:

(1) Randomly generate particles and the particle velocities; Initialize particle positions and velocities.

(2) Evaluate its degree of adaptability to the particle swarm. Calculate all the fitness values of each particle.

(3) Compare each individual's evaluation value with its *pbest*. The best evaluation value among the *pbest*'s is denoted as *gbest*. Update *pbest* and then *gbest* based on the values. If the new value is better than the previous *pbest* , the new value is set to be *pbest*.

(4) Modify the member velocity of each individual P_g according to Eqs.(1) and Eqs. (2).

(5) If the stopping criteria are met, the positions of particles represented by *gbest* are the optimal solution. If the result is satisfied with the number of iteration or precision, then terminate the calculation, or else return to the step (2).

2.3 Analysis on inertia weight in particle swarm optimization

Particle Swarm Optimization (PSO) is simple and easy to implement, and it can generate high-quality solutions within shorter calculation time and more stable convergence characteristic than other stochastic methods. There are not many parameters which need to be tuned in PSO. Although the dimension of particles and the range of particles is important factors in particle swarm optimization. Here, we focus on the influence of the number of particles and the inertia weight.

The inertia weight is the crucial parameter of the particle swarm optimization. It controls the impact of the previous history of velocities on the current velocity. Suitable selection of inertia weight w provides a balance between global and local explorations, thus requiring less iteration on average to find a sufficiently optimal solution. The inertia weight is an important parameter of the PSO choice which can be divided into two categories: fixed inertia weight and time variant inertia weight. The choice of a fixed inertia weight is to choose a constant value as a weight in the process, and it is unchanged in the optimization.

The time variant inertia weight is selected a certain range. For example, Here, a linearly iteration decreasing w is used, which is usually employed as a trade off between the global and local exploration abilities of the swarm. The general inertia weight linearly descend with the number of iterations in common. It is better to initially set the inertia weight to a large value, in order to promote global exploration of the search space, and gradually decreases it to get more refined solutions. Thus the inertia weight can be described as

$$w = wmax - [(wmax - wmin)/itermax] \cdot iter \qquad (3)$$

In the equation, where w_{max} and w_{min} denote the maximum and minimum value of the inertia weight; $iter_{max}$ denote the maximum number of iterations, and $iter$ is the current number of iterations. For example, the parameters of PSO algorithm are selected as w=1.5~0.2, which means that w starts from 1.5 and gradually decreases to 0.2.

In this paper, we supposed the weights are fixed inertia weight. Usually, larger w is expected to facilitate the global exploration at the beginning of the run, while smaller w is expected to facilitate the local exploration near the end of the run. The initial positions and velocities of all particles are generated randomly in the search space. Then, the processes from Eqs. (1) to Eqs. (3) are repeated until a user defined stopping criterion is reached.

2.4 Artifical experiments

2.4.1 Establish index system of an environmental assessment of agricultural project

In this paper, we use an agricultural project as an example and utilize PSO-BP to evaluate environmental-quality assessment. We establish the index system of an environmental assessment of agricultural project. The index we select here is mainly reflect environmental benefits ,environmental pollution and ecological degradation (Chen li ,Zhu Weidong, 2008) (Table 1.).

Table 1. Index system of an environmental assessment of agricultural project

Index	Contents
Environmental benefits	Grassplot area rate Nursery stock area
Environmental pollution	The agricultural product contaminates Soil contaminates Water body pollution
Ecocide	The vegetation covers lessening rate Apply chemical Loss of soil and water area

2.4.2 Application and Particle Swarm learning algorithm based on adjustment of parameter

To demonstrate the good performance of PSO, PSO-BP is also used to find the optimal parameter combination. However, the computation time using PSO-BP is much less than using BP. So, it is much faster using PSO-BP than BP. In this paper, we do not compare the computation time using PSO-BP and BP.

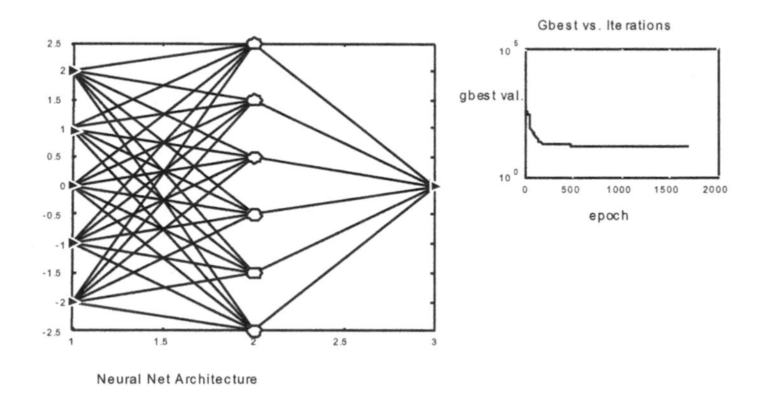

PSO: 1/2000 iterations, GBest = 10091.998215630534.
PSO: 500/2000 iterations, GBest = 17.466520066108341.
PSO: 1000/2000 iterations, GBest = 17.433778425823558.
PSO: 1500/2000 iterations, GBest = 17.433623293802363.
PSO: 1600/2000 iterations, GBest = 17.433623293802359.
PSO: 1691/2000 iterations, GBest = 17.433623293802359.
SSE = 2.3187e+003
--> Solution likely, GBest hasn't changed by at least 1e-009 for 400 epochs.

Fig. 1. Neural Net Architecture and Optimize process

The proposed PSO based approach was implemented by using the MATLAB language and the experiments have been carried on two infrared images using MATLAB7.0 on 2.0 GHz Pentium 4 PC. Initially, several runs have been done with different values of the PSO. Meanwhile, several experiments have been done in order to obtain the penalty parameters.

In the neural network algorithm trained by particle swarm optimization: Parameters in PSO algorithm are described as follows: the maximum allowed iteration number is 2000 ; the acceleration constants are selected as the default values $c_1 = c_2 = 2$.SSE is equivalent to 2.3187e+003; The segmentation results are shown in Fig1.

Tables 2 ~4 show the preliminary results of the comparisons. Each test is run 10 times and each run Loops 2000 generations.

Table 2. Comparison of experimental result

No.	Particle number	5			30		
	Weight	0.4	0.6	0.8	0.4	0.6	0.8
1	ESS	662.613	1860.8	201.626	0.235	0.423	0.340
2		1594.7	1.701	1242.0	0.332	0.248	0.507
3		129.401	1.826	2212.8	0.234	0.273	0.206
4		824.866	3819.1	748.385	0.186	0.256	0.210
5		654.370	1072.4	590.781	0.240	0.408	0.288
6		83.344	160.128	15.950	0.464	0.276	0.800
7		2231.7	1.423	19597	0.340	0.533	0.669
8		81.685	16.514	4.512	0.426	0.241	0.629
9		2318.7	41.596	267.348	0.353	0.246	0.508
10		11.648	5.275	0.777	0.467	0.496	0.506
No.	Particle number	50			90		
	Weight	0.4	0.6	0.8	0.4	0.6	0.8
1	ESS	0.233	0.644	0.205	0.224	0.215	0.373
2		0.286	0.303	0.283	0.265	0.242	0.399
3		0.458	0.398	0.255	0.289	0.285	0.238
4		0.130	0.523	0.270	0.150	0.174	0.340
5		0.279	0.250	0.344	0.236	0.159	0.219
6		0.307	0.202	0.349	0.122	0.436	0.256
7		0.154	0.139	0.302	0.495	0.323	0.218
8		0.347	0.504	0.481	0.420	0.358	0.388
9		0.346	0.266	0.308	0.288	0.378	0.375
10		0.390	0.358	0.226	0.258	0.413	0.220

Sufficient experiments have been done on inertia weight that it is an important parameter in the algorithm. The particle swarm, which optimizes neural networks, has overcome its disadvantage of slow convergent speed and shortcoming of local optimum. The parameter that the particle swarm optimization relates to is not much. But the particle swarm optimization is strongly sensitive to the parameter.

When the particle number is 5, 30, 50, 90 separately, and the inertia weight is 0.4, 0.6, 0.8 separately, calculate 10 times under the condition of each same parameter. Analyze the influence of the parameter. The acceleration constants are selected as the default values $c_1 = c_2 = 2$.The inertia weight w is 0.4, 0.6, and 0.8, respectively. The SSE of optimization with PSO by changing inertia weight is listed in Table 2.

The Optimum value and the poorest value of the target satellite are listed in Table 3. Results of optimization with PSO by changing inertia weight are listed in Table 4.

Table 3. Results of optimization with PSO

Particle Number	Inertia Weight 0.4		Inertia Weight 0.6		Inertia Weight 0.8	
	Optimum alue	Poorest value	Optimum alue	Poorest value	Optimum value	Poorest value
5	11.648	2231.7	1.423	3819.1	0.7778	19597
30	0.1865	0.4674	0.2412	0.5337	0.2062	0.8000
50	0.1308	0.4582	0.1390	0.6446	0.2052	0.4817
90	0.1223	0.4957	0.1599	0.4367	0.2182	0.3979

Table 4. Results of optimization with PSO by changing weight

Weight	Particle number 20			Particle number 25			Particle number 28		
				ESS					
0.4	0.630	0.6364	0.4818	0.2291	0.5244	0.6727	0.1326	0.7571	0.450
0.5	1.355	0.4219	0.3700	0.4812	0.429	0.3823	0.1926	0.5680	0.130
0.6	0.498	0.4624	0.3505	0.3489	0.3864	0.5889	0.3782	0.1982	0.347
0.7	0.595	0.3869	0.2984	0.6638	0.4863	0.4161	0.5681	0.1187	0.314
0.8	0.321	0.5116	0.3994	0.4028	0.2502	0.4951	0.4101	0.7096	0.156
0.9	0.905	0.6007	0.5599	0.8496	0.8480	1.2726	0.4405	0.6945	0.282

2.4.3 Experiments analysis

Analysis from Table 2, Table 3, and Table 4, we can conclude:

(1) We test each sample with 10 times for impartial results and treat the average of each sample as the final result. In the test, all the algorithms run 10 times with 2000 iterations. When the swarm size particle number is fewer, calculating under the same condition ten times, the SSE of optimization with PSO fluctuates greatly. When the number of the particle is 5, and the inertia weight is 0.8, and optimum value of the ten times is 0.7778.The worst value is 19597. When the number of the particle is greater than 20, calculating under the same condition, the SSE of optimization with PSO fluctuates not greatly (Table 2.).

(2) In Fig.1, the PSO algorithm converges fast at the beginning of the run but slows down when it gets close to the global optimum. It has overcome its disadvantage of local optimum when the inertia weight is greater, but it has the relatively slow convergent speed.

(3) Final optimization results of the algorithm are shown in Table 2. In the test, all the algorithms run 10 times with 2000 iterations.

From the result of PSO operation of a agriculture project, we find that 5-90 particles are used in PSO, which is a balance between the accuracy

From the result of the applications assessment of agricultural projects based on the PSO, we find that although only few parameters are required, they are important for the optimization efficiency of particle swarm optimization. The inertia weight and the

required in search of the global optimum and time consumed (Table 3.). Through being ten experiments, result is indicated that the number of particles is in 25 ~ 30 and the inertia weight is in 0.6 ~0.7, and the result of optimization is satisfied. When the inertia weight is too great, such as 0.9 , the result is not pretty good. (Table 4.).

The result of this paper has confirmed that the reliability of the PSO algorithm is more powerful in the aspects of an environmental assessment of agricultural project. PSO algorithm has been successfully applied to solve the environmental assessment of agricultural project.

3. CONCLUSION

From the result of the applications assessment of agricultural projects based on the PSO, we find that although only few parameters are required, they are important for the optimization efficiency of particle swarm optimization. The inertia weight and the number of PSO are the crucial parameters of the particle swarm optimization.

Results indicated that when the inertia weight is smaller, especially when the particle is fewer, it is fast convergent speed at this moment, but apt to fall into local optimum. The larger the number of particles adopted in PSO, the fewer the opportunities to be trapped in suboptima. It finds particle counts between 25-30, while weight in 0.6-0.7, and it optimizes that result is getting more ideal.

The results of this paper confirmed that the reliability of the PSO algorithm is more powerful in the aspects of an environmental assessment of agricultural project.

PSO-BP neural network can be used to predict PSO Applications assessment of agricultural projects based on adjustment of parameters, which not only overcomes the long time of network learning, but also avoids the failure of convergence. No more than 2.3187e+003 maximum errors showed that the model predicted the PSO Applications assessment of agricultural projects with highly accuracy.

The application field and the theory about PSO still need exploring. The results of this paper confirmed that the reliability of the PSO algorithm is more powerful in the aspects of an environmental assessment of agricultural project. Comparison functions adopted here are four benchmark functions used by many researchers. They are the Sphere, Griewank, Rastrigrin and Rosebrock . How to advance the accuracy of the results by inducing other optimization methods and how to make our method be applied in practice is future research for us. It should be noted that the validity of the proposed method in this study is limited within the cases where inertia weight is fixed. The time variant inertia weight and the four benchmark functions mentioned above should be addressed in the future.

REFERENCES

Chen li, Zhu Weidong,The enviromental quality assessment of neural network algorithm trained by particle swarm optimization, Acta Ecologica Sinica,Vol.28,No.3, 2008(3):1072-1079

Chen Li, Zhu Weidong. Application of rough neural network in agricultural engineering project evaluation . Transactions of the CSAE, 2006, 22 (7): 230- 232

Chen Li, Zhu Weidong. Modeling and Simulation of Post-evaluation Based on Rough Set-Neural Network. Journal of System Simulation, 2006,18 (8):2158-2161

Chui W F, Huo X X,Zhuang S H,etc. Appraise a model of agricultural high sci-tech agriculture projects base on BP. Jour. of Northwest Sci-Tech Univ. of Agri and For. (N at. Sci.Ed.), 2006,34(7) :160-164

Cui Guangzhou, Niu Yunyun, Wang Yanfeng, Zhang Xuncai and Pan Linqiang. A new approach based on PSO algorithm to find good computational encoding sequences, Progress in Natural Science, Vol. 17, No. 6, 2007(6):712-716

Li Qiang, Guo Fucheng, Zhou Yiyu. A New Satellite Passive Locating Algorithm Using Frequency-only Measurements Based on PSO, Journal of Astronautics, Vol. 28, No. 6,2007(6):1575-1582

Li Xiaoqing. Design optimization based on neural network and genetic algorithm, Computer Automated Measurement & Control,2006, 14 (2):253-255 (in Chinese)

Liu Yijian, Zhang Jianming, Wang Shuqing. Parameter estimation of cutting tool temperature nonlinear model using PSO algorithm, Journal of Zhejiang University Science, 2005,6(10):1026-1029

Pan Hongxia, Ma Qingfeng. Research on Gear-box Fault Diagnosis Method Based on Adjusting-learning-rate PSO Neural Network, Journal of Donghua University (Eng. Ed.) ,Vol . 23, No. 6, 2006(10):29-32

Yang Huachao, Zhang Shubi, Deng Kazhong, Du Peijun, Research into a Feature Selection Method for Hyperspectral Imagery Using PSO and SVM, Journal of China University of Mining & Technology, Vol.17, No.4, 2007(12):473-478

Zhao Bo, Guo Chuangxin, Cao Yijia. Optimal Power Flow Using Particle Swarm Optimization and Non-Stationary Multi-Stage Assignment Penalty Function, Transactions of China Electro Technical Society, Vol.19, No.5, 2004 (5):47-53

THE SCIENTIFIC AND TECHNICAL SERVICE FUNCTION OF AGRICULTURAL DIGITAL LIBRARY IN COUNTRYSIDE INFORMATIZATION DEVELOPMENT AND CONSTRUCTION UNDER NETWORK ENVIRONMENT

Xichuan Guo
The Agricultural Information Institute of the Chinese Academy of Agricultural Sciences, 100081,Beijing P. R. China, Tel: +86-010-82109886call2209,Email: guoxc@caas.net.cn

Abstract: By analyzing internet popularization in rural China and open questions in rural information service, together with service function of agricultural digital library in rural informatization construction, this article makes a study of how to give full play to agricultural digital library's various advantages, in order to make even greater contributions to our country's rural informatization construction.

Keywords: agricultural digital library, rural, informatization construction, information service, internet popularization

1. FOREWORD

21st century will be an era of information agriculture, computer and information technology will bring profound influence on modernization of traditional speciality. In point of countryside, the old conventional production awareness and agricultural structure have been smashed gradually and replaced by modern agriculture awareness and new pattern of agricultural economy; Ecological agriculture have been attached great importance to; Green food, economic planting, township enterprise and aquaculture have been developing at very fast speed; Township enterprises

Please use the following format when citing this chapter:

Guo, X., 2009, in IFIP International Federation for Information Processing, Volume 294, *Computer and Computing Technologies in Agriculture II, Volume 2*, eds. D. Li, Z. Chunjiang, (Boston: Springer), pp. 1389–1396.

are developing gradually towards industrialization, specialization, integration and socialization; A majority of agricultural science and technology workers, farmers, and farmer-enterpriser are seeking zealously for information in order to know the trends and master the skills. . Compared with the applications such as research, industry and trade in our country, the popularization of information technology and internet are relatively hysteretic in agricultural utilization, but as the fast increase of netizens in the rural China area, the implementation of construction of basic network facilities such as "extend to every village" project and "every town sets up their own website " project will hasten the improvement of rural internet environment. . Researches on agriculture informatization's development and how an agricultural digital library can play its role in development and construction of agriculture informatization is a topic worth exploring.

2. ACTUALITY OF RURAL INTERNET POPULARIZATION IN OUR COUNTRY

The number of netizens in China had increased rapidly to 210 million up to Dec, 2007. It was 48 million more than that of Jun. 2007, the number is 73 million for the year 2007 and the annual growth rate reached 53.3%. Namely, in the past year, the average growth of number of netizens is 200 thousand per day. Rural netizens（refer to the netizens dwelling in countryside for the moment）under swift growth is an importance component of newly increased netizens.. In 2007, the annual growth rate of rural netizens exceeded 100% and reached 127.7%; the number came up to 52.62 million. Among 7.3 million newly increased netizens, there are 2.917 million from countryside.

The number of rural netizens will keep growing in the future, which will be the major factor of the increase.. There are two reasons: Firstly, China's macroeconomic condition is very good, and will keep developing all along, what's more, the improvement of people's quality of life could be assured. Secondly, operators and service systems from different operators under Ministry of Information Industry (MII) such as “extend to every village" project and "every town has its own website" project would hasten the improvement of internet environmental in countryside, thus accelerate the increase of rural netizens. According to the survey, the utilization of rural netizens, for the moment, are mainly for information, minor for recreation.

The pinch of information in countryside may possibly become an important factor in enlarging the gap between developments of countryside and towns. Due to the geographic problem, the transportation in rural area is relatively inconvenient. Undoubtedly, the internet becomes the most convenient and prompt source to bring information to the area. Farmers could get access to advanced technology and information conveniently at a low price, and can obtain required information on planting, breeding and unclosing marketing channel, etc. What's more, the internet can play a great role in management, rural education, and medical treatment and so on. Information provided by internet will promote new socialism countryside construction.

3. ANALYSES OF OPEN QUESTION IN INFORMATION SERVICES IN COUNTRYSIDE

3.1 Frailty of Farmers' Information cell Consciousness

In a long time , because of small-scale production, farmers become accustom to listen to their experience on what to plant and what to breed, however, they are lack of awareness of the importance of information and not capable enough to get information.. Meanwhile, due to the restrictions of economic foundation ,cultural quality and geographic regions, they don't have the condition to obtain and put out information from and on internet in time. This kind of information consciousness impacts construction of modernization of agriculture. Every year, thousands of agricultural scientific payoffs come into being in our country, But those are converted to be used in agricultural production are only about one third. Network-based digital library, particularly digital library of agricultural colleges and universities can be justified in acting as the information medium and should be functional as information bridges.

3.2 Agricultural Information Resources Lacks Diversity

Countryside are always dotty area of economic development, so agricultural book materials, network information resources are lack of diversity.. Although most countryside have founded wireless or cable television network, broadcasting call system, the agricultural information service provided is limited. Actually, useful information on agriculture is very little; Scope of payoff generalization is finite. Our agronomist has quite a lot achievement, but as to vast rural area all over the country, agricultural production is not instructed under science. Utilization of agronomist system not yet draws forth farmers' response.

3.3 Issuance of Agricultural Information not Smooth

To spread agricultural information, grass roots in countryside are by dint of holding a meeting, giving agricultural lecture, reinfusion on wire, cable TV and so on. Apparently, it can not catch up with the request of turn of the market. Backwardness and passivity of agricultural information dissemination will certainly bring about the hysteresis of agricultural information, which will result in sharp swing of agricultural production, and crop surplus or shortage, and impact agricultural reorganization of an industry's process. According to statistics, information transfer between decentralized management farmer and National Bureau of Statistics will need at least half a month.

3.4 Lack of Agricultural Information Network Talents

Vast countryside are lack of compound senior talents who can collect and trim out network agricultural information, who can analyze trend of market, who can answer questions and provide in time precise agricultural products information for agricultural products dealers, who know information technology and is familiar with agro technique.. However, because of they pay little attention to agricultural information network talent, invest little fund, plus imperfect training mechanism, for the moment, useful agricultural information resources has not yet been exploited out, technological element of agricultural economy is little, economic efficiency is low.

4. CONCEPT OF AGRICULTURAL DIGITAL LIBRARY

In company with fast development of new techniques such as modern communication network technology, computer technology, highly dense storage technology, in 21st century, people are striding towards a truly information resources sharing frontier —digital library, This has became the best choice of development of world library, which is also the only way for libraries to occupy a "do some things "position on information super highway.

Agricultural digital library is to turn original agricultural book materials into digital storage, to precede format conversion and compression disposal to traditional information like text, image, and sounds and so on, in order to translate them into numerical information, then, to precede storage via computer technology, to precede information transfer by the aid of internet, to realize computer search, to build library holdings which can serve for local and remote users to access their OPAC, to inquire conventional nonnumeric information, to provide information service for readers bylibrary digitized foundation spermatic internal service in library and super-netization of connection between systems. Compared with traditional agricultural library, agricultural digital library can not only enable users to get summary of related main body, but also provide more original text information for users, to inquire digital resources inner and outer of libraries.

5. SERVICE FUNCTION OF AGRICULTURAL DIGITAL LIBRARY IN RURAL INFORMATIZATION CONSTRUCTION

In company with digitization of books and periodicals and other literary, services of library in high-tech agriculture turn to netization and modernization.To realize modernized service libraries need to make use of computer technology and network communication, to carry out scientific management on library bibliographic information, to enhance service efficiency and quality , to realize information resources sharing , realize robotization as well as to joint track with information super highway.Library digitization radically changed method to collect, store, spread and utilize literature information, reformed its service in high-tech agriculture, established user-oriented service pattern.

5.1 Intend to Train up Countryside High-tech Agricultural Talents out of Participators and Servants

As information center, digital library plays an irreplaceable role in high-tech agricultural training of qualified personnel. Digital library stockpiles prolific professional literature periodical of varied discipline, all manner of bibliographic information, electronic medium material, By reading, one can widen his visual field , widen his span of knowledge, perfect personal knowledge structure day by day and develop body and mind healthily.

5.2 To Provide Individuation Custom-tailor Service for Countryside Agricultural Research Production Personnel

Individuation customization service refers to custom-tailor interface for selected subscriber according to agricultural research production personnel, search and provide customizing messages meeting the users's demands, at the same time, to protect user's privacy by means of security authentication technology. This is also a major trend for digital library development.Individualization service is an important measure to enhance library service quality, to realize service modernization. Digital library make use of new technique, provide custom-built individualization services for all classes of users with specially designed computer system.Custom-built contents include system resources, system interface and regular or nonscheduled mails about relevant column. System resources include all digital resource in the library, network resources, and service information and so on; system interface then include page style and layouts of custom-tailored module. They sent latest agricultural scientific information and production information to registered users by email in order to provide them with worthy information in time.Agricultural scientific personnel acquire latest bibliographic information, information resources according to his needs from digital library so that they can get their job going..

5.3 To provide Interactive Intellectualization Reference Queries Service for Agricultural Research Production Personnel

Reference queries are to provide required information and support for agricultural research production personnel by librarian or information expert with professional method as much as possible. To retrieve among immense information resources from digital library, traditional idiomatic method is far from possible to solve recall ratio and precision ratio problems.Digitization reference queries is a new kind of service pattern establishing on traditional reference queries and network technique, that is to say , get going reference queries service on the internet by using new network technique and information technology .There are two main patterns

5.3.1 Email based Digitization Reference Queries Service

Email-based digitization reference queries service is the most simple and popular pattern.This kind of pattern generally sets Up "reference queries"or similar linkage window on homepage or some webpage of the library. Thus users could send consultation problem to relevant managerial staff by email, consultants could send answers to agricultural research production personnel from all over the country by email.

5.3.2 Digitization Reference Queries Service Based on Real-time Interactive Technology

The unique feature of this pattern is that agricultural research production personnel and librarian can interlocution in real time, the answer can be transferred immediately, nowadays, more popular methods are chatting room BBS, Network Conference, network call center and so on.Chatting rooms primarily use chatting software, both sides could proceed text pattern consultation, Intercourse and Transfer consultation results.We could get going reference queries service in type of network conference by Network Conference software, set up timed network conference connection between Library and agricultural research production personnel. Both sides could exchange ideas via text, image, and sounds. Take advantage of popular network client call center in electronic commerce , they can provide real-time and cooperation based consulting service for agricultural research production personnel, System can enable consultant to choose one-to-one or one-to-many consulting pattern, both sides could transmit various format files and information in real time. They could also browse webpage and demo operational process together, thus to realize telereference service.

5.4 To Provide Cooperative Digitization Reference Queries Service for Agricultural Research Production Personnel

Digital reference queries are now walking towards cooperation pattern based on combo form tone library activityCooperation digital reference queries based on netization join 2 or more libraries up, together provide network reference queries, together shoulder the duty, draw upon each other's strengths, so as to enhance quality of service by a wide margin, to incarnate the pursuit for collaboration and resource sharing for quite some time.In this pattern, 2 or more libraries take advantage of network technique to establish digitization reference queries service collaboration between systems. They are researching on agricultural research production personnel put forward consultation problem according to unified form by consulting service linkage on webpage, and then sent forms to supervisory servicer according to factors like nature of problem. Server will automatically pass them to most suitable cooperation member by email. Then answers will send back to library initially accepted problems by supervisory servicer and then to user.

6. CONCLUSION

In conclusion, agricultural library should realize that in an era when rural internet grows rapidly in china, they should pay attention to farmers' information requirements to realize supreme social benefit, to perfect agricultural digital library's agricultural information service function through indigenous effort, to strike up scientific agricultural information service system, to foster favorable agricultural information service market with a will, to serve to enhance farmers' scientific and cultural quality, to increase their yield and income, to build socialism new countryside.

REFERENCES

Chen Ai Min. Wang We Displays the library superiority Completes the countryside information service work. Journal of Library and Information Sciences in Agriculture, 2006, 6:84-86

CHEN Zhang jun. Information Service of the University Library and Rural Community, Journal of Library and Information Sciences In Agriculture, 2004, 10:109-111

CNNIC, 21st China internet development status report 2008

CNNIC, 21st China rural internet development status report 2008

Ma DeXing, Construction of Technology Communication Service System of Digital Agr iculture Platform. Journal of Anhui Agri. Sci. 2007，35（2）：614- 615

RESEARCH ON AGRICULTURE DOMAIN META-SEARCH ENGINE SYSTEM

Nengfu Xie*, Wensheng Wang
Agricultural Information Institute, The Chinese Academy of Agricultural Sciences, Beijing, China, 100081
* *Corresponding author, Address: Agricultural Information Institute, No.12 Zhongguancun South St., Haidian District Beijing 100081, P. R. China, Tel:+86-10-82109819, Fax:+86-10-82109819, Email:nf.xieg@caas.net.cn*

Abstract: The rapid growth of agriculture web information brings a fact that search engine can not return a satisfied result for users' queries. In this paper, we propose an agriculture domain search engine system, called ADSE, that can obtains results by an advance interface to several searches and aggregates them. We also discuss two key technologies: agriculture information determination and engine.

Keywords: meta search engine, agriculture information, web search engine.

1. INTRODUCTION

With the Web quick development, the information resources on the Web is gradually abundant. In the meantime, the agriculture web sites spread over China like mushrooms after rain. Recently China has about more than 14000 large agriculture web sites. The available enormous amount of Web information requires the use search engine tools such as google, baidu, whose aim is to match general users' requirements, but they suffer from many problems such as, a) low retrieval rate, b) freshness problem, c) poor retrieval rate, d) long list of result which consumes time and efforts, e) huge amount of rapidly expanded information which causes a storage problem, and finally, f) large number of daily hits which makes most search engines not able to provide enough computational power to satisfy each users information need (Eldesouky & Saleh, 2008).

Please use the following format when citing this chapter:

Xie, N. and Wang, W., 2009, in IFIP International Federation for Information Processing, Volume 294, *Computer and Computing Technologies in Agriculture II, Volume 2*, eds. D. Li, Z. Chunjiang, (Boston: Springer), pp. 1397–1403.

That leads a new generation of search engines called meta-search engines, which provides a united interface that send a user's query to multiple search engines, thus providing the means for a user to search a broader set of documents and potentially get a better set of results (Lawrence & Giles, 1999). A meta-search engine generally does not maintain its own index of web information instead of aggregates the certain search results into a unified result set that is re-ranked based on the relevance to the query. Metasearch engines increase the search coverage, solve the extendibility issues in searching eclectic information sources, facilitate the invocation of multiple search engines and improve information retrieval effectiveness (Meng et al., 2002).

In the paper, we propose an agriculture domain meta-search engine Architecture to build ADSE. ADSE uses different databases for searching information and the results are obtained at a unified interface.

The paper is organized as follows. Section 2 describes the related work. Architecture of the agriculture meta-search engine is proposed in the section 3. Section 4 discusses the domain Information determination algorithm in ADSE. Finally, Section 5 will conclude the paper.

2. RELATED WORK

A meta-search engine transmits user's search simultaneously to several individual search engines' interface, and gets results from all the search engines queried, and then re-ranked the results by relevance to the query. There has been a great deal of work done in making guided searches a reality. One of the best examples is GuideBeam (Guidebeam), which is the result of research work carried out by the DSTC (Distributed Systems Technology Centre) at the University of Queensland in Brisbane, Australia. GuideBeam works based on a principle called "rational monotonicity" that emerged from artificial intelligence research in the early nineties. In the context of GuideBeam, rational monotonicity prescribes how the user's current query can be expanded in a way which is consistent with the user's preferences for information. In other words it is a guiding principle of preferential reasoning (Peter & Bernd, 1999).

There are many meta-search engines in service. Dogpile and Vivisimo are commercial clustering engines. The Meta search engine Dogpile is a meta-search engine that searches multiple search engines at once (Dogpile, 1999). This site was used as it allows users to obtain results from some of the most popular search engines in one attempt, rather than having to perform the search at each of the sites individually. When a search was performed using the syntax e commerce security issues, results were found for most of the major search engines. The Dogpile meta-search engine is good, in that it shows the top search results for some of the top WWW search engines, web directories and Pay-for-Placements engines at one site. Vivisimo provides an innovative document clustering technology that acts as a metasearch engine. It transforms search engine outputs from long, tedious lists into crisply organized categories

(Vivisimo, 2006).One of the oldest meta search services, MetaCrawler began in July 1995 at the University of Washington. MetaCrawler was purchased by InfoSpace, an online content provider, in Feb. 97. Queries other search engines, organizes results into a uniform format, ranks them by relevance, and returns them to the user (Infospace, 1999).Delivers highly relevant results from a variety of different search engines and directories including Inktomi, GoTo, and the Open Directory.

3. NEURAL NETWORKS IDENTIFICATION ALGORITHM

In the web, agriculture information needs of users are stored in the database of multiple search engines. The rate of agriculture information explosion on the internet is much higher than the rate at which web search engines index the web. It is highly inefficient and inconvenient for an ordinary user to manually invoke multiple search engines and identify useful documents from the returned result sets. To support unified access to multiple search engines for getting agriculture information, an agriculture meta search engine can be constructed.

The key motivation behind the construction of our agriculture meta search mechanism is to increase the coverage and scope of search, handle the search of agriculture information efficiently, provide a single united access for several search interfaces. When a user issues a search query, the agriculture meta search engine fetches the results from various search engines, and the results relevant to the query. The results are unified into a individual result set with a single ranked list, and order them according to relevance.

Apart from the underlying search engines, a meta search hierarchy has four primary software components: a database selector (we select google, baidu, msn as our main databases), a document downloader, query dispatcher and a result merger. But ADSE, domain information determination is also a primary component. Reference ADSE architecture of an agriculture meta search engine is illustrated in Figure 1.

S1, S2, … Sn is the given search engine database from which the agriculture information will be fetched. The database selector is to identify one or more databases that are likely to contain useful documents for the query. The objective of performing database selection is to improve efficiency as by sending each query to only potentially useful search engines, network traffic and the cost of searching useless databases can be reduced.

The downloader component is responsible for establishing a connection with each selected search engine and passing the query to it, and downloads all URL obtained as result set.

The query dispatcher component focuses query management which parser the query, and specify options for each search provider, and finally send the result to the corresponding downloaders. Query dispatcher may also try to adjust the relative weights of query terms in the original query to fetch optimal results.

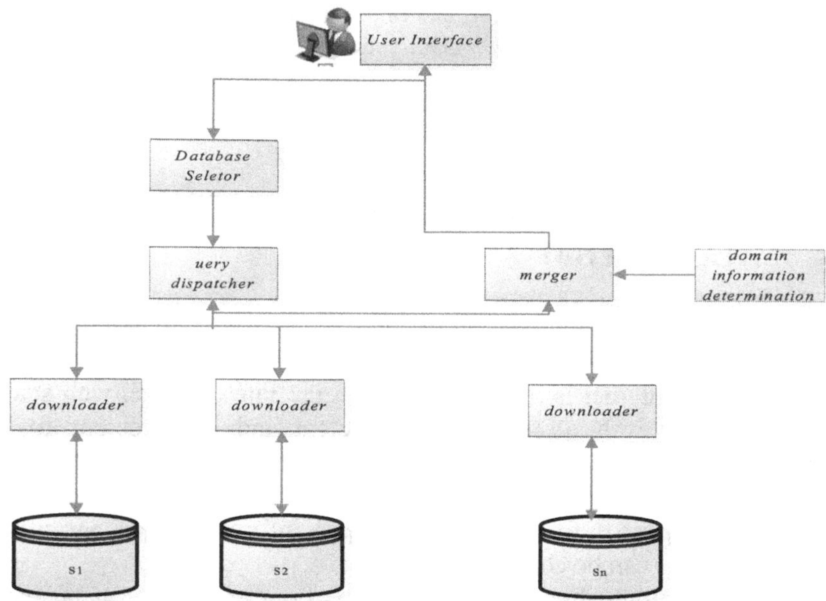

Fig.1. Agriculture Meta search Component Architecture

The merger component combines the results into a single ranked list by similarity between the query and documents in the result. The merger uses a ranking algorithm to re-rank the results.

The domain information determination algorithm component is to determine whether a document belongs to the agriculture domain based on a website url, which quickly collects the minimum closure of the domain-based web information, compared with other web page clustering algorithms in many domain search engines.

4. DOMAIN INFORMATION DETERMINATION ALGORITHM

For agriculture domain, one different point from other general met-search engines is to provide an interface for agriculture users to get agriculture information from the web. The general key technologies in meta search engines have been discussed in many papers. Here we proposed domain information determination algorithm based on website homepage content.

In a certain domain-specific website, whose homepage has a domain characteristic, and describes the abstract of the website's content, which is used to determine the website's domain. But some websites' homepage description is very brief or only contains login information. The following page' content is the main domain-determined information description after clicking the homepage.

A domain website homepage contains many links, whose descriptions can be extracted to form a agriculture word set according to the domain lexicons (Xie et al., 2008). The set contains enough semantic information to measure the website domain type. In our method, a website homepage can be defined by the frame description as following:

```
def category webHome
{
attribute: title
        :type string
    attribute: anchorInfo
        :type ArrrayAnchor
attribute: copyright
      : type string
}
```

The above website homepage description can be stored in XML format. The attribute "title" represents the title of the homepage, and the attribute "ArrrayAnchor" describes the information about an anchor, represented as 2-tuple <url, urlInfo>, the first coordinate url represents a link in a anchor, and the second coordinate urlInfo represents the responding link anchor text in the anchor, for example:

<a href="http://[URL goes here]">Link Anchor Text</a >

In which, the value of url is "http://[URL goes here]", and the value urlInfo is "Link Anchor Text".

In our method, the copyright information is not considered. So a domain website homepage can be represented as hInfo = <title,{<url, anchorInfo >}>. In order to compute the confidence of a website belonging to a certain domain, the content of url and anchorInfo is segmented to a set of words in the domain lexicons using our a fast algorithm for Chinese Word Segmentation. So hInfo can be also described as hInfo = <{ w_i },{<url, { w_j } >}>. For each word w_i, we define a word weight, called word rank (wr), so the vector of hInfo is divided two parts: 1) the vector of title as [wr_{t1}, wr_{t2},..., wr_{tn}] in which n is the vector dimension; 2) the anchor vector space (AVS), defined as:

$$AVS_{k \times n} = \left\{ \begin{bmatrix} wr_{l11} \\ wr_{l22} \\ \vdots \\ wr_{lk1} \end{bmatrix} \begin{bmatrix} wr_{l12} \\ wr_{l22} \\ \vdots \\ wr_{lk2} \end{bmatrix} \cdots \begin{bmatrix} wr_{l1n} \\ wr_{l2n} \\ \vdots \\ wr_{lkn} \end{bmatrix} \right\}$$

In which, n is the number of anchors in the homepage and k is the number of components for each anchor data. So the steps of the domain website determination algorithm are described as following:

1) Determining the weight of the title and anchor Information belonging to a certain domain, called W_t and W_a;

2) Given a website web address, a crawler was used to download the website homepage, and then store the homepage content (HC);

3) Applying the webpage parser to produce a DOM tree equivalent to HC, and then get the description of webpage (hInfo). For Chinese website, we use word Segmentation to split the content of the title and anchor Information into words;

4) Assigning the weight to each word;

$$\text{Confidenc} \quad \Pr = W_t \frac{\sum_i wr_i}{N_t} + W_a \, \text{Conf}_a(AVS_{k\times n})$$

5) Using the following formula to compute the confidence of a website belonging to a certain domain.

In which, $\text{Conf}_a(AVS_{k\times n}) = \sum_j^n \sum_l^k w_{jl}$, and Nt is the number of components in a title vector, n is the number of links in the homepage. For a given threshold confidence value (TCV) of a website belonging to a certain domain, if the confidencePr of a website is equal to or above TCV, we can conclude that the website belonging to the domain.

5. CONCLUSION

This paper proposes an agriculture domain search engine system, called ADSE. We have studied the agriculture domain search engine system hierarchy, in which we propose a domain information determination algorithm to solve agriculture information selection. The new approach significantly improved the domain-specific information closure, reducing the non agriculture information returned. The papers have also studied the domain Information determination algorithm in ADSE. Future work involves, building the agriculture domain meta search engine, incorporating more number of search engines in the study, studying the running performance of ADSE.

ACKNOWLEDGEMENTS

This work is supported by Special Fund Project for Basic Science Research Business Fee, AIIS, the Chinese Academy of Agricultural Sciences (Grant No 2008211) and The National Science & Technology Program (Grant No.2006BAD10A06).

REFERENCES

B. Peter and V.L.Bernd. Preferential Models of Query by Navigation. Chapter 4 in Information Retrieval: Uncertainty & Logics, The Kluwer International Series on Information Retrieval. Kluwer Academic Publishers, 1999. http://www.guidebeam.com/preflogic.pdf.

Dogpile Home Page. Internet. 16 Jun. 1999. Available: http://www.dogpile.com/

E. Eldesouky, A. Saleh, N. El Gendy.An Efficient Strategy for Mobile Focused Crawling (MFC) Based on Mobile Agent Technology. The Sixth International Conference on Informatics and Systems (INFOS2008), 2008.

Infospace, Inc. Infospace. http://www.infospace.com/, January 1999.

N.F. Xie, et al.. .Journal of Jiangxi Normal University (Natural Sciences Edition), 2008, Vol. 32:192-196.

S. Lawrence and C. L. Giles. the NECI meta search engine. Computer Networks and ISDN Systems, 1998,(30):95～105..

S.Lawrence, & C. L.Giles (1999b). Searching the Web: General and scientific information access. IEEE Communications, 37(1):116–122.

Vivisimo clustering engine, 2006. Available from: http://vivisimo.com/demos/ PubMed@NIH.html.

W. Meng, C. Yu, and K. Liu. Building Efficient and Effective Metasearch Engines. ACM Computing Surveys, 34(1), March 2002, pp.48-84.

Zheng Ye Lu.et all..The Strategies for Building Agriculture Web Sites. FITA2002, 2002. Guidebeam - http://www.guidebeam.com/aboutus.html.

PRACTICE AND EXPLORE ON AGRICULTURAL SCIENCE DATA CONSTRUCTION AND SHARING IN CHINA

Ruixue Zhao*

Agricultural Information Institute, Chinese Academy of Agricultural Sciences ,Beijing, R.P China 100081

* *Corresponding author, Address: Agricultural Information Institute,Chinese Academy of Agricultural Sciences, No 12 Zhongguancun South Avenue, Beijing, R.P China 100081, Tel: +86-10-82109885-2308, Email: zhaorx@mail.caas.net.cn*

Abstract: This paper analyzes the situation and the progress of agricultural science data construction and sharing in China. It also illustrates some building experiences, puts forward the necessary considerations and suggestions to the advanced development of agricultural science data sharing in China, which is expected to provide reference in advancing agricultural science data sharing in China.

Keywords: agricultural, science data, data resources, agricultural database, data sharing

1. INTRODUCTION

Science data are the basic data and materials created by science and technology activities of human society, as well as the data products which are systemic processed in accordance with different needs. They have clearly potential value and being developed value, and their value increases in the application process (Xu Guanhua,2003). Science data are the achievements and also the starting point of human research activities, as well as the kinds of important and influential strategic resources in information age. Now, the effect of science data is getting more and more focus, and the demand of them is more and more widespread. How to effectively preserved, in-depth excavate, and use these science data resources has become a new challenge that the worldwide science and technology development should face.

Please use the following format when citing this chapter:

Zhao, R., 2009, in IFIP International Federation for Information Processing, Volume 294, *Computer and Computing Technologies in Agriculture II, Volume 2*, eds. D. Li, Z. Chunjiang, (Boston: Springer), pp. 1405–1413.

International scientific organizations and some developed countries in Europe and America started earlier the science data collection, collation and data sharing work, they had made great progress in policy, law, management, technology, standards, etc. For example, the International Science and Technology Data Committee (CODATA) founded in 1966, had been pursuing the global science data sharing, assessment, conservation and distribution. The United States drew up and implemented a "complete and open" science data sharing policy. The EU established a set of science data sharing system in close connection with the actual operation, which integrated legislative system with policy guidelines (Lu Peng et al., 2007).

Affected by the international science data sharing wave, and also pushed by the increased domestic demand of science data sharing, The Ministry of Science and Technology (MOST) of China initiated the "science data sharing project" in 2001. The project is aimed at, under the national entire planning, applying modern information technology, integrating discrete science data resources, building a intellectualized network system of management and sharing service, which is geared to the needs of the whole society, and implementing standardized management and efficient use of science data resources, so as to provide strong science data resources support for scientific and technological progress, innovation in government decision making, economic growth, social development and national security(Yi Aining, 2007).

As an important part of national science data sharing project, the agricultural science data sharing centre (pilot) project was officially launched in 2003, and made the agricultural science data sharing in China into the rapid development period. The main task of this project is to collect, collate the scattered agricultural science data resources, and to shape the State's authoritative agricultural science database group. At the same time it will establish an agricultural science data resources management and sharing service system which is geared to the needs of the whole society, supply supporting mechanisms and personnel, realize effective protection, sharing and use of agricultural science data resources, and provide support and indemnify, via agricultural science data information resources, for agricultural science and technology innovation and decision making in agricultural science and technology management(Meng Xianxue et al., 2006).

2. STATUS OF AGRICULTURAL SCIENCE DATA SHARING

With the features of dispatch, complexity, and dynamic, the collection and processing of agricultural science data are very difficult. Also due to the system and mechanism reason, the accumulating and sharing of data are in great straits.

Since 2000, with the support of the government, sustainable science projects such as the agricultural scientific fundamental database, pilot projects of sharing of agricultural science data and national agricultural science data sharing center have been started one by one. These projects drive forward the construction and sharing of agricultural science data resources sharing work.

During the period of the building and practicing in these years, remarkable accomplishments have been achieved in the aspects of the integration of agricultural science data resources, the establishment of standards for sharing data, the construction of network system for sharing servers, training talented personnel, sharing services etc.

2.1 Integrated about 560 databases or data sets

Reorganized and integrated kernel agricultural science data resources including 12 major subjects covering crop science, animal and veterinary sciences, tropical crop science, agricultural resources and environmental science, agricultural zonation science, grassland science, food science, agricultural biotechnologies, agricultural microbiology, agricultural information technology and management, and other basic agricultural information and data. Over 560 databases or data sets have been integrated and can be accessed through the Internet at the Agricultural Science Data Center website. These data resources become the biggest agricultural science database group in China.

2.2 Developed Necessary standards and criterion

Critical and necessary standard for database construction, data integration, data management, and data sharing have been developed and used in the project. These standards and criterion ensured the smooth implementation of the project.

2.3 Established agricultural science data sharing system

The agricultural science data sharing center, which including 1 main center and 7 sub-centers and 28 data nodes has been established, and the agricultural science data sharing site groups have also been built up. Based on those work, the agricultural science data sharing union has been founded and the web based data sharing service has been provided.

2.4 Explored diversity service pattern

Actively explored and practiced diversified science data integration and service pattern which covering union and collaboration and sharing of data sources, and built up a multiple data service system that involves subject service sub-center, regional information service sub-center, institutions and university service nodes. The sharing service has achieved obviously progress, and also brought positive influence on the society.

2.5 Promote effective sharing union

Promote effective data sharing union among the various parties including the central government and the local governments, research institutes and colleges, administrative department and scientific research institutes and technical service departments. It also brought out the interactions between scientists, research and production, research and decision-making.

3. EXPERIENCES ON AGRICUTURAL SCIENCE DATA CONSTRUCTION AND SHARING

3.1 Scientific planning, Clear objectives

If you fail to plan, you plan to fail. Scientific planning is very important for those things which are achieved through longer period of efforts, it's a kind of systemic and comprehensive plan , and its quality will directly affect the success or failure of the whole event.

The project of agricultural science data sharing is a long-term pioneering project which has a significant impact and far-reaching significance. Many units and personnel participated in the project, and technical factors and management factors which relate to this project are very complex, so it is necessary to formulate scientific and rational goals and possible ways to achieve these goals, analysis and identify the critical success factors of project implementation according to needs of the major agricultural scientific and technological innovation in the beginning of the project to provide guidance for following various stages to prevent the arbitrary, and maintain consistency of action to ensure the effectiveness of projects.

3.2 Strengthen management, effective organization

The project of agricultural science data sharing not only relates to a simple technical problem, but also relates to a lot of problems about the organization and coordination including the organization and coordination between units which participate in the project, between data producers and data providers, between statistics and submit of data and so on. If these relations are not handled properly, the project will not be able to proceed smoothly. To this point, the project team set up a project organization and management system which is composed of a leading group, project expert group, project management office and project technical group, and thus developed a good working atmosphere and working platform and ensure the smooth implementation of the agricultural science data sharing.

In addition, the project is also thinking of the introduction of process management to control strictly and manage production of data, statistics and submit of data and applications of data, and the introduction of relevant technical standards to ensure that agricultural science data sharing can be implemented on the platform that is scientific and orderly, standardized and consistent.

3.3 Detailed analysis, rational design

Analysis is to investigate, research, and anatomize the status and then further define objectives, tasks and needs to draw up a detailed implementation plan to provide a scientific basis. Analysis equals to "what to do", and design equals to "how to do", and the latter relates to the specific technical solutions.

In the process of the implementation of agricultural science data sharing project, the status of existing agricultural science data resources , which including the number, status, distribution and so on in order to develop viable science data integration scheme, should be understood firstly. At the same time, it is also necessary to investigate the requirement information including information amounts, states and distribution and so on. On the basis, considering the economic, technical and operational environmental conditions, particular technical programme including data processing and integrating, organizing and storage resources, data sharing and so on, could be developed. In accordance with its situation, the agricultural science data sharing centre identified the data resources integrating framework on 12 major subjects, developed the solution on distributing data processing and preserving and integrated accessing strategy based on metadata, designed a technology programme of sharing networking system that was moderate dispersion in physics and consistent highly in logic, and also designed mechanisms of data exchange and data sharing based on metadata.

3.4 Applicable technology, uniform standards

Construction of agricultural science data sharing center is complex and systematic project that is a trans-specialty, trans-departmental, trans-region project, and set acquisition, collection, storage, management and distribution of data and activities of data sharing in one. Construction of agricultural science data sharing center needs to integrate theory, methods and techniques of agriculture, information science, systems science, management science, and other multi-disciplinary and also needs to unitize standards and technical specifications which are necessary in the course of agricultural science data sharing. Above all, these details have become an important guarantee for the smooth implementation of the project.

The development of modern information technology provided a powerful technological support for the sharing of science data. Because of dazzling variety of hardware and software systems, wc must carefully select the appropriate technologic platform, identify the authenticity of merchant publicity, and pay attention to cultivating their own power in research and maintenance. The formulation of standards can not be formalized, and it is necessary to integrate its own needs closely, comply with the practical principles and improve gradually in practical applications.

3.5 Quantity-oriented, service-first

The collecting and arranging of the science data and sources is for the purpose of better utilizing, so as to realize their strategic value. So the key issues include: whether the collected data and sources are necessary, whether they are acquired, whether they are of considerate scale, and whether they are dependable. If not, the collecting and arranging work does not only make no sense, but also has negative impact on scientific research, management and decision making because of the unreliability, incompletion and inaccuracy of the information, which results in severe losses. In order to ensure the quality of science data in the process of constructing and integrating the agricultural science data and resources, the control and management measures for the data and sources quality are constituted, and the four-class responsibility system is constructed including data suppliers, data processors, data publishers and data centers, by which the work is checked at each level, taken charge of class by class. Meanwhile, the effective technologies used for data checking and quality control are explored actively.

High-quality science data and resources are the premise of shared services, while the shared services are the final purpose. But in the social environment nowadays, there are still many problems on the shared services of the science data because of the lack of sharing mechanisms at macroscopic level and protection rules of science data property rights. It is critical for the sustainable development of the science data sharing project that how to construct feasible science data sharing mode, and supply effective service. The agricultural science data sharing center has taken active exploration and practice, and constructed kinds of service, such as the integration of on-line(web portals) and off-line(telephone, e_mail and CD) services on the network, utilize the mature information service measures of the library, for example, reference consultation, SDI(directed service), document delivery(data pushing), and active services through introduction meeting, training class.

4. FURTHER CONSIDERATIONS

Even we have achieved a great progress in the agricultural science data sharing, we still have a long distance to go towards the requirements from the country and the society. First, there are still lots of existing agricultural science data need to be integrated, and also new data are generated every day. It becomes more and more urgent to find a stable, expandable and efficient way to integrate and share those agricultural science data. Second, how to maintain the integrated data is another important challenge, which decides how to keep the results achieved. Third, these data sharing service currently provided are still have some limitations, how to improve the data sharing level, provide more value added service to fulfill the new requirement from agriculture study and innovation needs more work. Forth, the state level data sharing related policies have not issued yet, the issue of intelligent protection in data sharing is still open(Qin Dahe, 2004;Liu Runda et al.,2007).

In order to take full advantage of agriculture science data, a lot of works are required to push the agricultural science data sharing in a wide, deep area, following the science data sharing strategy of the country:

4.1 Change the cooperation model, to improve the integration level of agricultural science data

Currently the stakeholder of the agricultural science data sharing center only includes the related departments, institutes and universities. The data integration and data sharing are conducted via project based method. This model limits the range and the depth of the data sharing. It is necessary to build up a new cooperate model, which can include more stakeholders, to issue the data connection and data sharing processes and policies in a state level, to build up more closely cooperation relationships between the stakeholders, for the purposes of integrating more resources, provide more wide service area, and finally form the state level multilevel structured agriculture data collection and data sharing system, which can provide more strong support to the state agriculture science study and innovation research.

4.2 Change the focus from amount to quality of the data, improve the quality of databases

The purposes of agriculture data share center include not only the continuously enlarge the data sharing scale, but also the continuously improve the data quality, to develop the more and more high quality data sharing products, to integrate more and more data sources, and to focus on more and more hot points of agriculture science and study, to build up a batch of high quality database. The buildup of the agriculture data sharing platform will study how to coordinate the projects of national agriculture projects, to provide active database support and data share services to those project teams, it will also become the important data sources for the government to buildup and agriculture policies and develop the critical agriculture research plan and projects.

4.3 Build up a high Efficiency working model, improve the positive attitudes of different stakeholders

The agricultural science data has the features of dispatch, complexity, and dynamic. Due to those features it is necessary to buildup an efficiency working model among different data owners. The working model needs to ensure that each stakeholder can work in a positive attitude, avoiding the duplicate investment, and all parties can be benefited from the model. The model can also provide more support from policies, regulations, funds and human resources to expand the capability and the long term development of the data sharing.

4.4 Develop towards the digital science library

The agricultural science data sharing center is a product developed under the network environment, the original design of the center are on the basis of sharing and distribution, the different organizations and departments can cooperated over network from different locations. It is not only the country level agriculture data integration and sharing center, but also the important study and research center. The strategy of developing the data share center can refer to the strategy of developing the traditional library, it is possible that the model of data share center and the model of library can be integrated into one model, to ensue the new model can be the data share workspace for both science documents and science data which can provide more critical information and data to support the agricultural development strategy decision making, to ensure the model can provide strong support for China to improve the usage efficiency of agriculture resources, to improve the quality of agriculture products, to deduct the cost of agriculture productions, to protect the nature environment, and to improve the competency capability of the country in the international agriculture market.

REFERENCES

Jens Klump, Roland Bertelmann, Jan Brase, et al. Data Publication in the Open Access Initiative, Data Science Journal, 2006, 5:79-83

Liu Runda, Zhu Yunqiang. Explore Key Issues of Science data Sharing-Data Sharing Network of Earth System Science as an Example, Advances in Earth Science, 2007, 26(5):118-125

Lu Peng, Miao Liang-tian, Li Zhi-xiong, et al. Investigation on science data sharing situation and its analysis, Earthquake, 2007, 27(3):125-130

Meng Xianxue, Yang Congke. Science data: Preserving, Archiving and Sharing, Journal of Northeast Agricultural University, 2006, 13(2):174-177

Miehael Zgurovsky. Impact of Information Society on Sustainable Development: Global and Regional Aspects, Data Science Journal, 2007, 6:137-145

Qin Dahe. Process and Development on China Meteorological Sharing, Scientific Chinese, 2004, 9:18-19

Xu Guanhua. Implement Science data Sharing, Improve the Science and Technology Competitive Ability in China, China Basic Science, 2003, (1):63-68

Yi Aining. States and Development on Medical Science Data Sharing, Chinese Journal of Information on Traditional Chinese Medicine, 2007, 14(5):1-2

THE META-ONTOLOGY MODEL OF THE FISH-DISEASE DIAGNOSTIC KNOWLEDGE BASED ON OWL

Yongchang Shi [1], Wen Gao [2,3], Liang Hu [3], Zetian Fu [3,*]

[1] *Beijing Jiaotong University, No.3 of Shangyuan Residence Haidian Disrict in Beijing, P.R.CHINA 100044*

[2] *ShanDong Institute of Business and Technology, 191 BinHai Road Yantai, P.R.CHINA 264005*

[3] *College of Engineering, China Agricultural University, No.17 Qinghua Dong Lu, Haidian, Beijing 100083, P.R.CHINA*

[*] *Corresponding author, Address: College of Engineering, China Agricultural University, No.17 Qinghua Dong Lu, Haidian, Beijing 100083, P.R.CHINA, Tel: +86-010-62736717, Fax: +86-010-62736717, Email: gaowen_sd@163.com*

Abstract: For improving available and reusable of knowledge in fish disease diagnosis (FDD) domain and facilitating knowledge acquisition, an ontology model of FDD knowledge was developed based on owl according to FDD knowledge model. It includes terminology of terms in FDD knowledge and hierarchies of their class.

Keywords: Ontology; Fish Disease Diagnosis; OWL; knowledge base;

1. INTRODUCTION

Aquatic products are an important source of complementary food for improving people's living in China. Since many kinds of fishes have been affected by disease in deeply degree in the north of China, the knowledge engineer in earlier time tried to develop an expert system technology of the fish disease diagnosis in AI.

The designs of intelligent systems for medical diagnosis have been one of the most prolific areas from the very beginnings of AI. The experts in this

---

Please use the following format when citing this chapter:

Shi, Y., Gao, W., Hu, L. and Fu, Z., 2009, in IFIP International Federation for Information Processing, Volume 294, *Computer and Computing Technologies in Agriculture II, Volume 2*, eds. D. Li, Z. Chunjiang, (Boston: Springer), pp. 1415–1422.

applications, and software agents use the ontology and knowledge bases which based on the ontology as data. (Chen YanHong 2004)

2.2 OWL

OWL is a new ontology language for the Semantic Web developed by the World Wide Web Consortium (W3C)Web Ontology Working Group. OWL was primarily designed to represent information about categories of objects and how objects are interrelated the sort of information that is often called the ontology. OWL can also represent information about the objects themselves—the sort of information that is often thought as data.

3. FDD KNOWLEDGE MANAGEMENT FRAMEWORK

Ontology design is primarily a categorization process. Good categorizations can facilitate information retrieval. Studies on categorization that pertain to ontology design in the AI field Since the domain ontology of a knowledge-based system is an explicit specification of the objects, concepts, and other entities that are presumed to exist in some area of interest as well as the relationships that are held among them (Gruber 1993), it defines the set of terms and relations of a domain independent of any problem-solving method. Normally, such method-specific formulation of domain knowledge is difficult to reuse in a different application. Therefore, to separate the potentially reusable domain knowledge from the method-specific knowledge is a consideration that guided our structure of the domain ontology.(Chen and Chan)

3.1 Fish Disease Diagnostic Model

The fish disease diagnostic question concerning multi-elements is complex. According to our investigation, we identify three broad elements as disease, cause and manifestation, omit others. We specify the relations among the three elements as the category of fish disease diagnostic knowledge. Fig 1 shows the three broad elements and reflects relation among them.

Arrow from node of cause and node of disease reflects causality between cause and disease. Arrow from node of disease to node of manifestation also reflects similar causality between disease and manifestation. The arrow from cause to manifestation reflects pathology.

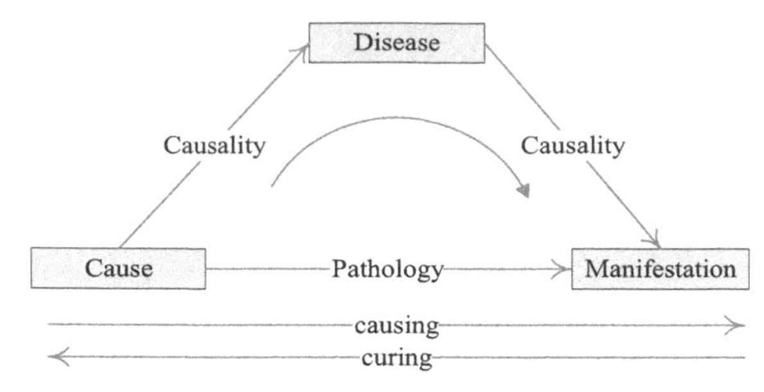

Fig 1 Conceptual Model of FDD

3.2 FDD Meta-Knowledge Categorizing

Knowledge acquisition is an important step in developing a knowledge-based system. The knowledge engineer acquires knowledge from one or more application experts who can explain the problem domain. In this process, detailed information on procedural problem-solving such as input and output, domain knowledge, and the entities and relations in that domain are obtained.(Chen and Chan)

We collect and construct the knowledge base for Fish Disease Diagnosis Acquisition System (FDDS). For doing this, we should make use of existed materials and knowledge elicitation conducted primarily through face-to-face interviews and textual cases provided by application experts. In addition, the knowledge engineer learned about the domain from previously published reports, related projects, and a commercial database on disease diagnostic case knowledge management. The strategy adopted for knowledge acquisition (KA) was based on teaching-learning and teaching-back. Teaching learning was used to obtain the knowledge from the experts including verbal data and references. Then, after the knowledge organized, teaching-back was used when the experts validated and clarified the knowledge presented to them by the knowledge engineer.

All this FDD case knowledge is collected and analyzed before stored in the knowledge database. We collect all case knowledge of FDD about ten species of the fish raised in the countryside of northern China.

3.2.1 FDD Knowledge Base Structure

In FDD system, the fish disease diagnostic knowledge was organized by knowledge representing scheme. There are two schemes: rule-based scheme and case-based scheme. So knowledge base consists of two parts, rule knowledge base and case knowledge base. They store rule knowledge and case knowledge respectively. Rule knowledge base store knowledge as if-then schema, and case knowledge base store knowledge as frame schema. For simplification, we only show the case base structure of our FDD model in Fig 2.

The case base is constituted by four parts: the new case base, the cleaned case base, the classified case base and the case knowledge base.

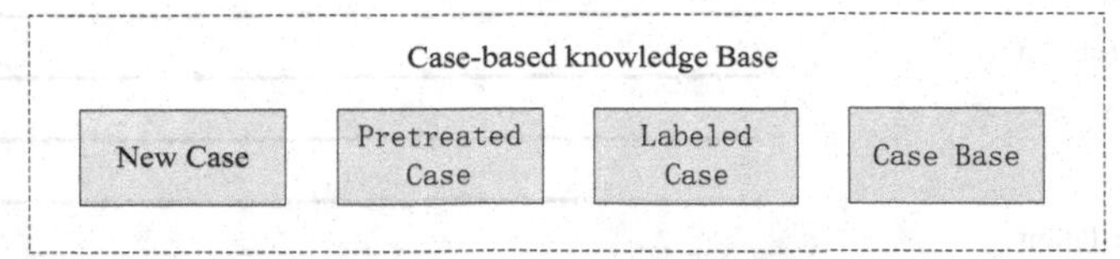

Fig 2 Structure of Case Base

3.2.2 Categorizing of FDD Case Knowledge Metadata

In terms of fish disease diagnosis model, there are two class sets which are nominal class subset, metadata for describing case knowledge and metadata for managing case knowledge in fish disease diagnostic case base. The FDD case knowledge class is named as case FDDK in our system.

Table 1 Class in fish disease diagnosis ontology

Class	1st 2nd 3rd Subclass
FDD Case knowledge	New case Labeled case Classified case Case knowledge
Role (User)	Knowledge engineer Domain expert Manager of knowledge base Other users
Fish	Structure of fish body Surface of fish body Head Body Tail Fin Internal organ Heart … Species grass carp carp … Growth Phase young fry luce ripe fish Size …

Water body	...
Feed program	...
Disease	...
Cause	...
Manifestation	...

3.2.3 Meta-Knowledge Frame of FDD

The metadata frame of our FDDK is shown in Fig 3 which shows the relationship of different sets in knowledge base.

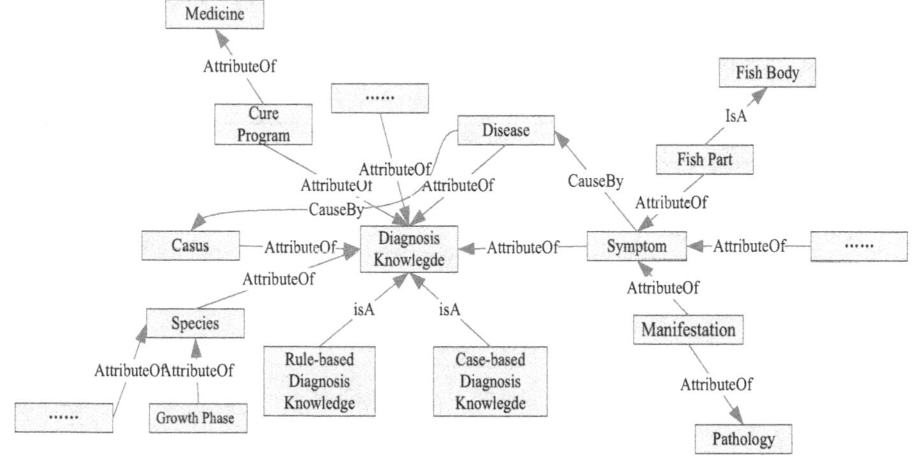

Fig 3 Metadata frame of FDD Knowledge

4. FORMALIZED FDD META-KNOWLEDGE FRAME BY USING OWL

OWL (Web Ontology Language) is designed to process the content of information instead of just presenting information to human. OWL can be used to explicitly represent the meaning of terms in vocabularies and the relationships between those terms. This representation and their interrelationships are called Ontology. OWL has more facilities for expressing meaning and semantics than XML, RDF, and RDFS, and thus OWL goes beyond these languages in its ability to represent machine interpretable content on the Web. OWL is a revision of the DAML+OIL web ontology language incorporating lessons learned from the design and application of DAML+OIL.

We give the formalize expression of FDD case knowledge metadata ontology. The formalized process makes the metadata easier to understand by computer. First we give the namespace by defining some entity and then start OWL ontology. We also import ontology; it will add what we have described into the whole knowledge base.

```
<?xml version="1.0"?>
<!DOCTYPE rdf:RDF [
  <!ENTITY owl "http://www.w3.org/2002/07/owl#" >
  <!ENTITY xsd "http://www.w3.org/2001/XMLSchema#" >
  <!ENTITY rdfs "http://www.w3.org/2000/01/rdf-schema#" >
  <!ENTITY rdf "http://www.w3.org/1999/02/22-rdf-syntax-ns#" >
]>
<rdf:RDF xmlns="http://www.owl-
ontologies.com/Ontology1176772345.owl#"
 xml:base="http://www.owl-ontologies.com/Ontology1176772345.owl"
 xmlns:xsd="http://www.w3.org/2001/XMLSchema#"
 xmlns:rdfs="http://www.w3.org/2000/01/rdf-schema#"
 xmlns:rdf="http://www.w3.org/1999/02/22-rdf-syntax-ns#"
 xmlns:owl="http://www.w3.org/2002/07/owl#">
```

Second we name all class of the system: FDD Knowledge class and subclass of FDD Knowledge, Case Knowledge Class, Role User, Fish class And sub-classes of fish Class and so on, then we define attribute of each class as below.

```
<owl:Ontology rdf:about="FDDK"/>
  <Bacteria rdf:ID="AeromonasPunctata0"/>
  <owl:Class rdf:ID="Bacteria">
    <rdfs:subClassOf rdf:resource="#Cause"/>
  </owl:Class>
  <PathologyDisplay rdf:ID="black">
    <rdfs:comment
rdf:datatype="&xsd;string">&#20307;&#33394;&#21457;&#40657;</rdfs:commen
t>
  </PathologyDisplay>
  <PathologyDisplay rdf:ID="blood">
    <rdfs:comment
rdf:datatype="&xsd;string">&#20986;&#34880;</rdfs:comment>
  </PathologyDisplay>
  <FishSpecies rdf:ID="Carp"/>
  <owl:Class rdf:ID="Cause">
    <owl:disjointWith rdf:resource="#Disease"/>
    <owl:disjointWith rdf:resource="#Manifestation"/>
    <owl:disjointWith rdf:resource="#PathologyDisplay"/>
    <rdfs:comment rdf:datatype="&xsd;string"
      >This is a conceptual specification of diseaseCause.</rdfs:comment>
  </owl:Class>
  <owl:ObjectProperty rdf:ID="causeBy">
    <owl:inverseOf rdf:resource="#Result_In"/>
```

```
</owl:ObjectProperty>
……
</owl:Class>
```

5. CONCLUSION

After the domain ontology of FDD Knowledge was constructed, we categorized the knowledge stored in FDD knowledge base. The terms in FDD case knowledge are classified and organized in taxonomies. All defined classes in FDD Knowledge formed the framework of FDD case Knowledge and associating classes formed a class hierarchy. We developed an ontology of FDD Case Knowledge. This ontology can then be used as a basis for some applications in a suite of expert-system tools.

ACKNOWLEDGEMENTS

This work was supported by VEGNET project (Contract No.CN/ASIAIT&C/005 (89099)), which was funded by the European Union.

REFERENCES

Chandrasekaran, B., J. R. Josephson, et al. (1999). What Are Ontologies, and Why Do We Need Them?, IEEE Computer Society: 20-26.

Chen YanHong. Component-based Software Reuse of the research and application [D],BeiJing: Capital University of Ecnomics and Business, 2004.

Chen, L. L. and C. W. Chan "Ontology Construction from Knowledge Acquisition." Proceedings of Pacific Knowledge Acquisition Workshop (PKAW 2000): 11-13.

Gruber, T. R. (1993). "A translation approach to portable ontology specifications." Knowledge Acquisition 5(2): 199-220.

Heijst, G. v., A. T. Schreiber, et al. (1997). Using explicit ontologies in KBS development, Academic Press, Inc. 46: 183-292.

Palma, J. and R. Marin (2002). "Modelling Contextual Meta-Knowledge in Temporal Model Based Diagnosis." Proceedings of ECAI: 407–411.

Perez, A. G. and V. R. Benjamins (1999). "Overview of Knowledge Sharing and Reuse Components: Ontologies and Problem-Solving Methods." IJCAI-99 workshop on Ontologies and Problem-Solving Methods (KRR5), Stockholm, Sweden, August 2: 1999.

BASED ON GENETIC ALGORITHM KNOWLEDGE ACQUISITION MODEL

Zetian Fu[1,*] , Le Chen[1] , Yonghong Guo[1] , Yongmei Guo[2]

[1] *Department of Mechanical Electrical Engineering,* China JiLiang *University, Hangzhou, Zhejiang Province, P. R. China 310018*

[2] *Department of Computer, Heilongjiang BaoQuanLing Farm Reclamation Industry School, HeGang, HeiLongJiang Province, P. R. China 154211*

[*] *Corresponding author, Address: China Agriculture University Key Laboratory of Modern Precison Agriculture System Integration Peking 100083*

Abstract: In this paper, genetic algorithms and machine learning theory and method of knowledge acquisition for expert system structure is proposed to solve the "bottleneck" problems in the traditional machine learning methods on the basis of AQ, the model of knowledge acquisition based on GA is proposed and its application is used in the Fish Disease Diagnosis reasoning system, the rules is accessed to fish disease diagnosis. Fish Disease Diagnosis and the problems of knowledge accessed into combinatorial optimization are solved.

Keywords: genetic algorithm, knowledge Acquisition, machine learning

1. INTRODUCTION

As expert system application areas continues to expand, the difficulties faced by the combinatorial explosion is increasingly obvious. Construction expert knowledge acquisition system is the "bottleneck" problems, and expertise will have a direct impact on the performance of the whole system. Diagnosis Expert System comes from the knowledge of experts in the field, which is summarized by the expert, it needs repeated exchange between knowledge engineers and experts in the field, and in many cases, often it is very difficult to experience their own knowledge and to make it clear for experts in the field that, especially intuitive problem, the lower the efficiency

Please use the following format when citing this chapter:

Fu, Z., Chen, L., Guo, Y. and Guo, Y., 2009, in IFIP International Federation for Information Processing, Volume 294, *Computer and Computing Technologies in Agriculture II, Volume 2*, eds. D. Li, Z. Chunjiang, (Boston: Springer), pp. 1423–1432.

of this method. In order to find a suitable for large-scale problems and have self-organizing, adaptive, self-learning ability of the algorithm ,it becomes a research subject goal, people has made a variety of different ways, machine learning is considered one of the most effective means. Traditional machine learning methods are based on pre-established rules and knowledge in the decision taken by the strategy. Therefore it is more difficult for the ever-changing environment of the problem.

Genetic Algorithm as a simulation of natural selection and evolution of the process of a random search algorithm and highly robust in recent years is gradually applied to machine learning system, which has a good performance of the intelligence in determining the coding schemes, fitness function and genetic operator, the algorithm will be used in the evolution of the information, such as self-organizing, adaptive characteristics. At the same time it has also given in accordance with changes in the environment, which can automatically discover the characteristics and environmental laws. Natural selection algorithm eliminates the process of designing one of the greatest obstacles: the need to advance the full description of the characteristics, and describe the different characteristics of the problem algorithm measures to be taken. Thus, we can resolve those structures no one can understand complex issues by using genetic algorithms.

2. BASED ON INDUCTIVE MACHINE LEARNING

The general machine learning framework is shown in Figure 1, the learning system seeks to provide teachers with the concept of a group of samples and background knowledge, identify the description of the concept.

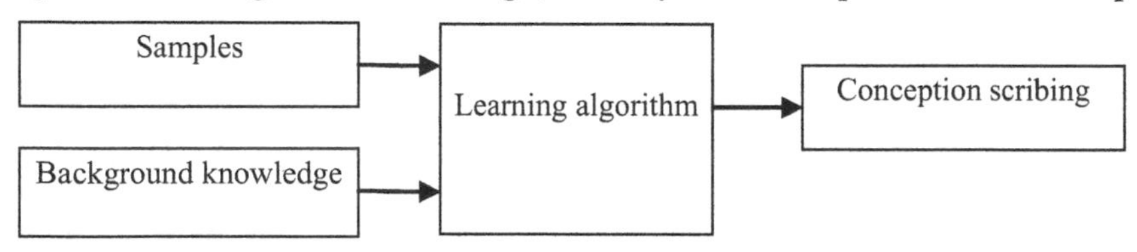

Fig.1 Machine learning framework

Concept can be understood as a group with some of the common nature of the object which is different from other objects. Semantic description of the concept is the basic unit of meaning, which is a significant feature set. Under normal circumstances, in accordance with the concept statement will be divided into the concept and the concept of reasoning. Presented the concept of reasoning is the basis of the analysis of the concept.

To set an example in the database collection, machine learning, known as background sets, $E=D_1 \times D_2 \times \ldots\ldots \times D_n$ is n-dimensional vector space limited, Dj for a limited collection, its base for mj = | Dj |. E attributes set $X = (x_1, x_2, \ldots\ldots, x_n)$, the first j attribute Xj is range D_j. $E = <e_1,e_2, \ldots\ldots, e_n>$ elements in $e = <v_1,v_2, \ldots\ldots, v_n>$ known as examples, $v_j \in D_j$, | E | E for the base said the collection includes examples numbers.

Cases are divided into E will be set of PE and anti-NE:

$PE=\{e^{1+},e^{2+},\ldots,,e^{k+}\},e^{i+}=<vi^{1+},vi^{2+},\ldots,vi^{n+}>,K_p \leqslant m$

$NE=\{e^{1-},e^{2-},\ldots,,e^{k-}\},ei-=<vi^{1-},vi^{2-},\ldots,vi^{n-}>,K_n \leqslant m$

NE PE and assumptions in n-dimensional attribute set for X, is the same, meet $PE \cup NE=E$, $PE \cap NE=\varphi$,$K_p+K_n=m$, $|PE|=K_p$, $|NE|=K_n$.

The concept of learning is a given set of background, the structure contains these statements into assertions, and under the guidance of bias , or the satisfaction of choosing the best summed up assertion that could explain the observed examples of a group concept.

Group related concepts are usually organized into the tree, which can be expressed by the plans or the level of generalization. In the hierarchical structure of a specific level, the concept does not usually intersect, but sometimes they are small differences between large. For example: the concept of a "rotten gill disease" is the "fish diseases," an example of "red skin disease" is a case in point.

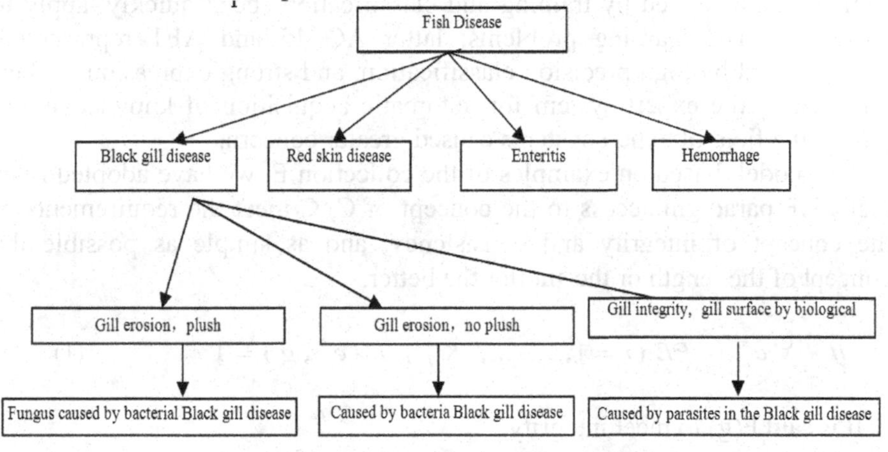

Fig.2 Fish disease concept framework

3. GENETIC ALGORITHMS IN THE APPLICATION OF KNOWLEDGE ACQUISITION

In recent years, many scholars at home and abroad have researched the concept of learning. In these ways, some have used heuristic approach to the rules to guide the search space, such as ID3 algorithm from the point of studying information theory decision tree; AQ heuristic algorithm using the Star algorithm learning the rules; YAI LS algorithm through an evaluation function, such as learning the rules , High operating efficiency of such methods, but not universal. Some also learn adaptive algorithm which is used to find the optimum rules. GA algorithm is based on Darwin's "natural selection" theory of genetic and biological populations and evolution of the theory of general search methods; it can be used in many areas. In practice, however, GA is not superior to specific areas of the specific algorithm. Therefore, to consider the efficiency and versatility from both sides, people often adopt a more strategic search mechanism, heuristic methods to be combined with the GA, to get a better overall performance of the learning algorithm.

Genetic algorithms for knowledge acquisition may constitute a GA at the core of machine learning system.

3.1 Concept of model

Examples of machine learning is a core areas, in accordance with the knowledge that it can be divided into two categories: decision tree into the decision-making rules and summarized, ID3 as the representative to the former, characterized by training and classification speed quickly, apply to large-scale The learning problems; latter AQ 15 and AE1 represented, characterized by high precision classification, and strong expression of their knowledge, the expert system for automatic acquisition of knowledge and thus in the field of expert systems caused greater concern.

The model, based on examples of the collection E, we have adopted rules that CNF paradigm access to the concept of C, C meet the requirements of the concept of integrity and consistency, and as simple as possible the concept of the length or the shorter the better.

$$if \quad \forall e_i^+ \in PE\,(i = 1,2,......,\ K_P),\ F(e_i^+, g) = 1 \qquad (1)$$

It is said F(g) to meet integrity.

$$if \quad \forall e_i^- \in NE\,(i = 1,2,......,\ K_N),\ F(e_i^-, g) = 0 \qquad (2)$$

It is said F (g) meet the consistency;

To meet the concept of integrity and consistency as binding conditions to simplicity as the goal, the establishment of an integer programming model are as follows:

$$MinZ = \sum_{j=1}^{n}\sum_{l=1}^{m_j} x_{jl} + \omega \sum_{j=1}^{n} x_j$$

$$s.t.\begin{cases} \sum_{j=1}^{n} x_{1l_j j} \ge 1 \\ \sum_{j=1}^{n} x_{il_j j} - \sum_{j=1}^{n} x_{1l_j j} < 0, i = 1,2,..., K_P \\ \sum_{j=1}^{n} x_{il_j j} - \sum_{j=1}^{n} x_{1l_j j} < 0, i = 1,2,..., K_N \\ x_j = x_{j1} \vee x_{j2} \vee ...... \vee x_{jm_j}, j = 1,2,..., n \end{cases} \quad (3)$$

$v_{ij} = d_{jl}, x_{il_j j} = x_{jl} (l = 1,2,..., m)$。 $x_j = x_{j1} \vee x_{j2} \vee ... \vee x_{jm_j}$

$if \quad \forall 1, x_{jl} = 0 \quad then \quad n_j = 0$

$if \quad \exists 1, x_{j1} = 1 \quad then \quad x_i = 1$

Where: ω is a big factor punished, the generalω≥2×max{mj,j=1,2,……n}.

3.2 Based on the example of learning the rules GA

The example: The grass carp Hemorrhage and Black gill disease in two cases the rules of study as an example, grass carp Hemorrhage Black gill disease and the symptoms and characteristics of cases Table 1 and shown in Table 2.

Hemorrhage in case of grass carp as are cases of Black gill disease cases as counter-examples to illustrate the concept of learning based on the GA the calculation process and characteristics. In practice, we have set examples of the range of variables and attributes such as shown in Table 3:

Table1 Grass carp Hemorrhage of the sample data Case

ID	Disease	Muscle	Surface	Abdominal	Scales	Head	Fin	Gill	Intestinal
1	Grass carp Hemorrhage	G03	000	000	000	B02	F02	C03	000
2	Grass carp Hemorrhage	G03	000	000	000	B03	F02	C03	000
3	Grass carp Hemorrhage	G03	000	000	000	B07	F02	C03	000
4	Grass carp Hemorrhage	000	A05	000	000	B02	F02	C03	000
5	Grass carp Hemorrhage	000	A05	000	000	B03	F02	C03	000
6	Grass carp Hemorrhage	000	A05	000	000	B07	F02	C03	000
7	Grass carp Hemorrhage	000	A01	000	000	B02	F02	C03	000
8	Grass carp Hemorrhage	000	A01	000	000	B03	F02	C03	000
9	Grass carp Hemorrhage	000	A01	000	000	B07	F02	C03	000

Table2 Black gill disease diagnosed cases of sample data

ID	Disease	Muscle	Surface	Abdominal	Scales	Head	Fin	Gill	Intestinal
1	Black gill disease	000	000	000	000	B01	000	C01	000
2	Black gill disease	000	000	000	000	B01	000	C05	000
3	Black gill disease	000	000	000	000	B01	000	C06	000

Table 3 Examples of existing disease and the collection of attributes Range

No.	x_{ij}	X1	X2	X3	X4	X5	X6	X7	X8
		Muscle	Surface	Abdominal	Scales	Head	Fin	Gill	Intestinal
1	x_{i0}	G03	000	000	000	B02	F02	C03	000
2	x_{i1}	000	A05	000	000	B01	F02	C01	000
3	x_{i2}	G03	000	000	000	B07	000	C03	000
4	x_{i3}	000	000	000	000	B03	000	C05	000
5	x_{i4}	000	A01	000	000	B03	F02	C06	000

Under (3), the issue of model rules for:

$$\begin{aligned}\min\ Z = & x_{10} + x_{12} \\ & + x_{20} + x_{21} + x_{25} \\ & + x_{50} + x_{51} + x_{52} + x_{53} + x_{54} \\ & + x_{60} + x_{61} + x_{64} \\ & + x_{70} + x_{71} + x_{72} + x_{73} + x_{74} \\ & + \omega(x_1 + x_2 + x_3 + x_4 + x_5 + x_6 + x_7 + x_8)\end{aligned} \tag{4}$$

$$s.t. = \begin{cases} x_{10} + x_{12} + x_{50} + x_{60} + x_{61} + x_{64} + x_{70} + x_{72} \geq 1 \\ x_{10} + x_{12} + x_{50} + x_{60} + x_{61} + x_{64} + x_{70} + x_{72} - x_{10} - x_{12} - x_{53} - x_{54} - x_{60} + x_{61} - x_{64} - x_{70} - x_{72} = 0 \\ x_{10} + x_{12} + x_{50} + x_{60} + x_{61} + x_{64} + x_{70} + x_{72} - x_{10} - x_{12} - x_{52} - x_{60} + x_{61} - x_{64} - x_{70} - x_{72} = 0 \\ x_{10} + x_{12} + x_{50} + x_{60} + x_{61} + x_{64} + x_{70} + x_{72} - x_{21} - x_{50} - x_{60} + x_{61} - x_{64} - x_{70} - x_{72} = 0 \\ x_{10} + x_{12} + x_{50} + x_{60} + x_{61} + x_{64} + x_{70} + x_{72} - x_{21} - x_{53} - x_{54} - x_{60} + x_{61} - x_{64} - x_{70} - x_{72} = 0 \\ x_{10} + x_{12} + x_{50} + x_{60} + x_{61} + x_{64} + x_{70} + x_{72} - x_{21} - x_{52} - x_{60} + x_{61} - x_{64} - x_{70} - x_{72} = 0 \\ x_{10} + x_{12} + x_{50} + x_{60} + x_{61} + x_{64} + x_{70} + x_{72} - x_{24} - x_{50} - x_{60} + x_{61} - x_{64} - x_{70} - x_{72} = 0 \\ x_{10} + x_{12} + x_{50} + x_{60} + x_{61} + x_{64} + x_{70} + x_{72} - x_{24} - x_{53} - x_{54} - x_{60} + x_{61} - x_{64} - x_{70} - x_{72} = 0 \\ x_{10} + x_{12} + x_{50} + x_{60} + x_{61} + x_{64} + x_{70} + x_{72} - x_{24} - x_{52} - x_{60} + x_{61} - x_{64} - x_{70} - x_{72} = 0 \\ x_{10} + x_{12} + x_{50} + x_{60} + x_{61} + x_{64} + x_{70} + x_{72} - x_{51} - x_{71} \geq 1 \\ x_{10} + x_{12} + x_{50} + x_{60} + x_{61} + x_{64} + x_{70} + x_{72} - x_{51} - x_{73} \geq 1 \\ x_{10} + x_{12} + x_{50} + x_{60} + x_{61} + x_{64} + x_{70} + x_{72} - x_{51} - x_{74} \geq 1 \end{cases} \tag{5}$$

$$\begin{cases} x_1 = x_{10} \vee x_{12} \\ x_2 = x_{21} \vee x_{24} \\ x_5 = x_{50} \vee x_{52} \vee x_{53} \vee x_{54} \\ x_6 = x_{60} \vee x_{61} \vee x_{64} \\ x_7 = x_{70} \vee x_{72} \end{cases} \tag{6}$$

3.3 Based on the binding plan for GA

In general, binding problems can be divided into a binding constraint satisfaction problem (CSP) and is bound by the optimization problem (COP).

area were considered as pioneers in the use of the deep causal models. The systems rapidly showed the advantages of modeling domain knowledge in this way, in contrast with the classical rule-based systems (Palma and Marin 2002). Recently the researches in this area have been paid increasingly attention to the use of ontology models, especially if they are considered as integrated in knowledge based Diagnosis (KBD) techniques, which have proved their efficiency in the design of intelligent diagnosis systems.

Ontology aims at capturing domain knowledge in a generic way and provides a commonly agreed understanding of a domain, which may be reused and shared across applications. and groups(Chandrasekaran, Josephson et al. 1999). Ontologies provide a common vocabulary of an area and define -with different levels of formality- the meaning of the terms and the relations between them. Ontologies are usually organized in taxonomies and typically contain modeling primitives such as classes, relations, functions, axioms and instances(Gruber 1993). Popular applications of ontology include knowledge management, natural language generation, enterprise modeling, knowledge-based systems, ontology-based brokers, and interoperability between systems. (Perez and Benjamins 1999)

The structure of the paper is as follows: the related work about Ontology and OWL in section 2, FDD Knowledge Management Framework is presented in section 3, the formalized expression of FDD Case Knowledge metadata frame using OWL was given in section 4, and we finally provide conclusions in section 5.

2. RELATED WORK

2.1 Ontology

In the past, ontological issues were investigated in such areas of AI as theoretical knowledge representation and natural language understanding. Recently, ontological issues are being widely used for the purposes of knowledge sharing and reuse, and object-oriented database design. Ontology can also be seen as the study of the organization and classification of knowledge. Ontological engineering in AI has the practical goal of constructing frameworks for "knowledge" that allow computational systems to tackle knowledge-intensive problems such as natural language processing and real-world reasoning. (Chen and Chan)

The ontology of the domain is not a goal in itself often. Developing the ontology is akin to defining a set of data and their structure for other programs to use. Problem-solving methods, domain-independent

For such problems, the issue of the search space does not belong to certain regions include the issue of the search point. Fish diseases from the above model of knowledge acquisition can be seen, it is bound by a non-linear, the problem of how to find such global optimal solution does not exist at present foolproof method.

Optimization of the penalties in the traditional function, and optimize the search process from the point to another point, according to structural constraints of the punishment, punishment will be added to the objective function, so that non-linear programming problems into a series of extreme value to the non-binding question , is the external their early problems that the optimal solution.

To function outside the penalty point method as an example, the model for knowledge acquisition of fish diseases such inequality constrained optimization problem, point penalty function outside the law for the following steps:

Construction penalty function, the original problem into a non-binding Optimization:

$$\phi(X, M^{(k)}) = f(X) + M^{(k)} \sum_{\alpha=1}^{m} \{\max[\ g\alpha(x), 0]\}^{\alpha} \tag{7}$$

Where: the right to punish the second; α punishment for the structural function of the index, its value will affect the function Φ (X, M (k)) in the contour of the binding nature of the general admission α = 2; M For the punishment factor is greater than 0 a progressive series, which should meet

$$0<M(0)<M(1)<\ldots<M(k)<M(k+1)<\ldots) \qquad \lim_{k\to\infty} M^{(k)} = +\infty \tag{8}$$

Where M(0) for the initial punishment factor, based on experience values (such as from $M^{(0)}=1$).

(2) fitness function is defined as

$$f(X)=C0-\Phi\,(X,M(k)) \tag{9}$$

where: C0 is a given number, to ensure that f (X) a non-negative, the optimization problem for the sake of fitness into the biggest problem.

Groups of individuals in accordance with the fitness of all sort of mechanism is follows: First, compare the binding of individual fitness Fcom, good adaptation of the top individual, if the fitness of equal value, compared to its optimized fitness Fopt, Good fitness top.

And punishment is usually based on the method, making it possible to the point of total points is better than not feasible. Thus enabling optimization of the process be feasible for the first point, then these potential points better not feasible, and the genetic operation be feasible optimal point. This will enter a viable area and has been optimized, unified, and without changing the optimum objective fitness to greater than zero. There's no need to set up Fcom and Fopt weight, the use of relatively simple.

Sort after i individuals for the survival chances

prob(i)=q(1-q)i-1 (10)

One q $\in$ (0,1) called the selection pressure, used to control the risk of the individual selected, usually for an average of several times the risk of individual choice.

In order to protect the weak offspring of individuals involved in reproductive capacity and prevent certain anomalies in the evolution of the individual to be disproportionately selected as the father of gene makes the convergence lead to premature convergence, add a counter pNum individual records were selected for the father of the frequency, Article i individuals for the survival chances.

prob(i)=q(1-q)i-1(fn)pNum (11)

One q $\in$ (0,1), can be used to control the individual had been selected as the father of the number of times, pNum bigger, prob (i) the smaller the risk.

(3) Canonical Genetic Algorithms CGA

$\max\{f(b):b\in IB^L\}$

where:0$\leqslant$f(b) $\leqslant\infty$,b$\in$IBl=(0,1)L,f(b)$\neq$const

Abstract speaking, CGA comes from the following seven components:

CGA=(λ ,L, P0,P',S,C,M)

Where

λ $\in$ N for the group in the total number of individuals;

L $\in$ N Binary coded for the length of string;

$$P^0 = (b_1^0, b_2^0, \ldots, b_\lambda^0) \in I^\lambda, I = IB^L = (0,1)^L \quad (12)$$

P' = (P_c, P_m), where P_c for cross-probability, P_m for the mutation rate;

S: I2→I for the selection;

C: I×IPc, I×I for the cross operator.

In general, Pc and Pm is set their own by the user.

(4) CGA process procedures

Initialization L, λ, Pc, Pm;

Randomly generated initial population, P_{old};

While the termination does not meet the conditions do;

The number of new individual P N = 0;

For j = 1 to λ do fj = f (bj) calculation of the individual fitness;

Optimal solution = max (fj) the corresponding individual;

while N <λ do;

To choose from the operator S P_{old} parents choose;

If PC ≥ Random (0,1) then use a C-operator of two generations;

If Pm ≥ Random (0,1) then use mutation operator M to change the two generations;

The two generations into the new P

N=N+1;

EndWhile;

Pnew=Pold;

EndWhile;

Output optimal values。

(5) Termination criteria

Three kinds of termination criteria can be used: a, using the generation to meet the average value of the older generation and to meet the average ratio of the value of the termination criteria; b, using the frequency of the cycle of the termination criteria; c, the optimal use of individual groups and more and the termination of the same criteria. All of the above criteria can be used alone, but also joint use.

GA solution can be reached through the following conclusions:

IF (Muscle =G03) ∧ (Head=B03) ∧ (Fins =F02) ∧ (Gill =C03) THEN Grass carp hemorrhage

IF (Muscle =G03) ∧ (Surface=A01) ∧ (Fins =F02) ∧ (Gill =C03) THEN Grass carp hemorrhage

Consistent with the actual situation, it can be seen that this method is effective。

4. SUMMARY

Construction expert knowledge acquisition system is the "bottleneck" problems, and expertise will have a direct impact on the performance of the whole system. As Fish Disease Diagnosis knowledge learning is, in fact, a combination of issues, the problem is transformed into knowledge acquisition portfolio optimization problem, fish disease diagnostic knowledge acquisition is used successfully in the model. Genetic Algorithm as a simulation of natural selection and evolution of the process of a random search algorithm, and highly robust in recent years gradually is applied to machine learning system, it has a good performance of the intelligence. A algorithm is proposed in this paper, which is used to solve the concept of access model, it is proved that the method of Fish Disease Diagnosis knowledge acquisition is effective.

ACKNOWLEDGEMENTS

Funding for this research was provided by China Agriculture University Key Laboratory of Modern Precision Agriculture System Integration (P. R. China). The first author is grateful to China Agriculture University for providing her with pursuing a PhD degree.

REFERENCES

D.E. Goldberg .P.Segresi. Finite Markov Chain Analysis of Genetic Algorithm. Genetic Algorithms and Their Applications: Proceedings of the Second international Conference on Genetic Algorithms. 1987, 1~8

G .Rudolph. Convergence Analysis of Canonical Genetic Algorithms. IEEE Trans. On Neural Network, 1994,5(1): 96~101

H. .Muhlenbein. How Genetic Algorithms Really Work: Mutation and Hillclimbing. in Parallel Problem Solving from Nature, 2, Amsterdam , North Holland,1992 , 15~25

J. H. Holland. Adaptation in natural and artificial systems. Ann Arbor: University of Michigan Press. 1975, 10

J. R. Koza. Genetic Programming.Cambridge , MA : MIT Press, 1992

J. R. Koza. Hierarchical Genetic Algorithms Operation on Populations of Computer Programs. Proceedings of 11th international Joint Conference on Artificial Intelligence, 1989

Joe Suzuki. A Further Result on the Markov Chain Model of Genetic Algorithms and Its Application to a Simulated Annealing-Like Strategy. IEEE transactions on System. Man and Cybernetics--Part B: Cybernetics. 1998, 28(1): 95~102T. M. Murdock. et al. Use of a Genetic Algorithm to Analyze Robust Stability Problems, Proceedings of American Control Conference. Boston , 1991, 886~889

Joe Suzuki. A Markov Chain Analysis on Simple Genetic Algorithms. IEEE transactions on System. Man and Cybernetics. 1995, 25(4): 655~659

M Srinivas, L.M. Patnaik. Adaptive Probability of Crossover and Mutation in Genetic Algorithms. IEEE Transactions on System. Man and Cybernetics. 1994, 24(4): 656~667

T .Back. The Interaction of Mutation Rate, Selection and Self-Adaption within a Genetic Algorithm. in Parallel Problem Solving from Nature , 2 , Amsterdam , Norh Holland,1992 , 84~94

X. Qi, F. Palmieri. Theoretical analysis of evolution algorithms with an infinite population size in continuous space. Part I: basic properties of selection and mutation. IEEE Trans on Neural Networks. 1994, 5(1): 102~119

CASE-BASED REASONING MODEL OF THE FISH DISEASE DIAGNOSIS

Zetian Fu[*] , Guoyong Hong[1] , Jian Sun[1]
[1] *Department of Mechanical Electrical Engineering, China JiLiang University,HangZhou, ZheJiang Province, P. R. China 310018*
[*] *Corresponding author, Address: China Agriculture University Key Laboratory of Modern Precison Agriculture System Integration Peking 100083*

Abstract: The system of Case Based Reason is researched in this paper, a kind of genetic algorithm is proposed to solve the problem of feature vector space, and proposes an approach of integration of CBR AND RBR, an example about fish disease diagnose inferring integration model is given.

Keywords: method, CBR , diagnose , reasoning

1. INTRODUCTION

In a general the knowledge of expert system have a direct impact on the performance of the traditional expert systems, because of the incomplete knowledge acquisition, to complete the construction and use of the reasoning rules for the system model is very difficult, construction expert knowledge acquisition system is the "bottleneck" problem. Since fish disease diagnosis system has the reasoning complexity and uncertainty, using case-based reasoning (CBR) approach can reduce the diagnosis in the process of man-made factors, the approach to solve the problem is very close to the human, the key is the method can memories the case, which is related to the most similar case when the system encountered a new problem, to a large extent the system is dependent on the quality of the case whether the reasoning ability of fish diseases can play its due role in this process, in order to get the better guarantee of objectivity and accuracy of the results, how to search

Please use the following format when citing this chapter:

Fu, Z., Hong, G. and Sun, J., 2009, in IFIP International Federation for Information Processing, Volume 294, *Computer and Computing Technologies in Agriculture II, Volume 2*, eds. D. Li, Z. Chunjiang, (Boston: Springer), pp. 1433–1442.

the expected result in a becoming huge case is a key part of reasoning in retrieval CBR system.

2. THE BASIC CONCEPT OF CASE-BASED REASONING

2.1 The concept of CBR

CBR to develop artificial intelligence in the rule-based reasoning is different from a mode of reasoning, it refers to the use of the old cases or experience to solve the problem and to evaluate solutions to explain the abnormal situation or understand the new situation. Case is contextual information with the knowledge, the knowledge of the reasoning machine can achieve its objectives in the process, which can play a key role in the experience. In CBR, the current situation or the problems facing the case is known as the goal, but the memory of the case is known as the source of the problem or case.

2.2 The general process of CBR

CBR to deal with all the main elements are: identification of new issues, and finding a source of a similar case, the case is to use this new source of the problem a solution, the evaluation of this programme, and through case study and revise Library. CBR is actually a solution to the problem; learn from experience, to resolve the issue of a new cycle and integration process. General CBR cycle process includes the following four: case retrieval, reuse case, case modification and case studies. Case retrieval is retrieved from the case with most new issues similar to the case; reuse is reusable case to case retrieval of information and knowledge to solve new problems; case for change is modification programme; case study is to learn new Experience and put it into the existing case library. Figure 1 can be used to describe a cycle of the CBR.

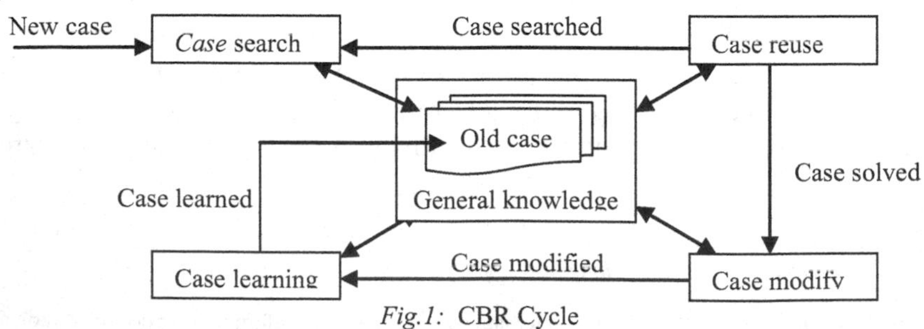

Fig.1: CBR Cycle

According to the above description, CBR can be found on the core issue includes five aspects: that case method, case retrieval method, case reuse methods, case modification method and case study methods.

3. THE CONTENT OF FISH DISEASE DIAGNOSIS

Fish Disease Diagnosis is through a variety of ways and means of observing and monitoring the fish's living conditions and in accordance with the known symptoms, were diagnosed to identify possible targets of fish or to be a disease, and led to the direct cause of these diseases, to ensure that Was diagnosed target fish in certain seasons and the living environment of healthy growth. Fish Disease inference system refers to the completion of a diagnostic function of the fish's computer system, and its main task is under observation by the user to show the fish and its related symptoms, such as the quality of information, identifying the fish living in water may exist disease and its related properties, and to determine the corresponding control measures or strategies.

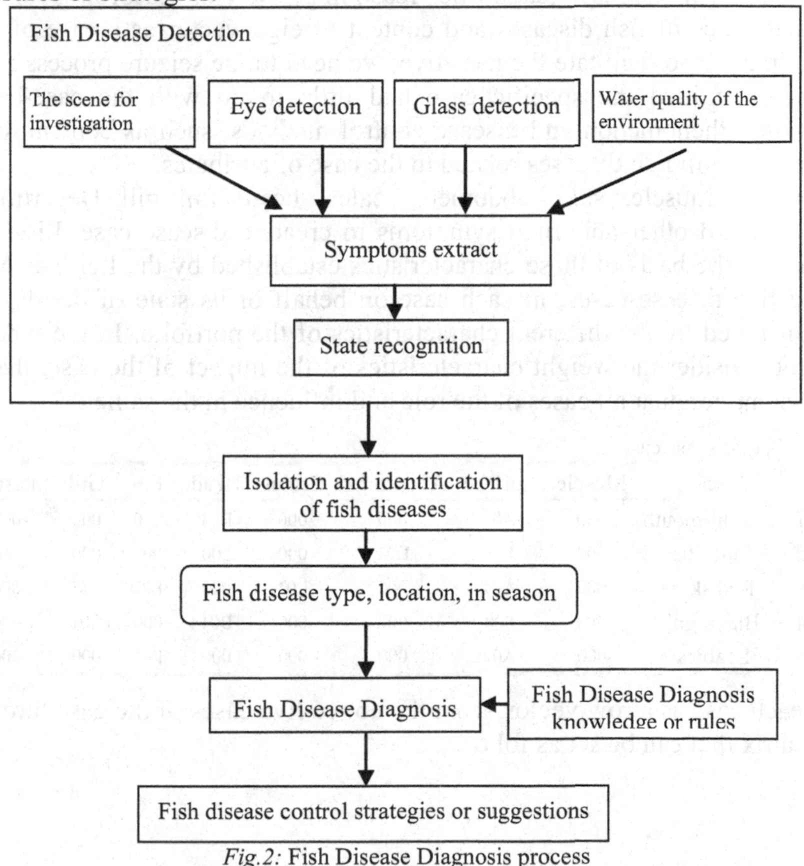

Fig.2: Fish Disease Diagnosis process

Fish Disease Diagnosis has three basic attributes, that is, symptoms, diseases and causes. One is the external manifestations of the disease or the results of the symptoms of the disease occurred before or during the course of fish in certain areas may show the abnormal phenomenon, which is the cause of diseases leading to the fundamental attributes collection. Therefore, a fish disease diagnosis system in the specific diagnosis of the process is shown in Figure 2.

4. CASE EXPRESSION OF FISH DISEASE DIAGNOSIS REASONING

Normally, a typical case should at least include a description of the problem and a description of the solution. For fish disease diagnosis system, the problem is demonstrated by the fish's symptoms, such as water quality parameters; solution is to identify the disease, causes, and gives the corresponding treatment.

In the fish disease diagnostic reasoning system, according to the characteristics of fish diseases and content to eigenvector as a case of that form, in order to facilitate the narrative, we head to the seizure process as an example, omitted the specific case had little to do with the case-based reasoning Phenomenon and disease control methods, such as content, only concerned with fish diseases related to the case of attributes.

We use muscle, skin, abdomen, scales, head, fin, gill Department, intestinal and other abnormal symptoms to create a disease case. Listed in Table 1 is the basis of these characteristics established by the Lei Yue Mun in the five disease cases, in each case on behalf of its state of the disease demonstrated by the different characteristics of the portfolio. In these cases did not consider the weight characteristics of the impact of the case, that is the assumption that all cases of the role and influence in the same.

Table1 Carp disease case

ID	Disease	Muscle	Surface	Abdominal	Scales	Head	Fin	Gill	Intestinal
Case1	Baekdu-mouth	000	A01	000	000	B04	000	000	000
Case2	Enteritis	000	000	D01	000	000	000	000	H03
Case3	Red skin	000	A04	000	E01	000	F01	C03	000
Case4	Black gill	000	000	000	000	B01	000	C01	000
Case5	Scabies	000	A07	000	000	000	F02	000	000

If each case as a row vector, from the above five cases of the case through the matrix that can be set as follows:

$$C = \begin{bmatrix} 000 & A01 & 000 & 000 & B04 & 000 & 000 & 000 \\ 000 & 000 & D01 & 000 & 000 & 000 & 000 & H03 \\ 000 & A04 & 000 & E01 & 000 & F01 & C03 & 000 \\ 000 & 000 & 000 & 000 & B01 & 000 & C01 & 000 \\ 000 & A07 & 000 & 000 & 000 & F02 & 000 & 000 \end{bmatrix} \quad (1)$$

In practice, the different characteristics of the role and impact of cases there is a certain degree of difference, but this difference in performance under different circumstances to different degrees and others. Table 1 listed in the case cited, if the consideration of the characteristics of eight cases of the impact and role of the differences, they give different weights, and you can get a new set of cases, as shown in table 2.

Table 2 Carp disease case

ID	Disease	Muscle	Surface	Abdominal	Scales	Head	Fin	Gill	Intestinal
Case1	Baekdu-mouth	000	A01	000	000	B04	000	000	000
	Weight	0	0.5	0	0	0.5	0	0	0
Case2	Enteritis	000	000	D01	000	000	000	000	H03
	Weight	0	0	0	0	0	0	0	0.5
Case3	Red skin	000	A04	000	E01	000	F01	C03	000
	Weight	0	0.25	0	0.25	0	0	0.25	0
Case4	Black gill	000	000	000	000	B01	000	C01	000
	Weight	0	0	0	0	0.5	0	0.5	0
Case5	Scabies	000	A07	000	000	000	F02	000	000
	Weight	0	0.5	0	0	0	0.5	0	0

As in the case of different features and characteristics of the weight may change, as in the example of Black gill disease symptoms and weight in the performance of the head and gills, and enteritis disease symptoms and weight are reflected in the abdomen and bowel, It is an extraordinary collection of configuration case, we use this case the largest collection of feature set to solve the unusual configuration problems.

We can organize the collection features the largest case characteristic matrix:

$$F = \begin{bmatrix} f_{1,1} & f_{1,2} & \cdots & f_{1,m} \\ f_{2,1} & f_{2,2} & \cdots & f_{2,m} \\ . & . & . & . \\ f_{n,1} & f_{n,2} & \cdots & f_{n,m} \end{bmatrix} \quad (2)$$

In characteristic matrix, each line represents a case, each column represents a case of different characteristics in case the characteristics of the matrix of every element of the value $f_{i,j}(1 \le i \le n, 1 \le j \le m)$, and we do the following:

$$f_{i,j} = \begin{cases} \text{the value of characteri stic j in case i}, & \text{when characteri stic } j \in F \\ \text{no definition} & , \text{when characteri stic } j \notin F \end{cases}$$

When the feature is a feature of the case, it said in the case of the characteristics of value, or no practical significance, can be used arbitrarily to fill the appropriate value of the element in the matrix in the position.

Similarly, we have the characteristics of the case in the weight also similar manner in accordance with the above organizations to become a matrix form, saying the matrix for the characteristics of the right matrix.

$$R = \begin{bmatrix} r_{1,1} & r_{1,2} & \cdots & r_{1,m} \\ r_{2,1} & r_{2,2} & \cdots & r_{2,m} \\ \cdot & \cdot & \cdot & \cdot \\ r_{n,1} & r_{n,2} & \cdots & r_{n,m} \end{bmatrix} \quad (3)$$

Where: R is the characteristics of the right matrix; n is the collection C features the largest number of features, the case for the number of cases in the collection.

Characteristics of the matrix R, each row represents a case of different characteristics in the case of the characteristics of the right values, the matrix R of every element of the value is $r_{i,j} (1 \le i \le n, 1 \le j \le m)$, we do the following:

$$f_{i,j} = \begin{cases} \text{the value of characteri stic j in case i}, & \text{when characteri stic } j \in F \\ \text{no definition} & , \text{when characteri stic } j \notin F \end{cases}$$

When the feature j is a feature of the case i, said $r_{i,j}$ is the right value of characteristics, otherwise $r_{i,j} = 0$.

In the above definition on the basis of a case that could set the case matrix collection C features and characteristics of the matrix right to buy the product matrix:

$$C = F \bullet R^T \quad (4)$$

For example, we can use matrix shown in table 2 from the collection of fish diseases of the case said:

$$\begin{bmatrix} C1 \\ C2 \\ C3 \\ C4 \\ C5 \end{bmatrix} = \begin{bmatrix} 000 & A01 & 000 & 000 & B04 & 000 & 000 & 000 \\ 000 & 000 & D01 & 000 & 000 & 000 & 000 & H03 \\ 000 & A04 & 000 & E01 & 000 & F01 & C03 & 000 \\ 000 & 000 & 000 & 000 & B01 & 000 & C01 & 000 \\ 000 & A07 & 000 & 000 & 000 & F02 & 000 & 000 \end{bmatrix} \bullet \begin{bmatrix} 0 & 0.5 & 0 & 0 & 0.5 & 0 & 0 & 0 \\ 0 & 0 & 0 & 0 & 0 & 0 & 0 & 0.5 \\ 0 & .25 & 0 & .25 & 0 & 0 & 0.5 & 0 \\ 0 & 0 & 0 & 0 & 0.5 & 0 & 0.5 & 0 \\ 0 & 0.5 & 0 & 0 & 0 & 0.5 & 0 & 0 \end{bmatrix} \quad (5)$$

5. FISH DISEASE DIAGNOSIS REASONING OF THE CASE RETRIEVAL SYSTEM

In general, CBR system is suitable for experienced, but the lack of knowledge of the field, while RBR system can handle rich and knowledge of the system on the field. Therefore, the CBR and RBR will be integrated into one of the two systems will have the strengths to overcome the shortage of both, so that the entire system to achieve a higher level of intelligence, is of great significance. It will not only greatly enhance the flexibility and comprehensive reasoning ability, but also significantly reduce the case retrieval and the burden of the case.

We will go into two tiers: the bottom of all cases; upper deck is a typical case, each case represents the typical one or several months and it is very similar to the case, typical case of the amended rules is essential, it should be the guidance of experts in the field. To be good to amend the rules, for a similar case semantic difference between the measures is essential, semantic word table can be defined by the index of keywords to describe, indexing vocabulary of keywords can be used in case the same function and the description of semantic network nodes. Application of amended rules also requires case assessment that the subsystem is the evaluation of the answer, because once the amendment is not likely to be the result of the establishment, but also further changes. Sometimes, the problem can not be with a certain rules of the first match, now need to separately image to the first of several rules, the final rules will be several interpretations of the results integrated into the whole question to answer.

5.1 Case index

The index case mechanism is an important issue, the objective is to provide a case for the search mechanism, to make the retrieval in the future in line with the need to quickly identify the cases or cases set. The index should be specific, clear and easy identification. Case of the index can have multiple, respectively, for different purposes. Search algorithms should be able to use the index case in time to meet the constraints to identify the premise of the case. This means that the case should be indexed; making the case in any need at all times can be found.

In this paper, C-means clustering algorithm to a polymerization center on behalf of a disease, according to the type of disease the number is divided into several polymerizations centre, the centre for polymerization index.

C means clustering model, established the objective function:

$$J = \min \sum_{r=1}^{P} \sum_{i=1}^{m_r} \| X_j^{(r)} - C_r \|^2 \tag{6}$$

Where: Cluster Center

$$C_r = \frac{1}{m_r}\sum_{i=1}^{m_r} X_i^{(r)} \quad (i = 1,2,\dots,\ r = 1,2,\dots P)\ ,\ \sum_{r=1}^{P} m_r = N \qquad (7)$$

where: m_r is the number of r sample; $X_i^{(r)}$ express sample X_i belong to r; N is the number of sample; P is the number of cluster center($2 \le P \le N-1$).

We use genetic algorithms to solve the problem dynamic clustering.

Step 1 chromosome structure

Genetic algorithms set up the relevant parameters, max_gen: the largest number of iterative; N: population size; length: the length of chromosomes; Pc: cross-probability; Pm: mutation probability; P: initial cluster centres; α: the fitness function of Parameters;

Step 2: groups initialization
```
    for i = 1 to N do
    for j = 1 to length do
      Xi the first-chromosome gene = random (0, P);
    Endfor
    Endfor
```
STEP 3: Calculation of fitness function
```
    for i = 1 to N do
     Xi chromosome calculation of the objective function Ji;
     endfor
       According to Xi chromosome target function of the value of Ji sort;
       for i = 1 to N do
       Individual fitness calculation
       endfor
       for j = 1 to N do
         An Introduction to calculate the cumulative
       endfor
```
Step 4: The Executive select Options

Step 5: the implementation of cross-operation

Step 6: The Executive mutation

Step 7: retain the elite strategy
```
  if (Minimum of NewGeneration <Minimum of OldGeneration) then
    Father and with the best of the individual offspring to replace the
  worst of the individual;
  endfor
```
Step 8: The dynamic changes in the number of cluster
```
    If (the best group of individuals in the M-generation, has not changed)
     then
     P = Popt + n, and to Step 2;
    else
     Running;
  endif
```

```
Step 9: termination of the conditions tested
   if (gen <max_gen) then
      Order gen = gen +1, and turn to Step 3;
   else
      Stop, the output results
   endif
```

5.2 Rule-based and case retrieval algorithms

Rule-based and case search algorithm is as follows:

1: input new case index generated in accordance with the relevant rules, create a new case description of the rules;

2: judge rules coding RULE_CODE the degree of similarity, if the same case were directly targeted. If not identical, the value of similar properties in the area threshold if the threshold value, then retrieves the corresponding case set; if the threshold, based on the similarity case match and then retrieves the corresponding The case. If the search to retrieve cases of unreasonable or no case, to 4;

3: In Search of the process, the preservation of relevant case study of key knowledge and interest in hobbies, credited to the inspiring knowledge base.

4: selected cases involving the target system (object to sort through), in accordance with our neighbors to improve the method of calculating targets under the similarity measure;

5: Does also involves other object, if so, the system for cases involving the next target, or else to 5;

6: calculate the case of the similarity system;

7: The similarity of the cases is the highest result;

8: The Case users or experts satisfied with.

If satisfied, such cases will be listed as a candidate case set, the end of reasoning.

6. CONCLUSION

The system based on the characteristics of fish diseases and content of case-based reasoning is studied in this paper, we can effectively solve the problems in the diagnosis of fish diseases, and improve the operating efficiency of the system. CBR has a crucial impact on the accuracy and practical application, but it is difficult to determine the right value, based on random genetic algorithm global search capability, a genetic algorithm to complete the space characteristics of the right to direct the search algorithm is presented, For people to solve practical problems is not simply relying on experience, not simply relying on theoretical knowledge, is often a combination of both the facts, it is a combination of methods, and practice shows the method is effective.

ACKNOWLEDGEMENTS

Funding for this research was provided by China Agriculture University Key Laboratory of Modern Precision Agriculture System Integration (P. R. China). The first author is grateful to China Agriculture University for providing her with pursuing a PhD degree.

REFERENCES

C. R. Sukumaran, B. P. N. Singh. Compression of bed of rapeseeds: the oil-point, Journal of Agricultural Engineering Research, 1989,42:77-84

Cheng Wang. Yinghong LI. Hengxi Zhang. The improvement of CBR technology in the diagnosis of failure. Computer Engineering and Applications. 2003（15）：44～46

E. Davion, A. G. Meiering, F.J. Middendof. A theoretical stress model of rapeseed, Canadian Agricultural Engineering, 1979, 21(1): 45-46

J. B. Noh. A case-based reasoning approach to cognitive map-driven tacit knowledge management. Expert Systems with Application. 2000(19)：249～259

J. M. Garrell. I. Guiu. Automatic diagnosis with genetic algorithms and case-based reasoning. Artificial Intelligence in Engineering. 13(1999)367～372

Pi Sheng Deng. Using case-based reasoning approach to the support of ill-structured decisions. European Journal of Operational Research. 1996 (93)511～521

S L.A Salzburg, Nearest Hyper rectangle Learning Method. Machine Learning. 1991,6:251-276.

S M.A Bryant Case-based Reasoning Approach to Bankruptcy Prediction Modeling

T. W. Liao. A case-based reasoning system for identifying failure mechanisms. Engineering Applications of Artificial Intelligence. 2000 (13)：199～213

REASONABLE SAMPLING SCALE OF MACROPORE BASED ON GEOSTATISTIC THEORY

Mingyao Zhou[1,2,*], Zhaodi Lin[2], Peng Wu[2], Susheng Wang[2], Fei Zhang[2]

[1] *State Key Laboratory of Hydrology - Water Resources and Hydraulic Engineering, Hohai University, Nanjing, Jiangsu Province, P. R. China 210098*

[2] *College of Hydraulic Science and Engineering, Yangzhou University, Yangzhou, Jiangsu Province, P. R. China 225009*

[*] *Corresponding author, Address: College of Hydraulic Science and Engineering, Yangzhou University, 31 middle Jiangyang Rord, Yangzhou 225009, Jiangsu Province, P. R. China, Tel: +86-514-87978640, Fax: +86-514-87978640, Email: myzhouyz@163.com*

Abstract: Spatial scale of soil character is forefront in soil-scientific research, and the analysis of macropore reasonable sampling size becomes significant in studyinig on soil spatial variability. From the micro-structure of the soil, using digital image preparation and analysis technology, this paper described the soil macropore with rations, discussed the spatial structure of soil macropore on both the vertical and horizontal directions by geologic geostatistic theory , analysed by the base value of variation function C0, structure variance C, variable function, model test parameter I and model accuracy rate of change k and so on. The results showed that the reasonable sampling interval of vertical depth was 20 mm, and the economic reasonable sampling diameter on horizontal was about 90 ~ 100 mm. The production has very important significance to ensuring the scientific of the soil macropore research, saving study time, and reducing the human and financial resources in the trial.

Keywords: digital image; soil macropore degree; geostatistic theory; scale sample

1. INTRODUCTION

Soil is a kind of uneven and variational continuum, its spatial scale problem is in the forefront of soil-scientific research. In the pedology,

---

Please use the following format when citing this chapter:

Zhou, M., Lin, Z., Wu, P., Wang, S. and Zhang, F., 2009, in IFIP International Federation for Information Processing, Volume 294, *Computer and Computing Technologies in Agriculture II, Volume 2*, eds. D. Li, Z. Chunjiang, (Boston: Springer), pp. 1443–1450.

configuration and variability of many regional variables usually exist in different scales, some structure just show under fixed observational measure, meanwhile, some observational measure just reveal relevant change rule or structural characteristic, the effect of spatial variability differs during different scales (Li Zizhong et at., 2001). Sampling is the first step in all macropore research, its scale effect can directly impact spatial variability rule of soil.

Geostatistical theory is a risingly fringe subject founded and developed in lately thirty years (Hou Jingru et at., 1982; Wang Renduo et at.,1988). Since it was taken into the spatial analysis of soil characteristic, geostatistical theory had largely applied to many fields, such as soil-water science, hydrology, and so on (Li Weiping et at., 2004; Shi Haibin et at.,1994).However, this theory was almostly used to study some characteristic and crop information in field scale presently. The depth of research was stressly about the spatial insert number and related interval of some soil characteristic and crop information, then drew isoline figure. Practical research showed that, in order to reflect macropore distributing in soil completely, the research of macropore spatial variability in both microcosmic and macrocosmic had become a necessarily solving problem, as the gradual development of macropore research.

Optimal sampling scale means mortal sampling scope、very tittle sampling interval and sampling volume (Westerna A W et at., 1999), but in fact, this is very difficult to achieve. This paper took geostatistical theory into microcosmic domain to research macropore distributing, analysis the spatial structural characteristic, obtain the reasonable sampling size of macropore study, accurately obtain macropore distributing rule, provide instruction for soil moisture, solute movement regulation and setting up exact predictive model for cropland moisture.

2. GEOSTATISTICAL THEORY

Geostatistical theory is based on regional variables、random function and balanced supposes, consider semi-variance function as its basic tool, studies the natural phenomena that both randomicity and configuration in the spatial distributing (Hou Jingru et at., 1993). Semi-variation function is the core during calculating regional variables, usually take following formula which Matheron commended:

$$\gamma(h) = \frac{1}{2N(h)} \sum_{i=1}^{N(h)} [Z(x_i) - Z(x_i + h)]^2 \qquad (1)$$

where $N(h)$ is experimentation logarithm at a distance of h on x axis; h is sampling interval; $\gamma(h)$ is the result of semi-variation function; $Z(x_i)$, $Z(x_i+h)$ is realization of observation value $Z(x)$, $Z(x+h)$.

3. MATERIAL AND METHOD OF EXPERIMENT

3.1 Prepare and dispose the soil column

The experiment took original soil as object, and the fetching soil containers included two kinds of UPVC pipe, that is: (1) Diameter 11cm, pipe thickness 3mm, length 60cm; (2) Diameter 20cm, pipe thickness 4mm, length 25cm. It fetched soil from the field, and original soil should avoid been disturbed during the whole process.

The original soil was then dried in nature air, took impregnant to marinate the soil columniation next, it must be slowly and separately immitted, to the greatest extent to be immerged into soil macropore. When there was 0.5cm impregnant higher than soil surface in the pipe, just stopped immerging and placed until impregnant completely solidified. The last step was skived the soil with electromotion thin grinding wheel, obtained many real multicolor surface every other 5mm by scanner。

3.2 Obtain digital images

Used the large-scale image disposal software ImageSys and Photoshop to pick-up macropore area (based on the precision of this experimental investigation and data processing level , considered the hole that larger than 100μm as macropore). The final data were following series macroporous degrees:(1) 600mm underground depth, diameter scope 104mm , every other 5mm interval ; (2) 230mm underground depth, diameter scope 25mm、40mm、60mm、80mm、100mm、125mm、150mm、 175mm, every other 10mm interval.

4. RESULTS AND ANALYSIS

4.1 Semi-variance analysis

4.1.1 Model choice

Semi-variance function was the foundation of spatial variability explaining by geostatistical theory. It was usually expressed by experimental semi-variogram. These curves can be fitted by curve equation. These curve equations were called of semi-variance function. There were spherical model, Gaussian model, the index model and the linear model. The theoretical model of choice depended on the specific cases. According to the studies of predecessors, theoretical variogram model of soil properties were generally the spherical model of transition (Masaru Hoshiya, 1995; Gotway C A. R B Ferguson,1996). When selecting the semi-variance model, first of all, used the equation (1) to calculate $\gamma(h)$ scatter plot, then fitted the curves with spherical model, that was :

$$\gamma(h)=\begin{cases}0 & h=0\\ C_0+C\left(\frac{3}{2}\frac{h}{a}-\frac{1}{2}\frac{h^3}{a^3}\right) & 0<h\le a\\ C_0+C & h>a\end{cases} \tag{2}$$

Where C_0 was the value for the block, meaned semi-variance when distance approached zero. It was the variability caused by random factors such as experimental error; C was structure variance. It was the variability caused by non-human factors such as soil parent material, terrain and so on.; a was the related socpe of the observation points.

4.1.2 Test of model precision

In order to make sure the theoretical models of semi-variance really describe the law of the changes, it must do the optimal test. The tests were usually cross-examination, the estimated variance test and I value test. The cross-examination was the theoretical variogram model of “the square of decrease between kriging estimates and measured value was minimum”. The estimated variance test was the method which do the test through the ratio of the practical variance and theoretical estimated variance. If the theoretical variation function properly fitted, then $\overline{(z^*-z)^2/(s^*)^2}$ should be fluctuated around 1. I value test integrated the above two indicators into a unified theory (Hou Jingru et at., 1982), formula was as follows:

$$I = \overline{(z^{*} - z)^{2}} \times \left[P \times \left| 1 - \frac{1}{\left[(z^{*} - z)/s^{*}\right]^{2}} \right| + (1 - P) \right] \quad (3)$$

Where P was parameters for the empirical.When $0 \le \overline{(z^{*} - z)^{2}} < 100$, P was 0.1. When $100 \le \overline{(z^{*} - z)^{2}}$, P was 0.2.The smaller I value, the higher fitted precision of variation function model. Three methods were used to test the model. The results were shown in Table1 and Table2. The corresponding map of semi-variogram was as Fig. 1.

In order to quantificationally describe how scale change impact model precision , change rate of model precision was introduced, that was:

$$k\ (‰) = |(I_2 - I_1)| / \delta * 1000 \quad (4)$$

The results were shown in Fig .2 and Fig.4.

Table 1. Theoretical model and model testing parameters of macroporous degree under the condition of the differernt minimum step

Minimum Step (mm)	C_0 (%)	C (%)	$\frac{C_0}{(C+C_0)}$ (%)	Variable a (mm)	$\overline{(z^{*} - z)^{2}}$	$\overline{((z^{*} - z)/s^{*})^{*}}$	P	I
5	0.23	0.43	34.8	201.6	0.0027	0.97	0.1	0.0025
10	0.29	0.47	38.2	199.8	0.0041	1.05	0.1	0.0037
15	0.40	0.35	53.3	202.7	0.0052	1.00	0.1	0.0047
20	0.42	0.48	46.7	199.1	0.0025	1.09	0.1	0.0048
25	0.59	0.47	55.7	210.8	0.0038	0.99	0.1	0.0058
30	0.55	0.55	50.0	231.6	0.0042	1.02	0.1	0.0055

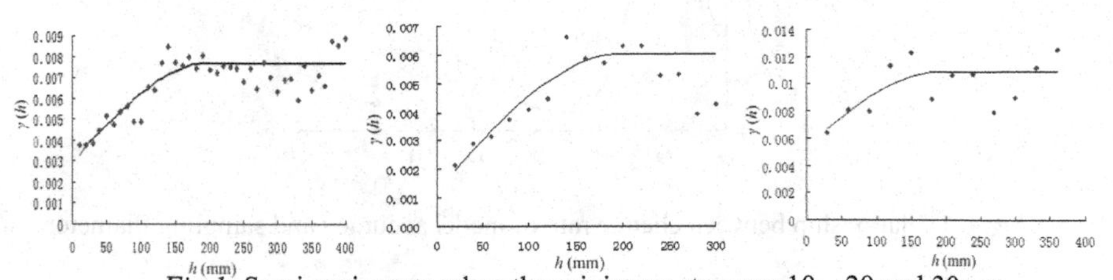

Fig.1: Semi-variogram when the minimum step was 10、20 and 30mm

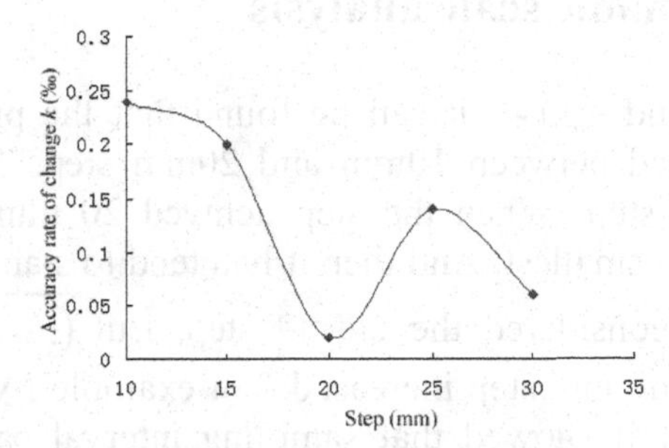

Fig.2 : Relationship between change rate of model accuracy and step

Table 2. Theoretical model and model testing parameters s of macroporous degree under the condition of differeent sampling diameter

Sampling Diameter (Mm)	C_0 (%)	C (%)	$\frac{C_0}{(C+C_0)}$ (%)	Variable a(mm)	$\overline{(z^*-z)^2}$	$\overline{((z^*-z)/s^*)^*}$	P	I
25	2.12	0.66	76.3	47.9	0.0268	0.96	0.1	0.0243
40	1.36	0.83	62.1	32.5	0.0237	0.94	0.1	0.0296
60	1.55	0.67	69.8	61.8	0.0197	1.02	0.1	0.0159
80	0.59	1.86	24.1	94.05	0.0128	1.19	0.1	0.0117
100	0.11	1.96	5.31	97.70	0.0084	1.28	0.1	0.008
125	0.28	3.95	6.62	93.24	0.0146	1.06	0.1	0.0132
150	0.10	2.41	3.98	89.44	0.0099	1.13	0.1	0.0093
175	0.02	1.54	1.28	97.92	0.0047	1.15	0.1	0.0044

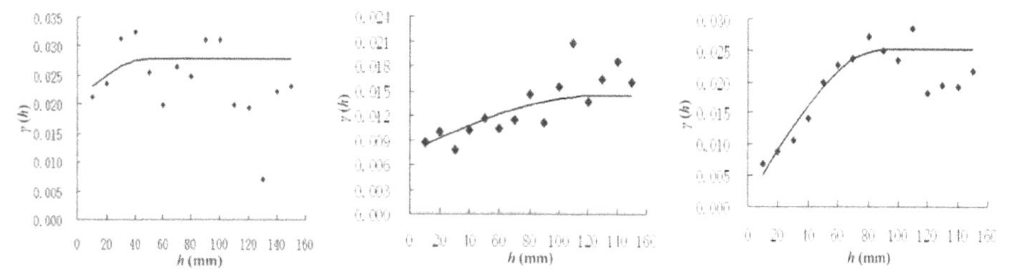

Fig. 3: Semi-variogram when the sampling diameter was 25、60、150mm

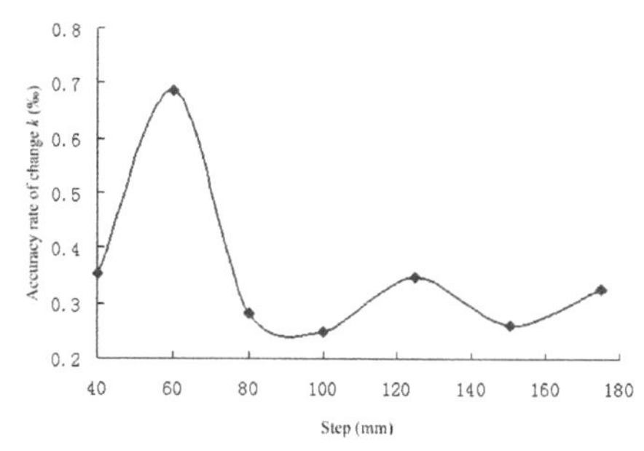

Fig.4: Rrelationship between change rate of model accuracy and sampling diameter

4.2 Reasonable scale analysis

From Table 1. and Fig.1~4, it can be found that the precision change rate gradually decreased between 10mm and 20mm step. The precision varied larger with these step. When the step achived 20 mm, the rate of model precision achieved smallest. And then it had tend to stabilize ups and downs. 20mm could be considered the critical step. But $\overline{(z^*-z)^2}$ and I increased along with the minimum step increased. For example I value increased from 0.0051 to 0.0025. It showed that sampling interval can not be too large. Because even if sample spacing increased, the workload reduced, and at the same time it caused academic model would not primely describe variable characteristics and special structure of macropore degree in soil. And the accuracy of estimates and simulation would also reduce.

Table 2. and Fig.5~8 said that when the sampling diameter was between 40 mm and 80 mm, model precision varied from 0.28‰ to 0.685‰, changed distinctly. Model precision change rate reached the smallest between 80mm and 100mm. In the context, the model precision achieved the highest point of local and then changed in the steady state. *I* value slightly reduced with the sampling increaced, concretely minished from 0.00243 to 0.0044(<10%). The larger the sampling diameter, the higher simulation accuracy of calculation and of variogram were. The sampling diameter went bigger, it could reflect the nature of the porosity more. But this would spend a significant amount of manpower, material. It said that 80~100mm could be taken for reasonable sample diameter.

5. CONCLUSION

This paper took geostatistical theory into microcosmic domain to research macropore distributing, analysis the spatial structural characteristic, broadened the application of goestatistical theory. It had two conclusions as follows:

(1) Used digital image analysis technology to visually describe the number of the soil macropore, size and other characteristics. It said that the macropore degree had roughly the same trend in different soil column. Max value basically concentrated lied underground 12cm to 18cm, that was cultivation layer. The maximum of macropore area changed little. But the maximum of macropore perimeter fluctuated in the apparent trend. This showed in the vertical section, the difference of macropore spatial distribution structure was different.

(2) Used geostatistical theory to research the spatial structure of the macropore of soil. Analysised the spatial structure of macropore on both vertical and horizontal directions. Through the parameters such as C_0, C, Variable, I, model precision change rate k and so on, it showed that 20mm was reasonable sampling interval in the vertical depth. 80～100mm was the economically reasonable sampling diameter in the level. This provided instruction for soil moisture, solute movement regulation and setting up exact predictive model for cropland moisture.

ACKNOWLEDGEMENTS

This research was funded by the Program of State Key Laboratory of Hydrology-Water Resources and Hydraulic Foundation, Hohai Universty(accession number 2006zd04).

REFERENCES

Gotway C A. R B Ferguson. G H Hergert and T A Peterson. Comparison of kriging and inverse-distance methods for soil parameters. Soil Sci，1996（60）：1237～1247

Hou Jingru, Guo Guangyu. Forecast of deposit Statistics and theory and application of Geostatistical theory [M]. Beijing: metallurgy and industry publishing company，1993，6～8

Hou Jingru, Huang Jingxian.Theory and method of geostatistics [M]. Beijing: Geological publishing company，1982

Hou Jingru, Hung Jingxian. Geostatistical theory and its application in calculation of mine rescourses [M]. Beijing: Geological publishing company，1982

Li Weiping, Shi Haibin, Huo Zailin. Spatial structure of the height and diameter of sunflower stem [J]. Transactions of The Chinese Society of Agricultural Engineering，2004，20(4)：30-33

Li Zizhong, Gong Yuanshi. Spatial variability of soil water content and electrical conductivity in field for different sampling scales and their nested models [J]. Plant Natrition and Fertilizen Science，2001，7(3): 255-261

Masaru Hoshiya. Kriging and Conditional simulation of Gaussian Field. Journal of Engineering Mechanics, 1995, 121 (2)

Shi Haibin，Chen Yaxin. Combination structure model of soil moisture spatial variability and regional information estimation[J]. Journal of hydraulic engineering, 1994, (7)：70-77

Wang Renduo, Hu Guangdao, Linear. Geostatistics. Beijing: Geological publishing company，1988

Westerna A W, Bloschl G. On the spatial scaling of soil moisture [J]. Journal of Hydrology, 1999, (17): 203-224

INTELLIGENT DESIGN OF VEHICLE PACKAGE USING ONTOLOGY AND CASE-BASED REASONING

Xiaoping Jin [1], Enrong Mao [2], Bo Cheng [1,*]
[1] *Department of Automotive Engineering, Tsinghua University, Beijing, P. R. China 100084*
[2] *College of Engineering, China Agricultural University, Beijing, P. R. China 100083*
[*] *Corresponding author, Address: Department of Automotive Engineering, Tsinghua University, Beijing, P. R. China, Tel: +86-10-62780355, Fax: +86-10-62780355, Email: chengbo@tsinghua.edu.cn*

Abstract: The similarity of varied vehicle package is a critical design feature that affects method selection, optimized design and driver performance. However there is limited understanding of what constitutes similarity in package design and limited computer-based support to identify this feature in a layout model. This paper contributes a case-based framework for representing and reasoning about layout similarity that builds on domain-specific ontological modeling and case-based reasoning techniques. Validation study of the system provides evidence that the framework is general and enables a more efficient package layout design process.

Keywords: vehicle package, CAD, ontology, case-based reasoning

1. INTRODUCTION

Designers of workplaces and products have three major tasks (Feyen et al. 2000): one, integrating information about processes, tools, machines, parts, tasks, and human operators; two, satisfying design constraints which often conflict; and three, generating a design acceptable to all parties involved. However, while completing these tasks, designers often have difficulty incorporating ergonomics information about the human operator into their designs. Although such information exists for use in the job design process,

---

Please use the following format when citing this chapter:

Jin, X., Mao, E. and Cheng, B., 2009, in IFIP International Federation for Information Processing, Volume 294, *Computer and Computing Technologies in Agriculture II, Volume 2*, eds. D. Li, Z. Chunjiang, (Boston: Springer), pp. 1451–1460.

one reason for the difficulties in using this information is that it is often poorly presented for use by designers.

The layout of the driver workstation is referred to as the vehicle package. Complete definitions of interior points can be found in SAE J1100 and associated practices (SAE 2001). Vehicle occupant packaging is the process of laying out the interior of a vehicle to achieve the desired levels of accommodation, comfort, and safety for the occupants. Creating a vehicle package should take the ergonomic factors into consideration, which include the requirements of driver seating comfort, operating convenience, visibility, etc. Theoretically there might be numerous layout design solutions. However, the following reasons have to be considered: (a) The vehicle cab interior design technology is quite mature and a lot of typical package layout designs have been created and proved to be feasible in iterative practice in different series of vehicle models; (b) Unproved novel design is unlikely to be adopted in fact due to the demands of high cost and high reliability; and (c) Most designers prefer reuse of previous designs as much as possible to reduce the workload because of large number of complicated design tasks and generally restricted time. Owing to these facts, the vehicle package layout design is usually achieved by referring to former design cases.

As one of AI technologies, case-based reasoning offers some advantages compared to other knowledge representation and reasoning formalisms. Cases represent specific knowledge of the domain, are natural and usually easy to obtain (Avramenko 2006). Case-based reasoning has been utilized to solve mechanical design problems since it well reflects the commonly used design methodology of consulting previous designs (designers rarely design an artifact from scratch). Structure-behavior function (SBF) device models are used to represent and comprehend specific design cases, and also to provide necessary knowledge for modifying a retrieved case to fit a new design problem (Han and Lee 2006). Some studies (Chen et al. 2006) applied the CBR for an automotive body assembly process design system in searching the identifying features. Some (Dan 1997) had proposed the framework of EDKBES (Ergonomic design knowledge base expert system) for vehicle interior design combining CAD with expert system technologies. Also RAMSIS was used for interior layout design (Vogt 2005). It worked effectively with a few components and small layout areas. However, serious errors occurred when it performed relatively complicated design tasks. In general, while AI technologies are widely applied in the fields of conceptual design, diagnostic and fault detection, little work has been done to automate and/or support the vehicle package design tasks with the ergonomic analysis.

Contemporary AI technologies can be used to represent knowledge and model reasoning processes. A system for intelligent design of interior layout can be developed on the basis of these technologies. Development of such system is a promising research field. In this paper, a case-based framework for vehicle package design was presented with the ontology employed in the development of knowledge representation model, which was a promising way to increase the quality and efficiency of vehicle interior ergonomic design. A coach package problem was selected as a test domain of layout design and analysis because of their importance and frequent interaction in driving practice.

2. ARCHITECTURE OF THE SYSTEM

To understand the subtleties for how practitioners think about the design in the context of assessing layout similarity, some researches have been performed to extract the human factor rules and principles (Jin et al. 2005), and approaches of quantifying the degree of comfort have been proposed.

Fig.1 graphically represents the framework that was developed for representing and reasoning about layout similarity:

a) Represent layout similarity: Designers specify their knowledge for defining layout similarity in a computer-interpretable template that is based on the ontology (Fig.1(1)). Users define layout similarity specifications according to their preferences. The system represents the instances of this feature generically, independent of a particular project. This project-independent knowledge is then utilized to compute similarity when the practitioner is ready to create an outcome for a domain-specific 3D design.

b) Identify layout similarity: A specific package configuration of layout similarity was created based on the practitioner's generic preferences defined in the previous research. Then the formal methods deduce out the geometric, topological, and symbolic similarities between components in the input 3D model and quantify the degree of similarity (Fig.1(2)). The result is a specific package configuration of layout similarity customized for the user.

3. KNOWLEDGE REPRESENTATION MODEL

According to the characteristics of layout design, analysis of the following types of knowledge was to be represented within the knowledge representation model:

a) Attributes of the vehicle package layout including: Vehicle types, Component parts, Properties of the layout and its parts, and Spatial and functional relations between component parts within a layout.

b) Human factor problems including: Types of design rules and layout constraints, Properties of the layout ergonomic problems.

c) Knowledge about solution procedures, including: Problem types, Algorithms and their parameters.

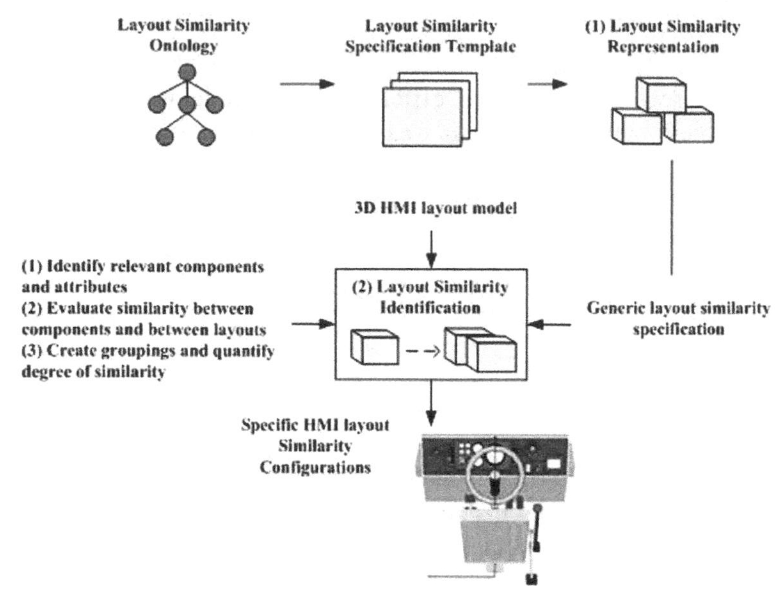

Fig.1: Framework for (1) representing and (2) reasoning about layout similarity to create specific package configurations of layout similarity

d) Dependencies between human factor problem types and properties and the layout design solutions.

e)Cases: a case's description includes definitions of: Physical components, Human factors problem set for this layout, Formulation of a corresponding layout reference, and Saved numerical solution routine specification.

Upon the comparative analysis, domain-specific formal ontology was selected as the knowledge representation model type. Knowledge representation model was developed which comprised: (a) ontology of whole package layout, (b) ontology of human factor principles, (c) ontology of solution techniques, (d) special concepts—subsets of the 'rule' concept, which represented rules and constraints reflecting dependencies between properties of ergonomic problems, layout and solutions.

So, the ontology-based domain knowledge representation model *OBM* can be structured according to the "human factor problems—layout—solution routine'' reasoning schema in the following way: *OBM = {HP, LP, SP, DE}*, where *HP* is the set of elements of description of human factor problems, *LP* the set of elements of components and layout problems, *SP* the set of elements of solution procedure descriptions and *DE* the set of dependencies between properties of ergonomic problems, layout problems and solutions.

This work aims to develop the general structure of the model so that it can be populated by a variety of ergonomic experts and be broadly applicable across a variety of vehicle designs and domains. The domain-specific ontology that was developed to represent layout similarity allowed designers to specify their varied preferences for what component properties needed to be similar and how much variation was acceptable for layout similarity to exist. Features were classified into the following types: a) Component Features were features that resulted from components in a 3D vehicle interior layout model; b) Spacial Features were features that resulted from the spacial relations and functions between components; and c) Macro Features were features that resulted from pre-specified combinations of other features. Layout similarity was a specific class of macro feature.

Fig.2 showed the features and attributes currently represented in the ontology. Each component feature represented the basic attributes of the components using the 'feature set' and 'property set'. Each spacial feature represented the emphasis of what special relation and function of the component would affect a component's location layout using the 'property set' attribute. Each macro feature represented the elements for defining layout similarity in terms of the component properties that need to be similar and the amount of variation that was allowed to exist according to the ergonomic rules. The attributes of the three feature types enabled the designers to represent their varied design preferences based on the layout similarity, and specifying their interior layout design characteristics by highlighting certain part of them.

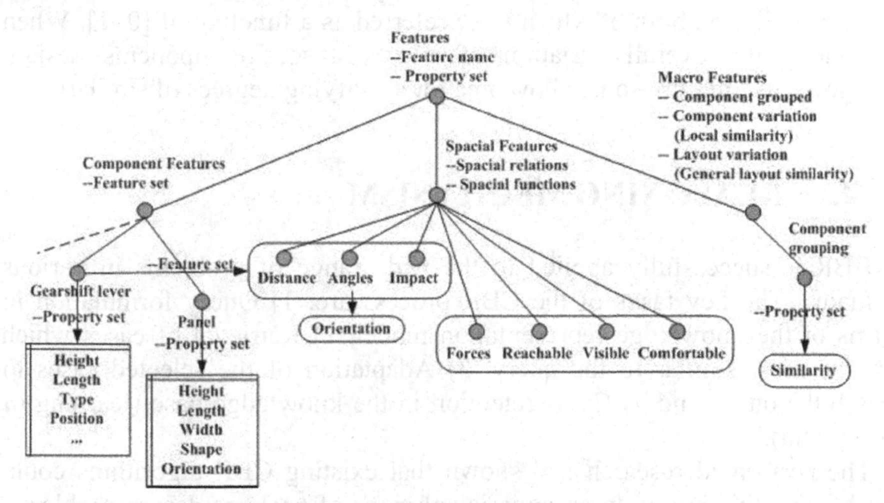

Fig.2: Domain-specific ontology that represented the attributes of package components

The attributes currently formalized in the ontology were to characterize the feature layout similarity. The attributes provided a formal way to specify the different types of component properties (e.g., geometric, symbolic and relational attributes) to be evaluated. The ontology also provided a way to characterize the degree of similarity at the component level (e.g., SeatHeight±10 cm) and at the system level (e.g., 0.1 variation).

The classification of component attributes facilitated the evaluation of layout similarity. Table1. showed the different component attributes currently implemented in the ontology based on the different case studies conducted to date. The component attributes listed were either represented explicitly in a 3D model or could be derived from a 3D model.

Table 1. Component attributes in the current ontology

Component	Component attributes			Component	Component attributes		
	Geometric	Symbolic	Relational		Geometric	Symbolic	Relational
Gearshift	Height	Type	Reachable	Panel	Height	Type	Visible
	Length	Shape	Comfortable		Length	Curvature	Reachable
	PositionX		Max force		Width	Shape	
	PositionY				PositionX	Orientation	
	PositionZ				PositionY		
					PositionZ		

The overall variation of the components allowed achieving layout similarity as a function of a maximum and minimum range. In the package cases, a minimum of 0.85 and a maximum of 1 were specified to represent an 'ideal' degree range of similarity for the similar case. The component variation corresponded to the local similarity and the layout variation to the overall similarity, both of which were referred as a function of [0, 1]. When evaluating the overall variation of all the layout components design, designers also require some allowance for specifying degrees of similarity.

4. REASONING MECHANISM

CBR is successfully applied to the wide range of problems in various domains. The key tasks of the CBR process are: 1) Query formulation in terms of the knowledge representation model, 2) Retrieval of cases which are the most similar to the query, 3) Adaptation of the selected cases to match the query and 4) Cases retention in the knowledge base (learning of the system).

The performed research had shown that existing CBR algorithms could not be directly applied to ergonomic solutions of package design problems, because this domain had some specific properties: layout problem (case) description was, as a rule, of complex structure, which could vary depending on the technical object and the problem type; values of qualitative and quantitative parameters are to be evaluated taking into account the context (values of other parameters).

Consequently, the following algorithms were developed to operate on the suggested domain model *OBM* in the framework of the CBR mechanism implementation: CBR-query formulation support algorithm; case retrieval algorithm which uses concept-based similarity (CBS) computation algorithm and slot-based similarity (SBS) computation algorithm; case adaptation algorithm, which used ergonomic rules and similarity paths.

The case retrieval algorithm, which was the key part of the CBR reasoner, was based on the similarity measure *Sim*. As the case indexes were represented in the formal ontology by features, the similarity of cases was reduced to the similarity between the features i1 and i2 of the ontology, which was local similarity. It could be represented as

$$Sim(i_1, i_2) = Func(CBS(i_1, i_2), SBS(i_1, i_2)) \tag{1}$$

where *Sim* is the overall similarity of the individuals, *CBS* the concept-based similarity, *SBS* the slot-based similarity, *Func* a real-valued composition function. The suggested case retrieval algorithm did not calculate the full *Sim* value for all the cases in the knowledge base. For most cases only the CBS value was calculated, which was much simpler to compute than the SBS, then the maximal possible value *Sm* of *Sim* for the given CBS was estimated, and SBS was computed only when *Sm* was greater than the current maximal *Sim* or greater than some given threshold.

The CBS computation algorithm used the vector space model, where each instance was represented by an n-dimensional vector. Components of this vector corresponded to user-defined ontology concepts. If the instance was subsumed by a concept *m*, the corresponding vector component was assigned the appropriate value of the weighting function *W(m)*, if not - zero. CBS was then computed using the well-known cosine measure:

$$CBS(i_1, i_2) = \frac{Vector_1 Vector_2}{\|Vector_1\| \|Vector_2\|} \tag{2}$$

The SBS computation algorithm was based on one-to-one comparison of individuals' relations, which could represent their parameters and/or structural relations. The SBS computation process was recursive; it started on the given features i_1 and i_2, compares one-to-one all the features related to them by ontology relations, and stopped on features with no relations, for which SBS was not computed. Use of weighting coefficients *W* for relations allowed flexible tuning of the algorithm and use of role similarity functions (tables) adequately handles ontologies with multiple similar relations:

$$SBS(i_1, i_2) = \sum_{j1=0}^{n1} \sum_{j2=0}^{n2} w_{j1} w_{j2} LocalSim(R_{j1}, R_{j2}) \tag{3}$$

where n_1 is the number of relations of the individual i_1, n_2 the number of relations of the individual i_2, w_{j1} the weight of relation j_1 of instance i_1, w_{j2} the weight of relation j_2 of the individual i_2, R_{j1} the j_1th relation of the individual i_1, R_{j2} the j_2th relation of the individual i_2, LocalSim the similarity function of the two given relations (local similarity). The general schema of the suggested similarity computation algorithm is presented in Fig.3.

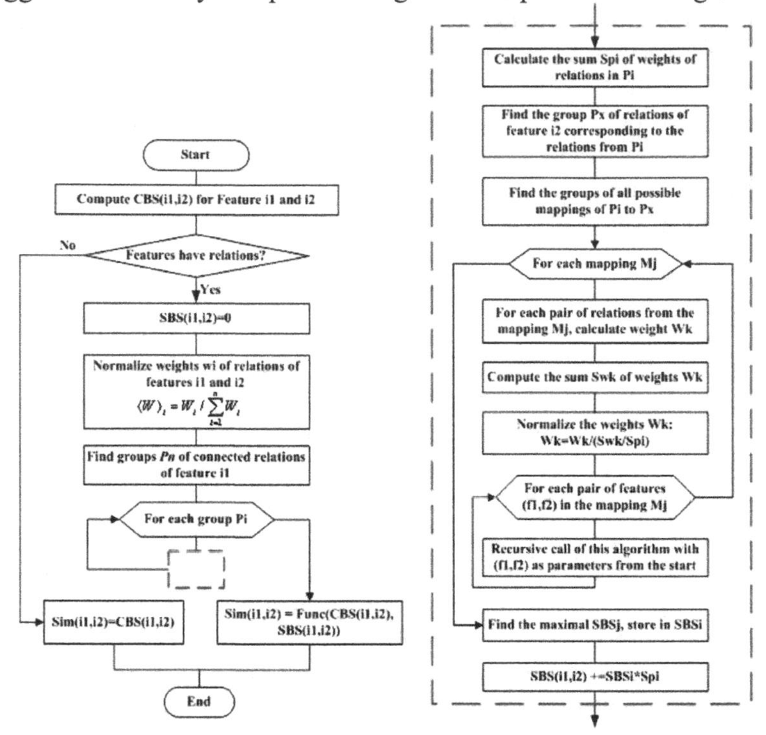

Fig. 3: Overall flowchart of CBS and SBS similarity computation algorithm

5. CASE STUDY

Based on the methodologies discussed above, a software tool called Vehicle Package Design Advisor had been developed. In this section, a case study was presented for the design of an automotive panel and instruments that could generate optimum layout considering user preference based on generic ergonomic knowledge base and feasible reasoning mechanisms by using this Advisor.

By the reasoning mechanism proposed above, several similar cases were successfully retrieved, and the basic layout information from the similar cases including the component grouped and detailed information were then displayed when one of the cases was highlighted along with the graphical representations of the components and layout. Fig.4 and Table 2. indicated the results of case retrieval for panel grouped, in which *Dist1* was the eye view distance, *Dist2*the distance from panel to floor, *Obliquity1*the angle of panel relative to floor， *Obliquity2* the eye sight angle relative to the horizontal, *Length* the effective panel length, W*idth* the effective panel width.

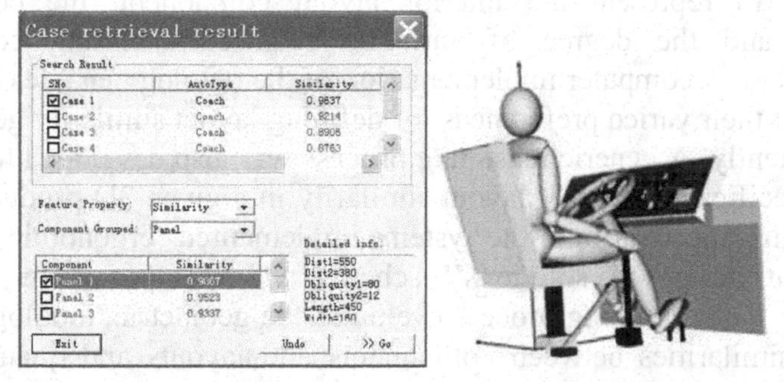

Fig. 4: Results of case retrieval and graphical representations

Table 2. Results of case retrieval for panel grouped

Items	Dist1	Dist2	Obliquity1	Obliquity2	Length	Width	Similarity
Expectation	440~710	350~450	60°~90°	10°			
Case 1	550	380	80°	12°	450	160	0.9867
Case 2	600	410	70°	13°	480	150	0.9523
Case 3	630	440	75°	10°	500	180	0.9337

After all feasible layout designs were found, the most similar layout case would be lightly revised and evaluated considering seating comfort. Since all the layout cases in this research used certain formerly adopted interior layouts as references, there were possibilities of comparing the manikin seating postures both in the new design and in the referred layout using empirical range of comfort angles (Judic 1993 and Kolich 2000) of driver joints. The grading evaluation method proposed in literature (Jin 2005) was used to estimate the seating comfort. It could be concluded from Table 3. that the newly designed coach interior layout was better than the referred layout case regarding the comfort quality for users of 50th percentile.

Table 3. Comparison of joint angles of manikin seating posture

Joint	Design sample	Grading	Referred case	Grading
Ankle	95.7°	better	90°	accepted
Knee	108°	better	97°	accepted
Elbow	89°	accepted	87°	accepted
Seat back angle	21°	accepted	15°	bad

6. CONCLUSION

This paper contributes a framework for vehicle package design by representing and reasoning about layout similarity that was built on ontological modeling and case-based reasoning techniques. It is presented that assumes use of ergonomic knowledge base. The ontology was formalized to represent the interior layout component, the component attributes, and the degree of variation required assessing component similarity, etc. A computer implementation of the ontology enables designers to represent their varied preferences for defining layout similarity generically and consistently. A generic reasoning process was also developed to identify domain-specific instances of layout similarity in a given 3D product model. The reasoning mechanism of the systems implemented "ergonomic problems <-> Layout design<-> solutions'' schema and was based on the CBR technology. The reasoning process evaluates the geometric, topological, and symbolic similarities between components and layouts and quantifies the degree of similarity.

Automating the detection of domain-specific design features, like layout similarity, has the potential to significantly improve the efficiency of the design process. Project teams could perform what-if analyses on different designs and explore a larger variety of design alternatives to identify the optimal design. Users could provide feedback to designers on the specific features that impacted the degree of comfort. Hence, project teams can adjust the case-based models to develop more comfortable designs in less time.

REFERENCES

Avramenko Y and Kraslawski A. Similarity concept for case-based design in process engineering, Computers & Chemical Engineering, 2006, 30: 548-557.

Chen, GL, Zhou, JQ and Cai, W. A framework for an automotive body assembly process design system, Computer-Aided Design, 2006, 38: 531-539.

Dan, MP. Using man modeling CAD system and expert systems for ergonomic vehicle interior design. Proceedings of the 13th Triennial Congress of the International Ergonomics Association, 1997, pp. 80-83.

Feyen R, Liu YL and Chaffin D. Computer-aided ergonomics: a case study of incorporating ergonomics analyses into workplace design, Applied Ergonomics, 2000, 31:291-300.

Han YH and Lee KW. A case-based framework for reuse of previous design concepts in conceptual synthesis of mechanisms, Computers in Industry, 2006, 57: 305–318.

Human Factors and Ergonomics in Manufacturing, 2005, 15(2): 197–212.

Jin XP, Song ZH and Mao ER. Research on the vehicle HMI qualitative and quantitive grading evaluation methods, Proceedings of Man-Machine-Environment System Engineering, 2005, 7: 232-236 (in Chinese).

Judic JM. More Objective Tools for the Integration of Postural Comfort in Automotive Seat Design, SAE Technical paper 930113.

Kolich M. Driver Selected Seat Position: Practical Applications, SAE Technical Paper 2000-01-0644.

Society of Automotive Engineers. Automotive engineering handbook. Warrendale, PA: Author. 2001.

Vogt C., Mergl C. and Bubb H. Interior Layout Design of Passenger Vehicles with RAMSIS,

APPLICATION OF VIRTUAL MANUFACTURING IN FIELDS CULTIVATE MACHINES

Jun Liu[1,*], Yan Li[1], Zhou Li[1]

[1] *College of Mechanical & Electrical Engineering, Henan Agriculture University, Zhengzhou Henan Province, P. R. China 450002*

[*] *Corresponding author, Address: College of Mechanical & Electrical Engineering, Henan Agriculture University, Zhengzhou 450002, Henan Province, P. R. China, Tel: +86-371-63554632, Fax: +86-371-63558040, Email: liujunshd@sina.com*

Abstract: The paper introduces virtual manufacturing application in the Fields Cultivate Machines and mainly discusses the model-building of rotary-cultivate part, physical model and mathematical model included. With the model, the rotary knife is optimized, the force on the bend part is decreased, and therefore, the service life is prolonged. At the same time, with dynamic analysis software Adams, the moment curve of the rotary knife is simulated, which provides a basis for the improvement of stability and farther research in future.

Keywords: Virtual Manufacturing, Dynamic analysis, Fields cultivate machine Rotary knife, Mathematical model, Physical model

1. INTRODUCTION

Virtual Manufacturing is a synthesized development and application of computer-simulation technology and virtual-reality technology. With VM, the actual manufacture can be realized essentially on computer. It is, based on information integrations, a new manufacture philosophy of enterprises. The extensive application of the virtual manufacturing technology will change the present manufacture pattern thoroughly, and will bring various kinds of changes into enterprise organization, business management and production methods and so on. It also will have the huge influence to the

Please use the following format when citing this chapter:

Liu, J., Li, Y. and Li, Z., 2009, in IFIP International Federation for Information Processing, Volume 294, *Computer and Computing Technologies in Agriculture II, Volume 2*, eds. D. Li, Z. Chunjiang, (Boston: Springer), pp. 1461–1466.

correlation profession, and will be one of important content of next generation manufacture technology. It is said surely, Virtual Manufacture technology will decide enterprise's future, and also will decide whether or not manufacturing industry can play an important role in the competition.

Pro/Engineer is a large-scale design software which integrates with CAD/CAM/CAE. The modules in common used by CAE are Mechanism Design Extension and Pro/MECHANICA. Comparing with high specialized analysis software, Pro/E is not a worthy opponent, however, up to now, Pro/E is still used most widely in all three-dimension software. Its accuracy rate of analysis reaches to 99%, and it is quite convenient and is popular to major users.

The fields cultivate machine has already applied commonly and mainly in some big awnings today. Although it has many functions, its work is not stable sometime, main reason is that various kinds of functions have not been considered completely in primary design. For having a more stable work, the construction and simulation must be done with some large-scale software. Nowadays, the higher quality cultivate machines are needed. Therefore, cultivating machines simulation is necessary so that model optimization can be realized, and intensive cultivation can be achieved.

2. MODEL ESTABLISHMENT

At first, with Pro/Engineer, the rotary-cultivate part of the machine should be built, for example in Fig.1.

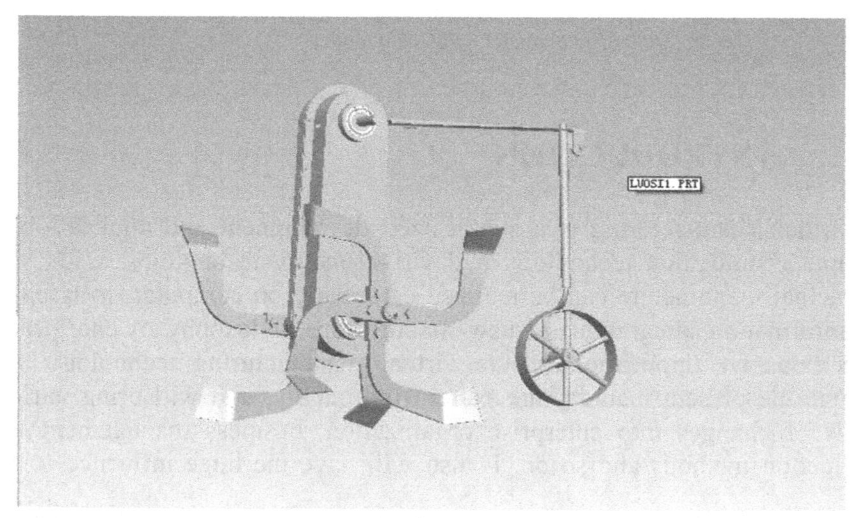

Fig.1 General Structure of model

The parameter method is adopted in the model establishment completely, the model can be revised easily, so that continued optimization can be realized conveniently, until the request is fulfilled.

The analysis result of a mechanical system, from initial geometry modeling to dynamics modeling, is obtained in the end, through the numerical value solution of the model, the analytical results can be obtained finally, and the whole process can be expressed in Fig.2.

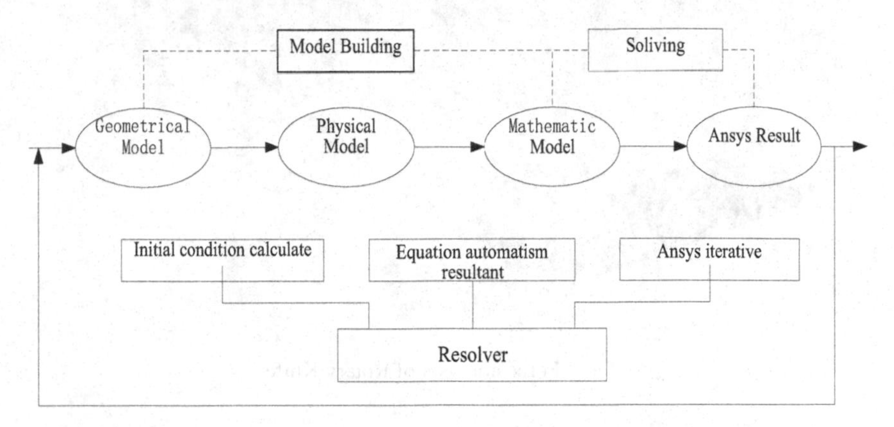

Fig.2 the general course of calculate many systematic dynamics modeling and solution

The whole process of polypore system of dynamic analysis includes two stages : constructing and solving models. Constructing model divides into both physical modeling and mathematical modeling, physical modeling is the model which can be established by geometry modeling; Mathematics modeling is anther model which can be constructed by physical model. The geometry modeling is formed of geometry-constructing modeling of analysis system of dynamics , or is introduced through general geometry-modeling software. Putting the kinematical restraint, the driving restraint, the force unit, external force or external moment into geometry model, the physical model may be formed, which is able to describe mechanics trait of the system. In process of physical modeling, Sometime the geometry model needs to be assembled in the light of the kinematics restraint and the initial position condition. With the physical model, Cartesian coordinates or the Lagrangian coordinates constructing-model method, and the application of automatically constructing-mold technology, various kinds of coefficient matrix can be built and the mathematical model of the system can be obtained. Using analytical and calculation methods of kinematics, dynamics, static equilibrium or restores dynamics. Analysis results needed are obtained with iterative solution . Combining with design goal, the result is analyzed repeatedly, which is returned the process of physical and geometry model, so that best result of the design is acquired.

3. OPTIMIZATION MODELS

It is necessary to optimize rotary knife for the sake of durative use, the process is shown on the following fig.3:

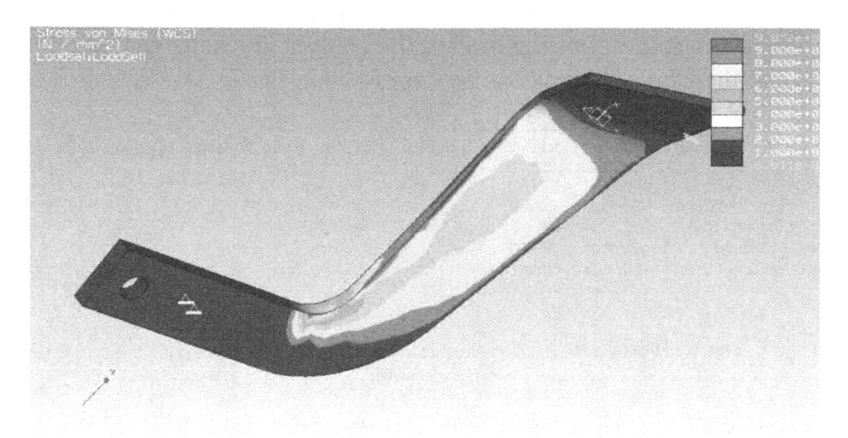

Fig.3 Force analysis of Rotary Knife

From Fig.3 , it is known that great force (about 9.072N/mm2)is put on the bent part of rotary knife. After optimization, although the force which applies on whole knife is equal, the force on the bent part becomes weak. The optimized knife is shown in following Fig.4.The force put on bent part of the knife which is optimized is about 8.313N/mm2 so that the knife is durable.

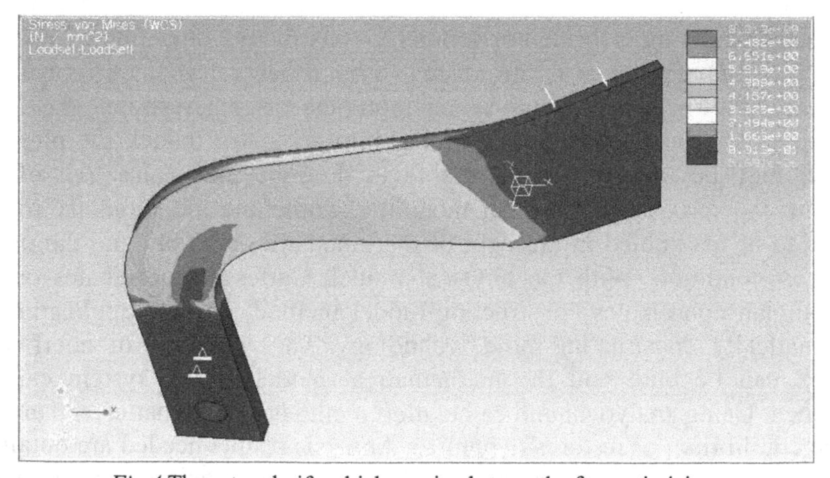

Fig.4 The rotary knife which received strength after optimizing

4. MODEL DYNAMICS ANALYSIS

After optimization, effective dynamics analysis can be adopted with Adams software. During work process, the movement curve of rotary knife is shown below fig.5. Through dynamics simulation of rotary knife by Adams, the speed ,acceleration , angular velocity and the force moment of rotary knife in the course of working are measured. Its certain shortage can be improved.

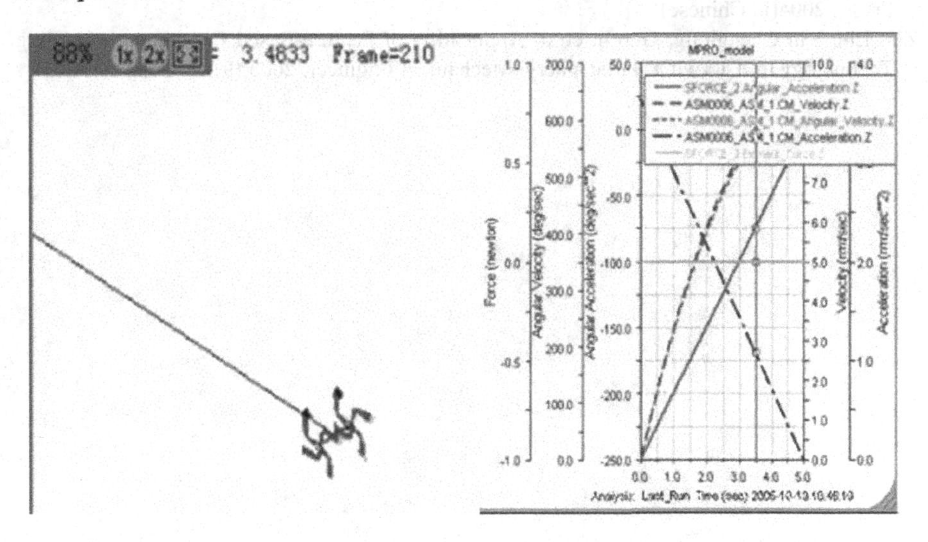

Fig.5 The movement trajectory of Rotary

5. CONCLUSION

Through using virtual manufacturing technology in fields cultivate machine, every components of the machine have been optimized, without question, feasibility of virtual manufacturing technology have been proved.

REFERENCES

Fu Gang, Dong Gang. The Cleaning Device of Rape Harvester Virtual Prototyping Modeling and Simulation Experiment, Journal of Agricultural Mechanization Research, 2006,(70):96-98(in Chinese)

K. Iwata, M. Onosato, et al. Virtual manufacturing systems as advanced information infrastructure for integrating manufacturing resources and activities. Annals of the CIRP 1997, 46(1):335-338

Kimura F. Product and Process Modeling as a Kernel for Virtual Manufacturing Environment. Annual of the CIRP. 2003, 52(1):85-93

Li Yong, Zeng Zhixin, Ye Mao, et al. Application of Virtual Prototyping Technology to Development of Small-sized Agricultural loader, Transactions of the CSAE, 2004, 20(5):134-137(in Chinese)

M. onosato, K.iwata. Development of a virtual manufacturing system by integrating Product models and factory models. Annals of the CIRP.1999, 48(l):475-478

Wang Xiao, Liu Huixia, Cai Lan. Virtual Product Development Technology in Automotive Development, Journal of Jiangsu University of Science and Technology, 2001,22(5):42-46(in Chinese)

Zan You Gang . Pro/E Chinese Wildfire edition study course. Beijin: TsingHua University Press , 2004(in Chinese)

Zhu Lin, Yin Chenglong, Guo li, et al. Application of Feature-Based Component Modeling Technology in Agricultural Machinery, Mechanical Engineer, 2005,(9):77-79(in Chinese)

VISUALIZATION OF VIRTUAL PLANT BASED ON 3D MODEL CONVERTER

Qian Wang [1] , Ying Zhang [1,*] ,Ji Liu [1]
[1] *Computer College of Chongqing University, Chongqing, P. R. China 400030*
[*] *Corresponding author, Address: 400030, Computer College of Chongqing University , Chongqing, P. R. China, Tel: +86-23-65102506, Email: linzhibinjelly@163.com*

Abstract: We present a method based on a 3D model converter to simulate the development of virtual plant. The converter is mainly used to import the plant organs which have fine details into the virtual plant development system, then communicating with the L-system to implement the simulation of the development of plants controlled by the physiology of plant. It improves the virtual effect that carried out by the former systems which only take into account the geometric model.

Keywords: modeling of virtual plant organs, L-system, 3D model converter, visualization of virtual plant

1. INTRODUCTION

Plant development is a dynamic process in which the topology and geometry change over time in a seemingly complex manner. We can simulate the developing process in the three-dimensional space on computer to reflect the morphological structure changing of the realistic scene, including the individual plant and the community (Ma Xinming et al.,2003).

Models of plant development can be implemented using a variety of methods. They contain the Particle Systems (Reeves,1985),A-System (Aono et al.,1984),the Realistic method for the Procedural Generation of Trees (Weber et al.,1995) and the AMAP (De Reffye et al., 1988;Jaeger et al.,1992) model etc, but they are all modeling plant organs through the geometrical parameters, not considering the fine detail on them, which leads to lacking of

---

Please use the following format when citing this chapter:

Wang, Q., Zhang, Y. and Liu, J., 2009, in IFIP International Federation for Information Processing, Volume 294, *Computer and Computing Technologies in Agriculture II, Volume 2*, eds. D. Li, Z. Chunjiang, (Boston: Springer), pp. 1467–1476.

the visual effect. An improved model Greenlab (De Reffye et al.,2003) based on the AMAP was proposed by the researchers, it's a function-structural model that uses the mathematical expression to represent the developing process which excels the AMAP in simulating the complicate morphological structure. Also it has limitation, it can only create the organs that have similar shapes and the surface detail was not incarnated. Moreover, there are plenty of modeling softwares that support the texture mapping that can express the detail on the surface which can enhance the virtual effect, but it can not implement the plant developing process which has relation with time, so they can not be used to simulate the developing process of plant directly.

To solve the above problems, we present a new method which models the plant elementary organs (like stem, petal, calyx, leaf et al.) by the outside modeling software and choose the L-system(Lindenmayer, 1968; Prusinkiew et al.,1990; Karwowski et al.,2003) to simulate the plant development. According to this formalism, a plant is viewed as a developing assembly of individual units, or modules. These modules are characterized by parameters such as length, width, and age, as well as parameters characterizing shape (Mundermann et al., 2005). Here we use a three-dimension model converter to import the plant organs designed by the outside modeling software to the L-system to control the plant growing. The experiment result shows this methodology can effectively solve the problem of lacking of the detail on the plant surface in real time and combine the physiology of plant seamlessly.

2. IMPLEMENTATION

2.1 Visual approach

A plant can be considered as a series of organs, each consists of many elementary components that array regularly in the three-dimensional space. Those are stems, leaves, flowers et al.. Now we will introduce our method to model the plant organs by the outside modeling software. All models are created in relative size not the actual size. All of the organs' centre must be in the origin of the world space. And all organs' texture comes from the photos taken by us. This will enhance the visual effect.

1) Leaves

The obvious characteristic of leaf is its geometrical shape and surface texture. So we design the leaf according to the photos taken from the plant of various view directions. First, we use a two dimension plane to simulate the leaf, then we extend it in the third dimension. (see Fig.1(a)).

As for the organs' texture, we will use the two-part texture mapping method (Bier et al.,1986) to transfer the texture space to the world space, we will use the plane as the intermediary surface, that is, first map the texture onto the plane, then use the texture image from the plane to map onto the leaf's surface.

2) Flowers

One of the most important parts of plant is the flower. It comprises by many components, namely, the petal, the pistil, the stamen, and the calyx after our simplifying. Here we illustrate our means to determine these organs.

Petal's shape is similar to the leaf, so we use the same way of the leaf's creating. And both of the pistil and the stamen (see Fig.1(b)) are gotten by metamorphosing the sphere. As to the calyx, its body can be constructed by two steps: an axis from the flower branch to the chaplet and the shape with different sizes circling the axis. So we can determine it by appointing the shape of the axis and the different radius of the shape infixed the axis.

We use gradual changing color as the texture of flower's all components except the calyx and petal (see Fig.1(c)). Only one color is used for the calyx (see Fig.1(d)). And the petal has the real texture from the photo of the dissected plant, we assemble them together for integration and saving space.

3) Branches and Stems

Main points to both of them are their radius and holistic bending state. Then we apply the mesh smoothing technique to make them bend naturally.

The final shape of each organ with real texture are listed (see Fig.1). Then we use the L-system to control the organs' appearing positions and sizes over different developing stages.

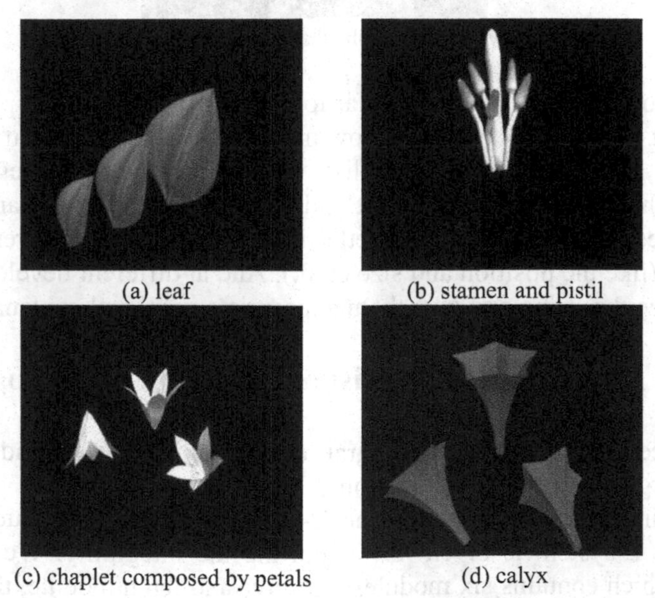

(a) leaf (b) stamen and pistil

(c) chaplet composed by petals (d) calyx

Fig.1. Some organs created by our method

2.2 Physiological approach

We choose the formalism of L-systems as growing three-dimensional structures. L-systems emphasize plant topology, namely, the neighborhood relations between cells or larger plant modules (Prusinkiewicz et al.,1990).

The central concept of L-systems is rewriting. In general, rewriting is a technique for defining complex by successively replacing parts of simple initial objects with a set of rewriting rules or productions. It consists of triple table G=<V,w,P>, V is the alphabet, and w is the initiator, P is the productions list. One begins with two components: an initiator and a generator. The iteration process is: The w is the first iterative string, then replacing the character appeared in the last string by finding in the P, repeating the step until all words have been replaced. Fig.2 gives the results evolved by the L-systems.

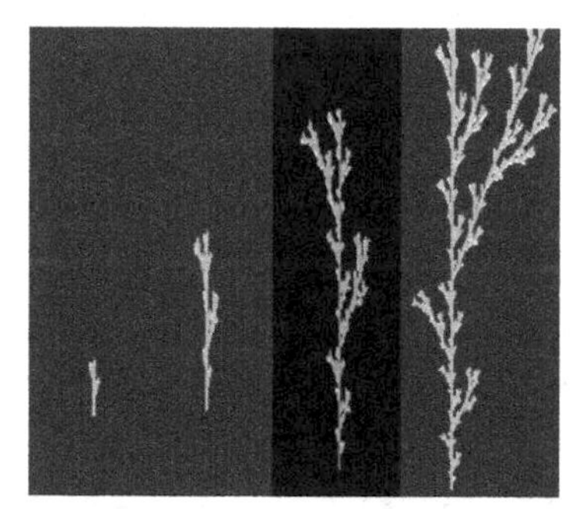

Fig. 2. The results that L-systems evolve

The evolving process is so similar to the development of the plant that we can use it to simulate the plant growing. For the L-systems can appoint the initiative state and productions list, we can apply these regulations to proceed the scheduled evolvement and form the complicate plant topology. Also we can appoint each of the letter in the alphabet the different graphical meaning (like the position and size et al.). And at different developing stages we analyze the string and give them appropriate graphical explanation.

2.3 Integration of the visual effect and physiology

This section illustrates the integration of the visual effect and L-systems. All the organs created in the section 2.1 have implemented the visual effect, then we import the organs into the L-systems. The whole structure of our improved L-system is given for supporting the integrality. We call it I-L-system which contains six modules: the save and fetch module, the grammar managing module, the organ transfer module, the alphabet appointing module, the evolving module and the rendering module. There are six steps to be proceeded, and Fig.3 shows the structure of our I-L-system:

1) import the organs from the model library, and draw them by visual technology in real time, including displaying the texture correctly;

2) appoint every organ a special character that presents different graphical meaning in the L-grammar through the alphabet appointing module;

3) write the plant developing grammar of special growing physiology which will control the arrangement of the organs;

4) appoint the number of the iteration step, and the result of each step can be shown in the system evolving module;

5) set the camera position and light condition, and draw the growing scene through the rendering module;

6) save all the things finished in the former five steps, including saving the organ models, the plant developing grammar and the evolving result.

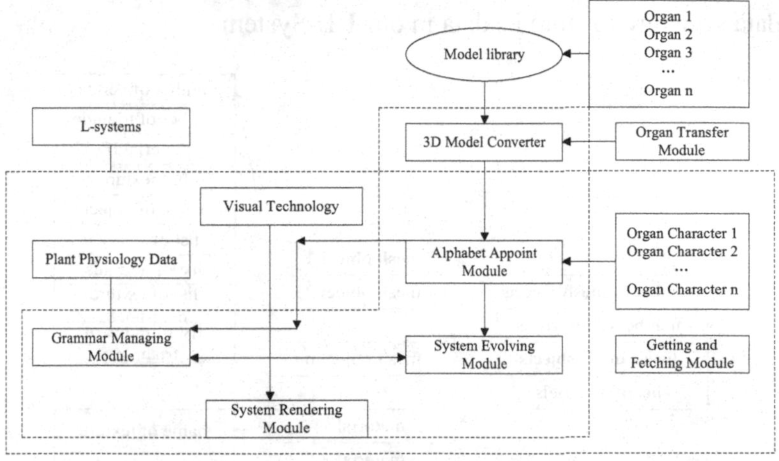

Fig. 3. The whole structure of the I-L-system

3. MAIN PROCEDURE OF THE 3D MODEL CONVERTER

We introduce the L-systems and analyze it in the second section, the findings show that L-systems are very good at exhibiting the plant topology due to its iteration process, and it also uses this way to produce the texture. To the realistic plant developing scene, there are various of texture of the organs, and these cost much iterative time. Compared to the real organ texture, the L-systems' are not so realistic for the former uses. Moreover when we want to remodel a new organ, the L-systems needs to recreate the organ, but our method support modifying the organs from the existing.

This paper use .3ds file as the input file format of the model library. Then we design a three dimension model converter to import these organs from the library into the L-systems.

3.1 Data structure extracted from the .3ds file

The .3ds is the native file format of the Autodesk 3D Studio. And it has promulgated the standard file structure. The file is constructed by various of blocks. Each block has a head and body which can contain other blocks that can also contain other blocks. All of them are nested. Each head of the block consists of an ID and the block length, and the body contain the different data. The models determined by this format are the mesh style, that is, all of the organ surfaces are made up of triangles (others are quads and so on). Through analyzing, we find all the data can be extracted to following structure. Fig.4 shows our extracted result for every organ, we will construct some data structure to store its data in our I-L-System.

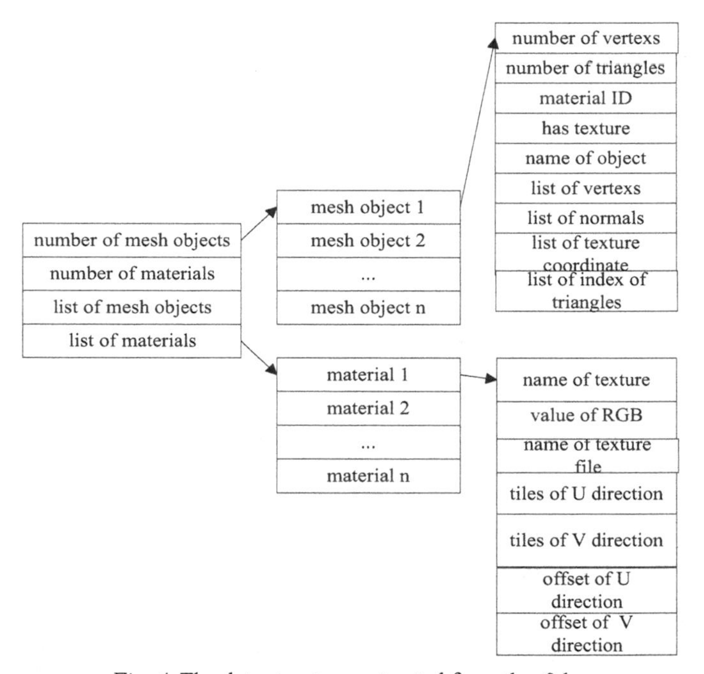

Fig. 4. The data structure extracted from the .3ds

3.2 Two-part procedure of the 3D model converter

In the last section, we have designed some data structure to store the organ data of the .3ds model file. There are three important structures which are the model object, mesh object and material object. The relationship between them is every model object may contain many mesh objects, many kinds of texture, and every mesh object has the vertex, surface and material data, every material object has different texture. After confirming their existing format, we can design a three dimension model converter to transfer the model data to the format recognized by the L-systems.

The converter has two parts. The first is to read the data of the block in the .3ds by iterative method to the appropriate data structure of the system, and the blocks we need to deal with are the main editing block, the material information block, the mesh object block and their interior subordinate block. Then do some relative computations including converting the coordinate of the model from the world space to the system space and computing the normal direction of each vertex of each mesh face. This can be done by dividing the number of the triangles sharing the vertex, and use it as the vertex's normal direction.

The second part is to make the imported model data correspond with the L-systems, that is, to let every organ has relevant character in the alphabet which may has the graphical meaning and let them be assembled by the developing grammar written according to the plant physiology. The process is to appoint a character to each model once we import it, when we start to draw the special developing scene, we read the string produced by the L grammar and analyse the character, if it is the organ character we call the draw module to display it. The two-part procedure is shown in Fig.5.

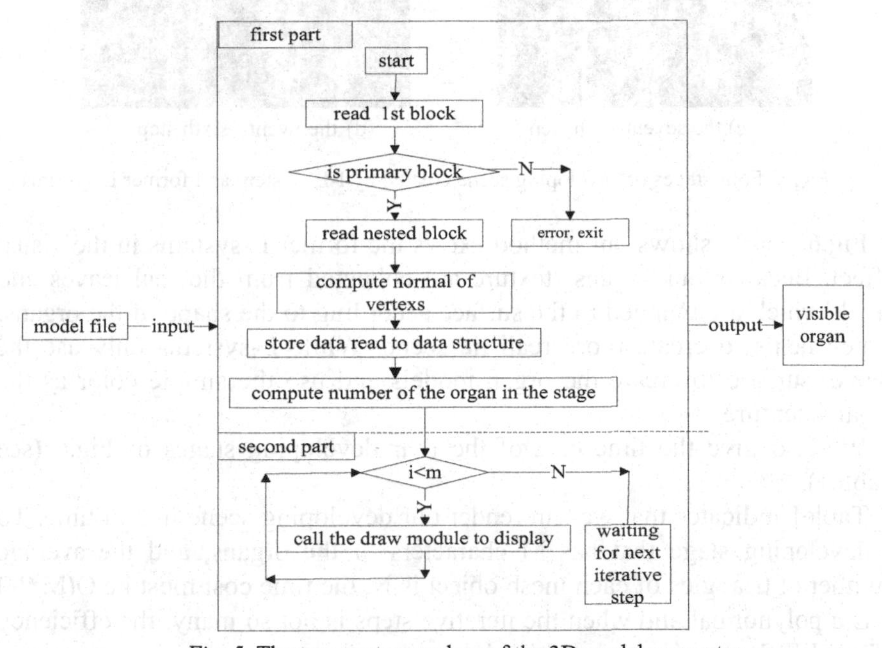

Fig. 5. The two-part procedure of the 3D model converter

4. EXPERIMENT AND ANALYSIS

We use the organ models created by our method mentioned in the first section (see Fig.1), and import them into the I-L-System we propose. There is comparison of the growing scene created by the former L-systems and our I-L-System while they use the same developing grammar. Fig.6 gives the result of our system and the former L-Systems.

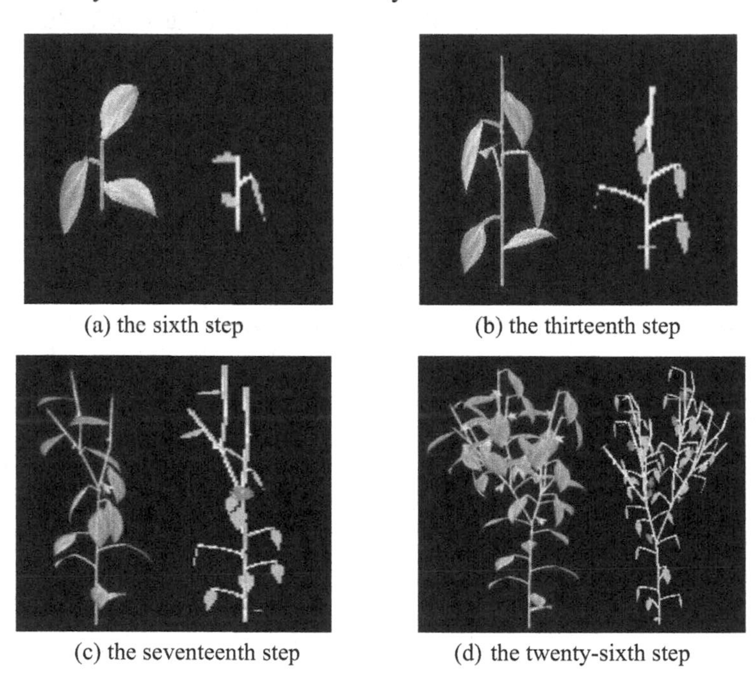

(a) thc sixth step (b) the thirteenth step

(c) the seventeenth step (d) the twenty-sixth step

Fig. 6. Four stages of developing scene created by I-L-System and former L-systems

Fig.6 clearly shows our method excels the former L-systems in the visual effect. Because our organs' texture are collected from the real leaves and petals which are mapped to the surface according to the shape of the organs, which helps to create more realistic scene. While L-systems only use the Bezier surface to create the organ models and use the simple color as the organs' texture.

We also give the time cost of the four developing stages of Fig.6 (scc Table1).

Table1 indicates that we can render our developing scene in real time. To a developing stage that has M characters of the organs, and the average number of triangles of each mesh object is N, the time cost must be O(M*N). It is a polynomial and when the iterative steps is not so many, the efficiency of our I-L-System is comparably high.

Table1. Time cost of the four developing stages of our I-L-System

Developing Stage	Iterative Steps	Cost(ms)
A	6	186
	8	232
	10	295
	12	373
B	13	435
	15	560
C	17	778
	19	1107
	22	1944
	24	2688
D	26	4085

5. CONCLUSION

The method based on the geometrical parameters leads to abstract and bad controllable modeling process, it also causes lacking of individual characteristics of the various plants' organs. The modeling method based on the image also has limitation that it has great dependence on the view point and needs plenty of photos taken from the plants. And the way based on the L-systems to create the organs can only support the Bezier surface to model them which have no realistic texture.

What we propose in this paper can solve the above problems. The results indicate that we can create the organs through outside modeling software which have a good revisability at any spatial position of the organ, it also helps us to enable them have abundant details on their surface. After finishing the models, we use the 3D model converter to import them and use the L-systems to assemble the whole organs created outside and call the rendering module to implement the visualization of the plant on computer in real time.

The environment is the living space for the plants, it has so big effect on the plants. While our I-L-System does not provide an interface for the environment to work on the plant. And the texture mapping method we use can only stick the smooth texture to the organ surface, however there must be rough on real organs, so we will concentrate on the two problems in future.

ACKNOWLEDGEMENTS

Funding for this research was provided by the National High Technology Research and Development Program of China (863 Program) (2006AA10Z233) and the Ph.D. Programs Foundation of Ministry of Education of China (20050611027). The second author is grateful to the Computer College of Chongqing University for providing her with pursuing a Master degree at the Chongqing University. The third author is also grateful to the Computer College of Chongqing University for providing him with pursuing a Doctor degree at the Chongqing University.

REFERENCES

A.Lindenmayer, Mathmatical models for cellular interaction in development: parts I and II, Theor boil, 1968,18:280-315

E.A.Bier, K.R.Sloan, Two-part texture mapping, IEEE computer graphics and applications, 1986, 6(9):20-53

J.Weber, J.Penn, Creation and rendering of realistic trees, Computer graphics proceedings, Annual conference series, 1995:119-128

L.Mundermann, Y.Erasmus, B.Lane, et al., Quantitative modeling of arabidopisis development, Plant physiology, 2005, 139(2):960-968

M.Aono, T.Kunii, Botanical tree image generation, IEEE computer graphics and applications, 1984, 4(5):10-34

M.Jaeger, P.De Reffye, Basic comcepts of computer simulation of plant growth, Journal of biosciences, 1992, 17(3):275-291

Ma Xinming, Yang Juan, Xiong Shuping, et al. Reality and prospect of the virtual plant, Research of crop, 2003, 17(3):48-151(in Chinese)

P.De Reffye, C.Edelin, F.Jetal, Plant models faithful to botanical structure and development, Computer graphics, 1988, 22(4):151-158

P.De Reffye, J.Leroux, Study on plant growth behaviors simulated by the function-structural plant moedel-Greelab, Plant growth modeling and applications proceedings, 2003:118-128

P.Prusinkiewicz, A.Lindenmayer, The algorithmic beauty of plants, Springer-verlag, 1990

P.Prusinkiewicz, J.Hannan, Lindenmayer systems, Fractals and plants, 1989

R.Karwowski, P.Prosinkiewicz, Design and implenmentation of the L+C modeling language, Electronic notes in theoretical computer science, 2003, 86:1-19

W.T.Reeves, Approximate and probabilistic algorithms for shading and rendering structured particle systems, Siggraph, 1985, 19(3):313-322

INTELLIGENT GROWTH AUTOMATON OF VIRTUAL PLANT BASED ON PHYSIOLOGICAL ENGINE

Qingsheng Zhu*, Mingwei Guo, Hongchun Qu, Qingqing Deng

College of Computer Science, Chongqing University, Chongqing 400044, P. R. China

* *Corresponding author, Address: College of Computer Science, Chongqing University, Chongqing 400044, P. R. China, Tel: +86-23-65105660, Fax: +86-23-65105660, Email: qszhu@cqu.edu.cn*

Abstract: In this paper, a novel intelligent growth automaton of virtual plant is proposed. Initially, this intelligent growth automaton analyzes the branching pattern which is controlled by genes and then builds plant; moreover, it stores the information of plant growth, provides the interface between virtual plant and environment, and controls the growth and development of plant on the basis of environment and the function of plant organs. This intelligent growth automaton can simulate that the plant growth is controlled by genetic information system, and the information of environment and the function of plant organs. The experimental results show that the intelligent growth automaton can simulate the growth of plant conveniently and vividly.

Keywords: intelligent growth automaton, intelligent physiological engine, two dimensional hierarchical automata, virtual plant

1. INTRODUCTION

In recent years, research on virtual plant has made great progress. Many methods and models for generating plant graphics has been proposed, for example, mended L-system (Prusinkiewice et al., 1990), IFS (Dekmo et al., 1985), reference axis technique (Blaise et al., 1998), branch matrix (Viennot et al., 1989), particle system (Reeves et al., 1985) and so on. However, it is very difficult to depict and reproduce the complicated growth of plant.

Please use the following format when citing this chapter:

Zhu, Q., Guo, M.,. Qu, H. and Deng, Q., 2009, in IFIP International Federation for Information Processing, Volume 294, *Computer and Computing Technologies in Agriculture II, Volume 2*, eds. D. Li, Z. Chunjiang, (Boston: Springer), pp. 1477–1486.

The growth and development is a kind of behavior of plant life activities. It includes two aspects. One is the increase of size and weight of plant which is the growth of plant. The other is the continuously occurring of new organs called morphogenesis, which is the development of plant (Yang Shijie et al., 2000). The growth and development of plant are controlled by the genetic information system and the environment information system. Genetic information system determines the potential pattern of growth and development of plant, while the environment information system impact on the specific performance of characters. The growth and development is a very complex dynamic process (Li Hesheng et al., 2002).

The existing plant growth models are focused on computer graphics and very limited knowledge of botany are employed to generate a beautiful plant models quickly and easily, excepting for L-system and reference axis technique. Dr. Zhao Xing et al. (Zhao Xing et al., 2001) proposed a two-scale automata model based on the growth mechanism of plant, and Kyle W. Tomlinson et al. (Kyle W. Tomlinson et al., 2007) proposed a functional-structural model for growth of clonal bunchgrasses. However, the parameters of two-scale automata model are acquired empirically. Dr. Qu Hongchun et al. (Hongchun Qu et al., 2007) proposed the intelligent physiological engine (IPE) which consists of virtual environment, two dimensional hierarchical automata (2DHA) controlling plant branch pattern, the carbohydrate balance model on the basis of ant colony system and some other key components. IPE controls and coordinates the interaction between individual agent virtual organs (IAVOs) which constitute the virtual plant, distribution and balance of carbohydrate inside plant, and the interaction of virtual plant and the virtual environment, to drive the growth and development of plant. However, 2DHA can't satisfy the demand that IPE drive the growth and development of plant. On the basis of this framework, a novel intelligent growth automaton (IGA) of virtual plant is proposed which extends the function of 2DHA and simulates the growth and development of plant controlled by genetic information system and environment information system.

2. FRAMEWORK DESCRIPTION

In this framework, 2DHA simulates the function of genetic information system while IPE simulates the function of environment information system and the function of IAVOs. IGA analyzes the 2DHA and uses the controlling information from 2DHA to produce the plant. At the same time, IGA sends the interaction request to IPE, and modifies the parameters of 2DHA according to the feedback information, to simulate the environment coordinating the plant growth.

Fig.1 describes the framework of the interaction of IGA (Intelligent Growth Automaton) with 2DHA and IPE. I represents the interface between IGA and 2DHA, while II represents the interface between IGA and IPE. ① represents the information of controlling the plant growth; ② represents the information of structure and physiology of plant which are computed by IGA, and the information is stored in the Bi-dimensional Hash Chain; ⑤ represents the information of current structure and physiology of plant; ⑥represents sending the information of environment and plant structure and physiology status to IPE through II. ⑦represents the information of environment and function of IAVOs' feedback from IPE; ⑧ represents the adjustment information of structure and physiology of plant which is computed by IGA; ③ represents the information to modify parameters of 2DHA and ④sending the modifying information to 2DHA.

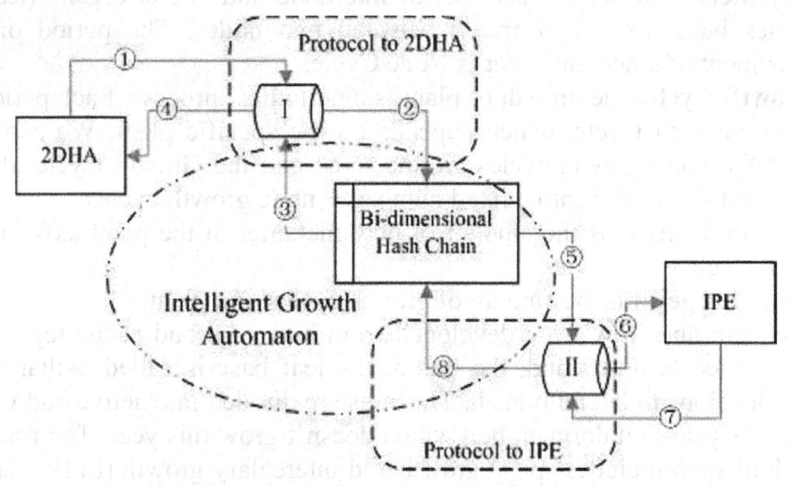

Fig.1 Framework of Intelligent Growth Automaton

3. THE DEVELOPMENT OF STEM

The appearance of plant includes the stem and the organs such as leaves, flowers and fruits which grow on it. In addition, the stem is the functional organ which links the root with leaves and transport water, inorganic salt and organic nutrition (Li Hesheng et al., 2002) (Yang Shijie et al., 2000). The part where leaf grows on is called node. The part between two nodes is called internode. As shown in Fig.2, the stems constitute the main structure of the plant, bearing some buds and leafage.

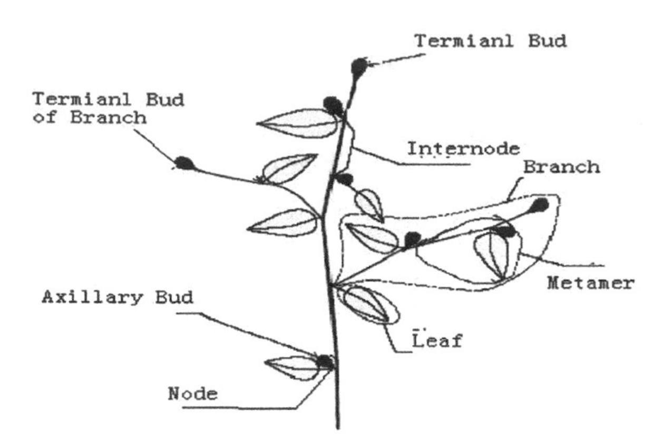

Fig.2 Form of Stem

Metamer: it is defined as a set of internode and lateral organs (leaves, axillaries bud, flowers or fruits) between two nodes. The period of the development of a new metamer is Node Cycle.

Growth Cycle: the growth of plant is a periodical process. Each period is called a Growth Cycle, which depends on the specific plant. We assumes that all the node growth cycles are the same and the Growth Cycle of one plant could be divided into a fixed number of node growth cycles.

Growth Unit: It is the amount of new metamer in the plant axis in one Growth Cycle.

Growth Age: It is the amount of growth cycles of a plant.

The stem and flower are developed from bud. The bud at the top of the axis is called terminal bud; the bud at the leaf base is called axillary bud which develops to lateral branch. The buds are divided into active bud which grows this year and dormant bud which doesn't grow this year. The primary growth of stem includes apical growth and intercalary growth (Li Hesheng et al., 2002) (Yang Shijie et al., 2000).

4. TWO DIMENSIONAL HIERARCHICAL AUTOMATA

2DHA samples the plant and then acquires the parameters by employing the statistical hidden markov tree (HMT) model and clustering. 2DHA is constituted of parent automaton and child automaton. Parent automaton is constituted of parent states which transfer mutually and child automaton is constituted of child states which transfer mutually too.

Definition 1(Parent automaton): the parent automaton is defined as a finite state automaton: $A_p ::=< Q_p, \pi_p, S_{GC}, \rho_p, F_p >$,where Q_p denotes the

active bud is 1, while the initial value of dormant bud is the negative of the dormant period. The value of *F* includes 0,1 and 2, that 0 represents the metamer is intercalary growth, while 1 represents the metamer is apical growth, and 2 represents the metamer and others in the branch is restrained.

Definition 2(Str-Hash Table): It is a hash table which stores the structure information of metamer. At present, there are three parameters --*POS*, *ORI* and *Mesh*. *POS* denotes the absolute position of the metamer and *ORI* denotes the absolute orientation in the plant. *Mesh* denotes the mesh graphics corresponding to an metamer, including color and size of the organs.

Definition 3(B-DHT): It is constituted of Phy-Hash Table and Str-Hash Table. If an internode has axillary buds, it is need to bracket the B-DHT of the axillary bud with '['and ']', and all B-DHTs of axillary buds are linked.

Definition 4(B-DHC): It is a chain which is constituted of all B-DHTs by link of a plant and mark Ω . Two denotation '{' and '}' are added at the head and the end of Ω which represents the start and the end of the plant. Obviously, all of the branches of the plant are bracketed by '['and ']'.

Definition 5(Sequential B-DHC): It is a chain which is constituted of B-DHTs without branch and mark δ . For instance, the Sequential B-DHC of main stem is constituted of all B-DHTs which are inside '{' and '}' but outside '['and ']'.

5.2 Modeling growth by intelligent growth automaton

Ω is initialized according to axiom, and updated according to controlling information from 2DHA continuously, to generate a plant. Meanwhile, IGA gets the information of environment and functional organs, and modifies parameters of 2DHA to make plant growth accustomed to the environment.

δ represents a stem without branch. It have four possibilities: The first is that GC is less than 1, which means there is only one dormant bud; The second is that all GC equal 1, which means that all metamers appearance only this year; The third is all GC are greater 1, which means that all metamers appearance only before this year and restrained; The fourth is that the metamers appears this year and before.

Main Algorithm: modeling growth by IGA.

Input: Axiom Q_A ,Growth Age *age* and 2DHA

Output: Ω representing the information of structure and physiology of plant

Ⅰ. Initialize Ω according to Q_A and compute the number of node cycle in a growth cycle, defined as T.

Ⅱ. The growth cycle loop from 1 to age. In a growth cycle, the node cycle loop from 1 to T. In a node cycle, do these operations:

(1). Establish a stack to partition Ω into a lot of δ ;

(2). Deal with every δ ;
(3). If the current Ω is processed successfully, plot Ω . Compute the absolute position and orientation, size, height and other geometry information of every organ;
(4). Modify parameters of 2DHA according to IPE.

The above main algorithm depicts the process of growth modeling by IGA. The following child algorithm ⅰ, ⅱ, ⅲ are the detailed description of operation (1), (2), (4)in the main algorithm.

Child Algorithm ⅰ **:** the description of (1)--partition Ω into one by one δ .

Input: Ω
Output: δ (initial δ is null)

①. Access every character from Ω , mark *ch* from left to right;
②. If *ch* is B-DHT, or '{','[', push *ch* and go to ①; otherwise, pop to *y*;
③. If *y* is B-DHT, *y* is added at the head of δ ; if *y* is '[', an entire δ is acquired, go to main algorithm(2). When return from (2), δ initialize null and access the character after *ch* from Ω , go to ②; if *y* is '{', Ω is processed successfully and go to main algorithm (3).

Child Algorithm ⅱ **:** the description of (2)--deal with δ .

Input: δ
Output: modified B-DHT

①. If the eldest metamer(the head) in δ is dormant, go to ②; if it appears this year, go to ③; if it appears before this year, go to ④;
②. The *NC* of this metamer plus 1, if *GC* is 0, wakens it. Return;
③. If the entire metamer sequence of this Growth Unit have already formed and this Growth Cycle is over, all *GC* of this Growth Unit plus 1, if Growth Cycle isn't over, the *NC* of all metamers of this Growth Unit plus 1. If the metamer sequence isn't completed, *n* and *NC* plus 1, *F* updated to 0, select a new metamer according to the transfer matrix of the Child Automaton, add to Ω , return;
④. If the youngest metamer (the end) in δ appears this year, partition δ into δ_1 (metamer sequence this year) and δ_2 (metamer sequence before this year). Process δ_1 with③ and process δ_2 with⑤. When δ_1 and δ_2 are processed successfully, return. If the youngest metamer appears before this year, go to⑤, after ⑤, return;
⑤. If the Growth Cycle is over, all *GC* of this Growth Unit plus 1, if the eldest metamer is restrained, return; if it isn't restrained, select a new metamer according to Parent Automaton and Child Automaton, add to Ω , return. If Growth Cycle isn't over, return.

Meanwhile, IGA modifies parameters of 2DHA by interacting with IPE, to make plant growth accustomed to environment dynamically. The function f_e depicts IPE, which is defined as $f_e = F(\vec{e}, \vec{y})$, where $\vec{e}$ denotes all kinds of environment factor, such as illumination, water, accumulated temperature and mineral. $\vec{y}$ denotes all kinds of factor of organ function, such as age, height, branch, photosynthesis outcome and so on. The following child algorithm iii describes IPE impact on plant growth and development.

Child Algorithm iii: the description of (4)--modifies parameters of 2DHA.

Input: f_e

Output: 2DHA modified parameters

①. If current Ω is processed successfully and Growth Cycle isn't over, modifies transfer matrix of Child Automaton according to function f_e. F can be updated to 2 to make a metamer restrained, and the dormant period of bud can be modified;

②. If Growth Cycle is over, modifies transfer matrix of Parent Automaton according to function f_e. And modifies π_c, L_c and transfer matrix of Child Automaton of the next parent state.

6. CONCLUSION AND DISCUSSION

The growth and development of plant are controlled essentially by genes, and plant must grow in feasible environment. Plant absorbs mineral nutrition and energy from environment, and then grows and develops by material metabolism and energy metabolism. Therefore, genes and environment are the driving force of the growth and development of plant.

In the framework, 2DHA simulates the function of genetic information to control the branch pattern of plant and drives the growth and development of plant. IPE controls and coordinates the interaction between IAVOs which constitute the virtual plant, distribution and balance of carbohydrate inside plant, and the interaction between virtual plant and virtual environment, to drive the growth and development of plant, and display 3D graphics of plant. Therefore, the IGA employs genes and environment to control the growth and development of plant, and is capable of demonstrate the drive of genes and environment to the growth and development of plant vividly.

On the basis of IGA, a software tool which can plot the topology of plant has been developed to validate this model. The different growth in the normal and abnormal environment is simulated successfully through the software. Normal environment is a feasible environment for plant growth. While abnormal environment is harmful to plant living and growth, such as drought, low temperature, insect pest, and so on. In the experiment, abnormal environment is drought and low temperature. Fig. 3 shows the topology of four continuous plant growth phases in normal environment, while Fig. 4 shows the topology of plant growing in abnormal environment.

Experimental results show that the potential pattern of plant is controlled by genes, while the specific character of plant is affected by environment. In normal environment, the potential branch and organ will grow and develop, while abnormal environment make some buds dormant, restrained, dead or transform to other kind of tissue, Fig.3 and Fig.4 can illustrate this effect.

The genetic information can't be acquired directly by 2DHA. At present, complex statistical and clustering method are employed. At first, the plant is sampled and statistically computed, and then the genetic information is clustered and recognized. Therefore, new method must be employed, for example, the gene control network which is on the basis of plant physiology can be employed. In addition, it is worthy to do thorough research on the functional organ model and environment model, and the latest achievements in plant physiology should be employed to control the development of plant more vividly and effectively.

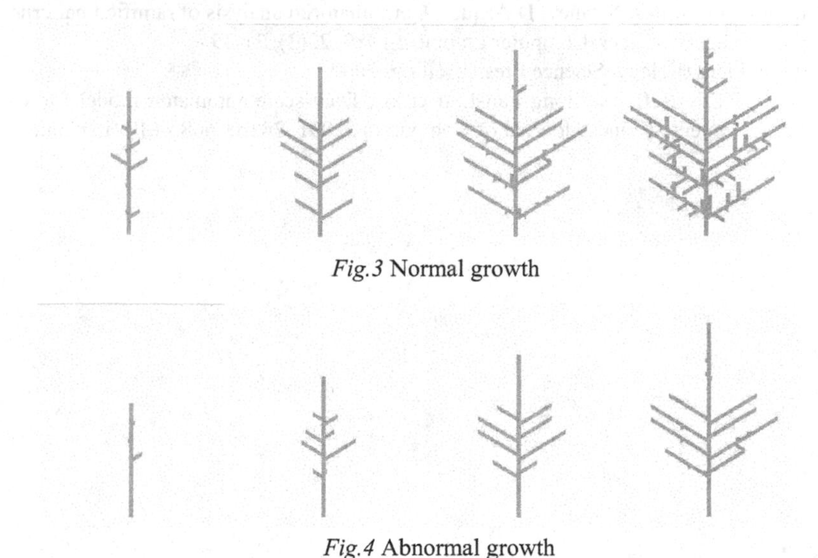

Fig.3 Normal growth

Fig.4 Abnormal growth

ACKNOWLEDGEMENTS

The authors are grateful to the National Natural Science Foundation of China (60773082) and the 863 program of China (2006AA10Z233). We are also grateful to Chongqing Academy of Agricultural Science of China who provided the plant physiological, morphological data for training and experimentations.

REFERENCES

F Blaise, J F Barczi, M Jaeger, P Dinouard, P De Reffye. Simulation of the growth of plants modeling of metamorphosis and spatial interactions in the architecture and development of plants. Cyberworlds, Tokyo: Springer-Verlag, 1998, 81-109

Hongchun Qu ,Qingsheng Zhu ,Qingqing Deng et al .Modelling and Consructing of Intelligent Physiological Engine Merging Artificial Life for Virtual Plants. Journal of Computational and Theoretical Nanoscience, 2007 (4) pp.1405-1411

Kyle W. Tomlinson et at. A functional-structural model for growth of clonal bunchgrasses. ecological modelling, 2007 (202) 243-264.

Li Hesheng. Modern Plant Physiology. Higher Education Press, Beijing, 2002

P Prusinkiewice, A Lindenmayer. The algorithmic beauty of plants. New York: Springer-Verlag, 1990

S Dekmo, L Hodges, B. C Naylor onstruction of fractal objects with iterated fuction systems. Computer Graphics, 1985, 19(3): 271-278

W Reeves. Approximate and probabilistic algorithms for shading and rendering structured Particle Systems. Computer graphics, 1985, 19(3): 313-322

X G Viennot, G Eyrolles, N Janey, D Arques. Combinatorial analysis of ramified patterns and computer imagery of trees. Computer graphics, 1989, 23(3): 31-39

Yang Shijie. Plant Biology. Science Press, Beijing, 2000.

Zhao Xing , Ph de Reffye , Xiong Fan-Lun et al . Dual-scale automaton model for virtual plant development . Chinese Journal of Computers , 2001 ,24 (6) :608 - 617 (in Chinese)

A VISUALIZATION MODEL OF FLOWER BASED ON DEFORMATION

Ling Lu[1,*], Lei Wang[1], Xuedong Yang[2]

[1] *School of Information Engineering, East China Institute of Technology, JiangXi Province, P. R. China 344000*

[2] *Department of Computer Science , University of Regina , Regina, Saskatchewan Canada S4S 0A2*

[*] *Corresponding author, Address: School of Information Engineering, East China Institute of Technology, FuZhou344000, JiangXi Province, P. R. China, Tel: +86-794-8258390, Fax: +86-794-8258390, Email: luling@ecit.edu.cn*

Abstract: We present a simple and effective modeling method for flowers. It starts with an initial geometric shape, such as ellipsoid, cylinder, or plane surface et al., and then simulates flower components (such as pedicel, receptacle, pistils, stamens, petals and sepals) by addition deformation to the basic geometric shape. The detailed geometry of flower component is defined by basic equation for the basic shape along with a deformation function. A variety of flower can be produced by varying the deformation parameters. A number of examples are given in the paper to demonstrate the effectiveness of the proposed model.

Keywords: plant flower, basic geometric shape, deformation

1. INTRODUCTION

Flowers pose an interesting and important challenge for three-dimensional computer graphics modeling. They have a great number of components, such as stems, torus, pistils, stamens, petals. They also have intricate structure and unique free-form. The L-System was first formulated by Lindenmayer (1968) and was introduced to the computer graphics community by Prusinkiewicz and Lindenmayer(1990). The L-System defines plant structures using a set of rewriting rules (Prusinkiewicz et al.,1990). Prusinkiewicz et al.

Please use the following format when citing this chapter:

Lu, L., Wang, L. and Yang, X., 2009, in IFIP International Federation for Information Processing, Volume 294, *Computer and Computing Technologies in Agriculture II, Volume 2*, eds. D. Li, Z. Chunjiang, (Boston: Springer), pp. 1487–1495.

(2001) also used the positional information to control parameters along a plant's axis. Prusinkiewicz et al. (1993) introduced dL-System by combining differential equations with the L-System. The dL-System provides a model that simulates plant growth in continuous time. These studies, however, mainly focused on branching structure or skeletal shape, and did not deal with the growth of surface tissues of plants, such as petals or leaves. Recently, some studies published methods dealing with plant surfaces. Runions et al. (2005) presented an algorithm for generating leaf venation patterns based on the canalization hypothesis. The algorithms can simulate many types of venation patterns. The initial leaf shape is specified interactively by the user, as a parametric curve that defines the leaf contour. Wang et al. (2004) considered leaf tissues to be viscous and simulated growth by expansion of incompressible fluids. Lintermann et al (1999) presented a modeling method that allows easy generation of many types of objects that have branching structures, including flowers, bushes, trees, and even some non-botanical objects. Flower's natural leaf was scanned and applied as a texture to the leaf component's surface. Their three-dimensional appearance was not satisfactory because of the use of two-dimensional texture. Recently, Wang et al (2005) presented a framework for real-time rendering of plant leaves with global illumination effects. Their leaf model can be captured from real leaves, which makes it easy to create highly realistic leaf appearance models. But their leaf model did not consider small features such as hairs on leaves. Ijiri et al. (2005) presented a system for modeling flowers in three dimensions quickly and easily while preserving correct botanical structures. This system is an application-customized sketch-based interface. The user needs to input the outline of petal and draws modifying strokes. The petal shapes that can be generated by this system are limited to elliptical outlines. Qin Peiyu et al. (2006) proposed a flower model using the L- system and Bezier surfaces. One of the drawbacks of using Bezier surfaces is the lack of ability to model fine concave or convex texture on petals and toothed borderlines.

We use our deformation technology (Lu Ling et al.2007) to form flower's components, such as stems, torus, pistils, stamens, and petals. A number of examples are given in the paper to demonstrate the effectiveness of the proposed model.

2. GEOMETRY MODELS FOR FLOWER COMPONENTS

Flower is composed of pedicel, receptacle, pistils, stamens, petals and sepals. Pistil consists of stigma, style and ovary. Stamen consists of anther and filament. They is shown in Fig.1(Hu Baozhonf et al.2002)

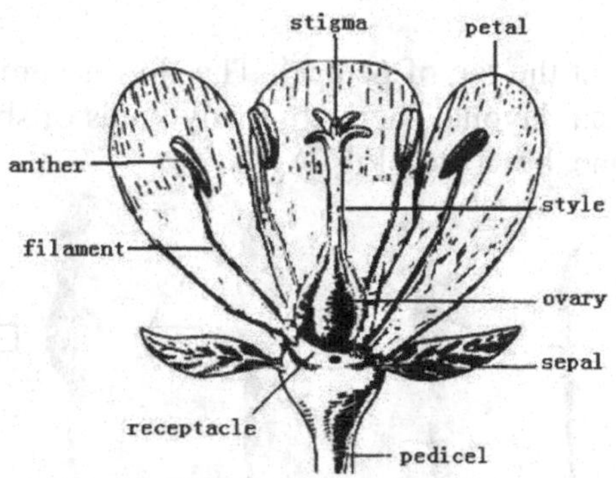

Fig.1: Flower's component.

2.1 Pedicel modeling

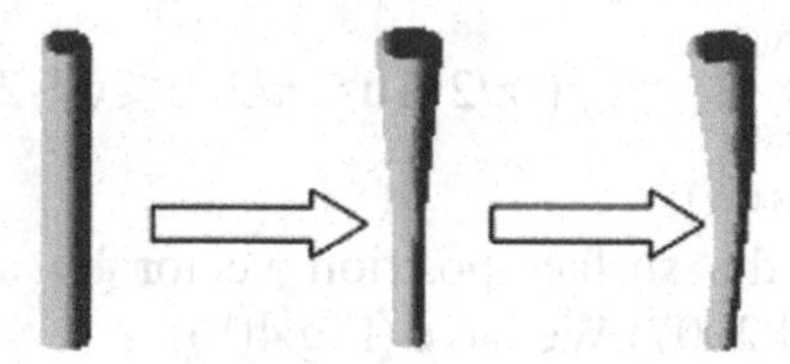

Fig.2: The simulator process of Pedicel

The pedicel is a small spray with flower and its structure is similar to stem. Its length and thickness is different from a plant. Since its shape looks likes cylinder. We may deform a cylinder to simulate pedicel (Fig.2). The cylinder is represented as the follows:

$$\begin{aligned} x(u,v) &= r\cos(u) \\ y(u,v) &= hv \\ z(u,v) &= r\sin(u) \end{aligned} \qquad (0 \leqslant u \leqslant 2\pi,\ 0 \leqslant v \leqslant 1) \qquad (1)$$

where r is radius of the cylinder, h is height of the cylinder. We add deformation function to make cylinder uneven and bend:

$$\begin{aligned} x(u,v) &= r\cos(u) + n_x a_1 \sin(\pi v - 1) + a_2 \sin(\pi v) \\ y(u,v) &= hv + n_y a_1 \sin(\pi v - 1) \\ z(u,v) &= r\sin(u) + n_z a_1 \sin(\pi v - 1) + a_2 \sin(\pi v) \end{aligned} \qquad (0 \leqslant u \leqslant 2\pi,\ 0 \leqslant v \leqslant 1) \qquad (2)$$

where a_1 is the extent of uneven deformation, a_2 is the extent of bend deformation, and (n_x, n_y, n_z) is unit surface normal at the point (x,y,z). In Fig.2, both a_1 and a_2 are equal 3.

2.2 Receptacle modeling

The receptacle is at the top of pedicel. The flower components grow on a receptacle with special layout. There are many kinds of shape for receptacle, such as cylinder, cone, bowl etc. (Fig.3)

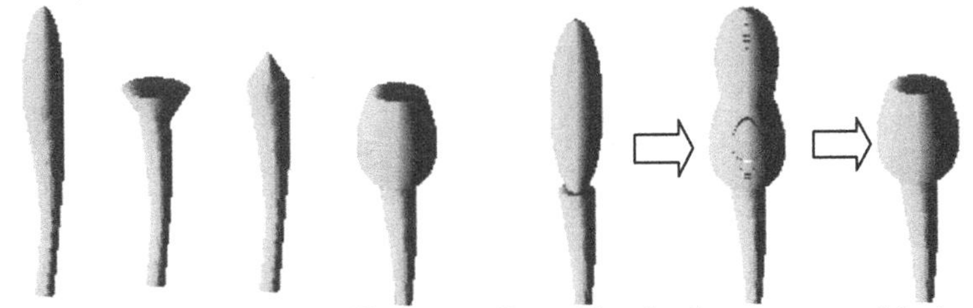

Fig.3: Receptacle shape of different *Fig4:* The simulator process of the bowl shape

For example, the bowl shape comes from ellipsoid deformation (Fig.4). First, the parametric ellipsoid equation is as the follows (Fig.4(a))

$$\begin{aligned} x(u,v) &= r\cos(u)\cos(v) \\ y(u,v) &= 3r\sin(u) \\ z(u,v) &- r\cos(u)\sin(v) \end{aligned} \qquad (-\pi/2 \leqslant u \leqslant \pi/2,\ 0 \leqslant v \leqslant 2\pi) \tag{3}$$

Second, we modify the surface-position vector by adding a deformation function (Lu ling et al.2007).We have (Fig.4(b))

$$\begin{aligned} x(u,v) &= r\cos(u)\cos(v) + n_x r/2\,|\sin(2u)| \\ y(u,v) &= 3r\sin(u) + n_y r/2\,|\sin(2u)| \\ z(u,v) &= r\cos(u)\sin(v) + n_z r/2\,|\sin(2u)| \end{aligned} \qquad (0 \leqslant u \leqslant \pi/2,\ 0 \leqslant v \leqslant 2\pi)$$

$$\begin{aligned} x(u,v) &= r\cos(u)\cos(v) + n_x r/1.4\,|\sin(2u)| \\ y(u,v) &= 3r\sin(u) + n_y r/1.4\,|\sin(2u)| \\ z(u,v) &= r\cos(u)\sin(v) + n_z r/1.4\,|\sin(2u)| \end{aligned} \qquad (-\pi/2 \leqslant u \leqslant 0,\ 0 \leqslant v \leqslant 2\pi) \tag{4}$$

Finally, when $0 \leqslant u \leqslant \pi/2$,we add a function $-r\cos(\pi/2-u)$ to $z(u,v)$(Eq(4)). The upper part of deform ellipsoid is changed to inside of bowl (Fig,4(c)).

2.3 Pistil and stamen modeling

(a) apocarpous gynoecium

(b)compound pistil

(c)compound pistil

Fig.5: Different kind of pistil

Pistil is formed by carpel intervolve. The type of pistil is determined by the number of carpel, and extent of separation. Simple pistil consists of a carpel. Many carpels construct a compound pistil. In some plant, a flower has a lot of carpel such that they are apart and form respective pistil. These pistils are named apocarpous gynoecium (Fig.5(a)).Every pistil is divided into three parts: stigma, style and ovary. We may use ellipsoid to simulate stigma, deformation cylinder to simulate style, and deformation ellipsoid to simulate ovary. In Fig.5(a), an ovary model follows

$$\begin{aligned} x(u,v) &= r\cos(u)\cos(v) \\ y(u,v) &= 2r\sin(u) \\ z(u,v) &= r\cos(u)\sin(v) - r/2\sin(\pi/2+u) \end{aligned} \qquad (-\pi/2 \leqslant u \leqslant \pi/2,\ 0 \leqslant v \leqslant 2\pi) \quad (5)$$

A style equation is

$$\begin{aligned} x(u,v) &= r\cos(u) + n_x\sin(\pi v - 1) \\ y(u,v) &= 10rv + n_y\sin(\pi v - 1) \\ z(u,v) &= r\sin(u) + 3\sin(\pi v) + n_z\sin(\pi v - 1) \end{aligned} \qquad (0 \leqslant u \leqslant 2\pi,\ 0 \leqslant v \leqslant 1) \quad (6)$$

A stigma may be expressed by ellipsoid. So we may combine a stigma, a style and an ovary to form a pistil. For apocarpous gynoecium, we use many several independent pistils with different rotation angles. The compound pistil is constructed by several pistils that are closely staggered (Fig 5(b)).

Every stamen consists of an anther and a filament. The anther, expanded to become a saccate, is at top of filament. We use a deformed ellipsoid to simulate anther. Since the shape of filament is long and thin, it may be made by deformation cylinder. Depending on the separation between the stamens and the different length of stamens, a stamen is divided into several types: monadelphous stamen, diadelphous stamens , polydelphous stamens,tetradynamous stamen , didynamous stamen and synantherous stamen. The methods for their simulation are similar to pistil.

2.4 Petal and sepal modeling

The shape of petal and sepal are similar. Therefore, we use petal as an example to describe the method. Because petal is a curve surface, we select the initial shape of a petal is a rectangle surface. We add deformation function ($\triangle x$ and $\triangle y$) to form free curve outline and modify the surface-position vector by adding a deformation function ($\triangle g$) . We have (Lu Ling et al., 2008)

$$\begin{aligned} x(u,v) &= a_x u + \Delta x + n_x \Delta g \\ y(u,v) &= b_y v + \Delta y + n_y \Delta g \\ z(u,v) &= c_z + n_z \Delta g \end{aligned} \qquad (7)$$

where a_x is the length with respect to the x direction, b_y is the length along the y direction, and c_z is the section distance.

For example, the deformation function of peony petal is as follows

$$\begin{aligned} \Delta x &= 0.8 a_x u \sin(\pi v / 2 - c) \\ \Delta y &= (v - 0.5)(1.5 a_x \sin(\pi (u + 0.5)) + |3v \sin(\pi a_x (u + 0.5) / 6) \\ \Delta g &= 0.3(1 - 2|v - 0.5|) \sin(20(u + 0.5)\pi)(1 - 2|u|) + 3(1 - v)^2 e^{-800u^2} + \\ &\quad dy \sin(\pi y / 40 - 1) + dx \sin(\pi x / 2 - 0.5) + 2v^3 \sin(\pi x / 8) \end{aligned} \tag{8}$$

The process from rectangle surface to a peony petal is showed in Fig.6.

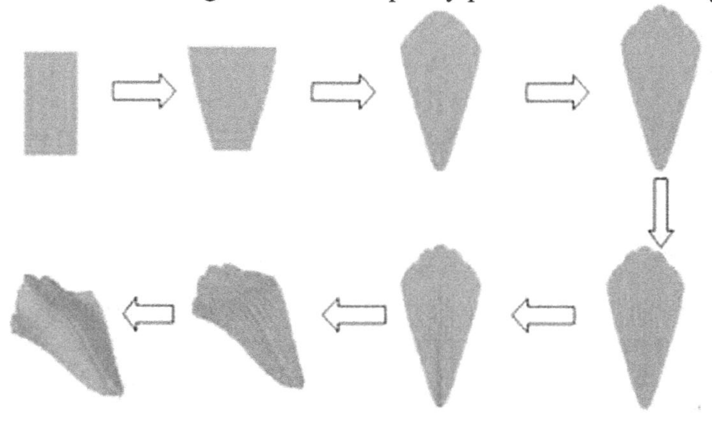

Fig.6: The process from rectangle surface to peony petal

3. FLOWER MODELING

3.1 Flower color

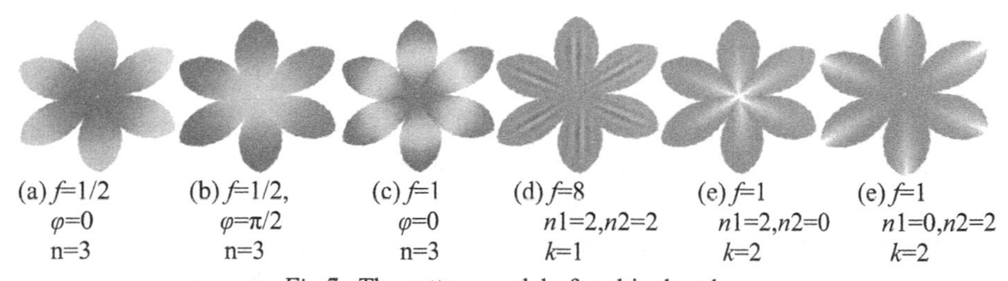

(a) f=1/2, φ=0, n=3 (b) f=1/2, φ=π/2, n=3 (c) f=1, φ=0, n=3 (d) f=8, $n1$=2, $n2$=2, k=1 (e) f=1, $n1$=2, $n2$=0, k=2 (e) f=1, $n1$=0, $n2$=2, k=2

Fig.7: The pattern model of multicolored

The color of flower is divided into two types (Zhao et al., 2005): one is single color, another is multicolored. Generally, a pedicel and a receptacle are single color. The composition of pistil and stamen, such as stigma, style and ovary, are also single color. But a petal is multicolored and its color model is more complex. In common, the pattern model of multicolored include flower centre, flower side, flower ring and flower rib etc (Zhou et al., 2007) (Fig 7). Maybe a kind of flower has many pattern models. For simplification, we use basic function to simulate a petal multicolored. For example, let the intensity vary from I_1 to I_2 ($I_1 < I_2$) with respect to v as the function of

$$I = I_1 + I_2 \sin^n(v\pi f + \varphi) \tag{9}$$

where n controls the change rate from I_1 to I_2. Fig 7(a-c) shows the result of different value in Eq.(9). For flower rib, we may create a function as follows

$$I = I_1 + I_2 \sin(f(u+0.5)\pi)(1-v)^{n1} v^{n2} (1-2|u|)^k \tag{10}$$

where f controls the number of rib,$n1$,$n2$ and k control the border rib of a petal. Fig 7(d-f) shows the result of different value in Eq.(10).

3.2 Synthesis flower components

After flower components have been modeled, we may build an entire flower model. Based on the characteristics of a flower, we may combine flower component (such as pedicel, receptacle, pistils, stamens, petals and sepals) accordingly. The number of individual component is different from one flower to another. For example, a lily has a pistil, six stamens and six petals. A peony has many stamens and many petals. Every petal is also not same. The inside petals are bigger and more bender than outside petals. The size and bent of a petal can be controlled easily by our deformation definition. Fig.8 show that we can use different parameter in Eq.(7) and Eq.(8) to construct different petal shape.

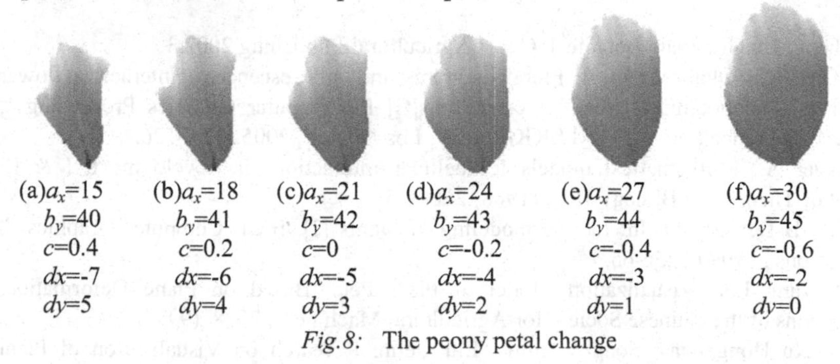

Fig.8: The peony petal change

We place geometric objects on the receptacle model. For peony petal, we may use a recursion iteration or L-system to simulate many stamens and petals (Fig.8).The position of stamens and petals are determined by random number.

set of parent states and a parent state represents a kind of Growth Unit. π_p is the initial probability vector and used to acquire the axiom. S_{GC} is the time signal to drive parent automaton and indicates time period requirement for growing one Growth Unit. ρ_p denotes the transfer matrix of all parent states. F_p denotes the set of terminal conditions, such as limit of plant age or a terminal parent state. It should be noted that a parent state corresponds to a child automaton.

Definition 2(Child automaton): a child automaton is a sequential automaton, it's defined as: $A_c ::= < Q_c, \pi_c, S_{NC}, B_c, L_c, \rho_c, F_c >$ .where Q_c denotes the set of child states and a child state represents a kind of Metamer. π_c denotes the initial probability of every child state when the child automaton begin to run. S_{NC} is the time signal to drive child automaton and indicates time period requirement for growing one Metamer. B_c denotes the set of axillary bud of every child state. L_c denotes the cycle number of the child automaton continuously locating in a child state. ρ_c denotes the transfer matrix of all child states. F_c denotes the set of terminal conditions, such as limit of Growth Cycle or a terminal child state.

When the child automaton of a parent state is established, Monte Carlo method is employed to make the sequence of internodes which is produced by child automaton to satisfy the statistical laws of parent state.

5. INTELLIGENT GROWTH AUTOMATON

The main data structure of IGA is bi-dimensional hash table (B-DHT) and bi-dimensional hash chain (B-DHC). The structure information of plant and physiology information of every organ are stored in B-DHT. IGA produces the plant and interacts with IPE by access and writing B-DHC.

5.1 Definition of bi-dimensional hash table and bi-dimensional hash chain

Definition 1(Phy-Hash Table): It is a hash table which stores the physiology information of metamer. At present, there are four parameters -- *GC*, *NC*, *n* and *F*. *GC* denotes the number of Growth Cycle from appearance to now. *NC* denotes the number of Node Cycle from appearance to now. *n* is the ordinal of child automaton in a cycle. The initial value of *GC*, *NC*, *n* of

Fig.9: Peony model

4. CONCLUSION

In this paper, we presented a modeling method for flowers using deformation. This method start with basic geometric shape, such as ellipsoid, cylinder and rectangle surface, and add deformation function, making a basic geometric shape to approach a desired flower component shape. The geometric model of a flower can be controlled through a small set of parameters. Because our models are simple mathematical expressions controlled by a few parameters, these parameters may be changed in a continuous manner, such that both the shape and color of the flower can vary continuously. A growth model based the current deformation technique will be our future research in the next step.

REFERENCES

Hu Paozhong, Hu Guoxuan. Botanical. China Agricultural Publishing 2002,3

Ijiri T Owada S Okabe M et al. Floral diagrams and inflorescences : interactive flower modeling using botanical structural constraints[C] PPComputer Graphics Proceedings , Annual Conference Series , ACM SIGGRAPH , Los Angles , 2005 :720 -726.

Lindenmayer A.: Mathematical models for cellular interactions in development, I & II. Journal of Theoretical Biology 18, 3 (1968),280-315.

Lintermann B Deussen O. Interactive modeling of plants [J]. IEEE Computer Graphics & Applications , 1999 (1):56-65.

Lu Ling Wang Lei. Visualization Model of Plant Petal Based on Plane Deformation. Transactions of the Chinese Society for Agricultural Machinery. 2008. (9)

Lu Ling, Xu Hongzhen, Song Wenlin, Liu Gelin. Research on Visualization of Plant Fruits Based on Deformation. New Zealand Journal of Agricultural Research.2007.11

Prusinkiewicz P., Hammel, M., Mjolsness E.: Animation of plant development. In Proc. IGGRAPH '93 , 351-360.

Prusinkiewicz P., Lindenmayer A.: The Algorithmic Beauty of Plants. Springer–Verlag, New York, 1990. With J. S. Hanan, F. D. Fracchia,D. R. Fowler, M. J. M. de Boer, and L. Mercer.

Prusinkiewicz P., Mündermann L., Karwowski R., Lane B.: The use of positional information in the modeling of plants. In Proc. ACM SIGGRAPH '01 (2001), 289-300.

Qin Peiyu, Chen Chuanbo, Lv Zehua. Simulation Model of Flower Using the Interaction of L-systems with Bezier Surfaces[J].Computer Engineering and application. 2006.16,pp:6-8.

Runions A., Fuhrer M., Lane B., Federl P., Rolland-Lagan A.-G.,and Prusinkiewicz P.: Modeling and visualization of leaf venation patterns. ACM Trans. Graph., 24, 3 (2005), 702-711.

Wang I., Wan J., Baranoski, G.: Physically-based simulation of plant leaf growth. Computer Animation and Virtual Worlds, 15, 3-4 (2004), 237-244.

Wang L Wang W Dorsey J et al. Real-time rendering of plant leaves [C] PPComputer Graphics Proceedings , Annual Conference Series , ACM SIGGRAPH , Los Angles , 2005:712-719.

Zhao Changling , Guo Weiming , Chen Junyu. Formation and regulation of flower color in higher plants [J].Chinese Bulletin of Botany , 2005 , 22 (1) : 70 - 81

Zhou Ning,Dong Weiming,Wang Jiaxin, Simulation of Flower Color Pattern[J], Journal of Computer-aided Design & Computer Graphics. 2007, 19(6):708-712

THREE-DIMENSIONAL COMPUTER AIDED DESIGN OF A VERTICAL WINNOWER

Yumei Bao*, Saijia Lin, Lijie Weng
The MOE Key Laboratory of Mechanical Manufacture and Automation, Zhejiang University of Technology, Hangzhou, Zhejiang Province, P.R.China 310032
** Corresponding author, Address: College of Mechanical & Electrical Engineering, Zhejiang University of Technology, Hangzhou310032,Zhejiang Province, P. R. China, Tel: +86-571-88320244, Fax: +86-571-88320837, Email: baoym@zjut.edu.cn*

Abstract: The research states home and abroad of the winnowing technology and winnowers are reviewed in brief. For the air duct, the core component of the winnower, the relevant technical parameters in the winnowing process are calculated based on the winnowing principle. The three-dimensional computer aided design (3D-CAD) software Solidworks is applied. The designed vertical winnower is able to separate different raw materials by adjusting the air speed and has been put into practical production to separate the Chinese traditional medicine with high separating effect.

Key words: Winnowing technology; Vertical winnower; 3D-CAD

1. INTRODUCTION

The winnowing technology has been used in agriculture for a long time as winnowing method, to separate chaff and broken rice stalk from the rice after harvest and threshing. Along with development of the industries such as mining and metallurgy industry, many kinds of winnowers have sprung up and were used in the field of mining, agriculture, pharmaceutical industry and urban construction, to separate such raw materials as Chinese traditional medicine, soybean and hops. Owing to the higher demands of separation, the winnowing technology and the structure design of the winnowers receive

---

Please use the following format when citing this chapter:

Bao, Y., Lin, S. and Weng, L., 2009, in IFIP International Federation for Information Processing, Volume 294, *Computer and Computing Technologies in Agriculture II, Volume 2*, eds. D. Li, Z. Chunjiang, (Boston: Springer), pp. 1497–1503.

more and more attention (Ma Jiguang,2001; Wu Jianzhang et al, 2002; Shapiro et al, 2005).

In the developed countries such as Europe and American , the production of the gravitational winnower is 1～15t/h .The shape of the table board may be triangle, rectangle and mixed shape .The air may be negative pressure or positive pressure. Mary series of gravitational winnowers have been developed featuring excellent technique, stable performance, low noise and high reliability. Among these manufacturing factory, WESTRUP in Denmark, HEID in Austria, PETKUS in Germany, OLIVER, LMC and CRIPPEN in American are the more famous. But in China the users depend mostly on pure import or reforming the imported machine (Hu Zhichao,2002; Liu Xiang, 2004).

A vertical winnower is designed based on three-dimensional computer aided design, the shape , characteristic , processing condition and material quality of winnower will display in computer with real size and shape, convenient for the data exchange and share. The three-dimensional virtual product model can be operated in several ways that cannot realize for the entity model, such as slitting of the parts or assemblies, assembling and moving simulation in dimension interference. In the earlier period of designing , all the parts, tool equipments and assemblies are digital pre-assembled to reduce the design alteration, mistakes and reworks obviously (Tang Rongxi, 2005; Li Yan, 2007). The winnower was designed and developed successfully and applied in winnowing of Chinese traditional medicine with a good effect.

2. THE PRINCIPLE OF WINNOWING

Different materials have different movement under the action of vertical and horizontal (or gradient) air flow due to different aerodynamical properties, and may be separated accordingly. The aerodynamical properties include suspending velocity and flight coefficient, and so on. The suspending velocity is the ratio of the horizontal pressure and the gravity, while the flight coefficient is used to reflect the ability of the horizontal air flow to take away the materials.

The material with small suspending velocity can be blown farther under the horizontal air flow, and vice versa. Under the action of the vertical or gradient air flow, if the air flow velocity is larger than the suspending velocity, the material will be carried away. But if the air flow velocity is smaller than the suspending velocity, the material will go down. Therefore the heavy material is separated from the light material (Guo RenNing et al, 2005).

When material is in the horizontal or gradient air-flow, considering the action of the gravity G and air-flow pressure P, it will move along the direction of resultant force T and the trail is a parabola, as shown in Fig.1 (Li Shijiu, 1992).

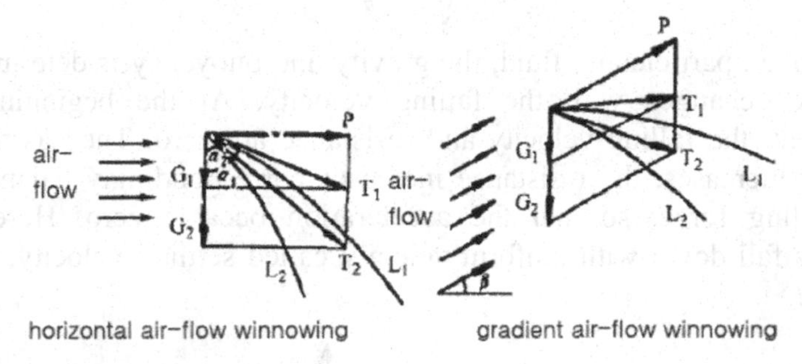

Fig.1: The movement of material in the horizontal and gradient air-flow

In horizontal or gradient air-flow, light materials is blown farther than the heavy one, i.e. the bigger the angle between gravity G and resultant force T (the direction of T is close to the horizontal), the materials is blown farther, and Vice versa. Included angle α depends on following equation,

$tg\alpha = \frac{P}{G}$, where $tg\alpha$ is the flight coefficient.

Different materials have different flight coefficient under the same horizontal or gradient air-flow. The larger gravity results to the smaller $tg\alpha$ so that the material will be blown closely. Whereas the smaller gravity leads to the larger $tg\alpha$ so that the will be blown far. So if only the suitable air flow velocity is chosen, the heavy and light materials can be separated accordingly.

In designing the vertical winnower, the gradient (about 30^0) air flow is used to winnow and remove impurities, since it owns farther flying distance thereby has better separating effect.

3. SETTLEMENT COMPUTATION AND AIR DUCT STRUCTURAL DESIGN

3.1 Settlement computation

If the flow rate in the air duct equals to the settling velocity, particles in the air will be taken away, this velocity is called carryover velocity. It is apparent that the carryover velocity is equal to settling velocity for single particle (Li Shijiu, 1992).

The carryover velocity is: $U = \sqrt{\frac{4gd_p(\rho_s - \rho_f)}{3C_p\rho_f}}$.

For given particle and fluid, the gravity and buoyancy is determinate, the resistance changes with the falling velocity. At the beginning of the settlement, the falling velocity and resistance are zero. Then as the falling velocity increases, the resistance increases correspondingly balancing with the settling force, so that the acceleration became zero. Hereafter the particles fall down with uniform velocity called settling velocity, as shown in Fig. 2.

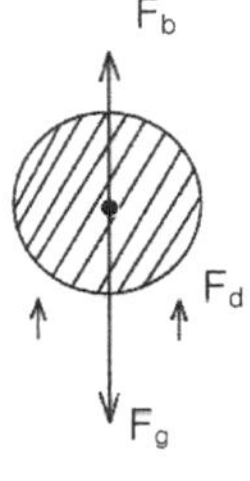

Fig.2: Forces balanced on a spherical particle

Considering a spherical particle with diameter d, the densityρ_s, settling down in the air with density ρ , there are gravity F_g, buoyancy F_b and resistance F_d along the settling direction (vertical):

$$F_g = \frac{\pi}{6} d^3 \rho_s g \quad F_b = \frac{\pi}{6} d^3 \rho g \quad F_d = \xi A \frac{\rho u^2}{2} = \frac{\xi \pi d^2 \rho u^2}{8}$$

Because the particle is in a balance, $F_g - F_b - F_d = 0$

The flow can be divided into three regions: laminar flow, transition flow and turbulent flow, depending on the Reynolds number, as shown in Table 1,

$$R_{et} = \frac{du_t\rho}{\mu}$$

Table 1 Computation of gravitational settling velocity

Flow regions	Reynolds number	Gravitational settling velocity
laminar flow	Re<2	$u_t = \frac{d^2(\rho s - \rho)g}{18\mu}$
transition flow	2<Re<500	$u_t = \sqrt[3]{\frac{4g^2(\rho_s - \rho)^2}{225\mu\rho}} d_p$
turbulent flow	500<Re<200000	$u_t = 1.74\sqrt{\frac{d(\rho s - \rho)g}{\rho}}$

Supposed the average density of Chinese traditional medicine is about 300 kg/m^{3}and its equivalent average diameter is about 3mm. Considering at the room temperature and atmospheric pressure, the density of the air 1.2 kg/m^3, the reasonable air flow velocity can be computed to be 4.71m/s by trial-and-error method.

3.2 Structural design of the air duct

The core structure of a vertical winnower is the air duct, the duct design will influence the efficiency and precision of winnowing .According to the air flow velocity, the corresponding dimensions can be computed, and then structure of the air duct is designed with 3D software- solidworks.

The vertical winnower has a vertical section and a horizontal winnowing trunk. The raw materials enter into the vertical duct and then the winnowing trunk with the blown air, as shown in Fig.3. The air is blown into the winnower at inlet 1. The raw materials are feed in 2 and then move toward the winnowing trunk. In this process, the heavy impurities such as small iron block and stone fall down from outlet 5. While the rest move into the winnowing trunk together with the air flow. The air flow rate and correspondingly the pressure must be adjustable to let most required materials come out from the outlet 4. The light materials (like hair, cotton yarn, paper scraps) will come out of 3. To ensure the airproof and good circulation, the flanges are welded to the air-inlet1 and outlet 3.

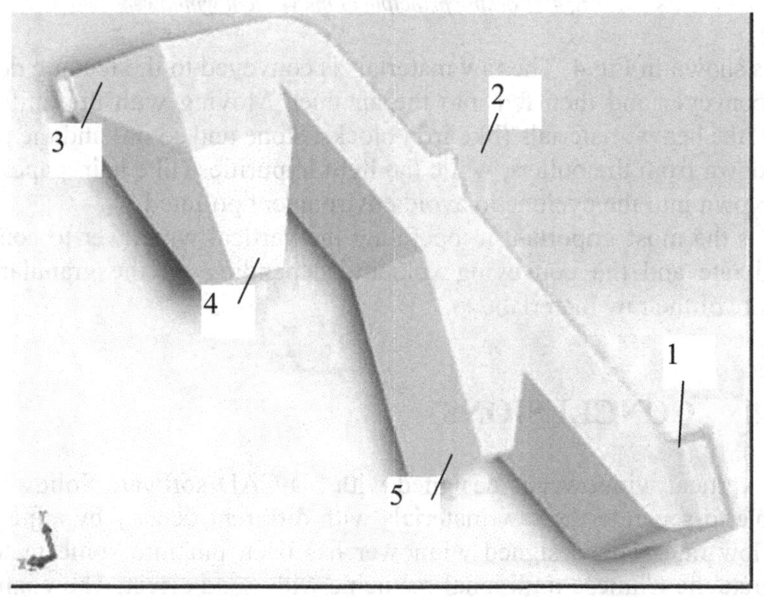

Fig.3: Air duct of the vertical winnower

1. air-inlet 2. raw materials inlet 3.air outlet 4.product outlet 5.heavy impurities outlet

Considering the designed vertical winnower is mainly applied to separate medicine, seed etc, therefore all the materials used are stainless steel 1Cr18Ni9.

3.3 Overall structure of winnower

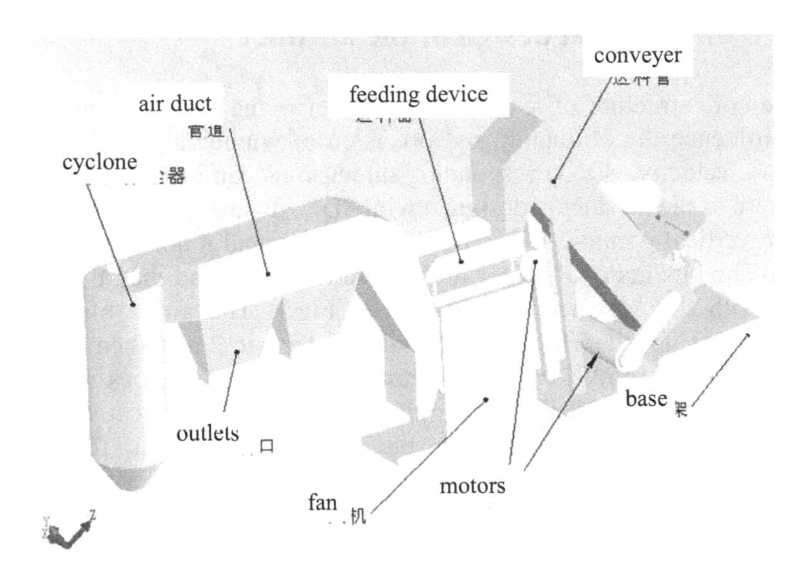

Fig.4: Working principle of the vertical winnower

As shown in Fig.4, The raw materials is conveyed to the feeding device by the conveyer and then fed into the air duct. Moving with the uniform air-flow, the heavy materials (like iron blocks, stone and so on) and the products fall down from the outlets, while the light impurities (like hair, paper scraps) are blown into the cyclone to avoid environment polluted.

It is the most important in operating the vertical winnower to control the wind rate and the conveying velocity, depending on the granularity and density of the raw materials.

4. CONCLUSIONS

A vertical winnower is designed with 3D-CAD software Solidworks and is able to separate the raw materials with different density by adjusting the air flow rate. The designed winnower has been put into some factories to separate the Chinese traditional medicine with good effect. The winnower is easy to maintenance, stable in operation and has healthful process.

The designed vertical winnower can also be applied to winnow other raw materials such as seed and soybean with a little modification and adjustment.

ACKNOWLEDGEMENTS

This research was supported by Hangzhou Chunjiang Institute of Pharmacy machinery.

REFERENCES

Guo RenNing, Gao Chunyu, Ji Junhong. Numerical Simulation and Analysis to Inner Flow Field of Horizontal Air Flow Winnower. Mining & Processing Equipment, 2005, 33(10): (57-59)

Hu Zhichao, Gu Renhong. On the nationalization of the winnowing equipment for the seeds. China Agriculture Mechanization, 2002,(3).

Li Shijiu, Zhou Xiaojun. pneumatic conveying theory and application. Beijing：China machine Press，1992.

Li Yan. Three-Dimensional Structure Design for Ships by CADDS. Chinese Journal of Ship Research, 2007, 2 (1): 14-18

Liu Xiang. Structure and working principle of air fiber sifter. China forest products industry. 2004,31(6):30-31,45

M. Shapiro, V. Galperin. Air classification of solid particles: a review. Chemical Engineering and Processing 44 (2005): 279–285

Ma Jiguang. Development trend of the gravitational winnowers abroad. World agriculture, 2001,(7): 32-33.

Tang Rongxi. Popularization and rapid rise of 3D SolidWorks. Manufacture information engineering of China, 2005, 4:54

Wu Jianzhang, Zhu Yongyi. Study on Cereal Winnowing by Gas - Solid Fluidization. Cereal & feed industry, 2002, 6: 11-13

METHOD ON VIRTUAL CROP SYSTEM MODELING BASED ON AGENT TECHNOLOGY

Ping Zheng, Zhongbin Su*, Jicheng Zhang, Yujia Zhang
Engineering Academy, Northeast Agricultural University, Harbin, Heilongjiang Province, P. R. China 150030
* *Corresponding author, Address: Engineering Academy, Northeast Agricultural University, Harbin, Heilongjiang Province, P. R. China 150030. Tel:+86-0451-55190170, Fax:+86-0451-55190170, Email: suzb001@163.com*

Abstract: On the basis of adequate understanding the significance of virtual crops research, according to the research state and characteristics of virtual crops' typical modeling methods at home and abroad, a new modeling thoughtway of virtual crops system was put forward based on Agent technology. This method established crops self-learning system based on neural network. It is helpful to understand crop's structural and physiological rule, and to simulate the interaction influnce both between crop and crop and between crop and environment effectively. This research provided a new test platform for Agent technology's verification and perfection which is of great significance academically as well as practically.

Keywords: Agent technology; neural network; virtual plant; interaction influence

1. INTRODUCTION

Virtual crops belongs to application basic research, involving agriculture, mathematics, computer graphics, and other related fields. It is of great significance to extract the rules of crops structural and physical information, high-yield plant type design, optimization of field management which is a hot issue in the forefront of agricultural research.

At present, virtual crops research methods include L-system, AMAP, particle systems and three-dimensional reconstruction system, with their own study objects and applictions. All these methods can be summed the

---

Please use the following format when citing this chapter:

Zheng, P., Su, Z., Zhang, J. and Zhang, Y., 2009, in IFIP International Federation for Information Processing, Volume 294, *Computer and Computing Technologies in Agriculture II, Volume 2*, eds. D. Li, Z. Chunjiang, (Boston: Springer), pp. 1505–1510.

law from the morphological characteristics, physical characteristics ,then crop modeling, and dynamic displaying its modeling. But under the existing method of modeling, it can show crop growth process in a specific environment , not the dynamic changes along with environment changes independently. Modeling with these mehods could not simulate the growth of groups crops taking into account less physical features, and plants can not simulate the interactive simulation both between plants, and plants and the environment.

So, according to the research state and characteristics of virtual crops' typical modeling methods at home and abroad, this paper puts forward a new modeling thoughtway of virtual crops system based on Agent technology. This method established crops self-learning system based on neural network. It is helpful to understand crop's structural and physiological rule, and to simulate the interaction influnce both between crop and crop and between crop and environment effectively. This research provided a new test platform for Agent technology's verification and perfection which is of great significance academically as well as practically.

2. THE ANALYSIS OF BIOLOGICAL SIMULATING RESEARCH BASED ON AGENT

Resently, agent and multi-agent systems research in the field of artificial intelligence is a very active issue, and has made great progress.

1987, C.W.Reynolds first proposed the animation of bird populations. He put forward a cluster of computing model, and each bird is an animated role to simulate the dynamic interaction between it and other birds and the environment. Then he put forward restriction simulation on the basis of his group bird simulation before.

MIT multimedia Laboratory designed and developed a virtual dog Silas T. Dog which is a typical representative in the virtual reality filed. Silas's T. Dog lived in an artificial enviornment ALIVE (Artificial Life Interactive Video Environment) system. The system uses hierarchical structure of organizational behavior. Users can directly interact with the virtual dog. The dog's behaviors including acts of self-interest, goal-driven behaviour and so on get more vivid along with the continuous interaction with users

In the study of artificial fish, the most influential one is the Xiaoyuan's Fish. Xiaoyuan's Fish is a computer simulation technology based on biophysical and intelligent behavior model. It also simulates a movable artificial fish social in the virtual ocean. In the society, each Xiaoyuan's Fish has a body driven by variable internal muscle ,fin and visual perceptible eye. It not only has vivid appearance image like natural fish, but has mental status、habit、perception ability to environment and life characteristic like intention、reaction、movement control、purposeful behavior and so on. All those actions are driven by the environmental perception and inner desire, not by key frames and scripts. The fish uses intention generation to produce action programme on the basis of judgement about the environment. Operators need only set the kind and their initial conditions when initialized, and the fish will move independently on their own intention.

Through the study of simulation artificial animals, it is helpful to divide crops growing system and the establish self-study mechanisms for crops model. But there is essential difference between crops and animals, the former is annual plants generally with short growth cycle, widely change range on morphologically and physiologically in one year. Moreover, complex changes in external factors will bring important impact on the crops morphology and physiology. Many factors also control the formation of crop yields, and lead to many key technologic issues exist in virtual crops simulation to be resolved. The study will play virtual crops advantage to support agricultural production.

3. ESTABLISHMENT VIRTUAL CROPS ARCHITECTURE

3.1 Establishment of architecture framework

Based on the characteristics of agent technology, the paper designs crop growth architecture including perception system、cognitive system、behavior system, see Fig.1. This architecture can not only reflect to outside environment real-time, make reasonable growing behavior, and feedback to the environment, but can carry out interactive simulation among the morphology, physiology and the environment effectively. Specific features are as follows:

The perception system: it is an information acquisition system, including receptor as to temperature, humidity, light intensity, perception processing module and information syncretize module. In a moment, all kinds of receptors receive and pre-treat the information received, then integrate them in order to prepare for the subsequent information processing and behavior decision.

The cognitive system: it is an information processing system, to process perception information with the output of perception system. It makes behavior decision, then pass its order to the behavior system to execute.

The behavior system: it is an behavior planning system. It plays a role in the environment directly, including a group of behavior program and a group of effectors. Behavior program is the high-level behavior, the effectors are specific executive actions, and each behavior program is decomposed with a series of low-level action to implement.

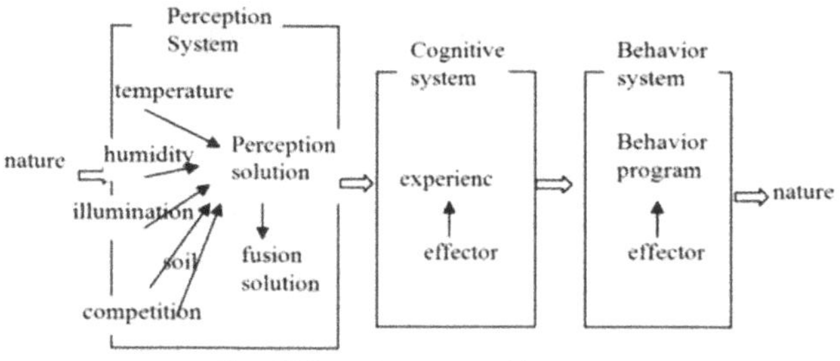

Fig.1 Virtual corps architecture

3.2 Virtual crops "self-learning" system based on the neural network

The plant grows in special environment. Different environmental impact will make crops change their growing status. As a result, crops model not only can real-time montor environmental change to make a reasonable growth; but change normal growth laws according to abnormality in the environment to achieve the purpose of reproduction and evolution. So the crop growth model wanted need to carry out "self-learning" in accordance with the outside environment, respond flexibly and reasonably, accord with the interaction between crops growth and the nature.

Virtual crops "self-learning" system based on the neural network is designed in perception system. The parameters of virtual farmland environment as input lead into the neural network to learn. Learned virtual crops can produce autonomic behavior under guide of cognitive model at the time of changes taken place in the external environment, then effect of action is passed on to neural work.. At the same time, fitness increase to some extent will lead to virtual crops better adapt to the changes in the external environment, see Fig.2.

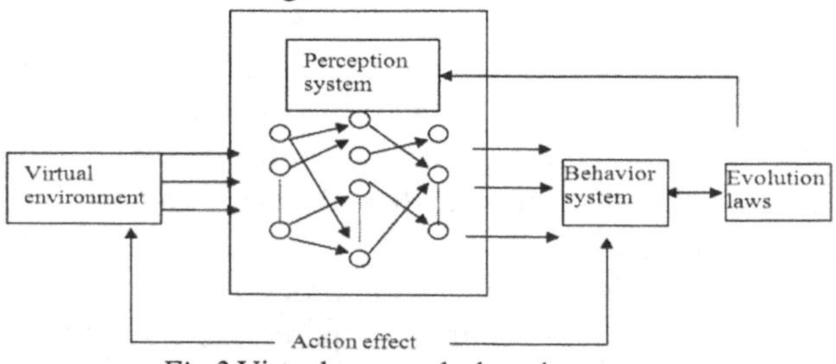

Fig.2 Virtual crops selp-learning system

3.3 Behavior simulation research based on crops population driven by resourse competition

The paper simulates crops goups by multi-agent involving cooperative law of crop growth on the basis of individual agent establishment and perfection. Usually crops growth belongs to selfish cooperative mode to compete solar energy、water and space among the individuals in order to survive and reproduce better. Agent cooperative behavior belongs to spontaneous cooperation, plan behavior in accordance with their respective objectives, the current state and long-term interests.

Each agent communicates through sending information and receiving coordination among them, which enable each agent can transfer information, ultimately implement cooperation. It is a reason device of communication and coordination mechanism impact on the cooperation efficiency, robustness and expansibility of entire system that it is quite important during devise multi-agent.

Crops make growth planning through the forecast of environmental resources because trigger factor of crops population from environment. In order to maintain the relative performance of agent, index will remain unchanged if the other agent is sharing the same resources. But when the lack of resources affect the survival of the agent, index have to be changed to adapt to the competitive environment. It will be a advantage that we apply agent technology to simulation of virtual crops growth, because the current research methods of virtual crops are all difficult to reach simulation results of groups.

4. CONCLUSION

Establish a architecture of virtual crops having learning mechanism, induct and abstract the common features and phenomena of crops growth and development from diversified natural features and phenomena, simulate functions and features of crops self-propagation、adaptive and self-optimizing as a result, manifest crops life characteristics of stability of internal、self-stabilization of adapt to the external environment、adaptive and self-coordination when agent technology is introduced into model construction of virtual crops. Simulate growth process of competition among crop groups based on distributed artificial intelligence theory, open up a new research method on simulation of virtual crop groups through researching proximity principle、quality principle and principle of adaptability of crops growth. Architecture of crop individual and groups are established, at the same time, based on simulation advantage of artificial intelligence in simulating crops growth aspects to deal with interactive impact between environmental and crops actively and effectively with environmental factors as integral part of simulation of crop growth, which contribute to training and strengthening self-learning system of crop model.

In this paper, we choose virtual crops research based on agent technology on the basis of understand research significance on virtual crops fully and combine with research status and characteristics of virtual crops at home and abroad, provide a new research platform for improving agent research technology because it expands research field of agent technology.

ACKNOWLEDGEMENTS

This study has been funded by Chinese 863 Plan (#2006AA10Z231), Program for Innovative Research Team of Northeast Agricultural University, "IRTNEAU"(CXZ010-2) and supported by "211 project".

REFERENCES

Ban Xiaojuan, Ai Dongmin et al. Artificial fish. Science Publisher. 2007(in Chinese)

Ban Xiaojuan, Liu Hongwei, et al. Study on Self-learning Method of "Artificial Fish". Computer Engineering. 2004.30(6),21-24(in Chinese)

Meng Xianyu, Yin Yixin et al. Research on virtual smelling perception system of artificial fish based on fuzzy reasoning. Journal of System simulation. 2007.19(20),4663-4780(in Chinese)

Su Zhongbin, Meng Fanjiang, Kang Li,et al.Virtual plant modeling based on Agent technology. Agricultural Engineering. 2005, 21 (8): 114- 116. (in Chinese)

Tu X. Artificial animals for computer animation: biomechanics, locomotion, perception and behavior. University of Toronto, Ph. D thesis, 1996

Zhang Shujun, Ban Xiaojuan, et al. Memory-based Cognitive Model of Artificial Fish. Computer Engineering. 2007.33(19),33-38(in Chinese)

THE RESEARCH OF ROUTE NAVIGATION BASED ON VISUAL NAVIGATION

Zhaobin Peng [1], Lijuan Wang [1], Yaru Zhang [1], Yuanyuan [1], Fangliang An [1], Rongjun Zhang [1], Yaoguang Wei [*]

[1] *Department of Information and Electrical Engineering, China Agricultural University, Beijing, P. R. China 100083*

[*] *Corresponding author, Address: Department of Information and Electrical Engineering, China Agricultural University, Beijing 100083, P. R. China, Tel: 13521591976, Email: wygzz@yahoo.com.cn*

Abstract: In order to solve the problems of Image pretreatment and the extraction of navigation line on agricultural machinery visual navigation, a route navigating system based on improved HOUGH transform is brought forward. First, extract the set of points from the original image based on the threshold segmentation and edge detection. Second, an improved HOUGH transform algorithm is used on detecting the route. This system is based on DM642 DSP and ARM9 Integrated Development Environment, which have achieved satisfactory experimental results.

Keyword: Visual navigation, threshold segmentation, edge extraction, improved HOUGH transform

1. INTRODUCTION

The visual navigation(Gan-Mor S et at., 2001) approach is one of the advanced navigation methods boomed up in recent years, which incorporates many advantages such as less noise, less harmful effects in comparison with non-visual sensors. In practical applications, drawing the leading path is all that's necessary, a robot can follow the path through visual navigation system all by itself (Pinto et at., 2000) (Toru et at., 2000).

Please use the following format when citing this chapter:

Peng, Z., Wang, L., Zhang, Y., Yuanyuan, An, F., Zhang, R. and Wei, Y., 2009, in IFIP International Federation for Information Processing, Volume 294, *Computer and Computing Technologies in Agriculture II, Volume 2*, eds. D. Li, Z. Chunjiang, (Boston: Springer), pp. 1511–1517.

Straight line detection is an important approach in image processing(Zhao Ying et at., 2006) , in which Hough Transform is commonly adopted. The Hough transform (HT) (Zhang Wei et at., 2005), in essence, is a method for grouping pixels in certain connections and searching for a parameterization to establish an equation by using the corresponding points of these pixels in the parameter space. But since this algorithm can not meet the real-time requirement(Illingworth J et at., 1988), the paper have proposed a visual navigation system, which provides visual-guiding for robot based on path guiding lines and marked identification(Wilson et at., 2000). In order to ameliorate the system, the real-time algorithm and robust are fully taken into account in each step from image preprocessing through path recognition and tracking.

We dig a little deeper into this question, and come up with the solution based on multi-known points Hough transform, increasing the processing speed of the real-time system.

2. IMAGE PRETREATMENT

2.1 Median filter

Median filter method was originally developed by J.W.Jukey in 1971 to process signal. Median filter is a non-linear image smoothing method, which is used to remove the high-frequency noise such as the Salt and Pepper noise and can better keep the edges from being blurred. The image is equipped with a small window, which of size with odd number of pixels, within the window, replacing each of them with the median value acquired from the middle pixel of the sequence rearranged in order of their gray value.

2.2 Edge detection (Sobel operator)

The edge detector inspect the neighborhood of each pixel, and quantify the gray rate, as well as the determination of the orientation. Most method employ the directive differentiation mask convolution.

Fig.1 shows the Sobel operator composed of two convolution nuclei, each pixel in the image is to be convoluted by these nuclei, one nucleus are sensitive to the vertical edge, the second nucleus, on the other hand, can detect the horizontal edge. The maximum value of the convolution results is output as the representative of this pixel. The effect of this algorithm is to get the contour of the image.

-1	-2	-1
0	0	0
1	2	1

-1	0	1
-2	0	2
-1	0	1

Fig.1: Template of Sobel

By using Sobel operator, the detected edge is clear. Sobel operator has always been chosen to detect the edge.

2.3 Threshold segmentation

After the edge detection, the threshold algorithm is done to the image, so the detail of the image is discernable.

3. EXTRACTION OF NAVIGATION LINE

The path recognition is the primary goal of making scientific researches on Agriculture autonomous walking robot visual navigation which usually represents as striated target. In the farmland environment, straight lines or lines slightly curved, the navigation line can be assumed to be straight. By processing the results of image analysis, navigation path can be isolated. Line detection is the important part in image analysis and computer vision. The most commonly used method is the least-squares and HOUGH transform. The least squares method is sensitive to the noise, and the classical HOUGH transform processing speed is very slow. A known point of HOUGH transform is faster, but is unable to the known point to carry on the accurate localization. For the above-mentioned problems, This article proposed based on multi-points HOUGH transform.

3.1 The Classical Hough transform

In 1962 Paul Hough first proposed the HOUGH transform, which realized one kind of mapping from the image space to the parameter space. The fundamental thought is duality between pixels and lines, namely the collinear point in image space corresponds to intersectant straight line in parameter space. Meanwhile, all straight line intersecting in a spot in parameter space corresponds to the collinear spot in the image space.

Since the straight line slope may be infinite, straight line represented by Polar coordinate equation is always used in the Hough transform, namely expressing the point of the straight line in the image space with the sine curve. The equation is as follows:

$$\rho = x * \cos\theta + y * \sin\theta \text{ , } \rho \in R \quad \theta \in [0, 2\pi] \tag{1}$$

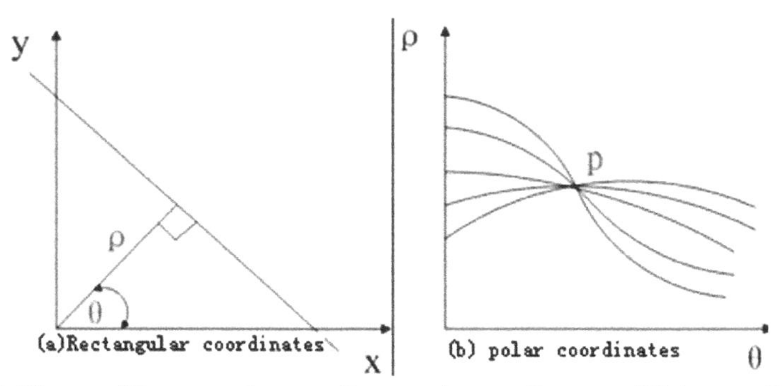

Fig.2: Change of Parameter between Rectangular coordinates and Polar coordinates

When all characteristic points in a straight line of a image have been carried out in this transformation, in the parameter space there will be many sine curves, all of which pass through (ρ,θ). So that the parameters on a straight line can be expressed by (ρ,θ) unit's coordinate, shown as picture b. The following two pictures are the results of a single image's HOUGH transform:

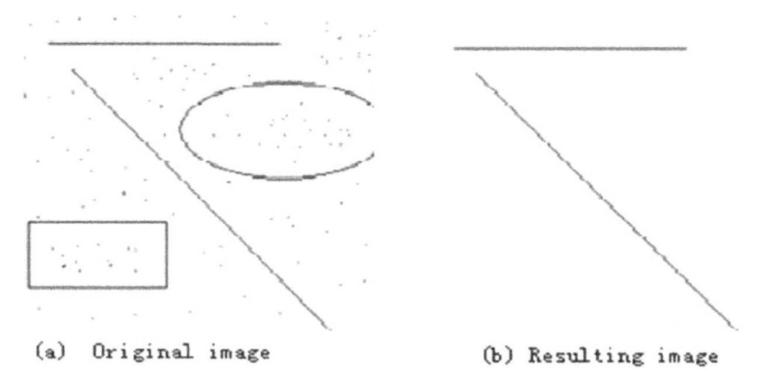

Fig.3: Result of Classic Hough transform

The extraction of straight line is accurate by using Hough transform, but it has a very serious shortcoming: It can't detect specific curve, and it's difficult to find the local maximum value of accumulator array, low precision, the large storage space. In order to solve these problems, the people proposed Hough transform with a known point.

3.2 The Hough transform with a known point

The Hough transform with a known point is a kind of improvement algorithm to the classical Hough transform .Firstly, finding a point on the line which will be defined as ρ_0 (the known point). Define this known point's coordinate as (x_0, y_0) and the slope of the straight line through ρ_0 as m, then the relationship between the slope and coordinate can be expressed by the following equation:

$$m_i = (y_i - y_0)/(x_i - x_0)\,. \qquad (2)$$

Map the slope value to a group of accumulators, making its corresponding accumulator's value to add 1 when obtains a slope each time. Each point on the same straight line has identical slope, so when there has straight line ingredient in the target area, its corresponding accumulator has the local maximum. The corresponding slope is taken as the straight line slope.

The Hough transform with a known point can greatly improve the computing speed, but the key issues is how to select the point correctly. It is random and the direction is uncertain in a real-time image. Taking into account the above issues, the method of Hough transform with multi-known points was adopted.

3.3 Improved HOUGH transform algorithm

From the above discussion we can see that the classical HOUGH transform and a known point HOUGH transform are not feasible to extract a straight line from real-time video clips. So the improved HOUGH transform algorithm based on multi-known points is adopted. Generally, the route may pass through the entire image. After several rounds of the image pretreatment, there are always bright spots on the 100th row for each image. After the pretreatment, very little noise is left. The following description tells the process of the whole experiment in detail.

4. THE APPROACH OF THE RECOGNITION ALGORITHM

The approach of the entire experiments are described as follows:

(1) Image pretreatment. Getting the binarization image with few noise

(2) Eliminate all the isolated bright spot of the 100th line of the binarization image

(3)Scan the entire image to get the total number C of the luminescent spots of the 100th line

(4) Divide the value of the slope into 10 parts

(5) Allocate an accumulator for each of the 10*C space

(6) Calculate slope for each of the C points with the other luminescent spots on the image and add 1 for the corresponding accumulator.

(7) Compare the values of the10*C accumulators with one another to get the maximum number, and make a record of the corresponding point's column coordinate

(8) Take the above determined coordinates as the known fiducial mark, figure out the slope with all the bright spots, if the slope of the line for the points fell within the defined area, they are the bright spots, otherwise set the brightness of this value to zero.

Through a number of experimental verification, various methods of straight line detection are compared as following:

Table 2 Contrast of different algorithm

	Classical HT	HT with kown point	Improved HT
Processing time per frame	3s	1.5ms	17.3ms
Effect	best	good	fair
Suitable for real-time image processing	NO	NO	YES

Using the improved HOUGH transformation to get the straight line, Fig.4 shows the original and effect image.

Fig.4 Result of improved HT

5. CONCLUSION

In order to meet the demand for the Agriculture vehicle visual navigation, this paper provide a IDE based on dual core (DM642 DSP and ARM9) to solve the problem of image pretreatment and the extraction of navigation line in the processing of image encountered in agricultural visual navigation. By adopting Improved HOUGH transform, this algorithm not only reduce the storage space but also it can meet the demand of real-time. All of the work above make a good foundation for the ARM9 MCU to precisely locate the robot position and control the navigation. Meanwhile, the embedded system has good flexibility to facilitate the transplantation of farmland in the experimental environment. A large number of experiments proved that the method is effective and practical.

ACKNOWLEDGEMENTS

This work is supported by China high tech development plan "863"(NO:2006AA10A304).

REFERENCES

Gan-Mor S, R L,Clark. 2001 DGPS-based on automatic guidance-implementation and economical analysis, ASAE Paper, No.01-1192

Illingworth J, Kittlor J. A survey of the Hough transform, CVGIP,1988,44:87～116

Pinto F A C ,Reid J F, Zhang Q. Vehicle guidance parameter determination from crop row images using principal component analysis, Journal of Agricultural Engineering Research,2000,75(3)：257～264

Toru Torii, Akirs Takamizawa. Crop row tracking by an autonomous vehicle using machine vision, JSAE,2000,62(5):37～42

Wilson J N. Guidance of agricultural vehicles-a historical perspective, Computers and Electronics in Agriculture,2000, 25 (1) : 3～ 9.

Zhang Wei, Du Shangfeng. Application of Hough transform in farmland machinery visual navigation, Instrumentation Journal, 2005,26(8)：706-707

Zhao Ying, Chen Bingqi, Wang Shumao, et al. Fast Detection of Furrows Based on Machine Vision on Autonomous Mobile Robot,Journal of Agricultural Machinery, 2006,37(4)：83～86.

A STUDY OF PADDYSTEM BORER (SCIRPOPHAGA INCERTULAS) POPULATION DYNAMICS AND ITS INFLUENCE FACTORS BASE ON STEPWISE REGRESS ANALYSIS

Linnan Yang[*], Lin Peng[1], Fei Zhong[2], Yinsong Zhang[3]

[1] *College of Basic Science and Information Engineering, Yunnan Agricultural University PO Box 650201, Kunming, Yunnan, China*

[3] *Faculty of Engineering and Technology, Yunnan Agricultural University PO Box 650201, Kunming, Yunnan, China*

[3] *International Exchange and Cooperation Department, Yunnan Agricultural University PO Box 650201, Kunming, Yunnan, China*

* *Corresponding author, Address: School of Life Science and Technology, University of Electronic Science and Technology PO Box 610054, Chengdu, Sichuan, China, lny5400@sina.com*

Abstract: Paddystem borer (Scirpophaga incertulas) is a serious rice pest. The damaged plants wither into dead tassel or white tassel. Such damage leads to decreased in rice production. In order to control the damages of paddystem borer efficiency, it is very important to analyze and study the regulation of population dynamics and the related factors affecting the development. This investigated the population dynamics of paddystem borer by means of light trap in JianShui County in Yunnan of China during 2004 to 2006, and analyzed the meteorological conditions affecting the population dynamics. The research suggests that: there exists a significant relationship between the population dynamics of paddystem borer and meteorological factors, among it, The most influenced are the average minimum temperature per month and relative humidity (RH).

KeyWords: paddystem borer (scirpophaga incertulas), stepwise regress analysis, population dynamics, meteorological factors

---

Please use the following format when citing this chapter:

Yang, L., Peng, L., Zhong, F. and Zhang, Y., 2009, in IFIP International Federation for Information Processing, Volume 294, *Computer and Computing Technologies in Agriculture II, Volume 2*, eds. D. Li, Z. Chunjiang, (Boston: Springer), pp. 1519–1526.

1. INTRODUCTION

Paddystem borer (Scirpophaga incertulas) is commonly known as borer, belong to Lepidoptera, Pyralidae, and a serious rice pest in tropical to sub tropical Asia. It is in the South Asian subcontinent, Southeast Asia , south of Japan and in most rice areas of south of the Yangtze in China. Paddystem borer feeding a single, designed to rice for food, Insect larvae bore the rice plant the damaged plants wither into dead tassel in stooling stage, and by boot stage to heading stage, it becomes death booting and white tassel, and heteroicous cause to the incidence of damaged rice plants. The "dead tassel" and "white tassel" are the main symptoms of rice seedlings damaged(Li Y R, 2002).

Paddystem borer (Scirpophaga incertulas) overwintering in rice stubble with mature larva, when temperature was about 16℃ in Spring, the pupation emergence and fly to paddy to spawn. There are 2-4 generations in a year, the occurrence period and the harm of each generation are as the followings: the first generation is in the first and second decade of June, do harm to the early rice and early medium rice, cause the plant to appear to dead tassel; the second generation is in July, do harm to the single late rice and Middle-Large Rice, cause the plant to appear to dead tassel, and do harm to the early rice and early medium rice, cause the plant to appear to white tassel; the third generation are from the middle ten days of August to the first ten-day of September, do harm to the Double-cropping Late Rice, cause the plant to appear to dead tassel, do harm to the Middle-Large Rice and single late rice, cause the plant to appear to white tassel; the forth generation is in September and October, do harm to the Double-cropping Late Rice and cause the plant to appear to white tassel(Lan X M,Yang F,Liang K Z,2002. Sun J,Wei G,Zhou X et al.,2003).

This paper utilize the pest population data which collected in JianShui County in Yunnan during the years 2004-2006, while the meteorological data in the corresponding period were acquired from JianShui County Meteorological Observatory, built the model of Stepwise Regress on Paddystem borer population Dynamics. First based on the 6 meteorological factors which effect Paddystem borer population, With the methods of regression, choose the most significant factor which influence the population, then establishes the corresponding models, and examine the pest population data in 2004-2006, finally obtain the most significant factor which influences the population.

2. MATERIALS AND METHODS

2.1 Datasets

The staff investigated the cardinal number of paddystem borer every November to December in JianShui County in Yunnan, in main rice producing area in JianShui County in Yunnan,. And observed the paddystem borer population in XiZhuang Town and recorded the data of the pest dynamics in next March to October.

The distribution of the paddystem borer in the field depends on the rice variety. The samples wouldn't correspond the actual field situation if the samples number is too small, because lots of different conditions in the field parcels affect pest population. Phototaxis of the paddystem borer is very strong. So black light trap for catching moth is useful because that the trapping area is large and fixed; the trapping data is representative. Though weather condition such as, raining or windy, is harmful for light trap catch of the moth, it effects the emergence period of the moth only, does not affect the quantity of the moth. Thus, the moth data from the light trap is effective for the primary data analysis.

Tab 1 Collecting Data of Paddystem Borer's Monthly Occurrence Quantityin Jianshui County from 2004-2006

Year	Jan	Feb	Mar	Apr	May	June	July	Aug.	Sep.	Oct	Nov	Dec
2004	0	0	15	190	2	0	0	1654	7456	0	0	0
2005	0	0	0	156	838	1342	677	18744	4206	0	0	0
2006	0	0	0	122	6	516	890	15748	838	0	0	0

The meteorological data such as the monthly average temperature, the high and the low temperature, Rainfall, evaporation and relative humidity are acquired from JianShui County Meteorological Observatory and were used for the primary data analysis.

2.2 Paddystem borer

There are 4 generations paddystem borer in a year in JianShui County. Usually the first generation lasts between last ten-days of March until the beginning of May. The pest devours the rice sprouts and the early transplanted rice seedling during this generation. The damaged plants wither in the actively growing shoots. The second generation often lasts between the first ten-days of June and mid July. Mid-season planted rice suffers damage from this generation and plants appear to have white tassels. The third generation lasts from the beginning of August until the last ten-days in

September. Late-growing plants suffer from the pest damage in this period. The last generation always begins in October and it over winters.

Tab 2 Meteorological Data in Jianshui County from 2004-2006

Years	Month	Jan	Feb	Mar	Apr	May	June	July	Aug	Sep	Oct	Nov	Dec
2004	Average Temperature (℃)	13.2	14.2	19.6	20.1	22.1	23.1	23.1	23.2	21.8	18	15.9	12.1
	Max-temperature (℃)	24.6	26.9	31.6	32.5	32.5	31.8	32.7	32	30.1	29	27.3	22.6
	Min-temperature (℃)	3	1.6	6	11.4	11.9	15.4	17.3	17.6	14.4	8.7	5.1	2.1
	Rainfall (mm)	8.6	24.3	0.2	104	133.5	151.5	162	96.8	196.1	13.3	22	4.6
	Evaporation (mm)	151.6	168	250.9	213.7	231.5	185.4	166.3	158.3	131.4	124.7	121.1	116.9
	relative humidity (%)	64	65	57	67	67	73	76	81	79	77	74	69
2005	Average Temperature (℃)	13.1	17.7	16.6	21.6	25	24.3	23.6	22.8	21.8	19.4	16.2	12.3
	Max-temperature (℃)	27.5	27.3	30	31.1	33.4	31.6	31.6	31.5	29.7	30.1	29	26.5
	Min-temperature (℃)	3.8	5	2.5	8.8	15.4	18.4	17.5	17.4	14.3	10.2	7.4	2.2
	Rainfall (mm)	26.6	0.3	44.5	37.9	22.3	165.7	103.5	163.6	39.8	89	46.9	36.3
	Evaporation (mm)	144	230.2	201.8	258	347.1	196.7	219.2	183.2	144.5	103.6	79.6	72.8
	relative humidity (%)	68	50	64	60	52	75	78	82	77	76	78	77
2006	Average Temperature (℃)	13.7	16.9	19.5	22.3	21.7	24.4	23.8	22.5	21.2	20.1	16.9	13
	Max-temperature (℃)	25.4	27.2	30.3	32.8	32.7	32.2	31.1	30.4	31.7	28.9	27.9	24.3
	Min-temperature (℃)	3.5	8.4	6.2	12.4	10.4	17	17.1	16.8	13.1	13.6	5.3	1.9
	Rainfall (mm)	0	5.9	1.3	85.7	88.1	94.4	205.9	75.4	65.9	143.6	9.2	0.9
	Evaporation (mm)	125.2	141.3	226.7	241.2	172.8	176.1	132.6	121.9	120.6	93.3	125.1	88.8
	relative humidity (%)	65	64	53	54	67	70	80	81	74	80	67	73

2.3 Stepwise Regress Analysis

This paper analyzes and studies the relationship between the paddystem borer population and meteorological factors by means of Stepwise Regress Analysis.

Stepwise Regress Analysis is to choose the Variables which has effective influence among the Variables related to Y(dependent variable) ,then use this variable to establish regression equation. Namely, include the factors which influenced Y as much as possible, and outstand some main factors at the same time. Mainly concludes the following steps: 1) Calculate the simple correlation coefficient matrix between each Variables, the Correlation of preliminary analysis, and the correlation between each Variables. and observe whether it has multicollinearity phenomenon. 2) Establish multiple linear regression equation with least square method, and do the Goodness-of-Fit Test for the equation. 3) If multiple linear regression equation has a good Goodness-of-Fit Test, it will calculate the T statistic of each regression parameter, Significance test on each Parameters, to reject non-significant factors under the level of significance a. 4) After rejecting non-significant factors, Re-establish multiple linear regression equation to the other variables, and do the significance test on each parameter till each factor is significance under the given level of significance a, under above conditions, the regression equation is the high Goodness multiple linear regression equation(WANG M C,SHENG H F, 1999).

3. RESULTS AND ANALYSIS

3.1 Data Selections

This paper get the monthly Occurrence Quantity of paddy borer adult by means of the traps quantity 10 areas per week in three years in Jianshui County in Yunnan Province of China from 2004-2006; choose 6 meteorological factors-monthly Average Temperature, monthly relative humidity, monthly rainfall, monthly evaporation, monthly max-temperature and min-temperature-as the influence factors to analyze. The reason to choose monthly max-temperature and min-temperature is because the pest will inhibit paddy borer larvae development and reproduction lower than 15℃ or higher than 35℃, this will effect the population of adult directly; the reason to choose monthly rainfall and monthly evaporation is because these two factors decide the humidity, it will effect the pest's development; the reason to choose monthly relative humidity and monthly rainfall is because the two factors effect the larvae survival

Law of Occurrence Quantity of paddy borer adult in 2004-2006

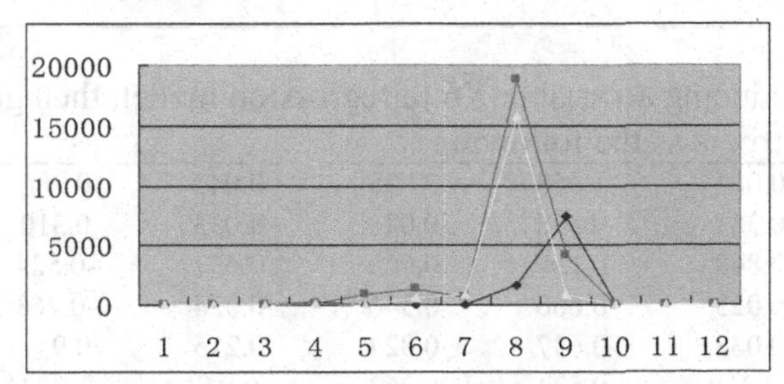

Fig 1 Occurrence Quantity of Paddy Borer Adult in 2004-2006

3.2 Stepwise Regress Analysis

Set up population of adult(Y), monthly Average Temperature(X1), monthly max-temperature(X2) , monthly min-temperature(X3), rainfall(X4), evaporation(X5), relative humidity(X6). To deal with these relative data by means of regression method, then get correlation coefficient matrix are as the following

1.000	0.825	0.927	0.675	0.493	0.146	0.299
0.825	1.000	0.735	0.525	0.443	0.062	0.161
0.926	0.735	1.000	0.773	0.232	0.428	0.414
0.675	0.525	0.773	1.000	0.000	0.545	0.332
0.493	0.443	0.232	0.000	1.000	-0.674	-0.060
0.146	0.062	0.428	0.545	-0.674	1.000	0.414
0.299	0.161	0.414	0.332	-0.060	0.414	1.000

partial regression square sum is:
V(1)= 0.089 V(2)= 0.026 V(3)= 0.171 V(4)= 0.110
V(5)= 0.004 V(6)= 0.171
F3(3)= 7.034

So we can get the important of each effected factors, the effect of X3 is equal to the effect of X6.

First introducing a variable X3 in regression model, then get correlation coefficient matrix is as the following

0.142	0.144	-0.927	-0.041	0.279	-0.250	-0.085
0.144	0.460	-0.735	-0.044	0.272	-0.253	-0.143
0.926	0.735	1.000	0.773	0.232	0.428	0.414
-0.041	-0.044	-0.773	0.403	-0.179	0.214	0.012
0.279	0.272	-0.232	-0.179	0.946	-0.773	-0.156
-0.250	-0.253	-0.428	0.214	-0.773	0.817	0.236
-0.085	-0.143	-0.414	0.012	-0.156	0.236	0.829

Table of Variance Analysis :

Source of variation	DF	SS	MS	F
Regression	1	0.171	0.171	7.034*
Off-regression	34	0.829	0.024	
Total	35	1		

Second introducing a variable X6 in regression model, then get correlation coefficient matrix is as the following

0.065	0.066	-1.058	0.025	0.042	0.306	-0.012
0.066	0.381	-0.868	0.023	0.033	0.310	-0.070
1.058	0.868	1.224	0.660	0.637	-0.524	0.290
0.025	0.023	-0.660	0.346	0.024	-0.263	-0.050
0.042	0.033	-0.637	0.024	0.215	0.947	0.067
-0.306	-0.310	-0.524	0.263	-0.947	1.224	0.289
-0.012	-0.070	-0.290	-0.050	0.067	-0.289	0.760

Table of Variance Analysis :

Source of variation	DF	SS	MS	F
Regression	2	0.24	0.12	5.204*
Off-regression	33	0.76	0.023	
Total	35	1		

Regression Equation:

$$Y= -9936.184+(211.567)*X(3)+(132.899)*X(6)$$

So the best regression equation of the model:

$$Y= -9936.184+(211.567)*X(3)+(132.899)*X(6)$$

Through the calculation we can drawn, after we introduce two factors, monthly min-temperature and relative humidity, standard deviation of the model of data simulation is 3720.721, multiple correlation coefficient is 0.49, it has already reached the acceptable results to over 75%.

According to the final regression equation, we know that the main factors which influence the Paddystem borer Population Dynamics are monthly min-temperature and relative humidity. That means along with the raise of monthly min-temperature and increase of relative humidity, the quantity of Paddystem borer adult has been increased. When enter into March in there, the average temperature may reach 16 ℃, Paddystem borer begins to propagation, and along with the coming of raining season and the further upturn of the temperature, the occurrence quantity of Paddystem borer rapid increment from July to September, and it will reach peak value in September, In October, along with the end of raining season, the quantity of Paddystem borer begins to decrease, till not appear any more. In Fig 1, we can see, the occurrence quantity in 2004 is different from the other two years, after comparing 3years temperature information, we found the raining season was later in 2004, the temperature was higher, the evaporation was also higher, and the rainfall amount was few obviously, these reason cause the difference of occurrence quantity of Paddystem borer between 2004 and other two yeas.

4. CONCLUSION

This paper analyze and study the paddystem borer population and meteorological factors such as the monthly average temperature, the high and the low temperature, Rainfall, evaporation and relative humidity in the same period by means of Stepwise Regress Analysis. Establish the relative regression model, and examine the data from 2004-2006, get the result that the occurrence quantity of Paddystem borer has a tight relationship with the monthly min-temperature and relative humidity in the same period. This research result laid a foundation of the further research work such as the research of the occurrence period of Paddystem borer, the occurrence quantity of Paddystem borer, the prediction of high peak period of Paddystem borer, etc.

ACKNOWLEDGEMENT

The findings and the opinions are partially supported by the Mega-projection of National Key Technology R&D Program for the 11th Five-Year Plan "Research on Information Technology of the ecological environmental protection in rural areas" (2006BAD10A14) and by the Technological Innovation projection of Yunnan Province "Research on Standard Producing Information Project of Organic Vegetables". This work is also supported by the Yunnan Agricultural University (China) and by the University of Electronic Science and Technology (China).

REFERENCES

CHEN Peng,YE Hui.Population dynamics of Bactrocera dorsalis(Diptera:Tephritidae) in LiuKu, Yunnan with an analysis of the influencing factors[J]. (in Chinese) Acta Entomologica Sinica, 2007,50(1):38-45.

Cheng G. H, Liu R. Q., Studies on forecasting the incidences of rice yellow stem borer by using muzzy close degree [J]. (In Chinese) Plant Protection Technology and Extension, 2002, 22(4):9-10.

He X.Q., The Modern Statistical Analysis Method and its Application [M]. (in Chinese) Beijing: China Renmin University Press, 2007.

Lan X. M, Yang F., Liang K.Z., Population Dynamics of Tryporyza incertulas and its Control Methods [J]. (in Chinese) Entomological Knowledge, 2002,39(2):113-115.

Lei X. S., Chen B. F., Forecasting of the Population Peak of the Paddy stem borer (Scirpophaga incertulas) based on meteorological factors in Jingdezheng City [J]. Journal of Anhui Agricultural Sciences. 2007, 35 (29):9307-9308.

Li Y R, Agricultural entomology [M]. Chinese Agriculture Press.2002.8,35-37.

Liu L.F., Feng D.Y., Prediction of amount of wheat aphid occurrence by application of principal component analysis. [J]. (in Chinese) Entomological Knowledge, 1997.34 (5): 260-263.

Sun J.M., Wei G., Zhou X.W et al., The Population Dynamics of the Yellow Rice Borer Causes of Outbreaks and Control Strategy [J]. (in Chinese) Entomological Knowledge, 2003,40(2):124-127.

WANG M C,SHENG H F. Probability theory and mathematical statistics [M]. Beijing: Higher Education Press, 1999,2:159-225.

Yazdan S., Mohsen H., Rostam M., Numerical solution of the nonlinear Schrodinger equation by feedforward neural networks [J]. Communications in Nonlinear Science and Numerical Simulation, 2008. 13, (10), 2132-2145.